T0254289

Theory of Statistical Inference

CHAPMAN & HALL/CRC
Texts in Statistical Science Series

Recently Published Titles

Modern Data Science with R, Second Edition
Benjamin S. Baumer, Daniel T. Kaplan, and Nicholas J. Horton

Probability and Statistical Inference
From Basic Principles to Advanced Models
Miltiadis Mavrakakis and Jeremy Penzer

Bayesian Networks
With Examples in R, Second Edition
Marco Scutari and Jean-Baptiste Denis

Time Series
Modeling, Computation, and Inference, Second Edition
Raquel Prado, Marco A. R. Ferreira and Mike West

A First Course in Linear Model Theory, Second Edition
Nalini Ravishanker, Zhiyi Chi, Dipak K. Dey

Foundations of Statistics for Data Scientists
With R and Python
Alan Agresti and Maria Kateri

Fundamentals of Causal Inference
With R
Babette A. Brumback

Sampling
Design and Analysis, Third Edition
Sharon L. Lohr

Theory of Statistical Inference
Anthony Almudevar

For more information about this series, please visit: https://www.crcpress.com/Chapman--Hall/CRC-Texts-in-Statistical-Science/book-series/CHTEXSTASCI

Theory of Statistical Inference

Anthony Almudevar

CRC Press is an imprint of the
Taylor & Francis Group, an **informa** business
A CHAPMAN & HALL BOOK

First edition published 2022

by CRC Press
6000 Broken Sound Parkway NW, Suite 300, Boca Raton, FL 33487-2742

and by CRC Press
4 Park Square, Milton Park, Abingdon, Oxon, OX14 4RN

CRC Press is an imprint of Taylor & Francis Group, LLC

Library of Congress Cataloging-in-Publication Data

Names: Almudevar, Anthony, author.
Title: Theory of statistical inference / Anthony Almudevar.
Description: Boca Raton : CRC Press, 2021. | Series: Chapman & Hall/CRC texts in statistical science | Includes bibliographical references and index.
Identifiers: LCCN 2021032159 (print) | LCCN 2021032160 (ebook) | ISBN 9780367488758 (hardback) | ISBN 9780367502805 (paperback) | ISBN 9781003049340 (ebook)
Subjects: LCSH: Mathematical statistics.
Classification: LCC QA276 .A46 2021 (print) | LCC QA276 (ebook) | DDC 519.5/4--dc23
LC record available at https://lccn.loc.gov/2021032159
LC ebook record available at https://lccn.loc.gov/2021032160

ISBN: 9780367488758 (hbk)
ISBN: 9780367502805 (pbk)
ISBN: 9781003049340 (ebk)

DOI: 10.1201/9781003049340

Publisher's note: This book has been prepared from camera-ready copy provided by the authors.

Contents

Preface

A typical graduate program in most departments of statistics has three core subjects: the theory of inference, probability theory and applied statistics. This volume is intended as a textbook for the first of these, and is based on lecture notes prepared by the author for this purpose.

Statistical science is concerned with the development of methods which transform data into inference statements, usually in the form of an estimate or a hypothesis test. The theory of statistical inference gives a mathematical foundation with which to characterize the properties of these methods, then to inform the choice of method. The first of these goals relies mainly on probability theory. The second, however, is unique to the theory of inference. This is the subject of our book.

Structure of the volume

This volume can be divided into four parts, each chapter ending with a section of practice problems. Chapters 1–2 define Part I, which covers the required distribution theory. If used as a textbook for a graduate course, these chapters would likely be a review for most students.

Chapters 3–6 define Part II, which introduces the problem of statistical inference. It begins with a formal definition of a statistical model in Chapter 3. Here, the core ideas of sufficiency and invariance are introduced. The problems of estimation and hypothesis testing are introduced in the next two chapters, with a separate chapter reserved for linear models (primarily multiple linear regression and ANOVA models). Some of the material in Part II would be covered in an undergraduate program in statistics, but the foundation of a formal theory is being laid, especially in Chapter 3. For many students, this will begin a transition from methodology to theory, with its reliance on formal proofs.

Part III (Chapters 7–10) constitutes the core of the volume, which presents a formal theory of statistical inference. Mathematically, this takes the form of optimization methods used to determine how to make the most efficient use of data. Decision theory (Chapter 7) is used to formulate the optimization problem precisely. Once this is done, a theory of inference is developed according to the canonical subdivisions adopted by most textbooks, in particular, *uniformly minimum variance unbiased estimation* (UMVU) in Chapter 8; *minimum risk equivariant estimation* (MRE) in Chapter 9; and the *Neyman-Pearson lemma* for *uniformly most powerful tests* (UMP) in Chapter 10. Of course, these are not entirely separate subjects, and throughout the volume the manner in which the notions of sufficiency and invariance complement each other will be emphasized. For this reason, sufficiency and invariance, usually associated with UMVU estimation and MRE estimation, respectively, are introduced earlier in Chapter 3.

Finally, Part IV (Chapters 11–14) introduces large sample, or asymptotic, theory. This describes approximate methods whose accuracy improves as the sample size of a data set approaches infinity. Chapter 11 covers stochastic limit theory, which allows us to characterize exactly what letting a sample size approach infinity represents. The ubiquitous central limit theorem and law of large numbers are introduced here. Chapter 12 introduces a number of large sample estimation methods (the δ-method and the Bahadur representation theorem, for example). The Cramér-Rao lower bound is also introduced, which places a quite definitive limit on the efficiency of large sample methods. For this reason, the theory of asymptotic inference assumes a somewhat different character than what is presented in Part III, and may be divided into two tasks. The first is to identify the most asymptotically efficient procedure, which is usually the maximum likelihood estimate, or likelihood ratio test (to be sure, this is related to the notion of sufficiency). The second task is to measure, and if possible improve, the accuracy of the approximation (a sample size will, after all, be finite). This is sometimes referred to as "small-sample asymptotics". This is a quite technical branch of mathematical statistics which usually relies on complex analysis (Field and Ronchetti, 1990; Brazzale *et al.*, 2006). We will, nonetheless, explore such problems using real analysis.

Chapter 13 applies large sample techniques to M-estimates, which covers a quite large class of estimation methods, including maximum likelihood estimation, nonlinear regression, generalized linear models (GLM) and generalized estimating equations (GEE). Here we emphasis the fact that these methods share essentially the same asymptotic theory, summarized in Theorem 13.1. Chapter 14 covers large sample hypothesis testing, and is structured around three widely used methods, the likelihood ratio test, the (Rao) score test and the Wald test. Each can be described as a χ^2 test, given the limiting distribution of the defining test statistic. Each is also an approximation of the other. However, all three are commonly used, and have relative advantages and disadvantages which depend on a particular application.

This volume is concerned not so much with methodology as with creating a single narrative based on a few themes, primarily sufficiency, invariance and decision theory. The material is therefore organized on that basis. For example, there is no separate chapter on nonparametric methods (i.e. rank-based methods). These methods are discussed, but since they can be derived by the same principles used for parametric models, they appear in the text according to their relationship to sufficiency (Chapter 8) or invariance (Chapter 9).

The examples considered rarely stray from independent sampling of univariate observations. Extension to more general models usually requires a broader range of mathematical techniques, but this process is largely independent of the themes of the book. A more thorough treatment of a smaller class of problems seems, arguably, the better approach.

Bayesian methods are discussed in Chapters 3, 4 and 7. Here the emphasis is not to develop an essentially separate branch of statistical methodology but to lend insight into the core themes of this book, especially decision theory in this case.

Mathematical background

It will be assumed that the reader is already familiar with commonly used statistical methods and elementary probability theory. This volume makes considerable use of matrix algebra. Appendix B summarizes most of the needed background on this subject. Some familiarity with multivariate calculus is also assumed, with Appendix C providing a review in this regard. More than most textbooks on inference, we make considerable use of group theory to express the notion of invariance (Appendix D). Of course, Lehmann and Casella (1998) and Lehmann and Romano (2005) are notable in this respect.

Throughout the volume (including some problems) reference is made to `R` programming. It is assumed that the reader has some familiarity with this computing platform, but this is not at all necessary to make use of this volume.

Acknowledgments

The author wishes to thank Jacob Almudevar for his insightful review of the manuscript, and David Grubbs of CRC Press for his patience and support.

1

Distribution Theory

1.1 Introduction

The first two chapters of this volume provide a review of those topics in probability theory important to statistical inference. Section 1.2 introduces the probability model, along with the related ideas of density, expected value and independence. Our discussion of conditional densities is targeted specifically to the theory of sufficiency (Chapters 3 and 8).

Section 1.3 gives a brief review of some important theorems of probability theory. Section 1.4 introduces the univariate densities which will appear in the examples and problems of this volume. Some attempt is made is to justify their common use in statistical practice. The notational conventions by which these distributions will be referred are also introduced here, while the analytic detail is reserved for Appendix A.

Section 1.5 covers stochastic order relations. If X and Y are two random variables, then the statement $X < Y$ can be interpreted as an event which occurs with probability p, but not as an ordering of real numbers (i.e. $0.5 < 2.6$). However, much of statistical methodology depends on our ability to order distributions (Y tends to be larger than X). What we will need in this respect is introduced here.

Statistical methods often concern quantiles or percentiles. If 5% of a population earns more than \$$M$ per year, then M is a 95th percentile. While these quantities have an intuitive meaning, their formal mathematical definition requires some care. This is the subject of Section 1.6.

Section 1.7 deals with a single technical problem which has important implications for the theory of inference. A standard result of probability theory is that if F is the cumulative distribution function of a random variable X (Equation (1.2)) and F is invertible, then $F(X)$ possesses a uniform distribution on $(0, 1)$. This predicts, for example, that the P-value under a null hypothesis is uniformly distributed (Chapter 5). It also provides a general method of constructing confidence intervals (Section 4.6.4). It is therefore important to determine if this result holds when an inverse $F^{-1}(p)$ is not defined for all $p \in (0, 1)$. It turns out that it holds for any continuous random variable. For discrete random variables, we can characterize the deviation of $F(X)$ from the uniform distribution in a way which serves our purpose in much the same manner.

Section 1.8 deals with the technical problem of determining the distribution of a transformed random variable $g(X)$ given the distribution of X. This problem arises frequently in the theory of inference. The final two sections introduce the moment generating function (MGF) and cumulant generating function (CGF), which have a number of important applications in inference.

DOI: 10.1201/9781003049340-1

1.2 Probability Measures

The axiomatic foundation of modern probability theory was established in 1933 by Andrey Kolmogorov (Kolmogorov, 1933) based on the concept of a measure, giving formal rules for the consistent assignment of probabilities (a brief review of basic measure theory is given in Appendix C). A probability measure models a single random outcome $\omega \in \Omega$, where Ω is the set of all possible outcomes (a probability cannot be calculated if Ω is not well defined). A probability measure assigns a number $P(E) \in [0,1]$ to a subset $E \subset \Omega$, which is interpreted as the probability that $\omega \in E$. This assignment must be consistent. If $E \subset F$, we expect $P(E) \leq P(F)$ (referred to as the monotonicity property). Or, if $E_1, E_2, \ldots$ is a countable collection of disjoint subsets of Ω, then we expect $P(\cup_{i=1}^{\infty} E_i) = \sum_{i=1}^{\infty} P(E_i)$ (referred to as the countable additivity property). The need for measure theory arises because it may not be possible to assign a probability to *every* subset of Ω in a manner consistent with these rules. In fact, this will be the case whenever Ω is an interval of real numbers.

The solution almost universally accepted (that proposed in Kolmogorov (1933)) is to restrict the class of subsets of Ω for which probabilities are defined to a σ-field $\mathcal{F}$ (Definition C.4), and to assume that P is a countably additive measure defined on $\mathcal{F}$ (Definition C.5). This is a very consequential restriction, since it imposes a continuity property on P. For example, suppose we are given an increasing sequence of sets $E_1 \subset E_2 \subset E_3 \subset \ldots \in \mathcal{F}$. A σ-field is closed under finite or countable union, so that $E = \cup_{i=1}^{\infty} E_i \in \mathcal{F}$. Thus, E can be interpreted as a limit of E_i (written $E_i \uparrow E$) and it follows directly from countable additivity that $\lim_{i \to \infty} P(E_i) = P(E)$. This assumption is suitable in support of statistical inference, although for some applications it is argued that the restriction of a probability measure to a σ-field is too limiting, and/or the continuity assumption is not necessary (Dubins and Savage, 1976).

For our purposes, we may assume that a random outcome is an element of a metric space $(\mathcal{X}, d)$ (Definition C.1). We reserve the "tilde" notation $\tilde{X}$ or $\tilde{x}$ for such a random or fixed observation. A metric leads to the definition of an open ball in $\mathcal{X}$ as $B_{\epsilon}(\tilde{x}) = \{\tilde{x}' \in \mathcal{X} \mid d(\tilde{x}', \tilde{x}) < \epsilon\}$, and an open set in $\mathcal{X}$ is defined as any union of open balls, which in turn defines a topology (Definition C.3). A closed set is any complement of an open set. Then $\mathcal{X}$ and $\emptyset$ are both open and closed under this definition. Note that this definition of an open set conforms to the conventional notion of an open set in $\mathbb{R}^m$.

We will usually refer to a metric space $\mathcal{X}$ without reference to the metric d. A measurable space is then the pair $(\mathcal{X}, \mathcal{F})$, where $\mathcal{F}$ is a σ-field. We will assume that $\mathcal{X}$ is also a metric space. The Borel sets $\mathcal{B}$ consist of all sets which are calculable as a countably infinite sequence of set operations (union, intersection and complement) performed on the open subsets of $\mathcal{X}$ (Example C.4). This, of course, includes all closed subsets. It may be shown that $\mathcal{B}$ is also the smallest σ-field containing all open subsets of $\mathcal{X}$ (or, equivalently, all closed subsets of $\mathcal{X}$). We refer to the σ-field $(\mathcal{X}, \mathcal{B})$ as a Borel space.

A probability measure is then a mapping of a σ-field to the unit interval which satisfies the following axioms:

Definition 1.1 (Axioms of a Probability Measure) Suppose we are given measurable space $(\Omega, \mathcal{F})$. Suppose we have a mapping $P : \mathcal{F} \to [0,1]$ which satisfies the following three axioms:

(i) $P(E) \geq 0$ for any $E \in \mathcal{F}$,
(ii) $P(\Omega) = 1$,
(iii) If $E_1, E_2, \ldots$ is a countable collection of disjoint subsets in $\mathcal{F}$ then $P(\cup_i E_i) = \sum_{i \geq 1} P(E_i)$.

Then P is a probability measure, and we refer to $(\Omega, \mathcal{F}, P)$ as a probability measure space. See also, Example C.3. ■

A random vector $\mathbf{X} \in \mathcal{X}$ is a measurable mapping (Section C.2) from a probability measure space $(\Omega, \mathcal{F}, P)$ to a Borel space $(\mathbb{R}^d, \mathcal{B})$. This construction is sometimes emphasized by writing $\mathbf{X} = \mathbf{X}(\omega)$, $\omega \in \Omega$, which is interpretable as a transformation of a random outcome ω. The marginal distribution of $\mathbf{X}$ is defined by the induced probability measure $(\mathcal{X}, \mathcal{B}, P_{\mathbf{X}})$, where $P_{\mathbf{X}}(B) = P(\mathbf{X}(\omega) \in B) = P(\mathbf{X}^{-1}(B))$ (the validity of this construction follows from the measurability of $\mathbf{X}$).

The term "almost everywhere" (Definition C.6) applied to a probability measure is equivalent to "with probability one", abbeviated as *wp*1.

1.2.1 Densities

Suppose ν and μ are two measures on a common measurable space $\mathcal{X}$. Recall from Definition C.7 that if $\mu(E) = 0 \Rightarrow \nu(E) = 0$ for all measurable $E \in \mathcal{X}$, then ν is absolutely continuous with respect to μ, written $\nu \ll \mu$ (equivalently, ν is dominated by μ). This is an idea of some practical significance in inference theory. The reader probably understands a normally distributed random variable to be a "continous random variable" and a random variable with a binomial distribution to be a "discrete random variable". What this means precisely is that the probability measure of a continuous random variable is absolutely continuous with respect to Lebesgue measure (Example C.5), and the probability measure of a discrete random variable is absolutely continuous with respect to some counting measure (Example C.6).

In inference theory probability measures are usually characterized by densities. A density is a function f on a measure space $(\mathcal{X}, \nu)$ which represents a probability measure by the evaluaton method

$$P(E) = \int_{x \in E} f(x) d\nu \tag{1.1}$$

(integration with respect to a measure is reviewed in Section C.3). In particular, a density function $f_{\tilde{X}}$ models the distribution of a random outcome $\tilde{X} \in \mathcal{X}$, and we accept the intepretation $P_{\tilde{X}}(E) = P(\tilde{X} \in E)$. We require that $f_{\tilde{X}}(x) \geq 0$, and $\int_{\tilde{x} \in \mathcal{X}} f_{\tilde{X}}(\tilde{x}) d\nu = 1$. If these conditions hold, then P satisfies the axioms of a probability measure.

A converse relationship exists. If a probability measure P is absolutely continuous with respect to ν then by the Radon-Nikodym theorem (Billingsley, 1995) there exists a density function f for which Equation (1.1) holds, and this density is unique *a.e.*

The support of a density f on $\mathcal{X}$ is the set $\bar{\mathcal{X}} = \{x \in \mathcal{X} : f(x) > 0\}$. If f is a density with respect to counting measure, then $\bar{\mathcal{X}}$ is unique. If f is a density with respect to Lebesgue measure μ then $\bar{\mathcal{X}}$ is not unique, since two densities need only be equal *a.e.* to define the same probability measure. It is useful, however, to define a support to be open. In fact, if E is a nonempty open subset of $\bar{\mathcal{X}}$, then we must have $P(X \in E) > 0$ if X has density f.

Our preference will be to express integrals over a sample space $\mathcal{X}$ in the general form of Equation (1.1) for both continuous and discrete random outcomes $\tilde{X} \in \mathcal{X}$. Of course, the more familiar evaluation methods will eventually emerge when needed.

Example 1.1 (Integrals *wrt* a Measure) If ν is Lebesgue measure on $\mathbb{R}$, then

$$P(E) = \int_{x \in E} f(x) d\nu = \int_{x \in E} f(x) dx.$$

That is, $P(E)$ can be evaluated using the conventional Riemann integral. If ν is counting measure on support $S = \{0, 1, 2, 3, \ldots\}$, then

$$P(E) = \int_{x \in E} f(x) d\nu = \sum_{x \in E \cap S} f(x) = \sum_{x=0}^{\infty} f(x).$$

That is, $P(E)$ can be evaluated using conventional summation. ∎

The cumulative distribution function (CDF) of a random variable $X \in \mathbb{R}$ is defined as

$$F_X(x) = P\{X \leq x\} = \int_{u=-\infty}^{x} f_X(u) d\nu. \tag{1.2}$$

We sometimes use the complement $\bar{F}_X(x) = 1 - F_X(x)$. Unlike the density function, a CDF is a unique representation of a distribution. The following theorem, given without proof, enumerates some of the important properties of a CDF.

Theorem 1.1 (Properties of the CDF) Suppose $P(X = -\infty) = P(X = \infty) = 0$. Then a CDF defined in Equation (1.2) possesses the following properties:

(i) $\lim_{x \to -\infty} F_X(x) = 0$.
(ii) $\lim_{x \to \infty} F_X(x) = 1$.
(iii) $F_X(x)$ is nondecreasing in x.
(iv) $F_X(x)$ is right-continuous. That is, $\lim_{\epsilon \downarrow 0} F_X(x + \epsilon) = F_X(x)$.
(v) Suppose, we denote the limit $\lim_{\epsilon \downarrow 0} F_X(x - \epsilon) = F_X(x-)$. Then $P(X = x) = F_X(x) - F_X(x-)$.
(vi) If the distribution of X is absolutely continuous *wrt* Lebesgue measure, then $F_X(x)$ is continuous over $x \in \mathbb{R}$.
(vii) If the distribution of X is absolutely continuous *wrt* Lebesgue measure the following relationship holds

$$dF_X(x)/dx = f_X(x), \tag{1.3}$$

wherever the derivative exists. ∎

Similarly, the definition of the CDF for a random vector $\mathbf{X} = (X_1, \ldots, X_n)$ is

$$F_{\mathbf{X}}(x_1, \ldots, x_n) = P\left(\cap_{i=1}^{n} \{X_i \leq x_i\}\right).$$

We finally note that the term density is sometimes reserved for densities defined with respect to Lebesgues measure. In this case, the term probability mass function (PMF) describes a density defined with respect to counting measure. We accept the more general definition of a density function, but we will also use the term "probability mass function".

1.2.2 Expected Values

The expected value of a random variable X defined on $(\Omega, \mathcal{F}, P)$ is taken to be

$$E[X] = \int_{\omega \in \Omega} X(\omega) dP,$$

that is, an integral evaluated with respect to the measure P. We then convert the integral to one which can be evaluated with standard techniques. If there is a measure ν for which $P \ll \nu$, then there exists a density f_X for X with respect to ν. Then

$$E[X] = \int_{x \in \mathbb{R}} x f_X(x) d\nu.$$

Usually, ν is Lebesgue or counting measure. If $\tilde{X} \in \mathcal{X}$ is any random outcome with density $f_{\tilde{X}}$ with respect to ν, then

$$E[g(\tilde{X})] = \int_{\tilde{x}\in\mathcal{X}} g(\tilde{x}) f_{\tilde{X}}(\tilde{x}) d\nu,$$

for any measurable function $g : \mathcal{X} \rightarrow \mathbb{R}$. The same logic of Example 1.1 follows for expected values.

Example 1.2 (Expected Values *wrt* a Measure) If ν is Lebesgue measure on $\mathbb{R}$, then

$$E[X] = \int_{x=-\infty}^{\infty} x f_X(x) d\nu = \int_{x=-\infty}^{\infty} x f_X(x) dx.$$

That is, $E[X]$ can be evaluated using the conventional Riemann integral. If ν is counting measure on support $S = \{0, 1, 2, 3, \ldots\}$, then

$$E[X] = \int_{x=-\infty}^{\infty} x f_X(x) d\nu = \sum_{x=0}^{\infty} x f_X(x),$$

so that $E[X]$ can be evaluated using conventional summation. ∎

1.2.3 Independence

Independence is an intuitive probabilistic notion, defining unrelatedness of random outcomes. However, it's formal definition requires some care.

Two events A, B on $(\Omega, \mathcal{F}, P)$ are independent if $P(AB) = P(A)P(B)$. We then say two collections of sets $\mathcal{F}$ and $\mathcal{G}$ are independent if $P(FG) = P(F)P(G)$ for all $F \in \mathcal{F}$ and $G \in \mathcal{G}$. Finally, two random variables X, Y are independent if $P(\{X \in E\} \cap \{Y \in F\}) = P(\{X \in E\} \cap \{Y \in F\})$ for all measurable sets E, F.

Independence of two sets, random variables or collections of sets is sometimes denoted as $A \perp B$, $X \perp Y$ or $\mathcal{F} \perp \mathcal{G}$, reflecting the fact that stochastic independence is in some ways analagous to geometric perpendicularity.

Suppose E_0, E_1 are any sets for which $P(E_0) = 0$ and $P(E_1) = 1$. Then for any set A, we must have $P(AE_0) = 0$, since $P(AE_0) \leq P(E_0) = 0$. This means $A \perp E_0$. Similarly, $P(AE_1) = P(A) - P(AE_1^c) = P(A)$, since $P(AE_1^c) \leq P(E_1^c) = 0$, and so $A \perp E_1$. In fact, $E \perp E$ implies $P(E) = P(E)^2$, that is $P(E)$ is 0 or 1.

A finite sequence of subsets $E_1, \ldots, E_n$ on $(\Omega, \mathcal{F}, P)$ is independent if

$$P(\cap_{i\in I} E_i) = \prod_{i\in I} P(E_i), \tag{1.4}$$

for all nonempty index subsets $I \subset \{1, \ldots, n\}$. In fact, any sequence of subsets E_i for which $P(E_i) \in \{0, 1\}$ is independent.

It is necessary to insist on the product rule for all non-empty selections from the sequence. To see this, consider three sets A, B, C. We may construct all joint probabilities by specifying suitable probability values for all eight intersections. Suppose we set $P(ABC) = 1/64$, $P(AB^cC^c) = P(A^cBC^c) = P(A^cB^cC) = 15/64$ and $P(A^cB^cC^c) = 18/64$. This gives $P(A) = P(B) = P(C) = 1/4$, and so $P(ABC) = P(A)P(B)P(C)$. On the other hand, $P(AB) = 1/64 \neq P(A)P(B) = 1/16$, and so A and B are not independent.

Conversly, pairwise independence does not imply independence. For example, if $A \perp B$, $P(A) = P(B) = 1/2$ and $C = AB \cup A^cB^c$, then it is easily verified that we also have $A \perp C$ and $B \perp C$, but that $P(ABC) \neq P(A)P(B)P(C)$.

Pairwise independence carries some implications. If $A \perp B$ then $A^c \perp B$, since $P(A^cB) = P(B) - P(AB) = P(B) - P(A)P(B) = P(B)(1 - P(A)) = P(B)P(A^c)$. This in turn implies $A^c \perp B^c$ and $A \perp B^c$. It is also true that $\Omega \perp A$ and $\emptyset \perp A$ for any set A.

It may be shown that random variables $\mathbf{X} = (X_1, \ldots, X_n)$ are independent if and only if,

$$F_{\mathbf{X}}(x_1, \ldots, x_n) = \prod_{i=1}^{n} F_{X_i}(x_i) or f_{\mathbf{X}}(x_1, \ldots, x_n) = \prod_{i=1}^{n} f_{X_i}(x_i). \tag{1.5}$$

It follows that if $E[|X_i|] < \infty$ for each i, independence implies

$$E\left\{\prod_{i=1}^{n} X_i\right\} = \prod_{i=1}^{n} E[X_i]. \tag{1.6}$$

1.2.4 Defining Conditional Distributions

The definition of a conditional probability is

$$P(A \mid B) = \frac{P(A \cap B)}{P(B)}, \tag{1.7}$$

assuming $P(B) > 0$. The conventional definition of a conditional observation $\mathbf{X} \mid \mathbf{Y}$, where $\mathbf{X}, \mathbf{Y}$ are random vectors is given by density

$$f_{\mathbf{X}|\mathbf{Y}}(\mathbf{x} \mid \mathbf{y}) = \frac{f_{\mathbf{X},\mathbf{Y}}(\mathbf{x}, \mathbf{y})}{f_{\mathbf{Y}}(\mathbf{y})} \tag{1.8}$$

where $f_{\mathbf{X},\mathbf{Y}}$ is the density of the combined observation $(\mathbf{X}, \mathbf{Y})$ and $f_{\mathbf{Y}}$ is the marginal density of $\mathbf{Y}$. If the probability measure of $(\mathbf{X}, \mathbf{Y})$ is absolutely continuous with respect to counting measure, then this definition follows from (1.7) where $f_{\mathbf{Y}}(\mathbf{y}) > 0$. Then an expectation $h(\mathbf{y}) = E[g(\mathbf{X}) \mid \mathbf{Y} = \mathbf{y}]$ is simply the expected value of $g(\mathbf{X})$ under density $f_{\mathbf{X}|\mathbf{Y}}$, and we may adopt the convention $E[g(\mathbf{X}) \mid \mathbf{Y}] = h(\mathbf{Y})$. Then $P(\mathbf{X} \in E \mid \mathbf{Y} = \mathbf{y})$ or $P(\mathbf{X} \in E \mid \mathbf{Y})$ is evaluated in the same way as the conditional expectation of $g(\mathbf{X}) = I\{\mathbf{X} \in E\}$. Informally, we use $P(\mathbf{X} \mid \mathbf{Y})$ as shorthand for the distribution of $\mathbf{X}$ given $\mathbf{Y}$.

However, if $(\mathbf{X}, \mathbf{Y})$ is absolutely continuous with respect to Lebesgues measure, then (1.8) is not directly comparable to (1.7), since $f_{\mathbf{Y}}(\mathbf{y})$ is not a probability, but a density. Essentially, we wish to condition on the event $\{\mathbf{Y} = \mathbf{y}\}$, but this event has probability zero. Of course, we may always define

$$P(\mathbf{X} \in E \mid \mathbf{Y} \in B_\epsilon(\mathbf{y})) = \frac{P(\mathbf{X} \in E \cap \mathbf{Y} \in B_\epsilon(\mathbf{y}))}{Y \in B_\epsilon(\mathbf{y})},$$

and accept the limit

$$P(\mathbf{X} \in E \mid \mathbf{Y} = \mathbf{y}) = \lim_{\epsilon \to 0} \frac{P(\mathbf{X} \in E \cap \mathbf{Y} \in B_\epsilon(\mathbf{y}))}{\mathbf{Y} \in B_\epsilon(\mathbf{y})}.$$

However, we will often need to define conditional distributions of the form $\mathbf{X} \mid S(\mathbf{X}) = \mathbf{s}$, where $\mathbf{X} \in \mathcal{X}$ is itself the complete observation. Suppose $\mathbf{X}$ possesses density $f(\mathbf{x})$ on $\mathcal{X}$ with respect to measure ν, and we wish to evaluate the distribution of $\mathbf{X}$ conditional on $A \subset \mathcal{X}$. Whatever measure theoretic issues arise, we assume that the conditional distribution is given by a density $f(\mathbf{x} \mid A)$ which is proportional to $f(\mathbf{x})$ on A. In particular, we may always find a normalizing constant H_A for which,

$$f(\mathbf{x} \mid A) = \frac{f(\mathbf{x})}{H_A} I\{\mathbf{x} \in A\}, \tag{1.9}$$

which may be expressed by

$$H_A = \int_{\mathbf{x} \in A} f(\mathbf{x}) d\nu_A. \tag{1.10}$$

If $P(A) > 0$, we use the original measure $\nu_A = \nu$. However, this construction may be used even when $P(A) = 0$, provided the correct measure ν_A on A is identified (this would not be ν).

We next consider conditions under which $H_A > 0$, otherwise the normalization cannot be constructed. Suppose $\bar{\mathcal{X}}$ is the support of f. If $A \subset \bar{\mathcal{X}}^c$, then clearly $H_A = 0$. Otherwise, basic measure theory may be used to verify under general conditions that if $\bar{\mathcal{X}}$ is an open set and $A \cap \bar{\mathcal{X}} \neq \emptyset$ then $H_A > 0$. If $\bar{\mathcal{X}}$ is not open, f can usually be replaced by another density with open support which possesses identical measure $P(E) = P\{\mathbf{X} \in E\}$. We then give the following definition.

Definition 1.2 (Existence of Conditional Density) For a density f on $\mathcal{X}$ with open support $\bar{\mathcal{X}} \subset \mathcal{X}$, the conditional density $f(\mathbf{x} \mid A)$ is defined *iff* $A \cap \bar{\mathcal{X}} \neq \emptyset$. ■

As discussed earlier, we may generally assume that the support of a density is an open set.

If we need to represent $P(\mathbf{X} \mid S(\mathbf{X}))$, we set $A_s = \{\mathbf{x} \in \mathcal{X} : S(\mathbf{x}) = s\}$, then let $H(s)$ denote the normalizing constant defined in (1.10), $H(s) = \int_{x \in A_s} f(\mathbf{x}) d\nu_{A_s}$. Following the previous comments, if $A_s \cap \bar{\mathcal{X}} \neq \emptyset$, we have $H(s) > 0$ and so we can write

$$f(\mathbf{x} \mid S(\mathbf{X}) = s) = \frac{f(\mathbf{x})}{H(s)} I\{S(\mathbf{x}) = s\}, \quad \text{when } A_s \cap \bar{\mathcal{X}} \neq \emptyset. \tag{1.11}$$

Furthermore, we may assume that if $\mathbf{x} \in \bar{\mathcal{X}}$ then $A_s \cap \bar{\mathcal{X}} \neq \emptyset$ for $s = S(\mathbf{x})$. This means we may substitute $S(\mathbf{x})$ for s in (1.11), so that

$$f(\mathbf{x} \mid S(\mathbf{X}) = S(\mathbf{x})) = f(\mathbf{x})/H(S(\mathbf{x})) \tag{1.12}$$

is always defined.

1.3 Some Important Theorems of Probability

A large part of probability theory relies on a relatively small number of classical theorems, which we briefly review. Boole's inequality is a direct consequence of countable additivity:

Theorem 1.2 (Boole's Inequality) For any countable collection of sets $\{E_i\}$ on a probability measure space $(\Omega, \mathcal{F}, P)$, it always holds that $P(\cup_i E_i) \leq \sum_i P(E_i)$. ■

Markov's inequality is quite simple to prove, while being indispensible to probability theory.

Theorem 1.3 (Markov's Inequality) If $X \geq 0$, $t > 0$, then $P(X \geq t) \leq E[X]/t$. ■

Proof. This follows from the inequality $tI\{x \geq t\} \leq x$ for $x \in [0, \infty)$, and the monotonicity of the expectation operator. □

Markov's inequality is widely used in the estimation of tail probabilities.

Example 1.3 (Markov's Inequality and Tail Probabilities) An exponential random variable with mean 1 has k-th order moment $\mu_k = k!$, and so using Markov's inequality gives a tail probability $P(X \geq t) \leq k!/t^k$, $t > 0, k = 1, 2, \ldots$. Since the inequality holds for all $k \geq 1$, we also have the inequality $P(X \geq t) \leq \inf_{k \geq 1} k!/t^k$, $t > 0$. See also Problem 11.2. ■

If we are given a countable sequence of events $E_1, E_2, \ldots$ on $(\Omega, \mathcal{F}, P)$ the following limits may be defined

$$\limsup_{n\to\infty} E_n = \cap_{n=1}^{\infty} \cup_{m\geq n} E_m = \lim_{n\to\infty} \cup_{m\geq n} E_m = \{E_n \text{ i.o.}\},$$

where i.o. means "infinitely often", that is, the event $\{E_n \text{ i.o.}\}$ occurs if an infinite number of the events E_n occur. A definition often accompanying this is

$$\liminf_{n\to\infty} E_n = \cup_{n=1}^{\infty} \cap_{m\geq n} E_m = \lim_{n\to\infty} \cap_{m\geq n} E_m = \{E_n \text{ a.f.}\},$$

where a.f. means "all but finitely often", that is, the event $\{E_n \text{ a.f.}\}$ occurs if for some finite n all E_m, $m \geq n$ occur. By De Morgan's law $\{E_n^c \text{ i.o.}\} = \{E_n \text{ a.f.}\}^c$ and $\{E_n^c \text{ a.f.}\} = \{E_n \text{ i.o.}\}^c$. The Borel-Cantelli lemmas apply to events of this type:

Theorem 1.4 (Borel-Cantelli Lemma I) If $\sum_{n\geq 1} P(E_n) < \infty$ then $P(E_n \text{ i.o.}) = 0$. ■

Proof. By Boole's inequality $P(\cup_{m\geq n} E_m) \leq \sum_{m\geq n} P(E_m)$. By hypothesis, this upper bound approaches 0 as $n \to \infty$ so the result holds by the continuity of P. □

Theorem 1.5 (Borel-Cantelli Lemma II) If the events E_n are independent and $\sum_{n\geq 1} P(E_n) = \infty$ then $P(E_n \text{ i.o.}) = 1$. ■

Proof. By independence we may write, for any $N \geq n$, $P(\cap_{m\geq n} E_m^c) \leq P\left(\cap_{m=n}^{N} E_m^c\right) = \prod_{m=n}^{N}(1 - P(E_m)) \leq \exp(-\sum_{m=n}^{N} P(E_m))$. By hypothesis, the upper bound approaches 0 as $N \to \infty$. Thus, $P(E_n^c \text{ a.f.}) = 0$, which concludes the proof. □

The Borel-Cantelli lemmas will be especially useful for the analysis of stochastic limits (Chapter 11), as suggested by the next example.

Example 1.4 (Application of the Borel-Cantelli Lemmas) On day n, $n = 1, 2, \ldots$, the numbers $\{1, \ldots, n\}$ are randomly ordered, and each ordering is independent of the previous orderings. The probability that 1 occurs first i.o. is 1, but the probability that 1 and 2 occur first and second i.o. is zero. The number of days on which any four consecutive integers appear consecutively anywhere in the random ordering is finite *wp*1. ■

1.3.1 Total Expectation and Total Variance Identities

Let $Y \in \mathbb{R}$ be a random variable, and $\tilde{X} \in \mathcal{X}$ be any other random observation. We will make frequent use of the total expectation identity:

$$E[Y] = E\left[E[Y \mid \tilde{X}]\right]. \tag{1.13}$$

The total variance identity also plays a crucial role in the theory of statistical inference. If Y is a random variable, and $\tilde{X}$ is any other random observation, we have

$$\text{var}[Y] = E[\text{var}[Y \mid \tilde{X}]] + \text{var}[E[Y \mid \tilde{X}]]. \tag{1.14}$$

In particular, $E[Y \mid \tilde{X}]$ has the same expected value as Y but strictly smaller variance, unless Y is a mapping of $\tilde{X}$, in which case $Y \sim E[Y \mid \tilde{X}]$. See Problem 1.7.

1.3.2 L^p Norms for Probability Measures

We may associate an L^p norm with a probability space $(\Omega, \mathcal{F}, P)$ as follows:

Definition 1.3 (L^p Norm) Let X be a random variable. The L^p norm is given by $\|X\|_p = E[|X|^p]^{1/p}$, $p \in [1, \infty)$, with the essential supremum L^∞ given by $\|X\|_\infty = \inf\{a : P\{X > a\} = 0\}$. ■

The L^p space is then the collection of random variables X for which $\|X\|_p < \infty$. It may be shown that L^p is a normed vector space (Definition C.8), where equality is defined *wp*1.

Two important theorems regarding L^p spaces follow.

Theorem 1.6 (Hölder Inequality) For any two random variables X, Y

$$\|XY\|_1 \leq \|XY\|_p \|XY\|_q$$

for any conjugate pairs $p^{-1} + q^{-1} = 1$. ■

Theorem 1.7 (Minkowski Inequality) For any two random variables X, Y

$$\|X + Y\|_p \leq \|X\|_p + \|Y\|_p$$

for $p \geq 1$. ■

That the triangle inequality of Definition C.8 is satisfied by L^p follows from the Minkowski inequality.

The Cauchy-Schwarz inequality takes various forms in various branches of mathematics. In probability theory, it states that

$$E[|XY|]^2 \leq E[X^2]E[Y^2], \tag{1.15}$$

with equality if and only if $X = Y = 0$ or $X = aY$ for some constant a. It's utility is extended by noting $E[XY]^2 \leq E[|XY|]^2$. It is actually a special case of the Hölder inequality with $p = q = 2$, but is usefully considered independently.

One consequence is that we must have

$$\begin{aligned} \text{cov}[X, Y]^2 &= E[(X - E[X])(Y - E[Y])]^2 \\ &\leq E[(X - E[X])^2]E[(Y - E[Y])^2] \\ &= \text{var}[X]\text{var}[Y], \end{aligned} \tag{1.16}$$

which is what guarantees that the correlation coefficient

$$\rho = \text{cov}[X, Y]/\sqrt{\text{var}[X]\text{var}[Y]}$$

is always in the interval $\rho \in [-1, 1]$. Then equality in (1.16), corresponds to (1.15), which implies $X = aY + b$ for some $a \neq 0$ (the covariance does not depend on b). In addition, if the means and variances of X and Y are equal it is easily verified that we must $a = 1$ and $b = 0$, that is $X = Y$.

1.3.3 Convexity and Jensen's Inequality

The concept of convexity emerges frequently in the theory of inference.

Definition 1.4 (Convexity) A subset $E \subset \mathbb{R}^m$ is a convex set if for any $x, x' \in E$ and $p \in (0,1)$, $px + (1-p)x' \in E$. That is, a convex set is closed under convex combinations.

If E is a convex set and $f : E \to \mathbb{R}$, then f is a convex function if and only if for all $x, x' \in E$ and $p \in (0,1)$ we have

$$f(px + (1-p)x') \leq pf(x) + (1-p)f(x'). \tag{1.17}$$

If in addition

$$f(px + (1-p)x') < pf(x) + (1-p)f(x') \tag{1.18}$$

whenever $x \neq x'$, then f is a strictly convex function. In addition f is (strictly) concave if -f is (strictly) convex. We assume that the domain of a convex function is a convex set. ■

For functions defined on $\mathbb{R}$, the following theorem is easily verified:

Theorem 1.8 (Convexity and Derivatives) Suppose f is a function on interval $E = (a,b)$ (possibly, $a, b = \mp\infty$) and possesses first derivative f' on E. Then f is convex *iff* $f'(x)$ is nondecreasing on E. In addition f is strictly convex *iff* $f'(x)$ is strictly increasing on E.

If f possesses second derivative f'' on E then f is convex *iff* $f''(x)$ is nonnegative on E. In addition f is stictly convex *iff* $f'(x)$ is positive on E. ■

The following theorem is essential to many arguments in probability theory.

Theorem 1.9 (Jensen's Inequality) Suppose f is a convex function on $E = (a,b)$ (possibly, $a, b = \mp\infty$). Let X be any random variable with $P(X \in E) = 1$. Then

$$E[f(X)] \geq f(E[X]).$$

In addition, if f is strictly convex then $E[f(X)] = f(E[X])$ *iff* $X = E[X]$ $wp1$. ■

Proof. For any $t \in E$ we may define a linear function $L(x) = a + bx$ for which $f(x) \geq L(x)$ for all $x \in E$, and for which $f(t) = L(t)$. Construct such a function $L(x)$ with $t = E[X]$. Then $E[f(X)] \geq E[L(X)] = a + bE[X] = L(E[X]) = f(E[X])$, which completes the first part of the proof.

Then set $h(x) = f(x) - L(x)$ for $x \in E$. If f is strictly convex, $h(E[X]) = 0$, and $h(x) > 0$ for any $x \neq E[X]$. Then $E[h(X)] = E[h(X)I\{X \neq E[X]\}]$. Therefore, $E[h(X)] > 0$ unless $P(X \neq E[X]) = 0$. □

A technical issue which often arises in the theory of inference is the existence of higher-order moments. Consider the following application of Jensen's inequality.

Theorem 1.10 (Jensen's Inequality and L^p Norms) Suppose $r > s > 0$. Then for any random variable X, $\|X\|_r \geq \|X\|_s$. ■

Proof. Apply Theorem 1.9 to the convex function $f(x) = |X|^{r/s}$. □

Thus, if the kth order moment does not exist, neither does any higher-order moment. Conversely, if the kth order moment does exist, so do all lower-order moments. See also Section C.7 for a discussion on role played by convexity for optimization problems.

1.4 Commonly Used Distributions

We can take a parametric class of densities on sample space $\mathcal{X}$ to be a family of density functions $\{f_\theta : \theta \in \Theta\}$ indexed by a finite dimensional parameter $\theta \in \Theta \subset \mathbb{R}^m$. We usually expect $f_\theta(x)$ to be expressible in some closed form. We reserve the term "parametric family" to define an actual inference problem (Section 3.2) and so is not what we refer to here as a parametric class. A probability density for a random variable $X \in \mathbb{R}$ can be any non-negative function $f_X(x)$ which integrates to one over $\mathbb{R}$. Yet much of statistical methodology seems to be concerned with a relatively small number of commonly used parametric classes. It is worth considering why this might be the case.

In this section, we will give brief descriptions of most of the parametric classes used in this volume. This includes a motivation for its use in real world applications, and also serves to introduce the notational convention. The actual form of each density is given in Appendix A, which also includes (when a closed form is known) the mean and variance, the moment generating function (Section 1.9), and moment formulas.

Thoughout the volume we will encounter the *iid* sequence $X_1, X_2, \ldots$ of random variables possessing identical distributions, and which are independent in the sense that Equation (1.5) holds for all n (the sequence may be finite or infinite). If X is a random variable with density f and CDF F, this may be denoted as $X \sim f$ or $X \sim F$, or by any of the variations of this notation introduced below.

Example 1.5 (The Central Limit Theorem and the Normal Distribution) The central limit theorem gives one reason for the ubiquitous appearance of the normal distribution in statistical methodology. Roughly, it states that sums of random quantities tend towards a normal distribution. Naturally occuring observations are typically perturbed by many stochastic inputs, and when these combine additively, we can often expect such observations to be approximately normally distributed. For a single random variable $X \in \mathbb{R}$, the normal distribution is defined by parameter $\theta = (\mu, \sigma^2)$, where $\mu = E[X]$ and $\sigma^2 = \text{var}[X]$, in which case we write $X \sim N(\mu, \sigma^2)$. By convention, the standard or unit normal distribution is $N(0, 1)$. The multivariate extension of the normal distribution is given in Example 2.3. ■

Observations used in statistical analysis which are not continuous are often integers representing either counts or frequencies. In the latter case, it is assumed that a population can be partitioned into categories, the frequencies then refering to the number of each category represented in a sample (Agresti, 2018). The next three examples describe distributions used for either type of inference problem.

Example 1.6 (Derivatives of Bernouli Processes) A Bernoulli random variable U attains value 0 or 1. It is defined entirely by the single parameter $P(U = 1) = p$, $p \in [0, 1]$, and is denoted as $U \sim bern(p)$. Suppose we have a Bernoulli process, which is an unbounded *iid* sequence $U_1, U_2, \ldots$, where $U_1 \sim bern(p)$. An observation $U_i = 1$ is conventionaly referred to as a "success" (as opposed to a "failure"). A number of important parametric classes are derived from this process. The binomial random variable is the number of successes in the first n observations, and is denoted as $X \sim bin(n, p)$. The multinomial distribution is a multivariate extension of the binomial (Example 2.1). The geometric random variable is variously defined as the number of observations (this volume), or number of failures, required to observe a success, and is denoted as $X \sim geom(p)$. Both definitions are used in the literature, so it should usually be made clear which convention is being used. Of course, two random variables defined by the two conventions always differ by one. The negative binomial random variable is the sum of $r \in \mathbb{I}_{>0}$ *iid* random variables $X_i \sim geom(p)$. It is therefore interpretable as the number of observations (or failures) required to observe r successes, and is denoted as $X \sim nb(r, p)$. Again, multiple definitions are used in the literature. Note that $nb(1, p)$ is equivalent to $geom(p)$.

For $X \sim bin(n, p)$, it is reasonable to allow $p = 0$ or $p = 1$, since the distribution of X is still well defined in either case. On the other hand, if $X \sim geom(p)$, the distribution of X is well defined for $p = 1$ but not $p = 0$. This issue will be discussed in Section 1.4.1. ■

Example 1.7 (The Hypergeometric Distribution) A bin contains m white and n black balls. A random selection of $k \leq m+n$ balls is made (this is referred to as sampling without replacement). Let X be the number of white balls among the k selected. This is known as a hypergeometric random variable, which we denote $X \sim hyper(k, m, n)$. We may temporarily label the balls, so that they are all distinct. The density is given by

$$p_X(i) = P(X = i) = \frac{N}{D} = \frac{\binom{m}{i}\binom{n}{k-i}}{\binom{m+n}{k}}, \tag{1.19}$$

with support

$$\mathcal{S}_X = \{i : \max(0, k-n) \leq i \leq \min(k, m)\}. \tag{1.20}$$

The hypergeometric and binomial distributions are similar in that they are both sums of Bernoulli random variables, and can both be used to model a random sample of size n from a population of two types. When the samples are replaced (or when the population is infinite), the sum has a binomial distribution. When the samples are not replaced (and when the population is finite), the sum has a hypergeometric distribution. See Problem 1.17 for more on this distinction.

The multivariate hypergeometric distribution is a multivariate extension of the hypergeometric distribution (Example 2.2). In this volume these distributions arise in conditional methods for the analysis of categorical data (for example, Sections 5.9 or 10.7). ■

Some distributions are used to model specific real world processes, so that their densities can be derived from some first principles argument. The following is an important example.

Example 1.8 (Pareto Distribution) The Pareto distribution is used in economics to model the distribution of wealth within a population, and belongs to a group of parametric classes which model resource allocation or size distributions. The Pareto principle (Vilfredo Pareto 1848–1923) states that the number of individuals with an income of at least x obeys the power law $N_x = Cx^{-\alpha}$ for constants $C > 0$, $\alpha > 0$. If we assume the existence of a minimum income $x_{min} > 0$, then $N_{x_{min}}$ is the total population size. The population proportion with incomes of at least x is therefore $N_x/N_{x_{min}} = (x/x_{min})^{-\alpha}$. Thus, if X is the income of a randomly selected individual it possesses CDF $F_X(x) = 1 - (x/\tau)^{-\alpha}$, which is denoted as $X \sim pareto(\tau, \alpha)$, setting $\tau = x_{min}$. The Pareto distribution is associated with the "80-20" law, which states that 80% of total wealth is owned by 20% of the population, but this holds only for a unique value of α (Problem 1.16). A first principles derivation is given in Wold and Whittle (1957), while overviews of the Pareto and other allocation distributions are given in Kleiber and Kotz (2003) or Newman (2005). ■

A distribution of a random time to event T may be defined by the memoryless property. These may be continuous or discrete.

Example 1.9 (The Exponential and Geometric Distribution) A positive random variable X possesses the memoryless property if we have

$$P(X > s + t \mid X > s) = P(X > t) \tag{1.21}$$

for all $s, t > 0$. Since $P(T > s + t \mid T > s) = P(T > s + t)/P(T > s)$, we can recognize the memorlyess property as a form of Cauchy's functional equation $\phi(s + t) = \phi(s)\phi(t)$. The only solution to this equation which models a probability distribution on $\mathbb{R}_{>0}$ is given by the CDF $F_T(x) = 1 - e^{-x/\tau}$, which defines the exponential density, denoted as $X \sim exp(\tau)$. Here $E[X] = \tau$.

It can also be shown that $X \sim geom(p)$ is the only random variable defined on $\mathbb{I}_{>0}$ which satisfies Equation (1.21) (Problem 1.15).

The sum X of α *iid* exponentially distributed random variables of mean τ possesses a gamma distribution, denoted as $X \sim gamma(\tau, \alpha)$. However, the gamma parametric class is also well defined when $\alpha > 0$ is any positive real number. It is related to the gamma function (Section C.5).

A random variable W has a Weibull distribution if there are two parameters $\alpha > 0$ and $\tau > 0$, such that

$$X = (W/\tau)^{\alpha} \tag{1.22}$$

has an exponential distribution with mean 1. This will be denoted as $W \sim weibull(\tau, \alpha)$.

The Rayleigh distribution is sometimes used in scientific applications, and can be derived as the radius $R = \sqrt{X_1^2 + X_2^2}$ of a random vector (X_1, X_2), where $X_i \sim N(0, \tau^2)$ are independent observations. This will be denoted as $R \sim rayleigh(\tau)$. It can be shown that $R \sim weibull(\tau, 2)$.

A good survey of parametric classes used to model survival times (positive random variables) can be found in Richards (2012). ■

While the normal distribution models the aggregation of additive noise, the aggregation of random counts is usually modeled as a Poisson process. The memoryless property plays an important role here.

Example 1.10 (The Poisson Process as an Aggregation of Arrival Processes) The Poisson distribution is often used to model random counts. Here $\mathcal{X} = \mathbb{I}_{\geq 0}$. For a single random variable $X \in \mathcal{X}$ the Poisson distribution is defined by a single parameter $\lambda = E[X]$, and is denoted as $X \sim pois(\lambda)$. Necessarily, $\lambda \geq 0$.

We can define an arrival process to be a nondecreasing right continuous integer-valued stochastic process $N(t)$, $t \in [0, \infty)$, for which $N(0) = 0$. Such a process models the time of occurrence of instantaenous arrivals in real time, so that $N(T)$ is the number of arrivals in the time period $[0, T]$. If the times between arrivals are *iid* then $N(t)$ is referred to as a renewal process. It can be shown (e.g. Section XI.3 of Feller (1971)) that superpositions of independent renewal processes tend towards a Poisson process, in which case $N(T) \sim pois(\lambda T)$, where λ is the arrival rate. Thus, when real world arrival processes are aggregations of many unrelated arrival processes, it is reasonable to expect that observed counts are approximately Poisson distributed.

It is important to note that the Poisson process $N(t)$ of rate λ can also be defined as a renewal process for which the interarrival times are $X_i \sim exp(1/\lambda)$. This means that the distribution of $N(t+s)$, $s > 0$, conditional on $N(t) = n$, does not depend on process history $N(t')$, $t' < t$, that is, a Poisson process is memoryless (Example 1.9). ■

Example 1.11 (Distributions Derived from the Normal Distribution) A number of important distributions can be derived from the normal distribution, and this relationship is important to understand. Let $X_1, \ldots, X_\nu$ be an *iid* sample from standard normal distribution $N(0, 1)$. If we let

$$W = \sum_{i=1}^{\nu} X_i^2 \tag{1.23}$$

then we say W possesses a χ^2 distribution with ν degrees of freedom, which we denote $W \sim \chi^2_\nu$. It also holds that $W \sim gamma(2, \nu/2)$.

Next, suppose $W_1 \sim \chi^2_{\nu_1}$ and $W_2 \sim \chi^2_{\nu_2}$, and that W_1 and W_2 are independent. If we set

$$F = \frac{W_1/\nu_1}{W_2/\nu_2}, \tag{1.24}$$

then we say F possesses an F distribution with ν_1 numerator degrees of freedom and ν_2 denominator degrees of freedom, which we denote $F \sim F_{\nu_1, \nu_2}$

The T distribution with ν degrees of freedom can be defined as

$$T = \frac{Z}{\sqrt{W/\nu}}, \tag{1.25}$$

where $Z \sim N(0,1)$, $W \sim \chi^2_\nu$ and Z,W are independent. This will be denoted as $T \sim T_\nu$.

It will sometimes be necessary to consider how these distributions are perturbed when $X_1, \ldots, X_\nu$ are independent with $X_i \sim N(\mu_i, 1)$, which defines noncentral variants of these distributions. This topic will be defered to Section 2.5. ∎

The beta and inverse gamma density (our next two examples) play an important role in Bayesian inference.

Example 1.12 (Beta Density) The beta density has a number of important applications in the theory of statistical inference. It defines a random outcome $X \in (0,1)$, and is sometimes used to model random probabilities p. It is denoted as $X \sim beta(\alpha, \beta)$, depending on two parameters $\alpha, \beta > 0$. It is related to the beta function (Section C.5). The Dirichlet distribution is a multivariate extension of the beta distribution (Example 2.4). ∎

Example 1.13 (Inverse Gamma Density) If $Y \sim gamma(\tau, \alpha)$ then $X = 1/Y$ possesses an inverse gamma distribution, denoted as $X \sim igamma(\xi, \alpha)$, where ξ is a scale parameter and α is a shape parameter. If X is constructed this way we have $\xi = 1/\tau$, and the shape parameter α is the same for X and Y. See Problem 1.29. ∎

Example 1.14 (Uniform Density) If $X \in \mathcal{X}$ is an observation on any bounded sample space $\mathcal{X} \subset \mathbb{R}^m$, then X possesses a uniform distribution on $\mathcal{X}$ if the density function is constant on $\mathcal{X}$. This is often what is meant by the phrase "at random". A dart lands on a dartboard "at random" if it is equally likely to land at any position. We will usually assume X is uniformly distributed on an interval (a,b), $a < b$, which we denote $X \sim unif(a,b)$. ∎

The final two examples introduce the location-scale parametric class, which will be discussed in more detail in Section 3.3.

Example 1.15 (Location-Scale Parametric Classes) Suppose $f(x)$ is a density function on $\mathbb{R}$. For any $\mu \in (-\infty, \infty)$ and $\tau \in (0, \infty)$ it is easily shown that $f_{(\mu,\tau)}(x) = \tau^{-1} f((x-\mu)/\tau)$ is also a density, and we have generated a location-scale parameter class. Similarly, if we fix $\tau = 1$, then we have location parameter class $f_{\mu,1}(x) = f(x-\mu)$, and if $\mu = 0$, then we have scale parameter class $f_{(0,\tau)}(x) = \tau^{-1} f(x/\tau)$.

Then suppose $X \sim f$. It may be verified that $Y = \tau X + \mu \sim f_{(\mu,\tau)}$, so that the location-scale class is suitable for observations which have a specific density form f, but are subject to an unknown affine transformation. In principle, any density can be used to induce a location-scale class, but interest is usually reserved for cases in which the transformation is natural to the problem.

First, note that the normal density $N(\mu, \sigma^2)$ is a location-scale class, with location parameter μ and scale parameter σ. In this volume, we will also encounter location-scale classes of the form $f_\theta(x) = \tau^{-1} f((x-\mu)/\tau)$, $\theta = (\mu, \tau)$, $x \in \mathbb{R}$, based on the Cauchy density $f(x) = \left[\pi(1+x^2)\right]^{-1}$, the double-exponential density $f(x) = \exp(-|x|)/2$ and the logistic density $f(x) = e^{-x}/(1+e^{-x})^2$.

These distributions will be denoted as $X \sim cauchy(\mu, \tau)$, $X \sim DE(\mu, \tau)$ and $X \sim logistic(\mu, \tau)$, respectively. If we wish to denote a location parameter class we will write, for example, $X \sim cauchy(\mu, 1)$, and the scale parameter class will be denoted as $X \sim cauchy(0, \tau)$. ∎

Example 1.16 (Exponential Density with Location Parameter) Suppose we have $X \sim exp(\tau)$. We may introduce a location parameter μ by the transformation $X + \mu$. We now have a location-scale parametric class defined by the density

$$f_{(\mu,\tau)}(x) = \tau^{-1}e^{-(x-\mu)/\tau}I\{x > \mu\}.$$

We may refer to this as the offset or shifted exponential density, which will be denoted as $X \sim exp(\mu, \tau)$. ∎

1.4.1 Natural Parameter Spaces

A parametric class $\{f_\theta : \theta \in \Theta\}$ is defined by both a functional form f_θ and parameter space Θ. However, conventions differ regarding the definition of Θ. In Example 1.10, for the Poisson parametric class $pois(\lambda)$ we assumed $\lambda \geq 0$, whereas some references will adopt the convention that $\lambda > 0$. If $X \sim pois(0)$, we simply have $P(X = 0) = 1$.

In settling on a single convention for this volume, we will follow two principles. A parametric class $\{f_\theta : \theta \in \Theta\}$ is defined by densities rather than distribution functions. We therefore expect all densities to be absolutely continuous with respect to a single measure. Thus, the $pois(\lambda)$ distribution is absolutely continuous with respect to counting measure on $\mathbb{I}_{\geq 0}$, including the case $\lambda = 0$. On the other hand, the expression for the Poisson density function is not a true density for $\lambda < 0$ (for example, it can be negative). So the parameter space for the Poisson parametric class is $\lambda \in \Theta = \mathbb{R}_{\geq 0}$.

Next, consider the normal distribution class $N(\mu, \sigma^2)$. For any $(\mu, \sigma^2) \in \mathbb{R} \times \mathbb{R}_{>0}$ the density is well defined, and absolutely continuous with respect to Lebesgue measure. Similar to the Poisson class, it might be reasonable to interpret the case $\sigma^2 = 0$ as a point mass at μ, that is $P(X = \mu) = 1$. However, this distribution will not be absolutely continuous with respect to Lebesgue measure. So we take the parameter space of the normal parametric class to be $\Theta = \mathbb{R} \times \mathbb{R}_{>0}$. We then define a natural parameter space.

Definition 1.5 (Natural Parameter Space of a Parametric Class) Given a parametric class $\{f_\theta : \theta \in \Theta\}$, Θ will be the natural parameter space of a parametric class if it is the largest parameter space for which each f_θ defines a valid probability distribution which is absolutely continuous *wrt* a common measure ν. ∎

Density parameters are often given broad classifications. The location and scale parameters were introduced in Example 1.15. We expect the location parameter to be in the same unit as the random variable. The reciprocal of a scale parameter is a rate parameter, and so they are interchangeable. If a random variable represents a random arrival time, a scale parameter might be the expected arrival time (or be related to it), while the rate parameter would represent the arrival rate (or be related to it).

A shape parameter changes the shape of a density, beyond the translation induced by a location parameter or the change in scale induced by a scale or rate parameter. The parameter α of the gamma density is an example, whereas the beta density has two shape parameters α, β. The symbol ν used as a parameter will usually denote a degrees of freedom (the χ^2, F and T distributions of Example 1.11).

As a default, we will reserve the symbols μ, τ for the location and scale parameter. Then $\lambda = 1/\tau$ will usually denote be a rate parameter. Thus, for survival times, τ will be in time units, while λ will be a rate per time unit. In deference to convention, for $N(\mu, \sigma^2)$ the symbol σ will often be used as the scale parameter. Similarly, for the Poisson distribution λ will be a mean, and not a rate, although this quantity is related to the rate of a Poisson process (Example 1.10). Finally, for the inverse gamma density, the symbol ξ may be used to denote the scale parameter. Recall that if $Y \sim gamma(\tau, \alpha)$, then $X = 1/Y \sim igamma(\xi, \alpha)$, where $\xi = 1/\tau$. Thus, ξ is a scale parameter for $1/Y$ and a rate parameter for Y (Example 1.13).

Example 1.17 (Exponential Scale Parameter) If $X \sim exp(\tau)$, then $E[X] = \tau$. However, τ is a scale parameter. The symbol μ is commonly used to denote both a mean and a location parameter. However, the two conventions conflict in this case, so our preference will be to select one of the two conventions, and use μ to denote a location parameter. Of course, in a nonparametric context we will frequently write something like $E[X] = \mu$, when no single parametric class is specified. ■

1.5 Stochastic Order Relations

The notion of stochastic order relations is behind a great deal of statistical inference, although this is not always recognized explicitly. In particular, we will see that these ideas play a crucial role in the construction of confidence intervals (Section 4.6) and hypothesis tests (Chapters 5 and 10). Suppose X, Y are two random variables for which $E[X] < E[Y]$. We might say the Y is larger in some sense than X, but we do not expect $P(X < Y) = 1$. Thus, stochastic order is concerned with the comparison of the distributions of X and Y, and not their outcomes. We start with the following definition.

Definition 1.6 (Stochastic and Likelihood Ratio Ordering) Suppose we are given random variables $X, Y \in \mathbb{R}$. Assume $P\{X \in (-\infty, \infty)\} = P\{Y \in (-\infty, \infty)\} = 1$.

(i) Likelihood ratio ordering: $Y \geq_{lr} X$ if

$$P(c < X < d)P(a < Y < b) \leq P(a < X < b)P(c < Y < d) \tag{1.26}$$

whenever $a \leq c$ and $b \leq d$.

(ii) Stochastic ordering: $Y \geq_{st} X$ if

$$P(t < X) \leq P(t < Y) \tag{1.27}$$

for all $t \in \mathbb{R}$. We say Y is stochastically larger than X. ■

Definition 1.6 offers only two forms of stochastic order relations, which will suit our purposes. Others exist (Problem 1.43) and these form a hierarchy of implication (Ross, 1996). The following theorem establishes such a hierararchy between likelihood ratio and stochastic ordering.

Theorem 1.11 (Likelihood Ratio Ordering Implies Stochastic Ordering) Suppose we are given random variables $X, Y \in \mathbb{R}$. Then $Y \geq_{lr} X \implies Y \geq_{st} X$. ■

Proof. The theorem folllows by allowing $b, d \to \infty$ and $a \to -\infty$ in Equation (1.26). □

We will make use of both stochastic and likelihood ratio ordering in this volume. By Theorem 1.11 likelihood ordering implies stochastic ordering, and inference methods tend to have more definitive properties when it holds (see, for example, the monotone likelihood ratio property discussed in Section 10.4). However, stochastic ordering does not imply likelihood ordering, as is shown in the next example. Therefore, if that condition suffices for a particular application, it should be used in place of likelihood ordering.

Example 1.18 (Counterexample for Likelihood Ratio Ordering) It is not difficult to construct random variables X, Y for which $Y \geq_{st} X$ holds but $Y \geq_{lr} X$ does not. Suppose X is uniformly distributed on $\mathcal{X}_X = (0, 1)$ and Y is uniformly distributed on $\mathcal{X}_Y = (0, 1/2) \cup (1, 5/2)$. Then Equation (1.27) will hold for all $t \in \mathbb{R}$, but Equation (1.26) does not hold for $a = 0$, $b = c = 1/2$, $d = 1$. ∎

We will make use of the following theorem, the proof of which is left to the reader (Problem 1.25).

Theorem 1.12 (Alternative Characterization of Stochastic Ordering) Let X, Y be two random variables. Then $Y \geq_{st} X$ *iff* $E[h(Y)] \geq E[h(X)]$ for all nondecreasing functions h. ∎

In Definition 1.6 stochastic ordering is characterized entirely by the probability measure P. In statistical inference interest is usually in probability densities, so we will develop suitable alternative characterizations of likelihood ratio ordering, based on the following definition.

Definition 1.7 (Ratio Ordering of Functions) Suppose two real-valued functions f, g are defined on $\mathcal{X} \subset \mathbb{R}$. Then $g \geq_r f$ if for any $s, t \in \mathcal{X}$, $s \leq t \Rightarrow f(t)g(s) \leq f(s)g(t)$. ∎

In Definition 1.7 we do not assume any form of the subset $\mathcal{X}$. In particular, it need not be an interval, and may be countable. The stronger likelihood ratio ordering is equivalent to $f_Y \geq_r f_X$. We state the theorem without proof.

Theorem 1.13 (Alternative Characterization of Likelihood Ratio Ordering) Suppose we are given random variables $X, Y \in \mathbb{R}$. Let f_X, f_Y be the densities, and let $\mathcal{X}_{XY} = \{t \in \mathbb{R} : f_X(t) + f_Y(t) > 0\}$, be the union of the support of X and Y. Then $f_Y \geq_r f_X$ on $\mathcal{X}_{XY}$ *iff* $Y \geq_{lr} X$. ∎

Clearly, the ratio ordering $g \geq_r f$ of Definition 1.7 holds *iff* $g(t)/f(t)$ is a nondecreasing function of t. A typical application is given in the next example.

Example 1.19 (Likelihood Ratio Ordering of the Exponential Density) Suppose $X \sim exp(\tau_x)$, $Y \sim exp(\tau_y)$, and $\tau_y > \tau_x$. The ratio of the densities is

$$\frac{f_Y(t)}{f_X(t)} = (\tau_x/\tau_y)e^{-(\tau_y^{-1} - \tau_x^{-1})t},$$

which is an increasing function of t over the support of the distributions. Therefore, by Theorem 1.13 we may conclude $Y \geq_{lr} X$. ∎

1.6 Quantiles

Suppose a random variable X has CDF $F(x)$. If F is continuous and increasing, the median m is the unique solution to $1/2 = F(m)$. However, this condition need not hold even when X is itself continuous. In fact, this equation may have multiple, or no, solutions (as can be seen in Figure 1.1). We therefore need a formal definition for the median of X, or more generally, the p-quantile.

Definition 1.8 (Quantile) For any $p \in (0, 1)$, let M_p be the set of all numbers $m \in \mathbb{R}$ for which

$$P(X \leq m) \geq p \text{ and } P(X < m) \leq p \tag{1.28}$$

equivalently, the set of all numbers m for which

$$P(X > m) \leq 1 - p \text{ and } P(X < m) \leq p. \tag{1.29}$$

Then any $m \in M_p$ is a p-quantile of X (alternatively, the pth quantile or $100 \times p$th percentile). The median is the 0.5-quantile. ∎

Remark 1.1 Definition 1.8 can be, in a sense, inverted, and is equivalent to the following. The number m is a p-quantile *iff* $p \in [P(X < m), P(X \leq m)]$. The interval is always nonempty. This formulation will be more convenient from time to time. ∎

It would be possible to extend Definition 1.8 to $p = 0$ and $p = 1$. If X had support $\mathbb{R}$, then condition (1.28) would lead to $M_0 = \{-\infty\}$ and $M_1 = \{\infty\}$. But these values are not in the support of X, and so they should be interpreted with some caution. By assuming $p \in (0, 1)$, the properties of M_p can be more precisely given.

Theorem 1.14 (Properties of Sets of Quantiles) For $p \in (0, 1)$, let M_p be the set given in Definition 1.8. Then:

(i) M_p is the nonempty bounded closed interval $[m_L, m_U]$ where $m_L = \inf_m M_p^+$ and $m_U = \sup_m M_p^-$ for sets $M_p^- = \{m : P(X < m) \leq p\}$ and $M_p^+ = \{m : P(X \leq m) \geq p\}$.

(ii) If $m_U > m_L$, then $P(X \leq m)$ is constant in the interval (m_L, m_U). ∎

Proof. (i) We start with the equation $M_p = M_p^- \cap M_p^+$. The set M_p^+ cannot be empty, since $P(X \leq m)$ approaches 1 from below as $m \to \infty$, and it must be bounded from below since $P(X \leq m)$ approaches 0 from above as $m \to -\infty$. Furthermore, we must have $M_p^+ = [m_L, \infty)$, $|m_L| < \infty$, since $P(X \leq m)$ is right-continuous and nondecreasing *wrt* m. It follows from a similar argument that $M_p^- = (-\infty, m_U]$ where $|m_U| < \infty$.

To complete the proof, we need to verify that $M_p \neq \emptyset$. By construction, for any $m' < m_L$ we must have $P(X < m') \leq P(X \leq m') < p$, hence $m' \in M_p^-$. The proof is completed after noting that M_p^- is closed, therefore $m_L \in M_p^-$.

(ii) Let M_p^* be the set of solutions to $p = P(X \leq m)$. Suppose $M_p^* \neq \emptyset$. Then M_p is the closure of M_p^*. Otherwise, $m = \inf\{m' : P(X \leq m') \geq p\}$ is the unique quantile. □

Remark 1.2 Suppose a quantile is not unique. Then by Theorem 1.14, $P(X \leq m)$ is constant in the interior of $M_p = [m_L, m_U]$. More precisely, by right continuity, $P(X \leq m) = p$ on $[m_L, m_U)$, and on $[m_L, m_U]$ if $P(X = m_U) = 0$. Similarly, $P(X < m) = p$ on $(m_L, m_U]$, and on $[m_L, m_U]$ if $P(X = m_L) = 0$. In this case, $P(m_L < X < m_U) = 0$, that is, the support of X does not include (m_L, m_U), and we may assume that its density is zero on that interval. ∎

Example 1.20 (Quantiles of the Bernoulli Distribution) Suppose $P(X = 0) = 1 - P(X = 1) = p$. Then

$$P(X \leq x) = \begin{cases} 0 & ; \quad x < 0 \\ p & ; \quad x \in [0, 1) \\ 1 & ; \quad x \geq 1 \end{cases}$$

and

$$P(X < x) = \begin{cases} 0 & ; \quad x \leq 0 \\ p & ; \quad x \in (0, 1] \\ 1 & ; \quad x > 1 \end{cases}.$$

This means $M_p = [0, 1]$ for any $p \in (0, 1)$. ∎

There will be good reason to resolve the quantile to a unique canonical choice (for example, Sections 5.6 or 12.6). This is usually done using the quantile function, which is defined next.

Definition 1.9 (Quantile Function) The quantile function of X is

$$Q(p; F) = \inf\{x \in \mathbb{R} : F(x) \geq p\}, \tag{1.30}$$

where F is the CDF of X. If F is continuous and strictly increasing on support $\mathcal{S}_X$, then it possesses an inverse F^{-1} which maps $[0, 1]$ to $\mathcal{S}_X$, and we have

$$Q(p; F) = F^{-1}(p). \tag{1.31}$$

In this case the p-quantile is unique. When the context is clear, we may write $Q(p) = Q(p; F)$.

For a random variable X with CDF F_X, the α critical value is defined as $X_\alpha = Q(1 - \alpha; F_X)$. ∎

Example 1.21 (Quantile Function of a Binomial Distribution) Suppose $X \sim bin(3, 1/2)$. Then the CDF of X evaluates to $F(x) = 1/8, 1/2, 7/8, 1$ for $x = 0, 1, 2, 3$. To characterize the median, set $p = 1/2$ in Theorem 1.14. Then $M_p^+ = [1, \infty)$ and $M_p^- = (\infty, 2]$, so the median is not unique, and its definition is satisfied by any number $q \in [m_L, m_U] = [1, 2]$. Note that the definition of the quantile function is equivalent to the definition of m_L in Theorem 1.14, so $Q(1/2) = 1$. The value of the quantile function is largely to resolve the choice of quantile, and is in this sense arbitrary.

Then suppose $X \sim bin(2, 1/2)$. Then the CDF of X evaluates to $F(x) = 1/4, 3/4, 1$ for $x = 0, 1, 2$. Then $M_p^+ = [1, \infty)$ and $M_p^- = (\infty, 1]$, so the median is uniquely 1, and of course we also have $Q(1/2) = 1$. ∎

It is certainly more convenient when the p-quantile is unique and equal to the quantile function $Q(p)$. However, this convenient relationship will not hold under two conditions. First, suppose $P(X = q') = p' > 0$. In this case, q' will be a p-quantile for multiple values of p. Second, suppose $P(X \in (a, b)) = 0$ for some nonempty open interval (a, b). In this case, the value $p' = F_X(q')$ is constant for all $q' \in (a, b)$, and all q' in this interval satisfy the definition of a p'-quantile. These cases are illustrated in the next example.

Example 1.22 (Quantiles and CDFs) Consider the CDF illustrated in Figure 1.1. First, note that $P(X = q_3) = p_3 - p_2 > 0$. We also have $P(X < q_3) = p_2$ and $P(X > q_3) = 1 - p_3$. It is easily verified that the inequalities of Definition 1.8 are satisfied for $q = q_3$ by any $p \in [p_2, p_3]$, that is, q_3 is a p-quantile for any $p \in [p_2, p_3]$.

Next, we note that $P(X \in (q_1, q_2)) = 0$. It may also be verified that the inequalities of Definition 1.8 are satisfied for $p = p_1$ and any $q \in [q_1, q_2]$. This means any $q \in [q_1, q_2]$ is a p_1-quantile. ∎

1.7 Inversion of the CDF

There is a technical result related to quantiles which proves to be of some significance to statistical inference. Suppose a random variable X has CDF $F_X(x)$. Then the transformed random variable $U = F_X(X)$ has a distribution which is invariant to F_X, in the sense that $U \sim unif(0, 1)$ when F_X is a bijective mapping from $\mathcal{X}$ to $(0, 1)$, where $P(X \in \mathcal{X}) = 1$,

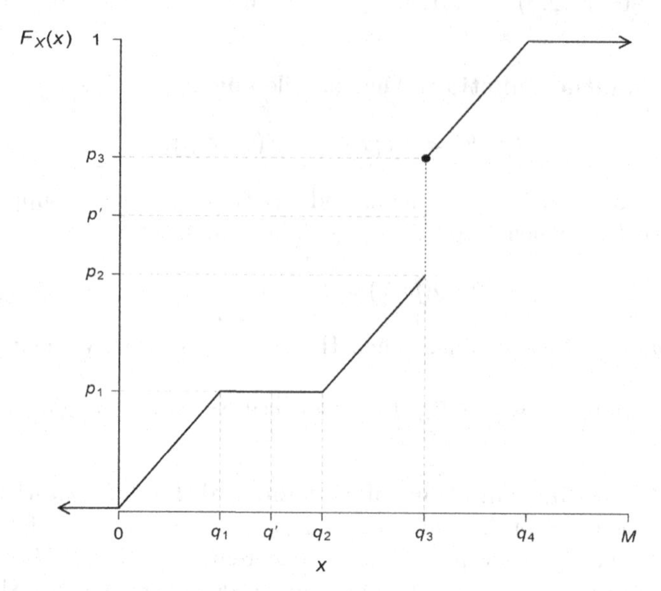

FIGURE 1.1
CDF for Example 1.22. Note that the function $F_X(x)$ extends to $\pm\infty$.

and is "uniform-like" otherwise. Here, the distinction is essentially between continuous and discrete random variables.

One of the consequences of this is that the P-value of a hypothesis test is, at least approximately, uniformly distributed on $(0, 1)$ when the null hypothesis is true, and so admits a universal interpretation. To be sure, opinions regarding the desirability of this vary. This author sees this as a good thing.

Theorem 1.15 (Inversion of the CDF (Continuous Case)) Suppose a random variable X has CDF $F_X(x)$, and that there exists a set $\mathcal{X}$ for which $P(X \in \mathcal{X}) = 1$, and for which $F_X : \mathcal{X} \to (0, 1)$ is a bijective mapping. Then $F_X(X)$ is uniformly distributed on $(0, 1)$. ■

Proof. Under the stated assumptions, $P(F_X(X) = 1) = P(F_X(X) = 0) = 0$. Then for $t \in (0, 1)$, $P(F_X(X) \leq t) = P(\{F_X(X) \leq t\} \cap \{X \in \mathcal{X}\}) = P(\{X \leq F_X^{-1}(t)\} \cap \{X \in \mathcal{X}\}) = F_X(F_X^{-1}(t)) = t$, where F_X^{-1} is taken to be the inverse mapping of $F_X : \mathcal{X} \to (0, 1)$. It follows that $F_X(X)$ is uniformly distributed on $(0, 1)$. □

A commonly stated assumption for this result is that F_X is strictly increasing between two values x_0, x_1 satisfying $F_X(x_0) = 0$, $F_X(x_1) = 1$, possibly $\pm\infty$. But this rules out examples with "probabilistic gaps". For example, suppose X is a continuous random variable, with $P(X \in (a, b)) > 0$ for any $a < b$, unless $c < a < b < d$ for two finite constants c, d (that is, $P(X \in (c, d)) = 0$). Then $F_X(x)$ is strictly increasing at all $x \notin [c, d]$ but is constant within $x \in (c, d)$, and so is not invertible on $\mathbb{R}$. However, the conditions of Theorem 1.15 will still be satisfied, so that $F_X(X)$ is uniformly distributed on $(0, 1)$.

Under Theorem 1.15 for continuous random variables, the transformation $U = F_X(X)$ satisfies $F_U(t) = t$, $t \in (0, 1)$. On the other hand, we have described $U = F_X(X)$ as "uniform-like" when X is discrete. The distribution of $F_X(X)$ is still invariant in that the deviation from the uniform distribution does not depend on F_X in any systematic way.

It will be convenient to define various modifications of the CDF. The *rotational image* of a CDF F may be taken as $F^*(x) = P(X \geq x)$. Then $F^-(x) = P(X < x) = 1 - F^*(x)$. Of course, this differs from the CDF complement $\bar{F} = 1 - F$.

Theorem 1.16 Suppose $F_X(x)$ is the CDF of a random variable X. Assume $P(-\infty < X < \infty) = 1$. Then the following inequalities hold:

$$P(F_X(X) \leq t) \leq t, \tag{1.32}$$
$$P(F_X^*(X) \leq t) \leq t, \tag{1.33}$$

for any $t \in (0, 1)$. ∎

Proof. For $t \in (0,1)$, define the set $E_t = \{x : F_X(x) \leq t\}$ Then $x_t = \sup E_t < \infty$, since $F_X(x)$ is nondecreasing, and $\lim_{x \to \infty} F_X(x) = 1$. We therefore have $(-\infty, x_t) \subset E_t$. If $x_t \notin E_t$, then $E_t = \cup_{n=1}^{\infty}\{x : x \leq x_t - 1/n\}$. By continuity of probability measures $P(E_t) = \lim_{n\to\infty} F_X(x_t - 1/n) \leq t$, by the definition of E_t. Otherwise $P(E_t) = F_X(x_t) \leq t$, so (1.32) follows. Then (1.33) is verified by applying (1.32) to $-X$. □

The phrase "uniform-like" can now be made precise.

Theorem 1.17 (Inversion of the CDF (General Case)) Suppose $F_X(x)$ is the CDF of a random variable X. Assume $P(-\infty < X < \infty) = 1$. Define $q(x) = P(X = x)$. Then the CDF of $U = F_X(X)$ satisfies the bounds

$$t - \delta \leq F_U(t) \leq t, \ \ t \in (0, 1), \tag{1.34}$$

where $\delta \geq \sup_{x \in \mathbb{R}} q(x)$. ∎

Proof. The assumptions of Theorem 1.16 apply here, so the upper bound of (1.34) is given directly by (1.32). Then subtracting both sides of the inequality (1.33) from 1 yields $P(F_X^-(X) \leq 1 - t) \geq 1 - t$, $t \in (0,1)$. Then $F_X(x) = F_X^-(x) + q(x)$, so that $1 - t \leq P(F_X(X) \leq 1 - t + q(X)) \leq P(F_X(X) \leq 1 - t + \delta)$, $t \in (0, 1)$. The lower bound of (1.34) is verified by setting $t' = 1 - t + \delta$ in the preceding inequality. □

Note that no assumption regarding the measure space on which X is defined has been made. This means that Theorem 1.15 becomes a special case of Theorem 1.17, obtained by noting that $q(x) \equiv 0$. Otherwise, deviation from the uniform distribution is limited by δ. A natural assumption of many models involving discrete random variables would be that δ approaches 0 as some measure of complexity (for example, sample size) increases. We demonstrate this in the following example.

Example 1.23 (Inversion of the CDF of a Binomial Distribution) Suppose $X \sim bin(n, 1/2)$, and let $U = F_X(X)$. Following Theorem 1.17 let $q(x)$ be the density function of X. It is easily verified that for even values of n $q(x)$ is maximized at $x = n/2$. Applying Stirling's approximation (Equation (C.7)) gives

$$\delta = q(n/2) = \binom{n}{n/2} 2^{-n} \approx \frac{1}{\sqrt{2\pi}} \frac{1}{\sqrt{n/2}} = \frac{1}{\sqrt{2\pi}} \frac{1}{\sqrt{E[X]}} \tag{1.35}$$

which bounds the error of the approximation of the CDF of U given by Equation (1.34). See also Problem 1.33. ∎

1.8 Transformations of Random Variables

Given a random variable X and a function g we easily obtain a new random variable $Y = g(X)$. This operation is often referred to as a transformation. As a general rule, distribution P_Y of $Y = g(X)$ can be directly obtained from the distribution P_X of X using the idea of the preimage: $g^{-1}(E) = \{x \in \mathbb{R} : g(x) \in E\}$ which means that $\{Y \in E\} = \{X \in g^{-1}(E)\}$, so that $P_Y(E) = P_X(g^{-1}(E))$. Note that the use of the preimage does not require that the function g actually have an inverse.

However, additional assumptions imposed on the transformation can yield simplified methods. We consider two such approaches. Transformations of random vectors will be considered in Section 2.3.

1.8.1 Monotone Transformations

Suppose $Y = g(X)$ and g is a strictly increasing function. Then g possesses a strictly increasing inverse function g^{-1}. In general, if h is any increasing function, and $x \leq y$ then we also have $h(x) \leq h(y)$. We may therefore write

$$\{Y \leq y\} = \{g(X) \leq y\} = \{g^{-1}(g(X)) \leq g^{-1}(y)\} = \{X \leq g^{-1}(y)\}. \tag{1.36}$$

We have the following result:

Theorem 1.18 (Monotone Transformations) Suppose X is a discrete or continuous random variable with CDF F_X. Let g be a strictly increasing function. Then $Y = g(X)$ possesses CDF $F_Y(y) = F_X(g^{-1}(y))$, $y \in (0,1)$. ■

Proof. The theorem follows from the equality $\{Y \leq y\} = \{X \leq g^{-1}(y)\}$ derived in Equation (1.36). □

Example 1.24 (The Weibull Random Variable as a Transformation of an Exponential Random Variable) The Weibull distribution for a random variable $W \sim weibull(\mu, k)$ was defined earlier in this chapter by the transformation of $X \sim exp(1)$ given by Equation (1.22). The CDF of X is $F_X(x) = 1 - \exp(-x)$. Then if for $\mu, k > 0$ we have $X = (W/\mu)^k$, by Theorem 1.18 the CDF of W is then $F_W(w) = F_X((w/\mu)^k) = 1 - \exp(-(w/\mu)^k)$. The density of W is therefore $f_W(w) = k\mu^{-1}(w/\mu)^{k-1}\exp(-(w/\mu)^k)$, equivalent to the Weibull density. ■

1.8.2 One-to-one transformations

Suppose the transformation $Y = g(X)$ is one-to-one. Then the transformation possesses an inverse g^{-1} between the support of X and the support of Y, which justifies its use. If X is discrete, the PMFs p_X and p_Y can be related by direct substitution:

$$p_Y(y) = P(Y = y) = P(g(X) = y) = P(X = g^{-1}(y)) = p_X(g^{-1}(y)).$$

If X is continuous, a similar, but more complex method is available. We'll return temporarily to the assumption that g is increasing. If we differentiate the expression of Theorem 1.18 using the chain rule we have

$$\frac{d}{dy}F_Y(y) = \frac{dg^{-1}(y)}{dy}\frac{d}{dx}F_X(g^{-1}(y)).$$

Recalling that the density f is the derivative of the CDF F we have

$$f_Y(y) = \left[\frac{dg^{-1}(y)}{dy}\right] f_X(g^{-1}(y)) = \left[\left.\frac{dg(x)}{dx}\right|_{x=x(y)}\right]^{-1} f_X(g^{-1}(y)).$$

However, it can be shown that this method does not depend on g being increasing. As long as it is a one-to-one transform, we have the transformation rule

$$f_Y(y) = \left|\frac{dg^{-1}(y)}{dy}\right| f_X(g^{-1}(y)) = \left|\frac{dg(x)}{dx}\right|_{x=x(y)}^{-1} f_X(g^{-1}(y)),$$

replacing the derivative of the transformation with its absolute value.

We summarize this section with the following theorem:

Theorem 1.19 (One-to-one Transformations) Suppose X is a random variable, and g is a function possessing inverse g^{-1}. Define random variable $Y = g(X)$. If X is discrete with PMF p_X then Y is discrete, and possesses PMF $p_Y(y) = p_X(g^{-1}(y))$. If X is continuous with density f_X then Y is continuous, and possesses density

$$f_Y(y) = \left|\frac{dg^{-1}(y)}{dy}\right| f_X(g^{-1}(y)) \tag{1.37}$$

over the support of Y. ■

Example 1.25 (The Weibull Random Variable as a Transformation of an Exponential Random Variable (Alternative Approach)) The Weibull density of Example 1.24 can also be obtained by Theorem 1.19. If $X \sim exp(1)$, $W \sim weibull(\mu, k)$, and X, W are related by the transformation $X = g(W) = (W/\mu)^k$, then by Theorem 1.19 the respective densities are related by

$$f_W(w) = \left|\frac{dg(w)}{dw}\right| f_X(g(w)) = k\mu^{-1}(w/\mu)^{k-1}\exp(-(w/\mu)^k),$$

as obtained in Example 1.24. Note that we are here intepreting $g(w)$ as the inverse of the transformation from W to X. ■

1.9 Moment Generating Functions

The moment generating function (MGF) of a random variable X is a type of functional transformation (the Fourier or Laplace transforms being well known examples). It maps the distribution of X to a function $m(t)$, $t \in \mathcal{T} \subset \mathbb{R}$, which is then a unique representation of that distribution. Like the Fourier or Laplace transforms, it is used because certain properties of the original object are more easily studied using the transformed representation. However, the MGF does not exist for all distributions, therefore conditions under which it does must be established. The formal definition follows.

Definition 1.10 (Moment Generating Function (MGF)) Let X be a random variable with distribution F_X. Define the transformation

$$m(t) = E[e^{tX}], \quad t \in \mathbb{R}. \tag{1.38}$$

Let $\mathcal{T}$ be the set of all $t \in \mathbb{R}$ for which $m(t) < \infty$. If $\mathcal{T}$ contains an open neighborhood of zero, we say that X (or the distribution F_X) possesses a MGF $m(t)$. We may conclude immediately that $m(0) = 1$, and $m(t) > 0$ for any $t \in (-\infty, \infty)$. ■

Recall the series expansion $e^x = \sum_{k \geq 0} x^k/k!$. What gives the MGF its utility is the power series expansion

$$m(t) = E\left[\sum_{k=0}^{\infty} \frac{(tX)^k}{k!}\right] = \sum_{k=0}^{\infty} E[X^k]\frac{t^k}{k!}, \tag{1.39}$$

so it is very important to derive conditions under which it is valid. We begin this investigation with the following theorem.

Theorem 1.20 (Existence of the MGF I) Let $m(t)$ be defined for some distribution F_X as in Equation (1.38). Suppose there exist values $t' < 0 < t''$ such that $m(t') < \infty$ and $m(t'') < \infty$. Then

(i) The distribution F_X possesses a MGF.
(ii) The power series Equation (1.39) converges for $t \in [-\tau, \tau]$ for some $\tau > 0$. ∎

Proof. By Jensen's inequality, if $m(t) < \infty$ for some $t \neq 0$, then $m(s) < \infty$ for all s between 0 and t (Problem 1.34). This proves statement (i).

It follows that $m(t) < \infty$ for $t = -\tau, \tau$, for some $\tau > 0$. Then $e^{|tx|} \leq e^{|\tau x|} \leq e^{\tau x} + e^{-\tau x}$, for all $t \in [-\tau, \tau]$, so we conclude that $E[e^{|tX|}] < \infty$, $t \in [-\tau, \tau]$. Then for each n

$$\left|\sum_{k=0}^{n} \frac{(Xt)^k}{k!}\right| \leq e^{|tX|}$$

for all $X \in \mathcal{X}$. The theorem follows from the dominated convergence theorem (Theorem 11.3). □

Theorem 1.20 gives a simple test for the existence of the MGF. It turns out that this condition is equivalent to a bound on the rate of growth of $|E[X^k]|$ and $E[|X|^k]$, $k \geq 1$. As a consequence, any bounded random variable possesses a MGF (Problem 1.42). The following theorem defines this bound.

Theorem 1.21 (Bound on Growth Rate of Moments) Suppose there exists $r > 0$ such that

$$\lim_{n \to \infty} r^n E[X^n]/n! = 0 \tag{1.40}$$

Then the following limit also holds

$$\lim_{n \to \infty} r^n E[|X|^n]/n! = 0. \tag{1.41}$$

for the same value r. ∎

Proof. By hypothesis the subsequence of Equation (1.41) defined by even n converges to zero. It remains to show the same for the subsequence defined by odd n. By the Cauchy-Schwarz inequality, we have $E[|X|^{2n+1}]^2 = E[|X|^{n+1}|X|^n]^2 \leq E[|X|^{2n+2}]E[|X|^{2n}]$. We may then write

$$\begin{aligned}
\frac{E[|X|^{2n+1}]r^{2n+1}}{(2n+1)!} &\leq \left[\frac{E[|X|^{2n+2}]r^{2n+2}}{(2n+1)!}\right]^{1/2} \left[\frac{E[|X|^{2n}]r^{2n}}{(2n+1)!}\right]^{1/2} \\
&\leq \left[\frac{E[|X|^{2n+2}]r^{2n+2}}{(2n+2)!}\right]^{1/2} \left[\frac{E[|X|^{2n}]r^{2n}}{(2n)!}\right]^{1/2} \left[\frac{2n+2}{2n+1}\right]^{1/2}
\end{aligned}$$

which completes the proof. □

We can now establish that the existence of a MGF is equivalent to the convergence of the power series given in Equation (1.39) for all t in an open neighborhood of zero, which in turn is equivalent to the bound given in Theorem 1.21.

Theorem 1.22 (Existence of the MGF II) A distributon F_X possesses a MGF if and only if condition (1.40) holds. ■

Proof. It may be verified that the power series expansion of Equation (1.39) converges in some neighbourhood $(-\tau, \tau)$, $\tau > 0$ *iff* (1.40) holds, which is therefore a necessary condition for the existence of an MGF, by Theorem 1.20.

Then suppose (1.40) holds. By Theorem 1.21 (1.41) also holds, and we may conclude that the power series $\sum_{k=0}^{\infty} E[|X|^k]t^k/k!$ converges in some neighbourhood of zero. By the monotone convergence theorem (Theorem 11.3) $E[e^{t|X|}] < \infty$ for all t in some open neighborhood of zero. We conclude that an MGF exists after noting that $0 < e^{tX} < e^{-t|X|} + e^{t|X|}$. □

The following two examples demonstrate how Theorem 1.22 may be used to test the existence of an MGF.

Example 1.26 (Moments of the Normal Density) Evaluation of the moments of a normal distribution can be carried out using the gamma function (Section C.5). Let $Z \sim N(0, 1)$. Consider the absolute moments:

$$E[|Z|^k] = \frac{\sqrt{2}}{\sqrt{\pi}} \int_0^{\infty} |z|^k e^{-z^2/2} dz = \frac{2^{k/2}}{\sqrt{\pi}} \int_0^{\infty} u^{(k-1)/2} e^{-u} du = \frac{2^{k/2}}{\sqrt{\pi}} \Gamma\left(\frac{k+1}{2}\right)$$

converting the integral to a gamma function using the variable transformation $u = z^2/2$. We then use Equations (C.5) and (C.6) to give

$$E[|Z|^k] = \begin{cases} (k-1)!! & ; \quad k \text{ is even} \\ \frac{2^{k/2}}{\sqrt{\pi}} \left(\frac{k-1}{2}\right)! & ; \quad k \text{ is odd} \end{cases} . \tag{1.42}$$

where $k!!$ is the double factorial, which evaluates to $k!! = k(k-2)(k-4)\cdots 4 \cdot 2$ for even k, and $k!! = k(k-2)(k-4)\cdots 3 \cdot 1$ for odd k.

That Z possesses an MGF is easily verified using Theorem 1.22 after noting that $E[|Z|^k]/k! < 2^{k/2}$. We may similarly test for the existence of an MGF for Z^2 by first noting that $E[|Z^2|^k] = E[|Z|^{2k}]$. Then, since $(2k-1)!!/k! \leq 2^k$, we have $E[|Z^2|^k]/k! < 4^k$, so that Z^2 possesses an MGF. Next, consider Z^4, for which $E[|Z^4|^k] = E[|Z|^{4k}]$. We then have $(4k-1)!!/k! > k^{k-1}$, and we may conclude $E[|Z^4|^k]/k! > k^{k-1}$. It follows that there can be no $r > 0$ for which the limit of Equation (1.41) holds, therefore, by Theorem 1.22 Z^4 cannot possess a MGF. ■

Example 1.27 (Nonexistence of an MGF for the Weibull Density with Shape Parameter $\alpha = 1/2$) Suppose $X \sim weibull(\mu = 1, \alpha = 1/2)$. We have $E[X^k] = (2k)!$. But there cannot exist a single constant θ for which

$$|E[X^k]|/k! = (2k)!/k! \leq \theta^k, \quad k = 1, 2, \ldots.$$

(Hint: $(2k)!/k! > k^k$). So X does not possess a MGF. ■

1.9.1 Derivatives of the MGF

The MGF is of value in part because of its tractability with respect to standard calculus. Much of this relies on the ability to exchange differentiation with respect to t and integration.

It suffices to prove this for $t \in (-\epsilon, \epsilon)$ for any $\epsilon > 0$, assuming the MGF exists on this interval. We may use Theorem C.7 to verify that the first order derivative can be evaluated as

$$dm(t)/dt = E[d\exp(tX)/dt] = E[X\exp(tX)], \quad t \in (-\epsilon, \epsilon).$$

We first require $E[X\exp(tX)] < \infty$, $t \in (-\epsilon, \epsilon)$. This can be verified by the Hölder inequality:

$$E[|X\exp(tX)|] \leq E[|X|^2]^{1/2} E[\exp(2tX)]^{1/2}.$$

We have already seen that the existence of the MGF implies that all order moments exist, thus we may assume that this upper bound is finite. Theorem C.7 applies after noting that $x\exp(xt)$ is monotone in t for fixed x. The argument can be extended to any order derivative in a straightforward manner.

This being the case, we have the following important identity. Let k be a positive integer. Then

$$\frac{d^k m_X(t)}{dt^k} = \frac{d^k}{dt^k} E[\exp(tX)] = E[\frac{d^k}{dt^k}\exp(tX)] = E[X^k \exp(tX)].$$

By evaluating this derivative at $t = 0$ we have the following theorem:

Theorem 1.23 (Evaluating Moments as Derivatives of the MGF) If X possesses MGF $m_X(t)$ then

$$d^k m_X(t)/dt^k \big|_{t=0} = E[X^k], \quad k \geq 1.$$

■

The MGF allows us to derive moments using differenation rather than integration, as the next example demonstrates.

Example 1.28 (MGF of a Normal Distribution) In Example 1.26 we showed indirectly that the MGF of $Z \sim N(0,1)$ exists by studying the moments $E[|Z|^k]$. This approach also allowed us to conclude that a MGF existed for Z^2 but not Z^4. Here, we will simply evaluate the MGF by the integral

$$m_Z(t) = \int_{-\infty}^{\infty} e^{tz}(2\pi)^{-1/2}e^{-z^2/2}dz = e^{\frac{t^2}{2}},$$

so that $m_Z(t) < \infty$ for all $t \in \mathbb{R}$ (this is evaluated by "completing the square" within the exponent of the integrand). Then using Theorem 1.23 we may obtain any order derivative of $m_Z(t)$, for example,

$$\begin{aligned}
dm_Z(t)/dt &= tm_Z(t), \\
d^2 m_Z(t)/dt^2 &= (1+t^2)m_Z(t), \\
d^3 m_Z(t)/dt^3 &= (3t+t^3)m_Z(t), \\
d^4 m_Z(t)/dt^4 &= (3+6t^2+t^4)m_Z(t).
\end{aligned}$$

Upon substituting $t = 0$ into the kth order derivative we obtain $E[Z^k]$, here, $E[Z^k] = 0$ for any odd k; $E[Z^2] = 1$ and $E[Z^4] = 3$, as predicted by Equation (1.42). ■

1.9.2 The Moment Problem

Probability distributions need not be uniquely determined by their moments (counter-examples of distinct distributions with identical and finite moments are given in, for example, Billingsley (1995) or Durrett (2010)). However, this will be the case under condition (1.40), as stated in the following theorem:

Theorem 1.24 (Uniqueness of Moments) Suppose X is a random variable with finite moments $E[X^n]$. If condition (1.40) holds then X possesses the unique distribution with those moments. ∎

Proof. See, for example, Billingsley (1995). □

Of course, by Theorem 1.22 the conditions of Theorem 1.24 also guarantee the existence of a MGF. Thus, when an MGF exists, it uniquely determines the distribution.

Example 1.29 (The Moments of a Constant are Unique) Suppose a random variable X possesses moments $E[X^k] = c^k$, $k = 1, 2, \ldots$ for some constant c. The conditions of Theorem 1.24 are satisfied, so only one random variable can possess those moments. That random variable is, of course, defined by $P(X = c) = 1$. ∎

Theorem 1.24 is important to statistical inference, since it allows us to deduce the distribution of a random variable by evaluating its MGF (examples of this type of problem will be seen in, for example, Section 2.6). Note, however, that existence of an MGF is not a necessary condition for the uniqueness of moments (Durrett, 2010).

1.10 Moments and Cumulants

The cumulant generating function (CGF) is defined as

$$c_X(t) = c(t) = \log(m_X(t)),$$

where the reference to X in the notation is omitted when there is no ambiguity. Clearly, a random variable X possesses a CGF *iff* it possesses a MGF, and when it does exist it will also possess an expansion

$$c_X(t) = \sum_{k=1}^{\infty} \kappa_k[X] \frac{t^k}{k!}, \tag{1.43}$$

in an open neighborhood of $t = 0$. We refer to $\kappa_k[X] = \kappa_k$ as the order k cumulant. In fact, this serves as their definition. Note that the order t^0 term in (1.43) is necessarily 0, since $m_X(0) = 1$. The cumulants can be obtained by essentially the same differentiation rule used for the MGF.

$$d^k c_X(t)/dt^k \big|_{t=0} = \kappa_k[X], \quad k \geq 1.$$

1.10.1 The MGF and CGF Under Linear Transformations

A particularly convenient transformation rule is available for the CGF. First, note that

$$m_{a+bX}(t) = E[e^{(a+bX)t}] = e^{at} m_X(bt). \tag{1.44}$$

Then

$$c_{a+bX}(t) = \log\left(m_{a+bX}(t)\right) = at + c_X(bt) = at + \sum_{k=1}^{\infty} b^k \kappa_k[X] \frac{t^k}{k!}.$$

Careful inspection of this modified expansion reveals the linear transformation rule for cumulants. It is particularly important to note that the location transformation parameter a affects only the first order cumulant. More generally, we have:

$$\kappa_1[a + bX] = a + b\kappa_1[X], \quad \kappa_k[a + bX] = b^k \kappa_k[X], \; k \geq 2, \tag{1.45}$$

so that the CGF is location invariant after the first order term, unlike the MGF, so that $\kappa_k[X] = \kappa_k[a + X]$ for any a, when $k \geq 2$. This can simplify calculation of higher-order cumulants considerably, since we can replace X with $X - E[X]$, or simply assume that $E[X] = 0$.

Example 1.30 (MGF and CGF for the Normal Distribution) Suppose $Z \sim N(0,1)$. From Example 1.28 we have $m_Z(t) = e^{t^2/2}$. Then $X = \mu + \sigma Z \sim N(\mu, \sigma^2)$, so by Equation (1.44) we have $m_X(t) = e^{\sigma^2 t^2/2+\mu t}$ and $c_X(t) = \sigma^2 t^2/2 + \mu t$. ∎

The CGF of random variable X is simply the transformed MGF $c_X(t) = \log(m_X(t))$. So what is the advantage of doing this? The fact that the CGF of a normal random variable is a 2nd order polynomial (Example 1.30) will prove to be a fact of some practical and theoretical significance. In fact, it can be shown that the CGF is a polynomial only for the normal distribution.

1.10.2 On the Relationship Between Various Classes of Moments

Let X be any random variable. The kth order (raw) moment, central moment, absolute moment, absolute central moment, standardized moment and standardized absolute moment are defined as

$$\begin{aligned}
\mu_k[X] &= \mu_k = E[X^k], \\
\bar{\mu}_k[X] &= \bar{\mu}_k = E[(X - E[X])^k], \\
\dot{\mu}_k[X] &= \dot{\mu}_k = E[|X|^k], \\
\dot{\bar{\mu}}_k[X] &= \dot{\bar{\mu}}_k = E[|X - E[X]|^k], \\
\gamma_k[X] &= \gamma_k = \frac{\bar{\mu}_k[X]}{\bar{\mu}_2[X]^{k/2}}, \\
\dot{\gamma}_k[X] &= \dot{\gamma}_k = \frac{\dot{\bar{\mu}}_k[X]}{\bar{\mu}_2[X]^{k/2}}, \quad k \geq 1,
\end{aligned} \tag{1.46}$$

where the reference to X in the notation is omitted when there is no ambiguity. The binomial theorem provides a transformaton rule for noncentral to central moments, setting $\mu_0 = 1$,

$$\bar{\mu}_n[X] = E[(X - \mu_1)^n] = \sum_{i=0}^{n} \binom{n}{i} \mu_i (-\mu_1)^{n-i}. \tag{1.47}$$

Many such formulae are described in the literature, and the algebraic manipulation of moment expressions sometimes represents a significant technical challenge (a system of tensors for multivariate moments is proposed in McCullagh (1984)).

It will be useful to write explicitly the first few transformations derived from (1.47):

$$\begin{aligned}
\bar{\mu}_2 &= \mu_1^2 - 2\mu_1^2 + \mu_2 & &= \mu_2 - \mu_1^2, \\
\bar{\mu}_3 &= -\mu_1^3 + 3\mu_1^3 - 3\mu_2\mu_1 + \mu_3 & &= \mu_3 - 3\mu_2\mu_1 + 2\mu_1^3, \\
\bar{\mu}_4 &= \mu_1^4 - 4\mu_1^4 + 6\mu_2\mu_1^2 - 4\mu_3\mu_1 + \mu_4 & &= \mu_4 - 4\mu_3\mu_1 + 6\mu_2\mu_1^2 - 3\mu_1^4.
\end{aligned} \tag{1.48}$$

The cumulants may be deduced by comparing the order t^k term in (1.43) to that obtained by substituting (1.39) into an expansion of $\log(x + 1)$. This is a straightforward (if cumbersome) procedure which is worth exploring to some degree. The expansion is given explictly by

$$\begin{aligned}\log(x+1) &= \sum_{i=0}^{\infty} \left.\frac{d^i \log(x+1)}{dx^i}\right|_{x=0} \times \frac{x^i}{i!} \\ &= \frac{x}{1} - \frac{x^2}{2} + \frac{x^3}{3} - \ldots \\ &= \sum_{i=1}^{\infty} (-1)^{i-1} i^{-1} x^i.\end{aligned}$$

By defintion $c(t) = \log(m(t))$, and $m(t) - 1 = \sum_{i=1}^{\infty} \mu_i x^i / i!$, so in order to relate the moments and the cumulants we "merely" have to substitute $m(t) - 1$ for x in the expansion of $\log(x+1)$:

$$\begin{aligned}c(t) &= \sum_{i=1}^{\infty} (-1)^{i-1} i^{-1} \left[\sum_{j=1}^{\infty} \mu_j \frac{t^j}{j!}\right]^i \\ &= \sum_{i=1}^{\infty} (-1)^{i-1} i^{-1} \left[\mu_1 t + \frac{\mu_2}{2} t^2 + \frac{\mu_3}{6} t^3 + \frac{\mu_4}{24} t^4 + \ldots\right]^i \\ &= \mu_1 t + \left[\frac{\mu_2}{2} - \frac{\mu_1^2}{2}\right] t^2 + \left[\frac{\mu_3}{6} - 2^{-1} 2\mu_1 \frac{\mu_2}{2} + 3^{-1} \mu_1^3\right] t^3 \\ &\quad + \left[\frac{\mu_4}{24} - 2^{-1} 2\mu_1 \frac{\mu_3}{6} - 2^{-1} \left(\frac{\mu_2}{2}\right)^2 + 3^{-1} 3\mu_1^2 \frac{\mu_2}{2} - 4^{-1} \mu_1^4\right] t^4 + \ldots\end{aligned} \tag{1.49}$$

By equating like ordered terms of (1.43) and (1.49) we get

$$\begin{aligned}\kappa_1[X] &= \mu_1, \\ \kappa_2[X] &= 2! \left[\frac{\mu_2}{2} - \frac{\mu_1^2}{2}\right], \\ \kappa_3[X] &= 3! \left[\frac{\mu_3}{6} - 2^{-1} 2\mu_1 \frac{\mu_2}{2} + 3^{-1} \mu_1^3\right] \\ \kappa_4[X] &= 4! \left[\frac{\mu_4}{24} - 2^{-1} 2\mu_1 \frac{\mu_3}{6} - 2^{-1} \left(\frac{\mu_2}{2}\right)^2 + 3^{-1} 3\mu_1^2 \frac{\mu_2}{2} - 4^{-1} \mu_1^4\right] \\ &\vdots \quad .\end{aligned} \tag{1.50}$$

The first order cumulant is simply equal to the first order moment. Recall, however that for $k \geq 2$ $\kappa_k[X] = \kappa_k[a+X]$ for any a. So set $\kappa_k[X] = \kappa_k[X - \mu_1]$. Of course, this transformation can be effected simply by replacing the noncentral moments with the central moments. It simplifies the analysis considerably by noting that the first central moment is always zero. We then have

$$\begin{aligned}\kappa_1[X] &= \mu_1, \\ \kappa_2[X] &= \bar{\mu}_2, \\ \kappa_3[X] &= \bar{\mu}_3 \\ \kappa_4[X] &= \bar{\mu}_4 - 3\bar{\mu}_2^2 \\ &\vdots \quad .\end{aligned} \tag{1.51}$$

The standardized moment $\gamma_k[X]$ is useful because it is invariant to linear transformation, that is $\gamma_k[aX + b] = \gamma_k[X]$ when $a \neq 0$. We will make use of the skewness and kurtosis, which are the 3rd and 4th order standardized moments of a random variable X:

$$SK = \gamma_3[X] = E[\frac{(X - \mu_1[X])^3}{\text{var}[X]^{3/2}}] = \frac{\bar{\mu}_3[X]}{\bar{\mu}_2[X]^{3/2}} = \frac{\kappa_3[X]}{\bar{\mu}_2[X]^{3/2}},$$
$$KURT = \gamma_4[X] = E[\frac{(X - \mu_1[X])^4}{\text{var}[X]^2}] = \frac{\bar{\mu}_4[X]}{\bar{\mu}_2[X]^2} = \frac{\kappa_4[X]}{\bar{\mu}_2[X]^{3/2}} + 3. \quad (1.52)$$

In large sample analysis, these quantities play a role in measuring the deviation of a distribution from the normal density (we will see this in Chapter 12). Recall that all cumulants of a normal random variable of order $k \geq 3$ are zero. Therefore, normal approximations can often be improved by reducing, or correcting for, skewness. Similarly, $KURT = 3$ for the normal distribution, therefore $KURT - 3$ is referred to as excess kurtosis. Consider the following example.

Example 1.31 (Kurtosis of the T Distribution) The T distribution with $\nu > 0$ degrees of freedom has density

$$f(x) = \frac{\Gamma((\nu + 1)/2)}{\sqrt{\nu\pi}\,\Gamma(\nu/2)} \left[1 + \frac{x^2}{\nu}\right]^{-(\nu+1)/2}, \quad x \in \mathbb{R}.$$

By examining the T distribution density, we can see that order k moments or cumulants exist only for $k = 1, \ldots, \nu - 1$ (Hint: The integrand of $E[X^k]$ is of order $x^k/x^{\nu+1}$). The T distribution is symmetric, so $SK = 0$, for $\nu > 3$. In addition, the T distribution has finite excess kurtosis for $\nu > 4$. This can be shown to be $KURT - 3 = 6/(\nu - 4)$. The excess kurtosis approaches 0 as $\nu \to \infty$. We would expect this, since the T distribution increasingly resembles the normal distribution as ν increases. Note that the T distribution does not possess a MGF or a CGF, but we may still evalute the cumulants κ_k up to order $\nu - 1$. ■

We will make use of the standardized absolute moment $\dot{\gamma}_k$ in Section 11.10.

1.11 Problems

Problem 1.1 Show that the axioms of Definition 1.1 imply each of the following statements.

(i) $P(\emptyset) = 0$.
(ii) $A \cap B = \emptyset$ implies $P(A \cup B) = P(A) + P(B)$ (finite additivity).
(iii) $P(A^c) = 1 - P(A)$.
(iv) $A \subset B$ implies $P(A) \leq P(B)$.

HINT: The set $\emptyset$ is disjoint to all other sets, including $\emptyset$ itself. This means $S = S \cup \{\cup_{i=1}^{\infty} \emptyset\}$. Also note that countable additivity and finite additivity (i.e. statement (ii) above) are distinct statements.

Problem 1.2 Let $E_1, E_2, \ldots$ be a countable collection of sets. It is sometimes useful to construct an associated collection of sets, denoted as

$$\bar{E}_1 = E_1, \quad \bar{E}_i = E_i \cap E_{i-1}^c \cap \ldots \cap E_1^c, \; i \geq 2.$$

(a) Verify the following properties of $\bar{E}_i$:

(i) $\bar{E}_i \subset E_i$ for all $i \geq 1$.
(ii) The sets $\bar{E}_1, \bar{E}_2, \ldots$ are mutually exclusive.
(iii) $\cup_{i=1}^{\infty} E_i = \cup_{i=1}^{\infty} \bar{E}_i$.

(b) Prove Theorem 1.2 (Boole's inequality).

(c) Suppose world records for a given sport are compiled annually. For convenience, label the years $i = 1, 2, \ldots$ with $i = 1$ being the first year records are kept. Define the events

$$E_i = \{ \text{ World record broken in year } i \ \}, \ \ i \geq 1.$$

Then, let Q_i be the probability that from year i onwards, the current world record is never broken. Prove that if $\sum_{i=1}^{\infty} P(E_i) < \infty$, then $\lim_{i \to \infty} Q_i = 1$. Verify that, in particular, if $P(E_i) \leq c/i^k$ for some finite constants $c > 0$ and $k > 1$ then $\lim_{i \to \infty} Q_i = 1$.

Problem 1.3 Prove Chebyshev's inequality: For any random variable with mean μ and variance σ^2 the inequality $P\left(|X - \mu| \geq t\sigma\right) \leq 1/t^2$ holds for any $t > 0$. **HINT:** This is a special case of Markov's inequality.

Problem 1.4 Two dice are tossed independently. Let X_1, X_2 be random variables representing the two outcomes, each from sample space $S_X = \{1, 2, 3, 4, 5, 6\}$. Derive the probability mass function of the following random variables:

(a) $X = X_1 - X_2$,

(b) $X = \begin{cases} X_1 & ; \quad X_1 = X_2 \\ 0 & ; \quad X_1 \neq X_2 \end{cases}$.

Problem 1.5 Consider the following CDF for a random variable X:

$$F_X(x) = \begin{cases} 0 & ; \quad x < 0 \\ x/2 & ; \quad x \in [0, 1/3) \\ 1/6 & ; \quad x \in [1/3, 2/3) \\ x/8 + 2/3 & ; \quad x \in [2/3, 2) \\ 1 & ; \quad x \geq 2 \end{cases}$$

Sketch the CDF, then determine the following probabilities:

(a) $P(X > 1)$.
(b) $P(X \in (3/7, 4/7))$.
(c) $P(X = 2/3)$.
(d) $P(X = 2)$.
(e) $P(X \leq 2)$.
(f) $P(X < 2)$.

Problem 1.6 A random variable X possesses the following density function for some constant c:

$$f_X(x) = \begin{cases} -c(x-2)(x-4) & ; \quad x \in [2, 4] \\ 0 & ; \quad \text{otherwise} \end{cases}.$$

(a) Determine c.
(b) Determine the CDF $F(x) = P(X \leq x)$. Give this as a function of $x \in (-\infty, \infty)$.

Problem 1.7 Conditional probabilities can be used to construct new conditional distributions of random variables in a natural way. Let X be a random variable and let A be an event. If we can evaluate $P(X \leq x \mid A)$ then we have a completely defined distribution for X *conditional on* A (this new random variable may be denoted as $X \mid A$). The CDF under this conditional distribution is typically denoted in the literature as $F_{X|A}(x)$ or $F_X(x \mid A)$, and the expected value conditional on A is denoted as $E[X \mid A]$.

Suppose we have a partition of the sample space $A_1, \ldots, A_n$. Then for any random variable X we have

$$E[X] = E[X \mid A_1]P(A_1) + E[X \mid A_2]P(A_2) + \ldots + E[X \mid A_n]P(A_n). \tag{1.53}$$

Calculating expected values by conditioning is often a very useful device. The following problem is a typical example.

A rat attempts to navigate a maze. It is possible for the rat to choose a path which will lead back to the starting point before exiting the maze. If this happens, the rat continues to attempt to exit the maze with no memory of its previous attempts. Let E be the event that the rat returns to the starting point before exiting the maze (it may do this more than once). Suppose $P(E) = p$. Also, let T_0 be the expected time to exit conditional on E^c, and let T_1 be the expected time to return to the starting point conditional on E. If W is the time taken to exit the maze, determine $E[W]$ as a function of p, T_0 and T_1.

Problem 1.8 Suppose X is a nonnegative random variable, that is, $P(X \geq 0) = 1$. Assume that X is a discrete random variable with sample space $S_X = \{0, 1, 2, \ldots\}$. Show that $E[X] = \sum_{n=0}^{\infty} 1 - F_X(n)$., where F_X is the CDF of X. **HINT:** Express X as a sum of Bernoulli random variable. See also Problem 1.25

Problem 1.9 What is the analogous statement of Problem 1.8 when X is a nonnegative random variable with a continuous CDF?

Problem 1.10 Prove that if $f(x)$ is a convex (concave) function on $E \subset \mathbb{R}$, and possesses an inverse f^{-1}, then that inverse is concave (convex). Then if $X > 0$ is a positive random variable with $E[X] < \infty$, prove that $E[\log(X)] \leq E[X]$, with strict inequality unless $X = E[X]$ *wp*1. **HINT:** Apply Jensen's inequality.

Problem 1.11 Let U_1 and U_2 be two independent random variables with a $unif[0,1]$ distribution. If $E \subset S$, where S is a unit square with sides $[0,1]$, then $P((U_1, U_2) \in E) = area(E)$. This holds whether or not E includes its own boundary.

(a) Determine the CDF $F_X(x)$ for $X = U_1 + U_2$.
(b) Determine the density function $f_X(x)$ for X.

Problem 1.12 Suppose $W \sim weibull(\mu, k)$. Derive an expression for the ith moment $E[W^i]$ expressed in terms of the gamma function Γ (Section C.5). Then determine the mean and variance.

Problem 1.13 Suppose $X \sim gamma(\tau, \alpha)$. Show that $E[\log X] = \log \tau + \psi(\alpha)$, where $\psi(\alpha) = \Gamma'(\alpha)/\Gamma(\alpha)$ is the digamma function, or the derivative of the logarithm of the gamma function $\Gamma(t)$. **HINT:** Evaluate the derivative of $\Gamma(t)$ by exchanging differentiation and integration within Equation (C.4), then compare to the gamma density.

Problem 1.14 We toss a fair coin (H or T) independently until a head H appears consecutively r times. What is the distribution of X (Example 1.6)?

Problem 1.15 Suppose you have been repeatedly playing a game of chance with a probability p of winning. After k games you have not yet won. We assume the outcomes are independent. Is the amount of time until you win, starting from that point, different from the time to win when you started to play? Many are tempted to believe that prior losses shorten the expected time to future wins, as though the number of losses is somehow fixed.

The issue is resolved by defining the memoryless property for a random variable X, which holds if $P(X > k+t \mid X > k) = P(X > k+t)/P(X > k)$ for all k, t. Verify that the geometric random variable is the only distribution on $\mathbb{I}_{>0}$ which possesses the memoryless property.

Problem 1.16 The "80-20" law states that 80% of total wealth is owned by 20% of the population. Show that this holds for the Pareto principle only if $\alpha = \log 5/\log 4$ (Example 1.8).

Problem 1.17 Consider the hypergeometric distribution of Example 1.7.

(a) Using principles of combinatorics, verify the density and support $\mathcal{S}_X$ given in Equations (1.19) and (1.20).
(b) Define a sequence of Bernoulli random variables $U_1, \ldots, U_k$, setting $U_i = 1$ if the ith selected ball is white. Expressing X as their sum, determine the mean and variance of X.
(c) Suppose we make a selection of k balls in the same manner, except that the balls are replaced immediately after being selected, and may be selected again (this is referred to as *sampling with replacement*). Let Y be the total number of white balls selected. What distribution does Y have? Show that $E[X] = E[Y]$ and determine the ratio $\text{var}[X]/\text{var}[Y]$. Verify that $\text{var}[X] \leq \text{var}[Y]$ for $k \geq 1$ and $\text{var}[X] < \text{var}[Y]$ for $k > 1$.
(d) Based on the comparisons of Part (c), under what conditions can the distribution of Y be used to approximate the distribution of X?

Problem 1.18 Show that if $Y \sim gamma(\tau, \alpha)$ then for any $k > 0$ $E[Y^k] = \tau^k \Gamma(k+\alpha)/\Gamma(\alpha)$ where $\Gamma(t)$ is the gamma function (Section C.5). Derive the mean and variance of the gamma distribution.

Problem 1.19 Show that if $Y \sim beta(\alpha, \beta)$ then for any $k > 0$

$$E[Y^k] = \frac{B(\alpha+k, \beta)}{B(\alpha, \beta)} = \frac{\Gamma(\alpha+k)/\Gamma(\alpha+k)}{\Gamma(\alpha+\beta+k)/\Gamma(\alpha+\beta+k)/\Gamma(\alpha+\beta)}.$$

where $B(a, b)$ is the beta function (Section C.5). Derive the mean and variance of the beta distribution.

Problem 1.20 Show that if $X \sim pois(\lambda)$ then $P(X \text{ is even}) = [1 + \exp(-2\lambda)]/2$, and as a consequence $P(X \text{ is even}) > P(X \text{ is odd})$ for any $\lambda > 0$. **HINT:** See Example 4.9.

Problem 1.21 For each of the following parametric classes, show that $E[Y] \geq E[X]$ implies $Y \geq_{lr} X$.

(a) $X, Y \sim bin(n, p)$, n is fixed.
(b) $X, Y \sim nb(r, p)$, r is fixed.
(c) $X, Y \sim pois(\lambda)$.
(d) $X, Y \sim unif(0, \theta)$, $\theta > 0$.
(e) $X, Y \sim gamma(\tau, \alpha)$, α is fixed.
(f) $X, Y \sim weibull(\tau, \alpha)$, α is fixed.

Problem 1.22 Suppose $X \sim N(\mu_x, \sigma_x^2)$, $Y \sim N(\mu_y, \sigma_y^2)$, and $\mu_y > \mu_x$. Prove that $Y \geq_{lr} X$ *iff* $\sigma_x^2 = \sigma_y^2$.

Problem 1.23 Suppose $X \sim bin(n, p)$ and $Y = 2X$.

(a) Does $Y \geq_{lr} X$ hold?
(b) Does $Y \geq_{st} X$ hold?

Problem 1.24 Suppose $X \sim pareto(\tau_x, \alpha_x)$, $Y \sim pareto(\tau_y, \alpha_y)$. For each of the following cases, determine whether or not a likelihood ratio ordering exists between X and Y.

(a) $\tau_y > \tau_x$, $\alpha_y = \alpha_x$.
(b) $\tau_y > \tau_x$, $\alpha_y > \alpha_x$.
(c) $\tau_y > \tau_x$, $\alpha_y < \alpha_x$.
(d) $\tau_y = \tau_x$, $\alpha_y > \alpha_x$.

Problem 1.25 Prove Theorem 1.12 by proving the following statements (Ross, 1996):

(i) If $Y \geq_{st} X$ and $X, Y \geq 0$ then $E[Y] \geq E[X]$. **HINT:** See Problem 1.8.
(ii) Let $X^+ = XI\{X > 0\}$, $X^- = -XI\{X < 0\}$, $Y^+ = YI\{Y > 0\}$, $Y^- = -YI\{Y < 0\}$. If $Y \geq_{st} X$ then $Y^+ \geq_{st} X^+$ and $X^- \geq_{st} Y^-$, and therefore $E[Y] \geq E[X]$.
(iii) Suppose $Y \geq_{st} X$ and h is a nondecreasing function. Then $h(Y) \geq_{st} h(X)$. **HINT:** Define $h^{-1}(a) = \inf\{x : h(x) \geq a\}$. Then $h(x) > a$ *iff* $x > h^{-1}(a)$.
(iv) If for all nondecreasing functions h we have $E[h(Y)] \geq E[h(X)]$ then $Y \geq_{st} X$ **HINT:** $h(x) = I\{x > a\}$ is a nondecreasing function.

Problem 1.26 Suppose $X \sim N(\mu, \sigma^2)$. Verify that $Y = aX + b$, $a \neq 0$, also possesses a normal distribution. Give the mean and variance.

Problem 1.27 Suppose $X \sim pareto(\tau, \alpha)$. Derive the distribution of $Y = \log(X/\tau)$. To what other parametric class is this related?

Problem 1.28 Suppose $X \sim N(0, 1)$, and we wish to determine the density of $Y = X^2$. Neither of the transformation methods of Section 1.8 is directly applicable to this problem, since the transformation $g(x) = x^2$ is neither monotone nor one-to-one on the support of X.

(a) Show how Theorem 1.18 can be modified for this example, and use this approach to determine the density of Y.
(b) Do the same for Theorem 1.19.
(c) Compare the density of Y to the exponential density.

Problem 1.29 If $Y \sim gamma(\tau, \alpha)$ then $X = 1/Y$ possesses an inverse gamma distribution. Show that this density is equal to $f(x) = x^{-\alpha-1}\theta^{-\alpha}/\Gamma(\alpha)e^{-1/(x\tau)}$, $x > 0$. Verify that the scale parameter is equal to $1/\tau$.

Problem 1.30 Derive the quantile function for the following distributions:

(a) $X \sim exp(\tau)$.
(b) $X \sim pareto(\tau, \alpha)$.
(c) $X \sim bin(4, 1/4)$.
(d) $X \sim unif(a, b)$.

Problem 1.31 Suppose we have a location-scale parametric class

$$f(x \mid \mu, \tau) = \tau^{-1} f((x - \mu)/\tau).$$

Let $Q_{0,1}(p)$ be the quantile function for density $f(x \mid 0, 1)$. Prove that $Q_{\mu,\tau}(p) = \mu + \tau Q_{0,1}(p)$ is the quantile function for density $f(x \mid \mu, \tau)$.

Problem 1.32 Suppose X has support $\mathcal{X} = (a, b) \cup (c, d)$, $b < c$, and that the CDF F_X is continuous. Prove directly (i.e. without using Theorem 1.15 or Theorem 1.17) that $F_X(X) \sim unif(0, 1)$.

Problem 1.33 Let $X \sim pois(\lambda)$, and $U = F_X(X)$. Suppose we wish to use Theorem 1.17 to approximate the distribution of U. Derive an estimate of the quantity δ in Equation (1.34). Show that your estimate is of the same form derived in Example 1.23. **HINT:** First identify the mode of f_X by considering the ratio $f_X(x+1)/f_X(x)$, then use Stirling's approximation (Equation (C.7)).

Problem 1.34 Find a mapping $\phi(e^a) = e^b$ that is convex, assuming $0 < a < b$. Use Jensen's inequality to verify that if $m(t) < \infty$, then $m(s) < \infty$ for any s between 0 and t.

Problem 1.35 Prove that the cumulant generating function (CGF) is convex. **HINT:** The CGF possesses a second derivative. Use the Hölder inequality to show that it is nonnegative.

Problem 1.36 For a random variable X the probability generating function (PGF) is defined as $\Pi(t) = E[t^X]$, $t \in \mathbb{R}$. Clearly, it shares the essential properties of a MGF, but is often more convenient when X is integer-valued. See Whittle (2000) for an excellent discussion of this topic.

(a) Show that if $m(t)$ is the MGF of X, then $\Pi(t) = m(\log(t))$.
(b) Show that

$$\left.\frac{d^k\Pi(t)}{dt^k}\right|_{t=1} = E[X^{(k)}].$$

(c) Use the PGF to evaluate $E[X^{(k)}]$ for a Poisson random variable.

Problem 1.37 Suppose $Z \sim N(0,1)$ and $Y = e^Z$. Then Y possesses a log-normal distribution. This distribution presents an interesting counter-example.

(a) Evalute the order k moments of Y. Verify that all order moments are finite. **HINT:** This can be evaluated directly from the MGF of Z.
(b) Determine whether or not Y possesses a MGF.

Problem 1.38 (Chernoff Bound) Let X be any random variable possessing *MGF* $m_X(t)$ and CGF $c_X(t)$. Prove the Chernoff bound

$$P(X \geq x) \leq \exp\left\{\inf_{t>0}\left(c_X(t) - tx\right)\right\}.$$

Then for $X \sim N(\mu, \sigma^2)$ derive the bound $P(X \geq x) \leq e^{-(x-\mu)^2/(2\sigma^2)}$, $x \geq \mu$.

Problem 1.39 Verify that Equation (1.42) may be written $E[|Z|^k] = c_k(k-1)!!$, where $c_k = 1$ if k is even and $c_k = \sqrt{2/\pi}$ if k is odd. Then determine for which positive integers k the random variable $W = |Z^k|$ possesses an MGF, where $Z \sim N(0,1)$.

Problem 1.40 Let X be an exponentially distributed random variable with mean 1. For which powers k does X^k possess an MGF?

Problem 1.41 Verify that the moment generating function of a gamma density $gamma(\tau, \alpha)$ is $m(t) = (1 - t/\tau)^{-\alpha}$.

(a) For what values of t is $m(t)$ finite?
(b) Show that if α is an integer, then $gamma(\tau, \alpha)$ is the density of the sum of independent random variables of density $exp(\tau)$.

Problem 1.42 Prove that if a random variable X satisfies $P(X \in [a, b]) = 1$ for finite constants a, b then it possesses an MGF. **HINT:** Either Theorem 1.20 or Theorem 1.22 may be used.

Problem 1.43 Two random variables X, Y satisfy MGF ordering $X \geq_{mgf} Y$ if $m_X(t) \geq m_Y(t)$ for all $t > 0$.

(a) Prove that if $X \geq_{mgf} Y$ and $E[X] = E[Y]$ then $\text{var}[X] \geq \text{var}[Y]$.
(b) Prove that if $X \geq_{mgf} Y$, $E[X] = E[Y]$ and $\text{var}[X] = \text{var}[Y]$, then $E[X^3] \geq E[Y^3]$.
(c) Prove that if $X \geq_{mgf} Y$ and $\kappa_j[X] = \kappa_j[Y]$ for $j = 1, \ldots, k$, then $\kappa_{k+1}[X] \geq \kappa_{k+1}[Y]$.

Problem 1.44 Verify that the standardized moments defined in Equation (1.46) are invariant with respect to linear transformations.

Problem 1.45 The standardized moments γ_3, γ_4 (Equation (1.46)) are related by various inequalities.

(a) Use the Cauchy-Schwarz inequality to prove the inequality

$$\gamma_3^2 \leq \gamma_4. \tag{1.54}$$

(b) Inequality (1.54) can be sharpened by noting that for a random variable X, $E[(X - a)^2(X - b)^2] \geq 0$ for any constants a, b (Rohatgi and Székely, 1989). In particular, prove the inequality

$$\gamma_3^2 + 1 \leq \gamma_4. \tag{1.55}$$

The essential part of the argument concerns the choice of a, b.
(c) Show that inequality (1.55) is sharp by verifying that equality holds for some random variable.

2

Multivariate Distributions

2.1 Introduction

Chapter 1 considered primarily univariate distributions, while this chapter considers distributions defined for random vectors. Section 2.2, following Section 1.4, introduces multivariate parameter classes which we will encounter in this volume. Section 2.3 deals with the technical problem of determining the distribution of a transformed random vector $g(\mathbf{X})$ given the distribution of $\mathbf{X}$. A number of important problems of inference will depend on its solution.

If we are given a random vector $\mathbf{X} = (X_1, \ldots, X_n)$, the order statistics are defined by the transformation $\mathbf{X}_{ord} = (X_{(1)}, \ldots, X_{(n)}) \in \mathbb{R}^n$, where $X_{(j)}$ is the jth largest element of $\mathbf{X}$. Some problems in statistical inference may be solved by determining the marginal or joint density of some subvector from $\mathbf{X}_{ord}$. This topic is reviewed in Section 2.4.

The χ^2 distribution plays a very important role in a large class of statistical methodology, along with the related T and F distributions, which are sometimes referred to collectively as sampling distributions (Example 1.11). Since these distributions are naturally interpreted as mappings of random vectors, they are best explored in the context of multivariate distribution theory. An understanding of these distributions, along with their noncentral variants, is essential to the practice of statistical methodology. To this end, this volume will emphasize the relationship between these distributions and the branch of matrix theory concerned with idempotent (or projection) matrices (Section B.4.5). This relationship is neatly summarized in a seminal contribution to statistical theory known as Cochran's theorem (Cochran, 1934). These topics are discussed in Section 2.5.

Our discussion of MGFs and CGFs from Chapter 1 continues here to the case of random sums (Section 2.6), and to the problem of determining the MGF for multivariate distributions (Section 2.7).

2.2 Parametric Classes of Multivariate Distributions

Suppose $\mathbf{X} = (X_1, \ldots, X_m)$ is a random vector. The mean vector can be written $\boldsymbol{\mu}_{\mathbf{X}} = E[\mathbf{X}] = (E[X_1], \ldots, E[X_m])^T$. In the context of matrix algebra $\boldsymbol{\mu}_{\mathbf{X}}$ will be interpreted as a column vector.

The $m \times m$ covariance matrix of $\mathbf{X}$ is defined elementwise as

$$\{\Sigma_{\mathbf{x}}\}_{i,j} = \text{cov}[X_i, X_j] = E[(X_i - E[X_i])(X_j - E[X_j])], \tag{2.1}$$

where we take $\text{var}[X_i] = \text{cov}[X_i, X_i]$. Note that by the Cauchy-Schwartz inequality $E[(X_i - E[X_i])(X_j - E[X_j])]$ is well defined and finite whenever $\text{var}[X_i], \text{var}[X_j] < \infty$. When the context permits we may write $\text{var}[\mathbf{X}] = \Sigma_{\mathbf{X}}$. Since $\text{cov}[X, Y] = \text{cov}[Y, X]$, $\Sigma_{\mathbf{X}}$ is always

DOI: 10.1201/9781003049340-2

symmetric. For any linear combination $Y = a_1X_1 + \cdots + a_mX_m$ based on constant coefficients a_i, it is easily shown that

$$\text{var}[Y] = \mathbf{a}^T \Sigma_{\mathbf{X}} \mathbf{a}, \tag{2.2}$$

where $\mathbf{a} = (a_1, \ldots, a_m)^T$ is taken to be a column vector. Since a variance is always nonnegative this must mean $\Sigma_{\mathbf{X}}$ is positive semidefinite, and is positive definite unless a subset of the elements of $\mathbf{X}$ are linearly dependent *wp*1.

Next, suppose $\mathbf{b} \in \mathcal{M}_{k,1}$ and $\mathbf{A} \in \mathcal{M}_{k,m}$ are constant matrices, and $\mathbf{X} \in \mathbb{R}^m$ is a random vector. Then

$$\mathbf{Y} = \mathbf{b} + \mathbf{A}\mathbf{X}$$

is an affine transformation yielding a random vector $\mathbf{Y} \in \mathbb{R}^k$. The mean and variance matrices of $\mathbf{X}$ and $\mathbf{Y}$ are always related by

$$E[\mathbf{Y}] = \mathbf{b} + \mathbf{A}E[\mathbf{X}] \text{ and } \text{var}[\mathbf{Y}] = \mathbf{A}\text{var}[\mathbf{X}]\mathbf{A}^T. \tag{2.3}$$

In particular, suppose $\text{var}[\mathbf{X}]$ is positive definite. Then there exists an invertible square root matrix $\text{var}[\mathbf{X}]^{1/2}$ which satisfies $\text{var}[\mathbf{X}] = [\text{var}[\mathbf{X}]^{1/2}]^T \text{var}[\mathbf{X}]^{1/2}$ (Section B.4.4). If $\mathbf{Y} = [\text{var}[\mathbf{X}]^{-1/2}]^T \mathbf{X}$, then

$$\begin{aligned}
\text{var}[\mathbf{Y}] &= [\text{var}[\mathbf{X}]^{-1/2}]^T \text{var}[\mathbf{X}] \text{var}[\mathbf{X}]^{-1/2} \\
&= [\text{var}[\mathbf{X}]^{-1/2}]^T [\text{var}[\mathbf{X}]^{1/2}]^T \text{var}[\mathbf{X}]^{1/2} \text{var}[\mathbf{X}]^{-1/2} \\
&= \mathbf{I}
\end{aligned}$$

Thus, any random vector with a positive definite covariance matrix possesses a linear transformation yielding independent coordinates of unit variance. Note that $\text{var}[\mathbf{X}]^{1/2}$ may be chosen to be symmetric, but for computational reasons other choices might be made.

We next introduce a number of important parametric classes of multivariate distributions.

Example 2.1 (The Multinomial Distribution) Suppose we are given a probability distribution $\mathbf{P} = (p_1, \ldots, p_m)$ on $\mathcal{S} = \{1, \ldots, m\}$. If we observe an *iid* sample of size n from $\mathbf{P}$, and we let $\mathbf{N} = (N_1, \ldots, N_m)$ be the vector of sample frequencies for each outcome, then $\mathbf{N}$ has a multinomial distribution with density given by

$$\begin{aligned}
f_{\mathbf{N}}(n_1, \ldots, n_m) &= P(N_1 = n_1, \ldots, N_m = n_n) \\
&= \frac{n!}{\prod_{i=1}^m n_i!} \prod_{i=1}^m p_i^{n_i}, \quad \min_i n_i \geq 0, \ n_1 + \cdots + n_m = n.
\end{aligned}$$

This is denoted as $\mathbf{N} \sim \textit{multinom}(n, \mathbf{P})$. The marginal distributions are $N_i \sim \textit{bin}(n, p_i)$. It may be verified that $E[N_iN_j] = n(n-1)p_ip_j$, $i \neq j$ (express N_i, N_j as sums of Bernoulli random variables). The covariance matrix of $\mathbf{N}$ therefore has elements:

$$[\text{var}[\mathbf{N}]]_{ij} = \begin{cases} np_j(1-p_j); & i = j \\ -np_ip_j; & i \neq j \end{cases}. \tag{2.4}$$

Note that this matrix is positive semidefinite, but not positive definite, and therefore not invertible. This follows from the linear constraint $n = N_1 + \cdots + N_m$. Nonetheless, it may still be used to calculate variances of linear combinations of $\mathbf{N}$. Note that we may write

$$n^{-1}\text{var}[\mathbf{N}] = \mathbf{D}_{\mathbf{P}} - \mathbf{P}\mathbf{P}^T \tag{2.5}$$

where $\mathbf{D}_{\mathbf{P}}$ is a diagonal matrix with elements $\{\mathbf{D}_{\mathbf{P}}\}_{jj} = p_j$. ∎

The following theorem will prove quite useful for the analysis of the multinomial distribution.

Theorem 2.1 (Linear Combinations of a Multinomial Vector) Suppose $\mathbf{N} = (N_1, \ldots, N_m) \sim multinom(n, \mathbf{P})$ is a mulitnomial vector of frequencies from probability distribution $\mathbf{P} = (p_1, \ldots, p_m)$. Then

$$E[\mathbf{c}^T\mathbf{N}] = n\mathbf{c}^T\mathbf{P}, \ n^{-1}\text{var}[\mathbf{c}^T\mathbf{N}] = \left[\sum_{j=1}^m c_j^2 p_j\right] - \left[\sum_{j=1}^m c_j p_j\right]^2 \tag{2.6}$$

for any vector of coefficients $\mathbf{c} \in \mathbb{R}^m$. ∎

Proof. The value of $E[\mathbf{c}^T\mathbf{N}]$ is easily verified. The variance follows directly by evaluating $\mathbf{c}^T\mathbf{D}_\mathbf{P}\mathbf{c} - \mathbf{c}^T\mathbf{P}\mathbf{P}^T\mathbf{c}$ from Equation (2.5). □

That var[$\mathbf{N}$] is not positive definite can be seen by applying Theorem 2.1 to the coefficient vector $\mathbf{c} = \mathbf{1}$.

Example 2.2 (The Multivariate Hypergeometric Distribution) Just as the multinomial distribution is an extension of the binomial distribution, the multivariate hypergeometric distribution is an extension of the hypergeometric distribution (Example 1.7). A bin contains n_i balls of a unique color i, $i = 1, \ldots, m$. There are total of $n = \sum_i n_i$ balls. A random selection of $k \leq n$ balls is made (equivalently, the balls are sampled without replacement). Let X_i be the number of each type of ball selected. The distribution of $\mathbf{X} = (X_1, \ldots, X_n)$ is

$$f_\mathbf{X}(x_1, \ldots, x_m) = \binom{n}{k}^{-1} \prod_{i=1}^m \binom{n_i}{x_i},$$

assuming the selection has nonzero probability. This distribution is denoted as $\mathbf{X} \sim multihyper(k, n_1, \ldots, n_m)$. See Problem 2.8. ∎

Example 2.3 (The Multivariate Normal Distribution) Suppose $\boldsymbol{\mu} = (\mu_1, \ldots, \mu_m)^T \in \mathcal{M}_{m,1}$ and $\Sigma \in \mathcal{M}_m$ is positive definite. The multivariate normal density function is defined on $\mathbb{R}^m$ by

$$\begin{aligned} f(\mathbf{x} \mid \boldsymbol{\mu}, \Sigma) &= (2\pi)^{-m/2} \det(\Sigma)^{-1/2} \exp(-Q/2), \quad \mathbf{x} \in \mathbb{R}^m, \text{ where} \\ Q &= (\mathbf{x} - \boldsymbol{\mu})^T \Sigma^{-1} (\mathbf{x} - \boldsymbol{\mu}). \end{aligned} \tag{2.7}$$

Then $\mathbf{X} = (X_1, \ldots, X_m)$ is a multivariate normal random vector if it possesses this density, in which case it may be shown that $E[\mathbf{X}] = \boldsymbol{\mu}$, var[$\mathbf{X}$] $= \Sigma$. This is denoted as $\mathbf{X} \sim N(\boldsymbol{\mu}, \Sigma)$. In addition, the marginal distributions are $X_i \sim N(\mu, \Sigma_{i,i})$. The $m = 2$ case is often referred to as the *bivariate normal* distribution.

It is important to note that a random vector with marginal normal densities is not necessarily multivariate normal. For example, if $X \sim N(0, 1)$ and $Y = SX$ where S is an independent random sign, then $Y \sim N(0, 1)$, cov[X, Y] $= \mathbf{0}$, but (X, Y) does not possess a multivariate normal density.

The definition of a multivariate normal random vector can be generalized to include any random vector of the form $\mathbf{X} = \boldsymbol{\mu} + \mathbf{A}\mathbf{Z}$, where $\boldsymbol{\mu} \in \mathcal{M}_{m,1}$, $\mathbf{A} \in \mathcal{M}_{m,k}$ and $\mathbf{Z}$ is a $k \times 1$ column vector of independent unit normal random variables. In this case var[$\mathbf{X}$] need not be positive definite, so (2.7) cannot be used directly. ∎

Example 2.4 (The Dirichlet Distribution) The Dirichlet distribution is a multivariate extension of the beta distribution (Example 1.12). Suppose $\mathbf{x} \in \mathbb{R}^m$, and let $\boldsymbol{\alpha} = (\alpha_1, \ldots, \alpha_m) \in \mathbb{R}^m$ be a parameter, $\alpha_i > 0$. Then,

$$f(\mathbf{x}) = B(\boldsymbol{\alpha})^{-1} \prod_{j=1}^{m} x_j^{\alpha_j - 1}, \ x_j \in (0,1), \ \sum_{j=1}^{m} x_j = 1, \tag{2.8}$$

where $B(\boldsymbol{\alpha}) = \Gamma\left(\sum_{j=1}^m \alpha_j\right)^{-1} \prod_{j=1}^m \Gamma(\alpha_j)$. If $\mathbf{X}$ possesses density (2.8), we write this as $\mathbf{X} \sim$ *dirichlet*$(\boldsymbol{\alpha})$. The marginal distribution of any component of $\mathbf{X}$ is $X_j \sim$ *beta*$(\alpha_j, \sum_{i \neq j} \alpha_i)$. See Problems 2.10 and 2.17. ∎

2.3 Multivariate Transformations

Suppose we are given a random vector $\mathbf{X} = (X_1, \ldots, X_n)$. Here we assume the distribution of $\mathbf{X}$ is absolutely continuous with respect to Lebesgue measure on $\mathbb{R}^n$. We wish to derive the density of the random vector $\mathbf{Y} = (Y_1, \ldots, Y_n)$, obtained from $\mathbf{X}$ by a transformation:

$$y_1 = g_1(x_1, \ldots, x_n), \quad \cdots \quad, \ y_n = g_n(x_1, \ldots, x_n).$$

If the transformation $g = (g_1(\mathbf{x}), \ldots, g_n(\mathbf{x}))$ is injective, we can define the inverse transformation $h = (h_1(\mathbf{y}), \ldots, h_n(\mathbf{y}))$:

$$x_1 = h_1(y_1, \ldots, y_n), \quad \cdots \quad, \ x_n = h_n(y_1, \ldots, y_n).$$

The Jacobian matrix of a multivariate transformation is the matrix of all partial first derivatives:

$$\mathbf{J} = \frac{\partial \mathbf{y}}{\partial \mathbf{x}} = \begin{bmatrix} \frac{\partial g_1(\mathbf{x})}{\partial x_1} & \cdots & \frac{\partial g_1(\mathbf{x})}{\partial x_n} \\ \vdots & \ddots & \vdots \\ \frac{\partial g_n(\mathbf{x})}{\partial x_1} & \cdots & \frac{\partial g_n(\mathbf{x})}{\partial x_n} \end{bmatrix}$$

The notation $\partial \mathbf{y}/\partial \mathbf{x}$ denotes this form of vector differentiation. The densities $f_{\mathbf{X}}$ and $f_{\mathbf{Y}}$ are related by the equation

$$f_{\mathbf{Y}}(y_1, \ldots, y_n) = |\mathbf{J}(y_1, \ldots, y_n)|^{-1} f_{\mathbf{X}}(h_1(y_1, \ldots, y_n), \ldots, h_n(y_1, \ldots, y_n)),$$

where $|\mathbf{J}| = |\det(\mathbf{J})|$.

In carrying out this derivation, we have a choice regarding $\mathbf{J}$. We could calculate the Jacobian matrix of the inverse transformation h:

$$\mathbf{J}' = \frac{\partial \mathbf{x}}{\partial \mathbf{y}} = \begin{bmatrix} \frac{\partial h_1(\mathbf{y})}{\partial y_1} & \cdots & \frac{\partial h_1(\mathbf{y})}{\partial y_n} \\ \vdots & \ddots & \vdots \\ \frac{\partial h_n(\mathbf{y})}{\partial y_1} & \cdots & \frac{\partial h_n(\mathbf{y})}{\partial y_n} \end{bmatrix}.$$

Then $|\mathbf{J}| = |\mathbf{J}'|^{-1}$.

The objective may be to derive the density of a subvector $\mathbf{Y}' = (Y_1, \ldots, Y_{n'})$ of $\mathbf{Y}$, where $n' < n$. This is obtained by integrating $f_{\mathbf{Y}}$ through the remaining coordinates:

$$f_{\mathbf{Y}'}(y_1, \ldots, y_{n'}) = \int_{y_{n'+1}} \cdots \int_{y_n} f_{\mathbf{Y}}(y_1, \ldots, y_{n'}, y_{n'+1}, \ldots, y_n)\, dy_{n'+1} \cdots dy_n. \tag{2.9}$$

Example 2.5 (Transformations Involving Polar Coordinates) One convenient method of simulating normal random variables is based on the following observation. Suppose $\mathbf{X} = (X_1, X_2)$ are two independent random variables with normal distribution $N(0,1)$. Then

$$f_{\mathbf{X}}(x_1, x_2) = \frac{1}{2\pi} e^{-\frac{x_1^2+x_2^2}{2}}, \quad (x_1, x_2) \in \mathbb{R}^2.$$

Then consider the transformation to polar coordinates

$$r = \sqrt{x_1^2 + x_2^2}, \quad \theta = \arctan(x_2/x_1). \tag{2.10}$$

The inverse transformation is $x_1 = r\cos(\theta)$, $x_2 = r\sin(\theta)$. The Jacobian of the latter transformation is

$$\mathbf{J} = \frac{\partial(x_1, x_2)}{\partial(r, \theta)} = \begin{bmatrix} \frac{\partial r\cos(\theta)}{\partial r} & \frac{\partial r\cos(\theta)}{\partial \theta} \\ \frac{\partial r\sin(\theta)}{\partial r} & \frac{\partial r\sin(\theta)}{\partial \theta} \end{bmatrix} = \begin{bmatrix} \cos(\theta) & -r\sin(\theta) \\ \sin(\theta) & r\cos(\theta) \end{bmatrix}.$$

Then $\det(\mathbf{J}) = r\cos(\theta)^2 + r\sin(\theta)^2 = r$. The density of (R, Θ) defined by the transformation (2.10) is therefore

$$f_{(R,\Theta)}(r, \theta) = r f_{\mathbf{X}}(x_1(r,\theta), x_2(r,\theta)) = \frac{1}{2\pi} r e^{-\frac{r^2\cos(\theta)^2 + r^2\sin(\theta)^2}{2}} = \frac{1}{2\pi} r e^{-\frac{r^2}{2}}.$$

The support of Θ is $[0, 2\pi)$, so by using (2.9), the marginal density of R is easily seen to be

$$f_R(r) = \int_0^{2\pi} \frac{1}{2\pi} r e^{-\frac{r^2}{2}} d\theta = r e^{-\frac{r^2}{2}}.$$

From Example 1.24, we conclude that $2R^2 \sim \exp(1)$, since $R \sim weibull(2, 1/\sqrt{2})$. Therefore, an independent pair of normal random variables is obtained from an exponentially distributed random variable R^2 and a uniformly distributed random variable Θ. It is easily verified that if $U \sim unif(0,1)$, then $\log(U) \sim exp(1)$, so that (X_1, X_2) is obtainable as a transformation of two independent uniformly distributed random variables.

It is important to note the role played by $\mathbf{J}$, which is to rescale a differential element used for integration. We can represent this as $dx_1 dx_2 = |\mathbf{J}| dr d\theta = r dr d\theta$. This relationship can be seen in Figure 2.1. Clearly, the size of $drd\theta$ is not constant. Thus, when intergrating with respect to Lebesgue measure, the role of the Jacobian is to standardize the measure. ■

The multivariate transformation method is often used to determine the density of real-valued mappings of several random variables. A typical example of this problem follows.

Example 2.6 (Deriving a Beta Distribution from a Gamma Sample) Suppose $X_i \sim gamma(1, \alpha_i)$ for $i = 1, 2$, and that X_1, X_2 are independent. We wish to derive the density of $X_1/(X_1 + X_2)$. Consider the injective transformation $(Y_1, Y_2) = \left(X_1 + X_2, \frac{X_1}{X_1+X_2}\right) \in (0, \infty) \times (0, 1)$. The joint density of $\mathbf{X} = (X_1, X_2)$ is

$$f_{\mathbf{X}}(x_1, x_2) = \frac{(x_1)^{\alpha_1 - 1}(x_2)^{\alpha_2 - 1}}{\Gamma(\alpha_1)\Gamma(\alpha_2)} \exp(-(x_1 + x_2)), \quad x_1, x_2 > 0.$$

The inverse transformation is $X_1 = Y_1 Y_2$, $X_2 = Y_1(1 - Y_2)$, and the Jacobian matrix of the (inverse) transformation is

$$\frac{\partial(x_1, x_2)}{\partial(y_1, y_2)} = \mathbf{J} = \begin{bmatrix} y_2 & y_1 \\ 1 - y_2 & -y_1 \end{bmatrix}$$

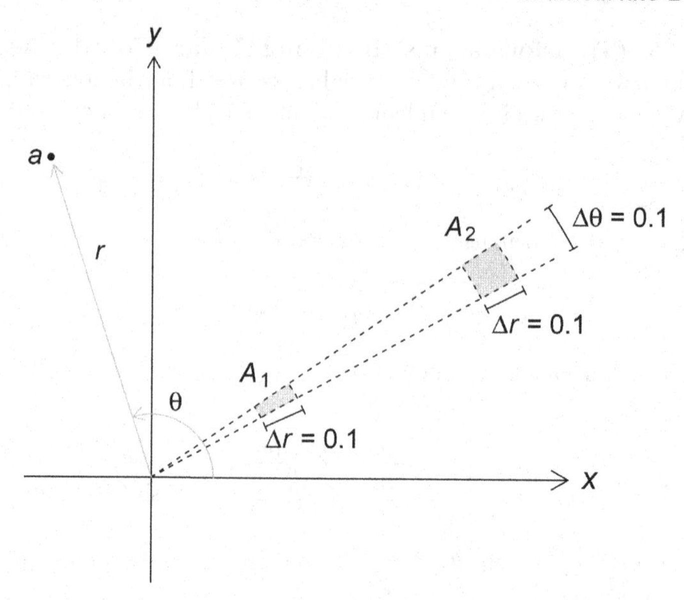

FIGURE 2.1
Comparison of differential elements for Euclidean and polar coordinates (Example 2.5).

with $\det(\mathbf{J}) = -y_1 y_2 - y_1(1 - y_2) = -y_1$. The joint density of $\mathbf{Y} = (Y_1, Y_2)$ is therefore

$$\begin{aligned} f_{\mathbf{Y}}(y_1, y_2) &= |\det(\mathbf{J})| f_X(y_1 y_2, y_1(1 - y_2)) \\ &= y_1 \frac{(y_1 y_2)^{\alpha_1 - 1}(y_1(1 - y_2))^{\alpha_2 - 1}}{\Gamma(\alpha_1)\Gamma(\alpha_2)} \exp(-y_1) \\ &= \frac{y_1^{\alpha_1 + \alpha_2 - 1} y_2^{\alpha_1 - 1}(1 - y_2)^{\alpha_2 - 1}}{\Gamma(\alpha_1)\Gamma(\alpha_2)} \exp(-y_1), \end{aligned}$$

where the support is $(y_1, y_2) \in (0, \infty) \times (0, 1)$.

Note that the density $f_{\mathbf{Y}}(y_1, y_2)$ is separable with respect to y_1 and y_2, so the marginal density of Y_2 is $f_{Y_2}(y_2) = K y_2^{\alpha_1 - 1}(1 - y_2)^{\alpha_2 - 1}$, where K does not depend on y_2. We then have $Y_2 \sim beta(\alpha_1, \alpha_2)$. ■

The Dirichlet distribution is a multivariate extension of the beta distribution, and can be constructed from independent *gamma* distributed random variables by an extension of the method used in Example 2.6 (Problem 2.17).

2.4 Order Statistics

Suppose we are given an *iid* sample $X_1, \ldots, X_n$ from density $f(x)$. It is sometimes necessary to consider the density of the kth highest element of the sample. Accordingly, we define the order transformation:

Definition 2.1 (Order Statistics) The order transformation of any $\mathbf{x} = (x_1, \ldots, x_n) \in \mathbb{R}^n$ is $\mathbf{x}_{ord} = (x_{(1)}, \ldots, x_{(n)}) \in \mathbb{R}^n$, where $x_{(k)}$ is the kth ranked element of $\mathbf{x}$. The order transformation is well defined in the presence of ties. The order transformation of a random vector $\mathbf{X} \in \mathbb{R}^n$ defines the order statistics. ■

Although interest is usually in $X_{(1)} = \min_i X_i$ or $X_{(n)} = \max_i X_i$, it is possible to characterize the distributions of any order statistic. In fact, a single method gives joint densities of any subvector of $\mathbf{X}_{ord}$, assuming X_1 is a continuous random variable. To see this, select indices $1 \leq k_1 < \cdots < k_m \leq n$, where $1 \leq m \leq n$. Select $t_1 < \cdots < t_m$. The joint density of subvector $\mathbf{X}'_{ord} = (X_{(k_1)}, \ldots, X_{(k_m)}) \in \mathbb{R}^m$ may be given as the limit

$$f_{\mathbf{X}'_{ord}}(t_1, \ldots, t_m) = \lim_{\epsilon \downarrow 0} \frac{P\left(\cap_{i=1}^m \{X_{(k_i)} \in (t_i - \epsilon, t_i]\}\right)}{\epsilon^m}. \tag{2.11}$$

In the following development, it will be convenient to define for any event $A \subset \mathbb{R}$ $N_A = \sum_{i=1}^n I\{X_i \in A\}$. At this point, consider the case $m = 1$, with $k_1 = k$. Then consider the event measured in the numerator of (2.11), for some positive ϵ, $F = \{X_{(k)} \in (t - \epsilon, t]\}$. We may then describe this event in terms of the *iid* sample. Clearly, if F occurs, we must have $N_{(t-\epsilon,t]} \geq 1$. More precisely, the interval $(t - \epsilon, t]$ contains the kth ranked observation $X_{(k)}$. So, if F occurs, and $N_{(t-\epsilon,t]} = 1$ we must also have $N_{(-\infty,t-\epsilon]} = k - 1$ and $N_{(t,\infty)} = n - k$. But F is not exclusive of the event $\{N_{(t-\epsilon,t]} = 2\}$, in which case we could also have $(N_{(-\infty,t-\epsilon]}, N_{(t,\infty)}) = (k-2, n-k)$ or $(N_{(-\infty,t-\epsilon]}, N_{(t,\infty)}) = (k-1, n-k-1)$, in addition to other possibilities.

The next step is to recognize that we have constructed a multinomial vector $\mathbf{N} = (N_1, N_2, N_3) = (N_{(-\infty,t-\epsilon]}, N_{(t-\epsilon,t]}, N_{(t,\infty)})$ with $\sum_{j=1}^3 N_j = n$ and probabilities $p_1 = P(X_1 \in (-\infty, t-\epsilon])$, $p_2 = P(X_1 \in (t-\epsilon, t])$, $p_3 = 1 - p_1 - p_2$. Furthermore, the event F can be expressed in terms of $\mathbf{N}$. That is, we may identify a set of multinomial vector outcomes, say $F_{\mathbf{N}}$, such that

$$P(F) = \sum_{(n_1,n_2,n_3) \in F_{\mathbf{N}}} P((N_1, N_2, N_3) = (n_1, n_2, n_3)),$$

where

$$P((N_1, N_2, N_3) = (n_1, n_2, n_3)) = \frac{n!}{n_1! n_2! n_3!} p_1^{n_1} p_2^{n_2} p_3^{n_3} \tag{2.12}$$

provided $n_j \geq 0$, $\sum_j n_j = n$.

Next, recalling our original goal of evaluating (2.11), we must investigate what happens when we divide any probability of the form (2.12) by ϵ, which is then allowed to approach 0. If f is a continuous density function, we have, given CDF F, $p_1 = F(t) + O(\epsilon)$, $p_2 = f(t)\epsilon + o(\epsilon)$, $p_3 = 1 - F(t) + O(\epsilon)$. Next, note that for any $(n_1, n_2, n_3) \in F_{\mathbf{N}}$ we must have $n_2 \geq 1$, in which case

$$\begin{aligned}
&\lim_{\epsilon \downarrow 0} \frac{1}{\epsilon} \frac{n!}{n_1! n_2! n_3!} p_1^{n_1} p_2^{n_2} p_3^{n_3} \\
&= \lim_{\epsilon \downarrow 0} \frac{1}{\epsilon} \frac{n!}{n_1! n_2! n_3!} \left(F(t) + O(\epsilon)\right)^{n_1} \left(f(t)\epsilon + o(\epsilon)\right)^{n_2} \left(1 - F(t) + O(\epsilon)\right)^{n_3} \\
&= \lim_{\epsilon \downarrow 0} \epsilon^{n_2 - 1} \frac{n!}{n_1! n_2! n_3!} \left(F(t) + O(\epsilon)\right)^{n_1} \left(f(t) + \frac{o(\epsilon)}{\epsilon}\right)^{n_2} \left(1 - F(t) + O(\epsilon)\right)^{n_3} \\
&= \left[\lim_{\epsilon \downarrow 0} \epsilon^{n_2 - 1}\right] \frac{n!}{n_1! n_2! n_3!} F(t)^{n_1} f(t)^{n_2} \left(1 - F(t)\right)^{n_3}.
\end{aligned}$$

But this limit is zero for any $n_2 > 1$, and there is only one outcome $(k-1, 1, n-k) \in F_{\mathbf{N}}$ for which $n_2 = 1$. We may then conclude

$$f_{X_{(k)}}(t) = \frac{n!}{(k-1)!(n-k)!} F(t)^{k-1} f(t) \left(1 - F(t)\right)^{n-k}.$$

Essentially the same argument may be used to give any joint density

$$f_{\mathbf{X}'_{ord}}(t_1,\ldots,t_m) = \frac{n!}{(n-k_m)!\prod_{j=1}^{m}(k_j-k_{j-1}-1)!}(1-F(t_m))^{n-k_m} \times \prod_{j=1}^{m}(F(t_j)-F(t_{j-1}))^{k_j-k_{j-1}-1}f(t_j), \tag{2.13}$$

where we set for convenience $k_0 = 0$, $t_0 = -\infty$ (Problem 2.26).

The maximum and minimum observations, in particular, have densities

$$f_{X_{(1)}}(t) = n(1-F(t))^{n-1}f(t), \quad f_{X_{(n)}}(t) = nF(t)^{n-1}f(t).$$

In addition, the order statitics themselves have the joint density

$$f_{\mathbf{X}_{ord}}(t_1,\ldots,t_n) = n!\prod_{j=1}^{n} f(t_j), \; t_1 < \cdots < t_n.$$

We will make use of the following example later in this volume.

Example 2.7 (Order Statistics for an Exponential Sample) Let $\mathbf{X} = (X_1,\ldots,X_n)$ be an *iid* sample with $X_1 \sim exp(1/\lambda)$. What is the density of $\mathbf{X}$ conditional on $\min_i X_i = X_{(1)} = t_1$?

First note that the CDF of f is $F(x) = 1 - e^{-x/\lambda}$. Then the conditional density is

$$f_{\mathbf{X}_{ord}|X_{(1)}}(t_1,\ldots,t_n \mid t_1) = (n-1)!\prod_{j=2}^{n} \lambda e^{-\lambda(t_j-t_1)}, \; t_1 < \cdots < t_n.$$

What this means is that conditional on $X_{(1)} = t_1$, the remaining order statistics $X_{(2)},\ldots,X_{(n)}$ have the same distribution as the order statistics of an *iid* sample $\mathbf{X}' = (X'_1,\ldots,X'_{n-1})$ of size $n-1$ from the density $f'(x) = \lambda e^{-\lambda(x-t_1)}I\{x > t_1\}$, which is a location shifted exponential density with the same rate parameter λ. This then completely characterizes the distribution of $\mathbf{X}$ conditional on $X_{(1)} = t_1$. Alternatively, $X_{(2)} - X_{(1)},\ldots,X_{(n)} - X_{(1)}$ have the same distribution as the order statistics of an *iid* sample of size $n-1$ from the density $f(x) = \lambda e^{-\lambda x}I\{x > 0\}$.

This is a consequence of the memoryless property of the exponentially distributed lifetime. ∎

In the next example, we evaluate the marginal distribution of a single order statistic.

Example 2.8 (Distribution of the Sample Median) Suppose we are given an *iid* sample of size n from a normal distribution $N(\mu,\sigma^2)$. Both the sample mean and sample median can be used to estimate μ. To fix ideas, suppose n is odd, so that the sample median is $M = X_{([n+1]/2)}$. The density of M is

$$f_M(x) = [([n-1]/2)!]^{-2}n!F(x)^{[n-1]/2}(1-F(x))^{[n-1]/2}f(x). \tag{2.14}$$

Densities for the sample mean and median for a sample of size $n = 11$ from $N(0,1)$ are shown in Figure 2.2. For this case, the sample mean is the more accurate estimate. The sample median might be preferred when the sample is potentially contaminated by outliers. In this case, the loss of efficiency can be bounded by comparing the respective densities. ∎

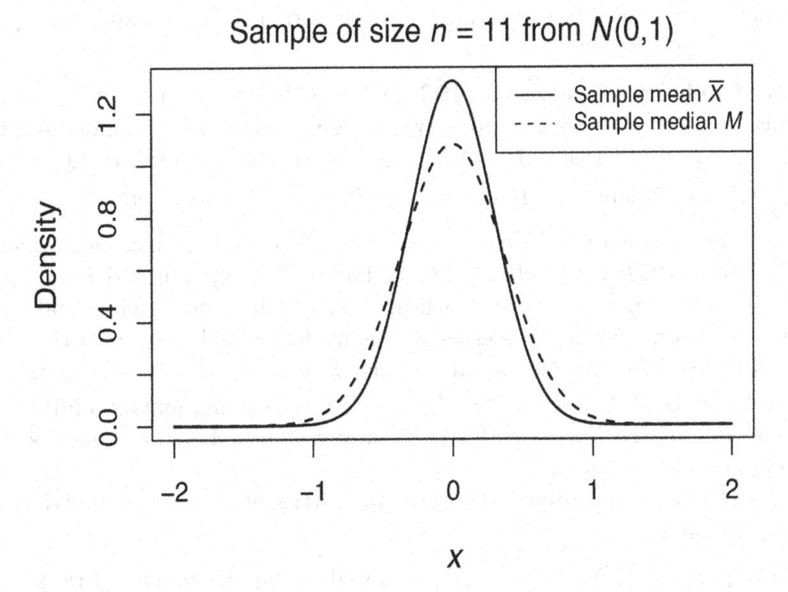

FIGURE 2.2
Densities of sample mean and median for a sample of size $n = 11$ from $N(0, 1)$ (Example 2.8).

2.5 Quadratic Forms, Idempotent Matrices and Cochran's Theorem

Recall from Example 1.11 that a random variable W possesses a χ^2 distribution with ν degrees of freedom if it may be constructed as $W = \sum_{i=1}^{\nu} Z_i^2$, where $Z_1, \ldots, Z_\nu$ are *iid* with $Z_1 \sim N(0, 1)$. This is denoted as $W \sim \chi^2_\nu$.

Formally, a quadratic form is any polynomial in m variables containing only second-order terms. It is represented in matrix form as $\mathbf{x}^T \mathbf{Q} \mathbf{x}$, where $\mathbf{x}$ is an $m \times 1$ column vector and $\mathbf{Q} \in \mathcal{M}_m$, so that the m variables are represented by $\mathbf{x}$.

If $\boldsymbol{\epsilon} \sim N(\mathbf{0}, \mathbf{I}_m)$ and $\mathbf{Q} = \mathbf{I}_m$, then, clearly, $W = \boldsymbol{\epsilon}^T \mathbf{Q} \boldsymbol{\epsilon} \sim \chi^2_m$. However, other quadratic forms $\boldsymbol{\epsilon}^T \mathbf{Q} \boldsymbol{\epsilon}$ commonly arise in statistical methodology which possess a χ^2_q distribution when $\mathbf{Q} \neq \mathbf{I}_m$. An understanding of this structure can simplify an analysis considerably, particularly when $\mathbf{Q}$ is an idempotent matrix (Section B.4.5).

The next theorem gives a direct relationship between the quadratic form and the χ^2 distribution.

Theorem 2.2 (χ^2 Distribution Derived from a Quadratic Form) Suppose $\mathbf{Y} \sim N(\mathbf{0}, \mathbf{Q})$ is an $n \times 1$ column matrix, and $\mathbf{Q} = \mathbf{Q}_{1/2}^T \mathbf{Q}_{1/2}$ is nonsingular. Then

(i) $[\mathbf{Q}_{1/2}^{-1}]^T \mathbf{Y} \sim N(\mathbf{0}, \mathbf{I}_n)$.

(ii) $\mathbf{Y}^T \mathbf{Q}^{-1} \mathbf{Y} \sim \chi^2_n$. ■

Proof. To prove (i), we note that the covariance matrix of $[\mathbf{Q}_{1/2}^{-1}]^T \mathbf{Y}$ is $[\mathbf{Q}_{1/2}^{-1}]^T [\mathbf{Q}_{1/2}]^T \mathbf{Q}_{1/2} \mathbf{Q}_{1/2}^{-1} = \mathbf{I}_n$. Then (ii) follows directly from (i). □

Cochran's theorem is of considerable importance in statistical theory, as it yields not only elegant solutions to some important problems of statistical methodology, but also characterizes precisely the analysis of variance, or the notion that variation can be decomposed into independent components based on its source. The theorem in its original form (Cochran, 1934) may be stated as follows:

Theorem 2.3 (Cochran's Theorem (1934)) Let $\mathbf{X}$ be an $n \times 1$ random vector of *iid* standard normal random variables. Suppose we are given m positive semidefinite matrices $\mathbf{Q}_j \in \mathcal{M}_n$, $j = 1, \ldots, m$ such that $\mathbf{I} = \sum_{j=1}^m \mathbf{Q}_j$. Let n_j be the rank of $\mathbf{Q}_j$. Then $q_j = \mathbf{X}^T \mathbf{Q}_j \mathbf{X} \in \chi^2_{n_j}$ and are independently distributed *iff* $n_1 + \cdots + n_m = n$. ■

The relationship of Theorem 2.3 to matrix algebra theory has since been a subject of some interest (James, 1952; Graybill and Marsaglia, 1957; Chipman and Rao, 1964; Tian and Styan, 2006). The proof offered in Cochran (1934) relies on showing that under the given sufficient condition each $\mathbf{Q}_j$ possesses n_j eigenvalues equal to 1, with the remaining eigenvalues equal to 0. We may also assume each $\mathbf{Q}_j$ is a real symmetric matrix, and so may be written $\mathbf{Q}_j = \mathbf{U}_j^T \Lambda_j \mathbf{U}_j$, where $\mathbf{U}_j \in \mathcal{M}_n$ is an orthogonal matrix, and $\Lambda_j \in \mathcal{M}_n$ is a diagonal eigenvalue matrix (Section B.4.4). However, it may be equivalently stated that the matrices $\mathbf{Q}_j$ are idempotent.

The connection between Theorem 2.3 and the properties of idempotent matrices depends on the following theorem.

Theorem 2.4 Suppose $\mathbf{A} = \sum_{j=1}^m \mathbf{A}_j \in \mathcal{M}_n$ is a real symmetric matrix of rank q, and $\mathbf{A}_j$ are real symmetric matrices of rank q_i. Consider the following four conditions:

(a) Each $\mathbf{A}_j$ is idempotent.
(b) $\mathbf{A}_i \mathbf{A}_j = \mathbf{0}$, $i \neq j$.
(c) $\mathbf{A}$ is idempotent.
(d) $q_1 + \cdots + q_m = q$.

Then the following statements hold:

(i) Any two of the conditions (a), (b) and (c) imply all four conditions (a)–(d).
(ii) Conditions (c) and (d) imply conditions (a) and (b). ■

Proof. See Theorem 1 of Graybill and Marsaglia (1957). □

This leads to a second version of Cochran's theorem.

Theorem 2.5 (Cochran's Theorem II) The following three statements regarding quadratic forms hold:

(i) Let $\mathbf{X}$ be an $n \times 1$ vector of uncorrelated random variables with zero mean and constant variance σ^2. If $\mathbf{H} \in \mathcal{M}_n$ is an idempotent matrix of rank ν then $E[Q] = \nu\sigma^2$, where $Q = \mathbf{X}^T \mathbf{H} \mathbf{X}$.
(ii) Suppose $\mathbf{X}$ is an $n \times 1$ vector of *iid* random variables of distribution $N(0, \sigma^2)$. If $\mathbf{H} \in \mathcal{M}_n$ is an idempotent matrix of rank ν then $Q = [\sigma^2]^{-1} \mathbf{X}^T \mathbf{H} \mathbf{X} \sim \chi^2_\nu$.
(iii) Suppose we are given m symmetric idempotent matrices $\mathbf{H}_j \in \mathcal{M}_n$, $j = 1, \ldots, m$ such that $\mathbf{H} = \sum_{j=1}^m \mathbf{H}_j$ is also idempotent. Then $Q_j = [\sigma^2]^{-1} \mathbf{X}^T \mathbf{H}_j \mathbf{X} \in \chi^2_{n_j}$ and are independently distributed. ■

Proof. By Theorems B.5 and B.8 we may write $\mathbf{H} = \mathbf{U}^T \Lambda \mathbf{U}$, where $\mathbf{U} \in \mathcal{M}_n$ is an orthogonal matrix, and $\Lambda \in \mathcal{M}_n$ is a diagonal matrix which contains exactly ν 1's and $n - \nu$ 0's on the diagonal. Then $\mathbf{Z} = \mathbf{U}\mathbf{X}$ is a vector of n uncorrelated random variables of mean zero and variance σ^2, since $\mathbf{U}$ is orthogonal. Then $Q = \mathbf{X}^T \mathbf{H} \mathbf{X} = \mathbf{Z}^T \Lambda \mathbf{Z}$ is the sum of squares of ν elements of $\mathbf{Z}$, therefore, $E[q] = \nu\sigma^2$.

(ii) The proof is essentially the same as for statement (i) after noting that $\mathbf{Z}$ is a vector of n *iid* random variables of distribution $N(0, \sigma^2)$.

(iii) The remainder of the theorem is a direct consequence of statement (ii) and Theorem 2.4. □

A large number of statistical methods are based on the T, χ^2 and F distributions, introduced in Example 1.11 as the "sampling distributions". They can be defined directly by their densities, but there is usually more interest in their relationship to an *iid* normal sample. Their importance to statistical inference can largely be explained by Cochran's theorem.

When a sampling distribution is constructed from vectors $\mathbf{Z}_0 \sim N(\mathbf{0}, \mathbf{I}_n)$ the distribution is qualified by the term "central". However, many statistical methods depend on understanding how a sampling distribution is perturbed when $\mathbf{Z}_0$ is replaced with $\mathbf{Z} \sim N(\boldsymbol{\mu}, \mathbf{I}_n)$, where $\boldsymbol{\mu} \neq \mathbf{0}$. The resulting distributions are referred to as the noncentral variants, which we define next.

Definition 2.2 (Noncentral χ^2 Distribution) Suppose $\mathbf{Z} \sim N(\boldsymbol{\mu}, \mathbf{I}_n)$ is an $n \times 1$ random column matrix, with mean vector $\boldsymbol{\mu} = [\mu_1, \ldots, \mu_n]^T$. Then $W = \sum_{i=1}^n Z_i^2$ possesses a noncentral χ^2 distribution with n degrees of freedom and noncentrality parameter $\Delta = \sum_{i=1}^n \mu_i^2$. The density of W may be completely specified using only n and Δ, so we denote the distribution $W \sim \chi^2_{\Delta;n}$. If $\Delta = 0$, we have the central χ^2 distribution $\chi^2_{0;n} = \chi^2_n$. The density is not expressible with a simple formula, but can be expressed either in series or integral form. ■

Definition 2.2 implicitly assumes that the noncentral χ^2 distribution depends on $\boldsymbol{\mu}$ only through a single noncentrality parameter $\Delta = \sum_{i=1}^n \mu_i^2$. This extraordinary fact must be verified, which we now do. This can be done using the MGF of the square of a normal random variable.

Example 2.9 (MGF of a Squared Normal Random Variable) Suppose $X \sim N(\mu, 1)$. The MGF of X^2 can be very informative of the sampling distributions:

$$m_{X^2}(t) = \int e^{tx^2} (2\pi)^{-1/2} e^{-(x-\mu)^2/2} dx = e^{-\mu^2 t/(1-2t)}/(1-2t)^{1/2}, \ t < 1/2.$$

By comparing MGFs, we can conclude from Example 2.9 that $W \sim gamma(2, n/2)$, when $W \sim \chi^2_n$ is a central χ^2 random variable. However, another important conclusion can be made. If $\mathbf{Z} \sim N(\boldsymbol{\mu}, \mathbf{I}_n)$, then the MGF of $W = \sum_{i=1}^n Z_i^2$ is, following Section 2.6,

$$m_W(t) = \prod_{i=1}^n m_{Z_i}(t) = \prod_{i=1}^n \frac{e^{-\mu_i^2 t/(1-2t)}}{(1-2t)^{1/2}} = \frac{e^{-\Delta t/(1-2t)}}{(1-2t)^{n/2}},$$

so that the distribution of W depends on $\boldsymbol{\mu}$ only through noncentrality parameter Δ. ■

We will need to evaluate the noncentrality parameter induced by various quadratic forms.

Theorem 2.6 (Noncentral Quadratic Form) Suppose $\mathbf{X} = (X_1, \ldots, X_n) \sim N(\boldsymbol{\mu}, \Sigma)$, where Σ is positive definite.

(i) $\mathbf{X}^T \Sigma^{-1} \mathbf{X} \sim \chi^2_{\Delta;n}$, where $\Delta = \boldsymbol{\mu}^T \Sigma^{-1} \boldsymbol{\mu}$ is the noncentrality parameter.

(ii) If $\Sigma = \mathbf{I}_n$, and $\mathbf{H} \in \mathcal{M}_n$ is an idempotent matrix of rank ν, then $\mathbf{X}^T \mathbf{H} \mathbf{X} \sim \chi^2_{\Delta;\nu}$, where $\Delta = \boldsymbol{\mu}^T \mathbf{H} \boldsymbol{\mu}$ is the noncentrality parameter. ■

Proof. The proof is left to the reader (Problem 2.30). □

The definition of the noncentral T and F distributions follow directly from the noncentral χ^2 following the construction method of Example 1.11.

Definition 2.3 (Noncentral T and F Distributions) The noncentral T distribution is defined by replacing $Z \sim N(0,1)$ in Equation (1.25) with $Z + \Delta$,

$$T = \frac{Z + \Delta}{\sqrt{W/\nu}}, \tag{2.15}$$

where Δ is the noncentrality parameter. Then $T_\Delta \sim T_{\Delta;\nu}$.

The noncentral F distribution is obtained from Equation (1.24) by replacing W_1 with $W_{\Delta,1} \sim \chi^2_{\Delta;\nu_1}$ (W_2 remains unchanged). Then $F \sim F_{\Delta;\nu_1,\nu_2}$. ■

In practice, we will encounter sums of independent squares of the form $W = \sum_{i=1}^n Z_i^2$, where $Z_i \sim N(\mu_i, \sigma^2)$. If σ^2 is known, it is a simple matter to scale by σ^{-2}, in which case $W/\sigma^2 = \sum_{i=1}^n (Z_i/\sigma)^2$, where $Z_i/\sigma_i \sim N(\mu_i/\sigma, 1)$. In this case $W/\sigma^2 \sim \chi^2_{\Delta;n}$, where the noncentrality parameter is $\Delta = \sigma^{-2}\sum_{i=1}^\infty \mu_i^2$.

Of course, it is more likely the case that σ^2 is unknown. In this case, there will usually be an unbiased estimate of σ^2 that is independent of W, say S^2 (the sample variance is a typical example). In many cases, we will have $S^2/\sigma^2 \sim \chi^2_m$. In this case, we may use the F distribution. Referring to Equation (1.24), we may construct an F statistic

$$F = \frac{W/n}{S^2/m} = \frac{(W/\sigma^2)/n}{(S^2/\sigma)/m} = \frac{W_{\Delta,1}/n}{W_2/m} \sim F_{\Delta;n,m},$$

which has a noncentral F distribution with noncentrality parameter Δ. In many cases (but not all), the role of the F statistic is to scale a χ^2 statistic.

An important fact regarding the sampling distributions is worth stating as a separate theorem.

Theorem 2.7 (Likelihood Ratio Ordering of the Noncentral Sampling Distributions) The noncentral T, χ^2 and F distributions all satisfy the likelihood ratio ordering property *wrt* the noncentrality parameter Δ. ■

Proof. See, for example, Ghosh (1970, 1973); Sun *et al.* (2010). □

2.5.1 The Sample Mean and Sample Variance

Suppose $X_1, \ldots, X_n$ is an *iid* sample from density f. The sample mean $\bar{X} = n^{-1}\sum_{i=1}^n X_i$ and sample variance $S^2 = (n-1)^{-1}\sum_{i=1}^n (X_i - \bar{X})^2$ are ubiquitous in statistical methodology. For any distribution, $\bar{X}$ and S^2 are unbiased estimates of $E[X_1] = \mu$ and $\mathrm{var}[X_1] = \sigma^2$, respectively.

If $X_1 \sim N(\mu, \sigma^2)$, then (as we will verify) the distributions of $\bar{X}$ and S^2 are easily determined. If the data is approximately normally distributed, these distributions will hold approximately. However, we will see in later chapters that these distributions will hold approximately even when f is not close to the normal distribution, as long as enough higher-order moments exist.

Suppose $\mathbf{X} = (X_1, \ldots, X_n)$ is an *iid* random vector with $X_1 \sim N(\mu, \sigma^2)$. First, note that when we evaluate the distribution of S^2 we may assume without loss of generality that $\mu = 0$ (here we are looking ahead to Section 3.3, but the reader who does not wish to rely on this device can consider Problem 2.29). Then the sample variance may be expressed in matrix form by

$$S^2 = (n-1)^{-1}\sum_{i=1}^n (X_i - [X_1/n + \cdots + X_n/n])^2 = (\mathbf{BX})^T(\mathbf{BX}) = \mathbf{X}^T\mathbf{B}^T\mathbf{BX}$$

where $\mathbf{B} = (n-1)^{-1/2}\left[\mathbf{I} - n^{-1}\mathbf{1}\right]$, $\mathbf{I}$ is the $n \times n$ identity matrix, and $\mathbf{1}$ is the $n \times n$ matrix with each element equal to 1. However, it is easily verified that the matrix $\mathbf{H} = \mathbf{I} - n^{-1}\mathbf{1}$ is symmetric, idempotent and of rank $n-1$ (the rank of an idempotent matrix is equal to its trace). This means $\mathbf{B}^T\mathbf{B} = (n-1)^{-1}\mathbf{H}^2 = (n-1)^{-1}\mathbf{H}$, so by Theorem 2.5 we have $(n-1)S^2/\sigma^2 \sim \chi^2_{n-1}$.

Furthermore, we may confirm that $\bar{X}$ and S^2 are independent. We may write $\bar{X} = n^{-1}\mathbf{1}^T_{n,1}\mathbf{X}$. Thus $\bar{X}$ is a mapping of $\mathbf{U} = \mathbf{1X}$ while S^2 is a mapping of $\mathbf{V} = \mathbf{BX}$. However, since $\mathbf{1B} = \mathbf{0}$, the covariance of $\mathbf{U}$ and $\mathbf{V}$ is zero, hence $\bar{X}$ and S^2 are independent.

The T statistic

Assume $X_1 \sim N(\mu, \sigma^2)$. For some fixed μ_0 the T-statistic is given by

$$T = \frac{\bar{X} - \mu_0}{S/\sqrt{n}} = \frac{\bar{X} - \mu_0}{S/\sqrt{n}} \times \frac{1/\sigma}{1/\sigma} = \frac{\frac{\bar{X}-\mu+(\mu-\mu_0)}{\sigma/\sqrt{n}}}{\sqrt{\frac{(n-1)S^2}{(n-1)\sigma^2}}} = \frac{Z + \Delta}{\sqrt{W/(n-1)}},$$

where $Z \sim N(0,1)$, $W \sim \chi^2_{n-1}$. However, we have just seen that $\bar{X}$ and S^2 are independent, therefore so are Z and W. This means $T \sim T_{\Delta;n-1}$, where $\Delta = \sqrt{n}(\mu - \mu_0)/\sigma$ is the noncentrality parameter.

The F statistic

The F statistic usually arises in statistical methodology as the ratio of two independent normalized quadratic forms. Let $W_N = \mathbf{X}^T\mathbf{Q}_N\mathbf{X}$ and $W_D = \mathbf{Y}^T\mathbf{Q}_D\mathbf{Y}$, where $\mathbf{X} \sim N(\boldsymbol{\mu}_N, \sigma^2_N\mathbf{I}_n)$, $\mathbf{Y} \sim N(\boldsymbol{\mu}_D, \sigma^2_D\mathbf{I}_m)$. Assume $\mathbf{Q}_N, \mathbf{Q}_D$ are idempotent matrices, with $\nu_N = \text{rank}(\mathbf{Q}_N)$, $\nu_D = \text{rank}(\mathbf{Q}_D)$. Then $W_N/\sigma^2_N \sim \chi^2_{\Delta_N;\nu_N}$, where $\Delta_N = \sigma_N^{-2}\boldsymbol{\mu}_N^T\boldsymbol{\mu}_N$. Similarly, $W_D/\sigma^2_D \sim \chi^2_{\Delta_D;\nu_D}$, where $\Delta_D = \sigma_D^{-2}\boldsymbol{\mu}_D^T\boldsymbol{\mu}_D$.

There are two common scenarios that lead to an F statistic. In the first, $\mathbf{X}$ and $\mathbf{Y}$ are mutually independent. We usually have $\Delta_N = \Delta_D = 0$. The F statistic is then

$$F = \frac{W_N/\nu_M}{W_D/\nu_D}.$$

In this case

$$F \times \frac{1/\sigma^2_N}{1/\sigma^2_D} \sim F_{\nu_N,\nu_D}.$$

In other words $F = (\sigma^2_N/\sigma^2_D)F^*$, where $F^* \sim F_{\nu_N,\nu_D}$. This type of F statistic is useful for inference problems concerning the ratio of two variances.

In the second scenario, we have one single sample $\mathbf{X} = \mathbf{Y}$, so that $\sigma^2_N = \sigma^2_D$. However, it may be the case that $\mathbf{Q}_N\mathbf{Q}_D = \mathbf{0}$, in which case the quadratic forms W_N, W_D are independent. In a typical application, we expect $\Delta_D = 0$, so that $F \sim F_{\Delta_N;\nu_N,\nu_D}$.

2.6 MGF and CGF of Independent Sums

Recall that the MGF, when it exists, is a unique representation of a distribution. This makes it useful for the analysis of distributions of random sums. If $X_1, \ldots, X_n$ are independent random variables then by Equation (1.6), we have

$$m_{X_1+\cdots+X_n}(t) = \prod_{i=1}^{n} m_{X_i}(t), \tag{2.16}$$

and for *iid* random variables this reduces to

$$m_{X_1+\cdots+X_n}(t) = m_{X_1}(t)^n. \tag{2.17}$$

In addition, the CGF of a random sum will be, following (2.16),

$$c_{X_1+\cdots+X_n}(t) = \sum_{i=1}^{n} c_{X_i}(t).$$

The density of the sum of two independent random variables is referred to as a convolution. It can be thought of as a functional analytic operation $*$ mapping two densities to a third. The convolution of densities f and g can generally be evaluated as

$$(f * g)(u) = \int_{u\in\mathbb{R}} f(x)g(u-x)d\nu(x)$$

where ν is the appropriate measure. The operation is associative, so we can refer precisely to the convolution of m densities, which can be used to evaluate the distribution of an independent sum $S = X_1 + \cdots + X_m$.

This is an important tool in statistical inference. Some parametric families are closed under convolution (normal, Poisson). In this case, the distribution of S is easily obtained by matching k parameters to the first k moments. For parametric families contructed as sums of independent Bernoulli random variables, closure under convolution only holds when the parameter p is common to all terms (the negative binomial and binomial, of which the geometric and Bernoulli distributions are special cases, respectively). Similarly, closure holds for the gamma density family only when the parameter λ is common (the exponential density is a special case of the gamma).

MGFs can be used to test this property. For example, the MGF of $X \sim exp(1/\lambda)$ is $m_X(t) = (1 - t/\lambda)^{-1}$. Using (2.16), the MGF of an *iid* sum from $exp(\lambda)$ is $m_S(t) = (1 - t/\lambda)^{-k}$. This is the MGF for $S \sim gamma(1/\lambda, k)$ which is therefore the density of S.

The specific constraints of closure under convolution can be made more precise, as shown in the following example.

Example 2.10 (MGF of the Sum of Binomials) Recall that MGFs uniquely define a distribution (Theorem 1.24). Here, they will be used to verify that if $X \sim bin(n, p)$ and $Y \sim bin(m, q)$, $X + Y$ is a binomial random variable if and only if $p = q$. Write

$$m_{X+Y}(t) = (pe^t + 1 - p)^n(qe^t + 1 - q)^m. \tag{2.18}$$

If $X + Y \sim bin(\alpha, s)$ for some α, s, we must have

$$m_{X+Y}(t) = (\alpha e^t + 1 - \alpha)^s. \tag{2.19}$$

Thus, $X + Y$ is binomial if and only if we can find α and s for which expressions (2.18) and (2.19) are equal. First, note that (2.18) is an order $n + m$ polynomial in e^t. Suppose (2.19) is equal to (2.18). Then (2.19) must also be an order $n + m$ polynomial in e^t. This means we must have $s = m + n$. Furthermore, the roots of the polynomials must be equal to each other. Then (2.18) has root $e^t = (p-1)/p$ of multiplicity n, and root $e^t = (q-1)/q$ of multiplicity m. But (2.19) has only one root $e^t = (\alpha - 1)/\alpha$ of multiplicity $s = n + m$. Thus, expressions (2.18) and (2.19) are equal if and only if $p = q = \alpha$. Therefore, $X + Y$ is binomial if and only if $p = q$. ■

We will see in Chapter 12 that cumulants of random sums and sample means play an important role in asymptotic theory. The following example suggests why this is the case.

Example 2.11 (Cumulants from *iid* Samples) Consider an *iid* sum $S_n = X_1 + \cdots + X_n$. Suppose $\mu = E[X_1]$, otherwise let κ_k be the kth cumulant of X_1, and let $\bar{\mu}_k$ be the central moment. Then the cumulants of S_n are $\kappa_k[S_n] = n\kappa_k$. Similarly, if we set $\bar{X}_n = n^{-1}S_n$, we have cumulants $\kappa_k[\bar{X}_n] = n^{1-k}\kappa_k$. We can easily get the central moments $\bar{\mu}_k[\bar{X}_n] = n^{1-k}\bar{\mu}_k$, $k = 2, 3$, and for $k = 4$ we have

$$\begin{aligned}\bar{\mu}_4[\bar{X}_n] &= \kappa_4[\bar{X}_n] + 3\bar{\mu}_2[\bar{X}_n]^2 \\ &= n^{-3}\kappa_4 + 3n^{-2}\bar{\mu}_2^2 \\ &= n^{-3}(\bar{\mu}_4 - 3\bar{\mu}_2^2) + 3n^{-2}\bar{\mu}_2^2 \\ &= 3\frac{\bar{\mu}_2^2}{n^2} + \frac{(KURT - 3)\bar{\mu}_2^2}{n^3}.\end{aligned}$$

Suppose we then normalize the sum as $Z_n = (S_n - \mu)/\sqrt{n}$. It will be a fact of some consequence that $\kappa_2[Z_n] = O(1)$, while $\lim_{n\to\infty} \kappa_k[Z_n] = 0$ for $k \geq 3$. ■

2.7 Multivariate Extensions of the MGF

Suppose we have a random vector $\mathbf{X} = (X_1, \ldots, X_m)$. Define vector $\mathbf{t} = (t_1, \ldots, t_m)$. We take the MGF of $\mathbf{X}$ to be

$$m_{\mathbf{X}}(\mathbf{t}) = E[e^{\mathbf{t}^T\mathbf{X}}]. \tag{2.20}$$

The relationship to the univariate MGF can be seen by definining the random variable $W_{\mathbf{t}} = \mathbf{t}^T\mathbf{X}$, and noting that $m_{\mathbf{X}}(\mathbf{t}) = E[e^{W_{\mathbf{t}}}] = m_{W_{\mathbf{t}}}(1)$. It follows that an MGF for $\mathbf{X}$ exists if an MGF exists for each $W_{\mathbf{t}}$. The observation that the distribution of $\mathbf{X}$ is implied by the collection of univariate distributions of all $W_{\mathbf{t}}$ is referred to as the Cramér-Wold device (see also Section 11.11).

The method of obtaining moments by differentiation of the MGF described in Section 1.9.1 is easily extended to joint moments by the rule

$$E[\prod_{j=1}^{n} X_j^{k_j}] = \left.\frac{\partial^K m_{\mathbf{X}}(\mathbf{t})}{\partial t_1^{k_1} \ldots \partial t_m^{k_m}}\right|_{\mathbf{t}=\mathbf{0}}, \quad K = k_1 + \cdots + k_m. \tag{2.21}$$

Example 2.12 (MGF of the Bivariate Normal Density) Suppose $\mathbf{X} \sim N(\boldsymbol{\mu}, \Sigma)$, where $\boldsymbol{\mu} = [\mu_1, \mu_2]^T$ and

$$\Sigma = \begin{bmatrix} \sigma_1^2 & \sigma_{12} \\ \sigma_{12} & \sigma_2^2 \end{bmatrix}.$$

If $W_{\mathbf{t}} = \mathbf{t}^T\mathbf{X}$, then $W_{\mathbf{t}} \sim N(\mathbf{t}^T\boldsymbol{\mu}, \mathbf{t}^T\Sigma\mathbf{t})$. Recall from Example 1.30 that the MGF of $Y \sim N(\mu, \sigma^2)$ is $m_Y(t) = \exp(\sigma^2 t^2/2 + \mu t)$. Then $m_{\mathbf{X}}(\mathbf{t}) = m_{W_{\mathbf{t}}}(1) = \exp(\mathbf{t}^T\Sigma\mathbf{t}/2 + \mathbf{t}^T\boldsymbol{\mu})$. By using Equation (2.21) the reader can verify, that, for example $E[X_1^2X_2] = \sigma_1^2\mu_2 + \sigma_{12}\mu_1 + \mu_1^2\mu_2$. ■

2.8 Problems

Problem 2.1 Prove the following:

(a) If X, Y are two random variables, then $\text{cov}[X, Y] = E[X(Y - E[Y])] = E[Y(X - E[X])]$.
(b) Let $\mathbf{x} = (x_1, \ldots, x_n)$, $\mathbf{y} = (y_1, \ldots, y_n)$ be two vectors. Then $\sum_i (x_i - \bar{x})(y_i - \bar{y}) = \sum_i x_i(y_i - \bar{y}) = \sum_i y_i(x_i - \bar{x})$.
(c) How does part (b) follow from part (a)? **HINT:** Consider the pairs (x_i, y_i), $i = 1, \ldots, n$. Construct a random vector (X, Y) by assigning it one of the pairs at random.

Problem 2.2 Suppose X_1, X_2 are independent observations from a geometric distribution with mean $1/p$, $p \in (0, 1)$.

(a) Derive the distribution of X_1 conditional on $\{X_1 + X_2 = s\}$ for any $s \geq 2$, in the form of the probability mass function (PMF) $p_X(x) = P(X_1 = x \mid X_1 + X_2 = s)$.
(b) How does $p_X(x)$ depend on x and p?

Problem 2.3 Suppose X and Y have joint density

$$f(x, y) = \begin{cases} 2; & 0 < x < y < 1 \\ 0; & \text{elsewhere} \end{cases}.$$

(a) Derive the marginal density of X.
(b) Derive the marginal density of Y.
(c) Derive the conditional density of X given Y.

Problem 2.4 Suppose $X_1, \ldots, X_n$ are independent random variables with a common CDF F_X.

(a) Show that the CDF of $Y = \max(X_1, \ldots, X_n)$ is given by $F_Y(t) = F_X^n(t)$.
(b) Suppose $X_1 \sim unif(0, 1)$. Derive the density function and mean of Y.

Problem 2.5 At a party, N men bring identical hats. At the end of the party each man brings home one of the hats chosen at random. Let S_N be the number of men who bring home their own hats. What is the mean and variance of S_N?

Problem 2.6 Suppose $X \sim N(\mu, \sigma^2)$. Then $Y = \exp(X)$ has a log-normal density. Let $X_1, \ldots, X_n$ be an *iid* sample with $X_1 \sim N(0, 1)$. Consider the power sum $S = \sum_{i=1}^n e^{X_i}$, which may be written $S = WV$, where $W = e^{X^*}$, $V = \sum_{i=1}^n e^{X_i - X^*}$, $X^* = \max\{X_1, \ldots, X_n\}$. Determine the cumulative distribution function of W, then determine $E[V \mid W = w]$ for any positive constant w.

Problem 2.7 Correlation between random variables can be modeled in the following way. Let X and ϵ be random variables with mean 0 and variance 1, and assume X and ϵ are independent. Then set, for constants $\beta_0, \beta_1, \beta_2$, a new random variable Y as $Y = \beta_0 + \beta_1 X + \beta_2 \epsilon$. Derive an expression for the correlation between X and Y as a function of β_1 and β_2 (verify that this expression will not depend on β_0).

Problem 2.8 Following Example 1.7, express precisely the support of the multivariate hypergeometric distribution of Example 2.2.

Problem 2.9 Suppose there are m types of coupons. A collector samples them one at a time. Whenever a collector samples another coupon, assume it is of each type with equal probability, and that the selections are independent. What is the expected number of coupon samples needed to have at least one of each type? **HINT:** Suppose T_j is the number of samples at which the jth new coupon is observed by the collector. What is the distribution of $T_{j+1} - T_j$, for any $j = 1, 2, \ldots, m - 1$?

Problem 2.10 Suppose $\mathbf{X} \sim dirichlet(\boldsymbol{\alpha})$ (Example 2.4).

(a) Verify that any component of $\mathbf{X}$ possesses marginal distribution $X_j \sim beta(\alpha_j, \sum_{i \neq j} \alpha_i)$.
(b) Derive the mean vector and covariance matrix of $\mathbf{X}$. **HINT:** See also Problem 1.19.

Problem 2.11 There are several methods of deriving the density of a χ^2_ν random variable. Obtain this density using a geometric argument, noting that the density of an *iid* sample from a $N(0,1)$ distribution is spherically invariant.

Problem 2.12 Suppose X and Y each have density $f \sim exp(1)$, and are independent. Consider the transformation $U = X + Y$, $V = X/Y$. Derive the joint density of (U, V). Are U and V independent? (See also Problem 2.19.)

Problem 2.13 Suppose (X, Y) has a bivariate normal distribution, and that U, V are two independent $N(0,1)$ random variables.

(a) Show that there exist constants a, b, c, d, e such that if $X^* = a + bU$, $Y^* = c + dU + eV$, then (X^*, Y^*) has the same density as (X, Y). Are the constants a, b, c, d, e unique?
(b) Prove that if $\mathrm{var}[X] = \mathrm{var}[Y]$, then $X + Y$ and $X - Y$ are independent.

Problem 2.14 Suppose random variables X_1 and X_2 each have uniform density $f(x) \sim unif(0,1)$, and are independent. Consider the transformation $Y_1 = X_1 + X_2$, $Y_2 = X_1 - X_2$. Derive the joint density of (Y_1, Y_2), and the marginal densities of Y_1 and Y_2.

Problem 2.15 Use the method of Section 2.3 to derive the convolution formula. Let X, Y be independent random variables with densities f_X, f_Y *wrt* Lebesgue measure. Then the density of $T = X + Y$ is given by

$$f_T(t) = \int_{\mathbb{R}} f_X(t-u) f_Y(u) du = \int_{\mathbb{R}} f_X(u) f_Y(t-u) du.$$

Problem 2.16 Suppose $X \sim unif(0,1)$ and $Y \sim unif(0,1)$ are independent random variables. Derive the density of $X - Y$ using two methods:

(a) Note that (X, Y) is a uniformly distributed point on a unit square. What is the area of $\{X - Y \leq t\}$?
(b) Use the convolution formula of Problem 2.15.

Problem 2.17 The Dirichlet distribution (Example 2.4) is commonly used to model random probability distributions $(p_1, \ldots, p_n)$. It can be derived as the random vector $\mathbf{Y} = (X_1/\sum_{i=1}^n X_i, \ldots, X_n/\sum_{i=1}^n X_i)$, where $X_1, \ldots, X_n$ are independent random variables with $X_i \sim gamma(\tau, \alpha_i)$. Use the method of Section 2.3 to verify that the distribution of $\mathbf{Y}$ is a Dirichlet distribution. **HINT:** Define transformation $\mathbf{V} = (V_1, \ldots, V_n)$, where $V_n = \sum_{i=1}^n X_i$ and $V_i = X_i/\sum_{i=1}^n X_i$, $i = 1, \ldots, n-1$. Determine the density of $\mathbf{V}$, then the marginal density of $(V_1, \ldots, V_{n-1})$.

Problem 2.18 A circle has radius R, circumference C and area A.

(a) If $R \sim exp(1)$ derive the density function for C and A. Which of these has an exponential distribution?
(b) If $R \sim unif(0,1)$ derive the density function for C and A. Which of these has a uniform distribution?

Problem 2.19 Suppose continuous random variables $X, Y \in \mathbb{R}$ have joint density $f_{XY}(x,y)$. Derive a general expression for the density of $U = X/Y$. Use this formula to derive the density of X/Y where X, Y are independent standard normal random variables (that is, derive the standard Cauchy density).

Problem 2.20 We are given *iid* observations X_1, X_2, where $X_1 \sim exp(\tau)$.

(a) Let $Y_1 = X_1 - X_2$, $Y_2 = X_2$. Determine the joint density of $\mathbf{Y} = (Y_1, Y_2)$.
(b) Are Y_1 and Y_2 independent? Justify your answer.
(c) Determine the density function of Y_1.

Problem 2.21 Suppose $U \sim exp(1/\lambda_1)$, $V \sim exp(1/\lambda_2)$ are independent random variables.

(a) Show that the density of $V - U$ is given by

$$f(\mathbf{x}) = \begin{cases} \frac{\lambda_1\lambda_2}{\lambda_1+\lambda_2}e^{\lambda_1 x}; & x < 0 \\ \frac{\lambda_1\lambda_2}{\lambda_1+\lambda_2}e^{-\lambda_2 x}; & x \geq 0 \end{cases}, \quad x \in \mathbb{R}.$$

(b) Verify that median$V - U \neq$ median$V -$ median$U$ unless $\lambda_1 = \lambda_2$.

Problem 2.22 Assuming n is odd, give explicitly the density of the median of an *iid* sample from a uniform distribution on $(0, 1)$.

Problem 2.23 This problem follows Example 2.8. Suppose we want to estimate the 0.75 quantile $Q(0.75)$ of a density using an *iid* sample $X_1, \ldots, X_n$. We can use as the estimator the appropriate order statistic: $\hat{Q}_1 = X_{(m)}$, where $m/n \approx 0.75$. However, suppose we know that $X_1 \sim N(0, \sigma^2)$, where σ is unknown. If $q_{0.75}$ is the 0.75 quantile for the standard normal distribution $N(0, 1)$, then we have $Q(0.75) = \sigma q_{0.75}$. An alternative estimator of $Q(0.75)$ is therefore $\hat{Q}_2 = S q_{0.75}$, where S^2 is the sample variance. Give, as far as possible, an analytic form for the densities of $\hat{Q}_1$ and $\hat{Q}_2$. Then plot the densities in a single graph. Use $n = 20$.

Problem 2.24 Let $\mathbf{X} = (X_1, \ldots, X_n)$ be an *iid* sample with $X_1 \sim exp(\tau)$. Suppose $X_{(1)}, X_{(2)}, \ldots, X_{(n)}$ are the order statistics of $\mathbf{X}$. For convenience let $U = X_{(1)}$, $V = X_{(2)}$.

(a) Derive the marginal density of U and the joint density of (U, V).
(b) Describe the relationship between the density of V conditional on U and the shifted exponential density $exp(\mu, \tau)$.

Problem 2.25 Suppose $\mathbf{X} = (X_1, \ldots, X_n)$ is an *iid* sample from a uniform density on $(0, 1)$, and let $X_{(1)}, \ldots, X_{(n)}$ be the order statistics of $\mathbf{X}$. Show that $E[X_{(j)}] = j/(n+1)$ for all $j = 1, \ldots, n$. **HINT:** You can make use of the beta function of Section C.5.

Problem 2.26 Carry out the argument required to derive Equation (2.13).

Problem 2.27 Let $X_1, \ldots, X_n$ be an *iid* sample from a uniform density on the unit interval $(0, 1)$.

(a) Derive the joint density of $(X_{(1)}, X_{(n)})$, where $X_{(1)} = \min_i X_i$ and $X_{(n)} = \max_i X_i$ are order statistics.
(b) Derive the marginal density of the range $R = X_{(n)} - X_{(1)}$.

Problem 2.28 Suppose $\mathbf{X} \sim N(\boldsymbol{\mu}, \sigma^2 \mathbf{I}_n)$, $\mathbf{X} \in \mathbb{R}^n$. Let $S = \sum_{i=1}^n (X_i - \bar{X})^2$. Show that $\sigma^{-2} S \sim \chi^2_{\Delta,\nu}$. Give Δ and ν explicitly.

Problem 2.29 Suppose we are given *iid* sample $X_1, \ldots, X_n$, $X_1 \sim N(\mu, \sigma^2)$. Let S^2 be the sample variance. In Section 2.5.1, the distribution $S^2/\sigma^2 \sim \chi^2_{n-1}$ was derived by arguing that the distribution will not depend on μ, which may therefore be set to zero. Suppose we attempt to derive the distribution of $S^2/\sigma^2 \sim \chi^2_{n-1}$ without constraining μ using Theorem 2.6. Verify that the noncentrality parameter will be zero.

Problem 2.30 Prove Theorem 2.6.

Problem 2.31 Use the MGF of the *gamma*(τ, α) density to derive the density function of a χ^2_ν random variable, defined as $Y = \sum_{i=1}^{\nu} Z_i^2$, where the Z_i's are *iid* standard normal.

Problem 2.32 Suppose two independent random variables X and Y have cumulant generating functions $C_X(t) = 6(e^t - 1)$ and $C_Y(t) = 2t^2$. Derive the mean, variance and standardized fourth order cumulant (excess kurtosis) of $X + Y$.

Problem 2.33 We are given n random variables $X_1, \ldots, X_n$. Suppose there exists a moment generating function $m(t)$ and a subset $\mathcal{T} \subset [0, \infty)$ such that $E[\exp(tX_i)] \le m(t)$ for all i and $t \in \mathcal{T}$.

(a) Show that, in general, $\max_i E[X_i] \le E[\max_i X_i]$.

(b) Show that

$$E\left[\max_i X_i\right] \le \inf_{t \in \mathcal{T}} t^{-1} \log(n \times m(t)).$$

(c) Use Part (b) to derive an upper bound for $E[\max_i X_i]$ when $X_i \sim N(\mu, \sigma^2)$, $i = 1, \ldots, n$.

3

Statistical Models

3.1 Introduction

A statistical model assumes the existence of a random observation $\tilde{X} \in \mathcal{X}$ sampled from a distribution with density p. It is assumed that p belongs to some family $\mathcal{P} = \{p_\theta : \theta \in \Theta\}$, indexed by a parameter $\theta \in \Theta$, where Θ is the parameter space. This defines a parametric family, which will be introduced formally in Section 3.2. Ideally, the analyst's understanding of the source of $\tilde{X}$ informs the choice of $\mathcal{P}$, and the smaller the family the more efficient can be the inference. Of course, this efficiency depends on correct model specification, here meaning that the assumption $p \in \mathcal{P}$ holds. Sections 6.6.1, 12.8.2 and 13.7 give examples in which this issue arises. In many cases the parametric family will be based on one of the parametric classes introduced in Section 1.4. However, the parameter space Θ may be strictly smaller than the natural parameter space (Definition 1.5), depending on the application.

The theory of inference includes a theory of statistical models which informs the choice of inference method. This must precede any development of inference itself, and is the subject of this chapter. Several general classes of statistical models are commonly defined. Location-scale models are introduced in Section 3.3. This class is a special case of the larger class of invariant models, which are models generated by a transformation group acting on the observation $\tilde{X}$. Section D.1 contains a background in group theory targeted towards this topic. There exists a theory of inference specialized to these models, to which Chapter 9 is devoted. However, these ideas are often relevant in other branches of inference, and may yield considerable insight, as well as simplification. For this reason, they will appear on a regular basis thoughout most of this volume.

The exponential family model is another general model class. Its prominence in the theory of inference can be explained, first, by the useful analytical properties it possesses, and second, by the fact that a very large number of commonly used statistical models belong to this class (Section 3.6).

An introduction to the regular family (Section 3.4) precedes introduction of the exponential family. The regular family is essentially defined by a number of technical assumptions which permit a definition, or at least a useful interpretation, of Fisher information (Section 3.5). Any exponential family model is a regular family, but the same need not be true of a location-scale model.

A discussion of statistical models leads to the problem of data reduction. While $\tilde{X}$ represents a complete observation, statistical methods usually involve some reduction to a statistic $T(\tilde{X}) \in \mathbb{R}^m$, usually of much lower dimension than $\tilde{X}$. The theory of sufficiency (Sections 3.7 and 3.8) allows us to determine in many cases the optimal reduction of data $T(\tilde{X})$, in the sense that all information of a parameter θ is retained, while spurious variation is excluded.

It will sometimes be advantageous to interpret an observation $\tilde{X} \in \mathcal{X}$ as conditional on a statistic $S(\tilde{X})$. This can be referred to as a conditional model, and is introduced in

DOI: 10.1201/9781003049340-3

Section 3.9. Conditional models typically address specific problems which may arise, but usually admit the same range of techniques as other models.

Bayesian models are introduced in Sections 3.10 and 3.11. It may be an oversimplification to say that statistical inference can be divided into Bayesian and frequentist (non-Bayesian) schools. Yet a fundamental difference exists, in particular, that in Bayesian models the parameter θ is regarded as a random outcome. Despite this, Bayesian and non-Bayesian methods can be to a large degree united by decision theory (Chapter 7), which is the point of view taken in this volume.

Section 3.12 considers the problem of nuisance parameters. This arises when a parameter $\theta = (\psi, \xi)$ is decomposed into a parameter of interest ψ, and a nuisance parameter ξ. Development of inference methods can be hampered by the fact that ξ is unknown, so the development of methods which do not depend on ξ is an important problem. A brief introduction to this topic is included here because two important methods of dealing with nuisance parameters are based on two concepts introduced in this chapter, in particular, sufficiency and group invariance.

The chapter ends with an enumeration of general principles of inference (Section 3.13). These include the sufficiency principle, the weak and strong likelihood principle, the conditionality (ancillarity) principle and the invariance principle. Any given inference method either does or does not conform to one or another of these principles. The section ends with a discussion of Birnbaum's theorem, which states that the sufficiciency principle and the conditionality principle together imply the strong likelihood principle.

3.2 Parametric Families for Statistical Inference

We are given an observation $\tilde{X}$ from a sample space $\mathcal{X}$. We assume $\mathcal{X}$ is a metric space, with the topology induced by the metric, and a measurable space defined by the Borel sets. A fixed value from the sample space will be denoted as $\tilde{x} \in \mathcal{X}$. We will usually associate a measure $\nu_{\mathcal{X}}$ with $\mathcal{X}$. Then $\tilde{X}$ possesses a density p_θ with respect to $\nu_{\mathcal{X}}$ from an indexed family $\mathcal{P} = \{p_\theta : \theta \in \Theta\}$. We refer to θ as the parameter and Θ as the parameter space. Where convenient we may use Θ to denote $\mathcal{P}$ itself. The probability measure associated with p_θ may be written P_θ. We may write $\nu = \nu_{\mathcal{X}}$ when the context is clear.

We will generally assume that for an indexed parametric family, the parameter θ is identifiable in the sense that $\theta \neq \theta'$ imples $p_\theta \neq p_{\theta'}$. This means that $\mathcal{P}$ and Θ are bijective. This avoids trivialities, since if this condition did not hold, there would be two distinct parameters $\theta \neq \theta'$ under which the observation $\tilde{X} \in \mathcal{X}$ would possess identical distributions. In this case the two parameters could not be distinguished by any statistical method. Normally, it is straightforward to verify identifiability. For most parametric families, for example, there will be a bijection between $\theta \in \Theta \subset \mathbb{R}^q$ and a set of q moments. For example, if $X \in \mathcal{X} = \mathbb{R}$, and $X \sim N(\mu, \sigma^2)$, then we have parameter $\theta = (\mu, \sigma^2) \in \Theta = \mathbb{R} \times \mathbb{R}_{>0}$. If $X \sim p_\theta$, $X' \sim p_{\theta'}$ and $\theta \neq \theta'$ then at least one of the statements $E_\theta[X] \neq E_{\theta'}[X']$ or $\mathrm{var}_\theta[X] \neq \mathrm{var}_{\theta'}[X']$ holds, so we must have $p_\theta \neq p_{\theta'}$. In this case, the model had q degrees of freedom, meaning that Θ defines q independent parameters. There will be cases, however, in which identifiability is at issue, so this definition should be kept in mind (for example, Sections 3.6.1, 6.2 or 6.7.1).

In our applications $\nu_{\mathcal{X}}$ will usually be either counting measure on a discrete set, or Lebesgue measure of appropriate dimension. We will generally assume $p_\theta(\tilde{x}) < \infty$, which need not be restrictive. This is because if we cannot replace any values $p_\theta(\tilde{x}) = \infty$ with zero without changing the model, then the density is not integrable. There is no reason, however, to insist that the density functions be bounded. Let $\bar{\mathcal{X}}_\theta = \{\tilde{x} \in \mathcal{X} : p_\theta(\tilde{x}) > 0\}$ be the support of p_θ. It will be convenient to define $\cup_\theta \bar{\mathcal{X}}_\theta = \bar{\mathcal{X}}$. Then for any $\tilde{x} \in \bar{\mathcal{X}}$, $p_\theta(\tilde{x}) > 0$ for at least one $\theta \in \Theta$. If $\bar{\mathcal{X}}_\theta$ does not depend on θ, we say $\mathcal{P}$ or Θ is a family of common support, which may be compactly expressed $\bar{\mathcal{X}}_\theta \equiv \bar{\mathcal{X}}$. In this case $p_\theta(\tilde{x}) > 0$ for all $\theta \in \Theta$ and $\tilde{x} \in \bar{\mathcal{X}}$. Although it might seem reasonable to assume that $\mathcal{X} = \bar{\mathcal{X}}$, we do not insist on this, since we may wish to compare distinct density families.

We write

$$E_\theta[g(\tilde{X})] = \int_{\tilde{x} \in \mathcal{X}} g(\tilde{x}) p_\theta(\tilde{x}) d\nu_{\mathcal{X}}(\tilde{x}) = \int_{\tilde{x} \in \mathcal{X}} g(\tilde{x}) p_\theta(\tilde{x}) d\nu_{\mathcal{X}},$$

where $\nu_{\mathcal{X}}$ is the appropriate measure. Adding θ to the subscript in $E_\theta[g(\tilde{X})]$ is a conventional approach, but it should be emphasized that this quantity will sometimes be interpreted as a function of θ.

3.2.1 The Likelihood Function

We have defined $\mathcal{P}$ formally as a class of real valued functions on $\mathcal{X}$ indexed by θ. We may also interpret $p_\theta(\tilde{X})$ as a function of θ for a fixed $\tilde{X} \in \mathcal{X}$. Viewed this way, the likelihood function is defined:

$$\mathcal{L}(\theta; \tilde{X}) = p_\theta(\tilde{X}).$$

It is often more tractable to use the log-likelihood function:

$$\ell(\theta; \tilde{X}) = \log p_\theta(\tilde{X}).$$

When there is no ambiguity, the term "likelihood" can refer to either form.

Usually, the likelihood $\mathcal{L}(\theta; \tilde{X})$ can be thought of as a general representation of the information about θ contained in observation $\tilde{X}$. Therefore, the likelihood is interpretable as an equivalence class of functions. Suppose we have

$$\mathcal{L}(\theta; \tilde{X}) = p_\theta(\tilde{X}) = u(\tilde{X}) f(\tilde{X}, \theta),$$

where $u(\tilde{x})$ may depend on $\tilde{x}$ but does not depend on θ. We may then also write

$$\mathcal{L}(\theta; \tilde{X}) = f(\tilde{X}, \theta).$$

Similarly, if we have

$$\ell(\theta; \tilde{X}) = g(\tilde{X}, \theta) + h(\tilde{X}),$$

where $h(\tilde{X})$ may depend on $\tilde{X}$ but not θ, we may choose to write $\ell(\theta; \tilde{X}) = g(\tilde{X}, \theta)$. However, some caution is needed here. For example, a statistical software application may report a value of the likelihood, so that the convention used should be understood.

Example 3.1 (Likelihood of a Binomial Parameter) Suppose $X \sim bin(n, \theta)$, $\theta \in [0, 1]$. Then the parametrized density is

$$p_\theta(x) = \binom{n}{x} \theta^x (1 - \theta)^{n-x}. \tag{3.1}$$

The log-likelihood is

$$\ell(\theta; \tilde{x}) = \log p_\theta(x) = x \log(\theta/(1 - \theta)) + n \log(1 - \theta) + \log \binom{n}{x}.$$

Note that $\log\binom{n}{x}$ does not depend on θ, so we could equivalently write the log likelihood $\ell(\theta; x) = x\log(\theta) + (n-x)\log(1-\theta)$.

Figure 3.1 shows plots of the likelihood function $\mathcal{L}(\theta; x) = \theta^x(1-\theta)^{n-x}$ for $(x, n) = (8, 10)$ and $(x, n) = (20, 100)$. We have not incorporated the binomial coefficient appearing in Equation (3.1). The commonly used estimate for θ is x/n, which gives the maximum value of $\mathcal{L}(\theta; x)$ ($8/10 = 0.8$ and $20/100 = 0.2$ respectively). This can be clearly seen in the plots. If we do not know the true value of θ, we might conclude that the maximum of the likelihood identifies the value of θ most compatible with the data, in some sense. But in each case, we can discern a neighborhood of the maximum likelihood value x/n representing models θ which are at least plausible. At these values the likelihood is "significantly greater than zero", a statement which will be made more precise in later chapters.

One further point of comparison should be noted. The likelihood for the case $(x, n) = (20, 100)$ is clearly more concentrated around x/n, so that the set of plausible models θ is smaller. This is meant in a relative sense, since the absolute value of the vertical scale plays no role. In eliminating a larger set of models, the inference will therefore be more precise, as we would expect given the larger sample size. Thus, the likelihood contains information about both the model most compatible with the data, and the degree of accuracy of any inference. ■

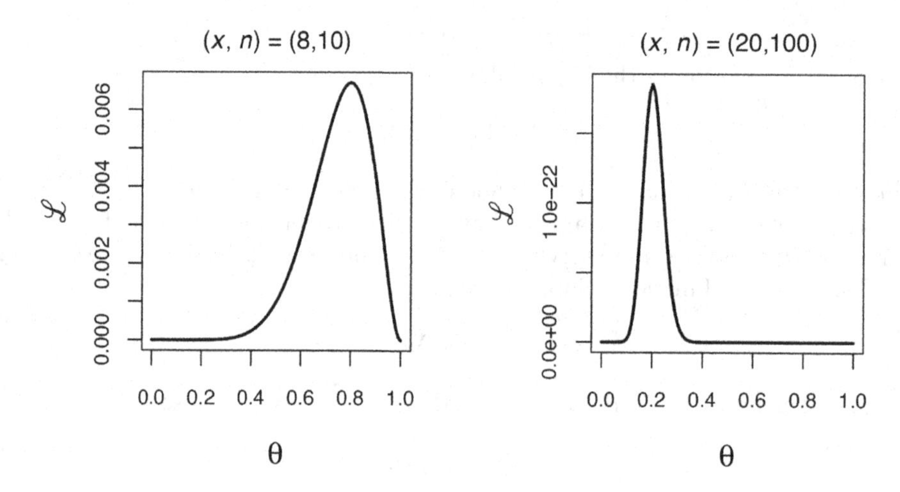

FIGURE 3.1
Binomial likelihood functions for Example 3.1.

3.2.2 Statistics and Their Structure

Of central concern to statistical inference is the following object:

Definition 3.1 (Statistic) Suppose we are given a random outcome $\tilde{X} \in \mathcal{X}$ from a parametric family Θ. A statistic is any mapping $S : \mathcal{X} \to \mathcal{S}$ which does not depend on θ, and is interpretable as a random outcome $S(\tilde{X})$. We will let $\mathcal{T}_{stat}$ denote the set of all statistics. ■

Any statistic $S \in \mathcal{S}$ induces a partition $\mathcal{X}_S$ of sample space $\mathcal{X}$ into equivalence classes of the form $\mathcal{X}_{S=s} = \{\tilde{x} \in \mathcal{X} : S(\tilde{x}) = s\}$ for any $s \in \mathcal{S}$.

If $S, T \in \mathcal{T}_{stat}$, the statement $S \mapsto T$ means that for any pair of observations $\tilde{x}, \tilde{y} \in \mathcal{X}$ $S(\tilde{x}) = S(\tilde{y}) \implies T(\tilde{x}) = T(\tilde{y})$. Alternatively, we may write $T(\tilde{x}) = T(S(\tilde{x}))$. If $S \mapsto T$

and $T \mapsto S$, then T and S are equivalent, which we denote $S \sim T$. In this case, S and T represent identical reductions of data in the sense that $\mathcal{X}_S = \mathcal{X}_T$.

It helps from time to time to recall that for *any* statistic S, we have $S \mapsto 1$. Equivalently, the partition induced by a mapping $S(\tilde{x}) \equiv 1$ is the singleton $\mathcal{X}_S = \{\mathcal{X}\}$.

We usually expect statistics to contain information about a parameter θ, however, we will also be interested in identifying statistics which, in contrast, have distributions which do not depend in θ.

Definition 3.2 (Ancillary Statistic) $V \in \mathcal{T}_{stat}$ is an ancillary statistic if its distribution *dnd* on θ. We denote the class of all ancillary statistics $\mathcal{T}_{anc}$. A statistic V is a first-order ancillary statistic if $E_\theta[V(\tilde{X})]$ *dnd* θ. ■

Example 3.2 (The Sample Variance is Ancillary *wrt* the Mean) Suppose $X_1, \ldots, X_n$ is an *iid* sample with $X_1 \sim N(\mu, \sigma^2)$. In Section 2.5.1, it was conjectured that the distribution of the sample variance does not depend on μ, which provides an example of ancillarity. This will be proven in Section 3.3. ■

3.2.3 Parametrized Families of Functions

Suppose we are given sample space $\mathcal{X}$ and parameter space Θ. We will make frequent use of parametrized families of mappings, say $h_\theta(\tilde{x})$ which map a subset $\bar{\mathcal{X}}_\theta \subset \bar{\mathcal{X}}$ to $\mathcal{H}$. This construction will be needed when $\bar{\mathcal{X}}_\theta$ varies with θ.

There will be some interest in statements of the form "$h_\theta(\tilde{x})$ *dnd* θ", which we take to mean that there is a mapping $h^* : \bar{\mathcal{X}} \to \mathcal{H}$ such that $h_\theta(\tilde{x}) = h^*(\tilde{x})$ *whenever* $\tilde{x} \in \bar{\mathcal{X}}_\theta$. Thus, $h_\theta(\tilde{x})$ may still depend on θ, in the sense that its domain may depend on θ. However, since for every $x \in \bar{\mathcal{X}}$, $h_\theta(x)$ is defined for at least one $\theta \in \Theta$, $h_\theta(\tilde{x}) = h^*(\tilde{x})$ can be evaluated without ambiguity. This idea will be applied below in Theorem 3.15.

3.3 Location-Scale Parameter Models

The location-scale model is an example of the larger class of invariant models, which will be introduced in Chapter 9. The reader may wish to review Section D.1 at this point. Suppose we have observation $\tilde{X} \in \mathcal{X}$. Then $g : \mathcal{X} \to \mathcal{X}$ is a bijective transformation on $\mathcal{X}$ if it possesses an inverse transformation g^{-1}, and if the range of g is $\mathcal{X}$. For example, suppose $\mathcal{X} = \mathbb{R}^n$. For a pair of numbers $\theta = (\mu, \sigma) \in \mathbb{R} \times \mathbb{R}_{>0} = \Theta$, we can define the transformation g_θ on $\mathbf{x} = (x_1, \ldots, x_n)$ by the evaluation method

$$g_\theta \mathbf{x} = g_\theta(x_1, \ldots, x_n) = (\sigma x_1 + \mu, \ldots, \sigma x_n + \mu) = \sigma \mathbf{x} + \mu, \tag{3.2}$$

where addition and multiplication by real numbers is applied to a vector element-wise. The next step is to define a class of transformations $G_{ls} = \{g_\theta : \theta \in \Theta\}$, which is an example of a transformation group (Section D.4).

If $\tilde{X} \sim p$ for some density p, then a parametric family $\mathcal{P} = \{p_\theta : \theta \in \Theta\}$ can be induced, where p_θ is the density of $g_\theta \tilde{X}$ for all $g_\theta \in G_{ls}$.

The notion of location-scale invariance is illustrated in Figure 3.2. The transformation of Equation (3.2) is referred to as a location-scale transformation (see Examples D.12 and D.13). Suppose we are given a sample $(X_1, \ldots, X_n)$ of $n = 100$ randomly generated observations, a histogram of which is shown in the top-left plot. The remaining three histograms show histograms of the same data following location-scale transformations defined by $(\mu, \sigma) = (-10, 1), (5, 0.25), (0, 3)$. The four histograms have the same shape, apart from shifts in location and scale factor. We might say that the four samples differ, trivially, only by a change in unit. Therefore, the inference should be the same for each, provided any change of unit is applied consistently.

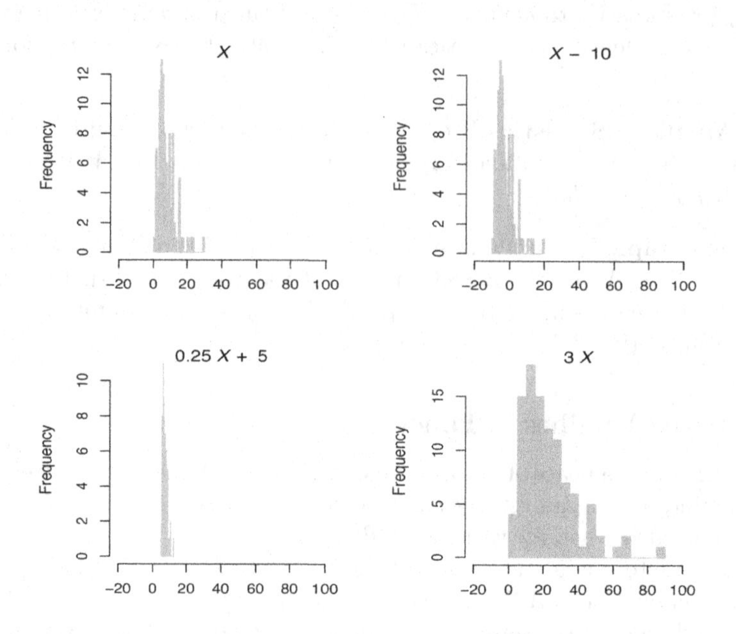

FIGURE 3.2
Location-scale transformations of a sample of size $n = 100$ in histogram form.

In this section we will consider the location, scale and location-scale models, which can be constructed in the manner described here. A theory of estimation for these models will be given later in Chapters 7 and 9. At this point we will not rely explicitly on group theoretic conventions, while a more comprehensive, and group theoretic, development of the more general invariant model will be given in Chapter 9.

3.3.1 Location Parameter Models

We first consider a special case of the location-scale transformation of Equation (3.2). Suppose $\mathbf{X} = (X_1, \ldots, X_n) \in \mathcal{X} \subset \mathbb{R}^n$ possesses density $p_0(x) = p_0(x_1, \ldots, x_n)$. Consider the location transformation

$$\mathbf{Y} = (Y_1, \ldots, Y_n) = g_\theta \mathbf{X} = (X_1 + \theta, \ldots, X_n + \theta) = \mathbf{X} + \theta. \tag{3.3}$$

Let $G_{loc} = \{g_\theta : \theta \in \Theta\}$, $\Theta = \mathbb{R}$, be the class of all location transformations. Then each $g_\theta \in G_{loc}$ is a bijection on $\mathcal{X}$. The Jacobian matrix of this transformation is the identity $\mathbf{J} = \mathbf{I}$, so the density of $\mathbf{Y}$ is

$$p_{\mathbf{Y}}(y_1, \ldots, y_n) = |\det(\mathbf{J})|^{-1} p_0(y_1 - \theta, \ldots, y_n - \theta) = p_0(y_1 - \theta, \ldots, y_n - \theta),$$

since $\det(\mathbf{J}) = 1$. Then define parametric family $\mathcal{P} = \{p_\theta : \theta \in \Theta\}$, $\Theta = \mathbb{R}$, where

$$p_\theta(\mathbf{x}) = p_\theta(x_1, \ldots, x_n) = p_0(x_1 - \theta, \ldots, x_n - \theta) = p_0(\mathbf{x} - \theta).$$

$\mathcal{P}$ is called a location parameter model or a location invariant model, in which case θ is a location parameter. As has just been shown, a location parameter model can be easily generated from any density $p_0(\mathbf{x})$ by setting $p_\theta(\mathbf{x}) = p_0(\mathbf{x} - \theta)$, $\theta \in \Theta = \mathbb{R}$.

Two examples of a location parameter model follow.

Example 3.3 (Normal Density as Location Parameter Model) Consider as an example an *iid* sample $\mathbf{X} = (X_1, \dots, X_n)$ from density $X_1 \sim N(\theta, 1)$. For parameter space $\Theta = \mathbb{R}$, the densities are given by

$$p_\theta(\mathbf{x}) = \frac{1}{(2\pi)^{n/2}} \exp\left(-\frac{1}{2}\sum_{i=1}^{n}(x_i - \theta)^2\right) = p_0(\mathbf{x} - \theta),$$

where $p_0(\mathbf{x})$ is the density when $X_1 \sim N(0, 1)$. This a location parameter model. ∎

There are two important classes of statistics, equivariant and invariant, associated with invariant families. Equivariant statistics preserve the effect of a transformation, and are usually used as estimators. In contrast, invariant statistics do not change value when the data is transformed. Both play an important and complementary role for invariant families. We first consider the location transformation G_{loc}.

Definition 3.3 (Location Invariant and Location Equivariant Statistics) Suppose we are given observation $\mathbf{X} \in \mathcal{X} = \mathbb{R}^n$. A statistic $S(\mathbf{X}) \in \mathbb{R}$ is (location) equivariant if $S(g_\theta \mathbf{X}) = S(\mathbf{X}) + \theta$ for all $g_\theta \in G_{loc}$. A statistic $T(\mathbf{X}) \in \mathcal{T}$ is (location) invariant if $T(g_\theta \mathbf{X}) = T(\mathbf{X})$ for all $g_\theta \in G_{loc}$. Then $M(\mathbf{X})$ is a maximal (location) invariant if $M(\mathbf{X})$ is invariant, and $M(\mathbf{X}) \mapsto T(\mathbf{X})$ for any other invariant statistic $T(\mathbf{X})$. ∎

If we regard a location transformation as a change in units, then an equivariant statistic is one that preserves this change.

Example 3.4 (The Sample Mean is Location Equivariant and the Sample Variance is Location Invariant) Consider the location transformation $\mathbf{Y} = \mathbf{X} + a$ defined in Equation (3.3). Then $\bar{\mathbf{Y}} = n^{-1}\sum_{i=1}^n Y_i = n^{-1}\sum_{i=1}^n (X_i + a) = \bar{\mathbf{X}} + a$, so that the sample mean is location equivariant. In contrast $S^2_{\mathbf{Y}} = (n-1)^{-1}\sum_{i=1}^n (Y_i - \bar{\mathbf{Y}})^2 = (n-1)^{-1}\sum_{i=1}^n (X_i - a - \bar{\mathbf{X}} + a)^2 = S^2_{\mathbf{X}}$ so that the sample variance is location invariant. ∎

The maximal invariant will play an important role throughout this volume, and it turns out to be simple to construct. Note that the definition of a maximal invariant in the context of group theory is given in Definition D.7.

Theorem 3.1 (Maximal Location Invariant) Let $S \in \mathcal{T}_{stat}$ be any location equivariant statistic. Then $M(\mathbf{X}) = (X_1 - S(\mathbf{X}), \dots, X_n - S(\mathbf{X}))$ is a maximal invariant. ∎

Proof. That $M(\mathbf{X})$ is location invariant is easily verified. Next, let T be any invariant statistic. Then $T(X_1, \dots, X_n) = T(X_1 - a, \dots, X_n - a)$ for any a. By setting $a = S(\mathbf{X})$ we conclude that $M \mapsto T$. □

The next two theorems motivate much of our interest in maximal invariants.

Theorem 3.2 Suppose $\mathbf{X} = (X_1, \dots, X_n) \in \mathcal{X}$ is an observation from a location parameter density $p_\theta(\mathbf{x}) = p_0(\mathbf{x} - \theta)$. Then the density of $\mathbf{Y} = \mathbf{X} - \theta = (X_1 - \theta, \dots, X_n - \theta)$ is $p_0(\mathbf{y})$. As a consequence, the distribution of $\mathbf{Y}$ does not depend on θ. ∎

Proof. Use the transformation rule $p_{\mathbf{Y}}(\mathbf{y}) = \det(\mathbf{J})^{-1} p_\theta(y_1 + \theta, \dots, y_n + \theta) = p_0(y_1 + \theta - \theta, \dots, y_n + \theta - \theta) = p_0(\mathbf{y})$, recalling that $\det(\mathbf{J}) = 1$. □

Theorem 3.2 has a very important consequence. In particular, an invariant statistic is usually an ancillary statistic.

Theorem 3.3 (Ancillarity of Location Invariant Statistics) Suppose $\mathbf{X} = (X_1, \ldots, X_n)$ is an observation from a location parameter density $p_\theta(\mathbf{x}) = p_0(\mathbf{x} - \theta)$. Then the distribution of any location invariant statistic does not depend on θ. ■

Proof. Suppose $T(\mathbf{X})$ is location invariant. Then $T(\mathbf{X}) = T(\mathbf{X} - \theta)$. But by Theorem 3.2 the distribution of $\mathbf{X} - \theta$ does not depend on θ. □

Example 3.5 (The Distribution of the Sample Variance *dnd* on a Location Parameter) The sample variance S^2 is location invariant, so its distribution will not depend on any location parameters. ■

Theorems 3.4 and 3.5 below give some practical methods for the construction of equivariant or invariant statistics.

Theorem 3.4 (Construction of Location Invariant Statistics) Let $S, S' \in \mathcal{T}_{stat}$ be any two location equivariant statistics. Then $T(\mathbf{X}) = S(\mathbf{X}) - S'(\mathbf{X})$ is location invariant. ■

Proof. We have $T(\mathbf{X}+a) = S(\mathbf{X}+a) - S'(\mathbf{X}+a) = S(\mathbf{X}) + a - S'(\mathbf{X}) - a = S(\mathbf{X}) - S'(\mathbf{X}) = T(\mathbf{X})$. □

Theorem 3.5 (Convexity and Location Equivariance) A convex combination of location equivariant statistics is location equivariant. ■

Proof. Suppose $S_1(\mathbf{X}), \ldots, S_m(\mathbf{X})$ are location equivariant statistics. Set constants $p_1 + \cdots + p_m = 1$, $p_i \geq 0$. Then define statistic $S(\mathbf{X}) = p_1 S_1(\mathbf{X}) + \cdots + p_m S_m(\mathbf{X})$. We then have

$$\begin{aligned} S(\mathbf{X} + a) &= p_1 S_1(\mathbf{X} + a) + \cdots + p_m S_m(\mathbf{X} + a) \\ &= p_1 S_1(\mathbf{X}) + \cdots + p_m S_m(\mathbf{X}) + p_1 a + \cdots + p_m a = S(\mathbf{X}) + a. \end{aligned}$$

□

The following two examples apply Theorems 3.4 and 3.5.

Example 3.6 (The Sample Mean as Convex Combination) The sample mean $\bar{\mathbf{X}}$ is a convex combination of observations X_i, which are location equivariant. Therefore $\bar{\mathbf{X}}$ is location equivariant. ■

Example 3.7 (Order Statistics and Location Parameter Models) Suppose $\mathbf{X} = (X_1, \ldots, X_n)$ is an observation from a location parameter density $p_\theta(\mathbf{x}) = p_0(\mathbf{x} - \theta)$. Consider location transformation (3.3). The transformation applies a single increasing transformation to each element of $\mathbf{X}$, therefore order is preserved. This means $Y_{(k)} = X_{(k)} + a$ so that the order statistics are location equivariant.

The sample median is usually taken to be

$$\hat{m} = \begin{cases} X_{(\frac{n+1}{2})}; & n \text{ is odd} \\ \frac{X_{(\frac{n}{2})} + X_{(\frac{n}{2}+1)}}{2}; & n \text{ is even} \end{cases}.$$

For any n, $\hat{m}$ is a convex combination of order statistics. So the sample median is location equivariant (Theorem 3.5).

Then consider the interquartile range (IQR) $IQR = Q_3 - Q_1$ where we have the quartiles $Q_i, i = 1, 2, 3$. These divide a sample into 4 quarters by rank, so that Q_1 is the 25th percentile and Q_3 is the 75th percentile. If the quartiles are convex combinations of order statistics, by Theorem 3.5 they are location equivariant, so IQR is location invariant (Theorem 3.4).

The sample range is $Range = \max_i X_i - \min_i X_i = X_{(n)} - X_{(1)}$ which is the difference of two equivariant statistics, and so is location invariant (Theorem 3.4). ■

3.3.2 Scale Parameter Models

Suppose $\mathbf{X} = (X_1, \dots, X_n) \sim \mathcal{X}$ possesses density $p_1(\mathbf{x}) = p_1(x_1, \dots, x_n)$. Consider the scale transformation

$$\mathbf{Y} = (Y_1, \dots, Y_n) = (\theta X_1, \dots, \theta X_n) = \theta \mathbf{X}, \;\; \theta > 0. \tag{3.4}$$

As for the location transformation, Equation (3.4) is a special case of the location-scale transformation of Equation (3.2). Let $G_{sc} = \{g_\theta : \theta \in \Theta\}$, $\Theta = \mathbb{R}_{>0}$, be the class of all scale transformations. For this model, we usually have $\mathcal{X} = \mathbb{R}^n$ or $\mathcal{X} = \mathbb{R}^n_{>0}$; in either case each $g_\theta \in G_{sc}$ is a bijection on $\mathcal{X}$.

The Jacobian matrix of this transformation is $\mathbf{J} = \theta^n \mathbf{I}$, so the density of $\mathbf{Y}$ is

$$p_{\mathbf{Y}}(\mathbf{y}) = p_{\mathbf{Y}}(y_1, \dots, y_n) = \theta^{-n} p_1(y_1/\theta, \dots, y_n/\theta),$$

since $\det(\mathbf{J}) = \theta^n$. Then define parametric family $\mathcal{P} = \{p_\theta : \theta \in \Theta\}$, $\Theta = \mathbb{R}_{>0}$, where

$$p_\theta(\mathbf{x}) = p_\theta(x_1, \dots, x_n) = \theta^{-n} p_1(x_1/\theta, \dots, x_n/\theta) = \theta^{-n} p_1(\mathbf{x}/\theta).$$

$\mathcal{P}$ is called a scale parameter model in which case $\theta > 0$ is a scale parameter. In fact, a scale parameter model can be easily generated from any density $p_1(\mathbf{x})$ by setting $p_\theta(\mathbf{x}) = \theta^{-n} p_1(\mathbf{x}/\theta)$, $\theta \in \Theta = \mathbb{R}_{>0}$.

Example 3.8 (Normal Density as Scale Parameter Model) Consider as an example an *iid* sample $\mathbf{X} = (X_1, \dots, X_n)$ from density $X_1 \sim N(0, \theta^2)$. For parameter space $\Theta = \mathbb{R}_{>0}$ the densities are given by

$$p_\theta(\mathbf{x}) = \frac{1}{\theta^n (2\pi)^{n/2}} \exp\left(-\frac{1}{2} \sum_{i=1}^n x_i^2/\theta^2 \right) = \frac{1}{\theta^n} p_1(\mathbf{x}/\theta),$$

where $p_1(\mathbf{x})$ is the density when $X_1 \sim N(0, 1)$. This defines a scale parameter model. ■

Example 3.9 (Pareto Density Family Defined by a Scale Parameter) The density of $X \sim pareto(\theta, \alpha)$ with $\theta = 1$, is given by $f(x) = \alpha x^{-(\alpha+1)} I\{x > 1\}$. We may generate a scale family by introducing a scale parameter $\theta \in \Theta = \mathbb{R}_{>0}$

$$f_\theta(x) = \theta^{-1} f(x/\theta) = \alpha\theta^{-1} (\theta/x)^{\alpha+1} I\{x/\theta > 1\} = \alpha\theta^{-1} (\theta/x)^{\alpha+1} I\{x > \theta\},$$

noting that the scale parameter must also be introduced into the support. But $f_\theta(x)$ is now a Pareto density with the conventional parametrization $pareto(\theta, \alpha)$, that is, θ is a scale parameter. ■

As for location equivariance, if we regard a scale transformation as a change in units, then an equivariant statistic is one that preserves this change. However, we need a more general definition for equivariance. In Example 3.8, θ is the standard deviation, while θ^2 is the variance, and both may be of interest. So to define scale equivariance we first fix the power θ^k of a scale parameter θ.

Definition 3.4 (Scale Invariant and Equivariant Statistics) A statistic $S(\mathbf{X}) \in \Theta$ is (scale) equivariant for estimand θ^k if $S(g_\theta \mathbf{X}) = \theta^k S(\mathbf{X})$ for all $g_\theta \in G_{sc}$, and $S(\mathbf{X}) \geq 0$. A statistic $T(\mathbf{X}) \in \Theta$ is (scale) invariant if $T(g_\theta \mathbf{X}) = T(\mathbf{X})$ for all $g_\theta \in G_{sc}$. Then $M(\mathbf{X})$ is a maximal (scale) invariant if $M(\mathbf{X})$ is invariant, and $M(\mathbf{X}) \mapsto T(\mathbf{X})$ for any other invariant statistic $T(\mathbf{X})$. ■

In Definition 3.4 it is assumed that $S(\mathbf{X}) \geq 0$, although the defining condition is still well-defined if $S(\mathbf{X})$ is allowed to be negative. However, we are following the convention of Lehmann and Casella (1998) (Section 3.3) that a scale equivariant statistic is interpreted as an estimator of a positive estimand.

A number of commoly used statistics satisfy Definition 3.4.

Example 3.10 (The Sample Variance is Scale Equivariant) Consider the scale transformation $\mathbf{Y} = a\mathbf{X}$ of Equation (3.4). Then for $a > 0$ $\bar{\mathbf{Y}} = n^{-1}\sum_{i=1}^n Y_i = n^{-1}\sum_{i=1}^n aX_i = a\bar{\mathbf{X}}$. This gives $S^2_{\mathbf{Y}} = (n-1)^{-1}\sum_{i=1}^n (Y_i - \bar{\mathbf{Y}})^2 = (n-1)^{-1}\sum_{i=1}^n (aX_i - a\bar{\mathbf{X}})^2 = a^2 S^2_{\mathbf{X}}$, so that the sample variance is a scale equivariant estimator of θ^2 while the sample standard deviation $\sqrt{S^2_{\mathbf{X}}}$ is a scale equivariant estimator of θ. ■

Example 3.11 (The T-statistic is Scale Invariant) Recall the T-statistic

$$T(\mathbf{X}) = \frac{\bar{\mathbf{X}}}{\sqrt{S^2_{\mathbf{X}}/n}}$$

(Section 2.5.1) and assume $S^2_{\mathbf{X}} > 0$. Consider the scale transformation $\mathbf{Y} = a\mathbf{X}$. From Example 3.13, we have $\bar{\mathbf{Y}} = a\bar{\mathbf{X}}$ and $S^2_{\mathbf{Y}} = a^2 S^2_{\mathbf{X}} > 0$ so that

$$T(\mathbf{Y}) = \frac{\bar{\mathbf{Y}}}{\sqrt{S^2_{\mathbf{Y}}/n}} = \frac{a\bar{\mathbf{X}}}{\sqrt{a^2 S^2_{\mathbf{X}}/n}} = \frac{\bar{\mathbf{X}}}{\sqrt{S^2_{\mathbf{X}}/n}} = T(\mathbf{X}).$$

Thus, T is scale invariant, assuming $S^2_{\mathbf{X}} > 0$. If the scale model is continuous, we will have $P(S^2_{\mathbf{X}} > 0) = 1$. ■

Derivation of a maximal scale invariant can follow closely that of the maximal location invariant given by Theorem 3.1. However, for this approach we will need to identify a scale equivariant statistic $S(\mathbf{x})$ which is strictly positive for $\mathbf{x} \in \mathcal{X}$. In practice, we may have a nonempty subset $\mathcal{X}_0 \subset \mathcal{X}$ on which $S(\mathbf{x}) = 0$ (the sample variance, for example). However, for continuous densities we will have probability $P_\theta(\mathbf{X} \in \mathcal{X}_0) = 0$, so this case can be disregarded. However, a more satisfactory approach would be to recognize that if $g_\theta \in G_{sc}$ is a bijection on $\mathcal{X}$, it is also a bijection on $\mathcal{X} - \mathcal{X}_0$ and $\mathcal{X}_0$ as well as $\mathcal{X}$. This formally justifies converting the sample space to $\mathcal{X} - \mathcal{X}_0$.

Theorem 3.6 Suppose each $g_\theta \in G_{sc}$ is a bijection on $\mathcal{X} \subset \mathbb{R}^n$. Let $S(\mathbf{x})$ be a scale equivariant statistic for θ^k, and define $\mathcal{X}_0 = \{\mathbf{x} \in \mathcal{X} : S(\mathbf{x}) = 0\}$. Then each $g_\theta \in G_{sc}$ is also a bijection on both $\mathcal{X} - \mathcal{X}_0$ and $\mathcal{X}_0$. ■

Proof. Suppose $S(\mathbf{x}) = 0$ for $\mathbf{x} \in \mathcal{X}$. Then $S(g_\theta \mathbf{x}) = \theta^k S(\mathbf{x}) = 0$, therefore g_θ maps $\mathcal{X}_0$ into $\mathcal{X}_0$. Similarly, if $S(\mathbf{x}) > 0$ for $\mathbf{x} \in \mathcal{X}$, then $S(g_\theta \mathbf{x}) = \theta^k S(\mathbf{x}) > 0$, since $\theta > 0$. Therefore, g_θ maps $\mathcal{X} - \mathcal{X}_0$ into $\mathcal{X} - \mathcal{X}_0$. But g_θ is bijective on $\mathcal{X}$, so it must be bijection on both $\mathcal{X} - \mathcal{X}_0$ and $\mathcal{X}_0$. □

Theorem 3.6 allows us to assume that $S(\mathbf{X}) > 0$, which does not really change the model if $P_\theta(S(\mathbf{X}) = 0) = 0$.

Theorem 3.7 (Maximal Scale Invariant) Let $S \in \mathcal{T}_{stat}$ be any statistic which is scale equivariant for θ. Assume $S(\mathbf{X}) > 0$ for all $\mathbf{X} \in \mathcal{X}$. Then $M(\mathbf{X}) = (X_1/S(\mathbf{X}), \ldots, X_n/S(\mathbf{X}))$ is a maximal (scale) invariant. ■

Proof. That $M(\mathbf{X})$ is scale invariant is easily verified. Next, let T be any invariant statistic. Then $T(X_1, \ldots, X_n) = T(X_1/a, \ldots, X_n/a)$ for any a. By setting $a = S(\mathbf{X})$, we conclude that $M \mapsto T$. □

That a scale invariant statistic is also an ancillary statistic follows essentially the same argument given in Theorems 3.2 and 3.3 for the location invariant statistic. This is given in the next two theorems.

Theorem 3.8 Suppose $\mathbf{X} = (X_1, \ldots, X_n) \in \mathcal{X}$ is an observation from a scale family density $p_\theta(\mathbf{x}) = \theta^{-n} p_1(\mathbf{x}/\theta)$. Then the density of $\mathbf{Y} = \mathbf{X}/\theta = (X_1/\theta, \ldots, X_n/\theta)$ is $p_1(\mathbf{y})$. As a consequence, the distribution of $\mathbf{Y}$ does not depend on θ. ■

Proof. Use the transformation rule

$$p_{\mathbf{Y}}(\mathbf{y}) = \det(\mathbf{J})^{-1} p_\theta(\theta y_1, \ldots, \theta y_n) = \theta^n \theta^{-n} p_1(\theta y_1/\theta, \ldots, \theta y_n/\theta) = p_1(\mathbf{y}),$$

noting that $\det(\mathbf{J}) = \theta^{-n}$. □

Theorem 3.9 (Ancillarity of Scale Invariant Statistics) Suppose $\mathbf{X} = (X_1, \ldots, X_n)$ is an observation from a scale family density $p_\theta(\mathbf{x}) = \theta^{-n} p_1(x/\theta)$. Then the distribution of any scale invariant statistic does not depend on θ. ■

Proof. Suppose $T(\mathbf{X})$ is scale invariant. Then $T(\mathbf{X}) = T(\mathbf{X}/\theta)$. But by Theorem 3.8 the distribution of $\mathbf{X}/\theta$ does not depend on θ. □

The next theorem shows how scale invariant statistics may be easily constructed from scale equivariant statistics.

Theorem 3.10 (Construction of Scale Invariant Statistics) Let $S, S' \in \mathcal{T}_{stat}$ be any two statistics which are scale equivariant for θ^k. Assume $S(\mathbf{X}), S'(\mathbf{X}) > 0$ for $\mathbf{X} \in \mathcal{X}$. Then $T(\mathbf{X}) = S(\mathbf{X})/S'(\mathbf{X})$ is scale invariant. ■

Proof. We may write

$$T(a\mathbf{X}) = S(a\mathbf{X})/S'(a\mathbf{X}) = (a^k S(\mathbf{X}))/(a^k S'(\mathbf{X})) = S(\mathbf{X})/S'(\mathbf{X}) = T(\mathbf{X}),$$

which completes the proof. □

Example 3.12 (Ratio of Quadratic Forms) We consider an important application of Theorem 3.10. Suppose $\mathbf{X} = (X_1, \ldots, X_n)$ and $\mathbf{Q}_1, \mathbf{Q}_2 \in \mathcal{M}_n$. Denote the quadratic form $W_i(\mathbf{X}) = \mathbf{X}^T \mathbf{Q}_i \mathbf{X}$ $i = 1, 2$, and assume $W_i(\mathbf{X}) > 0$. Consider the scale transformation $\mathbf{Y} = g_a \mathbf{X} = a\mathbf{X}$. Then

$$W_i(\mathbf{Y}) = \mathbf{Y}^T \mathbf{Q}_i \mathbf{Y} = a^2 \mathbf{X}^T \mathbf{Q}_i \mathbf{X} = a^2 W_i(\mathbf{X}).$$

Thus, $W_i(\mathbf{X})$ is a scale equivariant estimator of θ^2, so by Theorem 3.10 $W_1(\mathbf{X})/W_2(\mathbf{X})$ is scale invariant. ■

3.3.3 Location-Scale Invariant Models

Parametric families defined by a single location or scale parameter have convenient properties, but are unrealistic for many applications. It is, therefore, important to develop the location-scale invariant model, which can be generated by applying the location-scale transformations G_{ls} to an observation $\mathbf{X} \in \mathcal{X} \subset \mathbb{R}^n$ with density p (we assume each $g_\theta \in G_{ls}$ is a bijection on $\mathcal{X}$). We then have the location-scale parameter $\theta = (\mu, \sigma) \in \mathbb{R} \times \mathbb{R}_{>0} = \Theta$. The development is much the same, therefore, as for the location or scale invariant models. Suppose $\mathbf{X}$ possesses density $p(\mathbf{x}) = p(x_1, \ldots, x_n)$. Applying transformation $g_\theta \in G_{ls}$ (Equation (3.2)) gives

$$\mathbf{Y} = (Y_1, \ldots, Y_n) = (\mu + \sigma X_1, \ldots, \mu + \sigma X_n) = \mu + \sigma \mathbf{X}, \;\; \mu \in \mathbb{R}, \sigma \in \mathbb{R}_{>0}. \tag{3.5}$$

The Jacobian matrix of this transformation is $\mathbf{J} = \sigma^n \mathbf{I}$, so the density of $\mathbf{Y}$ is

$$p_{\mathbf{Y}}(\mathbf{y}) = p_{\mathbf{Y}}(y_1, \ldots, y_n) = \sigma^{-n} p\left((y_1 - \mu)/\sigma, \ldots, (y_n - \mu)/\sigma\right),$$

since $\det(\mathbf{J}) = \sigma^n$. Then define parametric family $\mathcal{P} = \{p_\theta : \theta \in \Theta\}$, where

$$p_\theta(\mathbf{x}) = \sigma^{-n} p\left((x_1 - \mu)/\sigma, \ldots, (x_n - \mu)/\sigma\right) = \sigma^{-n} p\left((\mathbf{x} - \mu)/\sigma\right). \tag{3.6}$$

$\mathcal{P}$ is called a location-scale parameter model, or a location-scale invariant model, in which case μ is the location parameter and σ is a scale parameter. In fact, a location-scale parameter model can be easily generated from any density $p(\mathbf{x})$ by setting $p_\theta(\mathbf{x}) = \sigma^{-n} p\left((\mathbf{x} - \mu)/\sigma\right)$, $(\mu, \sigma) = \theta \in \Theta$.

Recall the normal density $N(\mu, \sigma^2)$

$$f_{(\mu,\sigma)}(x) = \frac{1}{\sigma(2\pi)^{1/2}} \exp\left(-\frac{1}{2}(x - \mu)^2/\sigma^2\right).$$

We can see that μ is a location parameter, since $f_{(\mu,\sigma)}(x) = f_{(0,\sigma)}(x - \mu)$. However, unless $\mu = 0$, σ is not a scale parameter for a scale invariant model, since

$$\begin{aligned}
\frac{1}{\sigma} f_{(\mu,1)}(x/\sigma) &= \frac{1}{\sigma(2\pi)^{1/2}} \exp\left(-\frac{1}{2}([x/\sigma] - \mu)^2\right) \\
&= \frac{1}{\sigma(2\pi)^{1/2}} \exp\left(-\frac{1}{2}(x - \sigma\mu)^2/\sigma^2\right) \\
&= f_{(\mu,\sigma\mu)}(x) \neq f_{(\mu,\sigma)}(x).
\end{aligned}$$

So σ does not satsify the formal definition of a scale parameter. For this reason, we may consider the location and scale parameters jointly to define a parametric family of the form (3.6). Then for the normal density $N(\mu, \sigma^2)$

$$f_{(\mu,\sigma)}(x) = \frac{1}{\sigma(2\pi)^{1/2}} \exp\left(-\frac{1}{2}(x - \mu)^2/\sigma^2\right) = \frac{1}{\sigma} f_{(0,1)}\left(\frac{(x - \mu)}{\sigma}\right),$$

where $\theta = (\mu, \sigma)$ is the location-scale parameter.

The location-scale parameter model admits definitions of equivariance and invariance similar to those offered in Definitions (3.3) and (3.4). However, invariance and equivariance properties of a statistic S are defined relative to a specific parameter, so some care must be used with respect to terminology.

Definition 3.5 (Location-Scale Invariant and Equivariant Statistics) A statistic $S(\mathbf{X}) \in \mathbb{R}$ is location-scale equivariant if $S(g_\theta \mathbf{x}) = \mu + \sigma S(\mathbf{x})$ for all $g_\theta \in G_{ls}$. A statistic $T(\mathbf{X})$ is location-scale invariant if $S(g_\theta \mathbf{x}) = S(\mathbf{x})$ for all $g_\theta \in G_{ls}$. Then $M(\mathbf{X})$ is a maximal (location-scale) invariant if $M(\mathbf{X})$ is invariant, and $M(\mathbf{X}) \mapsto T(\mathbf{X})$ for any other invariant statistic $T(\mathbf{X})$. ■

The definitions of location equivariance and invariance Definition 3.3 and scale equivariance and invariance Definition 3.4 present no ambiguity when Definition 3.5 is added to the mix. In particular, we may still single out location or scale equivariant statistics for the estimation of location parameter μ or scale parameter estimand σ^r. However, it will be useful to impose on any scale equivariant estimator $\delta(\mathbf{X})$ the additional requirement of location invariance, that is, $\delta(a + b\mathbf{X}) = b^r \delta(\mathbf{X})$. Similarly, we may also impose on any location equivariant estimator the additional requirement of location-scale equivariance.

Example 3.13 (The Sample Variance is Scale Equivariant and Location Invariant) From Examples 3.4 and 3.10, the sample variance is scale equivariant (*wrt* σ^2) and location invariant. However, it is not location-scale invariant or location-scale equivariant. ∎

The characterization of the maximal location-scale invariant is similar to that of the maximal scale invariant, and is given by the following theorem. The proof is similar to those of Theorems 3.1 and 3.7 and is left to the reader (Problem 3.2).

Theorem 3.11 (Maximal Location-Scale Invariant) Let $S \in \mathcal{T}_{stat}$ be any scale equivariant/location invariant estimator of estimand σ. Suppose $S(\mathbf{X}) > 0$, $\mathbf{X} \in \mathcal{X}$. Let δ be any location-scale equivariant statistic. Define the mapping $M(\mathbf{X}) = ((X_1 - \delta(\mathbf{X}))/S(\mathbf{X}), \ldots, (X_n - \delta(\mathbf{X}))/S(\mathbf{X}))$. Then $M(\mathbf{X})$ is a maximal (location-scale) invariant. ∎

3.4 Regular Families

We may identify a set of regularity conditions imposed on a parametric family which may both simplify analysis and permit a richer theory. In practice, many (probably most) commonly used parametric familes are also regular families. These conditions vary between referencess, but are usually intended to permit the application of Fisher information (discussed in the following section) and the quite rich theory following from it. Our definition follows.

Definition 3.6 (Regular Families) Suppose we are given a parametric family $\mathcal{P} = \{p_\theta : \theta \in \Theta\}$ on sample space $\mathcal{X}$. Then $\mathcal{P}$ is a regular family if the following conditions hold:

(i) $\Theta \subset \mathbb{R}^p$ is open.
(ii) $\mathcal{P}$ is a family of common support.
(iii) All first and second-order partial derivatives of p_θ exist, and are continuous.
(iv) All first and second-order partial derivatives of p_θ exist, and are integrable.
(v) First and second partial differentiation of p_θ can be exchanged with integration. ∎

The main task of verifying Definition 3.6 is validating the exchange of integration and differentiation. A suitable form of Leibniz' rule (Section C.4) may be used for this purpose, and assumption (v) can be replaced with any set of conditions which, in addition to (i)–(iv), imply the hypothesis of that theorem.

For this reason, it is interesting to consider the assumptions of Theorem 6.2.6 of Lehmann and Casella (1998), which may also serve as a definition of a regular family. We adapt assumption (f) of that theorem. Suppose that for any $\theta^* \in \Theta$, and any two (possibly identical) coordinates θ_i, θ_j there exists $\epsilon > 0$ and integrable function $M(\tilde{x})$ for which

$$\left| \left\{ \left. \frac{\partial^2 p_\theta(\tilde{x})}{\partial\theta_i \partial\theta_j} \right|_{\theta'} \right\} \right| \leq M(\tilde{x}) \tag{3.7}$$

when $\theta'_i = \theta^*_i$, $i \neq j$ and $\theta'_j \in (\theta^*_j - \epsilon, \theta^*_j + \epsilon)$. The choice of $M(\tilde{x})$ and ϵ may depend on θ^* and the choice of coordinates θ_i, θ_j.

It is important to note that parametric families for which support $\bar{\mathcal{X}}_\theta$ varies with the parameter are not regular families.

3.5 Fisher Information

Definition 3.6 of the regular family mainly guarantees conditions which permit evaluation of Fisher information, which will play several important, and quite distinct roles. Suppose we have parameter space $\Theta \subset \mathbb{R}^p$. For convenience, we use the notation

$$\Psi_j(\tilde{x}, \theta) = \frac{\partial \ell(\theta; \tilde{x})}{\partial \theta_j}, \qquad \Psi'_{jk}(\tilde{x}, \theta) = \frac{\partial^2 \ell(\theta; \tilde{x})}{\partial \theta_j \partial \theta_k}, \quad j, k = 1, \ldots, p,$$

assuming the derivatives exist. Then the Fisher information matrix $\mathbf{I}(\theta)$ is a $p \times p$ matrix defined by $\{\mathbf{I}(\theta)\}_{jk} = E_\theta[\Psi_j(\tilde{X}, \theta)\Psi_k(\tilde{X}, \theta)]$. This quantity is especially useful when the following theorem holds:

Theorem 3.12 Suppose $\mathcal{P}$ is a regular parametric family (Definition 3.6). Then the following statements hold.

(i) For all $\theta \in \Theta$ and $j = 1, \ldots, p$, $E_\theta[\Psi_j(\tilde{X}, \theta)] = 0$.
(ii) $\mathbf{I}(\theta)$ equals the covariance matrix of $\Psi(\tilde{X}, \theta)$.
(iii) We have the identity $\mathbf{I}(\theta) = -E_\theta[\Psi'(\tilde{X}, \theta)]$. ■

Proof. The proof relies on the exchange of differentiation and integration, the validity of which follows from Definition 3.6. We may write

$$\begin{aligned}
E_\theta[\Psi_j(\tilde{X}, \theta)] &= \int_{\mathcal{X}} \left(\frac{\partial}{\partial \theta_j} \log p_\theta(\tilde{x}) \right) p_\theta(\tilde{x}) d\nu_{\mathcal{X}} \\
&= \int_{\mathcal{X}} \frac{1}{p_\theta(\tilde{x})} \frac{\partial p_\theta(\tilde{x})}{\partial \theta_j} p_\theta(\tilde{x}) d\nu_{\mathcal{X}} \\
&= \int_{\mathcal{X}} \frac{\partial p_\theta(\tilde{x})}{\partial \theta_j} d\nu_{\mathcal{X}} \\
&= \frac{\partial}{\partial \theta_j} \int_{\mathcal{X}} p_\theta(\tilde{x}) d\nu_{\mathcal{X}} \\
&= \frac{\partial 1}{\partial \theta_j} = 0,
\end{aligned}$$

which proves part (i). Part (ii) follows directly. To prove part (iii), first note that the following equality is easily verified:

$$\frac{\partial^2 \log p_\theta(\tilde{x})}{\partial \theta_j \partial \theta_{j'}} = \frac{1}{p_\theta(\tilde{x})} \frac{\partial^2 p_\theta(\tilde{x})}{\partial \theta_j \partial \theta_{j'}} - \left(\frac{\partial}{\partial \theta_j} \log p_\theta(\tilde{x}) \right) \left(\frac{\partial}{\partial \theta_{j'}} \log p_\theta(\tilde{x}) \right). \tag{3.8}$$

Then, we have

$$\begin{aligned}
E_\theta \left[\frac{1}{p_\theta(\tilde{x})} \frac{\partial^2 p_\theta(\tilde{x})}{\partial \theta_j \partial \theta_{j'}} \right] &= \int_{\mathcal{X}} \frac{1}{p_\theta(\tilde{x})} \frac{\partial^2 p_\theta(\tilde{x})}{\partial \theta_j \partial \theta_{j'}} p_\theta(\tilde{x}) d\nu_{\mathcal{X}} \\
&= \frac{\partial^2}{\partial \theta_j \partial \theta_{j'}} \int_{\mathcal{X}} p_\theta(\tilde{x}) d\nu_{\mathcal{X}} \\
&= \frac{\partial^2 1}{\partial \theta_j \partial \theta_{j'}} = 0.
\end{aligned}$$

The proof is completed by taking the expectation of Equation (3.8) and applying part (ii). □

Additivity of the Fisher Information Matrix

One important property of the Fisher information is that it is additive. Suppose $\ell(\theta; \tilde{X})$ and $\ell(\theta; \tilde{Y})$ are two log-likelihood functions for the same parameter based on independent observations (these need not come from the same density, as long as the parameter space itself is common to both). Suppose the respective Fisher information matrices are $\mathbf{I}_{\tilde{X}}(\theta), \mathbf{I}_{\tilde{Y}}(\theta)$. Under these conditions, the log-likelihoods of the pooled observations are additive, hence so is the Fisher information matrix, that is, for pooled data $(\tilde{X}, \tilde{Y})$ the log-likelihood and the Fisher information matrix are given by

$$\ell(\theta; \tilde{X}, \tilde{Y}) = \ell(\theta; \tilde{X}) + \ell(\theta; \tilde{Y}), \qquad \mathbf{I}_{\tilde{X},\tilde{Y}}(\theta) = \mathbf{I}_{\tilde{X}}(\theta) + \mathbf{I}_{\tilde{Y}}(\theta).$$

Fisher Information and Parametric Transformations

Suppose we have an injective parametric transformation $\lambda = g(\theta) \in \mathbb{R}^p$ of $\theta \in \Theta \subset \mathbb{R}^p$. Let $\mathbf{I}_\theta(\theta)$, $\mathbf{I}_\lambda(\lambda)$ be the Fisher information matrices evaluated with respect to the alternative parametrizations. The parametric family is invariant to this transformation, in the sense that we may write $p_\lambda(\tilde{x}) = p_{\theta(\lambda)}(\tilde{x}) = p_{g^{-1}(\lambda)}(\tilde{x})$. However, we cannot calculate the information with respect to λ with this type of simple substitution. The correct tranformation rule is

$$\mathbf{I}_\lambda(\lambda) = \mathbf{J}^T \mathbf{I}_\theta(\theta(\lambda))\mathbf{J} = \mathbf{J}^T \mathbf{I}_\theta(g^{-1}(\lambda))\mathbf{J} \tag{3.9}$$

where $\mathbf{J} = \partial\theta/\partial\lambda$ is the Jacobian matrix of the transformation $\theta(\lambda) = g^{-1}(\lambda)$. Then $\mathbf{I}_\lambda(\lambda)$ is the information that would be obtained starting from the parametrization λ.

For a one dimensional transformation $\lambda = g(\theta) \in \mathbb{R}$ the transformation rule is:

$$I_\lambda(\lambda) = I_\theta(g^{-1}(\lambda)) \left[dg^{-1}(\lambda)/d\lambda\right]^2 = I_\theta(g^{-1}(\lambda)) \left[dg(\theta)/d\theta\right]^{-2},$$

noting that we substitute $\theta = g^{-1}(\lambda)$ where needed.

Example 3.14 (Fisher Information Matrix for $N(\mu, \sigma^2)$) The log-likelihood for the $N(\mu, \sigma^2)$ density is

$$\ell(\theta; x) = -\frac{(x-\mu)^2}{2\sigma^2} - \log(\sigma) - \log(\sqrt{2\pi}).$$

We then have derivatives

$$\Psi_1(X, \theta) = \frac{\partial \ell(\theta; x)}{\partial \mu} = \frac{(x-\mu)}{\sigma^2}, \quad \Psi_2(X, \theta) = \frac{\partial \ell(\theta; x)}{\partial \sigma} = \frac{(x-\mu)^2}{\sigma^3} - \frac{1}{\sigma}.$$

To evaluate $\mathbf{I}_\theta$, we first use Theorem 3.12 (ii). We may write

$$\Psi(X, \theta) = (\Psi_1(X, \theta), \Psi_2(X, \theta)) = \left(\frac{(X-\mu)}{\sigma^2}, \frac{(X-\mu)^2}{\sigma^3} - \frac{1}{\sigma}\right).$$

Then $\mathbf{I}(\theta)$ is the covariance matrix of $\Psi(X, \theta)$. This means

$$\mathbf{I}(\theta) = \begin{bmatrix} \mathrm{var}[\frac{(X-\mu)}{\sigma^2}] & \mathrm{cov}[\frac{(X-\mu)}{\sigma^2}, \frac{(X-\mu)^2}{\sigma^3}] \\ \mathrm{cov}[\frac{(X-\mu)}{\sigma^2}, \frac{(X-\mu)^2}{\sigma^3}] & \mathrm{var}[\frac{(X-\mu)^2}{\sigma^3}] \end{bmatrix} = \begin{bmatrix} \frac{1}{\sigma^2} & 0 \\ 0 & \frac{2}{\sigma^2} \end{bmatrix}.$$

Alternatively, we may use Theorem 3.12 (iii). We need the 2nd order partial derivatives:

$$\Psi_{11}(x, \theta) = \frac{\partial^2 \ell(\theta; x)}{\partial \mu^2} = \frac{\partial}{\partial \mu}\left[\frac{(x-\mu)}{\sigma^2}\right] = -\frac{1}{\sigma^2},$$

$$\Psi_{12}(x, \theta) = \frac{\partial^2 \ell(\theta; x)}{\partial \mu \partial \sigma} = \frac{\partial}{\partial \sigma}\left[\frac{(x-\mu)}{\sigma^2}\right] = -2\frac{(x-\mu)}{\sigma^3},$$

$$\Psi_{22}(x, \theta) = \frac{\partial^2 \ell(\theta; x)}{\partial \sigma^2} = \frac{\partial}{\partial \sigma}\left[\frac{(x-\mu)^2}{\sigma^3} - \frac{1}{\sigma}\right] = -3\frac{(x-\mu)^2}{\sigma^4} + \frac{1}{\sigma^2}.$$

Note that $\Psi_{12}(X,\theta) = \Psi_{21}(X,\theta)$. This gives

$$\mathbf{I}(\theta) = \begin{bmatrix} -E[\Psi_{11}(X,\theta)] & -E[\Psi_{12}(X,\theta)] \\ -E[\Psi_{21}(X,\theta)] & -E[\Psi_{22}(X,\theta)] \end{bmatrix} = \begin{bmatrix} \frac{1}{\sigma^2} & 0 \\ 0 & \frac{2}{\sigma^2} \end{bmatrix}.$$

Then by additivity the Fisher information matrix for an *iid* sample $X_1,\ldots,X_n$, $X_1 \sim N(\mu,\sigma^2)$, *wrt* $\theta = (\mu,\sigma)$ is $n\mathbf{I}(\theta)$.

Finally, note that the Fisher information for the parameter σ is $I_\sigma(\sigma) = 2/\sigma^2$. What is $I_{\sigma^2}(\sigma^2)$? We have transformation $\sigma^2 = g(\sigma)$, where $g(x) = x^2$. For a one-dimensional transformation, the reparametrization takes the form:

$$I_{\sigma^2}(\sigma^2) = I_\sigma(g^{-1}(\sigma^2))\,[dg(\sigma)/d\sigma]^{-2} = \frac{2}{\sigma^2}\,[2\sigma]^{-2} = \frac{1}{2\sigma^4}.$$ ■

Example 3.15 (Fisher Information for Location and Scale Parameters) Fisher information possesses a distinct form for location or scale parameters. If θ is a location parameter then $I(\theta)$ is constant *wrt* θ. If θ is a scale parameter then $I(\theta) = K/\theta^2$, where K *dnd* θ. The proof is left to the reader (Problem 3.3). ■

3.6 Exponential Families

Exponential family densities allow an especially convenient and tractable definition of parametric models. These possess a common structure for which a unified theory of inference can be developed. Furthermore, many of the commonly used distributions (normal, binomial, Poisson, gamma, and so on) fall into this class.

Suppose we have sample space $\mathcal{X}$, parameter space $\theta \in \Theta \subset \mathbb{R}^p$, and we are given a statistic $\mathbf{T}(\tilde{x}) = (T_1(\tilde{x}),\ldots,T_m(\tilde{x})) \in \mathbb{R}^m$, for each $\tilde{x} \in \mathcal{X}$, mappings $B : \Theta \to \mathbb{R}$, $\eta_j : \Theta \to \mathbb{R}$, $j = 1,\ldots,m$ and a further mapping $h : \mathcal{X} \to \mathbb{R}_{>0}$. If we may define densities

$$p_\theta(\tilde{x}) = \exp\left[\sum_{i=1}^{m} \eta_i(\theta)T_i(\tilde{x}) - B(\theta)\right] h(\tilde{x}), \tag{3.10}$$

then the resulting parameteric family $\mathcal{P} = \{p_\theta(\tilde{x}) : \theta \in \Theta\}$ is known as an exponential family. We refer to $\mathbf{T}(\tilde{x})$ as the natural sufficient statistic (Definition 3.9). The existence of $\mathcal{P}$ depends on the integrability of (3.10). If this holds, then $B(\theta)$ is constrained by the requirement that $P_\theta(\mathcal{X}) = 1$. To see this, we write

$$\begin{aligned} 1 &= \int_{\mathcal{X}} p_\theta(\tilde{x})d\nu_{\mathcal{X}} \\ &= \int_{\mathcal{X}} \exp\left[\sum_{i=1}^{m} \eta_i(\theta)T_i(\tilde{x}) - B(\theta)\right] h(\tilde{x})d\nu_{\mathcal{X}} \\ &= \exp(-B(\theta)) \int_{\mathcal{X}} \exp\left[\sum_{i=1}^{m} \eta_i(\theta)T_i(\tilde{x})\right] h(\tilde{x})d\nu_{\mathcal{X}}. \end{aligned}$$

This means

$$\exp(B(\theta)) = \int_{\mathcal{X}} \exp\left[\sum_{i=1}^{m} \eta_i(\theta)T_i(\tilde{x})\right] h(\tilde{x})d\nu_{\mathcal{X}}, \tag{3.11}$$

so that when defining an exponential family, we don't need to specify $B(\theta)$ explicity. If the remaining components permit calculation of the preceding integral, the exact form of $B(\theta)$ follows.

Another important property can be given directly. It is easily seen that $\mathcal{P}$ is a family of common support $\mathcal{X}_\theta = \mathcal{X} = \{\tilde{x} : h(\tilde{x}) > 0\}$. Conversely, if the support for a family of densities depends on θ, it cannot be an exponential family.

The Canonical Parametrization

It is sometimes more convenient to represent an exponential family in the form

$$p_\eta(\tilde{x}) = \exp\left[\sum_{i=1}^{m} \eta_i T_i(\tilde{x}) - A(\eta)\right] h(\tilde{x}), \tag{3.12}$$

by applying the transformation $\eta = (\eta_1, \ldots, \eta_m) = (\eta_1(\theta), \ldots, \eta_m(\theta))$. Then $A(\eta)$ plays the same normalizing function as $B(\theta)$ in (3.10), while $\mathbf{T}$ and h remain unchanged. This is referred to as the canonical parametrization, and assumes considerable importance in the theory of this class of models. An immediate advantage is its simplicity. Thus, a useful strategy may be to analyze an exponential family density in its canonical form, then to translate the results back to any reparametrized form.

The Natural Parameter Space

Following (3.11), p_η is defined *iff*

$$\exp\left(A(\eta)\right) = \int_{\mathcal{X}} \exp\left[\sum_{i=1}^{m} \eta_i T_i(\tilde{x})\right] h(\tilde{x}) d\nu_{\mathcal{X}} \tag{3.13}$$

is finite, equivalently, $A(\eta) < \infty$. The set

$$\Xi = \{\eta : A(\eta) < \infty\} \tag{3.14}$$

is called the natural parameter space, and is the maximal parameter space (that is, any other parameter space must be a subset of it). It follows that Ξ must be a convex set (if $\eta, \eta' \in \Xi$, then $a\eta + (1-a)\eta' \in \Xi$ for any $a \in [0,1]$).

3.6.1 Dimensionality of an Exponential Family

Consider the general form of the exponential family density of Equation (3.10). It will be important to verify when this representation is minimal, in the sense that no equivalent representation of Equation (3.10) exists in which the summation in the exponent has fewer than m terms.

To fix ideas, consider the canonical parametrization of (3.12). Suppose the components of $\mathbf{T}$ satisfy a linear constraint $a_0 = a_1 T_1 + \ldots + a_m T_m$, where $a_j \neq 0$ for at least one $j \geq 1$. Without any loss of generality, we can assume $a_m \neq 0$ (exchanging labels if needed) then set $a_m = 1$. This gives

$$\begin{aligned}
\sum_{j=1}^{m} \eta_j T_j(\tilde{x}) &= \sum_{j=1}^{m-1} \eta_j T_j(\tilde{x}) + \eta_m (a_0 - a_1 T_1 - \ldots - a_{m-1} T_{m-1}) \\
&= \eta_m a_0 + \sum_{j=1}^{m-1} (\eta_j - \eta_m a_j) T_j(\tilde{x}).
\end{aligned}$$

The term $\eta_m a_0$ can be incorporated in $A(\eta)$ if $a_0 \neq 0$, so this gives parametric family

$$p_\eta(\tilde{x}) = \exp\left[\sum_{j=1}^{m-1}(\eta_j - \eta_m a_j)T_j(\tilde{x}) - A'(\eta)\right] h(\tilde{x}). \tag{3.15}$$

At this point, we need to remember the purpose of inference, which is to use data $\tilde{X}$ to make decisions regarding a parameter η. This is based on the assumption that if $\eta \neq \eta'$ we must have $p_\eta \neq p_{\eta'}$, otherwise $\tilde{X}$ has no information with which to distinguish the two parameters. In this case, we say that the parameters are unidentifiable.

This situation occurs in (3.15). Suppose we are given $\eta = (\eta_1, \ldots, \eta_m)$. Then construct $\eta' = \eta + (a_1, \ldots, a_{m-1}, 1)$. We can see that in (3.15) we must have $\eta_j - \eta_m a_j = \eta'_j - \eta'_m a_j$, $j = 1, \ldots, m-1$. Furthermore, since $A'(\eta)$ is constrained by the normalization condition, we would necessarily have $A'(\eta) = A'(\eta')$, so that $p_\eta = p_{\eta'}$, although $\eta \neq \eta'$. That means η is unidentifiable.

To make the idea precise, we introduce the condition.

Definition 3.7 A random vector $\mathbf{T} \in \mathbb{R}^m$ is of full rank if

$$P\left(\sum_{j=1}^{m} a_j T_j = a_0\right) < 1,$$

unless $a_j = 0$ for $j = 0, 1, \ldots, m$. ∎

When Definition 3.7 holds, each component T_j of $\mathbf{T}$ contains information independent of the remaining components.

The main definition of this section follows.

Definition 3.8 (Full Rank Exponential Family) Consider the general form of the exponential family density of Equation (3.10) with parameter space Θ. Suppose the following conditions hold;

(i) $\Xi_\Theta = \{(\eta_1(\theta), \ldots, \eta_m(\theta)) : \theta \in \Theta\} \subset \Xi$, where Ξ is the natural parameter space (Equation (3.14));
(ii) Ξ_Θ contains an open subset of $\mathbb{R}^m$;
(iii) $\mathbf{T} \in \mathbb{R}^m$ is of full rank for all $\theta \in \Theta$ (Definition 3.7).

Then $\{p_\theta : \theta \in \Theta\}$ is a full rank exponential family model. ∎

Condition (iii) of Definition 3.8 is usually assumed, and the model can be modified if it does not hold. For example, in the example considered earlier in this section, Equation (3.15) suggests that a full rank model might be constructed by reducing the dimension by one, redefining the parameter and statistic to be $(\eta_1 - \eta_m a_1, \ldots, \eta_{m-1} - \eta_m a_{m-1})$ and $(T_1, \ldots, T_{m-1})$ respectively. Thus, it is reasonable to assume that this process has already been carried out if needed. Condition (i) is largely a matter of ensuring that the parameter space of interest is well definined. We will see later in this chapter that condition (ii) is the one of most interest, and is sometimes referred to as the open set condition (OSC).

3.6.2 Distributional Properties of Exponential Family Statistics

Exponential family structures allow a unified treatment of the statistic $\mathbf{T}$, summarized in the next theorem.

Theorem 3.13 (Distribution of Natural Statistics) Given full rank exponential family of Equation (3.12), the following identities hold:

$$E_\eta[T_i] = \frac{\partial A(\eta)}{\partial \eta_i}, \qquad \text{cov}_\eta[T_i, T_j] = \frac{\partial^2 A(\eta)}{\partial \eta_i \partial \eta_j}. \tag{3.16}$$

In addition, the moment generating function (MGF) of $\mathbf{T}$ is

$$m_{\mathbf{T}}(u) = \exp\left(A(\eta + u) - A(\eta)\right), \tag{3.17}$$

therefore the cumulant generating function (CGF) is $c_{\mathbf{T}}(u) = A(\eta + u) - A(\eta)$. ■

Proof. The best place to start is with the MGF, which is given directly by

$$\begin{aligned}
m_T(u) &= E_\eta \left[\exp\left(\sum_{i=1}^m u_i T_i(\tilde{X})\right)\right] \\
&= \int_{\mathcal{X}} \exp\left[\sum_{i=1}^m (\eta_i + u_i) T_i(\tilde{x}) - A(\eta)\right] h(\tilde{x}) d\nu_{\mathcal{X}} \\
&= \exp\left(A(\eta + u) - A(\eta)\right) \\
&\quad \times \int_{\mathcal{X}} \exp\left[\sum_{i=1}^m (\eta_i + u_i) T_i(\tilde{x}) - A(\eta + u)\right] h(\tilde{x}) d\nu_{\mathcal{X}} \\
&= \exp\left(A(\eta + u) - A(\eta)\right),
\end{aligned}$$

since the integrand of the last integral is a density function. Then Equation (3.16) follows after applying standard methods of obtaining moments from the MGF. □

The statistic $\mathbf{T}$ also possesses an exponential family density, which follows directly from the original form (3.12). This is stated in the following theorem.

Theorem 3.14 Given full rank exponential family (3.12), the density of $\mathbf{T} = (T_1, \ldots, T_m)$ takes form

$$p_{\mathbf{T},\eta}(t_1, \ldots, t_m) = \exp\left[\sum_{i=1}^m \eta_i t_i - A(\eta)\right] h_{\mathbf{T}}(\mathbf{t}).$$ ■

Proof. The proof follows a change in variable argument. Suppose $\tilde{Y} = (T_1, \ldots, T_m, \tilde{U})$ is a one-to-one transformation of $\tilde{X}$. The density of $\tilde{Y}$ is

$$p_{\tilde{Y},\eta}(t_1, \ldots, t_m, \tilde{u}) = \exp\left[\sum_{i=1}^m \eta_i t_i - A(\eta)\right] h_{\tilde{Y}}(t_1, \ldots, t_m, \tilde{u}),$$

where $h_{\tilde{Y}}$ *dnd* η. Then

$$\begin{aligned}
p_{\mathbf{T},\eta}(t_1, \ldots, t_m) &= \int_{\tilde{u}} p_{\tilde{Y},\eta}(t_1, \ldots, t_m, \tilde{u}) d\nu_{\tilde{U}} \\
&= \exp\left[\sum_{i=1}^m \eta_i t_i - A(\eta)\right] \int_{\tilde{u}} h_{\tilde{Y}}(t_1, \ldots, t_m, \tilde{u}) \\
&= \exp\left[\sum_{i=1}^m \eta_i t_i - A(\eta)\right] h_{\mathbf{T}}(t_1, \ldots, t_m),
\end{aligned}$$

which completes the proof. □

3.6.3 The Exponential Family is a Regular Family

We will use the canonical parametrization to argue that the exponential family is a regular family (Defintion 3.6), therefore, Fisher information can be evaluated according to Theorem 3.12. For convenience, write

$$A_i' = \frac{\partial A(\eta)}{\partial \eta_i}, \quad A'' = \frac{\partial^2 A(\eta)}{\partial \eta_i \partial \eta_j}.$$

Then the second partial derivative of p_η can be written:

$$\begin{aligned}\frac{\partial^2 p_\eta}{\partial \eta_i \partial \eta_j} &= (T_i - A_i')(T_j - A_j')p_\eta - A_{ij}'' p_\eta \\ &= \left(T_i T_j - T_i A_j' - T_j A_i' + A_i' A_j' - A_{ij}''\right) e^{-A(\eta)} h(\tilde{x}) \exp\left[\sum_{i=1}^{m} \eta_i T_i\right].\end{aligned}$$

It is easily verified that all order moments of an exponential family density exist, and that $A(\eta)$ is twice continuously differentiable. This means we may find a single finite constant A^* such that in a neighborhood of η^*, we have

$$\sup_{\eta \in B_\epsilon(\eta^*)} \max\left\{1, |A_i'|, |A_j'|, |A_{ij}''|\right\} e^{-A(\eta)} \leq A^*.$$

Fix coordinate η_j. Then

$$\begin{aligned}\left|\left\{\left.\frac{\partial^2 p_\eta}{\partial \eta_i \partial \eta_j}\right|_{\eta_j}\right\}\right| &\leq \left(|T_i T_j| + |T_i| + |T_j| + 1\right) A^* h(\tilde{x}) \\ &\quad \times \left\{\exp\left[\sum_{i=1}^{m} \eta_i T_i + \epsilon T_j\right] + \exp\left[\sum_{i=1}^{m} \eta_i T_i - \epsilon T_j\right]\right\} \\ &= M(\tilde{x})\end{aligned}$$

when $\eta_i = \eta_i^*$, $i \neq j$ and $\eta_j \in (\eta_j^* - \epsilon, \eta_j^* + \epsilon)$. It can then be verified that $M(\tilde{x})$ is integrable for these parameters, and condition (3.7) holds. Following Definition 3.6 it follows that p_η defines a regular family.

Example 3.16 (Fisher Information for Exponential Family Models) Consider the exponential family in its canonical parametrization (Equation (3.12)). Clearly,

$$\Psi(\tilde{x}, \eta) = \mathbf{T}(\tilde{x}) - E_\eta[\mathbf{T}]$$

so that the Fisher information matrix $\mathbf{I}_\eta(\eta)$ in the canonical parametrization is simply the covariance matrix $\text{var}_\eta[\mathbf{T}(\tilde{X})]$. ■

3.6.4 Common Exponential Family Models

Table 3.1 gives the densities of some commonly used parametric families in their exponential family form using the notation of Equation (3.10) and in the canonical parametrization of Equation (3.12). Examples 3.17 and 3.18 below give the exponential family form for the bivariate normal and multinomial densities.

Example 3.17 (The Bivariate Normal Density as Exponential Family Density) The bivariate normal density can be written

$$f_\theta(x, y) = \frac{1}{2\pi\sigma_X\sigma_Y\sqrt{1-\rho^2}} \exp\{-Q/2\},$$

where the quadratic term is given by

$$Q = \frac{1}{(1-\rho^2)}\left[\frac{(x-\mu_X)^2}{\sigma_X^2} + \frac{(y-\mu_Y)^2}{\sigma_Y^2} - \frac{2\rho(x-\mu_X)(y-\mu_Y)}{\sigma_X\sigma_Y}\right],$$

and the full parameter is $\theta = (\mu_X, \mu_Y, \sigma_X, \sigma_Y, \rho)$. Then, for *iid* observations (X_i, Y_i) from f_θ, the density is given by

$$p_\theta(\tilde{x}, \tilde{y}) = \left[\frac{1}{2\pi\sigma_X\sigma_Y\sqrt{1-\rho^2}}\right]^n \exp\{-\sum_i Q_i/2\},$$

where Q_i is the quadratic term for pair (x_i, y_i). We then have $-\sum_{i=1}^n Q_i/2 = \sum_{j=1}^5 \eta_j(\theta)T_j(\tilde{x}, \tilde{y}) + B_1(\theta)$, where $(T_1, T_2, T_3, T_4, T_5) = (\sum_i x_i, \sum_i y_i, \sum_i x_i^2, \sum_i y_i^2, \sum_i x_i y_i)$, and

$$\eta_1(\theta) = -[2(1-\rho^2)]^{-1}\left[\frac{-2\mu_X}{\sigma_X^2} + \frac{2\rho\mu_Y}{\sigma_X\sigma_Y}\right],$$

$$\eta_2(\theta) = -[2(1-\rho^2)]^{-1}\left[\frac{-2\mu_Y}{\sigma_Y^2} + \frac{2\rho\mu_X}{\sigma_X\sigma_Y}\right],$$

$$\eta_3(\theta) = -[2(1-\rho^2)]^{-1}\left[\frac{1}{\sigma_X^2}\right],$$

$$\eta_4(\theta) = -[2(1-\rho^2)]^{-1}\left[\frac{1}{\sigma_Y^2}\right],$$

$$\eta_5(\theta) = -[2(1-\rho^2)]^{-1}\left[\frac{-2\rho}{\sigma_X\sigma_Y}\right],$$

$$B_1(\theta) = -n[2(1-\rho^2)]^{-1}\left[\frac{\mu_X^2}{\sigma_X^2} + \frac{\mu_Y^2}{\sigma_Y^2} - \left[\frac{2\rho\mu_X\mu_Y}{\sigma_X\sigma_Y}\right]\right].$$

Then $p_\theta(\tilde{x}, \tilde{y}) = \exp\left\{\sum_{j=1}^5 \eta_j(\theta)T_j(\tilde{x}, \tilde{y}) - B(\theta)\right\} h(\tilde{x}, \tilde{y})$ where $h(\tilde{x}, \tilde{y}) = (2\pi)^{-n}$ and $B(\theta) = n\log\left(\sigma_X\sigma_Y\sqrt{1-\rho^2}\right) - B_1(\theta)$. ■

Example 3.18 (The Multinomial Density as Exponential Family Density) The multinomial distribution for an integer vector $\mathbf{N} = (N_1, \ldots, N_m)$ is given by

$$p_\theta(n_1, \ldots, n_m) = P_\theta\left((N_1, \ldots, N_m) = (n_1, \ldots, n_m)\right) = \frac{n!}{\prod_{i=j}^m n_j!}\prod_{j=1}^m \theta_j^{n_j}$$

provided $n_i \geq 0$, $\sum_i n_i = n$. Then set

$$\theta_m = 1 - \theta_1 - \ldots - \theta_{m-1}, \quad n_m = n - n_1 - \ldots - n_{m-1}.$$

Then exponentiating gives

$$p_\theta(n_1, \ldots, n_m) = \exp\left\{\sum_{j=1}^{m-1} T_j\eta_j(\theta) - B(\theta)\right\} h(n_1, \ldots, n_m),$$

where $T_j = n_j$; $\eta_j = \log(\theta_j/\theta_m)$; $B(\theta) = -n\log(\theta_m)$; $h(n_1, \ldots, n_m) = n!/\prod_{i=j}^m n_j!$. We note that $(\eta_1(\theta), \ldots, \eta_{m-1}(\theta))$ is a 1-1 transformation of the $m-1$ dimensional set

$$\Theta^* = \left\{(\theta_1, \ldots, \theta_{m-1}) : \theta_j \geq 0, \ \sum_{j=1}^{m-1} \theta_j \leq 1\right\}.$$

TABLE 3.1
Representations of common exponential family densities using the notation $p_\theta(\tilde{x}) = \exp\{\sum_{j=1}^m \eta_j(\theta)T_j(\tilde{x}) - B(\theta)\}h(\tilde{x})$, as well as the canonical parametrizaton $p_\eta(\tilde{x}) = \exp\{\sum_{j=1}^m \eta_j T_j(\tilde{x}) - A(\eta)\}h(\tilde{x})$.

Binomial distribution, $X \in binom(n, \theta)$, $m = 1$
$\eta_1 = \log(\theta/(1-\theta))$, $T_1 = x$, $A(\eta) = n\log(1+e^\eta)$, $B(\theta) = -n\log(1-\theta)$
Negative binomial distribution, $X \in nb(r, \theta)$, $m = 1$
$\eta_1 = \log(1-\theta)$, $T_1 = x$, $A(\eta) = -r\log(e^{-\eta} - 1)$, $B(\theta) = -r\log(\theta/(1-\theta))$
Poisson distribution, $X \in pois(\theta)$, $m = 1$
$\eta_1 = \log(\theta)$, $T_1 = x$, $A(\eta) = e^\eta$, $B(\theta) = \theta$
Normal distribution, $X \in N(\mu, \sigma^2)$, $\theta = (\mu, \sigma^2)$, $m = 2$
$(\eta_1, \eta_2) = (\mu/\sigma^2, -1/(2\sigma^2))$, $(T_1, T_2) = (x, x^2)$, $A(\eta) = -\frac{\eta_1^2}{4\eta_2} - \frac{1}{2}\log(-2\eta_2)$, $B(\theta) = \frac{\mu^2}{2\sigma^2} + \log\sigma$
Gamma distribution, $X \in gamma(\tau, \alpha)$, $\theta = (\tau, \alpha)$, $m = 2$
$(\eta_1, \eta_2) = (-1/\tau, \alpha)$, $(T_1, T_2) = (x, \log(x))$, $A(\eta) = -\eta_2\log(-\eta_1) + \log\Gamma(\eta_2)$, $B(\theta) = \alpha\log(\tau) + \log\Gamma(\alpha)$
Beta distribution, $X \in beta(\alpha, \beta)$, $\theta = (\alpha, \beta)$, $m = 2$
$(\eta_1, \eta_2) = (\alpha, \beta)$, $(T_1, T_2) = (\log(x), \log(1-x))$, $A(\eta) = \log\Gamma(\eta_1) + \log\Gamma(\eta_2) - \log\Gamma(\eta_1 + \eta_2)$, $B(\theta) = \log\Gamma(\alpha) + \log\Gamma(\beta) - \log\Gamma(\alpha + \beta)$

In addition, it can be verified that $P_\theta\{\sum_{j=1}^{m-1} a_j n_j = a_0\} < 1$ for all θ, unless $a_j = 0$, $j = 0, 1, \ldots, m-1$. The exponential family is therefore of rank $m - 1$. ■

3.7 Sufficiency

The technical objective of inference is to discern the relationship between the parameter $\theta \in \Theta$ and the properties of the data $\tilde{X}$. Almost always, this is done through some reduction of $\tilde{X}$ to a statistic $S(\tilde{X})$ (a sample mean, for example). This process involves a type of balance. We don't want $S(\tilde{X})$ to depend on "noise", or variation unrelated to θ. Neither do we wish to throw away any part of the data that does contain information about θ. The concept of sufficiency allows us to deduce the optimal reduction of data from this point of view. In the following definition we may rely on the construction of conditional densities given in Section 1.2.4.

Definition 3.9 (Sufficient Statistic) A statistic $S \in \mathcal{T}_{stat}$ is sufficient if $P_\theta(\tilde{X} \mid S(\tilde{X}) = s)$ *dnd* on θ. The sufficiency principle holds that any inference should depend on the data only through $S(\tilde{X})$. The set of all sufficient statistics is denoted as $\mathcal{T}_{suff} \subset \mathcal{T}_{stat}$. A sufficient statistic always exists, since $\tilde{X}$ itself is sufficient. ■

This definition must be interpreted with some caution. First, consider the following example.

Example 3.19 (Sufficiency and the Uniform Scale Parameter) Let $\mathbf{X} = (X_1, \ldots, X_n)$ be *iid*, with X_1 uniformly distributed on $(0, \theta)$, $\theta \in \Theta = (0, \infty)$. Let $S(\mathbf{X}) = X_{(n)}$, the maximum of $\mathbf{X}$. It may be shown that the distribution of the elements of $\mathbf{X}$ other than $X_{(n)}$, conditional on $X_{(n)} = t$ is equivalent to an *iid* sample of $\mathbf{X}' = (X_1, \ldots, X_{n-1})$, where X_1 is uniformly distributed on $(0, t)$ (Problem 3.6). This means $X_{(n)}$ is sufficient for θ. ■

Here, the discussion in Section 3.2 regarding parametrized families of functions should be reviewed, since we are saying in the previous example that $P_\theta(\mathbf{X} \in E \mid X_{(n)} = t)$ *dnd* θ. However, what if $t > \theta$? Then $P_\theta(\mathbf{X} \in E \mid X_{(n)} = t)$ is not defined. Put another way, if for fixed t, $P_\theta(\mathbf{X} \in E \mid X_{(n)} = t)$ is defined for $\theta = \theta_1$ but not $\theta = \theta_2$, in what sense does the condition for sufficiency hold? (That the maximum in Example 3.19 is sufficient is reported in all textbooks that treat the subject.)

There are a number of ways to reconcile this apparent paradox. We can represent sufficiency using expectations. Suppose $g : \mathcal{X} \rightarrow \mathbb{R}$. If $S \in \mathcal{T}_{suff}$, then we expect the function $\eta(\theta, s) = E_\theta[g(\tilde{X}) \mid S(\tilde{X}) = s]$ to be constant for fixed s as θ is allowed to vary (a rich enough selection of functions g will completely define a distribution p_θ). If η is defined on a product space $\Theta \times I_S$, where I_S is the support of $S(\tilde{X})$ for all Θ, then the definition of sufficiency has no ambiguity, and we may write $\eta(\theta, s) = h(s)$ for some function h. However, in Example 3.19 the support of $S(\tilde{X})$, say $I_{S,\theta}$, depends on θ so we may only define $\eta(\theta, s)$ on the subset of form $\{(\theta, s) : s \in I_{S,\theta}\}$, which will not be a product space. However, where $\eta(\theta, s)$ is defined, it will still be possible to write $\eta(\theta, s) = h(s)$ for some function h.

Essentially what this says is that if, in Example 3.19, we observed $X_{(n)} = t$, we would know that $\theta > t$, and confine attention to a subset of the original parameter space, retaining the original purpose of sufficiency.

If we accept that the conditions for sufficiency hold only where $P_\theta(\tilde{X} \mid S(\tilde{X}) = s)$ is defined, even though this may depend on θ, we can regain the most important property of sufficiency, namely, that for $S \in \mathcal{T}_{suff}$, the mapping $G(\tilde{X}) = \eta(\theta, S(\tilde{X}))$ is always well defined, and does not depend on θ. In other words, $G(\tilde{X})$ is a statistic.

We summarize the preceding discussion in the following theorem.

Theorem 3.15 (Sufficiency for Nonconstant Support Models) Suppose $S \in \mathcal{T}_{suff}$. Define the set

$$I_{S,\theta} = \{s : p_\theta(\tilde{x} \mid S(\tilde{X}) = s) \text{ is well defined}\}.$$

Define range $R_S = \cup_\theta I_{S,\theta}$. Suppose we have mapping $g : \mathcal{X} \rightarrow \mathbb{R}$. Then there exists function $\bar{g} : R_S \rightarrow \mathbb{R}$ such that $\bar{g}(s) = E_\theta[g(\tilde{X}) \mid S(\tilde{X}) = s]$ whenever $s \in I_{S,\theta}$. Furthermore, $E_\theta[g(\tilde{X}) \mid S(\tilde{X})] = \bar{g}(S(\tilde{X}))$ is a well defined statistic *wp*1. ■

Proof. That $p_\theta(\tilde{x} \mid S(\tilde{X}) = s(\tilde{x}))$ is well defined for $\tilde{x} \in \mathcal{X}$ follows from the argument leading to Equation (1.12). Then note that the sufficiency condition need only hold where $p_\theta(\tilde{x} \mid S(\tilde{X}) = s)$ is well defined. Finally, under distribution p_θ, $\{\tilde{x} : S(\tilde{x}) = S(\tilde{X})\} \cap \mathcal{X}_\theta \neq \emptyset$ *wp*1. □

The order statistics are an important example of sufficiency. This fact will play an important role in nonparametric methods, and will in fact allow us to minimize the distinction between parametric and nonparametric estimators (Section 8.5.1).

Example 3.20 (Sufficiency of the Order Statistics for an *iid* Sample) Suppose $\mathbf{X} = (X_1, \ldots, X_n)$ is an *iid* sample from a distribution $p \in \mathcal{P}$ on $\mathbb{R}^n$. Then the order statistics $\mathbf{X}_{ord} = (X_{(1)}, \ldots, X_{(n)})$ are sufficient for $\mathcal{P}$. This will be proven in Theorem 8.8. ■

3.7.1 The Factorization Theorem

The theory of sufficiency is simplified analytically under the constant support assumption, but is not really strengthened. For this reason our own development will not require that assumption, although it will be singled out as an important special case. Theorem 3.15 supports this approach.

The main theorem characterizing sufficiency is the factorization theorem.

Theorem 3.16 (The Factorization Theorem) A statistic S is sufficient *iff* the densities $p_\theta \in \mathcal{P}$ may be written

$$p_\theta(\tilde{x}) = g(S(\tilde{x}), \theta)h(\tilde{x}). \tag{3.18}$$

where h *dnd* θ. ■

Proof. [(3.18) $\implies S \in \mathcal{T}_{suff}$] Set $I_h = \{\tilde{x} : h(\tilde{x}) > 0\}$, $I_{g,\theta} = \{s : g(s,\theta) > 0\}$ and $A_s = \{\tilde{x} \in \mathcal{X}_\theta : S(\tilde{x}) = s\}$. Then, from (1.11),

$$p_\theta(\tilde{x} \mid S(\tilde{x}) = s) = \frac{g(s,\theta)h(\tilde{x})I_{A_s}}{g(s,\theta)\int_{\tilde{x}\in A_s} h(\tilde{x})d\nu_{A_s}} = \frac{h(\tilde{x})I_{A_s}}{\int_{\tilde{x}\in A_s} h(\tilde{x})d\nu_{A_s}}$$

for $\tilde{x} \in I_h, s \in I_{g,\theta}$, which for fixed $s \in I_{g,\theta}$, *dnd* θ, therefore $S \in \mathcal{T}_{suff}$.

[(3.18) $\impliedby S \in \mathcal{T}_{suff}$] By hypothesis, the required factorization is given by Equation (1.12). □

Remark 3.1 As can be seen in the proof of Theorem 3.16, the support of p_θ as defined in (3.18) is equal to $\{\tilde{x} : S(\tilde{x}) \in I_{g,\theta}\} \cap I_h$. Thus, the while the support may depend on θ, it does so only through $S(\tilde{x})$. This is an essential part of the condition for sufficiency. ■

The sufficiency property is defined relative to the parameter space, which is itself a matter of definition. This is made clear in the following example.

Example 3.21 (The Shifted Location-Scale Invariant Model) Let $\mathbf{X} = (X_1, \ldots, X_n)$ be an *iid* sample with $X_1 \sim exp(\mu, \tau)$. The density of the sample is

$$\begin{aligned} p_\theta(\mathbf{x}) &= \prod_{i=1}^{n} \tau^{-1} \exp\{-(x-\mu)/\tau\} I\{x_i \geq \mu\} \\ &= \tau^{-n} \exp\left\{-\sum_{i=1}^{n}(x_i - \mu)/\tau\right\} I\{x_{(1)} \geq \mu\}, \end{aligned}$$

where $x_{(1)}, \ldots, x_{(n)}$ is the order transformation of $x_1, \ldots, x_n$. Consider the following three cases.

(a) Suppose the parameter is $\theta = \mu$, and τ is fixed. In this case $g(x_{(1)}, \mu) = \exp(n\mu) I\{x_{(1)} \geq \mu\}$. Then $p_\theta(\mathbf{x}) = g(x_{(1)}, \mu)h(\mathbf{x})$, where $h(\mathbf{x})$ does not depend on μ (but may depend on τ, which is fixed). Therefore, by the factorization theorem, $X_{(1)}$ is sufficient for μ.

(b) Suppose the parameter is $\theta = \tau$, and μ is fixed. Then set $s = \sum_i x_i$, $S = \sum_i X_i$. Let $g(s,\tau) = \tau^{-n}\exp\{-(s - n\mu)/\tau\}$. Then $p_\theta(\mathbf{x}) = g(s,\tau)h(\mathbf{x})$, where $h(\mathbf{x})$ does not depend on τ (but may depend on μ, which is fixed). Therefore, by the factorization theorem, S is sufficient for τ.

(c) Suppose $\theta = (\mu, \tau)$. We only need note that $p_\theta(\mathbf{x}) = g(s, x_{(1)}, \theta) = \tau^{-n}\exp\{-(s - n\mu)/\tau\} I\{x_{(1)} \geq \mu\}$. By the factorization theorem, $(S, X_{(1)})$ is sufficient for θ.

■

3.7.2 Minimal Sufficiency

Sufficient statistics are clearly not unique. If $S \in \mathcal{T}_{suff}$ and $S' \mapsto S$ then $S' \in \mathcal{T}_{suff}$. This follows from the factorizaton theorem (Theorem 3.16). In particular, a complete set of observations, say $\mathbf{X} = (X_1, X_2, X_3, X_4)$, is itself a statistic, and is clearly sufficient, so that a sufficient statistic always exists. Furthermore, if, for example, $S(\mathbf{X}) = X_1 + X_2 + X_3 + X_4$ is sufficient, then so is $S'(\mathbf{X}) = (X_1 + X_2, X_3 + X_4)$. It would therefore be useful to identify, if possible, a statistic that represents the greatest possible sufficient reduction. This statistic is referred to as a minimal sufficient statistic, which we define next.

Definition 3.10 (Minimal Sufficient Statistic) A sufficient statistic T is minimal sufficient if $S \mapsto T$ for every other $S \in \mathcal{T}_{suff}$. We denote the class of all minimal sufficient statistics $\mathcal{T}_{min}$. ■

Note that the definition requires that $T \in \mathcal{T}_{min}$ be sufficient. For example, the constant function is a function of all sufficient statistics but is itself not sufficient (unless $\mathcal{P}$ contains only one distribution). The following theorem gives a method of verifying that $S \in \mathcal{T}_{suff}$ is minimal. There are two technical details to keep in mind. We say that ratios $1/1$, $0/1$ and $1/0$ are well defined, but the ratio $0/0$ is not. Recall also that given two statistics $T(\tilde{x}), S(\tilde{x})$, we can say $S \mapsto T$ if $S(\tilde{x}) = S(\tilde{y}) \implies T(\tilde{x}) = T(\tilde{y})$.

Theorem 3.17 (Sufficient Conditions for Minimal Sufficiency) We are given parametric family $\mathcal{P} = \{p_\theta : \theta \in \Theta\}$. Suppose $T \in \mathcal{T}_{suff}$. Furthermore, suppose that for any pair $\tilde{x}, \tilde{y} \in \bar{\mathcal{X}}$ for which $p_\theta(\tilde{x})/p_\theta(\tilde{y})$, where defined, *dnd* θ we have $T(\tilde{x}) = T(\tilde{y})$. Then T is minimal sufficient for θ. ■

Proof. Suppose $S(\tilde{x})$ is a sufficient statistic. Suppose $p_\theta(\tilde{x}) + p_\theta(\tilde{y}) > 0$, so that the ratio $p_\theta(\tilde{x})/p_\theta(\tilde{y})$ is well defined. Then by the factorization theorem,

$$\frac{p_\theta(\tilde{x})}{p_\theta(\tilde{y})} = \frac{g(S(\tilde{x}), \theta)h(\tilde{x})}{g(S(\tilde{y}), \theta)h(\tilde{y})}.$$

Since $\tilde{x}, \tilde{y} \in \bar{\mathcal{X}}$ we must have $h(\tilde{x}) > 0$ and $h(\tilde{y}) > 0$. If $S(\tilde{x}) = S(\tilde{y}) = s$, then

$$\frac{p_\theta(\tilde{x})}{p_\theta(\tilde{y})} = \frac{g(s, \theta)h(\tilde{x})}{g(s, \theta)h(\tilde{y})} = \frac{h(\tilde{x})}{h(\tilde{y})},$$

so that $p_\theta(\tilde{x})/p_\theta(\tilde{y})$ *dnd*θ. But, by hypothesis, $T(\tilde{x}) = T(\tilde{y})$. This implies $S \mapsto T$. Since this is true for any sufficient statistic S, T must be minimal sufficient. □

The ratio $p_\theta(\tilde{x})/p_\theta(\tilde{y})$ is always well defined for families of constant support, and for this reason in many references, results such as Theorem 3.17 are given separately for that case. This avoids technical details which obscure the main point. However, we find that simply identifying when the ratio is well defined allows an argument of comparable simplicity. In fact, this is equivalent to the approach taken in Lehmann and Scheffé (1950), which offers a constructive characterization of the minimal sufficient statistic. Take any $\tilde{x}_0 \in \bar{\mathcal{X}}$. Let $D(\tilde{x}_0) \subset \bar{\mathcal{X}}$ be the set of $\tilde{x}$ for which there exists a function $k(\tilde{x}, \tilde{x}_0) > 0$, which *dnd* on θ, for which $p_\theta(\tilde{x}) = k(\tilde{x}, \tilde{x}_0)p_\theta(\tilde{x}_0)$ for all $\theta \in \Theta$. Then define the binary relation $\tilde{x} \sim \tilde{y}$ by the condition $\tilde{x} \in D(\tilde{y})$ (Section B.2). It is not hard to verify that $\tilde{x} \sim \tilde{y}$ is an equivalence relation (Definition B.1). It is then shown in Lehmann and Scheffé (1950) that the equivalence classes of $\tilde{x} \sim \tilde{y}$ define a partition of $\mathcal{X}$ that is equivalent to that induced by any minimal sufficient statistic. This is simply another way of expressing the argument made in Theorem 3.17. An alternate method of constructing minimal sufficient statistics, described in Lehmann and Casella (1998), is here given as Problem 3.15.

We next offer some applications of Theorem 3.17. The first demonstrates the approach for families of varying support.

Example 3.22 (Minimal Sufficient Statistic for Exponential Sample with Location Shift Parameter) This example follows Example 3.21. Let $\mathbf{X} = (X_1, \ldots, X_n)$ be an *iid* sample from density $f_\theta(x)$, $\theta \in \Theta = \mathbb{R}$, where $f_\theta(x) = \exp\{-(x-\theta)\} I\{x \geq \theta\}$. The density of the sample is

$$p_\theta(\mathbf{x}) = \exp\left\{-\sum_{i=1}^{n}(x_i - \theta)\right\} I\{x_{(1)} \geq \theta\},$$

Noting that $p_\theta(\mathbf{x}) > 0$ for $\theta < x_{(1)}$, the ratio

$$\frac{p_\theta(x)}{p_\theta(y)} = \exp\left[\sum_i y_i - \sum_i x_i\right] \frac{I\{x_{(1)} > \theta\}}{I\{y_{(1)} > \theta\}},$$

is well defined for $\theta < \max\{x_{(1)}, y_{(1)}\}$. It also depends on θ unless $x_{(1)} = y_{(1)}$, so that $T(X) = X_{(1)}$ is minimal sufficient for θ. ■

We will see below that sufficiency is an especially important idea for exponential family models, since the minimal sufficient statistic generally represents a considerable reduction in data (Section 3.8.2). For models of *iid* samples, the opposite extreme occurs when the order statistics are miminal sufficient. We will see that the optimal statistical methods for these two cases tend to be based on different principles. This can be seen especially by a comparison of Chapter 8 (UMVU estimation) and Chapter 9 (MRE estimation). Such an example is given next.

Example 3.23 (Cauchy Location Invariant Model) Let $\mathbf{X} = (X_1, \ldots, X_n)$ be an *iid* sample from density $f_\theta \sim cauchy(\theta, 1)$. Then $\mathbf{X}$ has density $p_\theta(\mathbf{x}) = \prod_i f_\theta(x_i)$. Clearly, the order statistic $\mathbf{X}_{ord} = (X_{(1)}, \ldots, X_{(n)})$ is sufficient for θ. We will show it is minimal sufficient using Theorem 3.17. Suppose for $\mathbf{x}, \mathbf{y} \in \mathcal{X}$ the ratio

$$\frac{p_\theta(\mathbf{x})}{p_\theta(\mathbf{y})} = \frac{\prod_{i=1}^{n}(1 + (y_i - \theta)^2)}{\prod_{i=1}^{n}(1 + (x_i - \theta)^2)}$$

dnd on θ. Note that this is a ratio of two polynomials in θ and the highest order term of each is θ^{2n}. The polynomials must therefore have equal coefficients, and therefore identical roots. However, it is easily verified that the $2n$ complex roots of the numerator are $y_j \pm i$, $j = 1, \ldots, n$, and those of the denominator are $x_j \pm i$, $j = 1, \ldots, n$ (note the convention $i^2 = -1$). Therefore, the order transformations of x and y must be equal. By Theorem 3.17, this implies that $\mathbf{X}_{ord}$ is minimal sufficient. See Problem 3.10 or 3.12 for a similar example. ■

3.8 Complete and Ancillary Statistics

There are two more classes of statistics which complement the theory of sufficiency. We have already defined ancillarity (Definition 3.2). Our goal later in this section will be to describe its relationship to sufficiency.

We first define the completeness property, which is of great importance to various branches of the theory of inference. When there exists a statistic which is a complete statistic as well as sufficient, the problem of determining optimal statistical procedures usually has a more definitive resolution, and the analysis is often simpler. This will be made clear in Chapters 8 and 9. While the completeness property is especially relevant to exponential family models, its domain of application extends to other models, notably, to nonparametric estimation (Section 8.5). We begin with the main definition.

Definition 3.11 (Complete Statistic) $T \in \mathcal{T}_{stat}$ is a complete statistic for $\theta \in \Theta$ if $E_\theta[g(T)] \equiv 0$ implies $g \equiv 0$ for any measurable function g, assuming g *dnd* θ. We denote the class of all complete statistics $\mathcal{T}_{comp}$.

If the completeness property is only required to hold for a class of functions $\mathcal{F}$, then T is $\mathcal{F}$ complete. In particular, if completeness holds on the space of bounded functions, then T is boundedly complete. The class of such statistics is denoted as $\mathcal{T}^b_{comp}$. Clearly, $\mathcal{T}_{comp} \subset \mathcal{T}^b_{comp}$. A statistic is L^2-complete if functions g are confined to those for which $g(T)$ possesses a finite second moment for all θ. ∎

Note that completeness equivalently requires that if $E_\theta[g(T)] \equiv c$ for some constant c, then $g \equiv c$. In this case c does not depend on θ, and so must be considered known. Then we may simply replace $g(T)$ with $g(T) - c$ to recover Definition 3.11.

Verifying the completeness property can often be challenging mathematically, so we will review a number of methods in this section. We begin with two (related) examples.

Example 3.24 (Completeness for the Bernoulli Distribution) Suppose $X \sim bern(p)$. For any function $g(X)$ can assume only two values. Suppose

$$E[g(X)] = g(0)(1-p) + g(1)p = g(0) + [g(1) - g(0)]p = 0$$

for all $p \in [0, 1]$. Clearly, we must have $g(1) - g(0) = 0$. But then we must also have $g(0) = 0$, which implies $g(1) = 0$. Thus, X is complete for p. ∎

A similar approach can be used for the binomial distribution.

Example 3.25 (Completeness for the Binomial Distribution) Suppose $X \in bin(n, \theta)$. Similar to Example 3.24, any function g of X only needs to be defined for $x = 0, 1, \ldots, n$. Accordingly, set $g(i) = g_i$ for $n + 1$ suitable constants. Then

$$E_\theta[g(X)] = \sum_{i=0}^{n} g_i \binom{n}{i} \theta^i (1-\theta)^{n-i} = \sum_{i=0}^{n} a_i \theta^i,$$

is a polynomial of (at most) order n. If $E_\theta[g(X)]$ for all $\theta \in (0, 1)$, then the coefficients a_i must each be zero. However, the only contribution to the 0 order term comes from the first term of the summation, that is, $a_0 = g_0$ (all remaining terms are at least order 1). Therefore, $g_0 = 0$. Applying a similar logic sequentially, we must have $g_i = 0$ for each i. Therefore, X is complete. ∎

We will see in Section 3.8.2 that completeness for the binomial distribution follows from the fact that it is an exponential family distribution.

One remarkable property of a complete sufficient statistic $T(\tilde{X})$ is that there can be only one unbiased estimator of any estimand of the form $\delta(T(\tilde{X}))$. This follows immediately from Definition 3.11. If two nonequivalent unbiased estimators δ_1, δ_2 of any estimand $\eta(\theta)$ exist which are both functions of a statistic $S(\tilde{X})$, then that statistic cannot be complete, since $E_\theta[\delta_1 - \delta_2] = \eta(\theta) - \eta(\theta) = 0$ for all θ, but $\delta_1 - \delta_2$ is not equivalent to zero. Consider the following example.

Example 3.26 (Noncompleteness of Order Statistics for Symmetric Location Invariant Models) Suppose $X_1, \ldots, X_n$ is an *iid* sample from any density which is symmetric about μ. Suppose $E_\mu[|X_1|] < \infty$. The sample mean $\bar{X}$ and sample median $\hat{m}$ are both unbiased estimators of μ. Both are functions of the order statistics, which cannot therefore be complete, since $E_\mu[\bar{X} - \hat{m}] = 0$, but $\bar{X} - \hat{m}$ is not equivalent to zero. ∎

It is important to note that completeness is a property not only of the density, but of the parameter space. Consider the following example.

Example 3.27 (Noncompleteness for the Binomial Parameter from a Reduced Parameter Space) Suppose $X \sim bin(n,p)$. We take the parameter to be $\theta = p$, but for this particular application the parameter space is limited to two densities, say, $p \in \Theta_0 = \{p_0, 1-p_0\}$ for some fixed $p_0 \in (0, 1/2)$. We then have $E_\theta[(X - n\theta)^2] = n\theta(1-\theta)$. But $E_{p_0}[(X-n\theta)^2] = E_{1-p_0}[(X-n\theta)^2]$, that is, $E_p[(X-n\theta)^2]$ is constant for all $p \in \Theta_0$, so X cannot be complete *wrt* Θ_0, even though is is complete *wrt* $\Theta = [0, 1]$. We may also reach this conclusion using the more general argument of Problem 3.24. ■

Example 3.27 demonstrates the dependence of the completeness property on the definition of the parameter space. We can say more, however. The completeness property is nested, in the sense that if it holds for a set of parameters Θ_0 it also holds for a superset $\Theta_1 \supset \Theta_0$. This is proven in the next theorem. We will make use of this fact in Section 8.5 (see also Problems 3.15 and 3.24).

Theorem 3.18 (The Completeness Property is Nested) We are given parametric family Θ, and nested subsets $\Theta_0 \subset \Theta_1 \subset \Theta$. If a statistic T is complete on Θ_0, it is also complete on Θ_1. ■

Proof. Suppose for function g, $E_\theta[g(T)] = 0$ for all $\theta \in \Theta_1$. This clearly holds on Θ_0, by hypothesis implying $g \equiv 0$, which completes the proof. □

A complete statistic need not be sufficient. Trivially, $T(\tilde{X}) \equiv c$, where c is a constant, is complete, since any function of $T(\tilde{X})$ is constant. Somewhat less trivially, if $T(\tilde{X}) = (T_1(\tilde{X}), T_2(\tilde{X}))$ is a complete minimal sufficient statistic, and $T_1(\tilde{X}), T_2(\tilde{X})$ are not equivalent, then $T_1(\tilde{X})$ will be complete but not sufficient (since any function of $T_1(\tilde{X})$ is a function of $T(\tilde{X})$). In practice, we are interested in complete statistics that are sufficient, but this must be verified. However, among sufficient statistics completeness and minimal sufficiency can be considered equivalent.

Theorem 3.19 (Completeness and Minimal Sufficiency) If a minimal sufficient statistic for Θ exists, then any complete sufficient statistic is minimal sufficient. ■

Proof. Given $S, T \in \mathcal{T}_{stat}$, suppose T is minimal sufficient and S is complete sufficient. Then we have $T = h(S)$ for some function h. Then set $\zeta(t) = E_\theta[g(S(\tilde{X})) \mid T(\tilde{X}) = t]$, which by sufficiency *dnd* θ. Suppose $\neg T \mapsto S$. Then we can choose g so that $g(S(\tilde{X})) \neq \zeta(h(S(\tilde{X}))$ with positive probability. However, we have $E_\theta\left[g(S(\tilde{X})) - \zeta(h(S(\tilde{X}))\right] = 0$ so that if S is complete we must have $g(S(\tilde{X})) \equiv \zeta(h(S(\tilde{X}))$. By contradiction we conclude $T \mapsto S$, from which is follows that S is minimal sufficient. □

3.8.1 Completeness and Functional Transformations

Suppose a statistic T has support $\mathcal{T} \subset \mathbb{R}^d$. Suppose we are given $g : \mathcal{T} \to \mathbb{R}$. The test for the completeness of g involves the evaluation of an expectation of g at all $\theta \in \Theta$, in particular

$$G(\theta) = \int_{t \in \mathcal{T}} g(t) p_{T,\theta}(t) d\nu_T, \tag{3.19}$$

where $p_{T,\theta}$ is the density of T with respect to ν_T under model p_θ. We can think of Equation (3.19) as a mapping $\mathcal{Q}$ between function spaces $\mathcal{G}$, $\mathcal{G}^*$, consisting of mappings $g : \mathcal{T} \to \mathbb{R}$ and $g^* : \Theta \to \mathbb{R}$, respectively. We then write $\mathcal{Q}g = G(\theta)$. Then $\mathcal{Q}$ is a linear operator, and we have an alternative characterization of completeness.

Theorem 3.20 (Completeness as an Injective Functional Transformation) Consider the mapping $\mathcal{Q}g = G(\theta)$ defined by (3.19). Then T is complete *iff* $\mathcal{Q}$ is an injective mapping. ∎

Proof. To prove sufficiency, suppose $\mathcal{Q}g = c$, which *dnd* θ. This equation is solved by $g(t) \equiv c$. Since $\mathcal{Q}$ is injective, $g \equiv c$ is the only solution. To prove necessity, suppose $\mathcal{Q}$ is not injective. Then there exist $g \neq g'$ such that $\mathcal{Q}g = \mathcal{Q}g'$. By the linearity of $\mathcal{Q}$ we have $\mathcal{Q}(g - g') = 0$. But by hypothesis $g - g'$ is not constant, therefore T is not complete. □

An $n \times n$ matrix is an injective operator on $\mathbb{R}^n$ if and only if it is invertible. By Theorem 3.20 this gives a method for testing the completeness property for distributions with finite support. Consider the following example.

Example 3.28 (Completeness and Matrix Transformations) Suppose X is a random variable with support $\mathcal{X} = \{1, \ldots, m\}$. Consider parameter $\theta \in \Theta = \{1, \ldots, m\}$. For any $\rho \neq 1/m$, define p_θ with PMF

$$p_\theta(x) = \begin{cases} \rho & ; \quad x = \theta \\ (1-\rho)/(m-1) & ; \quad x \neq \theta \end{cases}.$$

Then, any function g on $\mathcal{X}$ is equivalent to an m-dimensional vector $\mathbf{g} = (g_1, \ldots, g_m)$, essentially, $g(j) = g_j$. Then

$$G(\theta) = \int_{x \in \mathcal{T}} g(x) p_\theta(x) d\nu_{\mathcal{X}} = \sum_{j=1}^{m} g_i p_\theta(i).$$

In the same way, p_θ is also an m-dimensional vector $(p_\theta(1), \ldots, p_\theta(m)) = (p_{\theta,1}, \ldots, p_{\theta,m})$, as is $G(\theta) = \mathbf{G}_\theta = (G_1, \ldots, G_m)$. In effect, $\mathcal{Q}$ defines the linear transformation $\mathbf{G}_\theta = \mathbf{A}\mathbf{g}$, in particular,

$$\begin{bmatrix} G_1 \\ \vdots \\ G_m \end{bmatrix} = \begin{bmatrix} p_{1,1} & \cdots & p_{1,m} \\ \vdots & \ddots & \vdots \\ p_{m,1} & \cdots & p_{m,m} \end{bmatrix} \begin{bmatrix} g_1 \\ \vdots \\ g_m \end{bmatrix} = \begin{bmatrix} \rho & \cdots & \frac{1-\rho}{m-1} \\ \vdots & \ddots & \vdots \\ \frac{1-\rho}{m-1} & \cdots & \rho \end{bmatrix} \begin{bmatrix} g_1 \\ \vdots \\ g_m \end{bmatrix}.$$

The test for completeness requires that $\mathbf{G}_\theta = 0$ *iff* $\mathbf{g} = 0$. However, this condition is identical to the invertibility of the matrix $\mathbf{A}$ defining the transformation, which holds, provided $\rho \neq 1/m$. This argument is generalized in Problem 3.24. ∎

It suffices that $\mathcal{Q}$ be an injective mapping of positive functions.

Theorem 3.21 (Functional Transformations of Positive Functions) Suppose the mapping $\mathcal{Q}$ is defined by (3.19). Then T is complete if $\mathcal{Q}$ is an injective mapping on the space of positive functions. ∎

Proof. Suppose $\mathcal{Q}g = 0$. Suppose $g^+ = gI\{g > 0\}$ and $g^- = -gI\{g < 0\}$. Then $g^+, g^- \geq 0$, and $g = g^+ - g^-$. Since $\mathcal{Q}$ is a linear operator we must have $\mathcal{Q}g^+ = \mathcal{Q}g^-$. By hypothesis, $\mathcal{Q}$ is injective for positive functions, which implies $g^+ = g^-$, and hence $g \equiv 0$. □

Theorem 3.21 states that if $\mathcal{Q}$ is injective on the space of positive functions, then T is complete. But Theorem 3.20 states that if T is complete, then $\mathcal{Q}$ is injective in general. We may therefore conclude that the injective property on positive functions is sufficient for the general injective property, and Theorem 3.21 could be replaced by an alternative theorem making this (quite correct) claim.

3.8.2 Sufficiency and Completeness for Exponential Families

The theory of completeness is particularly important to exponential famly models, since the property holds for full-rank models (Definition 3.8). This will be seen especially in Chapter 8. The proof of this, however, is rather technical, and relies on the open set condition (ii) of Definition 3.8. Here, we use the functional transformation approach with Theorem 3.21.

Theorem 3.22 (Minimal Sufficiency and Completeness for Full-Rank Exponential Family Models) The natural sufficient statistic $\mathbf{T}$ of a full-rank exponential family model (Definition 3.8) is minimal sufficient and complete. ■

Proof. We will assume the canonical parametrization for an exponential family density. Then under Definition 3.8 the parameter space $\Xi_\Theta \in \Xi$ contains an open subset.

Sufficiency. That $\mathbf{T}$ is sufficient follows directly from the factorization theorem (Theorem 3.16).

Minimal sufficiency. We use Theorem 3.17. An exponential family is of constant support, so for any $\tilde{x}, \tilde{y} \in \mathcal{X}$, using Equation (3.12), the ratio

$$\frac{p_\eta(\tilde{x})}{p_\eta(\tilde{y})} = \exp\left[\sum_{i=1}^{m} \eta_i(T_i(\tilde{x}) - T_i(\tilde{y}))\right] \frac{h(\tilde{x})}{h(\tilde{y})}$$

is well defined. By the open set condition (ii) of Definition 3.8 the ratio $p_\eta(\tilde{x})/p_\eta(\tilde{y})$ is independent of η *iff* $\mathbf{T}(\tilde{x}) = \mathbf{T}(\tilde{y})$. Therefore, $\mathbf{T}$ is minimal sufficient.

Completeness. Consider a function $g(\mathbf{T}) = g(T_1, \ldots, T_m)$. Assume $g \geq 0$. Following Theorem 3.14, the expectation is

$$E_\eta[g(T_1, \ldots, T_m)] = \int_{\mathbf{t}} g(\mathbf{t}) \exp\left[\mathbf{t}^T \eta - A(\eta)\right] h_{\mathbf{T}}(\mathbf{t}) d\nu_{\mathbf{T}}. \tag{3.20}$$

Select one $\eta^* = (\eta_1^*, \ldots, \eta_m^*) \in \Xi_\Theta$. Then

$$\begin{aligned} E_\eta[g(T_1, \ldots, T_m)] &= \int_{\mathbf{t}} g(\mathbf{t}) \exp\left[\mathbf{t}^T \eta - A(\eta)\right] h_{\mathbf{T}}(\mathbf{t}) d\nu_{\mathbf{T}} \\ &= e^{-A(\eta)} \int_{\mathbf{t}} \exp\left[\mathbf{t}^T (\eta - \eta^*)\right] H(\mathbf{t}, \eta^*; g) d\nu_{\mathbf{T}}, \end{aligned}$$

where $H(\mathbf{t}, \eta^*; g) = g(\mathbf{t}) h_{\mathbf{T}}(\mathbf{t}) e^{\mathbf{t}^T \eta^*}$. Since η^* is in the natural parameter space, the normalization constant

$$H(\eta^*; g) = \int_{\mathbf{t}} g(\mathbf{t}) h_{\mathbf{T}}(\mathbf{t}) e^{\mathbf{t}^T \eta^*} d\nu_{\mathbf{T}}$$

is finite, and $\bar{H}(\mathbf{t}, \eta^*; g) = H(\mathbf{t}, \eta^*; g)/H(\eta^*; g)$ is a density function. Therefore

$$E_\eta[g(T_1, \ldots, T_m)] = e^{-A(\eta)} H(\eta^*; g) m_g(\eta - \eta^*),$$

where m_g is the MGF for density $\bar{H}(\mathbf{t}, \eta^*; g)$.

Now, suppose two function $g, g' \geq 0$ satisfy $E_\eta[g(T_1, \ldots, T_m)] = E_\eta[g'(T_1, \ldots, T_m)]$ for all η. Then

$$e^{-A(\eta)} H(\eta^*; g) m_g(\eta - \eta^*) = e^{-A(\eta)} H(\eta^*; g') m_{g'}(\eta - \eta^*).$$

Since $e^{-A(\eta)} > 0$ for all η, this implies

$$H(\eta^*; g) m_g(\eta - \eta^*) = H(\eta^*; g') m_{g'}(\eta - \eta^*).$$

We then note that any MGF satisfies $m(0) = 1$, which implies $H(\eta^*; g) = H(\eta^*; g')$, which in turn implies $m_g(\eta - \eta^*) = m_{g'}(\eta - \eta^*)$ for η in an open neighborhood of η^*. However, MGFs uniquely define distributions, so that $H(\mathbf{t}, \eta^*; g) = H(\mathbf{t}, \eta^*; g')$, which implies $g \equiv g'$. The proof is completed by applying Theorem 3.21. □

3.8.3 Sufficiency and Density Truncation

We present a second example of the functional transformation approach for verifying completeness. Suppose a parametric family is constructed in the following way. A fixed positive function $h(x) > 0$ of $x \in \mathcal{X} \subset \mathbb{R}$ is specified. Then define parameter $\theta \in \mathcal{X}$. Our intention is to construct a family of densities of the form

$$f_\theta(x) = c(\theta)h(x)I\{x > \theta\}, \tag{3.21}$$

where we have finite normalization constant

$$c(\theta)^{-1} = \int_{x=\mathcal{X}}^{\infty} h(x)I\{x > \theta\}dx = \int_{x>\theta} h(x)dx$$

satsfying $c(\theta) \in (0,\infty)$ (we simply require that $\int_{\mathcal{X}} h(x)dx < \infty$). This gives a parametric family with parameter space $\Theta = \mathcal{X}$.

We then wish to determine if $X \sim f_\theta$ is complete for θ. Suppose $g : \mathcal{X} \to \mathbb{R}$. Then

$$E_\theta[g(X)] = c(\theta)\int_\theta^\infty g(x)h(x)dx$$

But $c(\theta) \in (0,\infty)$ and $\int_\theta^\infty g(x)h(x)dx$ is an injective mapping of $g : \mathcal{X} \to \infty$ to $G : \Theta \to \mathbb{R}$, so by Theorem 3.20 X is complete for Θ.

However, it must be noted that for completeness to hold, the parameter space must extend to all $\mathcal{X}$. Suppose $\mathcal{X} = \mathbb{R}$, but $\Theta = (-\infty, 0)$. In this case, if $E_\theta[g(X)] = 0$ for all $\theta \in \Theta$, this implies only that $g(x) = 0$ for $x > 0$. We can construct a counter-example to completeness in the following way. Set

$$g(x) = I\{x \in (0,1)\} - I\{x \in (1,2)\}\frac{P_\theta\{X \in (0,1)\}}{P_\theta\{X \in (1,2)\}}.$$

It easily verified that the ratio $P_\theta\{X \in (0,1)\}/P_\theta\{X \in (1,2)\}$ does not depend on θ, provided $\theta < 0$, so that g is a statistic for this particular model. Clearly, $E_\theta[g(X)] = 0$ for all $\theta \in \Theta = (-\infty, 0)$, but $g(x) \neq 0$ on a set of positive probability.

Next, suppose $\mathbf{X} = (X_1, \ldots, X_n)$ is an *iid* sample from density (3.21), $\theta \in \Theta = \mathcal{X}$. The density of $\mathbf{X}$ is

$$p_\theta(\mathbf{x}) = c(\theta)^n \prod_i h(x_i)I\{x_{(1)} > \theta\}.$$

That $X_{(1)}$ is sufficient is a consequence of the factorization theorem. Then from Section 2.4 the density of $X_{(1)}$ is given by

$$f_\theta^*(t) = n(1 - F_\theta(t))^{n-1}f_\theta(t) = nc(\theta)^n H(t)^{n-1}h(t)I\{t > \theta\},$$

where $H(t), h(t)$ depend on t but not θ. By the preceding argument, we may conclude that $X_{(1)}$ is complete for Θ.

Example 3.29 (Examples of Complete $X_{(1)}$) Suppose we are given an *iid* sample from the Pareto density

$$f_\mu(x) = \mu^{-1}\alpha/(x/\mu)^{\alpha+1}I\{x > \mu\}, \quad \mu > 0,$$

or the location shifted exponential density

$$f_\mu(x) = \tau^{-1}e^{-(x-\mu)/\tau}I\{x > \mu\}, \quad \mu \in \mathbb{R}.$$

Then $X_{(1)}$ will be complete for μ. See also Problem 3.23. ■

3.8.4 Ancillary Statistics

Ancillarity (Definition 3.2) is a property which is complementary to sufficiency and completeness, in the sense that an ancillary statistic contains no information about $\theta \in \Theta$. This is made precise by Basu's theorem.

Theorem 3.23 [Basu's Theorem] Any statistic $T \in \mathcal{T}_{comp}^b \cap \mathcal{T}_{suff}$ is independent of any statistic $V \in \mathcal{T}_{anc}$. ∎

Proof. Write $p_A = P\{V \in A\}$, which by ancillarity *dnd* θ. Set $\xi_A(t) = P(V \in A \mid T = t)$. By sufficiency $\xi_A(t)$ *dnd* θ. By bounded completeness $E_\theta[\xi_A(T) - p_A] = 0$ implies $\xi_A(t) \equiv p_A$, that is, T and V are independent. □

Note that since a complete statistic is boundedly complete, Theorem 3.23 holds also for $\mathcal{T}_{comp}$.

We have already seen that location invariant and scale invariant statistics define two important classes of ancillary statistics (Theorems 3.3 and 3.9). An important application of Basu's theorem follows.

Example 3.30 (The Sample Mean and Variance of an *iid* Normal Sample are Independent) For an *iid* sample from $N(\mu, \sigma^2)$, $\bar{X}$ is a complete sufficient statistic for μ (Theorem 3.22), which is a location parameter. However, the sample variance S^2 is location invariant (Example 3.4), and is therefore ancillary *wrt* μ. By Theorem 3.23 $\bar{X}$ and S^2 are independent. ∎

3.9 Conditional Models and Contingency Tables

Suppose, we are given observation $\tilde{X} \in \mathcal{X}$ and density family $\mathcal{P} = \{p_\theta : \theta \in \Theta\}$. We sometimes have reason to interpret a model as conditional on some statistic $S(\tilde{X})$. Upon observing $S(\tilde{X}) = s$, we regard s as fixed, so that we now take $\tilde{X} \mid S(\tilde{X}) = s$ as the observation. The density family now becomes $\mathcal{P}_s$, with elements $p_\theta(\tilde{x} \mid S(\tilde{X}) = s)$ evaluated for each $p_\theta \in \mathcal{P}$. Motivations for this approach are given in Sections 3.12, 3.13.4 or Section 5.9. Usually, statistical methods can be applied to conditional models without any important modification.

As an example, consider the 2×2 contingency table of counts, based on classes (A, A^c) and (B, B^c):

	A	A^c	
B	N_{11}	N_{12}	R_1
B^c	N_{21}	N_{22}	R_2
	C_1	C_2	n

Here, $R_i = N_{i1} + N_{i2}$, $i = 1, 2$ are the row totals, and $C_j = N_{1j} + N_{2j}$, $j = 1, 2$, are the column totals, with total sample size $\sum_{ij} N_{ij} = n$. The observation $\mathbf{N} = (N_{11}, N_{12}, N_{21}, N_{22})$ is assumed to possess a multinomial density with probabilities $\boldsymbol{\theta} = (p_{11}, p_{12}, p_{21}, p_{22})$. Then $\mathbf{N}$ has density

$$p_{\boldsymbol{\theta}}(n_{11}, n_{12}, n_{21}, n_{22}) = \frac{n!}{\prod_{i,j} n_{ij}!} \prod_{i,j} p_{ij}^{n_{ij}}.$$

There is often interest in the odds ratio $\rho = (p_{11}p_{22})/(p_{12}p_{21})$. The hypothesis $H : \rho = 1$ is equivalent to the independence of A and B. More precisely, it may be shown that

$$\rho = \frac{P(A \mid B)/(1 - P(A \mid B)}{P(A \mid B^c)/(1 - P(A \mid B^c)} = \frac{P(B \mid A)/(1 - P(B \mid A)}{P(B \mid A^c)/(1 - P(B \mid A^c)},$$

which is interpretable as the effect of classification (B, B^c) on the odds of A in a manner which does not depend on the marginal probabilities $P(A), P(B)$.

We next consider the effect of conditioning on the marginal totals R_i, C_j. Introduce the substitutions

$$n_{11} = n_{11},\ n_{12} = r_1 - n_{11},\ n_{21} = c_1 - n_{11},\ n_{22} = n - r_1 - c_1 + n_{11}.$$

In effect, we will derive the density of N_{11} conditional on $R_1 = r_1$, $C_1 = c_1$. First, rewrite the density as

$$p_{\boldsymbol{\theta}}(n_{11}, n_{12}, n_{21}, n_{22}) = h(n_{11}; r_1, c_1, n) p_{11}^{n_{11}} p_{12}^{r_1 - n_{11}} p_{21}^{c_1 - n_{11}} p_{22}^{n - r_1 - c_1 + n_{11}},$$

where

$$h(n_{11}; r_1, c_1, n) = \frac{n!}{n_{11}!(r_1 - n_{11})!(c_1 - n_{11})!(n - r_1 - c_1 + n_{11})!}.$$

The distribution of N_{11} conditional on $R_1 = r_1, C_1 = c_1$ is then

$$\begin{aligned} p_{\boldsymbol{\theta}}(x \mid R_1 = r_1,\ C_1 = c_1) &= \frac{h(x; r_1, c_1, n) p_{11}^{x} p_{12}^{r_1 - x} p_{21}^{c_1 - x} p_{22}^{n - r_1 - c_1 + x}}{\sum_{x=0}^{\min(r_1, c_1)} h(x; r_1, c_1, n) p_{11}^{x} p_{12}^{r_1 - x} p_{21}^{c_1 - x} p_{22}^{n - r_1 - c_1 + x}} \\ &= \frac{h(x; r_1, c_1, n) \rho^x}{\sum_{x=0}^{\min(r_1, c_1)} h(x; r_1, c_1, n) \rho^x}. \end{aligned} \tag{3.22}$$

Therefore, by the factorization theorem, N_{11} conditional on $R_1 = r_1$, $C_1 = c_1$ is a sufficient statistic for the odds ratio ρ. This is the basis for Fisher's exact test for independence in contingency tables. See Problem 5.22.

3.10 Bayesian Models

Suppose we are given the parametric model $\mathcal{P}$ with parameter space Θ. In Bayesian inference, $\theta \in \Theta$ itself is taken to be a random variable or vector possessing a prior density $\pi(\theta)$ for θ with respect to measure ν_π. The model is hierarchical in structure. We can imagine θ being selected first, then $\tilde{X}$ sampled from $p_\theta \in \mathcal{P}$, θ now being considered fixed. This defines a joint distribution for θ and $\tilde{X}$, since $p_\theta(\tilde{x})$ is interpretable as the density of $\tilde{X}$ conditional on θ, that is,

$$f(\tilde{x} \mid \theta) = p_\theta(\tilde{x}).$$

Viewed this way, $\tilde{X}$ contains information about θ, so it is natural to define the posterior density

$$\pi(\theta \mid \tilde{x}) = \frac{f(\tilde{x} \mid \theta)\pi(\theta)}{f(\tilde{x})} = \frac{f(\tilde{x} \mid \theta)\pi(\theta)}{\int_\Theta f(\tilde{x} \mid \theta)\pi(\theta) d\nu_\pi}, \tag{3.23}$$

which is simply the density of θ conditional on $\{\tilde{X} = \tilde{x}\}$.

Prior densities are often selected as members of some parametric family. The parameters associated with the prior are referred to as hyperparameters. These may be selected on the basis of prior knowledge or belief, and can therefore be treated as known parameters. It is also possible, however, to model uncertainty regarding the hyperparameters by assuming they are randomly selected from a hyperprior distribution.

3.10.1 On the Choice of Prior Density

What distinguishes the Bayesian model is the existence of the prior density $\pi(\theta)$ (the parametric family $\mathcal{P}$ being defined as before) so we first offer a simple example of the role it may play. Suppose we are developing a test for the presence of a type of infection, the test outcome represented as a random variable $X \in \mathbb{R}$. The parameter space Θ may then consists of two models only, θ_0, θ_1, corresponding to negative and positive infection state. When we develop the test, we presumably have enough training data with which to estimate conditional densities $f(x \mid \theta_0)$ and $f(x \mid \theta_1)$. The posterior probability of infection is therefore given by (3.23):

$$\pi(\theta_1 \mid x) = \frac{f(x \mid \theta_1)\pi_1}{f(x \mid \theta_0)(1 - \pi_1) + f(x \mid \theta_1)\pi_1}, \tag{3.24}$$

where π_1 is the prior probability of infection (which sufficies to completely define the prior density in this example).

However, while the use of training data to estimate the conditional densities $f(x \mid \theta)$ would be entirely appropriate, it would usually not be appropriate to estimate the prior density by the proportions of each class in the training data. We might expect that π_1 would be much less than $1/2$, and so for the purposes of efficient estimation, the proportion of the infected class in any training data should be chosen to be much higher than π_1.

For this reason, the choice of prior probabilities is often made independently of the training data. To take an extreme example, suppose the infection in question is nonexistent (small pox, for example). In this case, it would be appropriate to set $\pi_1 = 0$. In this case, direct substitution into (3.24) yields posterior probability of infection $\pi(\theta_1 \mid \tilde{x}) = 0$, independent of the degree to which the test itself favors a prediction of infection.

3.10.2 Conjugate Priors

While the form of the posterior density (3.23) seems straighforward, its actual evaluation sometimes poses a considerable technical challenge. This issue usually depends on whether or not the denominator in (3.23) can be conveniently evaluated. It is therefore important to identify Bayesian inference problems which have an elegant analytic solution. An important example occurs when the prior and posterior densities are members of a single parameteric family, in which case we refer to $\pi(\theta)$ as conjugate prior. Two examples follow.

Example 3.31 (The Beta Distribution is a Conjugate Prior for the Binomial Parameter) Suppose $X \sim bin(n, p)$, and we assign prior density $p \sim beta(\alpha, \beta)$. We then have conditional density

$$f(x \mid p) = \binom{n}{x} p^x (1-p)^{n-x}, \quad x = 0, \ldots, n$$

and prior density

$$\pi(p) = \frac{1}{B(\alpha, \beta)} p^{\alpha-1}(1-p)^{\beta-1}, \; p \in [0, 1].$$

This leads to posterior density

$$\begin{aligned} \pi(p \mid x) &= \frac{P(X = x \mid p)\pi(p)}{\int_{p=0}^{1} P(X = x \mid p)\pi(p)dp} \\ &= \frac{\binom{n}{x} p^x (1-p)^{n-x} \frac{1}{B(\alpha,\beta)} p^{\alpha-1}(1-p)^{\beta-1}}{\int_{p=0}^{1} \binom{n}{x} p^x (1-p)^{n-x} \frac{1}{B(\alpha,\beta)} p^{\alpha-1}(1-p)^{\beta-1} dp}. \end{aligned} \tag{3.25}$$

Although this expression seems complicated, it is actually quite simple, as long as the objective is kept in mind, which is to derive a density of p. We can always express (3.25) in the form $\pi(p \mid x) = Kg(p)$, where K does not depend on p. This is conveniently written as a proportional relationship $\pi(p \mid x) \propto g(p)$. We can then renormalize to get

$$\pi(p \mid x) = \frac{g(p)}{\int_{p=0}^{1} g(p)dp},$$

noting that the denominator does not depend on p. Clearly, from (3.25) we have

$$\pi(p \mid x) \propto p^{x+\alpha-1}(1-p)^{n-x+\beta-1}.$$

In other words, the posterior density of p given observation $X = x$ is $beta(x+\alpha, n-x+\beta)$, so that the beta density defines a conjugate prior. As a technical matter, note that the quantities $\binom{n}{x}$ and $1/B(\alpha, \beta)$ in (3.25) play no role, since they do not depend on p (in fact, they appear in both the numerator and denominator, and so cancel). In particular, for an example such as this, we do not need to explicitly evaluate the integral in the denominator. ■

Example 3.32 (The Normal Distribution is a Conjugate Prior for the Normal Mean Parameter) Suppose we observe a normally distributed random variable $X \sim N(\mu, \sigma)$. Assume σ is known, and that μ has a prior density $\pi(\mu)$:

$$\mu \sim N(\mu_0, \sigma_0),$$

for some fixed μ_0, σ_0. We then have $\pi(\mu \mid x) \propto f(x \mid \mu)\pi(\mu)$ where $x \mid \mu \sim N(\mu, \sigma)$ and $\mu \sim N(\mu_0, \sigma_0)$. This means

$$\pi(\mu \mid x) = Ke^{-\frac{1}{2}Q_1}e^{-\frac{1}{2}Q_0}$$

where

$$Q_1 = \frac{(x-\mu)^2}{\sigma^2}, \quad Q_0 = \frac{(\mu-\mu_0)^2}{\sigma_0^2}.$$

We then have

$$Q_1 + Q_0 = \mu^2\left[\frac{1}{\sigma^2} + \frac{1}{\sigma_0^2}\right] - 2\mu\left[\frac{x}{\sigma^2} + \frac{\mu_0}{\sigma_0^2}\right] + \left[\frac{x^2}{\sigma^2} + \frac{\mu_0^2}{\sigma_0^2}\right].$$

This means

$$\pi(\mu \mid x) = Ke^{-\frac{1}{2}\left\{\mu^2\left[\frac{1}{\sigma^2}+\frac{1}{\sigma_0^2}\right]-2\mu\left[\frac{x}{\sigma^2}+\frac{\mu_0}{\sigma_0^2}\right]\right\}}$$

where K does not depend on μ, and that $\pi(\mu \mid x) \sim N(\mu_{post}, \sigma^2_{post})$ is a normal density function with mean and variance

$$\mu_{post} = \frac{\frac{x}{\sigma^2} + \frac{\mu_0}{\sigma_0^2}}{\frac{1}{\sigma^2} + \frac{1}{\sigma_0^2}}, \quad \sigma^2_{post} = \frac{1}{\frac{1}{\sigma^2} + \frac{1}{\sigma_0^2}}.$$ ■

3.11 Indifference, Invariance and Bayesian Prior Distributions

The notion that a parameter θ is random may seem unnatural, and one response to this is to attempt to select a prior density which has as little influence on the inference as possible, assuming such a goal can be precisely defined. The first step is to distinguish between two types of prior, the informative prior and the uninformative prior (we will see examples of both). An informative prior can be constructed by incorporating any available prior information in a manner which assigns greater weight to models compatible with that information (Example 4.18). However, the choice of informative prior can be based on any type of knowledge, and need not be supported by data. The construction of a prior density based on expert opinion is sometimes referred to as prior elicitation.

It then seems reasonable to refer to any prior that is not an informative prior as an uninformative prior. However, this does not yet reach the level of mathematical definition. If we try to be more precise, and accept that a prior distribution is a type of belief or plausibility weighting imposed on a model space prior to the examination of statistical evidence, then it is tempting to think that an uninformative prior should assign equal weight to each model. This is reasonable if all models are indistinguishable except for arbitrary labelling. We then take the unique uninformative prior π to be the uniform distribution on the model space Θ (we will discuss technical issues regarding the cardinality of Θ below). This is the simplest example of the principle of indifference.

When models are unlike in more fundamental ways, especially relating to complexity, then the principle of indifference as stated above becomes ambiguous, until we recognize it as a type of invariance principle, similar to that introduced in Section 3.3. This brings us back to the central issue regarding priors: that is, whether informative or uninformative, a choice must still be made. In particular, the application of the principle of indifference requires that we choose precisely with respect to what we are indifferent, and this choice is not always obvious.

3.11.1 Proper and Improper Priors

The principle of indifference is typically expressed using a uniform prior distribution. When the parameter space is of finite size m, the uniform prior is without ambiguity $\pi(\theta) = 1/m$.

For infinite parameter spaces a problem emerges. A uniform prior can only be formally defined if Θ is bounded. Clearly, we cannot define a uniform density on the entire real line, or on a countably infinite sample space. However, examining the definition of the posterior density in (3.23), we can see that the prior density appears only within a ratio. In particular, if a prior density is given as $\pi(\theta) = cg(\theta)$ then (3.23) can be evaluated without knowing normalization constant c. In fact, $g(\theta)$ need not even be integrable on Θ. This means we can construct a uniform prior simply by setting $\pi(\theta) \equiv 1$ for all $\theta \in \Theta$, for any Θ. Thus, we may define prior densities by proportionality statements such as $\pi(\theta) \propto g(\theta)$, for any nonnegative function $g(\theta)$. If the integral $\int_\Theta g(\theta) d\nu_\pi = \infty$ then $\pi(\theta)$ is called an improper prior distribution. Otherwise, $\pi(\theta)$ can be normalized if needed.

However, it must be noted that for some aspects of Bayesian inference the assumption that a prior is proper is crucial, as will be seen in Section 7.9.

3.11.2 The Jeffreys Prior and Invariance

There is a fundamental problem associated with the principle of indifference. The property of uniformity depends on the parametrization. For example, if we are interested in the variance parameter σ^2 of a normal density, then an improper uniform prior density for σ^2 will not be uniform for the transformed parameter σ. This leads to an inconsistency. For a uniform density on σ^2, the models $\sigma^2 = 1$ and $\sigma^2 = 4$ have equal prior weight. However, if we reparametrize the prior density with transformation $\sigma^2 \to \sigma$, we will not have equal prior weights for the equivalent models $\sigma = 1$ and $\sigma = 2$.

The Jeffreys prior distribution introduced in the seminal paper Jeffreys (1946) is based on an invariance principle which states that a measure of distance between two probability densities should not depend on any particular parametrization. This principle forces the choice of the Jeffreys prior, defined by

$$\pi(\theta) \propto [\det(\mathbf{I}(\theta))]^{1/2} \tag{3.26}$$

where $\mathbf{I}(\theta)$ is the Fisher information matrix (Section 3.5). For a one dimensional parameter it is given by the scalar

$$I(\theta) = E_\theta \left[\left(\frac{\partial}{\partial\theta} \log p_\theta(\tilde{X}) \right)^2 \right].$$

This is commonly described as invariant to reparametrization of θ. Of course, any density f_X can be reparametrized following a 1-1 transformation $x \mapsto y$ by the standard transformation rule

$$f_Y(y) = \left| \frac{dy}{dx} \right|^{-1} f_X(x), \tag{3.27}$$

so it should be made clear in what sense the Jeffreys prior is invariant.

The reader is encouraged to study the original paper (Jeffreys, 1946), written in a style which is technically informal, but in the essential features mathematically precise, and it is illuminating when understood. Very roughly, the argument is this. Suppose $D(P', P)$ is a distance between two distributions, for example, Jeffreys divergence

$$D(P', P) = \int_{\mathcal{X}} (p'(x) - p(x))(\log p'(x) - \log p(x))dx$$

where p, p' are the appropriate densities of distributions P, P'. The technical problem is to characterize the limiting behavior of $D(P', P)$ as P' approaches P. This can be done in a straightforward way if P', P are members of some parametric family $\mathcal{P}$, indexed by parameter $\theta \in \mathbb{R}$. We then set $P' = P_{\theta'} = P_{\theta+\Delta\theta}$, $P = P_\theta$, Δ denoting an arbitrarily small change. In Jeffreys (1946), it shown that

$$D(P_{\theta+\Delta\theta}, P_\theta) \approx I(\theta)[\Delta\theta]^2.$$

Taking the square root yields

$$D(P_{\theta+\Delta\theta}, P_\theta)^{1/2} \approx I(\theta)^{1/2}\Delta\theta,$$

so that $I(\theta)^{1/2}$ can be intepreted as a density with respect to $\theta \in \mathbb{R}$. But this is the Jeffreys prior for a one-dimensional parameter. The argument is extended to $\theta \in \mathbb{R}^m$ in Jeffreys (1946), yielding the form (3.26).

But this particular distance D can be shown to be independent of any particular parametrization. If $\eta = \eta(\theta)$ is a one-to-one reparametrization, we may write

$$\eta' = \eta(\theta') = \eta(\theta + \Delta\theta) \approx \eta(\theta) + \frac{d\eta(\theta)}{d\theta}\Delta\theta,$$

so that

$$\eta' - \eta = \Delta\eta \approx \frac{d\eta(\theta)}{d\theta}\Delta\theta. \tag{3.28}$$

Application of the principle of indifference with respect to reparametrization then forces the identity:

$$\begin{aligned} I(\theta)[\Delta\theta]^2 &\approx D(P_{\theta+\Delta\theta}, P_\theta) \\ &= D(P_{\theta'}, P_\theta) \\ &= D(P_{\eta'}, P_\eta) \\ &= D(P_{\eta+\Delta\eta}, P_\eta) \\ &\approx I_\eta(\eta)[\Delta\eta]^2, \end{aligned}$$

where $I_\eta(\eta)$ is the Fisher information with respect to η. Taking the square root then applying (3.28) yields the standard transformation rule

$$I(\theta)^{1/2} = \left| \frac{d\eta(\theta)}{d\theta} \right| I_\eta(\eta(\theta))^{1/2},$$

which was introduced in Equation (3.9).

Thus, we can think of the Jeffreys prior as possessing an evaluation method that happens to be invariant to parametrization. This is a consequence of the fact that $I(\theta)^{1/2}\Delta\theta$ measures a differential change in a probability distribution with respect to a distance which is itself invariant to parametrization, as is Jeffreys divergence (or rather, its square root). It is important to note, therefore, that the same result is obtained using Hellinger distance

$$H^2(P', P) = \frac{1}{2} \int_{\mathcal{X}} \left(\sqrt{p'(x)} - \sqrt{p(x)} \right)^2 dx$$

in place of Jeffreys divergence, which is also invariant to parametrization.

The next example will clarify the distinction between a uniform prior and an indifferent prior.

Example 3.33 (Dependence on Parametrization of the Uniform Prior) The uniform prior is not invariant to reparametrization in general. Suppose $X \sim N(0, \sigma^2)$, and we apply an (improper) uniform prior to $\sigma \in \Theta = (0, \infty)$:

$$\pi_\sigma(\sigma) \propto 1.$$

Under the usual transformation rule (which also applies to improper densities), we have

$$\pi_{\sigma^2}(\sigma^2) \propto \frac{1}{2\sqrt{\sigma^2}}.$$

This means a uniform prior assigned to σ^2 is not consistent with a uniform prior assigned to σ, in the sense that the conventional transformation rule does not hold. ■

The Jeffreys prior possesses a single form for any location or scale parameter, as shown in the next example.

Example 3.34 (The Jeffreys Prior for Location and Scale Parameters) Consider a location parameter family $f(x - \theta)$, $x \in \mathbb{R}$. We may show that the Jeffreys prior for θ is

$$\pi(\theta) \propto 1, \tag{3.29}$$

which is an improper prior. We can, however, characterize location invariance for π by the rule

$$P_\pi\{\theta \in A\} = P_\pi\{\theta \in A + c\}$$

where $A \subset \Theta$, and $A + c = \{\theta + c : \theta \in A\}$. The only prior satifying this invariance rule is the Jeffreys prior of Equation (3.29).

Next, consider a scale parameter family $\theta^{-1} f(x/\theta)$, $x \in \mathbb{R}$, $\theta > 0$. We may show that the Jeffreys prior for θ is

$$\pi(\theta) \propto \frac{1}{\theta}, \tag{3.30}$$

Suppose the prior probability of interval $(a, b) \subset \Theta$ is some number

$$\alpha = P_\pi\{\theta \in (a, b)\} = \int_a^b \pi(\theta) d\theta. \tag{3.31}$$

For the scale invariant model, we are observing data $X = \theta X_0$, where X_0 has a known density. If we decide we have no reason to favor either hypothesis $\theta = 1$ or $\theta = \theta' > 0$, this implies that (a, b) and $(\theta' a, \theta' b)$ should have the same prior probability. Set

$$\alpha' = P_\pi\{\theta \in (\theta' a, \theta' b)\} = \int_{\theta' a}^{\theta' b} \pi(\theta) d\theta. \tag{3.32}$$

We may write the Jeffreys prior for a scale parameter as $\pi(\theta) = K\theta^{-1}$, for some $K > 0$ (the exact value doesn't matter). Then

$$\alpha' = K \int_{\theta' a}^{\theta' b} \theta^{-1} d\theta = K \log(\theta' b) - K \log(\theta' a) = K \log(b) - K \log(a),$$

so that α' does not depend on θ'. Thus, the Jeffreys prior for any scale parameter is scale invariant. ∎

3.12 Nuisance Parameters

We introduce at this point a ubiquitous problem of statistical inference. This arises when parameters are multidimensional, but the inference problem does not concern the full parameter. This topic is often treated as a technically challenging extension of a methodology from one to multiple dimensions. However, in our view it is preferable to introduce the problem early, since many methods of dealing with nuisance parameters proceed by reducing the model to a single (or full) parameter problem. Once this is done, methods suitable for the full parameter case can be applied with little modification.

To be precise, suppose a multidimensional parameter can be decomposed into $\theta = (\psi, \xi)$, and we are interested in an inference concerning ψ alone, regarding ξ as a nuisance parameter. However, any statistic S used obviously cannot depend on ξ. In addition, a complete inference statement will usually require that the distribution of S also not depend on ξ.

We have already seen two methods which, if available, can transform the parameteric family $\mathcal{P}$ to one which does not depend on ξ. If the nuisance parameter ξ is a location or scale parameter, then by Theorem 3.3 or 3.9 any maximal invariant statistic will be an ancillary statistic with respect to ξ, and so can be used in place of the original data $\tilde{X} \in \mathcal{X}$. This idea is extended to other cases in Section 9.4.

Alternatively, from Section 3.7, if $S_\xi \in \mathcal{T}_{stat}$ exists which is sufficient for ξ, then we may base our inference on the conditional observation $X \mid S_\xi$. Then any suitable method may be used for inference on ψ. This type of conditional inference was already introduced in Section 3.9.

Both approaches will be used throughout this volume.

3.13 Principles of Inference

There are a number of general statistical principles that are widely employed. They do not describe specific inference methods, rather, they enumerate properties that such methods might possess. Of course, they have some theoretical justification, but we will find that if we are able to derive an optimal procedure from a decision theoretic formulation, the sufficiency or likelihood principle will usually be upheld (Sections 3.13.1 and 3.13.2). On the other hand, the invariance principle (Section 3.13.3) is more in the nature of a model assumption. The conditionality principle (Section 3.13.4) is not universally accepted, but an example in which it would be highly advisable is offered. According to Birnbaum's theory (Section 3.13.5) the sufficiency and conditionality principles jointly imply the strong likelihood principle. Again, this conclusion is not universally accepted.

3.13.1 The Sufficiency Principle

Suppose we are given an estimate $\delta(\tilde{X})$ of $\eta(\theta)$. Suppose a statistic $S(\tilde{X})$ is sufficient for θ. Then $\delta^*(\tilde{X}) = E_\theta[\delta(\tilde{X}) \mid S(\tilde{X})]$ is also a statistic. Nominally, the evaluation of the conditional expectation used to define $\delta^*(\tilde{X})$ depends on θ, but not in this case, as a consequence of the sufficiency property. Furthermore, by the total expectation and total variance identities, $\delta^*(\tilde{X})$ has the same expected value as δ, but strictly smaller variance, unless $S(\tilde{X}) \mapsto \delta(\tilde{X})$ (Problem 3.30). The Rao-Blackwell theorem is a refinement of this argument, and is given in this volume in two forms as Theorems 7.11 and 8.5.

The implications of this for inference are absolutely crucial, since any estimator $\delta(\tilde{X})$ that is not a function of a sufficient statistic $S(\tilde{X})$ can be improved by one that is, in the sense given. Furthermore, the improved statistic can be obtained simply by taking the expected value $\delta^*(\tilde{X}) = E_\theta[\delta(\tilde{X}) \mid S(\tilde{X})]$. By the same argument, a estimator that is a mapping of a sufficient statistic, but not a minimal sufficient statistic $T(\tilde{X})$, can be improved by conditioning on $T(\tilde{X})$. The leads to the sufficiency principle.

Definition 3.12 (Sufficency Principle) Suppose we are given observation $\tilde{X} \in \mathcal{X}$ from a density from the family $\mathcal{P}$. The sufficiency principle states that any inference concerning model $\mathcal{P}$ should depend on $\tilde{X} \in \mathcal{X}$ only through a sufficient statistic S. In particular, if we are given distinct observations $\tilde{X}, \tilde{X}' \in \mathcal{X}$, not necessarily equal, the respective inference should be identical if $S(\tilde{X}) = S(\tilde{X}')$. In addition, if a minimal sufficient statistic $T(\tilde{X})$ exists, any inference should depend on $\tilde{X} \in \mathcal{X}$ only through T. ■

3.13.2 The Weak and Strong Likelihood Principle

Given the factorization theorem (Theorem 3.16), the sufficiency principle provides a justification for the central role played in inference by the likelihood function.

Definition 3.13 (Weak Likelihood Principle) Suppose we are given family $\mathcal{P}$ for sample space $\mathcal{X}$. Define likelihood $\mathcal{L}(\theta; \tilde{X}) = p_\theta(\tilde{X})$. Suppose we are given distinct observations $\tilde{X}, \tilde{X}' \in \mathcal{X}$. The weak likelihood principle states that if the ratio $\mathcal{L}(\theta; \tilde{X})/\mathcal{L}(\theta; \tilde{X}')$, where defined, does not depend on θ, then the inference should be identical for $\tilde{X}$ and $\tilde{X}'$. ■

For both the weak likelihood principle and the sufficiency principle inference is contained within a single parametric family. However, we can always compare the likelihood function of two distinct inference models, provided they share a single parameter. This leads to the strong likelihood principle:

Definition 3.14 (Strong Likelihood Principle) Suppose we are given observations $\tilde{X} \in \mathcal{X}, \tilde{Y} \in \mathcal{Y}$ from densities $f_{\tilde{X}}(\tilde{x}; \theta)$, $f_{\tilde{Y}}(\tilde{y}; \theta)$ dependent on a common parameter $\theta \in \Theta$. Define likelihoods $\mathcal{L}(\theta; \tilde{X}) = f_{\tilde{X}}(\tilde{X}; \theta)$, $\mathcal{L}(\theta; \tilde{Y}) = f_{\tilde{Y}}(\tilde{Y}; \theta)$. The strong likelihood principle states that if the ratio $\mathcal{L}(\theta; \tilde{X})/\mathcal{L}(\theta; \tilde{Y})$, where defined, does not depend on θ, then the inference should be identical for $\tilde{X}$ and $\tilde{Y}$. ■

The wider applicability of the strong likelihood principle is demonstrated in the following example.

Example 3.35 (Comparison of Binomial and Negative Binomial Observation) Suppose we observe $X \sim bin(n, \theta)$ and $Y \sim nb(r, \theta)$. The respective densities are

$$p_X(x; \theta) = \binom{n}{x} \theta^x (1-\theta)^{n-x} \text{ and } p_Y(y; \theta) = \binom{y-1}{r-1} \theta^r (1-\theta)^{y-r}$$

for $x \in \{0, 1, \ldots, n\}$ and $y \in \{r, r+1, \ldots\}$. We would not usually regard X and Y as belonging to a common parametric family, and so the sufficiency principle and the weak likelihood principle cannot be applied. However, the likelihood functions can still be compared. For example, suppose we observe $X = r$ and $Y = n$. Then the likelihoods have the ratio

$$\frac{\mathcal{L}(\theta; r)}{\mathcal{L}(\theta; n)} = \frac{\binom{n}{r} \theta^r (1-\theta)^{n-r}}{\binom{n-1}{r-1} \theta^r (1-\theta)^{n-r}} = \frac{\binom{n}{r}}{\binom{n-1}{r-1}} = \frac{n}{r}.$$

Thus, the likelihoods are not generally equal, but the ratio does not depend on θ. So in this case, by the strong likelihood principle the inference regarding θ would be the same based on X or Y. ∎

3.13.3 The Invariance Principle

Suppose for model Θ an observation $\tilde{X} \in \mathcal{X}$ leads to an inferential statement regarding $\theta \in \Theta$. If the data is subjected to some injective transformation $g : \mathcal{X} \to \mathcal{Y}$, we now have observation $\tilde{Y} = g\tilde{X} \in \mathcal{Y}$. If $\tilde{X} \sim p_\theta$, we can identify the density of $\tilde{Y} = g\tilde{X}$ as a member $p^*_{\theta^*}$ of a parametric family $\mathcal{P} = \{p^*_{\theta^*} : \theta^* \in \Theta^*\}$. Thus, there exists a mapping $g^* : \Theta \to \Theta^*$ such that $\tilde{X} \sim p_\theta$ implies $g\tilde{X} \sim p^*_{g^*\theta}$, and Θ^* is the image $g^*\Theta$.

The invariance principle states that under this model, an inference about $\theta \in \Theta$ based on $\tilde{X}$ should be equivalent to an inference about $\theta^* \in \Theta^*$ based on $\tilde{Y} = g\tilde{X}$, in the sense that each inference can be obtained from the other by the mapping g^*. For example, estimates $\hat{\theta}$ of θ and $\hat{\theta}^*$ of θ^* should satisfy $\hat{\theta}^* = g\hat{\theta}$. The expectation that this transformation holds is referred to as the equivariance principle, while $\hat{\theta}$ is refered to as an equivariant statistic.

The intuitive notion that inference should not depend on measurement units is an example of the invariance principle. If we change units from feet to inches, this should not change an inference in any important way, and inference performed on data in one unit should be easily mapped to an inference performed on the same data expressed in a different unit. It turns out that imposing this quite reasonable constraint allows the application of powerful optimization methods, the topic covered in Chapter 9.

3.13.4 The Conditionality (Ancillarity) Principle

Recall that a statistic is ancillary if its distribution does not depend on parameter θ (Definition 3.2). However, this does not mean that an ancillary statistic should play no role in any inference. Suppose that T is a minimal sufficient statistic. It will sometimes be possible that if $\dim(T) \geq \dim(\theta) > 1$, we may write $T = (S, V)$, where V is ancillary. In this case, we have the following inference principle, which holds that inference should proceed with V considered fixed.

Definition 3.15 (Conditionality Principle) Suppose we are given family $\mathcal{P}$ for sample space $\mathcal{X}$. Suppose the minimal sufficient statistic may be written $T = (S, V)$, where V is an ancillary statistic. The conditionality principle, also referred to the ancillarity principle, states that inference should be conditional on V. ∎

Section 3.9 gives an important application of the conditionality principle. In addition, Example 3.36 below illustrates the perils of ignoring the conditionality principle.

Example 3.36 (Inference of Binomial Parameter with Ancillary Size Parameter) Suppose we observe $X \sim bin(N, \theta)$, where the size parameter N is random. Assume the density of N, $p_N(n)$ does not depend on θ. We can regard this as a hierarchical model, so that N is sampled first, then X is taken to be a binomial random variable with fixed sample size N. We assume N is observed. In this case, it may be verified that the minimial sufficient statistic for θ is $T = (X, N)$. However, N is clearly ancillary, so the conditionality principle states that once $N = n$ is observed, inference should be based on $X \sim bin(n, \theta)$, with n regarded as fixed.

This approach seems quite reasonable. Although N contains no information regarding θ, it helps interpret the observation X. So it's worth considering what happens if the conditionality principle is not followed. To fix ideas, suppose $p_N(10) = p_N(20) = 1/2$ and we observe $X = 5$. We'll consider three cases: (a) condition on $N = 10$; (b) condition on $N = 20$; (c) use the unconditional density of X. Then, in this order, we have likelihood functions:

$$
\begin{aligned}
\mathcal{L}(\theta; X = 5, N = 10) &= \binom{10}{5}\theta^5(1-\theta)^5, \\
\mathcal{L}(\theta; X = 5, N = 20) &= \binom{20}{5}\theta^5(1-\theta)^{15}, \\
\mathcal{L}(\theta; X = 5) &= \frac{1}{2}\left(\binom{10}{5}\theta^5(1-\theta)^5\right) + \frac{1}{2}\left(\binom{20}{5}\theta^5(1-\theta)^{15}\right).
\end{aligned}
$$

Thus, for the conditional procedure, we would use likelihood function $\mathcal{L}(\theta; X = 5, N = n)$ for either outcome $N = n$. The unconditional procedure would be based on likelihood $\mathcal{L}(\theta; X = 5)$, irrespective of the observed value of N.

These likelihood functions are ploted in Figure 3.3. Clearly, the likelihoods conditioned on $N = n$ yield maximum likelihood estimate $\hat{\theta} = X/n$, which would be reasonable. The unconditional likelihood is maximized at a single value slightly larger than the estimate $X/n = 5/20 = 0.25$ obtained conditional on $N = 20$. We further note that the conditional likelihoods are considerably more concentrated around their maxima, which, as we will see, is directly related to the accuracy of any inference. ■

3.13.5 Birnbaum's Theorem

Birnbaum's theorem states that the sufficiency and conditionality principles jointly imply the strong likelihood principle (Birnbaum, 1962). If one accepts the first two, one must also accept the latter. The proof is based on a random choice of two experiments, each yielding a distinct likelihood function on a common parameter space. Assume the randomization method is ancillary (it does note depend on the parameter). The conditionality principle is interpretable as meaning that once the choice is made, one should proceed as though only that experiment was considered. Under this model, a sufficient statistic S can be defined. The sufficiency principle requires that inference is based only on S. The strong likelihood principle follows from a mathematical argument (see, for example, Casella and Berger (2002) or Cox and Hinkley (1979)). The conclusion of Birnbaum's theorem is not universally accepted (Evans *et al.*, 2013; Mayo *et al.*, 2014), and the correct interpretation (or even the justification) of the principles enumerated above is still subject to debate.

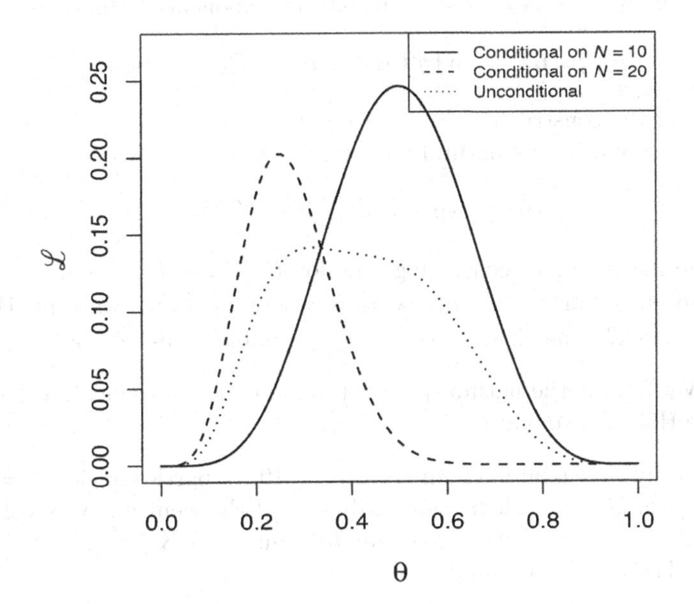

FIGURE 3.3
Plot of likelihood functions for Example 3.36.

3.14 Problems

Problem 3.1 For each model state whether or not the parameter $\theta \in \Theta$ is identifiable for the given observation. Justify your answer.

(a) The observation is $X \in \mathcal{X} = \mathbb{R}$, $X \sim N(0, \theta)$, $\theta \in \Theta = \mathbb{R}_{>0}$.
(b) The observation is $X \in \mathcal{X} = \mathbb{R}$, $X \sim N(\theta_1 - \theta_2, 1)$, $\theta = (\theta_1, \theta_2) \in \Theta = \mathbb{R}^2$.
(c) The observation is $X = \min\{n - Y, Y\} \in \mathcal{X} = \{0, 1, \ldots, \lfloor n/2 \rfloor\}$, $Y \sim bin(n, \theta)$, $\theta \in \Theta = [0, 1]$.
(d) The observation is $X \in \mathcal{X} = \mathbb{R}_{>0}$, $X \sim pareto(\tau, \alpha)$, $\theta = (\tau, \alpha) \in \Theta = \mathbb{R}^2_{>0}$.

Problem 3.2 Prove Theorem 3.11. **HINT:** The proof is similar to Theorem 3.7.

Problem 3.3 Prove the claims of Example 3.15:

(a) If θ is a location parameter then $I(\theta)$ is constant *wrt* θ. **HINT:** Make use of Theorem 3.2.
(b) If θ is a scale parameter then $I(\theta) = K/\theta^2$, where K *dnd* θ. **HINT:** Make use of Theorem 3.8.

Problem 3.4 In genetics, a genotype consists of two genes, each of which is one of (possibly) several types of alleles. Suppose we consider only two alleles, r and R. Furthermore, suppose the allele r exists in a population with frequency $\theta \in (0, 1)$. Under Hardy-Weinberg equilibrium, genotypes are essentially random samples of 2 alleles, one sampled from each parent. Since the genes are usually not ordered (because we don't know which is maternal and which is paternal), the probabilities of each possible genotype in an individual are $P(\mathtt{rr}) = \theta^2$, $P(\mathtt{rR}) = 2(1-\theta)\theta$, $P(\mathtt{RR}) = (1-\theta)^2$. Suppose we collect a random sample of n individual genotypes rr, rR, RR, observed with frequencies $\mathbf{N} = (n_{\mathtt{rr}}, n_{\mathtt{rR}}, n_{\mathtt{RR}}) = (n_1, n_2, n_3)$. Then $\mathbf{N}$ is a multinomial vector, with the class probabilities given above.

(a) Write the density p_θ of $\mathbf{N}$ as a one dimensional exponential family model in the form

$$p_\theta(\mathbf{n}) = \exp\{\eta(\theta)T(n_1,n_2) - B(\theta)\}\, h(n_1,n_2),$$

making use of the constraint $n_3 = n - n_1 - n_2$.

(b) Write the density using its natural parametrization:

$$p_\eta(\mathbf{n}) = \exp\{\eta T(n_1,n_2) - A(\eta)\}\, h(n_1,n_2).$$

Then derive the cumulant generating function (CGF) of $T(n_1,n_2)$.

(c) Consider an estimator $\hat{\theta} = cT(n_1,n_2)$ of θ, where c is some constant. For what value of c is $\hat{\theta}$ unbiased? What is the variance of the unbiased estimator?

Problem 3.5 Verify that the natural parameter space of an exponential family is convex. **HINT:** Use the Hölder inequality.

Problem 3.6 Verify the claim made in Example 3.19. In particular, let $\mathbf{X} = (X_1,\ldots,X_n)$ be *iid*, with $X_1 \sim unif(0,\theta)$. Then the distribution of the elements of $\mathbf{X}$ other than $X_{(n)}$, conditional on $X_{(n)} = t$ is equivalent to an *iid* sample of $\mathbf{X}' = (X_1,\ldots,X_{n-1})$, where $X_1 \sim unif(0,t)$. **HINT:** See Example 2.7.

Problem 3.7 Find the minimal sufficient statistic for θ from *iid* observations $X_1,\ldots,X_n$ where $X_1 \sim unif(-\theta,\theta)$.

Problem 3.8 Suppose X has a normal distribution with mean θ and variance 1. Is $T(X) = |X|$ sufficient for θ?

Problem 3.9 Suppose parametric family $\mathcal{P}$ on $\mathcal{X}$ is of constant support. Let Θ be the parameter space. Prove that S is a sufficient statistic *iff* $p_\theta(\tilde{x})/p_{\theta'}(\tilde{x})$, is a function of $S(\tilde{x})$ over $\tilde{x} \in \mathcal{X}$, for all pairs $\theta,\theta' \in \Theta$. **HINT:** If $\theta' \in \Theta$ is fixed, then $h(\tilde{x}) = p_{\theta'}(\tilde{x})$ is a function on $\mathcal{X}$ which *dnd* θ.

Problem 3.10 Suppose $\mathbf{X} = (X_1,\ldots,X_n)$ is an *iid* sample with $X_1 \sim logistic(\theta,1)$. Use Theorem 3.17 to verify that $\mathbf{X}_{ord}$ is minimal sufficient for θ. **HINT:** See Example 3.23.

Problem 3.11 Let $\mathbf{X} = (X_1,\ldots,X_n)$ be an *iid* sample from density $X_1 \sim N(\theta,1)$. Set $Y_i = g(X_i)$, where $g(x) = e^x$. Then Y_i has a log-normal distribution.

(a) Apply the factorization theorem directly to the log-normal density to verify that $T(\mathbf{Y}) = \prod_{i=1}^n Y_i$ is sufficient for θ, based on observation $\mathbf{Y} = (Y_1,\ldots,Y_n)$. Then verify that $T(\mathbf{Y})$ is minimal sufficient.

(b) The result of Part (a) seems reasonable, since $T(\mathbf{Y})$ is a 1-1 transformation of $\sum_{i=1}^n X_i$, which is minimal sufficient for θ *wrt* the parametric model $p_\theta(\mathbf{x})$ for $\mathbf{X}$. Suppose we are given parametric family p_θ for observation $\mathbf{X} = (X_1,\ldots,X_n) \in \mathbb{R}^n$ from Lebesgue measurable space $\mathcal{X}$. Suppose $g : \mathbb{R} \to \mathbb{R}$ is a differentiable injective transformation. Set $\mathbf{Y} = (Y_1,\ldots,Y_n) = (g(X_1),\ldots,g(X_n))$, and let p^*_θ be the density of $\mathbf{Y}$ when p_θ is the density of $\mathbf{X}$. Prove that if $S(X_1,\ldots,X_n)$ is sufficient for θ *wrt* parametric family p_θ, then $S^*(Y_1,\ldots,Y_n) = S(g^{-1}(Y_1),\ldots,g^{-1}(Y_n))$ is sufficient for θ *wrt* parametric family p^*_θ. In addition, if S is minimal sufficient, so is S^*. **HINT:** The technical problem is to verify that the transformation $\mathbf{X} \mapsto \mathbf{Y}$ preserves the lack of dependence of the relevant conditional density on θ. This may be easiest to do using the factorization theorem, exploiting the fact that the Jacobian of the transformation $\mathbf{X} \mapsto \mathbf{Y}$ does not depend on θ.

Problem 3.12 The T distribution with $\nu > 0$ degrees of freedom has density

$$f(x) = \frac{\Gamma((\nu+1)/2)}{\sqrt{\nu\pi}\,\Gamma(\nu/2)} \left[1 + \frac{x^2}{\nu}\right]^{-(\nu+1)/2}, \quad x \in \mathbb{R}$$

Let $p_0(\mathbf{x})$ be the density of an *iid* sample of size n from density $f(x)$. Then construct a location invariant model $p_\theta(\mathbf{x}) = p_0(\mathbf{x} - \theta)$, $\theta \in \Theta = \mathbb{R}$. Assume ν is known. Show that the order statistics are minimal sufficient for θ. **HINT:** See Example 3.23.

Problem 3.13 A random variable X is defined on sample space $\mathcal{X} = \{1, 2, 3\}$. Consider the following two parametric families for X. Determine for each whether or not X is a complete statistic.

	$P(X=1)$	$P(X=2)$	$P(X=3)$	
Family 1	$1/2 + p$	p^2	$1/2 - p - p^2$	$0 < p < 1/4$
Family 2	$p + p^2/2$	$p^2/2$	$1 - p - p^2$	$0 < p < 1/4$

Problem 3.14 Suppose $X \sim N(\mu, \sigma^2)$. We take the parameter to be $\theta = (\mu, \sigma^2)$, but for this particular application the parameter space is limited to two densities, say, $\Theta = \{(\mu_0, 1), (-\mu_0, 1)\}$. Is X complete for Θ?

Problem 3.15 Suppose parametric family $\mathcal{P}$ on $\mathcal{X}$ is of constant support. Let Θ_0 be a finite parameter space $\Theta = \{\theta_0, \theta_1, \ldots, \theta_m\}$. Define statistic

$$T(\tilde{x}) = \left(\frac{p_{\theta_1}(\tilde{x})}{p_{\theta_0}(\tilde{x})}, \ldots, \frac{p_{\theta_m}(\tilde{x})}{p_{\theta_0}(\tilde{x})}\right).$$

(a) Prove that $T(\tilde{x})$ is sufficient. **HINT:** Note that for any fixed θ', $h(\tilde{x}) = p_{\theta'}(\tilde{x})$ is a function on $\mathcal{X}$ which *dnd* θ.
(b) Prove that $T(\tilde{x})$ is minimal sufficient.
(c) Suppose now that $\Theta_0 \subset \Theta$. Prove that if $T(\tilde{x})$ is sufficient for Θ, it is also minimal sufficient for Θ. **HINT:** Use Theorem 3.18.

Problem 3.16 Let (X_i, Y_i), $i = 1, \ldots, n$ be an *iid* sample from a bivariate normal distribution with $E[X_i] = E[Y_i] = 0$ and $\text{var}[X_i] = \text{var}[Y_i] = 1$. That is, the only unknown parameter is the correlation coefficient $\rho = E[X_i Y_i]$.

(a) Derive the minimal sufficient statistic for ρ.
(b) Is this statistic complete?

Problem 3.17 Suppose X is a single normal observation from $N(\mu, 1)$. Express the density of X in exponential family form using the canonical parametrization. Use this form to derive the cumulant generating function (CGF) of the natural sufficient statistic for μ. Then suppose X is a single normal observation from $N(0, \sigma^2)$ Again, use the canonical parametrization of the exponential family form to derive the cumulant generating function of the natural sufficient statistic for σ^2.

Problem 3.18 Let $\mathbf{X} = (X_1, \ldots, X_n)$ and $\mathbf{Y} = (Y_1, \ldots, Y_n)$ be two sequences of *iid* random variables such that X_1 is uniformly distributed on $(0, \tau)$, $\tau > 0$, and $P(Y_1 = 1) = 1 - P(Y_1 = -1) = p$, $p \in (0, 1)$. Assume $\mathbf{X}$ and $\mathbf{Y}$ are independent of each other. Suppose we observe $Z_i = X_i Y_i$, $i = 1, \ldots, n$. Find the minimum sufficient statistic for the parameter $\theta = (p, \tau)$ based on $Z_1, \ldots, Z_n$.

Problem 3.19 Let $\mathbf{X} = (X_1, \ldots, X_n)$ and $\mathbf{Y} = (Y_1, \ldots, Y_n)$ be two sequences of *iid* random variables such that $X_1 \sim bern(\theta)$, $Y_1 \sim bern(\theta/2)$, $\theta \in [0, 1]$. Find the minimal sufficient statistic for θ.

Problem 3.20 Suppose we are given sample space $\mathcal{X} = \{1, 2, \ldots\}$, and parameter space $\Theta = \{1, 2, \ldots\}$. Define a family of densities

$$p_\theta(x) = \begin{cases} 1/\theta; & x = 1, \ldots, \theta \\ 0; & x > \theta \end{cases}.$$

We have one observation $X \in \mathcal{X}$ from some p_θ.

(a) Show that X is complete for Θ.
(b) Consider the same density class $p_\theta(x)$ for sample space, $\mathcal{X} = \{1, 2, \ldots\}$, but with reduced sample subspace $\Theta_0 = \{2, 4, 6, \ldots\}$ (i.e. θ is a positive even integer). Is X complete for Θ_0? Justify your answer.

Problem 3.21 We are given an *iid* sample $\mathbf{X} = (X_1, \ldots, X_n)$ from a normal distribution $X_1 \sim N(\theta, c\theta)$, where $c > 0$ is a known constant, and $\theta \in \Theta = (0, \infty)$ (that is, the mean and variance of X_1 are θ and $c\theta$, respectively).

(a) Show that $T(\mathbf{X}) = \sum_{i=1}^n X_i^2$ is a complete sufficient statistic for θ.
(b) Calculate the Fisher information for θ.

Problem 3.22 Suppose $\mathbf{X} = (X_1, \ldots, X_n)$ is an *iid* sample from the double exponential density $DE(\mu, \tau)$.

(a) Is the location-scale parametric class $DE(\mu, \tau)$ an exponential family model?
(b) Suppose we consider μ to be fixed. Show that the parametric family defined by $DE(\mu, \tau)$ on parameter space $\tau \in (0, \infty)$ is an exponential family.
(c) Suppose we now have $p_{\mu,\tau}(\mathbf{x})$, the density of $\mathbf{X}$. Show that $T(\mathbf{X}) = \sum_{i=1}^n |X_i - \mu|$ is a minimal sufficient statistic for τ, when μ is known.
(d) Find the mean and variance of $T(\mathbf{X})/n$ by expressing $p_{\mu,\tau}(\mathbf{x})$ in its canonical parametrization (again, assuming μ is known).

Problem 3.23 Let $X_1, \ldots, X_n$ be an *iid* sample with $X_1 \sim unif(0, \theta)$. Prove that $X_{(n)}$ is complete *wrt* θ. **HINT:** See Section 3.8.3.

Problem 3.24 Suppose $\mathcal{P}$ is a parametric family for observation X with finite support $\mathcal{X} = \{1, \ldots, m\}$. Let Θ be the parameter space.

(a) Prove that if $|\Theta| < m$ then X is not complete for Θ. **HINT:** See Example 3.28.
(b) Suppose $|\Theta| \geq m$. Select $\theta_j \in \Theta$, $j = 1, \ldots, m$, and define $\mathbf{A} \in \mathcal{M}_n$ with elements $\{\mathbf{A}\}_{i,j} = P_{\theta_i}(X = j)$. Prove that if there is any selection θ_j for which $\mathbf{A}$ is invertible, then X is complete for Θ. **HINT:** Use Theorems 3.18 and 3.20.

Problem 3.25 Let $\mathbf{X} = (X_1, \ldots, X_n)$ be an *iid* sample from density $X_1 \sim N(\mu, \sigma^2)$. Let $\mathbf{A}$ be an $n \times n$ matrix, and interpret $\mathbf{X}$ as an n-dimensional column vector.

(a) Show that if $\mathbf{A1}_n = \mathbf{0}$, where $\mathbf{1}_n$ is an n-dimensional column vector with each element equal to 1, and $\mathbf{0}$ is a vector of zeros, then the quadratic form $Q = \mathbf{X}^T\mathbf{A}\mathbf{X}$ is independent of sample mean $\bar{\mathbf{X}} = n^{-1}\sum_i X_i$.
(b) Give an $n \times n$ matrix $\mathbf{A}$ for which $S^2 = \mathbf{X}^T\mathbf{A}\mathbf{X}$, where $S^2 = (n-1)^{-1}\sum_i (X_i - \bar{X})^2$ is the conventional sample variance. Does $\mathbf{A}$ satisfy the condition $\mathbf{A1}_n = \mathbf{0}$?

Problem 3.26 Suppose $X_1, \ldots, X_n$ is a sample of independent random variables in which $X_i \sim pois(c_i\lambda)$, where λ is unknown and constants $c_1, \ldots, c_n$ are known.

(a) Write the model in exponential family form.
(b) What is the minimal sufficient statistic?

(c) Rewrite the model in the canonical parametrization.
(d) Derive the moment generating function of the minimal sufficient statistic.

Problem 3.27 For each of the following Bayesian models, show that the prior density $\pi(\theta)$ for parameter θ is conjugate for that model. Give the exact form of the posterior density $\pi(\theta \mid x)$. Assume only θ is unknown.

(a) $X \sim pois(\theta)$, $\pi(\theta) \sim gamma(\tau_0, \alpha_0)$.
(b) $X \sim nb(r, \theta)$, $\pi(\theta) \sim beta(\alpha_0, \beta_0)$.
(c) $X \sim N(\mu, \theta)$, $\pi(\theta) \sim igamma(\xi_0, \alpha_0)$.
(d) $X \sim weibull(\theta^{1/\alpha}, \alpha)$, $\pi(\theta) \sim igamma(\xi_0, \alpha_0)$.
(e) $X \sim pareto(\tau, \theta)$, $\pi(\theta) \sim gamma(\alpha_0, \beta_0)$.

Problem 3.28 Determine the Jeffreys prior of θ for each of the following models.

(a) $X \sim bin(n, \theta)$.
(b) $X \sim pois(\theta)$.
(c) $\theta = (\tau, \alpha)$, where $X \sim gamma(\tau, \alpha)$.

Problem 3.29 Suppose, conditional on parameter $\theta \in (0, 1)$, a random variable X has distribution $X \sim bin(n, \theta)$. Then suppose we assign a prior distribution $\pi(\theta)$ to θ of the form $\pi(1/4) = \pi(1/2) = \pi(3/4) = 1/3$. If $n = 10$, and we observe $X = 4$, give the posterior distribution $\pi(\theta \mid X = 4)$ of θ.

Problem 3.30 Verify the claim made in Section 3.13.1. Suppose $T(\tilde{X})$ is a sufficient statistic for $\theta \in \Theta$. Suppose an estimator $\delta(\tilde{X}) \in \mathbb{R}$ has finite variance for all θ. Then

(i) $\delta^*(\tilde{X}) = E_\theta[\delta(\tilde{X}) \mid T(\tilde{X})]$ is also a statistic.
(ii) $E_\theta[\delta^*(\tilde{X})] = E_\theta[\delta(\tilde{X})]$ for all $\theta \in \Theta$.
(iii) $\text{var}_\theta[\delta^*(\tilde{X})] \leq \text{var}_\theta[\delta(\tilde{X})]$ for all $\theta \in \Theta$, with equality only if $T(\tilde{X}) \mapsto \delta(\tilde{X})$.

HINT: Use the total expectation and total variance identities (Section 1.3.1). This is a version of the Rao-Blackwell theorem.

4

Methods of Estimation

4.1 Introduction

Suppose we are given observation $\tilde{X} \in \mathcal{X}$ from a density from a family $\mathcal{P}$. One of the central problems of statistical inference is the development of an estimator $\delta(\tilde{X})$ for $\eta(\theta)$ (the estimand). In this chapter, we introduce principles by which estimators can be derived. Of course, we would like to use the "best" estimator, the one which is closest to the estimand in some sense. However, because of the nature of random error, the problem of defining 'best' is a subtle one. This will be the topic of Chapter 7, which offers a decision theoretic formulation of the problem.

In this chapter, we also introduce the confidence interval (or confidence set more generally) (Section 4.6). A point estimate $\delta(\tilde{X})$ is intended to be a single value which is close to the true value of the estimand on average, or with high probability. In contrast, a confidence interval is a set of values, constructed from $\tilde{X}$, which contains the true value of the estimand with high probability. A confidence interval is often constructed from a point estimate and a margin of error ME, with end points defined by $\delta(\tilde{X}) \pm ME$. However, strictly speaking, the definition of a confidence interval contains no reference to an estimator, and may be viewed as more related to hypothesis testing than to estimation. As a crude rule of thumb, if the object is knowledge of a parameter's value, then the confidence interval or hypothesis test may be more relevant. If the object is to use an estimate in place of a true estimand in a subsequent application, then a point estimate, chosen using decision theoretic criterion, would be the more relevant approach.

However, before we consider the confidence interval, we review a number of methods and principles commonly used to obtain point estimates. It is worth noting at the very beginning that the sufficiency principle (Section 3.13.1) is quite relevant to the problem of estimation, so that where possible we expect estimators to depend on $\tilde{X}$ only through a minimal sufficient statistic $T(\tilde{X})$.

In Section 4.2, we introduce briefly the idea of unbiasedness (the expected value of an unbiased estimate equals its estimand). This is an important property, but by no means do we always expect estimators to be unbiased. First of all, in many important cases an unbiased estimator does not exist. Even when they do, they need not be the best choice. This will become clear in Chapter 7.

There will always be problems for which the derivation of *any* reasonable estimator would be a challenge. In such cases, the method of moments estimator provides a workable heuristic option (Section 4.3). Example 4.4 and Problem 4.8 concern the species sampling problem, which is an important example of this type (a somewhat less heuristic treatment of the problem is offered later in Section 9.5.3). We can identify some cases for which the method of moments estimator is equivalent to those derived by other principles (see, for example, Theorems 4.3–4.4).

Section 4.4 briefly covers sample quantiles. Our most important result concerning this topic will be the Bahadur representation theorem (Section 12.6) which gives an approximate large sample distribution. It seems reasonable to use, for example, a sample 0.75-quantile to

DOI: 10.1201/9781003049340-4

estimate a population 0.75-quantile. However, in the context of parametric inference, this would generally be inefficient. The extended example in Section 12.8.2 will make this clear. Of course, if the parametric model is not correctly identified, the more efficient procedure may then become a poor choice, while the sample quantile retains its advantage of being at least a reasonable choice under any distributional assumptions.

Section 4.5 on maximum likelihood estimation is necessarily a central topic in any volume on statistical inference. However, in this chapter we confine attention to its definition and evaluation, deferring any discussion of its properties as an inference method to later chapters. Certainly, a discussion confined to these issues is quite warranted. To see this, we look ahead to the definition of the maximum likelihood estimate (MLE), given by

$$\hat{\theta}_{MLE} = \text{argmax}_{\theta \in \Theta} \mathcal{L}(\theta; \tilde{X}),$$

where $\mathcal{L}(\theta; \tilde{X})$ is the likelihood function defined in Section 3.2.1. Section 4.5 largely considers the problem of maximizing $\mathcal{L}(\theta; \tilde{X})$. Much of the theory of likelihood estimation depends on the assumption that $\hat{\theta}_{MLE}$ is a stationary point, evaluated as a solution to a system of equations. But this assumption does not always hold. Figure 4.2 of Example 4.10 plots a likelihood function with respect to $\theta \in \mathbb{R}$ for which multiple solutions exist. In Example 4.7, $\hat{\theta}_{MLE}$ cannot be a stationary point. In contrast, Example 4.9 offers an example in which $\hat{\theta}_{MLE}$ may or may not be a stationary point, depending on the observed data $\tilde{X} \in \mathcal{X}$.

Our discussion of confidence sets begins in Section 4.6. We review a number of different methods, including pivots and CDF inversion. The question of optimal coverage of a confidence interval is discussed. The CDF inversion method includes the well-known Clopper-Pearson interval for the binomial parameter (Clopper and Pearson, 1934). This method is much easier to implement using the "beta-quantile" method proposed in that paper, which we give in a more general form.

Section 4.7 offers a brief look ahead to a theme which will recur throughout this volume. A shrinkage estimate can here be taken to be a convex combination of estimates. One of these is usually unbiased, while the other may be a constant (note that theory of inference allows estimators that are constants). In this case, the estimator 'shrinks' towards the constant, which is often (but need not be) zero. There are quite practical reasons for considering shrinkage estimators, as in Example 4.18 or Problem 6.36. However, shrinkage estimators also have important theoretical implications. First of all, they usually violate the invariance principle of Section 3.13.3. In addition, we will see that many commonly used estimators are 'inadmissible', in the sense that strictly better estimators exist, and those are usually shrinkage estimators (for example, the James-Stein estimator of Example 7.2). This topic will be covered in Chapter 7, however, it is worth alerting the reader to the issue at this stage.

The chapter finishes with an introduction to Bayesian estimation and credible regions (Section 4.8). In the examples given, a Bayesian estimate is the mean of a posterior distribution (Section 3.10). A more formal treatment of this topic is given in Sections 7.9–7.10.

4.2 Unbiased Estimators

A natural place to begin a discussion of estimation is with the notion of unbiasedness. We start with the following definition.

Definition 4.1 (Unbiased Estimator) Let $\delta(\tilde{X})$ be a statistic for estimand $\eta(\theta)$. The bias of $\delta(\tilde{X})$ is defined as

$$\text{bias}_\theta[\delta(\tilde{X}); \eta(\theta)] = E_\theta[\delta(\tilde{X})] - \eta(\theta).$$

If $\text{bias}_\theta[\delta(\tilde{X}); \eta(\theta)] = 0$ for all θ, then $\delta(\tilde{X})$ is an unbiased estimator. ∎

Everything else being equal, it seems preferable that an estimator be unbiased. However, in many important cases an unbiased estimator may not exist. Accordingly, we make this condition precise.

Definition 4.2 (U-Estimability) An estimand $\eta(\theta)$ is U-estimable if there exists statistic $\delta(\tilde{X})$ such that $E_\theta[\delta(\tilde{X})] = \eta(\theta)$ for all $\theta \in \Theta$. ∎

A lack of U-estimabilty exists in many commonly used models, as shown in the following example.

Example 4.1 (U-Estimability and the Binomial Distribution) Suppose $X \in bin(n, \theta)$. Any function g of X only needs to be defined for $x = 0, 1, \ldots, n$. Accordingly, set $g(i) = g_i$ for $n + 1$ suitable constants. Then

$$E_\theta[g(X)] = \sum_{i=0}^{n} g_i \binom{n}{i} \theta^i (1-\theta)^{n-i} = \sum_{i=0}^{n} a_i \theta^i. \tag{4.1}$$

Then any U-estimable estimand $\eta(\theta)$ must be a polynomial of at most order n. It may also be verified that any polynomial in θ of at most order n is U-estimable (Problem 4.1). Thus, the odds $\eta(\theta) = \theta/(1-\theta)$, or the log-odds $\eta(\theta) = \log(\theta/(1-\theta))$ are not U-estimable, since thay cannot be expressed as polynomials. ∎

We can also identify models for which U-estimabilty always exists, as in the following example.

Example 4.2 (U-estimability of the Location Parameter) Suppose θ is a location parameter and $\delta(\mathbf{X})$ is a location equivariant estimator (Definition 3.3) for which $E_\theta[|\delta(\mathbf{X})|] < \infty$. By definition $\delta(\mathbf{X}) - \theta = \delta(\mathbf{X} - \theta)$. By Theorem 3.2 the distribution of $\delta(\mathbf{X} - \theta)$ *dnd* θ. This means $E_\theta[\delta(\mathbf{X}) - \theta] = c$ where c *dnd* θ. Therefore $\delta(\mathbf{X}) - c$ is an unbiased estimator of θ, which is then U-estimable. ∎

It may also be shown that any scale parameter is U-estimable, following Example 4.2 (Problem 4.2).

4.3 Method of Moments Estimators

The method of moments is a heuristic principle sometimes used to construct estimators. Of course, estimators derived from other principles will sometimes be method of moments estimators (Theorems 4.3–4.4). Otherwise, the approach is generally applied to nonstandard problems for which application of other methods is not feasible, or otherwise poses some significant challenge. The species sampling problem of Example 4.4 below is such a problem.

Suppose $\Theta \in \mathbb{R}^m$ and we are given statistic $\mathbf{T}(\tilde{X}) = (T_1(\tilde{X}), \ldots, T_m(\tilde{X}))$. Then set

$$\eta_j(\theta) = E_\theta[T_j(X)], \; j = 1, \ldots, m.$$

This generates a system of equations

$$\eta_j(\theta) = T_j(X), \;\; j = 1, \ldots, m. \tag{4.2}$$

A solution to Equation (4.2) is a method of moments estimator There is no guarantee the solution exists, or is unique if it does. But where practical, it often yields a reasonable estimation method, particularly where $\mathbf{T}$ is complete or minimal sufficient.

Example 4.3 (Method of Moments Estimator for Gamma Parameters) The mean and variance of a gamma distribution $gamma(\tau, \alpha)$ are $\mu = \alpha\tau$, $\sigma^2 = \alpha\tau^2$. If we have sample mean and variance $\bar{X}$, S^2 from an *iid* sample of size n, we have approximation $\bar{X} \approx \alpha\tau$, $S^2 \approx \alpha\tau^2$. Solving for α and τ gives $\alpha \approx \bar{X}^2/S^2$, $\tau \approx S^2/\bar{X}$. However, these estimates violate the sufficiency principle, since from Theorem 3.22 $\mathbf{T} = (\sum_i X_i, \sum_i \log(X_i))$ is a complete sufficient statistic for (τ, α). ■

We next consider the species sampling problem.

Example 4.4 (The Species Sampling Problem (The Probability of Unseen Support)) This problem was considered in Good (1953). Let $\mathbf{P} = (p_1, \ldots, p_\nu)$ be a probability distribution on the set of categories $(1, \ldots, \nu)$. Let $X_1, \ldots, X_n$ be an *iid* sequence of observations from $\mathbf{P}$. There may be interest in the quantity

$$Z_n = \sum_{j=1}^{\nu} p_j I\{N_j(n) = 0\}, \;\; n \geq 1,$$

where $N_j(n) = \sum_{i=1}^{n} I\{X_i = j\}$ is the frequency of category j among the first n observations. The quantity Z_n can be interpreted as the probability that the next (the $n + 1$st) observation is a category unseen in the previous n observations (or the probability of the unseen support of $\mathbf{P}$). Let

$$\alpha_n = E[Z_n] = \sum_{j=1}^{\nu} p_j(1 - p_j)^n, \;\; n \geq 1. \tag{4.3}$$

We first note that α_n is not U-estimable based on $X_1, \ldots, X_n$, but α_{n-1} is, so we content ourselves with the related problem of estimating the latter estimand (compare to Example 4.1).

Then note that the terms of the sum in Equation (4.3) are related to the probability that a binomial random variable $X \sim bin(n, p)$ equals 1, that is, $P(X = 1) = np(1-p)^{n-1}$. Then let $a_1(n)$ be the number of categories observed exactly once in $X_1, \ldots, X_n$. We have directly, $E[a_1(n)/n] = \alpha_{n-1}$, so $a_1(n)/n$ is an unbiased estimator of α_{n-1}, and a still reasonable estimator of α_n. Finally, it was shown in Robbins (1968) that $E\left[(Z_{n-1} - a_1(n)/n)^2\right] < 1/n$. ■

4.4 Sample Quantiles and Percentiles

Suppose a random variable X has CDF F. A p-quantile of F is any value m for which $P(X \leq m) \geq p$ and $P(X < m) \leq p$ (Definition 1.8). By Theorem 1.14 for any $p \in (0, 1)$ the set $\mathcal{M}_p$ of p-quantiles is a nonempty closed and bounded interval. In particular, $\mathcal{M}_p$ need not be a singleton, but at least one p-quantile always exists. Because a p-quantile may not be unique, it will be useful to make some canonical selection. The quantile function for F, defined as $Q(p; F) = \inf\{x \in \mathbb{R} : F(X \leq x) \geq p\}$ is a common choice (Definition 1.9).

The empirical distribution of a sample is

$$F_n(x) = \frac{1}{n} \sum_{i=1}^{n} I\{X_i \leq x\}. \tag{4.4}$$

The sample quantile function can then be defined as $Q(p; F_n)$, and serves as an estimate of $Q(p; F)$ (the definition of a quantile function applies to the empirical distribution).

However, while the sample quantile function $Q(p; F_n)$ may be useful in theoretical developments, given the nonuniqueness of the p-quantile, a more intuitive definition of a sample quantile might be prefered. For example, the sample median is usually taken to be

$$\hat{m} = \begin{cases} X_{(\frac{n+1}{2})}; & n \text{ is odd} \\ \frac{X_{(\frac{n}{2})} + X_{(\frac{n}{2}+1)}}{2}; & n \text{ is even} \end{cases},$$

where $(X_{(1)}, \ldots, X_{(n)})$ are the order statistics. More generally, for n data points we may define the pairs: $(p_k, X_{(k)})$, $k = 1, \ldots n$, where we take the sample p_k-quantile to be $X_{(k)}$, with any other sample quantile evaluated by linear interpolation. Usually, $p_k = (k+a)/(n+b)$ for some constants a, b. Common choices include $p_k = (k-1)/(n-1)$ or $p_k = k/(n+1)$. The choice depends on the objective of the estimate, and any anticipated properties of F (for example, whether F is discrete or continuous). See Hyndman and Fan (1996) for a comprehensive discussion of this topic. Problem 4.10 gives an example of this approach.

4.5 Maximum Likelihood Estimation

Recall from Section 3.2.1 the likelihood function $\mathcal{L}(\theta; \tilde{X})$. The maximum likelihood estimate (MLE) is defined as

$$\hat{\theta}_{MLE} = \text{argmax}_{\theta \in \Theta} \mathcal{L}(\theta; \tilde{X}), \tag{4.5}$$

when well defined. We may equivalently substitute the log-likelihood function $\ell(\theta; \tilde{X})$ for $\mathcal{L}(\theta; \tilde{X})$ in (4.5). The definition is unchanged, but in practice computational and distributional methods tend to be simplified.

It is easily verified that maximum likelihood estimation satisfies the sufficiency principle (Section 3.13.1).

Theorem 4.1 (The MLE and the Sufficiency Principle) The MLE of Equation (4.5), when well defined, is a mapping of any sufficient statistic. ■

Proof. The theorem follows directly from the factorization theorem (Theorem 3.16). □

One straightforward fact is worth singling out as a separate theorem.

Theorem 4.2 (Invariance of the MLE to Reparametrization) The problem of evaluating the MLE of Equation (4.5) is invariant to a bijective reparametrization of Θ. ■

Proof. The argmax operation defining Equation (4.5) is invariant to bijective reparametrization. □

The MLE can often, but not always, be obtained by identifying a stationary point of $\ell(\theta; \tilde{X})$ via differentiation, as shown in the next example.

Example 4.5 (MLE for the Binomial Parameter) If $X \sim bin(n, \theta)$, we have log-likelihood $\ell(\theta, x) = x\log(\theta) + (n-x)\log(1-\theta)$. The first and second derivatives are

$$\frac{d\ell(\theta,x)}{d\theta} = \frac{x}{\theta} - \frac{n-x}{1-\theta}, \quad \frac{d^2\ell(\theta,x)}{d\theta^2} = -\frac{x}{\theta^2} - \frac{n-x}{(1-\theta)^2}.$$

The second derivative is strictly negative over $\theta \in (0,1)$, therefore $\ell(\theta, x)$ is strictly concave. Setting the first derivative equal to zero defines a stationary condition which is solved by $\hat{\theta} = x/n$ which is therefore the unique MLE of θ.

We should also note that by Theorem 4.2 the MLE of any estimand $\eta(\theta)$ is obtained by the substitution rule $\hat{\eta} = \eta(\hat{\theta})$. The MLE of $\text{var}_\theta[X] = n\theta(1-\theta)$ is $n\hat{\theta}(1-\hat{\theta})$, and the MLE of the odds $\theta/(1-\theta)$ is $\hat{\theta}/(1-\hat{\theta})$. ■

The method of Example 4.5 extends naturally to a multidimensional parameter $\boldsymbol{\theta} = (\theta_1, \ldots, \theta_m)$. A system of m equations, the likelihood equations, is constructed by setting each partial derivative $\partial\ell(\boldsymbol{\theta}, \tilde{X})/\partial\theta_j$ equal to 0. Of course, it must be verified that any solution to the likelihood equations is a global maximum. We demonstrate the method for a two parameter normal sample.

Example 4.6 (MLE for a Normal Sample) We are given an *iid* sample $\mathbf{X} = (X_1, \ldots, X_n)$, with $X_1 \sim N(\mu, \sigma^2)$. The log-likelihood for $\boldsymbol{\theta} = (\mu, \sigma^2)$ is

$$\ell(\boldsymbol{\theta}, \mathbf{x}) = -\sum_i (x_i - \mu)^2 / \left(2\sigma^2\right) - n\log\sigma^2/2.$$

The likelihood equations are then:

$$\begin{aligned} \partial\ell(\boldsymbol{\theta}, \mathbf{x})/\partial\mu &= \sum_i (x_i - \mu)/\sigma^2 = 0, \\ \partial\ell(\boldsymbol{\theta}, \mathbf{x})/\partial\sigma^2 &= \sum_i (x_i - \mu)/2(\sigma^2)^2 - n/\left(2\sigma^2\right) = 0. \end{aligned} \tag{4.6}$$

From the first equation of Equation (4.6) we have directly $\hat{\mu} = \bar{x}$. After substituting this solution into the second equation we have

$$\sum_i (x_i - \bar{x})/2(\sigma^2)^2 - n/\left(2\sigma^2\right) = 0.$$

which yields solution $\hat{\sigma}^2 = n^{-1}\sum_i (x_i - \bar{x})^2$. Note that the MLE for σ^2 is *not* the sample variance $S^2 = (n-1)^{-1}\sum_i (x_i - \bar{x})^2$. The two parameter density for an *iid* normal sample is a full-rank exponential family model. It will be shown in Theorem 4.4 for this class that any solution to the likelihood equations must be a global maximum. Clearly, the likelihood equation (4.6) always has a unique solution, which must therefore be the unique MLE.

Next, suppose μ is known. We now only have one likelihood equation:

$$d\ell(\boldsymbol{\theta}, \mathbf{x})/d\sigma^2 = \sum_i (x_i - \mu)/\left(2(\sigma^2)^2\right) - n/\left(2\sigma^2\right) = 0,$$

which leads to MLE $\hat{\sigma}^2 = n^{-1}\sum_i (x_i - \mu)^2$. Note that the MLE for μ, $\hat{\mu} = \bar{x}$, remains the same whether or not σ^2 is known. ■

Of course, we do not always evaluate an MLE using differentiation. An example of this case is given next.

Example 4.7 (MLE for Uniform Density Family with Location or Scale Parameter) Suppose we have an *iid* sample $X_1, \ldots, X_n$, $X_1 \sim unif(\theta, \theta+1)$, for location parameter $\theta \in \Theta = \mathbb{R}$. The likelihood function is $\mathcal{L}(\theta; x) = I\{\theta < x_{(1)} < x_{(n)} < \theta + 1\}$. A plot is shown in Figure 4.1 (left plot). An MLE is any value $\hat{\theta}_{MLE} \in \left(x_{(n)} - 1, x_{(1)}\right)$. Thus the MLE exists but is not unique.

Then suppose we have an *iid* sample $X_1, \ldots, X_n$, $X_1 \sim unif(0, \theta)$, for scale parameter $\theta \in \Theta = (0, \infty)$. The likelihood function is $\mathcal{L}(\theta; X) = \theta^{-n} I\{x_{(n)} \leq \theta\}$. A plot is shown in Figure 4.1 (right plot). The MLE is $\hat{\theta}_{MLE} = x_{(n)}$. Thus the MLE exists and is unique. See also Problem 4.29. ■

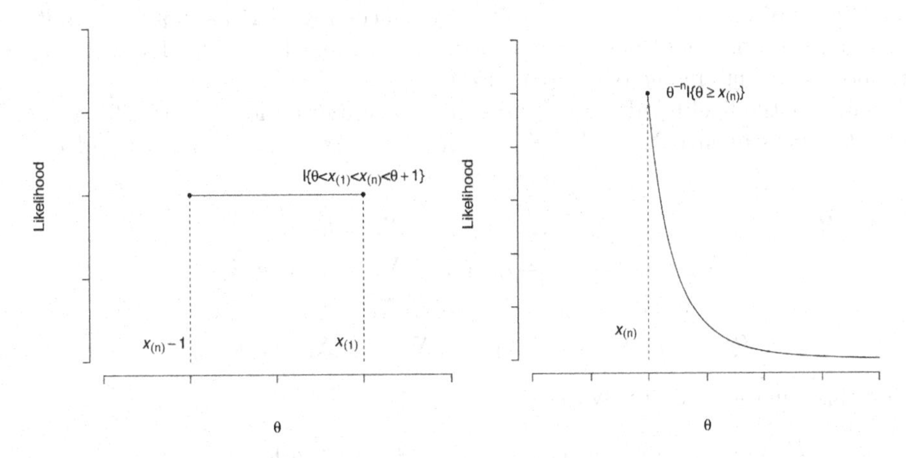

FIGURE 4.1
Likelihood functions for *iid* sample from a uniform density: location parameter model (left) and scale parameter models (right) Example 4.7).

There are, of course, a few technical issues underlying the definition of the MLE (Equation (4.5)). First, there is no guarantee that a maximum is unique (Example 4.7), or even exists. For some important cases regularity conditions under which a unique MLE exists can be derived (see, for example, Section 4.5.1). However, verifying the existence of a well defined MLE will often be a significant problem of its own, and will be one of the central issues considered in Chapter 13.

The next two examples together give an interesting hybrid case in which the method used to evaluate the MLE must depend on the observed data.

Example 4.8 (MLE for Markov Chains) The Markov chain is a discrete time stochastic process. The defining property is commonly known as the memoryless property, also the Markovian property. Suppose we are given a discrete time stochastic process $X_n \in \mathcal{X}$, $n = 0, 1, 2, \ldots$, which assumes values in a discrete state space $\mathcal{X}$. Without loss of generality we have either a finite state space $\mathcal{X} = \{0, 1, \ldots, m\}$ or countable state space $\mathcal{X} = \{0, 1, \ldots\}$. Then X_i is a Markov chain if the following memoryless property holds:

$$\begin{aligned} &P(X_{n+1} = j \mid X_n = i, X_{n-1} = i_{n-1}, \ldots, X_1 = i_1, X_0 = i_0) \\ &\quad = P(X_{n+1} = j \mid X_n = i) = P_{ij}. \end{aligned} \tag{4.7}$$

The quantity P_{ij} is called the transition probability from state i to state j. We also have transition probability matrix

$$\mathbf{P} = \begin{bmatrix} P_{00} & P_{01} & P_{02} & \cdots \\ P_{10} & P_{11} & P_{12} & \cdots \\ \vdots & \vdots & \vdots & \\ P_{i0} & P_{i1} & P_{i2} & \cdots \\ \vdots & \vdots & \vdots & \end{bmatrix}.$$

Row i of transition matrix $\mathbf{P}$ is equivalent to the conditional probability $P(X_{n+1} = j \mid X_n = i) = P_{ij}$, $j \in \mathcal{X}$. Note also that $\mathbf{P}$ will be a matrix of infinite dimension when $\mathcal{X}$ is countably infinite. We also have no difficulty conceiving of $\mathbf{P}$ as 'doubly infinite' when the state space is the set of positive and negative integers $\{\ldots, -2, -1, 0, 1, 2, \ldots\}$, which requires no important change of the definition.

To construct the likelihood, we must specify a distribution for X_0, say $P_0(x) = P(X_0 = x)$. Consider the sequence $\mathbf{X} = (X_0, X_1, \ldots, X_n) = (i_0, i_1, \ldots, i_n)$. Then by (4.7) we may write

$$\begin{aligned} &P(X_n = i_n, X_{n-1} = i_{n-1}, \ldots, X_1 = i_1, X_0 = i_0) \\ &\quad = P(X_n = i_n \mid X_{n-1} = i_{n-1}, \ldots, X_1 = i_1, X_0 = i_0) \\ &\qquad \times P(X_{n-1} = i_{n-1}, \ldots, X_1 = i_1, X_0 = i_0) \\ &\quad = P_{i_{n-1}, i_n} P(X_{n-1} = i_{n-1}, \ldots, X_1 = i_1, X_0 = i_0). \end{aligned}$$

Applying this equation iteratively gives

$$\begin{aligned} &P(X_n = i_n, X_{n-1} = i_{n-1}, \ldots, X_1 = i_1, X_0 = i_0) \\ &= P_0(i_0) \times \prod_{k=1}^{n} P_{i_{k-1}, i_k} \\ &= P_0(i_0) \times \prod_{(i,j) \in \mathcal{X} \times \mathcal{X}} P_{ij}^{N_{ij}} \\ &= \mathcal{L}(\mathbf{P}, P_0; \mathbf{X}) \end{aligned}$$

where N_{ij} is the observed number of transitions from states i to j. This yields directly the likelihood as a function of $\mathbf{P}$, P_0, based on the frequencies N_{ij}. ∎

The following example of an MLE evaluation continues from Example 4.8.

Example 4.9 (MLE for Markov Chains - Example) A stochastic process $X(t)$, $t \in [0, \infty)$, assumes one of two states 0 or 1. The switching times form a Poisson process on $[0, \infty)$ of rate λ (Example 1.10). Assume $X(0) = 0$ with probability 1/2. The process is monitored at equally spaced intervals to produce observations

$$X_i = X(\Delta i), \; i = 0, \ldots, n$$

for some known positive constant Δ. The objective is to estimate λ.

By the memoryless property of the Poisson process X_i is a Markov chain. The number of transitions in a time interval of length Δ is a Poisson random variable W with mean $\Delta\lambda$. This means

$$\begin{aligned} &P\left(X_n = X_{n-1} \mid X_{n-1} = i_{n-1}, \ldots, X_1 = i_1, X_0 = i_0\right) \\ &= P\left(X_n = X_{n-1} \mid X_{n-1} = i_{n-1}\right) = P(W \text{ is even}) = \sum_{j=0}^{\infty} \frac{(\Delta\lambda)^{2i}}{(2i)!} e^{-\Delta\lambda}. \end{aligned}$$

Recall the expansion $e^x = 1/0! + x/1! + x^2/2! + \ldots$. Then

$$e^{\Delta\lambda} + e^{-\Delta\lambda} = 2\sum_{j=0}^{\infty} \frac{(\Delta\lambda)^{2i}}{(2i)!} = 2P(W \text{ is even})e^{\Delta\lambda},$$

using the preceding identity. This gives

$$P(X_n = X_{n-1} \mid X_{n-1} = i_{n-1}) = P(X_n = X_{n-1}) = \frac{1 + \exp(-2\lambda\Delta)}{2}.$$

The probability transition matrix is therefore

$$\mathbf{P} = \begin{bmatrix} (1 + \exp(-2\lambda\Delta))/2 & (1 - \exp(-2\lambda\Delta))/2 \\ (1 - \exp(-2\lambda\Delta))/2 & (1 + \exp(-2\lambda\Delta))/2 \end{bmatrix}.$$

Then the number of transitions from 0 to 0 or from 1 to 1 is $N_e = \sum_{i=1}^n I\{X_i = X_{i-1}\}$, so that the likelihood for λ is

$$\mathcal{L}(\lambda; N_e) = \left[\frac{1 + \exp(-2\lambda\Delta)}{2}\right]^{N_e} \left[\frac{1 - \exp(-2\lambda\Delta)}{2}\right]^{n-N_e},$$

and the log-likelihood is

$$\ell(\lambda; N_e) = N_e \log[1 + \exp(-2\lambda\Delta)] + (n - N_e)\log[1 - \exp(-2\lambda\Delta)], \tag{4.8}$$

after eliminating constants that do not depend on λ. Note that the likelihood does not depend on P_0 or the initial state $X_0 = i_0$.

Conditions for a stationary point for (4.8) are easily derived as

$$\frac{N_e}{1 + \exp(-2\lambda\Delta)} - \frac{n - N_e}{1 - \exp(-2\lambda\Delta)} = 0$$

or equivalently,

$$N_e/n = [1 + \exp(-2\lambda\Delta)]/2. \tag{4.9}$$

It is tempting to think that this resolves the problem, after substituting the observed value of N_e/n into Equation (4.9) then solving for λ (after verifying that the stationary point is a local maximum). However, it can be seen that the right side of (4.9) is strictly larger than 1/2. What if $N_e/n \leq 1/2$? In this case, there is no stationary point, and the log-likelihood can be shown to be strictly increasing for all $\lambda > 0$. In other words, the MLE is

$$\hat{\lambda}_{MLE} = \begin{cases} \frac{-\log\left(2\frac{N_e}{n} - 1\right)}{2\Delta} & ; \quad \frac{N_e}{n} > 1/2 \\ \infty & ; \quad \frac{N_e}{n} \leq 1/2 \end{cases}.$$

This is quite reasonable. As $\lambda \to \infty$, the state transitions become more frequent, and the observations behave increasingly like an *iid* sequence of Bernouilli random variables of mean 1/2. Thus, $\lambda = \infty$ is interpretable as that limit, which is the maximum likelihood estimate for any observation $N_e/n \leq 1/2$. ■

In some cases, an MLE may exist, and be evaluated using likelihood equations. But these need not have a unique solution. An example follows.

Example 4.10 (Maximum Likelihood Function for a Cauchy Sample) Figure 4.2 shows the likelihood function for a Cauchy location parameter density $cauchy(\mu, 1)$, based on the (admittedly contrived) sample $\mathbf{X} = (-20, -15, -10, -5, 5, 10, 15, 20)$. The location of the individual observations are indicated by the '+' symbol. Note that the sample is symmetric about zero, which is therefore the value of the sample mean. Although $\mu = 0$ defines a stationary point, this is a local *minimum*. Of course, the Cauchy distribution does not have a first moment, so it is not surprising that the sample mean seems to play no role in determining the maximum likelihood solution. Recall that the order statistics are minimal sufficient for μ (Example 3.23). Accordingly, the maximum likelihood solution, roughly, favors values of μ near individual observations. ∎

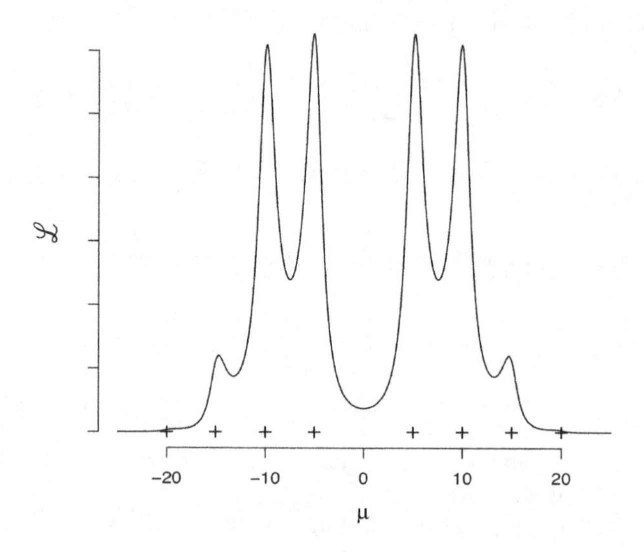

FIGURE 4.2
Likelihood function for a Cauchy density location parameter model based on sample $\mathbf{X} = (-20, -15, -10, -5, 5, 10, 15, 20)$ (Example 4.10). The location of the individual observations are indicated by the '+' symbol.

4.5.1 MLEs for Exponential Family Models

In contrast to Examples 4.9 and 4.10, the problem of evaluating the MLE of the full rank exponential family model is relatively well behaved. It will be most convenient to first consider the canonical parameter $\eta = (\eta_1, \ldots, \eta_m)$, which gives model

$$p_\eta(\tilde{x}) = \exp\left[\sum_{i=1}^m \eta_i T_i(\tilde{x}) - A(\eta)\right] h(\tilde{x}). \tag{4.10}$$

Recall from Section 3.6 the *natural parameter space* $\Xi = \{\eta : A(\eta) < \infty\}$, which is the maximal parameter space for this model, and is also convex. The log-likelihood is then

$$\ell(\eta; \tilde{x}) = \log p_\eta(\tilde{x}) = \sum_{i=1}^m \eta_i T_i(\tilde{x}) - A(\eta). \tag{4.11}$$

The likelihood equations are given by $\Psi(\tilde{x}, \eta) = (\Psi_1(\tilde{x}, \eta), \ldots, \Psi_m(\tilde{x}, \eta)) = \mathbf{0}$, where

$$\Psi_j(\tilde{x}, \eta) = \frac{\partial \log p_\eta(\tilde{x})}{\partial \eta_j} = T_j(\tilde{x}) - \frac{\partial A(\eta)}{\partial \eta_j} = T_j(\tilde{x}) - E_\eta[T_j(\tilde{X})], \tag{4.12}$$

$j = 1, \ldots, m$, where $\tilde{X} \sim p_\eta$. The MLE can be precisely characterized for this case.

Theorem 4.3 (MLE for Full-Rank Canonical Exponential Family Model) An MLE $\hat{\eta}_{mle} = \text{argmax}_{\eta \in \Xi} \ell(\eta; \tilde{X})$ exists for the full-rank canonical exponential family model if and only if there exists η^* such that

$$\mathbf{T}(\tilde{X}) = E_{\eta^*}[\mathbf{T}(\tilde{X})], \tag{4.13}$$

$\mathbf{T}(\tilde{X}) = (T_1(\tilde{X}), \ldots, T_m(\tilde{X}))$. In this case $\hat{\eta}_{mle} = \eta^*$. ■

Proof. The Hessian matrix (Section C.7) of $\ell(\eta; \tilde{x})$ exists and is given by $-A''(\eta)$ (Equation (4.11)). However, by Theorem 3.13, $A''(\eta)$ is the covariance matrix of $\mathbf{T}(\tilde{X})$, which is positive definite for the full-rank model. This implies $\ell(\eta; \tilde{x})$ is strictly concave on Ξ, and therefore possesses a global maximum if and only if the log-likelihood equation possesses a solution, in which case this solution is the MLE. □

This result extends to the noncanonical model

$$p_\theta(\tilde{x}) = \exp\left[\sum_{i=1}^{m} \eta_i(\theta) T_i(\tilde{x}) - B(\theta)\right] h(\tilde{x}). \tag{4.14}$$

Define the natural parameter space for θ, $\Theta_{nat} = \{\theta \in \mathbb{R}^m : \eta(\theta) \in \Xi\}$. The extension of Theorem 4.3 to Equation (4.14) follows.

Theorem 4.4 (MLE for Full-Rank General Exponential Family Model) Suppose Equation (4.14) is a full-rank exponential family model, with $\theta \in \Theta_{nat} \subset \mathbb{R}^m$, and assume $\eta : \Theta_{nat} \to \Xi$ is a bijective mapping. Then an MLE $\hat{\theta}_{mle} = \text{argmax}_{\theta \in \Xi} \ell(\theta; \tilde{X})$ exists if and only if there exists θ^* such that

$$\mathbf{T}(\tilde{X}) = E_{\theta^*}[\mathbf{T}(\tilde{X})], \tag{4.15}$$

$\mathbf{T}(\tilde{X}) = (T_1(\tilde{X}), \ldots, T_m(\tilde{X}))$. In this case $\hat{\theta}_{mle} = \theta^*$. ■

Proof. The theorem follows from Theorem 4.3. Note that $\eta(\theta)$ is bijective. Let $\theta(\eta) = \eta^{-1}(\eta)$ be the inverse transformation. If $\hat{\eta}$ is the unique solution to Equation (4.13), then $\hat{\theta} = \theta(\hat{\eta})$ will be the unique solution to $\mathbf{T}(\tilde{X}) = E_\theta[\mathbf{T}(\tilde{X})]$. Similarly, if $\hat{\eta}$ uniquely maximizes $\ell(\eta; \tilde{X})$, then $\hat{\theta} = \theta(\hat{\eta})$ will uniquely maximize $\ell(\theta; \tilde{X})$. □

We will use Theorem 4.4 to revisit the method of moments estimator of Example 4.3.

Example 4.11 (Method of Moments Estimator for Gamma Parameters (cont'd)) The natural sufficient statistic for an *iid* sample from $gamma(\tau, \alpha)$ is $\mathbf{T} = (\sum_i X_i, \sum_i \log(X_i))$. An alternative method of moments estimator of (τ, α) is given by the solution to the equations

$$n^{-1} \sum_{i=1}^{n} X_i = E_{(\tau,\alpha)}[X_1], \quad n^{-1} \sum_{i=1}^{n} \log(X_i) = E_{(\tau,\alpha)}[\log(X_1)],$$

wrt (τ, α). By Theorem 4.4 this is also the MLE. ■

One reason that exponential family models are so widely used in statistical modelling is that they may be constructed to match properties expected of a particular application. This is made clear by the idea of the quasi-likelihood function.

Example 4.12 (The Quasi-Likelihood Function) Suppose X has a one-parameter exponential family density $f(x) = \exp(\eta x - A(\eta))h(x)$. Then, following Equation (4.12), we have $\partial\ell(\eta; x)/\partial\eta = x - E_\eta[X]$. Suppose we reparametrize *wrt* $\mu = \mu(\eta) = E_\eta[X]$. This would lead to variance function $V(\mu) = \text{var}_\eta[X]$. It can then be shown that

$$\partial\ell(\mu; x)/\partial\mu = (x - \mu)/V(\mu). \tag{4.16}$$

What is interesting about Equation (4.16) is that (apart from the support of X) it depends only on $V(\mu)$, the relationship between the mean and the variance. This raises the possibility that in some cases parametric inference methods can be constructed by building the expression (4.16), without the need for an actual probability distribution. In this context, the integral of (4.16) was introduced as the quasi-likelihood function $K(x, \mu)$ in Wedderburn (1974) (that is, we may substitute $K(x, \mu)$ for $\ell(\mu; x)$ in (4.16)). See also McCullagh *et al.* (1983).

This raises the question as to whether or not Equation (4.16) suffices to define a probability distribution. We have already claimed that if X possesses a one-parameter exponential family density, then its log-likelihood satisfies (4.16). In Wedderburn (1974) it was shown that the converse holds, that is, if $E[X] = \mu$, $V(\mu) = \text{var}_\mu[X]$ and $\ell(\mu; x)$ satisfies (4.16)), then X has a one-parameter exponential family density. Thus, in this case, the quasi-likelihood function equals the log-likelihood function.

However, $K(x, \mu)$ need not be a log-likelihood function. If it is, it must correspond to a density of the form $f(x) = \exp(\eta x - A(\eta))h(x)$. Of course, in this case $A(\eta)$ would be completely determined by variance function $V(\mu)$. Note that by Theorem 3.13 $c(t) = A(\eta+t) - A(\eta)$ is a cumulant generating function, so that $A(\eta)$ must be an analytic function. This in turn places conditions on $V(\mu)$ under which the quasi-likelihood function is a log-likelihood function.

The main point, though, is that the quasi-likelihood function need not be an actual log-likelihood function in order to form the basis of an inference method, especially in a large-sample context. This point of view will be studied in some detail in Chapter 13. The formal proofs supporting this example are left to the reader (Problem 4.19). ∎

4.6 Confidence Sets

We have so far considered in this chapter the problem of point estimation. Of course, a point estimate by itself does not constitute a complete inference statement, which should include some assessement of accuracy. An error probability $P_\theta(\hat{\theta} \neq \theta)$ is not a useful number (for continuous densities it would always equal 1). A confidence set, on the other hand, replaces a point estimate with the claim $\theta \in C(\tilde{X}) \subset \Theta$, in such a way that $P_\theta(\theta \not\in C(\tilde{X})) = \alpha(\theta) \leq \alpha$, for some fixed number α. Intuitively, $C(\tilde{X})$ is a set of *plausible* values of θ based on an observation $\tilde{X}$, which contains the true parameter with a probability no less than $1 - \alpha$ for any $\theta \in \Theta$.

Of course, confidence sets and estimators are often related. A common approach is to define a standard error $S_{\hat{\theta}}$ for an estimator $\hat{\theta}$ (noting that $S_{\hat{\theta}} \in \mathcal{T}_{stat}$), with which a confidence interval $C(\tilde{X}) = (\hat{\theta} - cS_{\hat{\theta}}, \hat{\theta} + cS_{\hat{\theta}})$ can be defined. The value of α is determined by a critical value c, often based on a normal or T distribution. However, we will also see that confidence sets can be defined independently of any estimator, in fact their formal definition, given next, makes no reference to them.

Definition 4.3 (Confidence Sets) Suppose we are given family $\mathcal{P}$ for sample space $\mathcal{X}$.

Suppose $\tilde{X} \in \mathcal{X}$ is an observation from $p_\theta \in \mathcal{P}$. We have estimand $\eta(\theta) \in \Theta_\eta$. A confidence set is a mapping $C(\tilde{X})$ from $\mathcal{X}$ to the class of subsets of Θ_η which *dnd* θ. Then $C(\tilde{X})$ is of confidence level $1 - \alpha$ if $P_\theta(\eta(\theta) \notin C(\tilde{X})) \leq \alpha$ for all $\theta \in \Theta$. If $P_\theta(\eta(\theta) \notin C(\tilde{X})) < \alpha$ for some θ, then the confidence set is conservative for confidence level $1 - \alpha$, otherwise the confidence level is *exact.* The term confidence interval is used when $C(\tilde{X})$ is an interval on $\mathbb{R}$. If in addition $C(\tilde{X}) = (-\infty, u(\tilde{X})]$ or $C(\tilde{X}) = [l(\tilde{X}), \infty)$, then $u(\tilde{X})$ $(l(\tilde{X}))$ is an upper (lower) confidence bound. ∎

We next review a number of methods for constructing confidence sets.

4.6.1 The Pivot Method for Confidence Intervals

Many forms of confidence intervals can be obtained by pivots, defined below. This approach will also be relevant to hypothesis testing (Section 5.7).

Definition 4.4 (Pivots) Suppose we are given family $\mathcal{P}$ for sample space $\mathcal{X}$. Suppose $\tilde{X} \in \mathcal{X}$ has density $p_\theta \in \Theta$. We are given an estimand $\eta(\theta) \in \Theta_\eta$, $\theta \in \Theta$. Then a mapping $\xi : \mathcal{X} \times \Theta_\eta \to \mathbb{R}^m$, for some $m \geq 1$ is a pivot if the distribution of $\xi(\tilde{X}, \eta(\theta))$ does not depend on θ. ∎

We note that by the sufficiency principle we should confine attention to pivots which depend on $\tilde{X}$ through any minimal sufficient statistic. Then, since the distribution of $\xi(\tilde{X}, \theta)$ in Definition 4.4 does not depend on θ, we have a method of constructing a confidence set.

Theorem 4.5 (Confidence Sets via Pivots) Suppose we are given parametric family p_θ, $\theta \in \Theta$ on sample space $\mathcal{X}$. We are given an estimand $\eta(\theta) \in \Theta_\eta$, $\theta \in \Theta$. We have, in addition, a mapping $\xi : \mathcal{X} \times \Theta_\eta \to \mathbb{R}^m$. Finally, suppose there exists a subset $E \subset \mathbb{R}^m$ for which

$$P_\theta(\xi(\tilde{X}, \eta(\theta))) \in E) \geq 1 - \alpha, \quad \text{for all } \theta \in \Theta. \tag{4.17}$$

Then $C(\tilde{X}) = \{\eta \in \Theta_\eta : \xi(\tilde{X}, \eta) \in E\}$ is a confidence set of level $1 - \alpha$ (Definition 4.3). If (4.17) is replaced by equality, then the confidence level $1 - \alpha$ is exact. ∎

Proof. Fix any $\theta \in \Theta$. Then $\eta(\theta) \in C(\tilde{X})$ *iff* $\xi(\tilde{X}, \eta(\theta))) \in E$. But by assumption, the probability of this event is bounded below by $1 - \alpha$ (or equals $1 - \alpha$, in which case the confidence level is exact). □

It should be noted that in Theorem 4.5 it is not assumed that $\xi(\tilde{X}, \eta(\theta))$ is a pivot, and its distribution may depend on θ as long as the inequality of Equation (4.17) is satisfied. Clearly, though, constructing $\xi(\tilde{X}, \eta(\theta))$ to be a pivot is a natural approach to this method.

The next example gives two widely used examples of the pivot method.

Example 4.13 (Confidence Intervals for the Mean and Variance of a Normal Sample) Perhaps the best known example of a pivot is the T statistic (Section 2.5.1). Suppose $\bar{X}$, S^2 are the sample mean and variance for an *iid* sample $\mathbf{X} = (X_1, \ldots, X_n)$ of size n from $N(\mu, \sigma^2)$. Then $T = (\bar{X} - \mu)/(S/\sqrt{n})$ possesses a T distribution of $n - 1$ degrees of freedom. Therefore,

$$C(\mathbf{X}) = \left[\bar{X} - t_{n-1;\alpha/2}\frac{S}{\sqrt{n}}, \bar{X} + t_{n-1;\alpha/2}\frac{S}{\sqrt{n}}\right] \tag{4.18}$$

is a level $1 - \alpha/2$ confidence interval for μ, where $t_{n-1;\alpha/2}$ is the $\alpha/2$ critical value from a T_{n-1} distribution.

In addition, $W = (n-1)S^2/\sigma^2$ possesses a χ^2 distribution with $n-1$ degrees of freedom. Then

$$C(\mathbf{X}) = \left[\frac{S^2}{\chi^2_{n-1;\alpha/2}/(n-1)}, \frac{S^2}{\chi^2_{n-1;1-\alpha/2}/(n-1)} \right], \tag{4.19}$$

is a level $1-\alpha/2$ confidence interval for σ^2, where $\chi^2_{n-1;q}$ is the q critical value from a χ^2_{n-1} distribution. ∎

4.6.2 Confidence Sets for Location and Scale Parameters

Recall the location and scale parameter models of Section 3.3. Suppose $\mathbf{X} = (X_1, \ldots, X_n)$ is a sample from a model with location parameter μ. For any location equivariant estimator $\hat{\mu}(\mathbf{X})$ (Definition 3.3), $\xi(\mathbf{X}, \mu) = \hat{\mu}(\mathbf{X}) - \mu$ will be a pivot. Similarly, if $\mathbf{X} = (X_1, \ldots, X_n)$ is a sample from a model with scale parameter τ, then for any scale equivariant estimator $\hat{\tau}^k(\mathbf{X})$ of τ^k (Section 3.4), $\xi(\mathbf{X}, \tau^k) = \hat{\tau}^k(\mathbf{X})/\tau^k$ will be a pivot. The proofs are left to the reader (Problem 4.22). Consider the following example.

Example 4.14 (Confidence Intervals for an Exponential Scale Parameter) Suppose $\mathbf{X} = (X_1, \ldots, X_n)$ is an *iid* sample from an exponential density of mean μ, which is a scale parameter. Then $S(\mathbf{X}) = \sum_i X_i$ is scale equivariant, with $S \sim gamma(\tau, n)$. This means $\xi(S, \tau) = S(\mathbf{X})/\tau$ is a pivot, with distribution $S \sim gamma(1, n)$. Therefore,

$$C(\mathbf{X}) = \left[\frac{S(\mathbf{X})}{s_{\alpha/2}}, \frac{S(\mathbf{X})}{s_{1-\alpha/2}} \right],$$

is a level $1-\alpha$ confidence interval for τ, where s_q is the q critical value from a $gamma(1, n)$ distribution. ∎

Confidence sets for shape parameters can sometimes be constructed using this method, following a transformation that converts that parameter to a location or scale parameter. See Problem 4.25, for example.

4.6.3 Minimizing the Length of Confidence Intervals

Note that in Examples 4.13 and 4.14 the error probability α has been allocated equally to the lower and upper tails. Precisely, this means that if we write $C(\tilde{X}) = [L, U]$ for estimand θ, we get $P_\theta(L > \theta) = P_\theta(U < \theta) = \alpha/2$. However, if we select L and U so that $P_\theta(L > \theta) = q_1$, $P_\theta(U < \theta) = q_2$, with $q_1 + q_2 = \alpha$, then $C(\tilde{X})$ retains confidence level $1-\alpha$.

This generates an interesting optimization problem. It could be argued that, if we have a choice, we should use the confidence interval with the smallest length (although other criteria may exist). If we set $q_1 = \lambda\alpha$, $q_2 = (1-\lambda)\alpha$, $\lambda \in [0, 1]$, then we have defined a family of level $1-\alpha$ confidence intervals indexed by a single parameter λ. By symmetry, we might expect $\lambda = 1/2$ to be optimal in this sense for the confidence interval of Equation (4.18) (although it does no harm to verify this). However, $\lambda = 1/2$ will not be the optimal choice for the confidence interval of Equation (4.19) (Tate and Klett, 1959).

It is often the case that a confidence interval can be constructed from multiples of quantiles from a common distribution f, that is $C(\tilde{X}) = [KQ_{q_1}, KQ_{1-q_2}]$, where $q_1 + q_2 = \alpha$ and Q_p are quantiles from f. The confidence interval of Equation (4.19) is of this form, where the quantiles are from a single inverse-gamma density. In this case we may consider the problem of determining the subset $E \subset \mathbb{R}$ which minimizes $\int_E d\nu$ given constraint $\int_E f d\nu \geq 1-\alpha$. The following theorem states that under quite general conditions, any solution to this problem must be of the form $E_c = \{\tilde{x} \in \mathcal{X} : f(\tilde{x}) \geq c\}$. More specifically, if f is absolutely continuous with respect to Lebesgue measure, and is unimodel and continuous, then the optimal confidence set takes the form $C(\tilde{X}) = [KQ_{q_1}, KQ_{1-q_2}]$, where $f(Q_{q_1}) = f(Q_{1-q_2})$. The theorem follows.

Theorem 4.6 (Optimal Region Coverage) Suppose f is a density on Borel measure space $(\mathcal{X}, \nu)$, such that all measurable sets are continuity sets, and f is upper-semicontinuous. Assume ν is either Lebesgue or counting measure. For each $c > 0$ define $E_c = \{\tilde{x} \in \mathcal{X} : f(\tilde{x}) \geq c\}$, and let $\mathcal{E}$ be the class of all such sets. Then let E be any closed set with $q_E = \int_E f(\tilde{x})d\nu$ and $L_E = \int_E d\nu$. If $E \notin \mathcal{E}$, then there exists E' with $q_{E'} = \int_{E'} f(\tilde{x})d\nu$ and $L_{E'} = \int_{E'} d\nu$ such that $q_{E'} > q_E$ but $L_{E'} \leq L_E$.

Furthermore, if f is absolutely continuous *wrt* Lebesgue measure, then there exists closed E' such that $q_{E'} > q_E$ but $L_{E'} < L_E$. In this case, any closed E which minimizes L_E among all closed sets satisfying $q_E \geq q$ must be in $\mathcal{E}$. ∎

Proof. By assumption each E_c is closed. Then suppose $E \notin \mathcal{E}$ is a closed set satisfying $q_E = q$. By assumption, the interior of E, say E^o, also satisfies $q_{E^o} = q$. By assumption, we may find disjoint open balls $B_\epsilon(\tilde{x}_1)$, $B_\epsilon(\tilde{x}_2)$ such that $B_\epsilon(\tilde{x}_1) \subset E^o$, $B_\epsilon(\tilde{x}_2) \subset E^{oc}$, with $\sup_{\tilde{x} \in B_\epsilon(\tilde{x}_1)} f(\tilde{x}) < \inf_{\tilde{x} \in B_\epsilon(\tilde{x}_2)} f(\tilde{x})$. Then set $E' = [E \cap B_\epsilon(\tilde{x}_1)^c] \cup B_\epsilon(\tilde{x}_2)$. We then have $q_{E'} > q_E$ but $L_{E'} = L_E$. If f is absolutely continuous *wrt* Lebesgue measure, then we may replace $B_\epsilon(\tilde{x}_2)$ in the preceeding argument with $B_{\epsilon'}(\tilde{x}_2)$, such that $\epsilon' < \epsilon$ but $\int_{B_{\epsilon'}(\tilde{x}_2)} f(\tilde{x})d\nu > \int_{B_\epsilon(\tilde{x}_1)} f(\tilde{x})d\nu$, in which case we have $q_{E'} > q_E$ and $L_{E'} < L_E$. □

Remark 4.1 For our purposes, the assumption that all measurable sets are continuity sets means that f is absolutely continuous with respect to either Lebesgue measure or counting measure. Under the counting metric, if $\epsilon < 1$, then $B_\epsilon(\tilde{x})$ is either the singleton $\tilde{x}$, if $\tilde{x} \in \mathcal{X}$, or is the emptyset otherwise. ∎

The following example will help clarify the precise language of Theorem 4.6.

Example 4.15 (Counterexample to the Optimal Region Coverage Theorem) Theorem 4.6 addresses the problem of minimizing L_E among closed subsets $E \subset \mathcal{X}$ subject to $q_E \geq q$. Suppose f is a density *wrt* counting measure on $\{1, 2, 3, 4\}$, and we have density $f(1) = 0.02$, $f(2) = 0.04$, $f(3) = 0.47$, $f(4) = 0.47$. Set $q = 0.95$. The set $E = \{2, 3, 4\}$ is an member of the class $\mathcal{E}$, and has coverage $q_E = 0.98$. The set $E' = \{1, 3, 4\}$ has coverage $0.96 \geq q$, is of the same size as E, but is not a member of the class $\mathcal{E}$. So, both E, E' solve the stated problem. However, Theorem 4.6 states, at least, that a search for a solution to this problem can be confined to $\mathcal{E}$. In addition, if f is absolutely continuous with respect Lebesgue measure then any solution must be in $\mathcal{E}$. ∎

Next, we show how Example 4.13 can be refined in light of Theorem 4.6.

Example 4.16 (Confidence Intervals for the Variance of a Normal Sample - Optimal Coverage Problem) The confidence interval of Equation (4.19) can be generalized to $C(\tilde{X}) = [KQ_{q_1}, KQ_{1-q_2}]$, where $K = (n-1)S^2$, and Q_p are quantiles from density $f \sim igamma(1, n-1)$. For $n = 10$ and $q_1 = q_2 = 0.05/2$ we have $C(\tilde{X}) = K[0.069, 0.290]$, with length $K \times 0.221$. On the other hand, if we set $q_1 = 0.005$, $q_2 = 0.05 - q_1 = 0.045$, then we have $C(\tilde{X}) = K[0.0539, 0.217]$ with length $K \times 0.163$. For this solution $f(0.0539) \approx f(0.217)$. By Theorem 4.6 this yields the shortest length confidence interval. ∎

Theorem 4.6 will also play a role in Bayesian estimation (Section 4.8.1).

4.6.4 Inversion of CDFs

From Section 4.6.2, we can see that the pivot method gives a very natural approach for the construction of confidence itervals for location and scale parameters. The CDF inversion method is more generally applicable, and so is more often applied to the construction of confidence intervals for shape parameters, especially for discrete models.

Suppose $\tilde{X} \sim p_\theta$, $\theta \in \Theta$, is an observation from a parametric family $\mathcal{P}$. Then let $S(\tilde{X}) \in \mathcal{X}_S \subset \mathbb{R}$ be a statistic (here, we would generally use a minimal sufficient statistic). Then let $F_S(s;\theta)$ be the CDF for $S(\tilde{X})$. The CDF transformation $U = F_S(S;\theta)$ discussed in Section 1.7 provides what is essentially another form of pivot. By Theorem 1.17 we may conclude that the CDF of U, denoted as $F_U(u;\theta)$ satisfies the bounds

$$u - \delta \leq F_U(u;\theta) \leq u, \ \ u \in (0,1),$$

where $\delta = \sup_{(s,\theta)\in\mathcal{X}_S\times\Theta} P_\theta(S(\tilde{X}) = s)$. When $S(\tilde{X})$ is continuous, $\delta = 0$, so U is a pivot in the exact sense. In this case a confidence set of exact level can be constructed. If $S(\tilde{X})$ is discrete, then $\delta > 0$, and can be taken as a measure of the "coarseness" of the discrete structure. Generally, δ will be smaller when the support of $S(\tilde{X})$ is larger. For example, if $S \sim bin(n,p)$, then $\delta \to 0$ as $n \to \infty$. However, whatever the value of δ it will still be possible to construct at least a conservative confidence set of any level $1-\alpha$, with the sharpness of this level depending on δ. This is stated in the next theorem (note that, following Section 4.6.3, we do not assume that equal error probability is assigned to the upper and lower bound of the pivot).

Theorem 4.7 (Confidence Intervals by CDF Inversion) Suppose $\tilde{X} \sim p_\theta$, $\theta \in \Theta$, is an observation from a parametric family $\mathcal{P}$. Then let $S(\tilde{X}) \in \mathcal{X}_S \subset \mathbb{R}$ be a statistic with CDF $F_S(s;\theta)$. For real numbers $q_1, q_2 \in [0,1]$, $q_1 + q_2 \in [0,1]$, define the subset:

$$C(\tilde{X}) = \{\theta \in \Theta : F_S(S(\tilde{X});\theta) > q_1 \text{ and } F_S^*(S(\tilde{X});\theta) > q_2\}. \tag{4.20}$$

Then $C(\tilde{X})$ is a level $1 - q_1 - q_2$ confidence set for θ. ■

Proof. Fix $\theta \in \Theta$. Construct events $E_1 = \{F_S(S(\tilde{X});\theta) > q_1\}$, $E_2 = \{F_S^*(S(\tilde{X});\theta) > q_2\}$. Then $\theta \in C(\tilde{X})$ *iff* $\tilde{X} \in E_1 \cap E_2$. Then, an application of Theorem 1.16 gives $P_\theta(\tilde{X} \in E_1 \cap E_2) = 1 - P_\theta(\tilde{X} \in E_1^c \cup E_2^c) \geq 1 - q_1 - q_2$. The remainder of the proof follows Theorem 4.5. □

The remaining discussion concerns the exact form of $C(\tilde{X})$, the version appearing in Theorem 4.7 being of quite general structure. First, suppose $\Theta \subset \mathbb{R}^m$, $m > 1$, but we are interested in estimand $\eta(\theta) \in \mathcal{A} \subset \mathbb{R}$. The pivot $F_S(S;\theta)$ is defined on the full parameter space Θ, so a confidence set for $\eta(\theta)$ would have to take the form

$$C_\eta(\tilde{X}) = \{\eta' \in \mathcal{A} : \eta^{-1}(\eta') \subset C(\tilde{X})\}.$$

We can say more when $\theta \in \Theta \subset \mathbb{R}$ and $F_S(s;\theta)$ is monotone with respect to θ for a fixed $s \in \mathcal{X}_S$. First, if $F_S(s;\theta)$ is nonincreasing in θ, the confidence set (4.20) takes the form of an interval $C(\tilde{X}) = (\theta_L(\tilde{X}), \theta_U(\tilde{X}))$, where

$$\begin{aligned}\theta_L(\tilde{X}) &= \inf\{\theta \in \Theta : F_S^*(S(\tilde{X});\theta) > q_2\}\\ \theta_U(\tilde{X}) &= \sup\{\theta \in \Theta : F_S(S(\tilde{X});\theta) > q_1\}.\end{aligned} \tag{4.21}$$

If in addition $F_S(s;\theta)$ is continuous with respect to θ then $\theta_L(\tilde{X})$ and $\theta_U(\tilde{X})$ are the solutions to the equations

$$F_S^*(S(\tilde{X});\theta_L(\tilde{X})) = q_2 \quad \text{and} \quad F_S(S(\tilde{X});\theta_U(\tilde{X})) = q_1. \tag{4.22}$$

Conversely, if $F_S(s;\theta)$ is nondecreasing in θ, we similarly have $C(\tilde{X}) = (\theta_L(\tilde{X}), \theta_U(\tilde{X}))$, where

$$\begin{aligned} \theta_L(\tilde{X}) &= \inf\{\theta \in \Theta : F_S(S(\tilde{X});\theta) > q_1\} \\ \theta_U(\tilde{X}) &= \sup\{\theta \in \Theta : F_S^*(S(\tilde{X});\theta) > q_2\}, \end{aligned} \tag{4.23}$$

and under continuity conditions $\theta_L(\tilde{X})$ and $\theta_U(\tilde{X})$ are the solutions to the equations

$$F_S(S(\tilde{X});\theta_L(\tilde{X})) = q_1 \quad \text{and} \quad F_S^*(S(\tilde{X});\theta_U(\tilde{X})) = q_2. \tag{4.24}$$

Note that Equations (4.21) or (4.23) hold when the distribution of $S(\tilde{X})$ is stochastically increasing or decreasing with respect to θ, respectively (Definition 1.6). It is possible, of course, that these conditions hold for some but not all possible values of $s = S(\tilde{X}) \in \mathcal{X}_S$.

This type of procedure is especially useful for shape parameters, expecially for discrete distributions. This will be demontrated for the binomial parameter in Example 4.17 below. However, we will first discuss in the next section a method which can simplfy application of the CDF inversion method considerably.

4.6.5 CDF Inversion and the Quantile Method

It will sometimes be possible to express the quantities $\theta_L(\tilde{X})$ and $\theta_U(\tilde{X})$ defined by Equations (4.22) or (4.24) as quantiles of a density function defined on Θ. This, of course, is an idea natural to the Bayesian model (Section 3.10), and in Example 4.17 below we will see the appearance of what obviously resembles a conjugate density. This relationship will be made clearer in our discussion of Bayesian credible regions in Section 4.8.1.

We begin with the main theorem supporting the method.

Theorem 4.8 (The Quantile Method for Confidence Intervals) Suppose F_θ, $\theta \in \Theta \subset \mathbb{R}$ is a family of CDFs on support $\mathcal{S} \subset \mathbb{R}$. Let $\theta_L = \inf \Theta$ and $\theta_U = \sup \Theta$. Suppose for some $t \in \mathcal{S}$ the following conditions hold:

(i) The partial derivative

$$p_t(u) = \left. \frac{\partial P_\theta(X \geq t)}{\partial \theta} \right|_{\theta = u} \tag{4.25}$$

exists and is nonnegative for all $u \in \Theta$.

(ii) $\inf_{\theta\in\Theta} P_\theta(X \geq t) = 0$ and $\sup_{\theta\in\Theta} P_\theta(X \geq t) = 1$.

Then $p_t(\theta)$ is a probability density on Θ *wrt* Lebesgue measure. Furthermore

$$P_\theta(X \geq t) = \int_{\theta_L}^{\theta} p_t(u)du, \tag{4.26}$$

for $\theta \in \Theta$. Conversely, suppose for some $t \in \mathcal{S}$ the following conditions hold:

(i) The partial derivative

$$p_t(u) = \left. \frac{\partial P_\theta(X \leq t)}{\partial \theta} \right|_{\theta = u} \tag{4.27}$$

exists and is nonnegative for all $u \in \Theta$.

(ii) $\inf_{\theta\in\Theta} P_\theta(X \leq t) = 0$ and $\sup_{\theta\in\Theta} P_\theta(X \leq t) = 1$.

Then $p_t(\theta)$ is a probability density on Θ *wrt* Lebesgue measure. Furthermore

$$P_\theta(X \leq t) = \int_{\theta_L}^{\theta} p_t(u)du, \tag{4.28}$$

for $\theta \in \Theta$. ■

Proof. By the fundamental theorem of calculus

$$
\begin{aligned}
P_\theta(X \geq t) &= P_\theta(X \geq t) - P_{\theta_L}(X \geq t) \\
&= \int_{\theta_L}^{\theta} \left. \frac{\partial P_\theta(X \geq t)}{\partial \theta} \right|_{\theta = u} du \\
&= \int_{\theta_L}^{\theta} p_t(u) du.
\end{aligned} \tag{4.29}
$$

By hypothesis, $p_t(\theta) \geq 0$ for $\theta \in \Theta$, and $P_{\theta_U}(X \geq t) = 1$, therefore $p_t(\theta)$ is a density function on θ. The rest of the first part of the theorem follows directly from Equation (4.29). The remainder of the theorem uses essentially the same argument. □

Theorem 4.8 can be used directly to solve Equation (4.22). Suppose $S(\tilde{X}) = s$, and $F_S^*(s;\theta)$ is increasing in θ. Let $p_s(\theta)$ be the density on Θ given by Equation (4.25). Then by Equation (4.29), if

$$F_S^*(S(\tilde{X}); \theta_L(\tilde{X})) = q_2$$

then $\theta_L(\tilde{X})$ is the q_2 quantile of $p_s(\theta)$. If $S(\tilde{X})$ is a continuous random variable, we have from Equation (4.22),

$$q_1 = F_S(S(\tilde{X}); \theta_U(\tilde{X})) = 1 - F_S^*(S(\tilde{X}); \theta_U(\tilde{X})),$$

so that by Equation (4.29), $\theta_U(\tilde{X})$ is the $1 - q_1$ quantile of $p_s(\theta)$. If $S(\tilde{X})$ is integer-valued, then

$$q_1 = F_S(S(\tilde{X}); \theta_U(\tilde{X})) = 1 - F_S^*(S(\tilde{X}) + 1; \theta_U(\tilde{X})),$$

so that by Equation (4.29), $\theta_U(\tilde{X})$ is the $1 - q_1$ quantile of $p_{s+1}(\theta)$.

If, conversely, $F_S(s;\theta)$ is increasing in θ, then Equation (4.24) can be solved using essentially the same argument. In this case $p_s(\theta)$ is the density on Θ given by Equation (4.27). Then $\theta_L(\tilde{X})$ is the q_1 quantile of $p_s(\theta)$. If $S(\tilde{X})$ is continuous then $\theta_U(\tilde{X})$ is the $1 - q_2$ quantile of $p_s(\theta)$, and if $S(\tilde{X})$ is integer-valued then $\theta_U(\tilde{X})$ is the $1 - q_2$ quantile of $p_{s-1}(\theta)$. The procedure is demonstrated in the following example.

Example 4.17 (Quantile Form of the Clopper-Pearson Confidence Interval) The application of the ideas of Theorem 4.7 to the binomial parameter leads to the well-known Clopper-Pearson interval (Clopper and Pearson, 1934). Here we will use Theorem 4.8 to evaluate the confidence interval directly. If $X \sim bin(n, \theta)$, then for $t \in \{0, 1, \ldots, n\}$,

$$P_\theta(X \geq t) = \sum_{i=t}^{n} \binom{n}{i} \theta^i (1-\theta)^{n-i}.$$

Then for $t \geq 1$,

$$
\begin{aligned}
p_t(\theta) &= \sum_{i=t}^{n-1} \left[\binom{n}{i} i\theta^{i-1}(1-\theta)^{n-i} - (n-i)\binom{n}{i}\theta^{i-1}(1-\theta)^{n-i-1} \right] + n\theta^{n-1} \\
&= \sum_{i=t}^{n-1} \left[n\binom{n-1}{i-1}\theta^{i-1}(1-\theta)^{n-i} - n\binom{n-1}{i}\theta^{i}(1-\theta)^{n-i-1} \right] + n\theta^{n-1} \\
&= n\binom{n-1}{t-1}\theta^{t-1}(1-\theta)^{n-t},
\end{aligned}
$$

noting that all but one of the terms cancels. We can recognize the parametric form $p_t \sim beta(t, n-t+1)$.

Therefore for the Clopper-Pearson confidence interval $\theta_L(X)$ is the q_2-quantile of $beta(t, n-t+1)$, and $\theta_U(X)$ is the $1 - q_1$ quantile of $beta(t+1, n-t)$. See also Problem 4.24. ■

4.7 Equivariant Versus Shrinkage Estimation

If we accept the invariance principle of Section 3.13.3, then we would use equivariant estimators for location or scale parameters (the driving assumption of Chapter 9). Although we have included the invariance principle as a general principle of inference in Chapter 3, there will often be reason to choose statistical methods which violate it. We introduce here the class of shrinkage estimator as an important example of this type. The exact definition varies, and there are multiple principles by which they are derived. However, it will be useful to identify these estimators as a class of their own.

Shrinkage estimators are often modifications of other estimators. Suppose $\hat{\theta}$ is an estimator of $\theta \in \Theta$. Fix some other $\theta_0 \in \Theta$. The simplest form of shrinkage estimator is simply a convex combination of $\hat{\theta}$ and θ_0:

$$\hat{\theta}_\lambda = \lambda\hat{\theta} + (1-\lambda)\theta_0, \quad \lambda \in [0,1] \tag{4.30}$$

(it is sometimes assumed that $\lambda \in (0,1)$). We may also allow $\lambda = \lambda(\tilde{X})$. It is often assumed that $\theta_0 = 0$, but such estimators can usually be modifed to permit other values. In addition, the definition of a shrinkage estimator can be extended to mutlidimensional parameters.

The purpose of a shrinkage estimator appears to be to identify a "favored hypothesis" θ_0, and we would expect it to have better properties when the true parameter θ is at or near θ_0 (the concept of the "favored hypothesis" will be an important theme of Chapter 7). Alternatively, a shrinkage estimator is intended to force an estimate towards θ_0. This is usually the intention when $\theta_0 = 0$ (which suggests the term "shrinkage").

Of course, this is in violation of the principle of invariance, which implies that no hypothesis is to be favored. Yet both equivariant estimators and shrinkage estimators are well represented in conventional statistical methodology. The justification for equivariance was already offered in Section 3.13.3, so it is worth briefly discussing the motivation for using shrinkage estimators.

The notion of a favored hypothesis is of course quite natural to Bayesian models, and we will see in the next section that shrinkage estimators arise naturally in Bayesian estimation.

Shrinkage estimation also describes estimation methods which explicitly penalizes parameter values of large magnitude, in effect "shrinking" the estimates towards zero. Such methods need not rely on an explicit formulation such as Equation (4.30). Ridge regression is an example of such a procedure (Problem 6.36).

There is a third motivation for the study of shrinkage estimators which is primarily of theoretical interest. It sometimes happens that conventional statistical procedures are inadmissible, which means that they may be replaced by alternative procedures which are in some sense uniformly better (this topic will be covered in Chapter 7). Such alternatives tend to be shrinkage estimators, which therefore figure prominently in any discussion of the issue.

4.8 Bayesian Estimation

Recall the Bayesian model from Section 3.10. We are given a parametric model $\mathcal{P} = \{p_\theta : \theta \in \Theta\}$ for observation $\tilde{X} \in \mathcal{X}$, but we also assume that θ possesses a prior density $\pi(\theta)$ with respect to some measure ν_π on Θ. The joint density of $(\tilde{X}, \theta)$ is given by $f(\tilde{x}, \theta) =$

$f(\tilde{x} \mid \theta)\pi(\theta)$, where $f(\tilde{x} \mid \theta) = p_\theta(\tilde{x})$. The inference is then based on the posterior density

$$\pi(\theta \mid \tilde{x}) = \frac{f(\tilde{x} \mid \theta)\pi(\theta)}{f(\tilde{x})} = \frac{f(\tilde{x} \mid \theta)\pi(\theta)}{\int_\Theta f(\tilde{x} \mid \theta)\pi(\theta)d\nu_\pi}.$$

It will, of course, be convenient to simplify the inference to a single estimate $\hat{\theta} \in \Theta$, or, similar to a convidence interval, some subset $E \subset \Theta$ which contains the true value of θ with high probability. However, it must be remembered that, according to the Bayesian model, θ is an unobserved random outcome, which we must predict with some function $\hat{\theta}(\tilde{X})$. Of course, $\hat{\theta}$ must be a statistic, but can also depend on hyperparameters, which gives Bayesian inference much of its unique character. Suppose, to fix ideas, $\theta \in \mathbb{R}$. A reasonable choice for estimator might be $\hat{\theta}_B(\tilde{X}) = E[\theta \mid \tilde{X}]$, which is evaluated as the mean of the posterior density, that is,

$$\hat{\theta}_B(\tilde{x}) = E[\theta \mid \tilde{X} = \tilde{x}] = \int_{\theta\in\Theta} \theta \cdot \pi(\theta \mid \tilde{x})d\nu_\pi. \tag{4.31}$$

This is in fact a form of Bayes estimator, which will be given a more theoretical foundation in Chapter 7 (where we will see that other approaches exist). However, for the purposes of this chapter it will suffice to explore some of the characteristic features of the estimator defined by Equation (4.31).

Example 4.18 (Bayesian Estimation of the Binomial Parameter) Suppose we observe $X \sim bin(n, \theta)$. It was shown in Example 3.31 that the beta density $beta(\alpha, \beta)$ is a conjugate prior for θ, with $\theta \mid X = x \sim beta(x + \alpha, n - x + \beta)$. Suppose $U \sim beta(\alpha, \beta)$, and recall the moments

$$E[U] = \frac{\alpha}{\alpha + \beta}, \quad \text{var}[U] = \frac{\alpha\beta}{(\alpha + \beta)^2(\alpha + \beta + 1)}$$

(Problem 1.19). The Bayes estimator of θ is therefore the mean of posterior density $beta(x + \alpha, n - x + \beta)$, which is

$$\hat{\theta}_B = (x + \alpha)/\,(n + \alpha + \beta)\,. \tag{4.32}$$

It is instructive to reparametrize the beta density,

$$\rho = \alpha/\,(\alpha + \beta)\,, \quad T = \alpha + \beta. \tag{4.33}$$

Under the new parametrization we have $E[U] = \rho$ and $\text{var}[U] = \rho(1 - \rho)/\,(T + 1)$. Thus, ρ can be conceived of as a prior estimate of θ, while T is a measure of the certainty of the estimate. Then note that the Bayes' estimate (4.32) may be decomposed as

$$\hat{\theta}_B = q\hat{\theta}_{MLE} + (1 - q)\rho. \tag{4.34}$$

where $\hat{\theta}_{MLE} = X/n$ and $q = n/(n + T)$. Thus $\hat{\theta}_B$ is a convex combination of the (non-Bayesian) maximum likelihood estimate of θ and the prior estimate ρ, and is therefore a shrinkage estimator. Furthermore larger values of T relative to n place greater weight on the prior estimate.

This idea can be made more precise by thinking of the hyperparameter T as being itself a sample size. Suppose we have two independent binomial observations $X_0 \sim bin(m, \theta)$, $X_1 \sim bin(n, \theta)$. We would, of course, pool the observations into a single estimate $\hat{\theta}_{pooled} = (X_0 + X_1)/(m + n)$. Interestingly, this would be identical to first observing X_0, contructing a prior distribution for θ based on that observation, then evaluating the mean of $\pi(\theta \mid X_1)$. To see this, set hyperparameters $\rho = X_0/m$, $T = m$. Then from Equation (4.34) we have

$$\hat{\theta}_B = \frac{X_1}{n}\left(\frac{n}{n + m}\right) + \frac{X_0}{m}\left(\frac{m}{n + m}\right) = \hat{\theta}_{pooled}. \quad \blacksquare$$

In Section 3.11 the problem of defining an indifferent prior was introduced. We next consider this question in the context of Example 4.18.

Example 4.19 (Bayesian Estimation of the Binomial Parameter: Which Prior is Indiferrerent?) Although Example 4.18 presents a relatively straightforward example of Bayesian inference, it affords the opportunity to explore some of the subtleties of the idea of prior indifference. It seems reasonable in this model to express prior indifference by adopting a uniform prior on Θ, which assigns the same prior density to each $\theta \in \Theta$. This is equivalent to setting $\alpha = \beta = 1$ in the beta prior, and in our reparamatrization this is $\rho = 1/2$, $T = 2$. In this case Equation (4.34) becomes

$$\hat{\theta}_B = q\hat{\theta}_{MLE} + (1-q)/2, \tag{4.35}$$

where $q = n/(n+2)$. Thus, the Bayes' estimator under a uniform prior favors the estimate $\theta = 1/2$ (it could be said that this is the hypothesis favored by the symmetry about $\theta = 1/2$ imposed by the uniform prior).

Is there such a thing as a Bayesian estimator which is completely unaffected by the prior? It could be argued that this would occur when $T = \alpha + \beta = 0$, which would force the weight $(1-q) = 0$ in Equation (4.34). However, the beta density is conventionally defined only for $\alpha, \beta > 0$ (see Section 1.4.1). Nonetheless, we can still consider the limiting case as $\alpha, \beta \to 0$. In particular, in the limit the beta density becomes increasingly concentrated near $\theta = 0$ and $\theta = 1$, so we may at least reasonably claim that $P(\theta \in \{0,1\}) = 1$ under the prior distribution limit. Indeed, if we force ρ to remain constant as $\alpha, \beta \to 0$, then the limiting prior distribution is $bern(\rho)$. However, even in this case $\hat{\theta}_B$ will not depend on any particular choice of ρ. In other words, under the prior belief that θ is zero or one, our Bayesian estimate is equal to the MLE. This would seem to satisfy the notion of prior indifference at least as much as the uniform prior. This example will be further discussed in Example 7.16. ■

4.8.1 Bayesian Credible Regions

The formal definition of the confidence set (Definition 4.3) is not strictly interpretable in Bayesian inference. On the other hand, the simpler idea of a subset $\mathcal{C} \subset \Theta$ which contains the true parameter θ with high probability is well defined for the Bayseian model. This is formalized in the next definition.

Definition 4.5 (Bayesian Credible Region) Suppose we have a subset $\mathcal{C} \subset \Theta$ which is allowed to depend on $\tilde{X} \in \mathcal{X}$ but not θ. If under a posterior density $\pi(\theta \mid \tilde{x})$

$$P_\pi(\theta \in \mathcal{C} \mid \tilde{X} = \tilde{x}) = 1 - \alpha, \tag{4.36}$$

then $\mathcal{C}$ is referred to as a (level $1-\alpha$) Bayesian credible region, or credible interval, as appropriate. ■

To fix ideas, suppose a parameter $\theta \in \Theta \subset \mathbb{R}$ has posterior density $\pi(\theta \mid \tilde{x})$, and we wish to construct a level $1-\alpha$ credible interval (L, U). We must then have $P_\pi(L \leq \theta \leq U \mid \tilde{X} = \tilde{x}) = 1 - \alpha$. As was done for some types of confidence intervals we can do this by setting $L = Q_a$, $U = Q_b$, where Q_p is any p-quantile of $\pi(\theta \mid \tilde{x})$, and

$$b - a = 1 - \alpha, \tag{4.37}$$

or $a + (1-b) = \alpha$. A common approach is to set $a = 1 - b = \alpha/2$, which we will refer to as a symmetric credible interval (i.e. symmetric not geometrically, but in the tail probabilities).

However, we may wish to consider the problem of minimizing the interval length $K = U - L = Q_b - Q_a$ subject to the constraint of Equation (4.37). If the posterior density $\pi(\theta \mid \tilde{x})$ is symmetric about its mean, which is also the mode, we would expect the symmetric credible interval to have the smallest length K. When this does not hold, we can use the approach of Section 4.6.3, which leads to the following definition:

Definition 4.6 (Highest Posterior Density Region (HPDR)) A highest posterior density region (HPDR) is any subset $\mathcal{C} \subset \Theta$ containing the highest values of $\pi(\theta \mid \tilde{x})$. In other words, if $\theta' \in \mathcal{C}$, and $\theta'' \notin \mathcal{C}$, then $\pi(\theta' \mid \tilde{x}) > \pi(\theta'' \mid \tilde{x})$. If $\mathcal{C} = \mathcal{I} = (L, U) \subset \mathbb{R}$ is an interval, it is also a highest posterior density interval (HPDI). ■

By Theorem 4.6 an HDPR minimizes the total area among confidence regions of a fixed level $1 - \alpha$. The simplest way to construct an HPDR is to set $\mathcal{C} = \{\theta \in \Theta : \pi(\theta \mid \tilde{x}) \geq c\}$, selecting constant c to satisfy the requirement of Equation (4.36). Note that if $\pi(\theta \mid \tilde{x})$ is multimodal, $\mathcal{I}$ need not be an interval, but the mathematical argument is not affected by this.

We next given an example comparing the symmetric credible interval to the HPDI.

Example 4.20 (Credible Intervals for the Binomial Parameter) Recall from Example 4.18 that if $X \sim bin(n, p)$, and we assign conjugate prior density $\theta \sim beta(\alpha, \beta)$, we have posterior density $\theta \mid X = x \sim beta(x + \alpha, n - x + \beta)$ and Bayes estimate $\hat{\theta}_B(x) = (x + \alpha)/(n + \alpha + \beta)$. To take an example, use the values $\alpha = \beta = 1$, $x = 4$, $n = 20$. Note that for $\alpha = \beta = 1$, the beta density is equivalent to a uniform density, so in some sense we have an uninformative prior. The Bayes estimate is $\hat{\theta}_B(x) = 5/22$ (the sample frequency, however, is $\hat{p} = 4/20$). To construct a credible interval (L, U) for p, we set $L = Q_a$, $U = Q_b$, where Q_p is the p-quantile for posterior density and $a + (1 - b) = \alpha$. To construct a $1 - 0.05$ symmetric credible interval, we use posterior density $beta(5, 17)$, for which $L = Q_{0.025} \approx 0.082$, $U = Q_{0.975} \approx 0.419$.

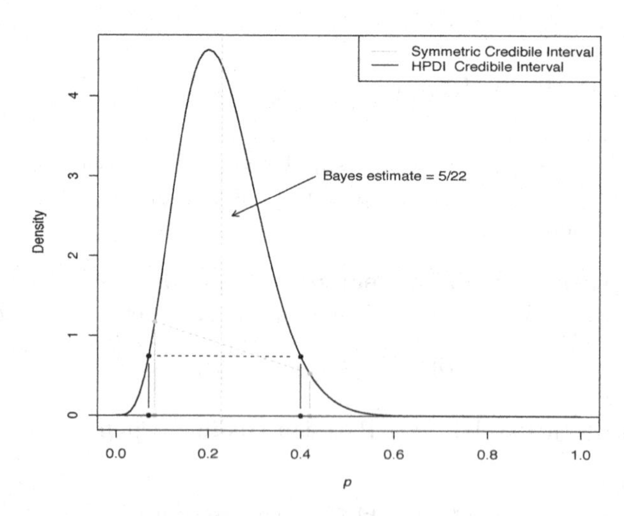

FIGURE 4.3
Plot of posterior density $beta(5, 17)$ for Example 4.20, with symmetric credible interval and HPDI indicated.

To construct a $1 - 0.05$ HPDI, we can write a simple program to determine c for which $\mathcal{I} = \{\theta : \pi(\theta \mid x) \geq c\}$ satisfies the level constraint $P_\pi(\theta \in \mathcal{I} \mid X = x) = 0.95$. We find that value is $c \approx 0.7413$, which yields interval bounds $L = Q_{0.0127} \approx 0.0692$, $U = Q_{0.0373} \approx 0.399$. For the symmetric credible interval we have $K = U - L = 0.419 - 0.082 = 0.337$. For the

HPDI we have $K = U - L = 0.399 - 0.0692 = 0.3298$. Figure 4.3 shows the $beta(5, 17)$ density, with the symmetric credible interval and the HDPI indicated. ■

4.9 Problems

Problem 4.1 Consider Equation (4.1) of Example 4.1. Construct a matrix $\mathbf{B}$ which maps the coefficients $(g_0, \ldots . g_n)$ to $(a_0, \ldots, a_n)$. Then verify the claim made in Example 4.1 that any polynomial in θ of at most order n is U-estimable given $X \sim bin(n, \theta)$.

Problem 4.2 Suppose for a scale parameter model there exists a scale equivariant estimator $\delta(X)$ for which $E_\theta[|\delta(X)|] < \infty$. Show that the scale parameter is U-estimable. **HINT:** Adapt the argument of Example 4.2, making use of Theorem 3.8.

Problem 4.3 Suppose we are given the sample mean $\bar{X}$ and sample variance S^2 of an *iid* sample of size n from a uniform distribution on the interval (a, b) where a and b are unknown. Derive a method of moments estimator for a and b.

Problem 4.4 An *iid* sample from a uniform distribution on (a, b) is partially observed through the two random variables U, V where

$$\begin{aligned} U &= \text{ the number in the sample below 1,} \\ V &= \text{ the number in the sample above 2.} \end{aligned}$$

Derive a method of moments estimator of a and b based on U and V.

Problem 4.5 Suppose $Y_1, Y_2, \ldots, Y_n$ is an *iid* sequence satisfying $Y_i = \beta x_i + \epsilon_i$, $i = 1 \ldots n$, where the x_i's are known constants, and $\epsilon_1, \epsilon_2, \ldots, \epsilon_n$ are *iid* with a normal distribution with mean 0 and variance σ^2.

(a) Calculate the maximum likelihood estimator of β. Is it unbiased?
(b) Calculate the method of moments estimator of β. In particular, derive an unbiased estimator of β of the form $a \sum_i Y_i$ for some a. Are there conditions on x_i?
(c) Which estimator has the smallest variance? Does this depend on x_i?

Problem 4.6 Suppose $X_1, \ldots, X_n$ is an *iid* sample with $X_1 \sim exp(\mu, \tau)$ for unknown parameters $\mu \in \mathbb{R}$ $\tau > 0$. Construct a method of moment estimators for (μ, τ) based on sample mean and variance $\bar{X}$, S^2.

Problem 4.7 Suppose $X_1, \ldots, X_n$ is an *iid* sample in which X_1 is sampled with equal probability from a uniform density on $(a, a+1)$ or a uniform density on $(b, b+1)$, where a and b are unknown. Construct a method of moments estimate of (a, b) based on the first two sample moments.

Problem 4.8 This problem is a continuation of the species sampling problem of Example 4.4. Let $\mathbf{P} = (p_1, \ldots, p_\nu)$ be a probability distribution on the set of categories $\{1, \ldots, \nu\}$. Let $X_1, \ldots, X_n$ be an *iid* sequence of observations from $\mathbf{P}$. We take n to be fixed. Previously, we consider the problem of estimating $z = \sum_{j=1}^{\nu} p_j I\{n_j = 0\}$, $n \geq 1$, where $n_j = \sum_{i=1}^{n} I\{X_i = j\}$ is the frequency of category j among the first n observations. The quantity z can be interpreted as the probability that a new observation (the $n+1$st) would be a category unseen in the previous n observations.

Here, we consider the problem of estimating $M = \sum_{j=1}^{\nu} I\{n_j = 0\}$, $n \geq 1$, the actual number of unseen categories (Chao, 1984). We assume ν is unknown, so that M is not observable. Let $a_k = \sum_{j=1}^{\nu} I\{n_j = k\}$, $n \geq 1$ interpretable as the number of categories observed exactly k times (commonly referred to as the *frequency of frequencies* in the literature). Note that $M = a_0$. We can observe a_k, $k \geq 1$, but we cannot observe a_0.

(a) Derive a general expression for the expected value $E[a_k]$, $k = 1, \ldots, n$.
(b) Use the Cauchy-Schwarz inequality to derive a lower bound of $E[a_0]$ that depends only on $E[a_1]$, $E[a_2]$ and n.
(c) Show how a lower bound for M can be estimated using the available data.

Problem 4.9 Suppose $X \in bin(n, 0.5)$, where n is odd. What is $Q(0.5)$, where $Q(p)$ is the quantile function? What is $E[\hat{m}]$, where $\hat{m}$ is the sample median of any sample from $bin(n, 0.5)$?

Problem 4.10 Suppose we estimate a quantile function $Q(p)$ from a sample $(X_1, \ldots, X_n)$ by setting $\hat{Q}(p_k) = X_{(k)}$, $k = 1, \ldots n$, where $p_k = k/(n+1)$, then using linear interpolation to evaluate $\hat{Q}(p)$ for any other $p \in (0, 1)$. If the observed sample is $(3.4, 3.9, 2.7, 1.9, 5.2, 0.6, 1.6)$, evaluate the estimates $\hat{Q}(1/4), \hat{Q}(1/2)$, and $\hat{Q}(3/4)$.

Problem 4.11 For some fixed m a random variable X_1 has density

$$f_\theta(i) = \frac{e^{-a_i\theta}}{\sum_{j=1}^{m} e^{-a_j\theta}}, \quad i = 1, \ldots, m$$

for some (distinct) constants $a_1, \ldots, a_m$. Suppose we observe a random sample $X_1, \ldots, X_n$ from f_θ.

(a) Derive an equation whose solution is the maximum likelihood estimate of θ.
(b) Is the solution to this equation unique? **HINT:** What is the variance of the random variable a_I where I has density f_θ?.

Problem 4.12 Define random variable $X \sim bin(n, p)$. Suppose (unlike the more common inference problem) we know that $p = 1/2$, and n is the unknown parameter.

(a) Derive the likelihood function L_n for parameter n.
(b) By deriving an expression for L_{n+1}/L_n, develop a method for determining the MLE of n. **HINT:** Determine when L_{n+1}/L_n is less than, equal to, or greater than 1.
(c) Determine the MLE of n when $X = 4$. If the MLE is not unique, determine all values which maximize L_n.

Problem 4.13 Suppose $(X_1, \ldots, X_n)$ is an *iid* sample with $X_1 \sim N(\mu, |\mu|)$, $\mu > 0$.

(a) Show that an MLE of μ always exists, and derive an explicit form.
(b) Suppose the parameter space is expanded to $\Theta = \{\mu \geq 0\}$ (but see Section 1.4.1). Here, $\mu = 0$ is interpreted as the point mass distribution $P(X_1 = 0) = 1$. Does this change the answer to Part (a)? What would be the value of the MLE derived in Part (a) if $\mu = 0$ was the true parameter? **HINT:** What is the limit of $\log(x)/x^{-1}$ as $x \to 0$?
(c) Suppose now the parameter space is expanded to $\Theta = \{\mu \in \mathbb{R}\}$, with the case $\mu = 0$ interpreted as before. If we accept maximum likelihood solutions of the type derived in Example 4.9, what would be the MLE for this case? Is it ever possible that the MLE would be strictly negative?

Problem 4.14 Suppose $X_1, \ldots, X_n$ is an *iid* sample with $X_1 \sim gamma(\tau, \alpha)$, $n = 15$, and we observe $\sum_{i=1}^{15} X_i = 14.42$ and $\prod_{i=1}^{15} X_i = 10.12$. What is the maximum likelihood estimate of $\theta = (\tau, \alpha)$ if the parameter space is $\Theta = \{\alpha \in \{2, 3, 4\}, \tau > 0\}$?

Problem 4.15 We are given an *iid* sample of size $n = 10$ from a $N(\mu, \sigma^2)$ distribution. Suppose for this particular sample $\sum_{i=1}^{10} X_i = 5.39$ and $\sum_{i=1}^{10} X_i^2 = 10.37$. Normally, the parameter space would be $\theta = (\mu, \sigma) \in \Theta = (-\infty, \infty) \times (0, \infty)$. However, suppose we have prior knowledge that μ is either 0 or 1, that is $\theta = (\mu, \sigma) \in \Theta^* = \{0, 1\} \times (0, \infty)$. Calculate the MLE of θ over the reduced parameter space Θ^*.

Problem 4.16 Suppose we are given paired random vectors $\mathbf{X} = (X_1, \ldots, X_n)$, $\mathbf{Y} = (Y_1, \ldots, Y_n)$, where (X_i, Y_i) has a bivariate normal density with means μ_x, μ_y, variances σ_x^2, σ_y^2 and correlation coefficient ρ. Assume all parameters are unknown. Derive the MLE of $\theta = (\mu_x, \mu_y, \sigma_x^2, \sigma_y^2, \rho)$. **HINT:** Make use of Theorem 4.4, then use a method of moments approach.

Problem 4.17 The MLE for $\theta = (\tau, \alpha)$ based in an *iid* sample $\mathbf{X} = (X_1, \ldots, X_n)$, $X_1 \sim$ *gamma*(τ, α) does not have a closed form solution. The generalized gamma distribution is defined by the power transformation X_1^γ, $\gamma > 0$, in which case we have the three parameter model $f(x) = (\tau^{\alpha\gamma})^{-1} \gamma x^{\alpha\gamma - 1} \exp\{-(x/\tau)^\gamma\}/\Gamma(\alpha)$ $x > 0$.

(a) Suppose we have an *iid* sample from the three parameter generalized gamma density. For the moment, assumed α is known. Give the likelihood equations for the remaining parameters τ and γ.
(b) Derive the values of α and τ for which the likelihood equations are solved when $\gamma = 1$. How do these compare to the MLEs of α and τ for the *gamma*(τ, α) density? See Ye and Chen (2017) for more on this problem.

Problem 4.18 Suppose the traffic flow rate on a certain highway is expressed as λ vehicles per hour. In other words, the number of vehicles which pass a given point in one hour is, on average, λ.

Then, suppose at one point on this highway there is a toll booth installation with $N = 10$ toll booths. When a car approaches the installation, any one of the tolls is chosen at random. As an approximation, we will assume that the number of vehicles which pass through a given toll in a time interval of length T has a Poisson distribution with mean $\lambda T/M$. We may also assume that the number of vehicles passing through distinct tolls are statistically independent. Let X be the number of toll booths through which at least one vehicle has passed within a one minute period.

(a) Derive the log-likelihood function of λ for given observation $X = x$. Plot this function against λ for $X = 6$.
(b) Give a closed form expression for the maximum likelihood estimate (MLE) of λ. Create a table of the MLEs for $x = 0, 1, \ldots, 9$.
(c) What happens when $x = 10$? Is it possible to give a MLE for λ, of any kind, for this case?

Problem 4.19 Prove Theorem 2 of Wedderburn (1974). The log-likelihood of a random variable $X \in \mathbb{R}$ satisfies Equation (4.16), where $E[X] = \mu$, $V(\mu) = \text{var}_\mu[X]$, *iff* its density is of the form $f_X(x) = \exp(\eta x - A(\eta))h(x)$.

Problem 4.20 A lake contains N fish. Suppose J fish are caught, tagged, then released. After a period of time, K fish are caught (assume these K fish are distinct). Suppose X of these have been previously tagged. Assuming both samples are random samples, we have $X \sim$ *hyper*$(J, N - J, K)$.

(a) Derive an expression using X, J, K, denoted $\hat{N}$, which can used as an estimate of total population size N. Use $E[X]$ as a guide. This method is known as *mark and recapture*, and is commonly used to estimate population sizes.

(b) We may construct a confidence set of confidence level $1-\alpha$ in the following way. Set x_{obs} to be the observed value of X. Then let N^* be a possible value of N. Then

$$N^* \in CS \textit{ iff } P(Y \leq x_{obs}) > \alpha/2 \text{ and } P(Y \geq x_{obs}) > \alpha/2, \tag{4.38}$$

where $Y \sim hyper(J, N^* - J, K)$. Then CS will consist of all integers between some lower and upper bounds, that is, $CS = \{N : N_L \leq N \leq N_U\}$ for some N_L, N_U. Then $P(N \in CS) \geq 1-\alpha$, so that the confidence set contains the true value of N with a probability of at least $1-\alpha$. Write an `R` function which accepts (X, J, K, α) as input, and outputs the bounds N_L, N_U for a confidence set CS for N of confidence level $1-\alpha$. To do this, set N_U, N_L to be the maximum and minimum values of N^*, respectively, that satisfy condition (4.38). If you use a search algorithm, a good starting point would be $\hat{N}$ (rounded off to the nearest integer), which would be within the bounds N_L, N_U.

(c) Use your function to determine a confidence set for N when $X = 13$, $J = 200$, $K = 100$, $\alpha = 0.05$.

Problem 4.21 Suppose $X_1, \ldots, X_n$ is an *iid* sample from a uniform density on $(0, \theta)$ for some $\theta > 0$, Show that the quantity $Z = X_{(n)}/\theta$ is a pivot, where $X_{(n)} = \max\{X_1, \ldots, X_n\}$. Use the pivot to construct an exact confidence interval for θ. Assume a 95% confidence level, and a sample size of $n = 10$. Use the quantities $0.025^{1/10} = 0.692$ and $0.975^{1/10} = 0.997$.

Problem 4.22 Recall the location and scale parameter models of Section 3.3.

(a) Suppose $\mathbf{X} = (X_1, \ldots, X_n)$ is a sample from a model with location parameter μ. Prove that for any location equivariant estimator $\hat{\mu}(\mathbf{X})$, $\xi(\mathbf{X}, \mu) = \hat{\mu}(\mathbf{X}) - \mu$ is a pivot. **HINT:** Use Theorem 3.2.

(b) Suppose $\mathbf{X} = (X_1, \ldots, X_n)$ is a sample from a model with scale parameter τ. Prove that for any scale equivariant estimator $\hat{\tau}^k = \hat{\tau}^k(\mathbf{X})$ of τ^k, $\xi(\mathbf{X}, \tau^k) = \hat{\tau}^k(\mathbf{X})/\tau^k$ is a pivot. **HINT:** Use Theorem 3.8

Problem 4.23 Suppose $X_1, \ldots, X_n$ is an *iid* sample with $X_1 \sim exp(\mu, \tau)$ where τ is known but μ is unknown.

(a) Construct a pivot function for μ based on $\sum_i X_i$. Adjust it so that it has a χ^2 distribution. (Recall that a gamma with scale parameter 2 and shape parameter n has a χ^2 distribution with $2n$ degrees of freedom).

(b) Show how you would use the pivot function and a χ^2 distribution table to obtain a lower confidence bound for μ of confidence level $1-\alpha$.

Problem 4.24 Suppose $\mathbf{X} = (X_1, \ldots, X_n)$ is an *iid* sample with $X_1 \sim pois(\lambda)$. Show how Theorem 4.7 and Theorem 4.8 can be used to construct a level $1-\alpha$ confidence interval for λ. **HINT:** See Example 4.17.

Problem 4.25 Suppose $\mathbf{X} = (X_1, \ldots, X_n)$ is an *iid* sample with $X_1 \sim pareto(1, \theta)$.

(a) What is the density of $\log X_1$?

(b) Following Section 4.6.2 show how a level $1-\alpha$ confidence interval for θ can be constructed.

(c) Suppose $\mathbf{X} = (2.96, 4.36, 4.32, 1.74, 1.10)$. Construct a symmetric level 95% confidence interval for θ (that is, $P_\theta(L > \theta) = P_\theta(U < \theta) = \alpha/2$), then determine the 95% confidence interval with the shortest length (see Theorem 4.6).

Problem 4.26 Suppose $X_1, \ldots, X_n$ is a random sample from a uniform density on $(0, \theta)$, with $\theta > 0$ unknown. Let $X_{(n)} = \max\{X_1, \ldots X_n\}$.

(a) Show that $X_{(n)}/\theta$ is a pivot, and give its distribution.

(b) Use this pivot to derive a family of confidence intervals for θ with confidence level $1-\alpha$ using two different methods:

 (i) By using the $\alpha/2$ and $1-\alpha/2$ quantiles of the distribution of the pivot to derive the upper and lower bound;

 (ii) By using $X_{(n)}$ as the lower bound.

 The confidence intervals should depend on $X_{(n)}, n$ and α.

(c) We may construct a level $1-\alpha$ confidence interval by using the $p\alpha$ and $1-(1-p)\alpha$ quantiles of the pivot distribution for any $p \in [0,1]$. Is there a unique value of p which gives the confidence interval of the shortest length for all α, n *wp*1?

Problem 4.27 In Tate and Klett (1959), it is shown that the confidence interval for the variance of a normal density given in Equation (4.19) is not the shortest possible. This can be shown by considering the class of confidence intervals allocating probabilities $p\alpha$ and $(1-p)\alpha$ to the lower and upper tails, for $p \in (0,1)$. Let $Q_\nu(p)$ be the quantile function for a χ^2_ν distribution. Create an R function which calculates $g_\nu(p) = Q_\nu(p\alpha)^{-1} - Q_\nu(1-(1-p)\alpha)^{-1}$, $p \in (0,1)$. For, say, $\alpha = 0.05$ and $\nu = 10$, plot the function over the interval $p \in (0,1)$. Then determine $\text{argmin}_p g_\nu(p)$ (in R, the function `optimize(...)` works well for this purpose). After suitable normalization, your results can be compared to those tabulated in Tate and Klett (1959).

Problem 4.28 This exercise is a continuation of Problem 4.27. Suppose the confidence interval of Equation (4.19) for σ^2 is modified to give a confidence interval for the standard deviation σ. Further optimize this confidence interval by adapting the procedure of Problem 4.27. Are the optimal values of p the same? In general, can the optimal confidence interval for σ be obtained by simply taking the square root of the optimal confidence bounds for σ^2? Finally, how does this inform the choice between estimating σ^2 and σ?

Problem 4.29 Consider the location parameter model of Example 4.7. How can the interval $I = (X_{(n)} - 1, X_{(1)})$ be interpreted? What do you expect to happen to I as the sample size n increases?

Problem 4.30 Suppose we observe a Poisson random variable $X \sim pois(\lambda)$, and that λ has a prior density $\pi(\lambda)$ given by $\lambda \sim gamma(1/\beta, \alpha)$ for some fixed β, α. It was shown in Problem 3.27 that the gamma density is a conjugate prior for this model.

(a) Suppose we observe a random sample $X_1, \ldots, X_n$ from a Poisson distribution with mean λ. Denote the sum $S = \sum_{i=1}^n X_i$. Define prior density $\lambda \sim gamma(1/\beta, \alpha)$. Show that the Bayesian estimate for λ (that is, the mean of the posterior density) can be given by $\hat{\lambda}_{Bayes} = q\bar{X} + (1-q)\alpha/\beta$ where $\bar{X} = S/n$ and $q = n/(n+\beta)$, assuming that the posterior density is conditioned directly on S.

(b) A study is to use this model, and a sample size n is planned. Suppose prior knowledge suggests that $\lambda = \lambda^*$ for some fixed value λ^*. Since sample size can be taken as a measure of precision or certainty, the confidence in the prior belief can be expressed as a fraction of the proposed sample size, say $n^* = n/10$. What values α, β would be appropriate for the prior density of λ in this situation?

Problem 4.31 Suppose $X \sim nb(r,p)$. Then $E[X] = \mu = r/p$. Suppose in the context of Bayesian inference, we fix r, and assign a $beta(\alpha, \beta)$ prior distribution to p. It was shown in Problem 3.27 that the beta density is a conjugate prior for this model.

(a) Show that given single observation $X \sim nb(r,p)$, the maximum likelihood estimate of p is $\hat{p}_{MLE} = r/X$, and that, therefore, the maximum likelihood estimate of μ is $\hat{\mu}_{MLE} = r/\hat{p}_{MLE} = X$.

(b) We then observe a single negative binomial random variable $X \sim nb(r, p)$. Assuming $\alpha > 1$, what is the expected value of $\mu = r/p$ under the prior distribution of p (say $\hat{\mu}_{prior}$) and under the posterior distribution of p given observation $X = x$ (say $\hat{\mu}_{post}$)? **HINT:** What is $E[Z^{-1}]$ if $Z \sim beta(a, b)$?

(c) Show that we can write $\hat{\mu}_{post} = q\hat{\mu}_{MLE} + (1 - q)\hat{\mu}_{prior}$ where q depends only on α and r.

5

Hypothesis Testing

5.1 Introduction

A statistical inference statement usually takes the form of a point estimate, a confidence set or a hypothesis test. If we are given a parametric family $\theta \in \Theta$ for an observation $\tilde{X} \in \mathcal{X}$, and an estimand $\eta(\theta)$, a point estimate is a value intended to be close to the true value of $\eta(\theta)$, while a confidence set is a set of values intended to contain the true value of $\eta(\theta)$ with high probability. In contrast, a hypothesis test is defined by a "yes or no" question about θ. For example, suppose μ_X, μ_Y are the population means of two samples $X_1, \ldots, X_n$, $Y_1, \ldots, Y_m$. We may be interested in determining whether or not the means are equal, their actual values not being important. We could always construct a level $1 - \alpha$ confidence interval $C(\tilde{X})$ for the estimand $\eta = \mu_Y - \mu_X$. If $C(\tilde{X})$ does not contain zero, we would report $\mu_X \neq \mu_Y$. By construction, if it were true that $\mu_Y = \mu_X$, then the probably of reporting otherwise in error would be α.

This approach is quite reasonable, and is commonly used, but it would be ultimately limiting to base a theory of hypothesis testing on it. For this reason, this theory relies on its own concepts and terminology, which are introduced in Sections 5.2 and 5.4, while Section 5.3 enumerates principles to which we would expect a test to adhere.

Sections 5.5 and 5.6 discusses the one- and two-sided hypothesis test, probably the most commonly used form. Conditions are given under which the principles of Section 5.3 will hold for these tests, and some examples for which they don't are offered. The pivot was already encountered in the context of confidence intervals (Section 4.6), and plays a similar role for hypothesis tests. This is discussed in Section 5.7.

The next two sections introduce a number of methods of constructing hypothesis tests. Section 5.8 introduces the likelihood ratio test (LRT), which relies on the important concept of the full and reduced model. For our two-sample problem, the full model would allow $(\mu_X, \mu_Y) \in \Theta_F = \mathbb{R}^2$, where Θ_F is the parameter space of the full model. We could then construct a reduced model by a constraint such as $\mu_X = \mu_Y = \mu_{XY}$, defining the parameter space to be $\mu_{XY} \in \Theta'_R = \mathbb{R}$. However, it helps to recognize the reduced model as a special case of the full model, and define its parameter space to be a subset of Θ_F, that is $\Theta_R = \{(\mu_X, \mu_Y) \in \Theta_F : \mu_X = \mu_Y\}$. We may also say that Θ_R and Θ_F are nested models, although that term is used in various other contexts in the statistical literature (i.e. nested factorial designs, (Box *et al.*, 1978; Kutner *et al.*, 2004)). The purpose of the LRT is to determine whether data are observed from a full or reduced model. The LRT will be revisited in Chapter 14 in the context of large sample theory.

Similar tests are introduced in Section 5.9, and are an important application of sufficiency (Section 3.7). Recall from Section 3.12 that a multidimensional parameter is often decomposed into $\theta = (\psi, \xi)$, where ψ is the parameter of interest and ξ is designated as nuisance parameter. It is simple enough to describe a similar test as one for which dependence on ξ is eliminated by conditioning on a statistic which is sufficient for ξ. This will largely be our approach, although a number of other theoretical properties of hypothesis tests, such

DOI: 10.1201/9781003049340-5

as unbiasedness (Section 5.3), can be approached by a study of similar tests (Lehmann and Scheffé, 1950, 1955; Lehmann and Romano, 2005).

5.2 Basic Definitions

We are given a parametric family of distributions $\mathcal{P}$ for observation $\tilde{X} \in \mathcal{X}$. Suppose we are given a strict subset of the parameter space $\Theta_o \subset \Theta$, and let $\Theta_a = \Theta - \Theta_o$, so that Θ_o and Θ_a partition Θ. This defines null hypothesis:

$$H_o : \theta \in \Theta_o$$

and alternative hypothesis:

$$H_a : \theta \in \Theta_a.$$

A hypothesis consisting of a single parameter is a simple hypothesis (that is, Θ_o or Θ_a are singletons) and one consisting of multiple parameters is a composite hypothesis.

The inference problem is to determine, based on the observed data, to which hypothesis the true value of θ belongs. It is possible to see this as a binary classification problem, since we either reject or accept H_o. However, this is not the point of view taken by the theory of hypothesis testing, since the choices are not really symmetric.

To see this, we introduce some basic definitions. A Type I error is a false rejection of H_o, and a Type II error is a failure to reject H_o when H_a is true. A common convention is to use the symbols α, β for Type I and II error, respectively. These errors play very different roles. The null hypothesis is regarded as a default state of affairs, which is accepted in the absence of evidence to the contrary. Mathematically, this idea is expressed by bounding the probability of a Type I error by a small number α (frequently set to 5%). The problem is then to minimize the probability of a Type II error β under this constraint.

The rejection region defines the decision rule for the test. Suppose $\mathcal{X}$ is partitioned into two subsets, rejection region R and acceptance region $A = \mathcal{X} - R$. The intention is that H_o is rejected if $\tilde{X} \in R$. If

$$P_\theta(\tilde{X} \in R) \leq \alpha \text{ for all } \theta \in \Theta_o$$

then α is a level of significance for the test or rejection region. If

$$\sup_{\theta \in \Theta_o} P_\theta(\tilde{X} \in R) = \alpha$$

then α is the size of the test or rejection region. We may express a hypothesis test as a decision rule

$$\delta(\tilde{X}) = \delta_R(\tilde{X}) = I\{\tilde{X} \in R\},$$

where H_o is rejected if $\delta(\tilde{X}) = 1$. Suppose R can be defined using a statistic S. If there is among these a statistic T which is a mapping of all other such statistics, then T is the test statistic.

For the moment, denote the rejection probability as a function of θ by $\gamma(\theta) = E_\theta[\delta_R(\tilde{X})]$. Mathematically, there is usually no reason to distinguish between $\theta \in \Theta_o$ and $\theta \in \Theta_o^c$ when evaluating $\gamma(\theta)$. But the interpretation obviously differs, and we will use the convention $\alpha = \sup_{\theta \in \Theta_o} \gamma(\theta)$ (size, or Type I error) and $\beta(\theta) = 1 - \gamma(\theta)$, $\theta \in \Theta_o^c$ (Type II error). We then have the power of a test $pow(\theta) = 1 - \beta(\theta) = \gamma(\theta)$, $\theta \in \Theta_o^c$, which is the probability of correctly rejecting H_o for all alternatives. A representation of $pow(\theta)$ is usually referred to as a power curve.

It is important to note at this point that the Type I error α is a single upper bound, while the Type II error $\beta(\theta)$ is allowed to depend on $\theta \in \Theta_o^c$. It is mainly in this sense that we say the problem of choosing to accept or reject H_o is not symmetric in the two choices.

Randomized Tests

In some cases, there will be some advantage in considering a randomized test. This is expressible as a mapping $\delta : \mathcal{X} \to [0,1]$. Then, given $\tilde{X}$, H_o is rejected with probability $\delta(\tilde{X})$. This means the deterministic hypothesis test just introduced may be written $\delta(\tilde{X}) = I\{\tilde{X} \in R\}$, and so is simply a special case of a randomized decision rule.

In this case, the probability of rejecting H_o for any θ is $E_\theta[\delta(\tilde{X})]$. The definition of size and level extend naturally to $E_\theta[\delta(\tilde{X})] \leq \alpha$ for all $\theta \in \Theta_o$ for a level α test, and $\sup_{\theta \in \Theta_o} E_\theta[\delta(\tilde{X})] = \alpha$ for a size α test.

Randomized tests play a necessary role in the theory of hypothesis testing, as will be made clear in Chapter 10. We may have reason to construct a hypothesis test with some fixed size α_0. When observations are discrete, this may be possible only with a randomized test. Beyond this, randomized tests can arise using conventional methods even when observations are continuous (Example 10.3).

5.3 Principles of Hypothesis Tests

The hypothesis test is defined simply by any partion into two subsets of both the parameter space and the sample space. The objective is easily defined as well, that is, to minimize the probability of a Type II error subject to an upper bound of α on the probability of a Type I error. However, a useful first step is to define some principles to which we would expect a hypothesis test to conform. It turns out that once this is done, the range of hypothesis tests δ considered is significantly restricted, in some cases, to a single possibility.

Principle 1 [Sufficiency] The principle of sufficiency (Definition 3.12) holds that any inference regarding a parameter θ should depend on the data only through a (minimal) sufficient statistic. If accepted, these principles are applicable to any decision rule δ, and in turn to any test statistic on which δ depends.

Principle 2 [Nested Rejection Regions] Suppose R_α and $R_{\alpha'}$ are two rejection regions of sizes $\alpha < \alpha'$. Then we should have $R_\alpha \subset R_{\alpha'}$. To see this, suppose $\tilde{X} \in R_\alpha$. The effect of increasing the size of a test is to make it less stringent with respect to the evidence required to reject H_o. It follows that X should also be in $R_{\alpha'}$. In terms of decision rules, this principle implies that $\delta'(\tilde{X}) \geq \delta(\tilde{X})$ if the size of δ' is greater than that of δ. This allows us to extend the notion of nested rejection regions to randomized tests.

Principle 3 [Unbiasedness] Suppose we are given a test $\delta(\tilde{X})$. Let $\gamma(\theta)$ be the rejection probability. If there exists some α for which $\gamma(\theta) \leq \alpha$ for all $\theta \in \Theta_o$ and $\gamma(\theta) \geq \alpha$ for all $\theta \in \Theta_o^c$ then $\delta(\tilde{X})$ is an unbiased test.

The definition of a hypothesis test is quite general, so it is always possible to construct counterexamples to these principles. The question is whether of not they are satisfied by a test derived from some methodological principle. We next consider the three principles in turn.

Principle 1 is a natural consequence of various methods used to construct hypothesis tests in which the likelihood function plays the central role. In particular, test statistics derived from these methods are usually expressible as ratios

$$T(\tilde{X}) = \frac{\mathcal{L}(\theta''(\tilde{X}), \tilde{X})}{\mathcal{L}(\theta'(\tilde{X}), \tilde{X})} = \frac{g(S(\tilde{X}), \theta''(\tilde{X}))h(\tilde{X})}{g(S(\tilde{X}), \theta'(\tilde{X}))h(\tilde{X})} = \frac{g(S(\tilde{X}), \theta''(\tilde{X}))}{g(S(\tilde{X}), \theta'(\tilde{X}))}, \tag{5.1}$$

where $S(\tilde{X})$ is any sufficient statistic, and $\mathcal{L}(\theta, \tilde{x}) = p_\theta(\tilde{x}) = g(S(\tilde{x}), \theta''(\tilde{x}))h(\tilde{x})$ follows from the factorization theorem (Theorem 3.16). In this case, if $S \mapsto \theta'(\tilde{X})$ and $S \mapsto \theta''(\tilde{X})$ then $T(\tilde{X})$ will be a mapping of $S(\tilde{X})$ as well. This will be the case for the likelihood ratio test of Section 5.8. It will also be true of the Neyman-Pearson test of Definition 10.2, for which the test statistic assumes the form of Equation (5.1) with $\theta'(\tilde{X})$ and $\theta''(\tilde{X})$ equal to known constants (recall that any statistic maps to a constant).

Most, if not all, hypothesis tests can be defined using a rejection region expressible as $R = \{T(\tilde{X}) > t\}$. This form is usually given by some methodological principle, with the critical value t left unspecified. In this case, $t = t_\alpha$ can be chosen to force a specific test size α. Clearly, if $\alpha' > \alpha$ we must have $t_{\alpha'} < t_\alpha$, so that the rejection regions are nested. However, the principle is not universal. The structure of the Neyman-Pearson test of Definition 10.2 is similar to $R = \{T(\tilde{X}) > t\}$, but is somewhat more complex, involving a second parameter. Although this parameter is associated with randomization, it is still possible to construct nonrandomized tests which are optimal but not nested. However, in this type of case it will generally be possible to construct a family of *randomized* tests which are nested in the sense given in the statement of Principle 2. Thus, Principle 2 can generally be assumed, and its violation is not usually of any consequence. See Problem 5.3.

In contrast to Principle 2, validation of Principle 3 is an important issue in the theory of hypothesis testing, and conditions under which it holds should be well understood. This will be shown in Section 5.5. Although the principle is stated as a comparision of the rejection probability $\gamma(\theta)$ within Θ_o and within Θ_o^c, it is often more useful to regard the principle of unbiasdness as a statement about $\gamma(\theta)$ without reference to Θ_o. For example, suppose $\theta \in \mathbb{R}$, and we are given hypotheses $H_o : \theta \leq \theta_0$ and $H_a : \theta > \theta_0$ for some fixed θ_0. Then the unbiasdness principle will hold for *all* θ_0 if and only if $\gamma(\theta)$ is nondecreasing in θ.

In hypothesis testing we usually interpret the likelihood function in ratios, as in Equation (5.1). The following example demonstrates the importance of this.

Example 5.1 (Counter-Example to the Likelihood Principle) Suppose $\mathbf{X} = (X_1, \ldots, X_n)$ is an *iid* sample from $N(\mu, 1)$. The density is then

$$p_\mu(\mathbf{x}) = (2\pi)^{-n/2} \exp\left\{-\frac{1}{2}\sum_{i=1}^n (x_i - \mu)^2\right\}.$$

Suppose we are given null hypothesis $H_o : \mu = \mu_0$. The likelihood principle is not concerned with likelihoods of a single model, but with likelihood ratios. Suppose, in contrast, we regarded small values of $p_\mu(\mathbf{x})$ as evidence against H_o. We would then use rejection region

$$\sum_{i=1}^n (X_i - \mu_0)^2 > t. \tag{5.2}$$

However, we can write

$$\sum_{i=1}^n (X_i - \mu_0)^2 = \sum_{i=1}^n X_i^2 - 2n\bar{X}\mu_0 + n\mu_0^2 = (n-1)S^2 + n(\bar{X} - \mu_0)^2.$$

As we might expect, the hypothesis is rejected when $\bar{X}$ differs significantly from μ_0, since, from the preceding equality, the rejection rule (5.2) will hold for all large enough values of $(\bar{X} - \mu_0)^2$. However, the test statistic also includes a term equal to $(n-1)S^2$. The distribution of this term does not depend on μ (Example 3.30), and so only serves to add spurious variation to the procedure, thus decreasing its efficiency. On the other hand, the likelihood ratio p_{μ_1}/p_{μ_0} depends on $\mathbf{X}$ only through $\bar{X}$, which is a complete sufficient statistic for μ. ∎

The point of this example is that the likelihood function $\mathcal{L}(\theta; \tilde{X})$ is a function of θ and not of $\tilde{X}$. Since it is invariant to positive scalar transformations, it is only interpretable as a ratio $\mathcal{L}(\theta'; \tilde{X})/\mathcal{L}(\theta; \tilde{X})$ for a fixed observation $\tilde{X}$. This would have the effect of eliminating any dependence of the decision rule on S^2.

5.4 The Observed Level of Significance (*P*-Values)

Although a hypothesis test is formally defined by a rejection region, and hence a binary decision, it is the usual practice to also report an observed level of significance or *P*-value. This serves as a type of index giving the degree to which the evidence contradicts the null hypothesis H_o (with smaller values tending to disprove H_o). Several definitions exist, and each must be interpreted carefully. However, they all should lead to the same number.

Definition 5.1 (Observed Level of Significance I) The observed level of significance or *P*-value is the probability that a new sample collected under identical conditions would be at least as contradictory of the null hypothesis H_o as the one observed, assuming H_o is true. If H_o is composite, then this number is the largest attainable among all $\theta \in \Theta_o$. ∎

The phrase "at least as contradictory of the null hypothesis" used in Definition 5.1 is, of course, subject to interpretation. Consider the following example.

Example 5.2 (The Unlikely Birthdays Problem) A restaurant currently has 20 customers, 3 of whom claim that today is their birthday. If we suspect subterfuge, the *P*-value might offer quantitative evidence. We can reason that, ignoring leap years, the number of birthday celebrations should be $X \sim bin(20, 1/365)$, which defines a null hypothesis H_o. Note that "at least as contradictory" as the observed evidence would here mean $X \geq 3$ (and not $X = 3$). This gives a *P*-value of $P_o(X \geq 3) \approx 2.3 \times 10^{-5}$. Our suspicions seem well-founded. ∎

The definition of the *P*-value can be made more precise when we have nested hypothesis tests. This means we have a family of hypothesis tests indexed by a level of significance $\alpha \in [0, 1]$ (for tests based on discrete observations α may not be a test size, but will be a level of significance). Now the phrase "at least as contradictory of the null hypothesis" is completely defined by the hypothesis test. This leads to our second definition.

Definition 5.2 (Observed Level of Significance II) Suppose we are given a family of nested hypothesis tests. The observed level of significance or *P*-value is the smallest level of significance under which the null hypothesis would be rejected given the observed data. ∎

The use of *P*-values in scientific literature is the subject of some disagreement (Halsey *et al.*, 2015). The protocol is well known. If $P \leq 0.05$, then a finding (in the form of an alternative hypothesis) can be reported as true, at least provisionally. There is nothing special about the number 0.05 (suggested by R.A. Fisher as a reasonable choice), but the adoption of a universal standard for statistical evidence seems to this author to have self-evident advantages. In practice, many researchers would be reluctant to report a finding with a P-value of 0.049. Ultimately, there is no escaping the decision theoretic character of the problem. To accept and report a finding is to make a binary decision. There are four possible outcomes, a true positive, a false positive, a true negative or a false negative. Conventional hypothesis testing is simply a way of balancing the probabilities of a false negative and a false positive, and it is difficult to see how an alternative approach would differ in any important way. The P-value is simply one tool used for this purpose. See also Wasserstein and Lazar (2016).

5.5 One- and Two-Sided Tests

Suppose $\mathcal{P} = \{p_\theta : \theta \in \Theta\}$, $\Theta \subset \mathbb{R}^m$, is a family of densities for observation $\tilde{X} \in \mathcal{X}$. A hypothesis test will often have the following structure. There exists a mapping $\Delta = \Delta(\theta) \in \mathbb{R}$, and a test statistic $T(\tilde{X}) \in \mathbb{R}$, the distribution of which depends on θ only through $\Delta(\theta)$. The test itself is defined as $\delta(\tilde{X}) = I\{T(\tilde{X}) \in R\}$ for some subset $R \subset \mathbb{R}$. Then suppose the null hypothesis can be expresses $H_o : \Delta(\theta) = \Delta_0$ for some fixed value $\Delta_0 \in \mathbb{R}$. Note that since $\Delta(\theta)$ is a mapping from $\mathbb{R}^m$ to $\mathbb{R}$, H_o need not be a simple hypothesis.

The One-Sided Upper-Tailed Test

A (one-sided) upper-tailed test compares null and alternative hypotheses $H_o : \Delta(\theta) \leq \Delta_0$ and $H_a : \Delta(\theta) > \Delta_0$. We assume $T(\tilde{X})$ is oriented so that large values tend to reject H_o, so we define the rejection region $R_\alpha = \{T(\tilde{X}) > t_\alpha\}$ for a size α test, where t_α satisfies

$$\alpha = \sup_{\{\theta:\Delta(\theta)\leq\Delta_0\}} P_\theta\left(T(\tilde{X}) > t_\alpha\right). \tag{5.3}$$

Note that a solution t_α to Equation (5.3) need not exist for all α, so it will suit our purposes for the moment to consider only α for which it does. In this case, the rejection regions are nested. Otherwise, this type of test can be randomized to attain any specified size. See Problem 5.2 or Theorem 10.1.

That the alternatives $\Delta(\theta) > \Delta_0$ can be indexed with a single parameter Δ allows us to clarify the unbiasedness principle introduced in Section 5.3 through the following example:

Example 5.3 (Power Curves and the Principles of Hypothesis Testing) Suppose we are given a single observation $X \sim N(\theta, v(\theta))$ for some function $v(\theta)$. To fix ideas, assume $v(1) = 1$, and take parameter space $\theta \in \Theta = (0, \infty)$. Then consider hypotheses $H_o : \theta \leq 1$ and $H_a : \theta > 1$. Our test statistic will be $T(X) = X$, so let F_θ be the CDF for $N(\theta, v(\theta))$. Let Z_p be the p-quantile for the $N(0,1)$ distribution, set $t_\alpha = 1 + Z_{1-\alpha}$, then define rejection region $R_\alpha = \{T(X) > t_\alpha\}$. The rejection probability as a function of θ is then $\gamma(\theta) = 1 - F_\theta(t_\alpha)$. We will consider three models: (i) $v(\theta) = \theta$; (ii) $v(\theta) = \theta^2$; (iii) $v(\theta) = \theta^4$, and set $\alpha = 0.05$. The rejection probability functions $\gamma(\theta)$, or power curves, for the three models are plotted in Figure 5.1. The limiting rejection probability $\gamma(\infty) = \lim_{\theta\to\infty} \gamma(\theta)$ is indicated for each model with a horizontal gray line, where $\Phi(x)$ is the CDF for the standard normal distribution $N(0,1)$ (see Problem 5.4).

For the given rejection region the test is unbiased for each model. In particular, for the null hypothesis the rejection probability is maximized at $\theta = 1$, so the size of the test is $\alpha = 0.05$. However, only for model (i) $v(\theta) = \theta$ can the hypothesis test be regarded as entirely satisfactory. For a one-sided test, we would expect larger values of $\theta > 1$ to yield larger rejecton probabilities, and for the rejection probability to approach 1 as θ becomes arbitrarily large. However, for models (ii) and (iii) the rejection probability does not approach 1, and for model (iii) the rejection probability is actually decreasing for all larger enough θ. Thus, while the test of this example is unbiased for all three models, we could construct a hypothesis $H_o : \theta \leq \theta_0$ and a test with rejection region $R = \{T(X) > t\}$ that was not unbiased for model (iii). ∎

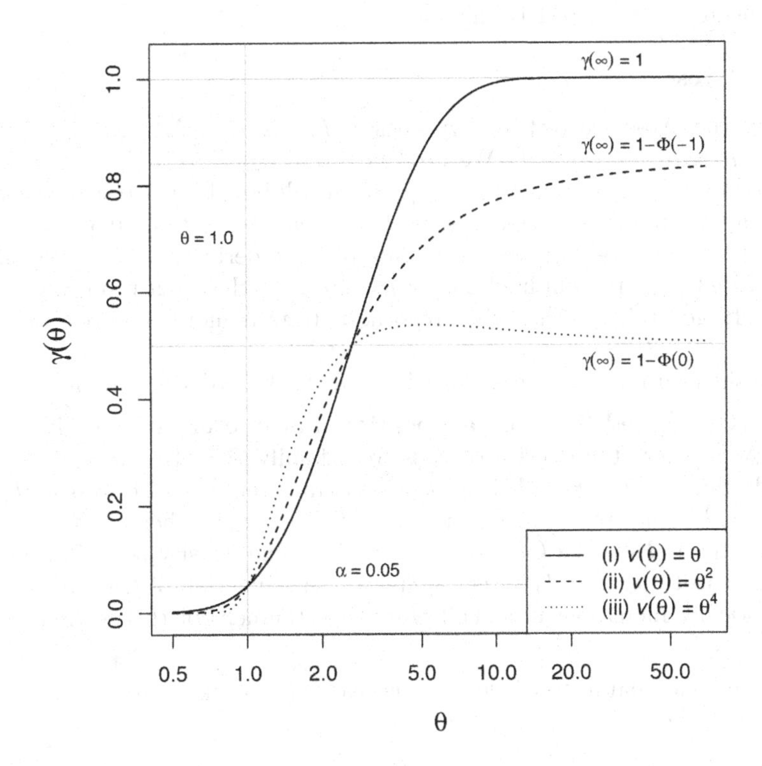

FIGURE 5.1
Power curves for Example 5.3. The limiting rejection probability $\gamma(\infty) = \lim_{\theta\to\infty} \gamma(\theta)$ is indicated for each model with a horizontal gray line, where $\Phi(x)$ is the CDF for the standard normal distribution $N(0, 1)$. The horizontal axis is on a logarithmic scale.

5.6 Unbiasedness and Stochastic Ordering

The reader may wish at this point to review the role played by stochastic ordering (Section 1.5) in the construction of confidence intervals (Section 4.6.4), since much the same role is played here. This may be stated concisely as the condition that the CDF of $T(\tilde{X})$, say F_Δ, is stochastically increasing in Δ. Suppose $T(\tilde{X}') \sim F_{\Delta'}$, $T(\tilde{X}) \sim F_\Delta$ and $\Delta \leq \Delta'$. Then $T(\tilde{X}') \geq_{st} T(\tilde{X})$. Equivalently, $F_\Delta(t)$ is a nonincreasing function of Δ for each t. In order to construct a level α test, set the critical value $t_\alpha = Q(1-\alpha)$, where $Q(p)$ is the quantile function of F_{Δ_0}. Then define the rejection region $R_\alpha = \{T(\tilde{X}) > t_\alpha\}$. Under the stochastic ordering assumption $pow(\Delta) = P_\Delta(T(\tilde{X}) > t_\alpha)$ is a nondecreasing function of Δ, and the size of the test is $\alpha^* = pow(\Delta_0)$. Note that $\alpha^* \leq \alpha$, but also that α^* is the maximum size of a level α test (Problem 5.1). Thus, since the power curve if nondecreasing, the test must be unbiased for one-sided upper-tailed tests.

The One-Sided Lower-Tailed Test

Clearly, the same properties hold for a (one-sided) lower-tailed test for hypotheses H_o : $\Delta(\theta) \geq \Delta_0$ and $H_a : \Delta(\theta) < \Delta_0$. In this case, the rejection region becomes $R_\alpha = \{T(\tilde{X}) < t_{1-\alpha}\}$. Note that the stochastic ordering F_Δ remains the same as for the upper-tailed test, and similarly ensures that the test is unbiased.

The Two-Sided Test

Finally, we have the two-sided test for hypotheses $H_o : \Delta(\theta) = \Delta_0$ and $H_a : \Delta(\theta) \neq \Delta_0$. Suppose $\alpha_L + \alpha_U = \alpha$, $\alpha_L, \alpha_U \geq 0$. We need not have $\alpha_L = \alpha_U$. Then we set rejection region $R_\alpha = \{T(X) < t_{1-\alpha_L}\} \cup \{T(X) > t_{\alpha_U}\}$, which will be a level α test by construction. Then the nesting and unbiased properties need not hold. In Section 10.5.2, we will make use of the fact that the two-sided test is unbiased *iff* the derivative $d\,pow(\Delta)/d\Delta = 0$ at $\Delta = \Delta_0$. The nested property will hold if α_L, α_U increase with α, but it obviously need not hold otherwise. In general, the theory of hypothesis tests is more definitive for one-sided tests.

We next consider an important case for which stochastic ordering will hold.

Example 5.4 (One-Sided Test for a Location Parameter) We are given a sample $\mathbf{X} = (X_1, \ldots, X_n)$, where the density of $\mathbf{X}$ is from family $\mathcal{P} = \{p_\theta : \theta \in \Theta\}$ defined on sample space $\mathcal{X}$. Suppose $\theta \in \Theta = \mathbb{R}$ is a location parameter. We wish to test $H_o : \theta \leq \theta_0$ against $H_a : \theta > \theta_0$ using rejection region $R = \{T(\mathbf{X}) > t_\alpha\}$, where $T(\tilde{X})$ is a location equivariant statistic. If F_0 is the CDF of $T(\tilde{X})$ under some density $p_0 \in \mathcal{P}$, then $F_\theta(t) = F_0(t - \theta)$ under density p_θ. But $F_0(t)$ is a nondecreasing function of t, which implies that $F_\theta(t)$ is stochastically increasing in θ. Following the argument of this section, the test is unbiased. ■

The case of the scale parameter is left to the reader (Problem 5.10).

5.7 Hypothesis Tests and Pivots

Recall the pivot of Definition 4.4, used in the context of confidence intervals. Suppose we are given a family of densities $\mathcal{P} = \{p_\theta : \theta \in \Theta\}$ for observation $\tilde{X} \in \mathcal{X}$, and we have estimand $\eta(\theta) \in \Theta_\eta$. Then a mapping $\xi : \mathcal{X} \times \Theta_\eta \to \mathbb{R}^m$, for some $m \geq 1$ is a pivot if the distribution of $\xi(\tilde{X}, \eta(\theta))$ does not depend on θ. This leads to the construction of a confidence interval.

A similar idea can be used to construct hypothesis tests, but with an important difference. Suppose the null hypothesis takes the form $H_o : \eta(\theta) = \eta_0$ for some fixed value $\eta_0 \in \Theta_\eta$. If we substitute the hypothetical value η_0 into the pivot, we get test statistic $T(\tilde{X}) = \xi(\tilde{X}, \eta_0)$. The distributon of $T(\tilde{X})$ will not depend on θ when $\theta \in \Theta_0$, but will depend on θ over Θ_0^c, and the degree to which this is true determines the efficiency of the test. This is the basis for the commonly used T test, which we consider in the next example.

Example 5.5 (Pivots Based on the T statistic) In Example 4.13, we discussed the T statistic in the context of confidence intervals. Suppose $\bar{X}$, S^2 are the sample mean and variance for an *iid* sample of size n from $N(\mu, \sigma^2)$. Then

$$T = \frac{\bar{X} - \mu}{S/\sqrt{n}} \tag{5.4}$$

possesses a T distribution of $n-1$ degrees of freedom, and is therefore a pivot of the form $\xi(\tilde{X}, \mu)$. For a hypothesis test we replace μ with a specific value μ_0 which will define the null hypothesis. This gives

$$T_0 = \frac{\bar{X} - \mu_0}{S/\sqrt{n}} \tag{5.5}$$

From Section 2.5, we have $T_0 \sim T_{\Delta;n-1}$, where $\Delta = \sqrt{n}(\mu - \mu_0)/\sigma$ is the noncentrality parameter of a noncentral T distribution. Then by Theorem 2.7 the distribution of T_0 is stochastically increasing in Δ. Since the hypotheses $H_o : \Delta \leq 0$ and $H_o : \Delta \geq 0$ equivalent to $H_o : \mu \leq \mu_0$ and $H_o : \mu \geq \mu_0$, respectively, we may conclude that the upper- and lower-tailed tests will be unbiased.

Similarly, the hypothesis $H_o : \Delta = 0$ is equivalent to $H_o : \mu = \mu_0$. It may be shown that $T_0^2 \sim F_{\Delta;1,n-1}$, so that by Theorem 2.7 the distribution of T_0^2 is stochastically increasing in Δ (Problem 5.12). This means that the test for hypotheses $H_o : \Delta = 0$ and $H_a : \Delta \neq 0$ based on rejection region $R = \{|T_0| > t\}$ is unbiased. However, the more general rejection region $R = \{T_0 \notin [t_0, t_1]\}$ will not be unbiased, unless $t_0 = -t_1$ (Problem 5.11). ■

We next consider the conventional tests for the variance of a normal distribution based on the sample variance.

Example 5.6 (Pivots Based on the Sample Variance) In Example 4.13 the pivot

$$W = (n-1)S^2/\tau$$

was introduced, which possesses a χ^2 distribution with $n-1$ degrees of freedom when S^2 is calculated from an *iid* sample of size n from $N(\mu, \tau)$. One- and two-sided tests with null hypothesis $H_o : \tau = \tau_0$ can be based on test statistic $W_0 = (n-1)S^2/\tau_0$. In this case, W_0 is scale equivariant, therefore the distribution of W_0 will be stochastically ordered *wrt* τ, and the one-sided test will be unbiased (Example 5.4, Problem 5.10).

We next consider the two-sided test. In Example 5.5 we were able to verify that for the two-sided hypotheses $H_o : \mu = \mu_0$ and $H_a : \mu \neq \mu_0$, a symmetric rejection region $R = \{|T_0| > t\}$ yielded an unbiased test. This, however, is due to the fact that the distribution of T_0 is symmetric about zero under H_0, a condition which does not hold here. Following Example 5.5, consider rejection region $R = \{W_0 \notin [t_0, t_1]\}$. The rejection probability as a function of τ is then

$$\gamma(\tau) = P_{\tau_0}(W_0 < t_0) + P_{\tau_0}(W_0 > t_1) = P(U < (\tau_0/\tau)t_0) + P(U > (\tau_0/\tau)t_1),$$

where $U \sim \chi^2_{n-1}$. Let $g(u)$ be the density function of U. Then the derivative of $\gamma(\tau)$ *wrt* τ is given by

$$\gamma'(\tau) = -\tau_0/\tau^2 g((\tau_0/\tau)t_0) + \tau_0/\tau^2 g((\tau_0/\tau)t_1).$$

If the test is unbiased, we must have $\gamma'(\tau_0) = 0$, which is equivalent to $g(t_0) = g(t_1)$. A common practice is to use a rejection region which is symmetric in the tail probabilites. Since under H_o we have $W_0 \sim \chi^2_{n-1}$, this would give critical values $t_0 = \chi^2_{n-1;1-\alpha/2}$ and $t_1 = \chi^2_{n-1;\alpha/2}$. However, for $n-1 = 10$ and $\alpha = 0.05$, this would give $t_0 \approx 3.247$ and $t_1 = 20.483$, as well as $g(t_0) \approx 0.0285$ and $g(t_1) \approx 0.00817$, so that this test is not unbiased. However, an unbiased test may be derived from the condition $g(t_0) = g(t_1)$, which we have earlier associated with the problem of determining the optimal confidence interval. Thus, by comparing this case to Example 4.16, we can see that the problem of determining a two-sided unbiased test is the same as determining the confidence interval of the smallest length. ■

5.8 Likelihood Ratio Tests

Suppose we are given observation $\tilde{X} \in \mathcal{X}$ from parametric family $\mathcal{P} = \{p_\theta : \theta \in \Theta\}$. We have likelihood function $\mathcal{L}(\theta; \tilde{X})$ and log-likelihood function $\ell(\theta; \tilde{X})$. For some subset $\Theta_0 \subset \Theta$ we wish to test hypotheses

$$H_o : \theta \in \Theta_0, \quad H_a : \theta \notin \Theta_0.$$

The likelihood ratio test statistic (LRT) is defined as

$$\Lambda(\tilde{X}) = \frac{\sup_{\theta \in \Theta_0} \mathcal{L}(\theta; \tilde{X})}{\sup_{\theta \in \Theta} \mathcal{L}(\theta; \tilde{X})}.$$

Necessarily, $\sup_{\theta \in \Theta_0} \mathcal{L}(\theta; \tilde{X}) \leq \sup_{\theta \in \Theta} \mathcal{L}(\theta; \tilde{X})$, so we must have $\Lambda(\tilde{X}) \in [0, 1]$. By construction, $\Lambda(\tilde{X})$ does not depend on any unknown parameters and is therefore a statistic.

Implicit in this test is a comparison of a full model to a reduced model, which forms the basis of much of the theory of hypothesis testing. For the full model, θ can take any value in Θ. The reduced model is defined by some constraint, expressible in the form $\theta \in \Theta_0$, which defines the null hypothesis. The models are nested models, in the sense that the reduced model is a special case of the full model, since $\Theta_0 \subset \Theta$. Then we can fit each model, that is, determine the full and reduced MLEs

$$\hat{\theta}_f = \text{argmax}_{\theta \in \Theta} \mathcal{L}(\theta; \tilde{X}), \quad \hat{\theta}_r = \text{argmax}_{\theta \in \Theta_0} \mathcal{L}(\theta; \tilde{X}).$$

We may write the LRT statistic

$$\Lambda(\tilde{X}; \hat{\theta}_r, \hat{\theta}_f) = \frac{\mathcal{L}(\hat{\theta}_r; \tilde{X})}{\mathcal{L}(\hat{\theta}_f; \tilde{X})}.$$

The LRT statistic satisfies the sufficiency principle (Section 3.13.1). Suppose $S(\tilde{X})$ is sufficient for $\theta \in \Theta$. Then it is also sufficienct for $\theta \in \Theta_0$, by the factorization theorem (Theorem 3.16). Then by Theorem 4.1 $S \mapsto \hat{\theta}_r$ and $S \mapsto \hat{\theta}_f$, so that by Equation (5.1) $S \mapsto \Lambda(\tilde{X}; \hat{\theta}_r, \hat{\theta}_f)$. Bear in mind that sufficiency over Θ_0 does not guarantee sufficiency over $\Theta \supset \Theta_0$.

The construction of the LRT statistic is demonstrated in the next example.

Example 5.7 (Full and Reduced Models: Difference of Means) We are given two independent samples $\mathbf{X} = (X_1, \ldots, X_n)$, $\mathbf{Y} = (Y_1, \ldots, Y_m)$, with respective means μ_X, μ_Y. Ignoring any nuisance parameters, we have full parameter space $\theta = (\mu_X, \mu_Y) \in \mathbb{R}^2$. Suppose we wish to test $H_o : \mu_X = \mu_Y$ against $H_a : \mu_X \neq \mu_Y$. The reduced model is defined by the parameter subset $\Theta_0 = \{(\mu_X, \mu_Y) \in \Theta : \mu_X = \mu_Y\}$. Suppose, to fix ideas, the two samples posess a common location parameter density, so that the joint density of $(\mathbf{X}, \mathbf{Y})$ is given by $p_\theta(\mathbf{x}, \mathbf{y}) = p_{\mu_X}(\mathbf{x}) p_{\mu_Y}(\mathbf{y}) = p_0(\mathbf{x} - \mu_X) p_0(\mathbf{y} - \mu_Y)$ for some density p_0. Under H_o we have $\mu_X = \mu_Y$ so the LRT statistic is

$$\begin{aligned} \Lambda(\mathbf{X}, \mathbf{Y}) &= \frac{\sup_{\theta \in \Theta_0} \mathcal{L}(\theta; \mathbf{X}, \mathbf{Y})}{\sup_{\theta \in \Theta} \mathcal{L}(\theta; \mathbf{X}, \mathbf{Y})} \\ &= \frac{\sup_{\mu \in \mathbb{R}} p_0(\mathbf{X} - \mu) p_0(\mathbf{Y} - \mu)}{\sup_{(\mu_X, \mu_Y) \in \mathbb{R}^2} p_0(\mathbf{X} - \mu_X) p_0(\mathbf{Y} - \mu_Y)}. \end{aligned}$$

■

Suppose $\Theta \in \mathbb{R}^m$ is a q-dimensional subset, whereas $\Theta_0 \subset \Theta$ is a p-dimensional subset, $p < q \leq m$. For example, this would be the case if Θ was defined by imposing $m - q$ functional constraints on some subset of $\mathbb{R}^m$ (if $q < m$) and Θ_0 was defined by imposing $q - p$ additional functional constraints on Θ. Let $\Lambda(\tilde{X})$ be an LRT statistic for testing $H_o : \theta \in \Theta_0$ against $H_a : \theta \in \Theta - \Theta_0$. We sometimes take the transformation

$$Dev = -2\log(\Lambda(\tilde{X})) \tag{5.6}$$

referred to as the model deviance between the full and reduced models. It is a fact of some consequence, known as Wilks' theorem, that under general assumptions, when H_o is true the distribution $Dev \sim \chi^2_{q-p}$ holds approximately for large sample sizes (Wilks, 1938). This gives a means of defining approximate critical values for a test of specific size α. However, it will also allow us to construct ANOVA decompositions of the type given in Theorem 6.7 (i), which are central to much of the theory of hypothesis testing. This topic will be covered more fully in Section 6.6. We will prove a version of Wilks' theorem in Chapter 14 for the general case (which for reference may be compared to the *iid* case in Theorems 10.3.1 and 10.3.3 of Casella and Berger (2002)).

We have stated earlier that most tests can be expressed as one- or two-sided tests. However, the LRT can be especially useful for constructing decision rules for other forms of test.

Example 5.8 (Null Hypothesis as a Bounded Interval) We have *iid* sample $\mathbf{X} = (X_1, \ldots, X_n)$, $X_1 \sim N(\mu, \sigma^2)$, where σ^2 is known. The density of the sample is

$$p(\mathbf{x}) = (2\pi\sigma^2)^{-n/2} e^{-\sum_{i=1}^n (x_i - \mu)^2/(2\sigma^2)}.$$

We wish to test $H_o : \mu \in [a, b]$ against $H_a : \mu \notin [a, b]$, where $a < b$. The unconstrained MLE is $\hat{\mu}^f = \bar{x}$, while the constrained MLE is

$$\hat{\mu}^r = \operatorname{argmax}_{\mu \in [a,b]} \ell(\mu) = \begin{cases} \bar{x}; & \bar{x} \in [a, b] \\ a; & \bar{x} < a \\ b; & \bar{x} > b \end{cases}.$$

We make use of the identity

$$\begin{aligned} \sum_{i=1}^n (x_i - \bar{x})^2 - \sum_{i=1}^n (x_i - a)^2 &= \sum_{i=1}^n x_i^2 - 2n\bar{x}^2 + n\bar{x}^2 - \sum_{i=1}^n x_i^2 + n2a\bar{x} - na^2 \\ &= -n\bar{x}^2 + n2a\bar{x} - na^2 \\ &= -n(\bar{x} - a)^2. \end{aligned}$$

Noting that $\hat{\mu}^r = \hat{\mu}^f$ when $\bar{x} \in [a, b]$, the LRT statistic is defined by

$$\Lambda(x) = \frac{\mathcal{L}(\hat{\mu}^r)}{\mathcal{L}(\hat{\mu}^f)} = \begin{cases} 1; & \bar{x} \in [a, b] \\ e^{\frac{-n(\bar{x}-a)^2}{2\sigma^2}}; & \bar{x} < a \\ e^{\frac{-n(\bar{x}-b)^2}{2\sigma^2}}; & \bar{x} > b \end{cases}.$$

The LRT rejects H_o with rejection region $R = \{\bar{X} < a - t\} \cup \{\bar{X} > b + t\}$ for some constant $t > 0$. The rejection probability for R is therefore

$$\gamma_t(\mu) = P_\mu\left(\bar{X} < a - t\right) + P_\mu\left(\bar{X} > b + t\right).$$

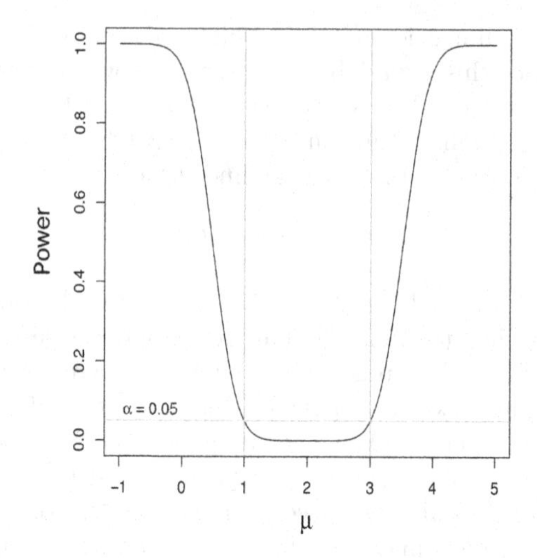

FIGURE 5.2
Power curve for Example 5.8.

Recall that for a test with null hypothesis $H_o : \theta \in \Theta_0$ and rejection region R, the size of a test is given by $\alpha = \sup_{\theta \in \Theta_0} P_\theta(X \in R)$. Thus, we need to identify the parameter $\theta^* \in \Theta_0$ at which the rejection probability is maximized. It is easily checked that $\gamma_t(\mu)$ is maximized for $\mu \in [a, b]$ by either $\gamma_t(a)$ or $\gamma_t(b)$. Therefore, to determine the critical value t_α for a size α test we need to solve $\gamma_{t_\alpha}(a) = \alpha$. Suppose $n = 10$, $\sigma^2 = 1$, and $[a, b] = [1, 3]$, and set size $\alpha = 0.05$. We can determine numerically the critical value $t_{0.05} \approx 0.52016$. The resulting power curve is shown in Figure 5.2. ∎

5.8.1 LRT for the Multinomial Model

Suppose we are given a probability distribution $\mathbf{P} = (p_1, \ldots, p_{q+1})$ on $\mathcal{S} = \{1, \ldots, q+1\}$. If we are given an *iid* sample of size n from $\mathbf{P}$, and we let $\mathbf{N} = (N_1, \ldots, N_{q+1})$ be the vector of sample frequencies for each outcome, then $\mathbf{N}$ has a multinomial distribution (Example 2.1) with density given by

$$f_{\mathbf{N}}(n_1, \ldots, n_{q+1}) = P(N_1 = n_1, \ldots, N_{q+1} = n_{q+1}) = \binom{n}{n_1, \ldots, n_{q+1}} \prod_{i=1}^{q+1} p_i^{n_i}.$$

The log-likelihood is then $\ell(\mathbf{P}; \mathbf{N}) = \sum_{i=1}^{q+1} N_i \log(p_i)$. The unconstrained hypothesis is

$$\mathbf{P} \in \Theta = \left\{ (p_1, \ldots, p_{q+1}) \in \mathbb{R}^{q+1} : \sum_{i=1}^{q+1} p_i = 1, \;\; p_i \geq 0 \right\},$$

so that Θ is an q dimensional subset of $\mathbb{R}^{q+1}$. The null hypothesis is then p dimensional subset $\Theta_0 \subset \Theta$, usually defined by specifying $q - p$ functional constraints on $\mathbf{P}$. The unconstrained (full) MLE is easily shown to be

$$\mathbf{P}_f = (\hat{p}_1^f, \ldots, \hat{p}_{q+1}^f) = (N_1/n, \ldots, N_{q+1}/n).$$

The constrained (reduced) MLE is then

$$\mathbf{P}_r = (\hat{p}_1^r, \ldots, \hat{p}_{q+1}^r) = \mathrm{argmax}_{\mathbf{P} \in \Theta_0} \ell(\mathbf{P}; \mathbf{N}),$$

the exact form of which will depend on the constraints. The model deviance becomes

$$dev = 2\left(\ell(\mathbf{P}_f;\mathbf{N}) - \ell(\mathbf{P}_r;\mathbf{N})\right) = 2\sum_{i=1}^{q+1} N_i \log\left(\frac{\hat{p}^f}{\hat{p}^r}\right),$$

where under H_o by Wilk's theorem we have, approximately, $dev \sim \chi^2_{q-p}$. It is often the practice to express the model deviance in terms of expected and observed counts, where

$$O_i = n\hat{p}_i^f = N_i$$

is the observed count for outcome i, and

$$E_i = n\hat{p}_i^r$$

is the expected count under H_o. The deviance then becomes

$$\begin{aligned} dev &= 2\left(\ell(\mathbf{P}_f;\mathbf{N}) - \ell(\mathbf{P}_r;\mathbf{N})\right) \\ &= 2\sum_{i=1}^{q+1} N_i \log\left(\frac{\hat{p}^f}{\hat{p}^r}\right) \\ &= 2\sum_{i=1}^{q+1} O_i \log\left(\frac{O_i}{E_i}\right). \end{aligned}$$

Suppose we have a simple null hypothesis

$$\begin{aligned} H_o &: p_1,\ldots,p_{q+1} \text{ are the population frequencies,} \\ H_a &: \text{At least one of the hypothetical frequencies is incorrect.} \end{aligned}$$

The observed and expected counts are $O_i = N_i$ and $E_i = np_i$. The null hypothesis Θ_0 has dimension $p = 0$, so by Wilk's theorem the model deviance has an approximate χ^2_q distribution under the null hypothesis. The following example is a typical application (Weir (1996) offers an excellent development of this method applied to statistical genetics).

Example 5.9 (A Goodness-of-Fit Test for Statistical Genetics) Gregor Mendel hypothesized that a certain pattern of breeding would yield a generation of pea plants consisting of four types: RY (round, yellow); WY (wrinkled, yellow); RG (round, green); WG (wrinkled, green). Furthermore, these types would occur in approximately the ratios:

$$RY{:}\,WY{:}RG{:}\,WG = 9:3:3:1. \tag{5.7}$$

Suppose an experiment yields a generation of pea plants with observed type frequencies given in the following table. Here $n = 332$. The observed counts are observations from a

	RY	WY	RG	WG	Total
Observed counts O_i	201	45	51	35	332

multinomial distribution with probabilities $\mathbf{P} = (p_1, p_2, p_3, p_4)$ corresponding to the four plant types. We can formulate Equation (5.7) as a simple null hypothesis using the normalization constant $9 + 3 + 3 + 1 = 16$, which yields

$$H_o : (p_1, p_2, p_3, p_4) = (9/16, 3/16, 3/16, 1/16). \tag{5.8}$$

Then the expected counts are $E_j = np_j$. Here we have $(E_1,\ldots,E_4) = (186.75, 62.25, 62.25, 20.75)$, which gives model deviance $dev = 16.62$. Given critical value $\chi^2_{3,0.95} = 7.815$ we would reject H_o with a significance level of $\alpha = 0.05$. See also Problem 5.19. ∎

The problem of detecting independence in contingency tables was introduced in Section 3.9. The LRT for the multinomial model can also be applied to this problem. In this case, large sample χ^2 approximations are usually employed, which will be further developed in Chapter 14 (see also Example 12.12).

5.9 Similar Tests

Recall that the size of a test is given by $\alpha = \sup_{\theta \in \Theta_o} P_\theta(\tilde{X} \in R)$. Usually, when Θ_o defines a composite null hypothesis, we expect that the rejection probability within the null hypothesis is maximized on the boundary with the alternative hypothesis (as in Example 5.8). We have seen that this will be the case if $\theta \in \mathbb{R}$ and the relevant test statistic is stochastically ordered with respect to θ.

The situation is somewhat more complex if a multidimensional parameter can be decomposed into $\theta = (\psi, \xi)$, and we are interested in a hypothesis test concerning ψ alone, regarding ξ as a nuisance parameter. Then Θ_o defines a composite hypothesis. In this case, it may be possible to construct a similar test, which is formaly defined as a test for which $P_\theta(\tilde{X} \in R) = \alpha$ for all $\theta \in \Theta_0$ (Lehmann and Scheffé, 1950; Lehmann and Romano, 2005).

There is a simple approach to the development of similar tests. Our test statistic obviously cannot depend on ξ. Then a rejection region R_α is called a similar region of size α if

$$P_{(\psi_0,\xi)}(\tilde{X} \in R_\alpha) = \alpha, \text{ for all } \xi.$$

If R_α is nested in α, then R_α defines a similar test against null hypothesis $H_o : \psi = \psi_0$.

Such a test can be constructed if there exists a statistic S_ξ which is sufficient for ξ, at least when $\psi = \psi_0$. Suppose we can construct a level α test against $H_o : \psi = \psi_0$ based on conditional observation $\tilde{X} \mid S_\xi = s$, defined by rejection region $R_\alpha(s)$, which is nested with respect to α. Then

$$P_{(\psi_0,\xi)}(\tilde{X} \in R_\alpha(s) \mid S_\xi = s) \leq \alpha, \text{ for all } \xi. \tag{5.9}$$

The rejection region is now $R_\alpha = \{\tilde{X} : \tilde{X} \in R_\alpha(S_\xi(\tilde{X}))\}$. The test itself is level α since

$$P_{(\psi_0,\xi)}(\tilde{X} \in R_\alpha(S_\xi)) = E_{(\psi_0,\xi)}\left[P_{(\psi_0,\xi)}(\tilde{X} \in R_\alpha(S_\xi) \mid S_\xi)\right] \leq \alpha,$$

based on (5.9). Furthemore, if (5.9) is replaced with equality, then R_α will define a size α test.

Section 3.9 gives an example of this approach applied to inference for the odds ratio in the context of contingency tables. The next example is, in a sense, a simpler version. But it also offers an example of a statistic S_ξ sufficient for ξ only for some values of ψ.

Example 5.10 (Conditional Test for Arithmetic Difference of Two Binomial Parameters) Suppose X_1, X_2 are independent binomial random variables $X_i \sim bin(n_i, p_i)$, $i = 1, 2$. Suppose we write the parameter $\theta = (\psi, \xi)$, where $\psi = p_1 - p_2$, $\xi = p_1 + p_2$. Define $S_\xi = X_1 + X_2$. Of course, S_ξ is sufficient for ξ only if $\psi = 0$. But this defines the null hypothesis $H_o : p_1 = p_2$, which is of interest in many applications. Suppose it holds that $p_1 = p_2 = p$. Then $S_\xi \sim bin(n_1 + n_2, p)$. The distribution of $X = (X_1, X_2)$ given S_ξ is equivalently characterized by the distribution of X_1 given S_ξ which is, for permitted values,

$$\begin{aligned} p_\theta(x_1 \mid S_\xi = s) &= \frac{\binom{n_1}{x_1} p^{x_1}(1-p)^{n_1-x_1}\binom{n_2}{s-x_1} p^{s-x_1}(1-p)^{n_2-s+x_1}}{\binom{n_1+n_2}{s} p^s(1-p)^{n_1+n_2-s}} \\ &= \frac{\binom{n_1}{x_1}\binom{n_2}{s-x_1}}{\binom{n_1+n_2}{s}}. \end{aligned}$$

This is a hypergeometric distribution, and it does not depend on ξ. A similar test against null hypothesis $H_o : \psi = 0$ can be constructed on this basis. ∎

The method of the previous example can be refined.

Example 5.11 (Conditional Test for Odds Ratio of Two Binomial Parameters) To continue Example 5.10, suppose we derive the conditional density of $X_1 \mid S_\xi = s$ in its general form. First, re-express the probabilities p_i as odds $\rho_i = p_i/(1-p_i)$, $i = 1, 2$. We then have

$$\begin{aligned} p_\theta(x_1 \mid S_\xi = s) &= \frac{\binom{n_1}{x_1} p_1^{x_1}(1-p_1)^{n-x_1}\binom{n_2}{s-x_1} p_2^{s-x_1}(1-p_2)^{n_2-s+x_1}}{\sum_{x=0}^{\min\{s,n_1\}} \binom{n_1}{x} p_1^{x}(1-p_1)^{n-x}\binom{n_2}{s-x} p_2^{s-x}(1-p_2)^{n_2-s+x}} \\ &= \frac{\binom{n_1}{x_1} \rho_1^{x_1}(1-p_1)^{n_1}\binom{n_2}{s-x_1} \rho_2^{s-x_1}(1-p_2)^{n_2}}{\sum_{x=0}^{\min\{s,n_1\}} \binom{n_1}{x} \rho_1^{x}(1-p_1)^{n_1}\binom{n_2}{s-x} \rho_2^{s-x}(1-p_2)^{n_2}} \\ &= \frac{\binom{n_1}{x_1} \rho_1^{x_1}\binom{n_2}{s-x_1} \rho_2^{s-x_1}(1-p_1)^{n_1}(1-p_2)^{n_2}}{\sum_{x=0}^{\min\{s,n_1\}} \binom{n_1}{x} \rho_1^{x}\binom{n_2}{s-x} \rho_2^{s-x}(1-p_1)^{n_1}(1-p_2)^{n_2}} \\ &= \frac{\binom{n_1}{x_1} \rho_1^{x_1}\binom{n_2}{s-x_1} \rho_2^{s-x_1}}{\sum_{x=0}^{\min\{s,n_1\}} \binom{n_1}{x} \rho_1^{x}\binom{n_2}{s-x} \rho_2^{s-x}}. \end{aligned} \tag{5.10}$$

This form suggests how the range of possible null hypotheses for which a similar test exists can be expanded. We need a injective mapping $(\rho_1, \rho_2) \mapsto (\psi, \xi)$ such that the form (5.10) depends on ψ but not ξ. Suppose we set $\psi = \rho_1/\rho_2$, and let ξ be any other function of (ρ_1, ρ_2) defining the required injective mapping ($\xi = p_1 + p_2$ would work). Then substititing $\rho_1 = \psi\rho_2$ into (5.10) gives

$$\begin{aligned} p_\theta(x_1 \mid S_\xi = s) &= \frac{\binom{n_1}{x_1} \rho_1^{x_1}\binom{n_2}{s-x_1} \rho_2^{s-x_1}}{\sum_{x=0}^{\min\{s,n_1\}} \binom{n_1}{x} \rho_1^{x}\binom{n_2}{s-x} \rho_2^{s-x}} \\ &= \frac{\binom{n_1}{x_1} (\psi\rho_2)^{x_1}\binom{n_2}{s-x_1} \rho_2^{s-x_1}}{\sum_{x=0}^{\min\{s,n_1\}} \binom{n_1}{x} (\psi\rho_2)^{x}\binom{n_2}{s-x} \rho_2^{s-x}} \\ &= \frac{\binom{n_1}{x_1} \psi^{x_1}\binom{n_2}{s-x_1} \rho_2^{s}}{\sum_{x=0}^{\min\{s,n_1\}} \binom{n_1}{x} \psi^{x}\binom{n_2}{s-x} \rho_2^{s}} \\ &= \frac{\binom{n_1}{x_1}\binom{n_2}{s-x_1} \psi^{x_1}}{\sum_{x=0}^{\min\{s,n_1\}} \binom{n_1}{x}\binom{n_2}{s-x} \psi^{x}}. \end{aligned}$$

Since this conditional distribution does not depend on ξ, we can construct a similar test for any null hypothesis based on the odds ratio $H_o : \rho_1/\rho_2 = \psi_0$. It is easily verified that for $\psi_0 = 1$ this test reduces to that given in Example 5.10. ∎

5.10 Problems

Problem 5.1 Suppose a null hypothesis H_o has a rejection region of the form $R = \{T(\tilde{X}) > t\}$ for some fixed t. Suppose test statistic $T(\tilde{X})$ has CDF F_T under H_o. Prove that the size of the test is maximized among all level α tests by setting $t = Q(1-\alpha)$, where $Q(p)$ is the quantile function of F_T.

Problem 5.2 Suppose $X \sim bin(4, \theta)$. We wish to test null hypothesis $H_o : \theta \leq 0.5$ against alternative hypothesis $H_a : \theta > 0.5$. Define a randomized decision rule $\delta : \{0, 1, \ldots, 4\} \rightarrow [0, 1]$ which satisfies the following conditions.

(i) δ is nondecreasing.
(ii) The size of the test is $\alpha = 0.05$.

Problem 5.3 This example is given in Problem 3.17 of Lehmann and Romano (2005). A random variable X has support $\mathcal{X} = \{1, 2, 3\}$, and we are given two densities $p_0(x) = 0.85, 0.1, 0.05$ and $p_1(x) = 0.7, 0.2, 0.1$, $x = 1, 2, 3$. Let p be the true density of X. We wish to test hypothesis $H_o : p = p_0$ against alternative $H_a : p = p_1$.

(a) Define nonrandomized tests which have size $\alpha = 0.05$, $\alpha' = 0.1$ and $\alpha'' = 0.15$. Are these tests nested?
(b) Let $T(X) = p_1(X)/p_0(X)$. Show that the three nonrandomized tests of Part (a) can be interpreted as randomized tests based on $T(X)$ which are nested. **HINT:** What is $P(X = 3 \mid X \geq 2)$ under H_o?

Problem 5.4 Consider the model of Example 5.3, and let $v(\theta)$ be any variance function for which $\lim_{\theta \to \infty} v(\theta) = \infty$, and for which the limit $t = \lim_{\theta \to \infty} \theta / v(\theta)^{1/2}$ exists. Show that the limiting rejection probability it given by $\gamma(\infty) = \lim_{\theta \to \infty} \gamma(\theta) = 1 - \Phi(-t)$, where Φ is the CDF of the $N(0, 1)$ distribution.

Problem 5.5 A company has developed a predictive model for the screening of applicants based on a questionnaire. The responses are converted to 3 components:

$$\begin{aligned} X_1 &= \text{Leadership skills} \\ X_2 &= \text{Communication skills} \\ X_3 &= \text{Level of expertise.} \end{aligned}$$

Each component is normally distributed, and has been standardized to have zero mean and standard deviation $\sigma_X = 25$. A composite score believed to be especially predictive of success is given by

$$T = \frac{1}{2}X_1 + \frac{1}{6}X_2 + \frac{1}{3}X_3.$$

The company wishes to use T for screening job applicants. If an applicant's score exceeds a threshold $T \geq t$ they are selected for further interviews. The company wishes to select 10% of applicants for further screening, so it sets the threshold at the value

$$t = \sigma_T \times z_{0.1}$$

where σ_T is the standard deviation of T, and $z_{0.1}$ is the 10% critical value of a standard normal distribution $N(0, 1)$. If σ_T is correctly calculated, and $E[T] = 0$ as expected, we would have

$$P(T > \sigma_T \times z_{0.1}) = 0.1.$$

It is then noted that in order to calculate σ_T, the correlations between X_1, X_2 and X_3 must be known. Following this, two points of view emerge, which we'll refer to as the null and alternative hypotheses.

H_o Scales from psychometric questionnaires are designed to measure independent constructs. So, although we might expect, say, leaderships skills and communication skills to be positively correlated in everyday life, the scales X_1, X_2 and X_3 are designed to measure these qualities in a manner that is independent of the others. Therefore, we should expect zero correlation between X_1, X_2 and X_3.

H_a A statistical analysis has estimated the following correlations, and these should therefore be used to calculate σ_T.

$$\begin{aligned} \rho_{X_1,X_2} &= 0.56 \\ \rho_{X_1,X_3} &= 0.18 \\ \rho_{X_2,X_3} &= 0.21. \end{aligned}$$

In order to test these hypotheses, a sample of n test scores T is to be collected. Design a size $\alpha = 0.05$ hypothesis test for null and alternative hypotheses H_o and H_a. Note that only the scores T will be available, and not the underlying scores X_1, X_2 and X_3. Construct a plot of power against sample size n, for $n = 2, 3, \ldots, 199, 200$. Superimpose a horizontal line at $power = 90\%$. What is the minimum sample size needed to attain at least 90% power?

Problem 5.6 In Diaconis *et al.* (2007) it was conjectured that there is a bias in coin tossing in favor of the side facing up at the start of the toss. The biased proportion was estimated to be 50.8% in place of the commonly expected 50%.

(a) What sample size is need to test $H_o : p = 1/2$ with power 0.9 at alternative $p = 0.508$? Do the calculation for both a one-sided and two-sided test.
(b) What will be the margin of error for these samples sizes? Base your answer on an approximate 95% confidence interval?
(c) In 2009 two students each flipped a coin 20,000 times, one starting with Heads facing up, the other with Tails facing up. See `http://www.stat.berkeley.edu/~aldous/Real-World/coin_tosses.html` for details. The two students attained 10231 Heads, and 10014 Tails according to the respective starting conditions. The pooled estimate of the proportion is $\hat{p} = 50.6\%$, rather close to the value predicted in Diaconis *et al.* (2007). What are the P-values for the one- and two-sided tests? Is the conjectured hypothesis $p = 0.508$ within the margins of error of Part (b)?

Problem 5.7 Suppose a population of measurements is claimed to be normally distributed with mean no larger that $\mu = 150$ and standard deviation $\sigma = 10$. To test this claim, a random sample of n components is to be collected to do a hypothesis test for $H_o : \mu = 150$ against $H_a : \mu > 150$.

(a) Draw a power curve, that is plot $Power(\mu) = 1 - \beta(\mu)$ as a function of μ over a suitable range of alternative hypotheses, say $\mu \in (150, 175)$, to generate the values of μ for your plot). Do this for a Type I error of $\alpha = 0.05$. Superimpose on the same plot power curves for $n = 5, 10, 15, 20, 25, 30$. Label the appropriate axes μ and $Power(\mu)$. Include a horizontal line at level 0.05, labelled $\alpha = 0.05$. Also, indicate the positions of the $n = 5$ and $n = 30$ curves.
(b) Create a table giving the power for each combination of $n = 5, 10, \ldots, 30$ and $\mu = 155, 156, 157, 158, 159, 160$. For each of these values of μ, give the minimum sample size (from those considered) required to attain a power of 80%.

Problem 5.8 A study is planned that will involve a one-sample upper-tailed T test for hypotheses $H_o : \mu = 100$ against $H_a : \mu > 100$ based on an *iid* sample from a $N(\mu, \sigma^2)$ distribution. The significance level will be 5%. Assume $n = 20$.

(a) Construct a power curve in which the vertical axis is $pow = 1 - P(\text{ Type II Error })$, and the horizontal axis is $\delta = (\mu - 100)/\sigma$. Note that σ doesn't have to be known.

(b) For a given sample size n we can calculate the power for rejecting an alternative for which the effect size is $\delta = (\mu - 100)/\sigma = 0.25$. Construct a plot in which the horizontal axis is sample size $n = 100, 101, \ldots, 150$ and the vertical axis is the power for an effect size $\delta = 0.25$ as a function of sample size n. Determine the smallest sample size needed to attain a power of 0.8.

Problem 5.9 An industrial process produces a component with a diameter which is, due to process variability, normally distributed with mean $\mu = 35.5$ mm and standard deviation $\sigma = 0.043$ mm. After some time it is suspected that the process mean has lowered, so a random sample of n components is to be collected to do a hypothesis test for $H_o : \mu = 35.5$ against $H_a : \mu < 35.5$.

(a) Draw a power curve, that is plot $Power(\mu) = 1 - \beta(\mu)$ as a function of μ over a suitable range of alternative hypotheses, say $\mu \in (35.4, 35.5)$, to generate the values of μ for your plot). Do this for a Type I error of $\alpha = 0.05$. Superimpose on the same plot power curves for $n = 5, 10, 15, 20, 25, 30$. Label the appropriate axes μ and $Power(\mu)$. Include a horizontal line at level 0.05, labelled $\alpha = 0.05$. Also, indicate the positions of the $n = 5$ and $n = 30$ curves. What is the value of $Power(35.5)$ for each curve?

(b) Create a table giving the power for each combination of $n = 5, 10, \ldots, 30$ and $\mu = 35.47, 35.48, 35.49$. For each of these values of μ, give either the minimum sample size (from those considered) required to attain a power of 80% or state that the largest sample size considered is not sufficient.

Problem 5.10 We are given a sample $\mathbf{X} = (X_1, \ldots, X_n)$, where the density of $\mathbf{X}$ is from family $\mathcal{P} = \{p_\theta : \theta \in \Theta\}$ defined on sample space $\mathcal{X}$. Suppose $\theta \in \Theta = \mathbb{R}$ is a scale parameter. We wish to test $H_o : \theta \leq \theta_0$ against $H_a : \theta > \theta_0$. Following Example 5.4, define a class of hypothesis tests which are unbiased and have nested rejection regions.

Problem 5.11 Verify the claim of Example 5.5, that the rejection region $R = \{T_0 \notin [t_0, t_1]\}$ for the two-sided test will not be unbiased, unless $t_0 = -t_1$. **HINT:** Evaluate the derivative of the rejection probability $\gamma(\Delta) = P_\Delta(R)$ at $\Delta = 0$. See also Example 5.6.

Problem 5.12 Verify the claim of Example 5.5, that if $T \sim T_{\Delta;n-1}$, then $T^2 \sim F_{\Delta;1,n-1}$.

Problem 5.13 Suppose $(X_1, \ldots, X_{n_1})$ and $(Y_1, \ldots, Y_{n_2})$ are independent *iid* samples with $X_1 \sim N(\mu_1, \sigma_1^2)$ and $Y_1 \sim N(\mu_2, \sigma_2^2)$. Let $\bar{X}$, $\bar{Y}$, S_1^2, S_2^2 be the respective sample means and variances. We define the pooled variance

$$S_p^2 = \frac{(n_1 - 1)S_1^2 + (n_2 - 1)S_2^2}{n_1 + n_2 - 2}.$$

Assume $\sigma_1^2 = \sigma_2^2 = \sigma^2$. Setting $\Delta = \mu_2 - \mu_1$, consider the statisitic

$$T = \frac{\bar{Y} - \bar{X} - \Delta}{S_p\sqrt{\frac{1}{n_1} + \frac{1}{n_2}}}.$$

Prove that when $\Delta = \mu_2 - \mu_1$ we have $T \sim T_{n_1+n_2-2}$. This is the well known pooled T statistic for testing the difference between two normal means.

Problem 5.14 Consider the pooled T statistic of Problem 5.13 for two independent samples with respective sample sizes n_1, n_2, assuming equal variances $\sigma_1^2 = \sigma_2^2 = \sigma^2$. We wish to test null hypothesis $H_o : \mu_1 \geq \mu_2$ against alternative hypothesis $H_a : \mu_1 < \mu_2$. Then H_o is rejected for large values of the pooled T statistic:

$$T = \frac{\bar{Y} - \bar{X}}{S_p\sqrt{\frac{1}{n_1} + \frac{1}{n_2}}},$$

where S_p^2 is the pooled variance.

(a) Suppose $\mu_2 - \mu_1 = \Delta \neq 0$. Show that T_{obs} has a noncentral T distribution with noncentrality parameter:

$$ncp = \frac{\Delta}{\sigma}\sqrt{\frac{n_1 n_2}{n_1 + n_2}},$$

assuming that the populations are normally distributed.

(b) Given the form of the noncentrality parameter ncp, show that if the total sample size $N = n_1 + n_2$ is fixed, the most powerful test is obtained by the balanced design $n_1 = n_2 = N/2$, assuming N is even.

(c) Construct power curves for the one-sided pooled variance T test just described, for $\alpha = 0.05$, with the following features:

   (i) Assume a balanced design $n = n_1 = n_2$. The plot will superimpose power curves for $n = 5, 10, 15, 20$ on the same plot.

   (ii) The power curve will plot power $1-\beta$ for one-sided alternatives $\Delta = \mu_2 - \mu_1 > 0$ against Δ/σ, over the range $0 \leq \Delta/\sigma \leq 3$ (use increments of 0.1).

   (iii) A grid should be superimposed with grid size 0.05 for the vertical axis and 0.125 for the horizontal axis.

(d) Suppose we need to determine a per-sample sample size n for a one-sided two-sample pooled variance T test, testing null hypothesis $H_o : \mu_1 \geq \mu_2$ against alternative hypothesis $H_a : \mu_1 < \mu_2$. A power of 90% is needed for an alternative $\mu_2 - \mu_1 = 5.85$, assuming standard deviation $\sigma = 5.2$ and using $\alpha = 0.05$. Using your power curves, what value of n would you recommend (select from 5,10,15,20)?

Problem 5.15 Suppose $(X_1, \ldots, X_n)$ and $(Y_1, \ldots, Y_m)$ are independent *iid* samples with $X_1 \sim N(\mu_x, \sigma_x^2)$ and $Y_1 \sim N(\mu_y, \sigma_y^2)$. Let $\bar{X}$, $\bar{Y}$, S_X^2, S_Y^2 be the respective sample means and variances. Consider the pooled T statistic of Problem 5.13.

(a) Does the answer to Problem 5.13 hold when $\sigma_x^2 \neq \sigma_y^2$?

(b) If σ_x^2 and σ_y^2 were known (and not necessarily equal), how would $\bar{X} - \bar{Y}$ be standardized so that

$$Z = \frac{\bar{X} - \bar{Y} - a}{b} \sim N(0,1)$$

under the null hypothesis $H_o : \mu_x - \mu_y = \Delta$?

(c) When $\sigma_x^2 \neq \sigma_y^2$ Welch's T test is often used in place of the pooled T test (Welch, 1947). The statistic may be constructed by replacing σ_x^2 and σ_y^2 in Z with estimates S_X^2, S_Y^2. The resulting statistic has an approximate T_{ν_W} distribution, where

$$\nu_W = \frac{\left(\frac{S_X^2}{n} + \frac{S_Y^2}{m}\right)^2}{\frac{(S_X^2/n)^2}{n-1} + \frac{(S_Y^2/m)^2}{m-1}},$$

which is a special case of the Satterthwaite equation (Satterthwaite, 1946). Prove that $\min(n-1, m-1) \leq \nu_W \leq n+m-2$ for any S_X^2, S_Y^2, and that the maximum is attained if and only if $S_X^2 = S_Y^2$. **HINT:** Show that the problem of maximizing ν_W over all S_X^2, S_Y^2 is equivalent the problem of maximizing $g(u) = (1+u^2)^2/(c+u^2)$ over $u \geq 1$, where $c = (m-1)/(n-1)$.

(d) Show that if $S_X^2/n = S_Y^2/m$ then ν_W is twice the harmonic mean of $n-1$ and $m-1$.

Problem 5.16 The following table gives data from two samples $(X_1, \ldots, X_n)$ and $(Y_1, \ldots, Y_n)$, $n = 10$. Suppose the data are paired, in the sense that the pair (X_i, Y_i) are sampled from a single source (a common household, for example). We therefore have respective means μ_i^x, μ_i^y, and interest is in the differences $\Delta_i = \mu_i^y - \mu_i^x$, and the null hypothesis $H_o : \Delta_i = 0$ for all i.

	1	2	3	4	5	6	7	8	9	10
X	95.14	101.09	89.52	97.73	104.51	103.63	106.25	104.13	104.23	97.88
Y	96.76	104.39	93.22	98.81	109.72	106.13	105.36	106.04	104.99	100.41

(a) For a paired t-test, we perform a one-sample T test on the differences $D_i = Y_i - X_i$. Evaluate for P-value for this test againt the two-sided alternative.
(b) Repeat part (a), but assume the samples are independent, and use the pooled T test of Problem 5.13. Does your conclusion change? How might we interpret Δ_i in this case?

Problem 5.17 Suppose $X_1, \ldots. X_m$ are independent observations, with $X_j \sim pois(\lambda_j)$, $j = 1, \ldots, m$. We wish to test hypothesis $H_o : \lambda_1 = \ldots = \lambda_m$ against alternative $H_a : \lambda_i \neq \lambda_j$ for some pair i, j.

(a) Derive the LRT statistic for these hypotheses. Express the statistic as model deviance Dev. What is the approximate distributon of Dev?
(b) The entropy of a probability distribution $\mathbf{P} = (p_1, \ldots, p_m)$ is defined as $H(\mathbf{P}) = -\sum_{j=1}^m p_j \log(p_j)$, where we take $0 \times \log(0) = 0$. Show that $0 \leq H(\mathbf{P}) \leq H(\mathbf{P}_e)$, where $\mathbf{P}_e = (1/m, \ldots, 1/m)$ is the uniform distribution on $\{1, \ldots, m\}$. **HINT:** This problem can be solved using the Lagrange multiplier method.
(c) Show that $Dev = 2T(H(\mathbf{P}_e) - H(\hat{\mathbf{P}}))$, where $T = \sum_{j=1}^m X_j$ and $\hat{\mathbf{P}} = (X_1/T, \ldots, X_m/T)$.
(d) If we accept the approximation Dev of Part (a), we would reject H_o for $Dev > t_\alpha$ for some critical value. Noting that H_o is a compound hypothesis, the intention is that the distribution of the test statistic is (approximately) the same for all parameters in H_o. In this case, the critical value t_α can be chosen in this manner. To test this idea, create a program which simulates $X_j \sim pois(\lambda_j)$, $j = 1, \ldots, m$, then calculates Dev. Use this program to estimate the rejection probabilty for a size $\alpha = 0.05$ test for a range of parameters in H_o, based on the approximate distribution of Part (a). Use $\lambda_1 = 1, 2, 5, 10, 20, 50$. Do the rejection probabilities vary significantly? What, roughly, is the true size of this test? How would we have to modify the test to ensure that it's size was actually $\alpha = 0.05$? Use $m = 10$.
(e) What is the distribution of $(X_1, \ldots. X_m)$ conditional on $T = t$, where t is a fixed integer? How can we modify our test to ensure that is it a similar test?

Problem 5.18 Show that the two-sided pooled T test of Problem 5.13 is equivalent to a likelihood ratio test.

Problem 5.19 Suppose in Example 5.9 the observed counts are replaced by $(O_1, \ldots, O_4) = (183, 60, 59, 19)$. What would be the appropriate conclusion?

Problem 5.20 Let $X \in \{1, \ldots, K_1\}$, $Y \in \{1, \ldots, K_2\}$, $Z \in \{1, \ldots, K_3\}$, be three discrete random variables possessing joint density $f_{XYZ}(x, y, z) = P(X = x, Y = y, Z = z)$. Assume K_1, K_2, K_3 are positive integers larger than one. Suppose we observe an independent sample of triplets $(X_1, Y_1, Z_1), \ldots, (X_n, Y_n, Z_n)$ sampled from f_{XYZ}.

(a) Use the LRT for the multinomial model to develop a test for the null hypothesis

$$H_o : X \text{ and } Y \text{ are independent}$$

against all alternatives. Use Wilk's theorem to derive an approximate rejection rule for a size α test.

(b) Repeat Part (a) for the null hypothesis

$$H_o : X \text{ and } Y \text{ are independent conditional on } Z,$$

that is, under H_o we always have $P(X = x, Y = y \mid Z = z) = P(X = x \mid Z = z)P(Y = y \mid Z = z)$.

HINT: Let N_{ijk} be the number of times the triplet $(X, Y, Z) = (i, j, k)$ is observed in the sample. Then the collection of all such frequencies define a multinomial random vector. We can express marginal totals as $N_{ij\cdot} = \sum_{k=1}^{K_3} N_{ijk}$ or $N_{i\cdot\cdot} = \sum_{j=1}^{K_2} \sum_{k=1}^{K_3} N_{ijk}$. Express the model deviance using this notation.

Problem 5.21 Suppose $(X_{11}, \ldots, X_{1n})$, $(X_{21}, \ldots, X_{2n})$, and $(X_{31}, \ldots, X_{3n})$, are three *iid* samples from normal $N(\mu_1, \sigma^2)$, $N(\mu_2, \sigma^2)$ and $N(\mu_3, \sigma^2)$ densities respectively. Assume that the three samples are independent of each other. Develop a likelihood ratio test (LRT) to test the hypothesis $H_o : \mu_1 + \mu_2 = \mu_3$ against $H_1 : \mu_1 + \mu_2 \neq \mu_3$ assuming σ^2 is known. How would you calculate a specific size α rejection region.

Problem 5.22 Consider the contingency table model of Section 3.9. Fisher's exact test is based on the conditional distribution of Equation (3.22).

(a) Show that under hypothesis $H_o : \rho = 1$ the conditional density $p_\theta(x \mid R_1 = r_1, C_1 = c_1)$ is equivalent to a hypergeometric distribution. Give the exact parameters.
(b) For $\rho \in \mathbb{R}$, this distribution is referred to as Fisher's noncentral hypergeometric distribution. Show that the family $p_\theta(x \mid R_1 = r_1, C_1 = c_1)$ is stochastically increasing *wrt* ρ.
(c) Suppose we observe table entries $(N_{11}, N_{22}, N_{12}, N_{21}) = (13, 12, 4, 5)$ What is the P-value for the hypotheses $H_o : \rho = 1$ against $H_a : \rho > 1$. What about $H_o : \rho = 2$ against $H_a : \rho > 2$?

Problem 5.23 Suppose X_1, X_2 are independent Poisson random variables of means λ_1, λ_2. Define $\psi = \lambda_1/\lambda_2$. We can write the parameter as $\theta = (\psi, \xi)$ where $\xi = \lambda_1 + \lambda_2$. Develop a similar test for $H_o : \psi = \psi_0$ against $H_a : \psi \in \Theta_a$, where Θ_a depends only on ψ. **HINT:** Condition on $S_\xi = X_1 + X_2 \sim pois(\lambda_1 + \lambda_2)$.

Problem 5.24 A goodness-of-fit test seeks to determine how well a set of observations conforms to a model. For example, given a sample $X_1, \ldots, X_n$ from a CDF F we may wish to test hypotheses $H_o : F = F_0$ against $H_a : F \neq F_0$. Here, we are not assuming that F is in a parametric family containing F_0. Rather, we may wish to determine whether or not an assumption of normality is justified. We therefore need a testing procedure based on a more general distance measure $d(F_1, F_2)$ between two distributions. F_1, F_2. Then suppose $\hat{F}$ is a nonparametric estimate of F based on the observations. We would then use $d(\hat{F}, F_0)$ as a test statistic. Either the distributional properties of the statistic under H_o are known by theory, or the null distribution of $d(\hat{F}, F_0)$ can be estimated by simulating from the (known) distribution F_0.

However, the choice of distance measure has some practical implications. The Glivenko-Cantelli (or Kolmogorov) distance between two CDFs F_1, F_2 on $\mathbb{R}$ is

$$d_{GC}(F_1, F_2) = \sup_x |F_1(x) - F_2(x)|,$$

while the total variation distance is

$$d_{TV}(F_1, F_2) = \sup_A |P_{F_1}(A) - P_{F_2}(A)| \tag{5.11}$$

the supremum being taken over all measurable sets.

(a) Prove that $d_{GC}(F_1, F_2) \leq d_{TV}(F_1, F_2)$ for any F_1 and F_2.

(b) Prove that if F_1 is absolutely continuous *wrt* some counting measure, and F_2 is absolutely continuous with respect to Lebesgue measure, then $d_{TV}(F_1, F_2) = 1$

(c) Prove that if F_1 and F_2 are absolutely continuous *wrt* some common measure ν, then $d_{TV}(F_1, F_2) = 2^{-1} \int |f_1(t) - f_2(t)| d\nu$, where f_1, f_2 are the densities of F_1, F_2, respectively.

(d) Suppose $X_1, \ldots, X_n$ is an *iid* sample from continuous CDF F, and we define the empirical density $\hat{F}(x) = n^{-1} \sum_{i=1}^n I\{X_i \leq x\}$. The Kolmogorov-Smirnov goodness-of-fit statistic for testing hypotheses $H_o : F = F_0$ against $H_a : F \neq F_0$ is $S_{KS} = d_{GC}(\hat{F}, F_0)$. An approximation of the null distribution was derived in Smirnov (1948). Why would the distance measure d_{GC} but not d_{TV} be appropriate for this test?

6

Linear Models

6.1 Introduction

Linear models warrant a separate chapter, both for their central role in statistical methodology, and because of the opportunity it affords to demonstrate some of the ideas we have encountered up to this point. We will rely as much as possible on matrix algebra, making use of the ideas associated with Cochran's theorem (Section 2.5). This will allow a certain unification of the inference theory for these models. The reader may wish to use Appendix B as a reference.

The linear model is formally defined in Section 6.2, where we introduce two important examples, simple linear regression and one-way analysis of variance (ANOVA). The best linear unbiased estimator (BLUE) and the Gauss-Markov theorem are introduced in Section 6.3, which provides the framework for our theory of estimation for linear models. In Section 6.4, we show that the BLUE can also be derived using two other approaches, least squares minimization and projection onto vector subspaces.

Section 6.5 then interprets the BLUE as a least squares estimator. We consider both the ordinary and the generalized least squares estimator. Section 6.6 introduces the idea of the analysis of variance (ANOVA), which is the decomposition of the total variation of a set of observations into identifiable sources. Here, the interpretation of the BLUE as a projection onto a vector subspace leads to the F test for linear models. One and two-way ANOVA and multiple linear regression are then considered in more detail in Sections 6.7 and 6.8.

The chapter ends with consideration of two more specialized problem, constrained least squares in Section 6.9 and simultaneous confidence intervals in Section 6.10.

6.2 Linear Models – Definition

Suppose $\mathbf{Y} = [Y_1, \ldots, Y_n]^T$ is a column vector of random variables, and $\boldsymbol{\beta} = (\beta_1, \ldots, \beta_q) \in \mathbb{R}^q$ is a q-dimensional parameter that is related to $\mathbf{Y}$ by the equation

$$E[\mathbf{Y}] = \mathbf{X}\boldsymbol{\beta} = \boldsymbol{\mu}, \tag{6.1}$$

where $\mathbf{X} \in \mathcal{M}_{n,q}$, or, equivalently,

$$\mathbf{Y} = \mathbf{X}\boldsymbol{\beta} + \boldsymbol{\epsilon}, \tag{6.2}$$

where $\boldsymbol{\epsilon} = \mathbf{Y} - E[\mathbf{Y}]$, so that $E[\boldsymbol{\epsilon}] = \mathbf{0}$. The distribution of $\boldsymbol{\epsilon}$ is general. We only assume that $E[\boldsymbol{\epsilon}] = \mathbf{0}$ and that $\Sigma_{\boldsymbol{\epsilon}}$ is finite and invertible (and therefore positive definite). Necessarily $\Sigma_{\mathbf{Y}} = \Sigma_{\boldsymbol{\epsilon}}$, which we refer to as the model covariance. We assume $n > q$ (the case $q \geq n$ does arise, particularly in the analysis of high-throughput data (Gentleman *et al.*, 2006)).

It will also be assumed that $\text{rank}(\mathbf{X}) = q$, equivalent to the condition that the column vectors of $\mathbf{X}$ are linearly independent. A matrix is of full rank if the rank equals the smallest

DOI: 10.1201/9781003049340-6

dimension, here q. In this case, the degrees of freedom of the model is q. Understanding this assumption is important, and a number of rank identities are given in Section B.4.1. Interestingly, this condition is not required for a portion of our analysis. This is because when it does not hold, it can be replaced by an equivalent model for which it does. Note that there are really two versions of the estimation problem, and our development will depend on an understanding of these multiple viewpoints. Let $\mathcal{V}_{\mathbf{X}}$ be the vector space spanned by the column vectors of $\mathbf{X}$ (Section B.3). Instead of estimating $\boldsymbol{\beta}$, we may instead estimate $\boldsymbol{\mu}$ (Equation (6.1)) by the fitted values $\hat{\mathbf{Y}} \in \mathcal{V}_{\mathbf{X}}$ closest to $\mathbf{Y}$, without the need to refer to $\boldsymbol{\beta}$. In this case, the model is really defined by $\mathcal{V}_{\mathbf{X}}$, so that $\mathbf{X}$ can be replaced by any matrix with the same column space, which need not have the same number of columns.

However, the objective in linear models is usually an estimate $\hat{\boldsymbol{\beta}} \approx \boldsymbol{\beta}$, which we expect to conform to the fitted values by the relation $\hat{\mathbf{Y}} = \mathbf{X}\hat{\boldsymbol{\beta}}$. However, for any given $\hat{\mathbf{Y}}$ the relation is satisfied by a unique $\hat{\boldsymbol{\beta}}$ *iff* $\mathbf{X}$ is of full rank. Under these conditions the parameter $\boldsymbol{\beta}$ is identifiable. See also Problem 6.2.

We next introduce two important examples of linear models.

6.2.1 Simple Linear Regression

Suppose we observe pairs $(Y_1, x_1), \ldots, (Y_n, x_n)$ which are related by

$$Y_i = \beta_0 + \beta_1 x_i + \epsilon_i, \quad i = 1, \ldots, n, \tag{6.3}$$

where $\boldsymbol{\beta} = [\beta_0, \beta_1]^T$ is an unknown parameter, Y_i is the response (or dependent) variable, x_i is a known predictor (or independent) variable, and $\epsilon_i \sim N(0, \sigma^2)$ is an unobserved error term. It is often assumed that the error terms are independent, but this is not required for the general theory of linear models. Note that Y_i is interpreted as a random variable, while x_i is a real number.

The model (6.3) can be written in the form of Equation (6.2) as

$$\mathbf{Y} = \mathbf{X}\boldsymbol{\beta} + \boldsymbol{\epsilon}, \tag{6.4}$$

where $\boldsymbol{\epsilon} = (\epsilon_1, \ldots, \epsilon_n)^T$ and

$$\mathbf{X} = \begin{bmatrix} 1 & 1 & \cdots & 1 \\ x_1 & x_2 & \cdots & x_n \end{bmatrix}^T. \tag{6.5}$$

An important assumption is that $\mathbf{X}$ is of full rank, which will hold unless all predictor values are equal.

The clear interpretation of Equation (6.3) is that a functional relationship of the form $Y = \beta_0 + \beta_1 x$ exists between Y and x, but that Y is observed with error ϵ. The inference problem is to make use of the data to construct estimate $\hat{\boldsymbol{\beta}}$ of $\boldsymbol{\beta}$ (the responses and predictors are often referred to collectively as the training data). Figure 6.1 presents an example of an estimate of model (6.3), commonly refered to as a "fit". The conventional least squares method was used to derive $\hat{\boldsymbol{\beta}}$, which will be introduced below. This approach also yields an estimate $\hat{\sigma}^2$ of the variance σ^2.

6.2.2 One-Way Analysis of Variance (ANOVA)

We introduce here the simplest of a class of models known by the acronym ANOVA (analysis of variance). Like many statistical methods, its prevalence today is largely due to R.A. Fisher (Fisher, 1925, 1935). See, for example, Yates (1951), Box (1980) or Pearce (1992) for historical commentaries. A good modern introduction to the subject would be Box *et al.* (1978) or Kutner *et al.* (2004).

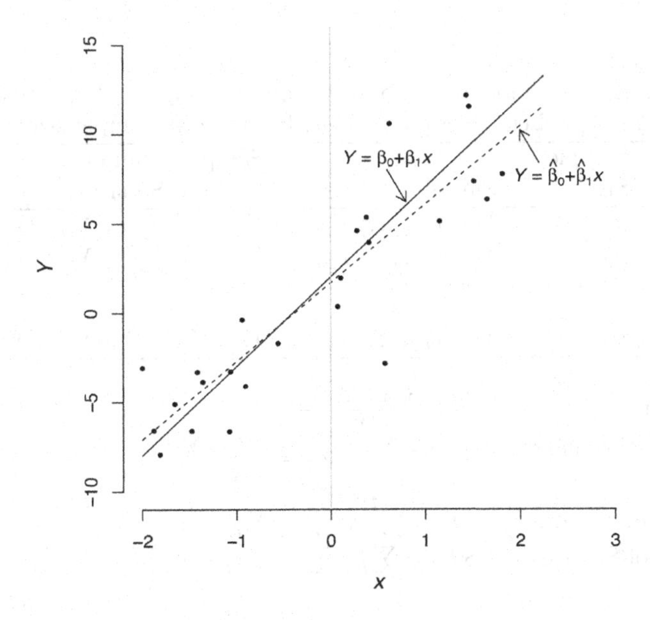

FIGURE 6.1
Example of a simple linear regression model and fit (Section 6.2.1). Data was simulated from the model of Equation (6.3) with $\boldsymbol{\beta} = (2.0, 5.0)$, $\sigma^2 = 4$, $n = 25$. The predictor variables were simulated from a $unif(-2, 2)$ distribution. The estimated regression coefficients were $\hat{\boldsymbol{\beta}} = (1.67, 4.39)$ and the estimate of the variance was $\hat{\sigma}^2 = 2.84$. Observations (Y_i, x_i) are plotted individually.

Suppose we are given independent *iid* samples from k distinct populations $N(\beta_j, \sigma^2)$, $j = 1, \ldots, k$. The population label j is referred to as a treatment. Table 6.1 summarizes the structure of the data.

Analysis of this model usually begins with a test of the hypotheses

$$H_o : \beta_1 = \beta_2 = \ldots = \beta_k \text{ against } H_a : \beta_i \neq \beta_j \text{ for some } i,j, \tag{6.6}$$

so that interest is in whether or not there is variation among the treatment means. This inference problem would appear to be different from the estimation problem associated with simple linear regression. However, from a more general point of view, the two problems are simply different examples of a linear model. We may always combine the responses Y_{ji} into a single vector

$$\mathbf{Y} = (Y_{11}, Y_{12}, \ldots, Y_{1n_1}, Y_{21}, Y_{22}, \ldots, Y_{2n_2}, \ldots, Y_{k1}, Y_{k2}, \ldots, Y_{kn_k})^T.$$

Then the data of Table 6.1 can be concisely expressed as the linear model

$$\mathbf{Y} = \mathbf{X}\boldsymbol{\beta} + \boldsymbol{\epsilon}, \tag{6.7}$$

where $\boldsymbol{\beta} = [\beta_1, \ldots, \beta_k]^T$, $\boldsymbol{\epsilon} = (\epsilon_1, \ldots, \epsilon_n)^T$ is a vector of *iid* error terms with $\epsilon_1 \sim N(0, \sigma^2)$, and $\mathbf{X} \in \mathcal{M}_{n,k}$ is a matrix for which $\{\mathbf{X}\}_{ij}$ equals 1 if the ith element of $\mathbf{Y}$ is from treatment j, and 0 otherwise. This means that each row of $\mathbf{X}$ contains exactly one element equal to 1, and each column contains at least one element equal to 1. Thus, the column vectors are linearly independent and $\mathbf{X}$ is of full rank k. For example, for $k = 3$, $n_1 = 3$, $n_2 = 4$, $n_3 = 2$ we would have

$$\mathbf{X} = \begin{bmatrix} 1 & 1 & 1 & 0 & 0 & 0 & 0 & 0 & 0 \\ 0 & 0 & 0 & 1 & 1 & 1 & 1 & 0 & 0 \\ 0 & 0 & 0 & 0 & 0 & 0 & 0 & 1 & 1 \end{bmatrix}^T. \tag{6.8}$$

TABLE 6.1
Structure of one-way ANOVA model. The observations $Y_{j1}, Y_{j2}, \ldots, Y_{jn_j}$ are an *iid* sample from $N(\beta_j, \sigma^2)$, $j = 1, \ldots, k$. Let $n = n_1 + \ldots + n_j$ be the total sample size.

Pop	Pop Mean	Sample Size	Sample	Sample Mean	Sum of Squares
1	β_1	n_1	$Y_{11}, Y_{12}, \ldots, Y_{1n_1}$	$\bar{Y}_1$	$\sum_{i=1}^{n_1} (Y_{1i} - \bar{Y}_1)^2$
2	β_2	n_2	$Y_{21}, Y_{22}, \ldots, Y_{2n_2}$	$\bar{Y}_2$	$\sum_{i=1}^{n_2} (Y_{2i} - \bar{Y}_2)^2$
$\vdots$	$\vdots$	$\vdots$	$\vdots$	$\vdots$	$\vdots$
k	β_k	n_k	$Y_{k1}, Y_{k2}, \ldots, Y_{kn_k}$	$\bar{Y}_k$	$\sum_{i=1}^{n_k} (Y_{ki} - \bar{Y}_k)^2$

Total Mean	$\bar{Y} = \frac{n_1\bar{Y}_1 + n_2\bar{Y}_2 + \ldots + n_k\bar{Y}_k}{n_1 + n_2 + \ldots + n_k}$
Treat. Sum of Squares	$SSTr = \sum_{i=1}^{k} \sum_{j=1}^{n_i} (\bar{Y}_i - \bar{Y})^2 = \sum_{i=1}^{k} n_i (\bar{Y}_i - \bar{Y})^2$
Error Sum of Squares	$SSE = \sum_{i=1}^{k} \sum_{j=1}^{n_i} (Y_{ij} - \bar{Y}_i)^2$
Total Sum of Squares	$SSTo = \sum_{i=1}^{k} \sum_{j=1}^{n_i} (Y_{ij} - \bar{Y})^2$

6.3 Best Linear Unbiased Estimators (BLUE)

Our first approach to the estimation of $\boldsymbol{\beta}$ for the linear model (6.2) is to define in advance properties of an estimate $\hat{\boldsymbol{\beta}}$ which we hope to attain. This is the basis of the Gauss-Markov theorem and results in the best linear unbiased estimator (BLUE), which we define next.

Definition 6.1 (Best Linear Unbiased Estimator (BLUE)) Let $\mathbf{Y} = [Y_1, \ldots, Y_n]^T$ be a column vector of random variables, with a distribution that depends on parameter $\boldsymbol{\eta} = (\boldsymbol{\beta}, \xi)$, where ξ is a nuisance parameter. An estimator $\hat{\boldsymbol{\beta}} \in \mathbb{R}^q$ of estimand $\boldsymbol{\beta}$ is linear and unbiased if there exists a known matrix $\mathbf{A} \in \mathcal{M}_{q,n}$ such that $\hat{\boldsymbol{\beta}} = \mathbf{AY}$, with $E_{\boldsymbol{\eta}}[\hat{\boldsymbol{\beta}}] = \boldsymbol{\beta}$.

An linear unbiased estimator $\hat{\boldsymbol{\beta}} \in \mathbb{R}^q$ of estimand $\boldsymbol{\beta}$ is a best linear unbiased estimator (BLUE) if for any constant vector of coefficients $\mathbf{c} \in \mathcal{M}_{q,1}$, $\text{var}_{\boldsymbol{\eta}}[\mathbf{c}^T\hat{\boldsymbol{\beta}}] < \infty$, and $\text{var}_{\boldsymbol{\eta}}[\mathbf{c}^T\hat{\boldsymbol{\beta}}] \leq \text{var}_{\boldsymbol{\eta}}[\mathbf{c}^T\hat{\boldsymbol{\beta}}']$ for any other linear unbiased estimate $\hat{\boldsymbol{\beta}}'$. ∎

Remark 6.1 There is a subtle point that arises from the definition of a BLUE, which takes the form $\hat{\boldsymbol{\beta}} = [\hat{\beta}_1, \ldots, \hat{\beta}_q]^T$. The BLUE property states that $\hat{\eta} = \mathbf{c}^T\hat{\boldsymbol{\beta}}$ minimizes the variance among estimators of $\eta = \mathbf{c}^T\boldsymbol{\beta}$ of the form $\mathbf{c}^T\hat{\boldsymbol{\beta}}'$, where $\hat{\boldsymbol{\beta}}' = [\hat{\beta}_1', \ldots, \hat{\beta}_q']^T$ is any other linear unbiased estimate of $\boldsymbol{\beta}$. Can we simply say that $\hat{\eta}$ minimizes the variance among all linear unbiased estimators of η? Conveniently, the answer is yes. First, suppose $\eta = \beta_1$, and that δ is a linear unbiased estimate of η for which $\text{var}[\delta] < \text{var}[\hat{\beta}_1]$. Then $\hat{\boldsymbol{\beta}}^* = (\delta, \hat{\beta}_2, \ldots, \hat{\beta}_q)$ is a linear unbiased estimator of $\boldsymbol{\beta}$, which contradicts the assumption that $\hat{\boldsymbol{\beta}}$ is a BLUE. The argument can be extended to $\mathbf{c}^T\boldsymbol{\beta}$ by applying a suitable bijective linear transformation to $\boldsymbol{\beta}$. ∎

Given the linear model (6.2), the Gauss-Markov theorem begins with an inspired guess of the BLUE, in particular,

$$\hat{\boldsymbol{\beta}}_{BLUE} = \left[\mathbf{X}^T\Sigma_{\mathbf{Y}}^{-1}\mathbf{X}\right]^{-1}\mathbf{X}^T\Sigma_{\mathbf{Y}}^{-1}\mathbf{Y}, \tag{6.9}$$

where $\Sigma_{\mathbf{Y}} = \text{var}[\mathbf{Y}]$ is the model covariance. That $\hat{\boldsymbol{\beta}}_{BLUE}$ is unbiased can easily be seen by substituting $E_{\beta}[\mathbf{Y}] = \mathbf{X}\beta$ into Equation (6.9). By Equation (2.3) the covariance matrix is easily evaluated as

$$\Sigma_{\hat{\beta}_{BLUE}} = \left[\mathbf{X}^T \Sigma_{\mathbf{Y}}^{-1} \mathbf{X}\right]^{-1} \tag{6.10}$$

Of course, we need to assume that $\hat{\beta}_{BLUE}$ does not depend on nuisance parameter ξ. In practice, $\Sigma_{\mathbf{Y}}$ will be unknown. However, if we may assume that $\Sigma_{\mathbf{Y}} = \sigma^2 \mathbf{R}$, where σ^2 is unknown, but $\mathbf{R}$ is a known correlation matrix, then the scale factor σ^2 will cancel in Equation (6.9), yielding

$$\hat{\beta}_{BLUE} = \left[\mathbf{X}^T \mathbf{R}^{-1} \mathbf{X}\right]^{-1} \mathbf{X}^T \mathbf{R}^{-1} \mathbf{Y}. \tag{6.11}$$

We will see that this type of assumption plays an important role in the more practical aspects of statistical modeling (for example, generalized estimating equations (GEE), Section 13.7). It should be noted that the proof of the Gauss-Markov theorem, which we give next, makes use of Loewner ordering (Section B.4.6).

Theorem 6.1 (Gauss-Markov Theorem) Suppose, following Definition 6.1, Equation (6.2) holds. Assume $\mathbf{X}$ is full rank and $\Sigma_{\mathbf{Y}}^{-1}$ exists. Then the estimator defined by Equation (6.9) is the unique BLUE, provided it does not depend on any unknown parameters. ∎

Proof. Any linear estimator of β can be written

$$\hat{\beta}^* = \mathbf{A}\mathbf{Y} = \hat{\beta}_{BLUE} - \hat{\beta}_{BLUE} + \mathbf{A}\mathbf{y} = \left[\left[\mathbf{X}^T \Sigma_{\mathbf{Y}}^{-1} \mathbf{X}\right]^{-1} \mathbf{X}^T \Sigma_{\mathbf{Y}}^{-1} + \mathbf{D}\right] \mathbf{Y}.$$

If $\hat{\beta}^*$ is unbiased, we must have $E[\mathbf{D}\mathbf{Y}] = \mathbf{0}$, which implies $\mathbf{D}\mathbf{X}\beta = \mathbf{0}$. This must hold for all β, so we must have $\mathbf{D}\mathbf{X} = \mathbf{0}$. The covariance matrix of $\hat{\beta}^*$ is then

$$\begin{aligned} \Sigma_{\hat{\beta}^*} &= \left[\left[\mathbf{X}^T \Sigma_{\mathbf{Y}}^{-1} \mathbf{X}\right]^{-1} \mathbf{X}^T \Sigma_{\mathbf{Y}}^{-1} + \mathbf{D}\right] \Sigma_{\mathbf{Y}} \left[\left[\mathbf{X}^T \Sigma_{\mathbf{Y}}^{-1} \mathbf{X}\right]^{-1} \mathbf{X}^T \Sigma_{\mathbf{Y}}^{-1} + \mathbf{D}\right]^T \\ &= \Sigma_{\hat{\beta}_{BLUE}} + \mathbf{D}\Sigma_{\mathbf{Y}}\mathbf{D}^T, \end{aligned}$$

where we make use of the unbiasedness constraint $\mathbf{D}\mathbf{X} = \mathbf{0}$. This means $\Sigma_{\hat{\beta}^*} - \Sigma_{\hat{\beta}}$ is equal to a positive semidefinite matrix, so that for any set of coefficients $\mathbf{c}^T = [c_1, \ldots, c_q]$ we must have

$$\text{var}[\mathbf{c}^T \hat{\beta}^*] = \mathbf{c}^T \left[\Sigma_{\hat{\beta}_{BLUE}} + \mathbf{D}\Sigma_{\mathbf{Y}}\mathbf{D}^T\right] \mathbf{c} = \text{var}[\mathbf{c}^T \hat{\beta}_{BLUE}] + \mathbf{c}^T \mathbf{D}\Sigma_{\mathbf{Y}}\mathbf{D}^T \mathbf{c},$$

which proves that $\hat{\beta}_{BLUE}$ is a BLUE after noting that $\mathbf{c}^T \mathbf{D}\Sigma_{\mathbf{Y}}\mathbf{D}^T \mathbf{c} \geq 0$. That the BLUE is unique follows after noting that we must have $\mathbf{c}^T \mathbf{D}\Sigma_{\mathbf{Y}}\mathbf{D}^T \mathbf{c} > 0$ for some $\mathbf{c}$, unless $\mathbf{D} = \mathbf{0}$. □

Although the theory of linear models anticipates that β is multidimensional, the Gauss-Markov theorem has some important implications for the one-dimensional case, which we describe in the next two examples.

Example 6.1 (Reciprocal Variance Weighting) Suppose we are given a sample $X_1, \ldots, X_n$ for which $E[X_i] = \theta$ and $\text{var}[X_i] = \sigma_i^2$. If the observations are independent then from Equation (6.9) the BLUE of θ is given by $\hat{\theta}_{BLUE} = \sum_{i=1}^n b_i X_i$, where

$$b_i = \sigma_i^{-2} / \sum_{i=1}^n \sigma_i^{-2}, \; i = 1, \ldots, n. \tag{6.12}$$

The contribution of X_i to the BLUE is thus weighted by the reciprocal of its variance. The proof is left to the reader (Problem 6.1). ∎

The next example is a simple application of Example 6.1.

Example 6.2 (Pooled Estimates and BLUEs) Equation (6.12) confirms the best way to pool several sample means into a single estimate. Suppose $\bar{X}_j$, $j = 1, \ldots, m$ are independent sample means formed from *iid* samples of sizes n_j from a common distribution of mean μ and variance σ^2. The variances are therefore $\text{var}[\bar{X}_j] = \sigma^2/n_j$. The BLUE of μ is is then

$$\bar{X}^* = \frac{\sum_{j=1}^m n_j \bar{X}_j}{\sum_{j=1}^m n_j}.$$

Essentially, the sample means are disassembled, the data pooled, then a single sample mean calculated. ∎

We next apply Theorem 6.1 to the simple linear regression model.

Example 6.3 (BLUE for Simple Linear Regression) Given Equations (6.4)–(6.5) we can see that the simple linear regression model of Section 6.2.1 is a linear model as defined by the Gauss-Markov theorem. Suppose $\Sigma_{\mathbf{Y}} = \sigma^2 \mathbf{I}_n$, where σ^2 is unknown. By Equation (6.11) the BLUE of $\boldsymbol{\beta}$ is

$$\hat{\boldsymbol{\beta}} = \left[\mathbf{X}^T \mathbf{X}\right]^{-1} \mathbf{X}^T \mathbf{Y}. \tag{6.13}$$

This gives

$$\begin{aligned}
\hat{\boldsymbol{\beta}} &= \begin{bmatrix} n & \sum_{i=1}^n x_i \\ \sum_{i=1}^n x_i & \sum_{i=1}^n x_i^2 \end{bmatrix}^{-1} \begin{bmatrix} \sum_{i=1}^n Y_i \\ \sum_{i=1}^n x_i Y_i \end{bmatrix} \\
&= \begin{bmatrix} n & S_x \\ S_x & S_{xx} \end{bmatrix}^{-1} \begin{bmatrix} S_Y \\ S_{xY} \end{bmatrix} \\
&= \frac{1}{n S_{xx} - S_x^2} \begin{bmatrix} S_{xx} & -S_x \\ -S_x & n \end{bmatrix} \begin{bmatrix} S_Y \\ S_{xY} \end{bmatrix}
\end{aligned} \tag{6.14}$$

using an obvious representation of summations. We first simplify $\hat{\beta}_1$,

$$\hat{\beta}_1 = \frac{n S_{xY} - S_x S_Y}{n S_{xx} - S_x^2} = \frac{\sum_{i=1}^n (x_i - \bar{x})(Y_i - \bar{Y})}{\sum_{i=1}^n (x_i - \bar{x})^2} = \frac{\sum_{i=1}^n (x_i - \bar{x}) Y_i}{\sum_{i=1}^n (x_i - \bar{x})^2}.$$

Then

$$\hat{\beta}_0 = \frac{S_{xx} S_Y - S_x S_{xY}}{n S_{xx} - S_x^2} = \frac{S_{xx} S_Y - S_x^2 S_Y/n + S_x^2 S_Y/n - S_x S_{xY}}{n S_{xx} - S_x^2} = \bar{Y} - \hat{\beta}_1 \bar{x},$$

so that the BLUE has a simple closed form expression. ∎

In the next example it will be shown that the BLUE for the one-way ANOVA model consists simply of the within-treatment sample means. However, there will be some advantage to recognizing this as an optimal solution in the sense of Definition 6.1.

Example 6.4 (BLUE for One-Way ANOVA) The one-way ANOVA model of Equation (6.7) is a linear model as defined by the Gauss-Markov theorem, so we may derive the BLUE of the parameter $\boldsymbol{\beta}$. Suppose we have k treatments with sample sizes $n_1, \ldots, n_k$. We will make use of the zero and one matrices $\mathbf{0}_{n,m} \in \mathcal{M}_{n,m}$, $\mathbf{1}_{n,m} \in \mathcal{M}_{n,m}$ (Section B.4). The matrix $\mathbf{X}$ can be written in block form

$$\mathbf{X} = \begin{bmatrix} \mathbf{1}_{n_1,1} & \mathbf{0}_{n_1,1} & \cdots & \mathbf{0}_{n_1,1} \\ \mathbf{0}_{n_2,1} & \mathbf{1}_{n_2,1} & \cdots & \mathbf{0}_{n_2,1} \\ \vdots & \vdots & \vdots & \vdots \\ \mathbf{0}_{n_k,1} & \mathbf{0}_{n_k,1} & \cdots & \mathbf{1}_{n_k,1} \end{bmatrix}.$$

If we assume $\Sigma_{\mathbf{Y}} = \sigma^2 \mathbf{I}_n$, then the BLUE of $\boldsymbol{\beta}$ is

$$\hat{\boldsymbol{\beta}} = \left[\mathbf{X}^T\mathbf{X}\right]^{-1}\mathbf{X}^T\mathbf{Y} = \begin{bmatrix} n_1 & 0 & \cdots & 0 \\ 0 & n_2 & \cdots & 0 \\ \vdots & \vdots & \vdots & \vdots \\ 0 & 0 & \cdots & n_k \end{bmatrix}^{-1} \begin{bmatrix} \sum_{i=1}^{n_1} Y_{1i} \\ \sum_{i=1}^{n_2} Y_{2i} \\ \vdots \\ \sum_{i=1}^{n_k} Y_{ki} \end{bmatrix} = \begin{bmatrix} \bar{Y}_1 \\ \bar{Y}_2 \\ \vdots \\ \bar{Y}_k \end{bmatrix}.$$

In other words, the obvious estimates for the treatment means are also the BLUEs. ■

While the Gauss-Markov theorem is most prominently associated with linear regression models, it is equally applicable to models not normally regarded as of this type. Consider the following example.

Example 6.5 (Estimation of Mixtures and BLUEs) A natural application of BLUEs is in mixed sampling. Suppose we have m populations, with population means and variances β_j, σ^2, $j = 1, \ldots, m$. A mixed estimate is formed by sampling n_j observations from populations $j = 1, \ldots, m$. Suppose only the sample mean of the pooled observations is available, denoted as Y. This means population j contributed a proportion $p_j = n_j/n$ of the sample, where $n = \sum_{i=1}^m n_i$. Assuming observations are independent, this means

$$E[Y] = p_1\beta_1 + \ldots + p_m\beta_m, \quad \text{var}[Y] = n^{-1}\sigma^2.$$

Next, suppose we have n such estimators Y_i, $i = 1, \ldots, n$, based on population sample sizes $(n_1[i], \ldots, n_m[i])$, with $n[i] = \sum_{j=1}^m n_j[i]$. As before, population j contributed a proportion $p_j[i] = n_j[i]/n[i]$ of the observations to the ith estimator. If we let $\mathbf{Y} = [Y_1, \ldots, Y_n]^T$, define $\mathbf{X} \in \mathcal{M}_{n,m}$ by $\{\mathbf{X}\}_{i,j} = p_j[i]$ and set $\boldsymbol{\beta} = [\beta_1, \ldots, \beta_m]^T$, then we have an example of Equation (6.2), $\mathbf{Y} = \mathbf{X}\boldsymbol{\beta} + \boldsymbol{\epsilon}$, where $E[\boldsymbol{\epsilon}] = 0$, and $\Sigma_{\mathbf{Y}} = \sigma^2\mathbf{R}$, where $\mathbf{R} \in \mathcal{M}_n$ is a diagonal matrix with $\{R\}_{ii} = n[i]^{-1}$. Then by Equation (6.11) the BLUE of $\boldsymbol{\beta}$ becomes

$$\hat{\beta}_{BLUE} = \left[\mathbf{X}^T\mathbf{R}^{-1}\mathbf{X}\right]^{-1}\mathbf{X}^T\mathbf{R}^{-1}\mathbf{Y},$$

which does not depend on any unknown parameters (note that the sample sizes $n_j[i]$ must be chosen so that $\mathbf{X}$ is a full rank matrix). ■

6.4 Least Squares Estimators, BLUEs and Projection Matrices

The BLUE is the unique solution to a well defined optimization problem. However, the BLUE is also the solution to two other other problems. As such, this means that the three problems are simply reformulations of a single one, and an understanding of this equivalence offers considerable insight, as well as some simplification and unification of the theory.

All three problems concern estimation of $\boldsymbol{\beta}$ for the linear model of Equation (6.2). Although there is no restriction on the model covariance matrix $\Sigma_{\mathbf{Y}}$ other than invertibility, in Section 6.5.2 we will see that little generality is lost by assuming $\Sigma_{\mathbf{Y}} = \sigma^2\mathbf{I}_n$ (this idea is made precise in Section 9.5.1). If we do so, then $\mathbf{Y} \sim N(\boldsymbol{\mu}, \sigma^2\mathbf{I}_n)$, where $\boldsymbol{\mu} = \mathbf{X}\boldsymbol{\beta}$. Then as $\boldsymbol{\beta}$ ranges over $\boldsymbol{\beta} \in \mathbb{R}^q$, $\boldsymbol{\mu}$ ranges over $\mathcal{V}_{\mathbf{X}}$. Since a bijection exists between $\mathbb{R}^q$ and $\mathcal{V}_{\mathbf{X}}$, the problem of estimating $\boldsymbol{\beta} \in \mathbb{R}^q$ can be replaced by the problem of estimating $\boldsymbol{\mu} \in \mathcal{V}_{\mathbf{X}}$.

We take the least squares estimate of $\boldsymbol{\mu}$ to be

$$\hat{\mathbf{Y}}_{LSE} = \text{argmin}_{\mathbf{y} \in \mathcal{V}_{\mathbf{X}}} (\mathbf{Y} - \mathbf{y})^T(\mathbf{Y} - \mathbf{y}). \tag{6.15}$$

The solution is easy to characterize. Let $\mathbf{y}^*$ be the orthogonal projection of $\mathbf{Y}$ onto $\mathcal{V}_\mathbf{X}$ (Section B.4.5). By the Pythagorean theorem, for any $\mathbf{y} \in \mathcal{V}_\mathbf{X}$

$$(\mathbf{Y}-\mathbf{y})^T(\mathbf{Y}-\mathbf{y}) = |\mathbf{Y}-\mathbf{y}|^2 = |\mathbf{Y}-\mathbf{y}^*+\mathbf{y}^*-\mathbf{y}|^2 = |\mathbf{Y}-\mathbf{y}^*|^2 + |\mathbf{y}^*-\mathbf{y}|^2,$$

since $\mathbf{Y}-\mathbf{y}^*$ is orthogonal to $\mathcal{V}_\mathbf{X}$ while $\mathbf{y}^*-\mathbf{y} \in \mathcal{V}_\mathbf{X}$. Clearly, $|\mathbf{Y}-\mathbf{y}|^2$ is minimized by setting $\mathbf{y}=\mathbf{y}^*$, so that $\hat{\mathbf{Y}}_{LSE} = \mathbf{y}^*$, which is the orthogonal projection of $\mathbf{Y}$ onto $\mathcal{V}_\mathbf{X}$.

Next, note that the BLUE is $\hat{\boldsymbol{\beta}}_{BLUE} = (\mathbf{X}^T\mathbf{X})^{-1}\mathbf{X}^T\mathbf{Y}$, which gives fitted values

$$\hat{\mathbf{Y}}_{BLUE} = \mathbf{X}\hat{\boldsymbol{\beta}}_{BLUE} = \mathbf{X}(\mathbf{X}^T\mathbf{X})^{-1}\mathbf{X}^T\mathbf{Y} = \mathbf{H}_\mathbf{X}\mathbf{Y}.$$

But $\mathbf{H}_X$ is the projection matrix onto $\mathcal{V}_\mathbf{X}$ (Section B.4.5), so that

$$\hat{\mathbf{Y}}_{BLUE} = \hat{\mathbf{Y}}_{LSE},$$

the orthogogonal projection of $\mathbf{Y}$ onto $\mathcal{V}_\mathbf{X}$.

We then refer to the residuals as

$$\mathbf{e} = \mathbf{Y} - \hat{\mathbf{Y}}_{LSE} = (\mathbf{I}_n - \mathbf{H}_\mathbf{X})\mathbf{Y}. \tag{6.16}$$

By the preceding argument $\mathbf{e}$ must be orthogonal to $\mathcal{V}_\mathbf{X}$. An important related quantity is the error sum of squares (SSE):

$$SSE = (\mathbf{Y}-\hat{\mathbf{Y}}_{LSE})^T(\mathbf{Y}-\hat{\mathbf{Y}}_{LSE}) = \mathbf{e}^T\mathbf{e} = \mathbf{Y}^T(\mathbf{I}_n - \mathbf{H}_\mathbf{X})\mathbf{Y}, \tag{6.17}$$

noting that $(\mathbf{I}_n - \mathbf{H}_\mathbf{X})$ is symmetric and idempotent, so that $(\mathbf{I}_n - \mathbf{H}_\mathbf{X}) = (\mathbf{I}_n - \mathbf{H}_\mathbf{X})(\mathbf{I}_n - \mathbf{H}_\mathbf{X})$ (Section B.4.5).

It is important to ensure that the least squares criterion is precisely interpreted, as the next example makes clear.

Example 6.6 (Alternative Least Squares Criterion for Simple Linear Regression) The estimated regression line $\hat{\beta}_0 + \hat{\beta}_1 x$ shown in Figure 6.1 was calculated using the least squares criterion of Equation (6.15), that is, the estimates $\hat{\beta}_0, \hat{\beta}_1$ minimize the sum of squares

$$SSE = (\mathbf{Y}-\hat{\mathbf{Y}})^T(\mathbf{Y}-\hat{\mathbf{Y}}) = \sum_{i=1}^n (Y_i - \hat{\beta}_0 - \hat{\beta}_1 x)^2. \tag{6.18}$$

Figure 6.2 shows estimated regression lines calculated using three variations of the least squares criterion applied to the same data set. When we regress $\mathbf{Y}$ on $\mathbf{x}$ we minimize the sum of squares of the *vertical* distance between Y_i and $\hat{\beta}_0 + \hat{\beta}_1 x_i$ (Δy of Figure 6.2). This is the method described above, expressed mathematically by Equations (6.15) and (6.18), and used in Figure 6.1.

Suppose we then exchange the roles of $\mathbf{Y}$ and $\mathbf{x}$, and regress $\mathbf{x}$ on $\mathbf{Y}$. Mathematically, at least, there is no reason we can't do this. Should we expect to get the same estimated regresson line? This is now equivalent to minimizing the sum of squares of the *horizontal* distance between Y_i and $\hat{\beta}_0 + \hat{\beta}_1 x_i$ (Δx of Figure 6.2). This is also shown in the plot and results in a different fit.

Given that the two preceding criterion yield different estimates, it might be asked why we don't simply minimize the sum of squares of the *perpendicular* distance between Y_i and $\hat{\beta}_0 + \hat{\beta}_1 x_i$ ($\Delta(x,y)$ of Figure 6.2). This procedure, refered to as total regression, yields a third fit, also shown in the plot.

The choice of (horizontal) least squares criterion (6.15) is determined by the form of Equation (6.2), in particular by the fact that we are trying to model the distribution of the response vector $\mathbf{Y}$, while the independent variables $\mathbf{x}$ are known constants. Error is therefore in the direction of what we are trying to estimate. The statistical properties of the estimation problem would depend quite significantly on the type of least squares criterion used, and the fact that the BLUE estimate is equivalent to (6.15) suffices for our purposes to determine this choice.

It should be noted, however, that if x_i is observable only with error then the estimation problem changes in character, and new techniques must be introduced. The term errors-in-variables model or measurement error model is commonly used to describe this problem (Cochran, 1972; Fuller, 2009). For example, a laboratory assay method (Y) might be calibrated to a second (X) by collecting paired measurements (Y_i, X_i) from each and estimating the relation $Y = \beta_0 + \beta_1 X$. Both X_i and Y_i are random, so we cannot meaningfully designate the role of dependent and independent variable. This means the BLUE criterion of Definition 6.1 no longer captures the objectives of the problem. In this case, total regression would be more appropriate (Deming, 1943). ∎

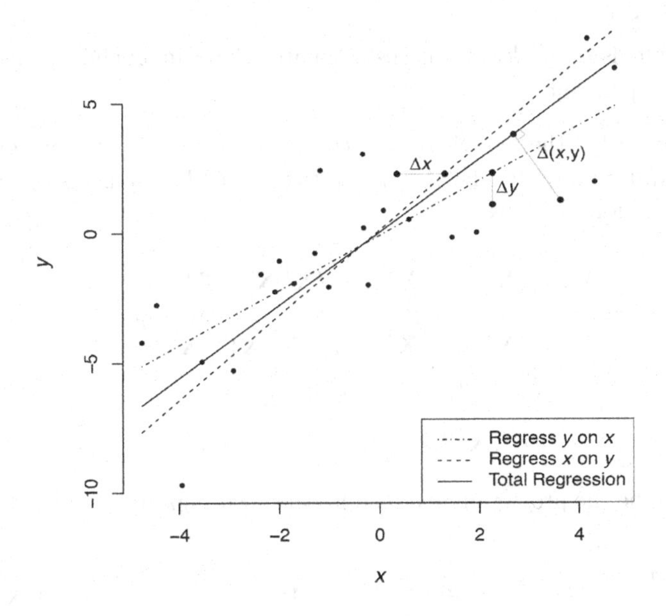

FIGURE 6.2
Three regression estimates using alternative least squares criterion (Example 6.6). Observations (Y_i, x_i) are plotted individually.

6.5 Ordinary and Generalized Least Squares Estimators

The analysis of the linear model is clearly simplified for the case $\Sigma_{\mathbf{Y}} = \sigma^2 \mathbf{I}_n$. Although the Gauss-Markov theorem in its most general form does not impose this constraint, it is referred to widely in the literature as part of the Gauss-Markov conditions. This defines an important special case of linear model, which we single out in the next theorem. Note that we are making use of the ideas surrounding Cochran's theorem (Theorem 2.5).

Theorem 6.2 (Ordinary Least Squares Estimator (OLS)) Suppose for the linear model (6.2), with $\boldsymbol{\beta} \in \mathbb{R}^q$, the error terms ϵ_j satisfy the Gauss-Markov conditions:

$$E[\epsilon_j] = 0, \ \ \text{var}[\epsilon_j] = \sigma^2, \ \ \text{cov}[\epsilon_j, \epsilon_k] = 0, \ \ j \neq k. \tag{6.19}$$

Assume that $\mathbf{X}$ is of full rank. Let $\mathbf{e}$ be the residuals defined in Equation (6.16), and let SSE be the error sum of squares defined in Equation (6.17). Then the following statements hold:

(i) The BLUE of $\boldsymbol{\beta}$ is given by $\hat{\boldsymbol{\beta}} = (\mathbf{X}^T\mathbf{X})^{-1}\mathbf{X}^T\mathbf{Y}$.
(ii) We have $E[\hat{\boldsymbol{\beta}}] = \boldsymbol{\beta}$ and $\Sigma_{\hat{\boldsymbol{\beta}}} = \sigma^2(\mathbf{X}^T\mathbf{X})^{-1}$.
(iii) $E[SSE] = (n-q)\sigma^2$.
(iv) The residuals $\mathbf{e}$ and least squares estimates $\hat{\boldsymbol{\beta}}$ are uncorrelated.
(v) If, in addition, $\boldsymbol{\epsilon} \sim N(\mathbf{0}, \sigma^2\mathbf{I}_n)$, then
   (a) $\hat{\boldsymbol{\beta}} \sim N(\boldsymbol{\beta}, \sigma^2(\mathbf{X}^T\mathbf{X})^{-1})$;
   (b) $\sigma^{-2}SSE \sim \chi^2_{n-q}$;
   (c) The residuals $\mathbf{e}$ and least squares estimates $\hat{\boldsymbol{\beta}}$ are independent, hence SSE and $\hat{\boldsymbol{\beta}}$ are independent. ■

Proof. By Equation (6.19) the model covariance matrix is $\Sigma_{\mathbf{Y}} = \sigma^2\mathbf{I}_n$. Statement (i) then follows directly from Equation (6.11). Regarding (ii) any BLUE is unbiased. Noting that $\hat{\boldsymbol{\beta}}$ is a linear transformation of $\mathbf{Y}$, we have

$$\begin{aligned}
\Sigma_{\hat{\boldsymbol{\beta}}} &= (\mathbf{X}^T\mathbf{X})^{-1}\mathbf{X}^T\Sigma_{\mathbf{Y}}\left[(\mathbf{X}^T\mathbf{X})^{-1}\mathbf{X}^T\right]^T \\
&= (\mathbf{X}^T\mathbf{X})^{-1}\mathbf{X}^T\sigma^2\mathbf{I}_n\left[(\mathbf{X}^T\mathbf{X})^{-1}\mathbf{X}^T\right]^T \\
&= \sigma^2(\mathbf{X}^T\mathbf{X})^{-1}\mathbf{X}^T\mathbf{X}(\mathbf{X}^T\mathbf{X})^{-1} \\
&= \sigma^2(\mathbf{X}^T\mathbf{X})^{-1}.
\end{aligned}$$

To prove (iii) let $\mathbf{H}_{\mathbf{X}}$ be the projection matrix onto the vector space $\mathcal{V}_{\mathbf{X}}$. Then from Equation (6.17)

$$SSE = \mathbf{Y}^T(\mathbf{I}_n - \mathbf{H}_{\mathbf{X}})\mathbf{Y} = [\boldsymbol{\mu} + \boldsymbol{\epsilon}]^T(\mathbf{I}_n - \mathbf{H}_{\mathbf{X}})[\boldsymbol{\mu} + \boldsymbol{\epsilon}]^T = \boldsymbol{\epsilon}^T(\mathbf{I}_n - \mathbf{H}_{\mathbf{X}})\boldsymbol{\epsilon}^T,$$

since, by assumption, $\mathbf{H}_{\mathbf{X}}\boldsymbol{\mu} = \boldsymbol{\mu} \in \mathcal{V}_{\mathbf{X}}$. The proof is completed by applying Theorem 2.5.

Note that $\mathbf{e} = (\mathbf{I}_n - \mathbf{H}_{\mathbf{X}})\mathbf{Y}$ and $\boldsymbol{\beta} = \mathbf{H}_{\mathbf{X}}\mathbf{Y}$. Statement (iv) follows from the fact that $(\mathbf{I}_n - \mathbf{H}_{\mathbf{X}})\mathbf{H}_{\mathbf{X}} = \mathbf{0}$.

(v)-(a) follows from (ii) and the fact that $\hat{\boldsymbol{\beta}}$ is a linear transformation of a multivariate normal random vector $\mathbf{Y}$. (v)-(b) follows the proof of (iii), and is an application of Theorem 2.5. To prove (v)-(c) note that $\mathbf{e}$ and $\hat{\boldsymbol{\beta}}$ are both linear transformations of $\mathbf{Y}$ and therefore possess a joint multivariate normal density. By (iv) they are uncorrelated and therefore independent. The proof is completed by noting that SSE is a mapping of $\mathbf{e}$. □

6.5.1 Inference for σ^2 and $\boldsymbol{\beta}$

Consider a linear model with $\boldsymbol{\epsilon} \sim N(\mathbf{0}, \sigma^2\mathbf{I}_n)$, $\boldsymbol{\beta} \in \mathbb{R}^q$. Suppose we are interested in the estimand $\eta = \mathbf{c}^T\boldsymbol{\beta}$. If $\hat{\boldsymbol{\beta}}$ is a BLUE for $\boldsymbol{\beta}$, then by the Gauss-Markov theorem $\hat{\eta} = \mathbf{c}^T\hat{\boldsymbol{\beta}}$ minimizes the variance among linear unbiased estimators of η (see Remark 6.1). In particular, $\text{var}[\hat{\eta}] = \mathbf{c}\Sigma_{\hat{\boldsymbol{\beta}}}\mathbf{c}^T = \sigma^2\mathbf{c}(\mathbf{X}^T\mathbf{X})^{-1}\mathbf{c}^T = \sigma^2 C$, noting that the constant C does not depend on any unknown parameters. Then, since $\hat{\eta}$ is a linear combination of the responses Y_i, we have

$Z = (\hat{\eta} - \eta)/\sigma\sqrt{C} \sim N(0,1)$. We next note that $\sigma^{-2}SSE \sim \chi^2_{n-q}$ (Theorem 6.2 (v)-(b)), so we define the mean sum of squares (MSE) $MSE = SSE/(n-q)$, which is an unbiased estimate of σ^2, and is independent of $\hat{\boldsymbol{\beta}}$, and hence $\hat{\eta}$. This means

$$T = \frac{Z}{\sqrt{MSE}/\sigma} = \frac{\hat{\eta} - \eta}{\sqrt{C \times MSE}} \sim T_{n-q}$$

has a T distribution with $n - q$ degrees of freedom, thus forming a pivot for $\mathbf{c}^T\boldsymbol{\beta}$, while $(n-q)MSE/\sigma^2$ forms a pivot for σ^2 (see Sections 4.6.1 and 5.7).

We next specialize this approach to the simple linear regression model.

Example 6.7 (Confidence Intervals for Simple Linear Regression) Consider the simple linear linear regression model of Sections 6.2.1 and 6.3 and define the estimand $\mu_x = \beta_0 + \beta_1 x$. The BLUE of μ_x is then $\hat{\mu}_x = \hat{\beta}_0 + \hat{\beta}_1 x$. The reader can verify that the values of $\sigma^2_{\hat{\beta}_i}$ and $\sigma^2_{\hat{\mu}_x}$ are

$$\sigma^2_{\hat{\beta}_1} = \frac{\sigma^2}{\sum_{i=1}^n (x_i - \bar{x})^2} \tag{6.20}$$

and

$$\sigma^2_{\hat{\mu}_x} = \sigma^2 \left[\frac{1}{n} + \frac{(x - \bar{x})^2}{\sum_{i=1}^n (x_i - \bar{x})^2} \right] \tag{6.21}$$

where we have mean value of the predictor $\bar{x} = n^{-1}\sum_{i=1}^n x_i$. Since $\beta_0 = \mu_0$ we can obtain directly from (6.21) the variance of $\hat{\beta}_0$ by substituting $x = 0$:

$$\sigma^2_{\hat{\beta}_0} = \sigma^2 \left[\frac{1}{n} + \frac{\bar{x}^2}{\sum_{i=1}^n (x_i - \bar{x})^2} \right]. \tag{6.22}$$

If we substitute MSE for σ^2 in the preceding expressions we obtain the standard error

$$S_{\hat{\beta}_1} = \frac{MSE^{1/2}}{\sqrt{\sum_{i=1}^n (x_i - \bar{x})^2}}, \tag{6.23}$$

$$S_{\hat{\mu}_x} = MSE^{1/2} \sqrt{\frac{1}{n} + \frac{(x - \bar{x})^2}{\sum_{i=1}^n (x_i - \bar{x})^2}} \tag{6.24}$$

and

$$S_{\hat{\beta}_0} = MSE^{1/2} \sqrt{\frac{1}{n} + \frac{\bar{x}^2}{\sum_{i=1}^n (x_i - \bar{x})^2}}. \tag{6.25}$$

Then from Theorem 6.2 the $1 - \alpha$ confidence intervals for any estimate $\hat{\eta}$ of η of this type is given by

$$CI = \hat{\eta} \pm t_{n-2;\alpha/2} S_{\hat{\eta}}.$$

See Problem 6.20 for the related problem of prediction. ■

6.5.2 Generalized Least Squares Estimators

Next, suppose the model covariance is given by $\Sigma_{\mathbf{Y}} = \sigma^2 \mathbf{R}$, where σ^2 is unknown and $\mathbf{R}$ is a correlation matrix known but not assumed to equal the identity matrix. Then the BLUE is given by Equation (6.11). In this context, $\hat{\boldsymbol{\beta}}_{BLUE}$ is also known as the generalized least squares (GLS) estimate of $\boldsymbol{\beta}$, as opposed to the ordinary least squares (OLS) ordinary estimate of $\hat{\boldsymbol{\beta}} = [\mathbf{X}^T\mathbf{X}]^{-1}\mathbf{X}\mathbf{Y}$ given in Theorem 6.2.

Of course, both estimators are BLUEs for linear models with specific model covariances.

However, it will be useful to distinguish the two estimators by their forms rather than by their relationship to any actual model covariance, so we write

$$\hat{\beta}_{GLS} = \left[\mathbf{X}^T\mathbf{R}^{-1}\mathbf{X}\right]^{-1}\mathbf{X}\mathbf{R}^{-1}\mathbf{Y}, \quad \hat{\beta}_{OLS} = \left[\mathbf{X}^T\mathbf{X}\right]^{-1}\mathbf{X}\mathbf{Y}.$$

If $\mathbf{R}$ is the true correlation matrix then $\hat{\beta}_{GLS}$ will be the BLUE. However, even in this case $\hat{\beta}_{OLS}$ will be an unbiased estimate of $\boldsymbol{\beta}$, and has the advantage of not requiring knowledge of $\mathbf{R}$ (Example 6.8).

It should be noted that for the GLS estimate the fitted value vector $\hat{\mathbf{Y}}$ is no longer the projection of $\mathbf{Y}$ onto the vector space $\mathcal{V}_{\mathbf{X}}$, so that the conclusions of Theorem 6.2 will not in general hold. However, the orthogonality structure of Section 6.4 on which Theorem 6.2 depends can be easily recovered by a linear transformation of the model.

To see this, let $\mathbf{R}_{1/2}$ be the symmetric square root matrix of $\mathbf{R}$ (Section B.4.4). The original model is given by $\mathbf{Y} = \mathbf{X}\boldsymbol{\beta} + \boldsymbol{\epsilon}$ where the covariance matrix of $\boldsymbol{\epsilon}$ is $\sigma^2\mathbf{R}$. We can premultiply the entire expression by $\mathbf{R}_{1/2}^{-1}$ to get

$$\mathbf{R}_{1/2}^{-1}\mathbf{Y} = \mathbf{R}_{1/2}^{-1}\mathbf{X}\boldsymbol{\beta} + \mathbf{R}_{1/2}^{-1}\boldsymbol{\epsilon}, \text{ or equivalently } \mathbf{Y}_{\mathbf{R}} = \mathbf{X}_{\mathbf{R}}^T\boldsymbol{\beta} + \boldsymbol{\epsilon}_{\mathbf{R}}, \tag{6.26}$$

where $\mathbf{X}_{\mathbf{R}} = \mathbf{R}_{1/2}^{-1}\mathbf{X}$, $\mathbf{Y}_{\mathbf{R}} = \mathbf{R}_{1/2}^{-1}\mathbf{R}$, $\boldsymbol{\epsilon}_{\mathbf{R}} = \mathbf{R}_{1/2}^{-1}\boldsymbol{\epsilon}$, so that the transformed model (6.26) now has covariance matrix $\Sigma_{\mathbf{Y}_{\mathbf{R}}} = \Sigma_{\boldsymbol{\epsilon}_{\mathbf{R}}} = \sigma^2\mathbf{R}_{1/2}^{-1}\mathbf{R}\mathbf{R}_{1/2}^{-1} = \sigma^2\mathbf{I}_n$. Now the model of Equation (6.26) satisfies the Gauss-Markov conditions of Theorem 6.2, the conclusions of which therefore apply to the OLS estimate for that model. But this estimate is identical to $\hat{\beta}_{GLS}$, that is

$$\begin{aligned}
\hat{\beta}_{\mathbf{R}} &= \left[\mathbf{X}_{\mathbf{R}}^T\mathbf{X}_{\mathbf{R}}\right]^{-1}\mathbf{X}_{\mathbf{R}}^T\mathbf{Y}_{\mathbf{R}} \\
&= \left[\mathbf{X}\mathbf{R}_{1/2}^{-1}\mathbf{R}_{1/2}^{-1}\mathbf{X}\right]^{-1}\mathbf{X}\mathbf{R}_{1/2}^{-1}\mathbf{R}_{1/2}^{-1}\mathbf{Y} \\
&= \left[\mathbf{X}\mathbf{R}^{-1}\mathbf{X}\right]^{-1}\mathbf{X}\mathbf{R}^{-1}\mathbf{Y} = \hat{\beta}_{GLS}.
\end{aligned}$$

This means the fitted values of the two models are related by $\hat{\mathbf{Y}}_{\mathbf{R}} = \mathbf{X}_{\mathbf{R}}\hat{\beta}_{\mathbf{R}} = \mathbf{X}_{\mathbf{R}}\hat{\beta}_{GLS} = \mathbf{R}_{1/2}^{-1}\mathbf{X}\hat{\beta}_{GLS} = \mathbf{R}_{1/2}^{-1}\hat{\mathbf{Y}}_{GLS}$. We can then define the generalized error sum of squares $SSE_{\mathbf{R}} = (\mathbf{Y}_{\mathbf{R}} - \hat{\mathbf{Y}}_{\mathbf{R}})^T(\mathbf{Y}_{\mathbf{R}} - \hat{\mathbf{Y}}_{\mathbf{R}}) = (\mathbf{Y} - \hat{\mathbf{Y}}_{GLS})^T\mathbf{R}^{-1}(\mathbf{Y} - \hat{\mathbf{Y}}_{GLS})^T$. In this way, Theorem 6.2 can be applied to $\hat{\beta}_{GLS}$ after substituting $\mathbf{X}_{\mathbf{R}}$ and $SSE_{\mathbf{R}}$ for $\mathbf{X}$ and SSE. Therefore, given a linear model with correlated errors, we may base our inference either on the GLS estimator, or the OLS estimator of the transformed model (6.26) (both being the same).

We illustrate the distinction between ordinary and generalized least squares estimation with a simple version of the mixed effects model.

Example 6.8 (Simple Linear Regression with Random Effects) We will modify the simple linear regression model by including a random effect, which is a hidden or latent variable which contributes to the response and induces correlation. To implement a simple example, suppose we observed pairs of dependent and independent variables (Y_i, x_i), $i = 1, \ldots, n$ similar to the training data for a simple linear regression model. We next impose a partition on the index set $\{1, \ldots, n\}$ into classes $A_1, \ldots, A_k$, assuming that if indices i, i' fall into different classes then Y_i and $Y_{i'}$ will be independent, otherwise they share a common random effect. As a result the pairwise correlation of the responses is given by

$$\mathrm{cor}[Y_i, Y_{i'}] = \begin{cases} 1; & i = i' \\ \rho; & i \neq i' \text{ are in the same class} \\ 0; & i \neq i' \text{ are in different classes} \end{cases}, \tag{6.27}$$

where $\rho > 0$ (note that we are assuming ρ is the same for each class, which can be generalized, with some care).

Given this partition structure, it makes sense to adopt the double subscript notation of Table 6.1. Suppose the jth class includes n_j observations, so that $n = \sum_{j=1}^k n_j$. Then Y_{ji} is the ith response from the jth class, $i = 1, \ldots, n_j$, and we can define the model as

$$Y_{ji} = \beta_0 + \beta_1 x_{ji} + U_j + \epsilon_{ji}, \ i = 1, \ldots, n_j, \ j = 1, \ldots, k,$$

the double subscript notation extended to the predictor and error terms in an obvious way. What is new is the inclusion of the random effect U_j. The error terms ϵ_{ji} remain an *iid* sample from $N(0, \sigma^2)$, while the random effects $U_1, \ldots, U_k$ are an *iid* sample from $N(0, \sigma_U^2)$, assumed to be independent of the error terms.

This allows us to evaluate the correlation ρ in Equation (6.27) as $\rho = \sigma_U^2/(\sigma^2 + \sigma_U^2)$. The model correlation matrix can then be expressed in the block form

$$\mathbf{R} = \begin{bmatrix} \mathbf{R}_1 & \mathbf{0}_{n_1,n_2} & \cdots & \mathbf{0}_{n_1,n_k} \\ \mathbf{0}_{n_2,n_1} & \mathbf{R}_2 & \cdots & \mathbf{0}_{n_2,n_k} \\ \vdots & \vdots & \vdots & \vdots \\ \mathbf{0}_{n_k,n_1} & \mathbf{0}_{n_k,n_2} & \cdots & \mathbf{R}_k \end{bmatrix}$$

where $\mathbf{R}_j$ is an $n_j \times n_j$ matrix equal to

$$\mathbf{R}_j = (1 - \rho)\mathbf{I}_{n_j} + \rho \mathbf{1}_{n_j,n_j}, \ j = 1, \ldots, k.$$

We may invert or factorize $\mathbf{R}$ by block. Suppose we given two $m \times m$ matrices of the form $\mathbf{A} = a\mathbf{I}_m + b\mathbf{1}_{m,m}$, $\mathbf{B} = c\mathbf{I}_m + d\mathbf{1}_{m,m}$, for some constants a, b, c, d. Then the following identity is easily verified

$$\mathbf{AB} = ac\mathbf{I}_m + (ad + bc + mbd)\mathbf{1}_{m,m}, \tag{6.28}$$

and can be used for the necessary matrix operations. Since $\mathbf{R}_j$ is positive definite (given suitable restrictions on ρ), there is a unique positive definite square root matrix $\mathbf{R}_j^{1/2}$, which, following Equation (6.28), is given by

$$\mathbf{R}_j^{1/2} = \sqrt{1-\rho}\left[\mathbf{I}_{n_j} + \frac{\sqrt{1 + n_j\rho/(1-\rho)} - 1}{n_j}\mathbf{1}_{n_j,n_j}\right],$$

with inverse

$$\mathbf{R}_j^{-1/2} = \frac{1}{\sqrt{1-\rho}}\left[\mathbf{I}_{n_j} - \frac{\kappa_j}{n_j}\mathbf{1}_{n_j,n_j}\right],$$

where

$$\kappa_k = 1 - \frac{1}{\sqrt{1 + n_j\rho/(1-\rho)}}, \ j = 1, \ldots, k,$$

Similarly, for any vector $\mathbf{z} \in \mathbb{R}^{n_j}$ we have

$$\mathbf{R}_j^{-1/2}\mathbf{z} = \frac{1}{\sqrt{1-\rho}}[z_1 - \kappa_j\bar{z}, \ldots, z_{n_j} - \kappa_j\bar{z}]^T,$$

where $\bar{z} = n_j^{-1}\sum_{i=1}^{n_j} z_i$. The transformed responses $\mathbf{Y_R}$ are therefore

$$Y_{ji}^{\mathbf{R}} = \frac{1}{\sqrt{1-\rho}}Y_{ji} - \kappa_j\bar{Y}_j, \ i = 1, \ldots, n_j, \ j = 1, \ldots, k,$$

and the transformed matrix of predictors is

$$\mathbf{X_R} = \frac{1}{\sqrt{1-\rho}} \begin{bmatrix} 1-\kappa_1 & x_{11} - \kappa_1 \bar{x}_1 \\ 1-\kappa_1 & x_{12} - \kappa_1 \bar{x}_1 \\ \vdots & \vdots \\ 1-\kappa_1 & x_{1n_1} - \kappa_1 \bar{x}_1 \\ \vdots & \vdots \\ \vdots & \vdots \\ 1-\kappa_k & x_{k1} - \kappa_k \bar{x}_k \\ 1-\kappa_k & x_{k2} - \kappa_k \bar{x}_k \\ \vdots & \vdots \\ 1-\kappa_k & x_{kn_k} - \kappa_k \bar{x}_k \end{bmatrix}$$

where $\bar{Y}_j$ and $\bar{x}_j$ are the sample means of the response and predictor variables within class j. Then if we consider the transformed model $\mathbf{Y_R} = \mathbf{X_R}\boldsymbol{\beta} + \boldsymbol{\epsilon}_\mathbf{R}$ we have $\boldsymbol{\epsilon}_\mathbf{R} \sim N(\mathbf{0}, (\sigma^2 + \sigma_U^2)\mathbf{I}_n)$, so that the Gauss-Markov conditions of Equation (6.19) are satisfied. Of course, we have already established that the OLS estimate of $\boldsymbol{\beta}$ for the transformed model will equal the GLS estimate obtained for the untransformed model, the advantage being that inference can be based on Theorem 6.2, using $\mathbf{Y_R}$ and $\mathbf{X_R}$ in place of $\mathbf{Y}$ and $\mathbf{X}$.

This example represents a rather ideal case in that we are assuming that ρ is known. However, in Section 13.7 we will see how generalized estimating equations (GEE) can be used to simultaneously estimate ρ and approximate the GLS of $\boldsymbol{\beta}$. In that approach, ρ would be treated as a nuisance parameter (also, the assumption that $U_j \sim N(0, \sigma_U^2)$ is not crucial). It is possible, on the other hand, to estimate all parameters $\boldsymbol{\beta}, \sigma^2$ and ρ using maximum likelihood estimation, or some suitable modification, for example, the restricted maximum likelihood (REML) method (McCulloch *et al.*, 2008).

Linear models for which the error terms have a predictable dependence structure define an important branch of statistical methodology, which can be quite technical, and is somewhat beyond the scope of this volume. Of course, such models arise quite naturally. For example, multiple responses may be observed from each subject (or class of homogenous subjects), refered to as repeated measures. When repeated measures are observed over time, this yields longitudinal data. This can be modeled by including random effects in a linear model. For example, in the current model, U_j represents a subject-level contribution. Of course, we could also have included a subject-level parameter $\boldsymbol{\beta}_j$, which could be estimated using a BLUE. However, this increases the complexity of a model, and experience suggests that the subject-level contribution can usually be more effectively modelled as a random effect. See also Verbeke (1997); Pinheiro and Bates (2009); Gałecki and Burzykowski (2013). ∎

6.6 ANOVA Decomposition and the F Test for Linear Models

We have seen that a linear model $\mathbf{Y} = \mathbf{X}\boldsymbol{\beta} + \epsilon$ can be defined entirely by the vector space $\mathcal{V}_\mathbf{X}$. Alteratively, if we define only a vector space, we may always find a matrix $\mathbf{X}$ for which $\mathcal{V} = \mathcal{V}_\mathbf{X}$. If we are given any such vector space $\mathcal{V}$, then the fitted values for the ordinary LSE will be $\hat{\mathbf{Y}} = \text{argmin}_{\mathbf{y} \in \mathcal{V}_\mathbf{X}} (\mathbf{Y} - \mathbf{y})^T (\mathbf{Y} - \mathbf{y})$. Then

$$SSE = \min_{\mathbf{y} \in \mathcal{V}_\mathbf{X}} (\mathbf{Y} - \mathbf{y})^T (\mathbf{Y} - \mathbf{y}) = (\mathbf{Y} - \hat{\mathbf{Y}})^T (\mathbf{Y} - \hat{\mathbf{Y}}) = |\mathbf{Y} - \hat{\mathbf{Y}}|^2.$$

Then recall from our discussion of the likelihood ratio test (Section 5.8) the notion of a full and reduced model. Here, we only need observe that vector spaces $\mathcal{V}_f$ and $\mathcal{V}_r$ define full and reduced linear models if $\mathcal{V}_r \subset \mathcal{V}_f$. To avoid trivialities, we will assume that $\mathcal{V}_r$ is of strictly smaller dimension than $\mathcal{V}_f$. This means that if SSE_r, SSE_f are the respected error sums of squares for the least squares fit, we necessarily have $SSE_f \leq SSE_r$. This consequence of the nesting property is important for the ANOVA decomposition, since it allows the definition of the extra sum of squares $SSR_{f,r} = SSE_r - SSE_f$, interpretable at the reduction in the sum of squares attributable to the full model relative to the reduced model, and we have ANOVA decomposition

$$SSE_r = SSR_{f,r} + SSE_f. \tag{6.29}$$

Much of the theory of hypothesis testing relies on this nesting assumption. An important consequence is given in the following theorem.

Theorem 6.3 (Projections of Full and Reduced Linear Models) Let $\mathcal{V}_r \subset \mathcal{V}_f \subset \mathbb{R}^n$ be vector subspaces of $\mathbb{R}^n$, with bases of dimension $q_r < q_f$ respectively. Let $\mathbf{H}_r, \mathbf{H}_f \in \mathcal{M}_n$ be the projection matrices from $\mathbb{R}^n$ to $\mathcal{V}_r, \mathcal{V}_f$, respectively. Then

(i) $\mathbf{H}_r\mathbf{H}_f = \mathbf{H}_f\mathbf{H}_r = \mathbf{H}_r$.
(ii) $\mathbf{H}_f - \mathbf{H}_r$ is idempotent.
(iii) $\text{rank}(\mathbf{H}_f - \mathbf{H}_r) = \text{rank}(\mathbf{H}_f) - \text{rank}(\mathbf{H}_r)$. ■

Proof. Suppose $\mathbf{y} \in \mathbb{R}^n$. Let $\hat{\mathbf{y}}_f = \mathbf{H}_f\mathbf{y}$, $\mathbf{e}_f = \mathbf{y} - \hat{\mathbf{y}}_f$ and $\hat{\mathbf{y}}_r = \mathbf{H}_r\mathbf{y}$, $\mathbf{e}_r = \mathbf{y} - \hat{\mathbf{y}}_r$. Then $\mathbf{H}_r\mathbf{H}_f\mathbf{y} = \mathbf{H}_r\hat{\mathbf{y}}_f = \mathbf{H}_r(\mathbf{y} - \mathbf{e}_f)$. But $\mathbf{e}_f$ is orthogonal to all vectors in $\mathcal{V}_f$ and therefore, by assumption, all vectors on $\mathcal{V}_r$. This implies $\mathbf{H}_r\mathbf{e}_f = \mathbf{0}$, and therefore $\mathbf{H}_r\mathbf{H}_f\mathbf{y} = \hat{\mathbf{y}}_r = \mathbf{H}_r\mathbf{y}$ for all $\mathbf{y}$. It follows that $\mathbf{H}_r\mathbf{H}_f = \mathbf{H}_r$. Similarly, $\mathbf{H}_f\mathbf{H}_r\mathbf{y} = \mathbf{H}_f\hat{\mathbf{y}}_r = \hat{\mathbf{y}}_r = \mathbf{H}_r\mathbf{y}$ since $\hat{\mathbf{y}}_r \in \mathcal{V}_r \subset \mathcal{V}_f$.

Statement (ii) follows from statement (i) and the fact that projection matrices are idempotent, that is, $(\mathbf{H}_f - \mathbf{H}_r)^2 = \mathbf{H}_f^2 - \mathbf{H}_f\mathbf{H}_r - \mathbf{H}_r\mathbf{H}_f + \mathbf{H}_r^2 = \mathbf{H}_f - \mathbf{H}_r$.

Statement (iii) follows from the fact that the rank of an idempotent matrix is equal to its trace. □

Theorem 6.4 (ANOVA Decomposition for Full and Reduced Models) Let $\mathbf{Y} \in \mathbb{R}^n$ be any response vector. Let $\hat{\mathbf{Y}}_f = \mathbf{H}_f\mathbf{Y}$, $\hat{\mathbf{Y}}_r = \mathbf{H}_r\mathbf{Y}$ be the projections of $\mathbf{Y}$ onto vector spaces $\mathcal{V}_r \subset \mathcal{V}_f \subset \mathbb{R}^n$. Then

$$\begin{aligned} SSE_f &= |\mathbf{Y} - \hat{\mathbf{Y}}_f|^2 = \mathbf{Y}^T(\mathbf{I} - \mathbf{H}_f)\mathbf{Y}, \\ SSE_r &= |\mathbf{Y} - \hat{\mathbf{Y}}_r|^2 = \mathbf{Y}^T(\mathbf{I} - \mathbf{H}_r)\mathbf{Y}, \\ SSR_{f,r} = SSE_r - SSE_f &= |\hat{\mathbf{Y}}_f - \hat{\mathbf{Y}}_r|^2 = \mathbf{Y}^T(\mathbf{H}_f - \mathbf{H}_r)\mathbf{Y}. \end{aligned}$$ ■

Proof. Any projection matrix is idempotent. By Theorem 6.3 $\mathbf{H}_f - \mathbf{H}_r$ is also idempotent. The theorem follows directly. □

Example 6.9 (Decomposition for One-Way ANOVA) Conventional one-way ANOVA is based on the three sums of squares

$$SSTr = \sum_{i=1}^{k}\sum_{j=1}^{n_i}(\bar{Y}_i - \bar{Y})^2, \qquad \text{Treatment Sum of Squares}$$

$$SSE = \sum_{i=1}^{k}\sum_{j=1}^{n_i}(Y_{ij} - \bar{Y}_i)^2, \qquad \text{Error Sum of Squares}$$

$$SSTo = \sum_{i=1}^{k}\sum_{j=1}^{n_i}(Y_{ij} - \bar{Y})^2, \qquad \text{Total Sum of Squares} \tag{6.30}$$

related by the identity

$$SSTo = SSTr + SSE. \tag{6.31}$$

This follows from Theorem 6.4, once we interpret the one-way ANOVA hypotheses of Equation (6.6) as defining reduced and full models. As before, we may vectorize the data:

$$\mathbf{Y} = (Y_{11}, Y_{12}, \ldots, Y_{1n_1}, Y_{21}, Y_{22}, \ldots, Y_{2n_2}, \ldots, Y_{k1}, Y_{k2}, \ldots, Y_{kn_k})^T.$$

The full model can be expressed as $E[\mathbf{Y}] = \boldsymbol{\mu} = \mathbf{X}_f\boldsymbol{\beta}$, where $\boldsymbol{\beta} \in \mathbb{R}^k$, and $\{\mathbf{X}_f\}_{ij}$ equals 1 if the ith element of $\mathbf{Y}$ is from treatment j, and 0 otherwise (an example of this type of matrix is given in Equation (6.8)). This means that each row of $\mathbf{X}_f$ contains exactly one element equal to 1, and each column contains at least one element equal to 1. The column vectors of $\mathbf{X}_f$ are therefore linearly independent and span a vector space of dimension k. In Example 6.4 it was shown that the BLUE of $\boldsymbol{\beta}$ is given by $\hat{\boldsymbol{\beta}}_{BLUE} = (\bar{Y}_1, \ldots, \bar{Y}_k)^T$, where $\bar{Y}_j$ is the sample mean for the jth treatment. The fitted values are $\hat{\mathbf{Y}}_f = \mathbf{X}_f\hat{\boldsymbol{\beta}}_{BLUE}$, so that the ith fitted value is the sample mean for the treatment class corresponding to that observation. Then

$$SSE = \sum_{i=1}^{k}\sum_{j=1}^{n_i}(Y_{ij} - \bar{Y}_i)^2 = (\mathbf{Y} - \hat{\mathbf{Y}}_f)^T(\mathbf{Y} - \hat{\mathbf{Y}}_f) = SSE_f.$$

On the other hand, the reduced model is defined by setting $\boldsymbol{\mu} = \mathbf{X}_r\theta$, where $\mathbf{X}_r = \mathbf{1}_{n,1}$ and $\theta \in \mathbb{R}$, in which case the fitted values obtained by the BLUE can be easily shown to be $\hat{\mathbf{Y}}_r = \bar{Y}\mathbf{1}_{n,1}$. We similarly have

$$SSTo = \sum_{i=1}^{k}\sum_{j=1}^{n_i}(Y_{ij} - \bar{Y})^2 = (\mathbf{Y} - \hat{\mathbf{Y}}_r)^T(\mathbf{Y} - \hat{\mathbf{Y}}_r) = SSE_r.$$

Of course, given our multiple intrepretations of the BLUE, $\hat{\mathbf{Y}}_f$ and $\hat{\mathbf{Y}}_r$ are also the orthogonal projections of $\mathbf{Y}$ onto the vector spaces $\mathcal{V}_f = \mathcal{V}_{\mathbf{X}_f}$ and $\mathcal{V}_r = \mathcal{V}_{\mathbf{X}_r}$, respectively. Furthermore, $\mathcal{V}_r \subset \mathcal{V}_f$. This follows from the fact that the sum of the column vectors of $\mathbf{X}_f$ equals $\mathbf{1}$, therefore $\mathbf{1} \in \mathcal{V}_f$. Thus, by Theorem 6.4,

$$SSTo - SSE = SSE_r - SSE_f = SSR_{f,r} = (\hat{\mathbf{Y}}_f - \hat{\mathbf{Y}}_r)^T(\hat{\mathbf{Y}}_f - \hat{\mathbf{Y}}_r),$$

and it is easily checked that

$$SSTr = \sum_{i=1}^{k}\sum_{j=1}^{n_i}(\bar{Y}_i - \bar{Y})^2 = (\hat{\mathbf{Y}}_f - \hat{\mathbf{Y}}_r)^T(\hat{\mathbf{Y}}_f - \hat{\mathbf{Y}}_r) = SSR_{f,r},$$

which verifies the correctness of Equation (6.31) (using a minimal amount of algebra).

This decomposition relies on the properties of idempotent matrices, but the analysis of problems involving ANOVA decompositions such as (6.31) generally does not involve the explicit construction of such matrices. However, it may clarify matters to do so here. By Theorem 6.4 we may express (6.31) explicitly using idempotent matrices:

$$\mathbf{Y}^T\mathbf{H}_{SSTo}\mathbf{Y} = \mathbf{Y}^T\mathbf{H}_{SSTr}\mathbf{Y} + \mathbf{Y}^T\mathbf{H}_{SSE}\mathbf{Y}.$$

The simplest term to construct is the total sum of squares, the method for which has already been developed in Example 2.5.1, here giving

$$SSTo = \mathbf{Y}^T\mathbf{H}_{SSTo}\mathbf{Y}, \quad \mathbf{H}_{SSTo} = \left(\mathbf{I}_n - \frac{1}{n}\mathbf{1}\right). \tag{6.32}$$

Examing the form of SSE in Equation (6.30), we can see that it the sum of quadratic terms of the form (6.32) applied separately to each treatment, which leads to

$$SSE = \mathbf{Y}^T \mathbf{H}_{SSE} \mathbf{Y}, \text{ where}$$

$$\mathbf{H}_{SSE} = \begin{bmatrix} I_{n_1} - \frac{1}{n_1}\mathbf{1} & \mathbf{0} & \cdots & \mathbf{0} \\ \mathbf{0} & I_{n_2} - \frac{1}{n_2}\mathbf{1} & \cdots & \mathbf{0} \\ \vdots & \vdots & \cdots & \vdots \\ \mathbf{0} & \mathbf{0} & \cdots & I_{n_k} - \frac{1}{n_k}\mathbf{1} \end{bmatrix}$$

$$= \mathbf{I}_n - \begin{bmatrix} \frac{1}{n_1}\mathbf{1} & \mathbf{0} & \cdots & \mathbf{0} \\ \mathbf{0} & \frac{1}{n_2}\mathbf{1} & \cdots & \mathbf{0} \\ \vdots & \vdots & \cdots & \vdots \\ \mathbf{0} & \mathbf{0} & \cdots & \frac{1}{n_k}\mathbf{1} \end{bmatrix}.$$

Finally, since $SSTr = SSTo - SSE$, it follows that $\mathbf{H}_{SSTr} = \mathbf{H}_{SSTo} - \mathbf{H}_{SSE}$, giving

$$SSTr = \mathbf{Y}^T \mathbf{H}_{SSTr} \mathbf{Y}, \text{ where}$$

$$\mathbf{H}_{SSTr} = \mathbf{H}_{SSTo} - \mathbf{H}_{SSE} = \begin{bmatrix} \frac{1}{n_1}\mathbf{1} & \mathbf{0} & \cdots & \mathbf{0} \\ \mathbf{0} & \frac{1}{n_2}\mathbf{1} & \cdots & \mathbf{0} \\ \vdots & \vdots & \cdots & \vdots \\ \mathbf{0} & \mathbf{0} & \cdots & \frac{1}{n_k}\mathbf{1} \end{bmatrix} - \frac{1}{n}\mathbf{1}. \quad \blacksquare$$

The next theorem states the distributional properties of the various sums of squares, which lead to the F test. It follows directly from Theorems 2.6 and 6.4.

Theorem 6.5 (Nested Models and the F Test) Let $\mathbf{Y} \in N(\boldsymbol{\mu}, \sigma^2\mathbf{I}_n)$ be a multivariate normal random vector. Suppose $\mathcal{V}_0 \subset \mathcal{V}_1 \subset \mathcal{V}_2 \subset \mathbb{R}^n$ are nested vector subspaces of $\mathbb{R}^n$ of dimensions $q_0 < q_1 \leq q_2$, respectively, and let $\mathbf{H}_0, \mathbf{H}_1, \mathbf{H}_2 \in \mathcal{M}_n$ be the respective projection matrices. Finally, assume $\boldsymbol{\mu} \in \mathcal{V}_2$, and denote $\boldsymbol{\mu}_i = \mathbf{H}_i\boldsymbol{\mu}$, $i = 0, 1$. Then the following statements hold.

(i) $\sigma^{-2}SSE_2 \sim \chi^2_{n-q_2}$.
(ii) $\sigma^{-2}SSE_i \sim \chi^2_{\Delta_i;n-q_i}$, $\Delta_i = \sigma^{-2}(\boldsymbol{\mu} - \boldsymbol{\mu}_i)^T(\boldsymbol{\mu} - \boldsymbol{\mu}_i)$.
(iii) $\sigma^{-2}SSR_{1,0} \sim \chi^2_{\Delta_{1,0};q_1-q_0}$, $\Delta_{1,0} = \sigma^{-2}(\boldsymbol{\mu}_1 - \boldsymbol{\mu}_0)^T(\boldsymbol{\mu}_1 - \boldsymbol{\mu}_0)$.
(iv) SSE_2 and $SSR_{1,0}$ are independent.
(v) As a consequence,

$$F = \frac{SSR_{1,0}/(q_1 - q_0)}{SSE_2/(n - q_2)} \sim F_{\Delta_{1,0};q_1-q_0,n-q_2}. \quad \blacksquare$$

Theorem 6.5 is relevant for testing reduced and full models $\mathcal{V}_r \subset \mathcal{V}_f$ of dimensions $q_r < q_f$. We would set $q_0 = q_r$, $q_1 = q_2 = q_f$, in which case $\mathcal{V}_1 = \mathcal{V}_2$. Suppose under the conditions of Theorem 6.5 we wanted to test $H_o : \boldsymbol{\mu} \in \mathcal{V}_r$ against $H_a : \boldsymbol{\mu} \in \mathcal{V}_f$. We could then use the F statistic

$$F = \frac{SSR_{f,r}/(q_f - q_r)}{SSE_f/(n - q_f)} \sim F_{\Delta;q_f-q_r,n-q_f},$$

where $\Delta = \sigma^{-2}(\boldsymbol{\mu} - \boldsymbol{\mu}_r)^T(\boldsymbol{\mu} - \boldsymbol{\mu}_r)$, $\boldsymbol{\mu}_r = \mathbf{H}_r\boldsymbol{\mu}$. However, there will be some advantage to considering the intermediate model $\mathcal{V}_1$ given in the theorem's hypothesis. See also Problem 6.7.

TABLE 6.2
ANOVA table for a sequence of models.

Source of Variation	Sum of Squares	df
$M_1 \mid M_0$	$SSR_{1,0} = SSE_0 - SSE_1$	$q_1 - q_0$
$M_2 \mid M_1$	$SSR_{2,1} = SSE_1 - SSE_2$	$q_2 - q_1$
$\vdots$	$\vdots$	$\vdots$
$M_m \mid M_{m-1}$	$SSR_{m,m-1} = SSE_{m-1} - SSE_m$	$q_m - q_{m-1}$
Error	SSE_m	$n - q_m$
Total	SSE_0	$n - q_0$

6.6.1 Decomposition into Multiple Sources of Variation

Cochran's theorem concerns the partition of a quadratic term of the form

$$\mathbf{Y}^T\mathbf{H}\mathbf{Y} = \sum_{j=1}^{q} \mathbf{Y}^T\mathbf{H}_j\mathbf{Y},$$

given a matrix sum $\mathbf{H} = \sum_{j=1}^{q} \mathbf{H}_j$. An important consequence of this theorem is that the terms of the decomposition are independent. Sufficient conditions given in the literature vary (see, for example, Theorem 2.3), but by Theorem 2.4 it suffices that $\mathbf{H}$ and $\mathbf{H}_1, \ldots, \mathbf{H}_q$ are idempotent. This gives a direct relationship between Cochran's theorem and nested linear models.

The R statistical computing environment includes the generic function `anova()` which produces an ANOVA table from a list of models. This table uses various conventions which depend on the type of model input, and the manner in which the models are input. But an ANOVA table usually captures the features given in Table 6.2. Suppose linear model M_j, defined by a projection $\hat{\mathbf{Y}}_j = \mathbf{H}_j\mathbf{Y}$ of response vector $\mathbf{Y} \in \mathbb{R}^n$ onto vector subspace $\mathcal{V}_j$, yields error sums of squares SSE_j, $j = 0, 1, \ldots, m$. Let q_j be the degrees of freedom of model, so we may assign $n - q_j$ degrees of freedom to SSE_j. The extra sum of squares is given by $SSR_{j,j-1} = SSE_{j-1} - SSE_j$.

We may then relate SSE_0 to SSE_m using the telescoping series

$$SSE_0 = SSR_{1,0} + SSR_{2,1} + \ldots + SSE_m. \tag{6.33}$$

Equation (6.33) is always correct, but for it to be interpretable we must have $SSR_{j,j-1} \geq 0$, which holds under the nesting conditon $\mathcal{V}_{j-1} \subset \mathcal{V}_j$. In this case the ANOVA table summarizes the process of building a model by successively introducing additional parametric complexity. A crucial problem then becomes to decide when to stop building.

At this point, we should discuss the special roles played by the first and last model in any nested sequence $M_0, M_1, \ldots, M_m$. Model M_0 represents the simplest model, and in a complete ANOVA table this will be a null model, such as the null hypothesis of (6.6), in which no predictors or treatments play any role. This model typically reduces to $Y_i = \mu + \epsilon_i$, $i = 1, \ldots, n$, for a single parameter $\mu \in \mathbb{R}$, but more complex models can also play this role. Then SSE_0 is commonly referred to as the total variation of the observations, which is to be decomposed into m sources of variation following Equation (6.33).

The model M_m also plays a specific role. It is the most complex model, and can be referred to as the full model. An important underlying assumption is that M_m is correctly specifed, that is, if $\mathbf{Y} \sim N(\boldsymbol{\mu}, \sigma^2\mathbf{I}_n)$ then $\boldsymbol{\mu} \in \mathcal{V}_m$. A reduced model may also be correct, but it is a special case of the full model. An important consequence of this is that under the conditions of Theorem 6.5 $\sigma^{-2}SSE_m \sim \chi^2_{n-q_m}$, so that $MSE_m = SSE_m/(n-q_m)$ is always an unbiased estimate of σ^2. Unless σ^2 is known, such an estimate is needed for any inference based on an ANOVA table.

Suppose that $\boldsymbol{\mu}^*$ is the true mean vector. By assumption $\boldsymbol{\mu}^* \in \mathcal{V}_m$, but we would like to determine the smallest value of j, say j^*, for which $\boldsymbol{\mu}^* \in \mathcal{V}_j$. If we set $\boldsymbol{\mu}_j = \mathbf{H}_j\boldsymbol{\mu}^*$, then $\boldsymbol{\mu}_j = \boldsymbol{\mu}^*$ for $j = j^*, j^*+1, \ldots, m$, but $\boldsymbol{\mu}_j \neq \boldsymbol{\mu}^*$ for $j = 0, 1, \ldots, j^*-1$. By Theorem 6.5 $\sigma^{-2}SSR_{j,j-1} \sim \chi^2_{\Delta_{j,j-1};q_j-q_{j-1}}$, where $\Delta_{j,j-1} = \sigma^{-2}(\boldsymbol{\mu}_j - \boldsymbol{\mu}_{j-1})^T(\boldsymbol{\mu}_j - \boldsymbol{\mu}_{j-1})$. Thus, the noncentrality parameter of $\sigma^{-2}SSR_{j,j-1}$ is zero if and only if $j > j^*$. Statistically, the problem is to find the simplest model compatible with the observed responses, and this fact yields a procedure to effect this. Refering to Table 6.2, we can consider each extra sum of squares term $SSR_{j,j-1}$, using the F test of Theorem 6.5 to determine whether or not the associated noncentrallity parameter is zero. With well defined error, this sufficies to identify j^*.

What if M_m is not correctly specified? Suppose we take the full model to be $\mathbf{Y} = \mathbf{X}\boldsymbol{\beta} + \boldsymbol{\epsilon}$, $\boldsymbol{\beta} \in \mathbb{R}^q$, but in reality there are important predictors missing from the model. The correct model may be, for example, $\mathbf{Y} = \mathbf{X}'\boldsymbol{\beta}' + \boldsymbol{\epsilon}$, where $\boldsymbol{\beta}' \in \mathbb{R}^{q+s}$, and the first q columns of $\mathbf{X}'$ are those of $\mathbf{X}$. Then M_m and the true model are nested. From the argument of Theorem 6.5 we would now have $\sigma^{-2}SSE_m \sim \chi^2_{\Delta;n-q_m}$, with noncentrality parameter

$$\Delta = \sigma^{-2}(\boldsymbol{\mu} - \mathbf{H}_m\boldsymbol{\mu})^T(\boldsymbol{\mu} - \mathbf{H}_m\boldsymbol{\mu}),$$

where $\mathbf{H}_m$ is the projection matrix of M_m. Then $\Delta = 0$ *iff* model M_m is correct. Otherwise MSE_m is a biased estimate of σ^2 (Gilmour, 1996; Mallows, 2000). The bias is more likely to be large when missing predictors are highly correlated with observed predictors. Otherwise, a missing predictor will, in effect, be additively subsumed into the error terms $\boldsymbol{\epsilon}$ with little effect on the model assumptions. If a model is correctly specified, then a plot of the residuals against the fitted values should reveal no functional structure. If this is not the case, this suggests that the model is not correct. Techniques have been developed for fitting regression models which anticipate this problem (Robins *et al.*, 1994).

6.6.2 Orthogonal Models

The intention of the ANOVA table given in Table 6.2 is usually to decompose the total variation SSE_0 into multiple sources. Suppose we have a full model $\mathbf{Y} = \mathbf{X}\boldsymbol{\beta} + \boldsymbol{\epsilon}$, $\boldsymbol{\beta} \in \mathbb{R}^q$, for $q = 4$. We can partition the design matrix as $\mathbf{X} = [\mathbf{1} \ \ \mathbf{x}_1 \ \ \mathbf{x}_2 \ \ \mathbf{x}_3]$, where $\mathbf{1}$ is the intercept, and $\mathbf{x}_j \in \mathcal{M}_{n,1}$, $j = 1, 2, 3$ are predictors representing some variable expected to induce variation in the mean response (say, *Age*, *Height* and *Weight*, in order). To construct the ANOVA table, we then fit the following models in succession:

$$\begin{aligned}
\mathbf{Y} &= \beta_0 + \boldsymbol{\epsilon}, && \text{Model } M_0\\
\mathbf{Y} &= \beta_0 + \beta_1 Age + \boldsymbol{\epsilon}, && \text{Model } M_1\\
\mathbf{Y} &= \beta_0 + \beta_1 Age + \beta_2 Height + \boldsymbol{\epsilon}, && \text{Model } M_2\\
\mathbf{Y} &= \beta_0 + \beta_1 Age + \beta_2 Height + \beta_3 Weight + \boldsymbol{\epsilon}, && \text{Model } M_3.
\end{aligned} \tag{6.34}$$

The design matrices for the successive models are $\mathbf{X}_0 = \mathbf{1}$, $\mathbf{X}_1 = [\mathbf{1} \ \mathbf{x}_1]$, $\mathbf{X}_2 = [\mathbf{1} \ \mathbf{x}_1 \ \mathbf{x}_2]$, $\mathbf{X}_3 = [\mathbf{1} \ \mathbf{x}_1 \ \mathbf{x}_2 \ \mathbf{x}_3]$, so that the models are clearly nested. The resulting ANOVA table is shown in Table 6.3.

Note, however, that SSR_{Height} is the reduction in the sum of squares of a reduced model $\mathbf{Y} = \beta_0 + \beta_1 Age + \boldsymbol{\epsilon}$ realized by fitting a full model $\mathbf{Y} = \beta_0 + \beta_1 Age + \beta_2 Height + \boldsymbol{\epsilon}$. Suppose we changed the order in which predictors were added to the model, say *Height*, *Age* then *Weight*, so that M_1 is now $\mathbf{Y} = \beta_0 + Height + \boldsymbol{\epsilon}$ while M_2 remains the same. The new

TABLE 6.3
ANOVA table for a sequence of four models of Equation (6.34).

Source of Variation	Sum of Squares	df
Age	$SSR_{Age} = SSE_0 - SSE_1$	1
Height	$SSR_{Height} = SSE_1 - SSE_2$	1
Weight	$SSR_{Weight} = SSE_2 - SSE_3$	1
Error	SSE_3	$n-4$
Total	SSE_0	$n-1$

extra sum of squares SSR^*_{Height} is now reduction in the sum of squares of a reduced model $\mathbf{Y} = \beta_0 + \boldsymbol{\epsilon}$ realized by fitting a full model $\mathbf{Y} = \beta_0 + \beta_1 Height + \boldsymbol{\epsilon}$.

This begs the question as to whether or not $SSR_{Height} = SSR^*_{Height}$, or more generally, whether or not the ANOVA decomposition of Tables 6.2 or 6.3 depend on the order in which the predictors are added to the model. In fact, there is a specific condition under which an ANOVA table wil not depend on the predictor order, in which case we would have $SSR_{Height} = SSR^*_{Height}$. This occurs when $\mathbf{X}$ can be partitioned as $\mathbf{X} = [\mathbf{X}_1 \cdots \mathbf{X}_m]$ and the matrices defining the partition are mutually orthogonal. This leads to the next definition.

Definition 6.2 (Orthogonal Linear Models) A linear model $\mathbf{Y} = \mathbf{X}\boldsymbol{\beta} + \boldsymbol{\epsilon}$ possesses orthogonal predictors if $\mathbf{X}^T\mathbf{X}$ is a diagonal matrix. Suppose for a common response vector $\mathbf{Y}$ we consider two linear models $\mathbf{Y} = \mathbf{X}\boldsymbol{\beta} + \boldsymbol{\epsilon}$ and $\mathbf{Y} = \mathbf{X}'\boldsymbol{\beta}' + \boldsymbol{\epsilon}$. The two models are orthogonal if $\mathbf{X}^T\mathbf{X}' = \mathbf{0}$. ∎

An important consequence follows from the orthogonality of predictors. Under the Gauss-Markov conditions (6.19) the covariance matrix of the ordinary LSE of $\boldsymbol{\beta}$ of a linear model $\mathbf{Y} = \mathbf{X}\boldsymbol{\beta} + \boldsymbol{\epsilon}$ is given by

$$\Sigma_{\hat{\boldsymbol{\beta}}} = \sigma^2(\mathbf{X}^T\mathbf{X})^{-1},$$

which will be a diagonal matrix for diagonal predictors. This means the elements $\hat{\beta}_j$ are uncorrelated, so that the contribution of the associated predictors to the model can be assessed independently.

This idea can be generalized by the following theorem. The proof is left to the reader (Problem 6.14).

Theorem 6.6 (Orthogonal Linear Models) Suppose $\mathbf{X} = [\mathbf{X}_1 \cdots \mathbf{X}_m] \in \mathcal{M}_{n,q}$ where $\mathbf{X}_j \in \mathcal{M}_{n,q_1}$, $q_1 + \ldots q_m = q < n$. Assume each matrix is of full rank. If the matrices $\mathbf{X}_j$ are mutually orthogonal, that is $\mathbf{X}_j^T\mathbf{X}_k = \mathbf{0}$, $j \neq k$, then $\mathbf{H}_{\mathbf{X}} = \sum_{j=1}^m \mathbf{H}_{\mathbf{X}_j}$. ∎

6.7 One- and Two-Way ANOVA

In Example 6.9 we saw that for the data of Table 6.1, the hypotheses H_a and H_o of Equation (6.6) define full and reduced linear models which are nested. This suffices to construct an F test for these hypotheses, given in the following theorem.

Theorem 6.7 (F Statistic for One-Way ANOVA) Suppose we are given the ANOVA model summarized in Table 6.1. The following statements hold:

(i) $SSTr + SSE = SSTo$.
(ii) Under the null hypothesis of Equation (6.6), $\sigma^{-2}SSTr$, $\sigma^{-2}SSE$ and $\sigma^{-2}SSTo$ are χ^2_ν random variables with $\nu = k-1, n-k$ and $n-1$ degrees of freedom, respectively.
(iii) Under the alternative hypothesis of Equation (6.6), $\sigma^{-2}SSTr$, $\sigma^{-2}SSE$ and $\sigma^{-2}SSTo$ are independent random variables, where $\sigma^{-2}SSE \sim \chi^2_{n-k}$ and $\sigma^{-2}SSTr \sim \chi^2_{\Delta,k-1}$, where $\Delta = \sigma^{-2}\sum_{j=1}^k n_j(\beta_j - \bar{\beta})^2$ and $\bar{\beta} = n^{-1}\sum_{j=1}^k n_j\beta_j$.

As a consequence

$$F = \frac{SSTr/(k-1)}{SSE/(n-k)} \sim F_{\Delta;k-1,n-k}, \tag{6.35}$$

for the given noncentrality parameter. ∎

The noncentrality parameter Δ of the statistic F of Equation (6.35) equals zero *iff* H_o holds. By Theorem 2.7, the noncentral F distribution satisfies the likelihood ratio ordering property *wrt* Δ, and therefore also the stochastic ordering property. Thus, following the discussion of Section 5.6, the rejection region $F > t$ defines an unbiased test for H_o against H_a.

The important components of Theorem 6.7 are commonly summarized in the ANOVA table:

Source	SS	df	MSS	F
Treatment	$SSTr$	$k-1$	$MSTr = \frac{SSTr}{k-1}$	$F = \frac{MSTr}{MSE}$
Error	SSE	$n-k$	$MSE = \frac{SSE}{n-k}$	
Total	$SSTo$	$n-1$		

The total variation is decomposed into two sources, that due to treatments (variation of the means β_j), and that due to error (variation of a response about its own mean). Within each source is given a sum of squares (SS) degrees of freedom (df) and mean sum of squares (MS). The table also includes the F statistic described in Theorem 6.7. The sums of squares are those given in Table 6.1. The degrees of freedom is that of the χ^2 distribution for each sum of squares given by Theorem 6.7. We can also equate this number with the dimension of the minimal reduction of the data needed to evaluate the sum of squares. For example, to evaluate $SSTo$, we may replace each response Y_{ji} with $Y'_{ji} = Y_{ji} - \bar{Y}$, then use only the transformed observations Y'_{ji}. However, the sum of the transformed responses is zero, introducing one linear constraint which reduces their dimension by one. The other degrees of freedom can be similarly deduced. The mean sum of squares (MSS) is simply the sum of squares divided by the degrees of freedom. Inference methods are more naturally expressed in terms of the MSS, for example, MSE is an unbiased estimator of σ^2 under H_o or H_a, while the F statistic of Section 6.7 is the ratio of $MSTr/MSE$ (this is also commonly included in an ANOVA table). The structure of the table emphasizes the decomposition $SSTr + SSE = SSTo$ (Theorem 6.7 (i)). Note that the degrees of freedom decompose into sources in a similar manner.

The following straightfoward example demonstrates the basic ANOVA test of Theorem 6.7.

Example 6.10 (Example of a One-Way ANOVA Test) Independent samples for $k = 3$ treatments are summarized in the table below. Assume sample j is from a normally distributed population with mean β_j and fixed variance σ^2.

This data can be used to construct the ANOVA table below. For example $SSE = SS_1 + SS_2 + SS_3 = 33.18 + 7.18 + 39.94 = 80.3$. Then $\bar{Y} = (5 \times 14.66 + 5 \times 8.99 + 5 \times 18.68)/(5+5+5) = 14.11$, so that

$$SSTr = 5(14.66 - 14.11)^2 + 5(8.99 - 14.11)^2 + 5(18.68 - 14.11)^2 = 237.0.$$

	1	2	3	4	5	$\bar{Y}_j$	SS_j	n_i
Treatment 1	13.13	15.16	10.60	16.41	17.99	14.66	33.18	5
Treatment 2	9.04	6.94	9.30	9.00	10.69	8.99	7.18	5
Treatment 3	20.55	16.37	20.20	14.39	21.89	18.68	39.94	5

Then $SSTo = SSTr + SSE = 317.3$. This gives ANOVA table:

ANOVA Table

Source	SS	df	MSS	F
Treatment	237.0	2	118.42	17.70
Error	80.3	12	6.69	
Total	317.3	14		

The F statistic is $F = 17.702$. We reject H_o at significance level $\alpha = 0.05$ if

$$F \geq F_{k-1,n-k;0.05} = 3.885.$$

We therefore reject the null hypothesis (P-value ≈ 0.00026). ■

In the next example we use the noncentrality parameter of Theorem 6.7 to estimate the sample size needed to test for a specific alternative hypothesis.

Example 6.11 (Power Analysis for One-Way ANOVA) For ANOVA, a specific alternative hypothesis is characterized entirely by the noncentrality parameter Δ, which depends on the distribution of the total sample size among the treatments. The probability of a Type II error for an alternative with noncentrality parameter Δ is

$$\beta(\Delta) = P(F \leq f_{k-1,n-k;\alpha}), \quad F \sim F_{\Delta;k-1,n-k}.$$

Then for fixed n, k and α, β we may solve for $\beta(\Delta) = \beta$ by determing the value of Δ, say Δ_β, such that $f_{k-1,n-k;\alpha}$ is the β-quantile of the $F_{\Delta;k-1,n-k}$ distribution.

To estimate a sample size, a number of assumptions must be made. First, we may assume that the total sample size n is going to be equally allocated to each treatment. That is, $n_i = n/k$, so that

$$\Delta = \sigma^{-2}\frac{n}{k}\sum_{i=1}^{k}(\beta_i - \bar{\beta})^2, \text{ where } \bar{\beta} = \frac{\sum_{i=1}^{k}\beta_i}{k}.$$

We may then make a conjecture regarding the treatment means $\beta_1, \ldots, \beta_k$. We would then determine the smallest value of Δ compatible with that conjecture. This would result in a conservative procedure, unless the conjecture consisted of a specification of $\sum_{i=1}^{k}(\beta_i - \bar{\beta})^2$ itself.

For example, we might wish to have a large enough sample size to reject any alternative hypothesis for which

$$\max_{i,j} |\beta_i - \beta_j| \geq m,$$

for some specfic value m, with the stated error probabilities α, β. It can be shown that $\sum_{i=1}^{k}(\beta_i - \bar{\beta})^2$ is minimized by setting $\beta_1 = \beta_2 + m$, then $\beta_i = \bar{\beta}$, $i = 3, \ldots, k$ (Problem 6.9). We can assume without loss of generality that $\bar{\beta} = 0$. This gives

$$\sum_{i=1}^{k}(\beta_i - \bar{\beta})^2 = (\beta_1 - 0)^2 + (\beta_2 - 0)^2 = (m/2 - 0)^2 + (-m/2 - 0)^2 = \frac{m^2}{2}.$$

We then solve $\beta = \beta(\Delta_\beta) = P(F \leq f_{\alpha;k-1,n-k})$, $F \sim F_{\Delta,k-1,n-k}$. Then set

$$\Delta_\beta = \sigma^{-2}(n/k)\sum_{i=1}^{k}(\beta_i - \bar{\beta})^2 = \sigma^{-2}(n/k)\left(m^2/2\right).$$

Solving for n gives $n \geq \sigma^2\Delta_\beta\left(2k/m^2\right)$, or the smallest integer n satisfying the preceding inequality. ■

6.7.1 Two-Way ANOVA

Under our framework, we can build the two-way ANOVA model using successive refinments of the one-way model. Suppose we conduct an experiment to determine the effect of two factors (or categorical variables) on crop yield Y. The first is seed variant, of which there are I types, the second is crop density, of which there are J levels (for example, low or high). Possible factor values are refered to as levels. We can more generally refer to factor A and B.

Suppose we are given a response vector $\mathbf{Y} \sim N(\boldsymbol{\mu}, \sigma^2\mathbf{I}_n)$. We may associate with each individual response Y a predictor variable $x = (i,j)$ giving the associated pair of factor levels. Each factor pair (i,j) is associated with n_{ij} responses, for which the common mean is μ_{ij}. For the moment, we assume that $n_{ij} \geq 1$ for each possible (i,j). Let $\bar{Y}_{ij}$ be the sample means of the responses for (i,j). This defines the full model.

Note that we have just defined a one-way ANOVA model where each (i,j) defines a treatment, therefore the model degrees of freedom is $q_f = IJ$. Of course, we are interested in identifying the separate effects of each factor on the response, and the manner in which the factors interact. Thus, while we could write the model $Y = \mu_{ij} + \epsilon$, the two-factor ANOVA model relies on the reparametrized form

$$Y = \mu + \alpha_i + \beta_j + \gamma_{ij} + \epsilon, \;, \; i = 1,\ldots,I, \; j = 1,\ldots,J, \tag{6.36}$$

where $\boldsymbol{\epsilon} \sim N(0,\sigma^2)$. The mean for factor pair (i,j) is therefore $\mu_{ij} = \mu + \alpha_i + \beta_j + \gamma_{ij}$.

Note that the full model has $q_f = IJ$ degrees of freedom, while Equation (6.36) introduces $1 + I + J + IJ$ parameters. Therefore, for the model to be identifiable we must introduce $1 + I + J$ constraints. This is accomplished by setting

$$\sum_{i=1}^{I}\alpha_i = 0, \; \sum_{j=J}^{J}\beta_j = 0, \; \sum_{j=1}^{J}\gamma_{ij} = 0, \; i = 1,\ldots,I, \; \sum_{i=1}^{I}\gamma_{ij} = 0, \; j = 1,\ldots,J. \tag{6.37}$$

Nominally, Equation (6.37) defines $I + J + 2$ constraints, but the number of constraints involving γ_{ij} reduces to $I + J - 1$ (it may be shown that any one of those constraints is implied by the remaining $I + J - 1$).

The parameters α_i and β_j are referred to as the main effects of factors A and B, respectively, while the parameters γ_{ij} are the factor interations. Hypothesis tests commonly concern the parameters grouped in these ways, leading to the following null hypotheses:

$$\begin{aligned} &H_o : \alpha_i = 0, \; \forall i && \text{Main effect of factor } A \\ &H_o : \beta_j = 0, \; \forall j && \text{Main effect of factor } B \\ &H_o : \gamma_{ij} = 0, \; \forall i,j && A \times B \text{ factor interations.} \end{aligned} \tag{6.38}$$

Example 6.12 (Examples of Two-Way Anova Models) We will express the parameters of Equation (6.36) as vectors $\boldsymbol{\alpha} = (\alpha_1, \ldots, \alpha_I)^T$, $\boldsymbol{\beta} = (\beta_1, \ldots, \beta_J)^T$ and matrix

$$\boldsymbol{\gamma} = \begin{bmatrix} \gamma_{11} & \gamma_{12} & \cdots & \gamma_{1J} \\ \gamma_{21} & \gamma_{22} & \cdots & \gamma_{2J} \\ \vdots & \vdots & \vdots & \vdots \\ \gamma_{I1} & \gamma_{I2} & \cdots & \gamma_{IJ} \end{bmatrix}.$$

We will consider four examples with $I = 2$, $J = 3$. When interpreting each model it will be useful to refer to the marginal response means for each factor:

$$\mu_{i\cdot} = J^{-1} \sum_{j=1}^{J} \mu_{ij} = \mu + \alpha_i + \beta_\cdot + \gamma_{i\cdot} = \mu + \alpha_i,$$
$$\mu_{\cdot j} = I^{-1} \sum_{i=1}^{I} \mu_{ij} = \mu + \alpha_\cdot + \beta_j + \gamma_{\cdot j} = \mu + \beta_j,$$

where $\alpha_\cdot = \beta_\cdot = \gamma_{i\cdot} = \gamma_{\cdot j} = 0$, after applying the constraints of Equation (6.37).

(a) In our first example we will have **two main effects and no interactions**, in particular, $\mu = 4.5$, $\boldsymbol{\alpha} = (-1.5, 1.5)^T$, $\boldsymbol{\beta} = (-2.0, -1.0, 3.0)^T$, with $\boldsymbol{\gamma} = \mathbf{0}$. Figure 6.3, Plot (a), gives a diagrammatic representaton of the treatment means μ_{ij} against factor B levels $j = 1, 2, 3$ for each fixed level i of factor A. Under this model, the main effects are additive in the sense that $\mu_{ij} = \mu + \alpha_i + \beta_j$, so that the plot of μ_{ij} against j has the same shape for each factor A level i.

(b) In this example we will have **one main effect and no interactions**. Set parameters $\mu = 4.5$, $\boldsymbol{\alpha} = (-1.5, 1.5)^T$, $\boldsymbol{\beta} = (0, 0, 0)^T$, with $\boldsymbol{\gamma} = \mathbf{0}$. The model is represented in Figure 6.3, Plot (b). There is a main effect for factor A but not B, and no interactions. This means $\mu_{i\cdot} = \mu + \alpha_i = \mu_{ij}$, while $\mu_{\cdot j} = \mu$ for all j. As a result, μ_{ij} varies by i but not j.

(c) In this example we will have **one main effect with interactions**. Set parameters $\mu = 4.5$, $\boldsymbol{\alpha} = (-0.5, 0.5)^T$, $\boldsymbol{\beta} = (0, 0, 0)^T$, with

$$\boldsymbol{\gamma} = \begin{bmatrix} 0.5 & -1.0 & 0.5 \\ -0.5 & 1.0 & -0.5 \end{bmatrix}.$$

Thus, we have main effects for factor A but not B. The model is represented in Figure 6.3, Plot (c). Unlike the model of Plot (b), both factors play a role in determing μ_{ij}. Thus, if there are no factor B main effects, then μ_{ij} may vary with j if interactions are present, but the marginal response mean $\mu_{\cdot j} = \mu$ will not.

(d) In this example we will have **two main effects with interactions**. Set parameters $\mu = 4.5$, $\boldsymbol{\alpha} = (-1.4, 1.4)^T$, $\boldsymbol{\beta} = (-0.8, 0, 0.8)^T$, with

$$\boldsymbol{\gamma} = \begin{bmatrix} -0.8 & 0.0 & 0.8 \\ 0.8 & 0.0 & -0.8 \end{bmatrix}.$$

The model is represented in Figure 6.3, Plot (d). For a full model, either series μ_{1j} or μ_{2j} may assume any shape independently, since $6 = IJ$ model degrees of freedom are available. Clearly, the main effects are not additive. ■

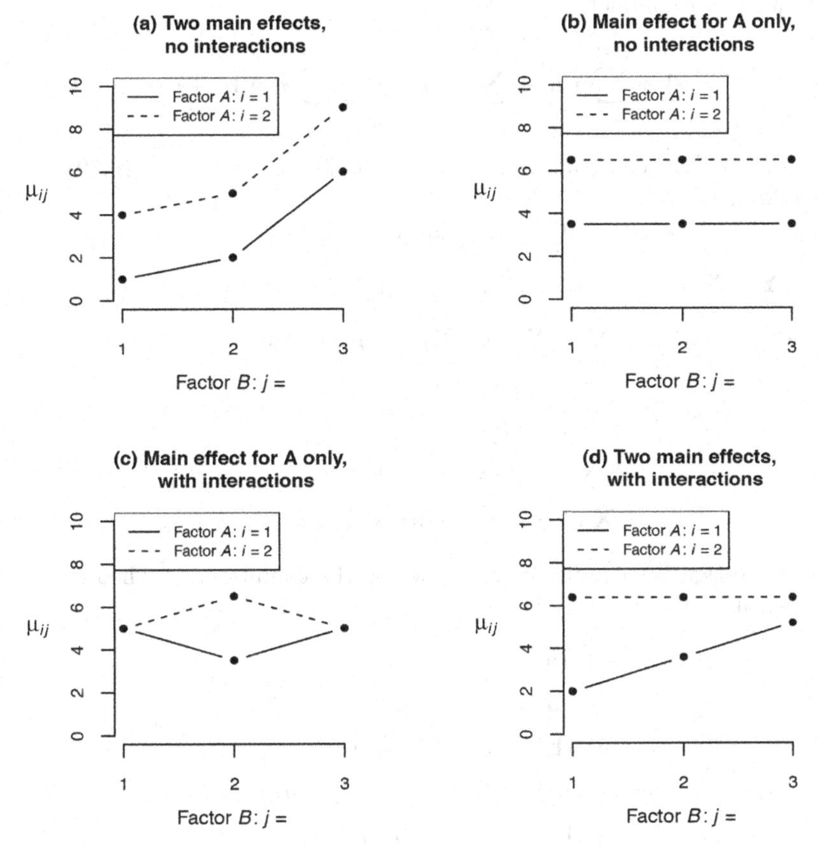

FIGURE 6.3
Examples of two-way ANOVA models from Example 6.12.

To represent the two-way ANOVA model as a linear model, let $\mathbf{a}_i, \mathbf{b}_j \in \mathcal{M}_{n,1}$ be the indicator functions for levels i, j of factors A and B. If we include only parameters μ, α_i, β_j, γ_{ij}, for factor levels $i = 1, \ldots, I-1$, $j = 1, \ldots, J-1$, we attain the correct model degrees of freedom. To do this write the model in full as

$$\mathbf{Y} = \mu\mathbf{1} + \sum_{i=1}^{I} \alpha_i \mathbf{a}_i + \sum_{j=1}^{J} \beta_j \mathbf{b}_j + \sum_{i=1}^{I}\sum_{j=1}^{J} \gamma_{ij}\mathbf{a}_i\mathbf{b}_j + \boldsymbol{\epsilon}, \tag{6.39}$$

where $\bar{\mathbf{a}}_i\bar{\mathbf{b}}_j \in \mathcal{M}_{n,1}$ is an elementwise product, or interaction, then reduced the number of parameters by applying the contraints of Equation (6.37).

It is worth seeing the process in some detail. We will first consider the model with factor A main effects only:

$$\mathbf{Y} = \mu\mathbf{1} + \alpha_1\mathbf{a}_1 + \ldots + \alpha_I\mathbf{a}_I + \boldsymbol{\epsilon}. \tag{6.40}$$

We have one linear constraint for the parameters, which can take the form $\alpha_I = -\alpha_1 - \ldots - \alpha_{I-1}$. After substitution we have

$$\mathbf{Y} = \mu\mathbf{1} + \alpha_1\bar{\mathbf{a}}_1 + \ldots + \alpha_{I-1}\bar{\mathbf{a}}_{I-1} + \boldsymbol{\epsilon}, \tag{6.41}$$

where $\bar{\mathbf{a}}_i = \mathbf{a}_i - \mathbf{a}_I$. The mth element of $\bar{\mathbf{a}}_i$ equals 1 if the factor A level of the mth response is i, equals -1 if the level is I, and equals 0 otherwise. If we similarly define $\bar{\mathbf{b}}_j = \mathbf{b}_j - \mathbf{b}_J$, we may then write the full model

$$\mathbf{Y} = \mu\mathbf{1} + \sum_{i=1}^{I-1} \alpha_i \bar{\mathbf{a}}_i + \sum_{j=1}^{J-1} \beta_j \bar{\mathbf{b}}_j + \sum_{i=1}^{I-1}\sum_{j=1}^{J-1} \gamma_{ij} \bar{\mathbf{a}}_i \bar{\mathbf{b}}_j + \boldsymbol{\epsilon}. \tag{6.42}$$

It may be verified that if the constraints of Equation (6.37) hold, equations (6.39) and (6.42) are equivalent (Problem 6.8).

We can then express Equation (6.42) as a linear model

$$\mathbf{Y} = \mathbf{X}[\mu, \alpha_1, \ldots, \alpha_{I-1}, \beta_1, \ldots, \beta_{J-1}, \gamma_{11}, \ldots, \gamma_{I-1,J-1}]^T \boldsymbol{\epsilon}.$$

We can construct the partition $\mathbf{X} = [\mathbf{X_1}\ \mathbf{X}_A\ \mathbf{X}_B\ \mathbf{X}_{AB}]$, where

$$\begin{aligned}
\mathbf{X_1} &= \mathbf{1}, \\
\mathbf{X}_A &= [\bar{\mathbf{a}}_1 \cdots \bar{\mathbf{a}}_{I-1}], \\
\mathbf{X}_B &= [\bar{\mathbf{b}}_1 \cdots \bar{\mathbf{b}}_{J-1}], \\
\mathbf{X}_{AB} &= [\bar{\mathbf{a}}_1 \bar{\mathbf{b}}_1 \cdots \bar{\mathbf{a}}_{I-1} \bar{\mathbf{b}}_{J-1}].
\end{aligned} \tag{6.43}$$

It is then worth investigating conditions under which the conditions of Theorem 6.6 hold. The following identities can be easily verified:

$$\begin{aligned}
\mathbf{1} \circ \bar{\mathbf{a}}_i &= n_{i\cdot} - n_{I\cdot}, \\
\mathbf{1} \circ \bar{\mathbf{b}}_j &= n_{\cdot j} - n_{\cdot J}, \\
\bar{\mathbf{a}}_i \circ \bar{\mathbf{b}}_j &= n_{ij} - n_{iJ} - n_{Ij} + n_{IJ}, \\
\bar{\mathbf{a}}_i \circ (\bar{\mathbf{a}}_i \bar{\mathbf{b}}_j) &= n_{ij} - n_{iJ} + n_{Ij} - n_{IJ}, \\
\bar{\mathbf{a}}_i \circ (\bar{\mathbf{a}}_{i'} \bar{\mathbf{b}}_j) &= n_{Ij} - n_{IJ}, \quad i \neq i', \\
\bar{\mathbf{b}}_j \circ (\bar{\mathbf{a}}_i \bar{\mathbf{b}}_j) &= n_{ij} - n_{iJ} + n_{Ij} - n_{IJ}, \\
\bar{\mathbf{b}}_j \circ (\bar{\mathbf{a}}_i \bar{\mathbf{b}}_{j'}) &= n_{iJ} - n_{IJ}, \quad j \neq j'.
\end{aligned}$$

The inner product of any two column vectors from different matrices of the partition listed in Equation (6.43) is of one of these types. Note also that each of these expressions is a contrast *wrt* the cell sizes n_{ij}, that is, a linear combintation $\sum_{i=1}^{I}\sum_{j=1}^{J} c_{ij} n_{ij}$ for which $\sum_{i=1}^{I}\sum_{j=1}^{J} c_{ij} = 0$. This means that if the ANOVA design is balanced, meaning that all cell sizes $n_{ij} = m$ are equal, then each of the inner products equal zero. This in turn implies that the matrices of Equation (6.43) are mutually orthogonal, so that

$$\mathbf{H_X} = \mathbf{H_{X_1}} + \mathbf{H_{X_A}} + \mathbf{H_{X_B}} + \mathbf{H_{X_{AB}}},$$

which may be arranged as

$$\mathbf{I} - \mathbf{H_{X_1}} = \mathbf{H_{X_1}} + \mathbf{H_{X_A}} + \mathbf{H_{X_B}} + \mathbf{H_{X_{AB}}} + (\mathbf{I} - \mathbf{H_X}),$$

leading to

$$SSTo = SSR_A + SSR_B + SSR_{AB} + SSE,$$

and the ANOVA table given in Table 6.4.

For the balanced design SSR_A and SSR_B are equivalent to $SSTr$ for a one-way ANOVA model, that is

$$SSR_A = mI \sum_{i=1}^{I} (\bar{Y}_{i\cdot} - \bar{Y}_{\cdot\cdot})^2, \quad SSR_B = mJ \sum_{j=1}^{J} (\bar{Y}_{\cdot j} - \bar{Y}_{\cdot\cdot})^2.$$

TABLE 6.4
ANOVA table for the two-way ANOVA model.

Source of Variation	Sum of Squares	df
Main effect A	SSR_A	$I-1$
Main effect B	SSR_B	$J-1$
Interactions AB	SSR_{AB}	$(I-1)(J-1)$
Error	SSE	$n-IJ$
Total	SST_0	$n-1$

Then note that $\mathbf{H}_{\mathbf{X}_{AB}} = (\hat{\mathbf{Y}}_f - \hat{\mathbf{Y}}_r)^T(\hat{\mathbf{Y}}_f - \hat{\mathbf{Y}}_r)$, which gives

$$SSR_{AB} = m\sum_{i=1}^{I}\sum_{i=1}^{J}(\bar{Y}_{ij} - \bar{Y}_{i\cdot} - \bar{Y}_{\cdot j} + \bar{Y}_{\cdot\cdot})^2.$$

However, this elegant decomposition is not available if the design is not balanced. In this case an ANOVA table such as Table 6.3 can be constructed by considering a sequence of nested linear models, for example, a model with only factor A main effects, then both main effects, then the full model. Of course, the values of the ANOVA table would then depend on the order in which the models parameters were introduced.

6.8 Multiple Linear Regression

The multiple linear regression model is given by

$$Y_i = \beta_0 + \beta_1 x_{i,1} + \ldots + \beta_{q-1}x_{i,q-1} + \epsilon_i, \quad i = 1, \ldots, n, \tag{6.44}$$

where β_j are unknown parameters, x_{ij} are known predictor (or independent) variables, Y_i is the observed response (or dependent) variable, and ϵ_i are unobserved error variables with full-rank covariance matrix $\Sigma_{\boldsymbol{\epsilon}}$. We usually have $q < n$. We refer to β_0, with which no predictor is associated, as the intercept term. The special case of simple regression, written $Y_i = \beta_0 + \beta_1 x_i + \epsilon_i$, $i = 1, \ldots, n$, has already been introduced in Section 6.2.1.

We can also represent Equation (6.44) using matrix algebra. The responses are expressed as the $n \times 1$ matrix $\mathbf{Y}$ and the predictors are represented as the $n \times q$ matrix $\mathbf{X} = \{x_{i,j}\} = [\mathbf{x}_1 \cdots \mathbf{x}_q]$. The jth column $\mathbf{X}$ is the jth predictor variable $\mathbf{x}_j$, which may be of independent interest. Then Equation (6.44) becomes

$$\mathbf{Y} = \beta_1\mathbf{x}_1 + \ldots + \beta_q\mathbf{x}_q + \boldsymbol{\epsilon} = \mathbf{X}\boldsymbol{\beta} + \boldsymbol{\epsilon} \tag{6.45}$$

where $\boldsymbol{\epsilon}$ is the column vector of error terms ϵ_i. It is important to note that the intercept term of Equation (6.44) has not been removed. It is now a column vector $\mathbf{1}_{n,1}$ in $\mathbf{X}$, the remaining columns corresponding to the $q-1$ predictors of the original model. In practice, linear regression models include an intercept term by default, and there are good reasons for this. However, when the model is expressed in the form of Equation (6.45) the intercept term is just another predictor variable $\mathbf{x}_j = \mathbf{1}_n$, which may or may not be included in the model. See, for example, Problems 6.17 and 6.21.

The reader may be wondering if anything new is being introduced here, since the model (6.45) is identical to the linear model originally introduced in Section 6.2, except for the notation (the one- and two-way ANOVA models are special cases of a linear model, while model (6.45) is not constrained in any way). Indeed, all the theory of the previous sections of this chapter are all applicable, and have already been demonstrated for the simple linear regression model (Examples 6.3, 6.7 and 6.8).

Regression models can be regarded as an attempt to model a functional relationship of the form

$$Y = g(\mathbf{x}; \boldsymbol{\beta}) + \epsilon$$

where $\mathbf{x} = (x_1, \ldots, x_q)$ is a vector of predictor variables, $\boldsymbol{\beta} \in \mathbb{R}^q$ is an unknown parameter and ϵ is an error term. Given observations from the model $(Y_1, \mathbf{x}_1), \ldots, (Y_n, \mathbf{x}_n)$ we may fit the model with the least squares estimator

$$\hat{\boldsymbol{\beta}}_{LSE} = \text{argmin}_{\boldsymbol{\beta}} \sum_{i=1}^{n} (Y_i - g(\mathbf{x}_i; \boldsymbol{\beta}))^2,$$

the fitted values given by $\hat{Y}_i = g(\mathbf{x}_i; \hat{\boldsymbol{\beta}}_{LSE})$. Linear regression is simply a special case of this problem, for which the class of models $g(\mathbf{x}; \boldsymbol{\beta})$ is defined by Equation (6.44).

Suppose $\boldsymbol{\epsilon} \sim N(0, \sigma^2 \mathbf{I}_n)$ then Theorem 6.2 may be applied.

Theorem 6.8 (Estimation of Multiple Linear Regression Coefficients) Suppose for model Equation (6.45) $\boldsymbol{\epsilon} \sim N(0, \sigma^2 \mathbf{I}_n)$, and $\mathbf{X}$ is of full rank. Then,

(a) The BLUE of $\boldsymbol{\beta}$ is given by $\hat{\boldsymbol{\beta}} = (\mathbf{X}^T\mathbf{X})^{-1}\mathbf{X}^T\mathbf{Y}$.
(b) $\hat{\boldsymbol{\beta}} \sim N(\boldsymbol{\beta}, \sigma^2(\mathbf{X}^T\mathbf{X})^{-1})$;
(c) $\sigma^{-2} SSE \sim \chi^2_{n-q}$;
(d) The residuals $\mathbf{e}$ and $\hat{\boldsymbol{\beta}}$ are independent, hence SSE and $\hat{\boldsymbol{\beta}}$ are independent. ■

Inference for specific parameters can proceed as shown in Section 6.5.1, and is demonstrated for the simple linear regression model in Example 6.7. For estimand $\eta = \mathbf{c}^T\boldsymbol{\beta}$, $\hat{\eta} = \mathbf{c}^T\hat{\boldsymbol{\beta}}$ will be the best linear unbiased estimator. Inference can be based on the pivot

$$T = \frac{\hat{\eta} - \eta}{S_{\hat{\eta}}} \sim T_{n-q},$$

where

$$S_{\hat{\eta}} = \sqrt{MSE \mathbf{c}^T(\mathbf{X}^T\mathbf{X})^{-1}\mathbf{c}}$$

is the standard error for the estimator $\hat{\eta}$. A level $1 - \alpha$ confidence interval for η is then

$$CI = \hat{\eta} \pm t_{n-q;\alpha/2} S_{\hat{\eta}}.$$

As for one-way ANOVA, there is a conventional format for linear regression fits adopted by most statistical software. Suppose we fit the model

$$Y = \beta_0 + \beta_1 x_1 + \beta_2 x_2 + \epsilon.$$

The output might look something like the following table:

Coefficients	Estimate	Standard Error	T statistic	P-value
(Intercept)	-0.07389	0.20174	-0.366	0.72500
x1	1.33538	0.26547	5.030	0.00151
x2	0.16825	0.19259	0.874	0.41128

There are 3 coefficients to estimate. The intercept is usually included by default, and is specifically identified. It is followed by a single row for each remaining predictor. The estimate column lists the least squares estimates $\hat{\beta}_j$. For example, we have $\hat{\beta}_2 \approx 0.16825$. The next column lists the standard error ($S_{\hat{\beta}_1} \approx 0.26547$). There is usually interest in whether or not a particular predictor x_j is "significant", and contributes to the predictive value of the model. Clearly, if $\beta_j \neq 0$ this will be the case. Therefore, linear regression summaries usually include for each x_j the T statistic $T = \hat{\beta}_j / S_{\hat{\beta}_j}$ with which to test the null hypothesis $H_o : \beta_j = 0$ against the two-sided alternative. This is given in a separate column of the summary. The final column contains the P-value for this test. In our example, for β_1 we have $P \approx 0.00151$, but for β_2 we have $P \approx 0.41128$. This suggests that the predictor x_1 belongs in the model, but x_2 might not. Note that although the same test is given for the intercept β_0, there are good reasons to retain it in the model even is it is not significantly different from zero. This will be made clear when we discuss the ANOVA decomposition for the regression model.

The following example demonstrates how analysis can proceed directly from a regression summary table.

Example 6.13 (Regression Summary Table) A client hires a consulting firm to conduct a study of two types of mutual funds (we'll call them simply Type A and Type B). It uses a regression model

$$Y = e^{\beta_0 + \beta_1 x + \epsilon}$$

where $x = 1$ for a Type A mutual fund, and $X = 0$ otherwise; $\epsilon \sim N(0, \sigma^2)$; and Y is the value of an original investment of \$1 after a year (that is, if $Y = 1.05$, the yearly rate of return is 5%). The model is first log-transformed, giving

$$\log(Y_i) = \beta_0 + \beta_1 x_i + \epsilon_i, \quad i = 1, \ldots, n. \tag{6.46}$$

A sample of $n = 62$ paired observations (Y_i, x_i), $i = 1, \ldots, 62$ is collected. A simple least squares regression model is used to fit the model (6.46), using transformed responses $\log(Y_i)$, producing the following summary table:

Coefficients	Estimate	Standard Error	T statistic	P-value
$\hat{\beta}_0$	0.0745	0.0040	18.8376	2.43×10^{-34}
$\hat{\beta}_1$	0.0333	0.0056	5.9514	4.13×10^{-8}

The consultant believes Type A mutual funds have a higher average yield, but the client currently purchases mutual funds of Type B, and there would be a significant cost to switching to Type A. Therefore, the consultant will only recommend switching to Type A if there is significant statistical evidence that $\beta_1 > 0.015$ (approximately, that the rate of return of Type A mutual finds exceeds Type B mutual funds by more than 1.5%). The hypotheses are therefore

$$H_o : \beta_1 \leq 0.015 \text{ against } H_a : \beta_1 > 0.015.$$

The appropriate T statistic is given by

$$T = \frac{\hat{\beta}_1 - 0.015}{SE_{\hat{\beta}_1}} = \frac{0.0333 - 0.015}{0.0056} \approx 3.268.$$

Since $T > t_{60;0.05} = 1.671$ we reject H_o at significance level $\alpha = 0.05$. ■

6.8.1 ANOVA Decomposition for Multiple Linear Regression

Theorems 6.4 and 6.5 give a general characterization of the ANOVA decomposition based on nested full and reduced models $\mathcal{V}_r \subset \mathcal{V}_f \subset \mathbb{R}^n$. In Section 6.7 this approach led to various F tests commonly used for one- and two-way ANOVA models.

The same approach can be used for linear regression models, although we would like to allow somewhat more flexibility in formulating the problem. As usual, we begin with the linear regression model $\mathbf{Y} = \mathbf{X}\boldsymbol{\beta} + \epsilon$, where $\epsilon \sim N(0, \sigma^2\mathbf{I}_n)$, $\boldsymbol{\beta} \in \mathbb{R}^q$ and $\text{rank}(\mathbf{X}) = q$. Then the full model is defined by vector space $\mathcal{V}_f = \mathcal{V}_\mathbf{X}$. In regression applications it will usually be more natural to define the reduced model $\mathcal{V}_r$ by constraining the parameter $\boldsymbol{\beta}$. To see this, we first expand the linear model

$$\mathbf{Y} = \beta_1\mathbf{x}_1 + \ldots + \beta_q\mathbf{x}_q + \boldsymbol{\epsilon}, \tag{6.47}$$

where $\mathbf{x}_j$ is the jth column of $\mathbf{X}$. Then suppose for some $p < q$ we define a reduced model

$$\mathbf{Y} = \beta_1\mathbf{x}_1 + \ldots + \beta_p\mathbf{x}_p + \boldsymbol{\epsilon}, \tag{6.48}$$

which is equivalent to setting $\beta_j = 0,\ j = p+1, \ldots, q$ in model (6.47). The problem of variable selection, for example, then reduces to the hypotheses

$$H_o : \beta_j = 0,\ \ j = p+1, \ldots, q \text{ and } H_a : \beta_j \neq 0 \text{ for some } j = p+1, \ldots, q. \tag{6.49}$$

The reduced model can be written $\mathbf{Y} = \mathbf{X}_r\boldsymbol{\beta}_r + \epsilon$, where $\boldsymbol{\beta}_r = (\beta_1, \ldots, \beta_p)$ (compared to full parameter $\boldsymbol{\beta} = (\beta_1, \ldots, \beta_q)$ and $\mathbf{X}_r$ is constructed from the first p columns of $\mathbf{X}$. The columns of $\mathbf{X}_r$ span vector space $\mathcal{V}_r \subset \mathcal{V}_f$, and Theorems 6.4 and 6.5 can be applied.

It is not much more work, however, to define a reduced model by applying any type of linear constraints to $\boldsymbol{\beta}$, of which the variable selection problem will be a special case. Suppose we define a reduced model by the null hypothesis

$$H_o : \mathbf{C}\boldsymbol{\beta} = \mathbf{0}, \tag{6.50}$$

where $\mathbf{C}$ is a $c \times q$ matrix. Here, we assume $c < q$, and that the rows of $\mathbf{C}$ are linearly independent, so that $\text{rank}(\mathbf{C}) = c$. We may always complete $\mathbf{C}$ to create an invertible $q \times q$ matrix

$$\mathbf{C}_q = \begin{bmatrix} \mathbf{C}_p \\ \mathbf{C} \end{bmatrix}, \tag{6.51}$$

where $\mathbf{C}_p$ is a $p \times q$ matrix of rank $p = q - c$. The full model can be written

$$\mathbf{Y} = \mathbf{X}\boldsymbol{\beta} + \epsilon = \mathbf{X}\mathbf{C}_q^{-1}\mathbf{C}_q\boldsymbol{\beta} + \epsilon = \mathbf{X}'\boldsymbol{\beta}' + \epsilon \tag{6.52}$$

where $\mathbf{X}' = \mathbf{X}\mathbf{C}_q^{-1}$ and $\boldsymbol{\beta}' = \mathbf{C}_q\boldsymbol{\beta} = (\beta'_1, \ldots, \beta'_q)$. Next, note that the vector space spanned by the columns of $\mathbf{X}'$ is equal to $\mathcal{V}_\mathbf{X}$, and the null hypothesis of Equation (6.50) can be equivalently expressed as

$$H_o : \beta'_j = 0,\ \ j = p+1, \ldots, q. \tag{6.53}$$

Under the null hypothesis of Equation (6.53) the reduced model becomes $\mathbf{Y} = \mathbf{X}'_r\boldsymbol{\beta}'_r + \epsilon$, where $\boldsymbol{\beta}'_r = (\beta'_1, \ldots, \beta'_p)$, and $\mathbf{X}'_r$ is constructed from the first p columns of $\mathbf{X}'$. This means the reduced model is fit by projecting $\mathbf{Y}$ onto the vector space $\mathcal{V}_{\mathbf{X}'_r} \subset \mathcal{V}_f$. In effect, following the transformation we can express the original problem as a variable selection problem. We can summarize the argument in the following theorem.

Theorem 6.9 (Goodness-of-Fit Test (F Test) for Multiple Linear Regression) Suppose $\mathbf{Y} = \mathbf{X}\boldsymbol{\beta} + \boldsymbol{\epsilon}$ is a linear model, where $\boldsymbol{\beta} \in \mathbb{R}^q$, $\mathbf{X} \in \mathcal{M}_{n,q}$ and $\boldsymbol{\epsilon} \in N(\mathbf{0}, \sigma^2\mathbf{I}_n)$. Let

$\hat{\mathbf{Y}}_f$ be the fitted values from the full model, and let $\hat{\mathbf{Y}}_r$ be the fitted values of the reduced model defined by the null hypothesis (6.50), where $\mathbf{C}$ is a $c \times q$ matrix of rank $c < q$. Set $p = q - c$. Let $SSE_F = |\mathbf{Y} - \hat{\mathbf{Y}}_f|^2$, $SSE_r = |\mathbf{Y} - \hat{\mathbf{Y}}_r|^2$ be the error sums of squares of the full and reduced models, respectively. Then the following statements hold.

(i) The ANOVA decomposition $SSE_r = SSR_{F,R} + SSE_F$ holds, where $SSR_{F,R} = |\hat{\mathbf{Y}}_f - \hat{\mathbf{Y}}_r|^2$

(ii) The F statistic for the hypothesis of Equation (6.50) is given by

$$F = \frac{SSR_{F,R}/(q-p))}{SSE/(n-q)} \sim F_{\Delta;q-p,n-q}.$$

where $\Delta = \sigma^{-2}(\boldsymbol{\mu} - \mathbf{H}_r\boldsymbol{\mu})^T(\boldsymbol{\mu} - \mathbf{H}_r\boldsymbol{\mu})$, and $\mathbf{H}_r$ is the projection marix for the reduced model. ■

Proof. We have shown above that the reduced model defined by null hypothesis (6.50) is equivalent to the projection onto a $c = q - p$ dimensional vector subspace $\mathcal{V}_r \subset \mathcal{V}_f$, where $\mathcal{V}_f$ is the q dimensional vector space defining the full model. The theorem follows directly from Theorems 6.4 and 6.5. □

Example 6.14 (Example of a Variable Selection Problem) Suppose we wish to compare two models

$$Y_i = \beta_0 + \beta_1 x_{i1} + \epsilon_i, \text{ and}$$
$$Y_i = \beta_0 + \beta_1 x_{i1} + \beta_2 x_{i2} + \beta_3 x_{i3} + \epsilon_i, \quad i = 1, \ldots, n.$$

This is an example of the classic variable selection problem. The second model is clearly interpretable as the full model, while the first is the reduced model defined by a null hypothesis of the form (6.50), setting constraint matrix

$$\mathbf{C} = \begin{bmatrix} 0 & 0 & 1 & 0 \\ 0 & 0 & 0 & 1 \end{bmatrix}.$$

The two rows of $\mathbf{C}$ are linearly independent, so $\text{rank}(\mathbf{C}) = 2$. ■

Example 6.15 (Hypothesis Test for Equality of Regression Parameters) Suppose we define a full model

$$Y_i = \beta_0 + \beta_1 x_{i1} + \beta_2 x_{i2} + \beta_3 x_{i3} + \epsilon_i, \quad i = 1, \ldots, n, \tag{6.54}$$

and we wish to test against the null hypothesis

$$H_o : \beta_1 = \beta_2 = \beta_3. \tag{6.55}$$

A reasonable approach would be to simply consolidate the three predictor variables into a single sum, then fit the resulting simple linear regression model

$$Y_i = \beta_0 + \beta_1'(x_{i1} + x_{i2} + x_{i3}) + \epsilon_i, \quad i = 1, \ldots, n. \tag{6.56}$$

We would like to know that we can test against H_o by comparing the sums of squares of the full and reduced models using the F test of Theorem 6.9. To do this, express H_o in the form (6.50), setting constraint matrix

$$\mathbf{C} = \begin{bmatrix} 0 & -1 & 1 & 0 \\ 0 & 0 & -1 & 1 \end{bmatrix}.$$

The two rows of $\mathbf{C}$ are linearly independent, so rank($\mathbf{C}$) = 2. We can complete $\mathbf{C}$ as in Equation (6.51) by setting

$$\mathbf{C}_q = \begin{bmatrix} 1 & 0 & 0 & 0 \\ 0 & 1 & 0 & 0 \\ 0 & -1 & 1 & 0 \\ 0 & 0 & -1 & 1 \end{bmatrix}.$$

It is not hard to verify that

$$\mathbf{C}_q^{-1} = \begin{bmatrix} 1 & 0 & 0 & 0 \\ 0 & 1 & 0 & 0 \\ 0 & 1 & 1 & 0 \\ 0 & 1 & 1 & 1 \end{bmatrix}. \tag{6.57}$$

so that, following Equation (6.52), the design matrix $\mathbf{X}_r'$ of the reduced model is given by the first two columns of $\mathbf{X}' = \mathbf{X}\mathbf{C}_q^{-1}$. Given matrix (6.57), this is equal to

$$\mathbf{X}_r' = \begin{bmatrix} 1 & x_{11} + x_{12} + x_{13} \\ 1 & x_{21} + x_{22} + x_{23} \\ \vdots & \vdots \\ 1 & x_{n1} + x_{n2} + x_{n3} \end{bmatrix},$$

which is equivalent to the reduced model (6.56). Thus, we may test against H_o by using the F test of Theorem 6.9 based on the error sums of squares obtained by full and reduced models (6.54) and (6.56). ∎

It is worth reviewing the process of completing the matrix $\mathbf{C}$ in Example 6.15. Since $\mathbf{C}$ was a $2 \times q$ of rank 2, all that was needed to complete $\mathbf{C}$ was to select *any* $2 \times q$ matrix $\mathbf{C}_p$ such that the completed matrix $\mathbf{C}_q$ (Equation (6.51)) is invertible (equivalently, of rank q). However, as long as this simple rule is applied, the vector space of the design matrix for the reduced model will always be the same. This is a consequence of the following theorem, the proof of which is left to the reader (Problem 6.38).

Theorem 6.10 (Vector Space of Reduced Model Induced by Completion of the Constraint Matrix) Let $\mathbf{C}_q$ be the completed matrix of Equation (6.51). Then the first p column vectors of $\mathbf{C}_q^{-1}$ are linearly independent, and orthogonal to each row vector of the constraint matrix $\mathbf{C}$. ∎

To see the relevance of Theorem 6.10 to this problem, suppose $\mathcal{V}_{\mathbf{C}} \subset \mathbb{R}^q$ is the c dimensional vector subspace spanned by the row vectors of $\mathbf{C}$. By Theorem 6.10 the first p column vectors of $\mathbf{C}_q^{-1}$ are linearly independent elements of the p dimensional vector subspace $\mathcal{V}_{\mathbf{C}}^{\perp} \subset \mathbb{R}^q$ which is orthogonal to $\mathcal{V}_{\mathbf{C}}$ (Section B.3). But $\mathcal{V}_{\mathbf{C}}^{\perp}$, and therefore the vector space spanned by the first p column vectors of $\mathbf{X}\mathbf{C}_q^{-1}$, are uniquely determined by $\mathbf{C}$.

Example 6.16 (Goodness-of-Fit Test *versus* Marginal Inferences) For this example, data ($n = 10$) was simulated from the model

$$Y = \beta_0 + \beta_1 x_1 + \beta_2 x_2 + \epsilon$$

and is summarized in the following table.

Considered separately neither β_1 or β_2 differs significantly from zero ($P = 0.605, 0.259$, respectively). We then construct an ANOVA table similar to that introduced in Section 6.7.

Coefficients	Estimate	Standard Error	T statistic	P-value
$\hat{\beta}_0$	0.4784	0.4349	1.100	0.308
$\hat{\beta}_1$	0.5945	1.0983	0.541	0.605
$\hat{\beta}_2$	1.3003	1.0588	1.228	0.259

Source	SS	df	MSS	F
Regression	36.79	2	18.40	10.40
Error	12.37	7	1.77	
Total	49.16	9		

As before, the means sum of squares (MSS) equals the sum of squares (SS) divided by the degrees of freedom. Here, $SSR = 36.79$, $SSE = 12.37$, then $MSR = 36.79/2 = 18.40$, $MSE = 12.37/7 = 1.77$, and finally $F = MSR/MSE = 10.40$. The null hypothesis is $H_o : \beta_1 = \beta_2 = 0$, under which $F \sim F_{2,7}$, leading to P-value $P = 0.008$. This means we can reject the null hypothesis, even though both β_1 and β_2 were found, marginally, to be insignificant.

The reason for this counterintuitive behavior has to do which the fact that, in this example, the predictors x_1 and x_2 are highly correlated ($r^2 = 0.92$) (in the context of linear regression this is referred to as collinearity) . Suppose, to fix ideas, the true coefficient values are $\beta_1 = \beta_2 = 1$. Suppose also that $x_1 \approx x_2$. Then

$$\beta_0 + x_1 + x_2 \approx \beta_0 + 2x_1 \approx \beta_0 + 2x_2 \approx \beta_0 + 1.5x_1 + 0.5x_2,$$

and so on. This means that the coefficient estimate $\hat{\beta}_1$ can be changed significantly without significantly changing the SSE, as long as $\hat{\beta}_2$ is changed in a complementary manner. This results in a wide confidence interval for each coefficient, but it also means that the estimates $\hat{\beta}_1$ and $\hat{\beta}_2$ will be negatively correlated. See Problem 6.26. ■

6.9 Constrained Least Squares Estimation

The null hypothesis of Equation (6.50) defines a reduced linear model defined by a vector subspace constraint $\boldsymbol{\mu} \in \mathcal{V}_r$. Suppose we consider instead the affine constraint:

$$H_o : \mathbf{C}\boldsymbol{\beta} - \mathbf{d} = \mathbf{0}, \tag{6.58}$$

where $\mathbf{C}$ is a $p \times q$ matrix, $\mathbf{d}$ is a $q \times 1$ column matrix, and $\text{rank}(\mathbf{C}) = p$. If $\mathbf{d} \neq \mathbf{0}$ the reduced model will no longer be defined by a vector subspace, and the Gauss-Markov theorem cannot be used.

Example 6.17 (Example of Affine Constraint) Suppose we wish to compare two models

$$\begin{aligned} Y_i &= \beta_0 + \beta_1 x_{i1} + (\beta_1 + 1)x_{i2} + +(\beta_1 + 2)x_{i3} + \epsilon_i, \text{ and} \\ Y_i &= \beta_0 + \beta_1 x_{i1} + \beta_2 x_{i2} + \beta_3 x_{i3} + \epsilon_i, \ \ i = 1, \ldots, n. \end{aligned}$$

This leads to hypotheses

$$H_o : \beta_2 = \beta_1 + 1, \ \ \beta_3 = \beta_1 + 2 \text{ against } H_a : \beta_2 \neq \beta_1 + 1 \text{ or } \beta_3 \neq \beta_1 + 2.$$

In this case we have parameter $\boldsymbol{\beta} = [\beta_0, \beta_1, \beta_2, \beta_3]^T$, and we would define a null hypothesis (6.58) using the components

$$\mathbf{C} = \begin{bmatrix} 0 & -1 & 1 & 0 \\ 0 & -1 & 0 & 1 \end{bmatrix} \text{ and } \mathbf{d} = \begin{bmatrix} 1 \\ 2 \end{bmatrix}.$$

The two rows of $\mathbf{C}$ are linearly independent, so $\text{rank}(\mathbf{C}) = 2$. ■

While the Gauss-Markov theorem is only available if $\mathbf{d} = \mathbf{0}$, a natural approach to testing hypothesis (6.58) would be to minimize SSE subject to contraint $\mathbf{C}\boldsymbol{\beta} - \mathbf{d} = \mathbf{0}$. We may then denote the SSE of the reduced (constrained) model by SSE_r, then SSE_f is the SSE of the full (unconstrained) model, as before. We necessarily have $SSE_r \geq SSE_f$, since the model space of the reduced optimization problem is a subset of the full model space.

We can show (Problem 6.29) that the least squares estimates of $\boldsymbol{\beta}$ under constraint $H_o : \mathbf{C}\boldsymbol{\beta} - \mathbf{d} = \mathbf{0}$ is given by

$$\hat{\boldsymbol{\beta}}_c = \hat{\boldsymbol{\beta}}_u + (\mathbf{X}^T\mathbf{X})^{-1}\mathbf{C}^T \left[\mathbf{C}(\mathbf{X}^T\mathbf{X})^{-1}\mathbf{C}^T\right]^{-1} (\mathbf{d} - \mathbf{C}\hat{\boldsymbol{\beta}}_u), \tag{6.59}$$

where $\hat{\boldsymbol{\beta}}_u = (\mathbf{X}^T\mathbf{X})^{-1}\mathbf{X}^T\mathbf{Y}$ is the unconstrained least squares solution. Let $\mathbf{B} = \mathbf{X}(\mathbf{X}^T\mathbf{X})^{-1}\mathbf{C}^T$. It follows that $\mathbf{B}^T\mathbf{B} = \mathbf{C}(\mathbf{X}^T\mathbf{X})^{-1}\mathbf{C}^T$. Note that $\mathbf{B}^T\mathbf{B}$ is an invertible $p \times p$ matrix. Then premultiply $\hat{\boldsymbol{\beta}}_c$ by $\mathbf{X}$ to obtain

$$\begin{aligned} \mathbf{X}\hat{\boldsymbol{\beta}}_c &= \mathbf{X}\hat{\boldsymbol{\beta}}_u - \mathbf{X}(\mathbf{X}^T\mathbf{X})^{-1}\mathbf{C}^T \left[\mathbf{C}(\mathbf{X}^T\mathbf{X})^{-1}\mathbf{C}^T\right]^{-1} (\mathbf{C}\hat{\boldsymbol{\beta}}_u - \mathbf{d}) \\ &= \mathbf{H}\mathbf{Y} - \mathbf{B}\left[\mathbf{B}^T\mathbf{B}\right]^{-1} (\mathbf{C}\hat{\boldsymbol{\beta}}_u - \mathbf{d}). \end{aligned}$$

We will make use of the identity

$$\mathbf{H}\mathbf{B} = \mathbf{X}(\mathbf{X}^T\mathbf{X})^{-1}\mathbf{X}^T\mathbf{X}(\mathbf{X}^T\mathbf{X})^{-1}\mathbf{C}^T = \mathbf{X}(\mathbf{X}^T\mathbf{X})^{-1}\mathbf{C}^T = \mathbf{B}. \tag{6.60}$$

We can now evaluate SSE_r:

$$\begin{aligned} SSE_r &= (\mathbf{Y} - \mathbf{X}\hat{\boldsymbol{\beta}}_c)^T(\mathbf{Y} - \mathbf{X}\hat{\boldsymbol{\beta}}_c) \\ &= \left([\mathbf{I}_n - \mathbf{H}]\mathbf{Y} + \mathbf{B}\left[\mathbf{B}^T\mathbf{B}\right]^{-1}(\mathbf{C}\hat{\boldsymbol{\beta}}_u - \mathbf{d})\right)^T \\ &\quad \times \left([\mathbf{I}_n - \mathbf{H}]\mathbf{Y} + \mathbf{B}\left[\mathbf{B}^T\mathbf{B}\right]^{-1}(\mathbf{C}\hat{\boldsymbol{\beta}}_u - \mathbf{d})\right) \\ &= \mathbf{Y}^T[\mathbf{I}_n - \mathbf{H}]\mathbf{Y} + (\mathbf{C}\hat{\boldsymbol{\beta}}_u - \mathbf{d}))^T \left[\mathbf{B}^T\mathbf{B}\right]^{-1} \mathbf{B}^T\mathbf{B}\left[\mathbf{B}^T\mathbf{B}\right]^{-1} (\mathbf{C}\hat{\boldsymbol{\beta}}_u - \mathbf{d})) \\ &= SSE_f + (\mathbf{C}\hat{\boldsymbol{\beta}}_u - \mathbf{d})^T \left[\mathbf{B}^T\mathbf{B}\right]^{-1} (\mathbf{C}\hat{\boldsymbol{\beta}}_u - \mathbf{d}), \end{aligned}$$

equivalently,

$$Q = SSE_r - SSE_f = (\mathbf{C}\hat{\boldsymbol{\beta}}_u - \mathbf{d})^T \left[\mathbf{B}^T\mathbf{B}\right]^{-1} (\mathbf{C}\hat{\boldsymbol{\beta}}_u - \mathbf{d}),$$

where we make use of Equation (6.60). Then write

$$\begin{aligned} \mathbf{C}\hat{\boldsymbol{\beta}}_u - \mathbf{d} &= \mathbf{C}(\mathbf{X}^T\mathbf{X})^{-1}\mathbf{X}^T\mathbf{Y} - \mathbf{d} \\ &= \mathbf{C}(\mathbf{X}^T\mathbf{X})^{-1}\mathbf{X}^T[\mathbf{X}\boldsymbol{\beta} + \boldsymbol{\epsilon}] - \mathbf{d} \\ &= (\mathbf{C}\boldsymbol{\beta} - \mathbf{d}) + \mathbf{B}^T\boldsymbol{\epsilon} \\ &= \mathbf{B}^T\left[\mathbf{B}(\mathbf{B}^T\mathbf{B})^{-1}(\mathbf{C}\boldsymbol{\beta} - \mathbf{d}) + \boldsymbol{\epsilon}\right], \end{aligned}$$

which gives quadratic form

$$Q = \left[\mathbf{B}(\mathbf{B}^T\mathbf{B})^{-1}(\mathbf{C}\boldsymbol{\beta} - \mathbf{d}) + \boldsymbol{\epsilon}\right]^T \mathbf{M} \left[\mathbf{B}(\mathbf{B}^T\mathbf{B})^{-1}(\mathbf{C}\boldsymbol{\beta} - \mathbf{d}) + \boldsymbol{\epsilon}\right],$$

where $\mathbf{M} = \mathbf{B}(\mathbf{B}^T\mathbf{B})^{-1}\mathbf{B}^T$ is a symmetric idempotent matrix of rank rank$(\mathbf{B}) = p$ (Theorem B.9). We can conclude that

$$\frac{SSE_r - SSE_f}{\sigma^2} \sim \chi^2_{\Delta,p}, \tag{6.61}$$

where Δ is the noncentrality parameter

$$\begin{aligned}\Delta &= \sigma^{-2}\left[\mathbf{B}(\mathbf{B}^T\mathbf{B})^{-1}(\mathbf{C}\boldsymbol{\beta} - \mathbf{d})\right]^T \mathbf{M}\left[\mathbf{B}(\mathbf{B}^T\mathbf{B})^{-1}(\mathbf{C}\boldsymbol{\beta} - \mathbf{d})\right] \\ &= \sigma^{-2}(\mathbf{C}\boldsymbol{\beta} - \mathbf{d})^T(\mathbf{B}^T\mathbf{B})^{-1}\mathbf{B}^T\mathbf{B}(\mathbf{B}^T\mathbf{B})^{-1}\mathbf{B}^T\mathbf{B}(\mathbf{B}^T\mathbf{B})^{-1}(\mathbf{C}\boldsymbol{\beta} - \mathbf{d}) \\ &= \sigma^{-2}(\mathbf{C}\boldsymbol{\beta} - \mathbf{d})^T(\mathbf{B}^T\mathbf{B})^{-1}(\mathbf{C}\boldsymbol{\beta} - \mathbf{d}).\end{aligned}$$

If σ^2 is unknown then Equation (6.61) does not define a statistic, so the next step is to construct an F statistic,

$$F = \frac{(SSE_r - SSE_f)/p}{SSE_f/(n-q)}$$

noting that $SSE_f/\sigma^2 \sim \chi^2_{n-q}$. However, we must verify that $SSE_r - SSE_f$ and SSE_f are independent. We have

$$SSE = \mathbf{Y}^T(\mathbf{I}_n - \mathbf{H})\mathbf{Y}, \quad SSE_r - SSE_f = \mathbf{Y}^T\mathbf{M}\mathbf{Y},$$

so we may verify independence by showing that $(\mathbf{I}_n - \mathbf{H})\mathbf{M} = \mathbf{0}$. First evaluate

$$\begin{aligned}\mathbf{H}\mathbf{M} &= \mathbf{X}(\mathbf{X}^T\mathbf{X})^{-1}\mathbf{X}^T \\ &\quad \times \left\{\mathbf{X}(\mathbf{X}^T\mathbf{X})^{-1}\mathbf{C}^T\left[\mathbf{C}(\mathbf{X}^T\mathbf{X})^{-1}\mathbf{C}^T\right]^{-1}\mathbf{C}(\mathbf{X}^T\mathbf{X})^{-1}\mathbf{X}^T\right\} \\ &= \mathbf{X}(\mathbf{X}^T\mathbf{X})^{-1}\mathbf{C}^T\left[\mathbf{C}(\mathbf{X}^T\mathbf{X})^{-1}\mathbf{C}^T\right]^{-1}\mathbf{C}(\mathbf{X}^T\mathbf{X})^{-1}\mathbf{X}^T \\ &= \mathbf{M}\end{aligned}$$

and as a consequence $(\mathbf{I}_n - \mathbf{H})\mathbf{M} = \mathbf{0}$, so that SSE_f and $SSE_r - SSE_f$ are independent.

We have proved our main theorem:

Theorem 6.11 (F Test for Constrained Least Squares) Suppose we are given a multiple linear regression model with q coefficients and $n \times q$ predictor matrix $\mathbf{X}$ of rank q. Assume $\boldsymbol{\epsilon}$ is an *iid* vector from distribution $N(0, \sigma^2)$. We wish to test

$$H_o : \mathbf{C}\boldsymbol{\beta} - \mathbf{d} = \mathbf{0} \text{ against } H_a : \mathbf{C}\boldsymbol{\beta} - \mathbf{d} \neq \mathbf{0},$$

where $\mathbf{C}$ is a $p \times q$ matrix, $1 \leq p \leq q$, such that rank$(\mathbf{C}) = p$. Let SSE_f be the error sum of squares for the unconstrained model, and let SSE_r be the error sum of squares for the model under constraint H_o. Then

$$F = \frac{(SSE_r - SSE_f)/p}{SSE_f/(n-q)} \sim F_{\Delta;p,n-q},$$

where Δ is the noncentrality parameter

$$\Delta = \sigma^{-2}(\mathbf{C}\boldsymbol{\beta} - \mathbf{d})^T(\mathbf{B}^T\mathbf{B})^{-1}(\mathbf{C}\boldsymbol{\beta} - \mathbf{d}) = (\mathbf{C}\boldsymbol{\beta} - \mathbf{d})^T\Sigma^{-1}_{\mathbf{C}\boldsymbol{\beta}_u}(\mathbf{C}\boldsymbol{\beta} - \mathbf{d})$$

where $\mathbf{B} = \mathbf{X}(\mathbf{X}^T\mathbf{X})^{-1}\mathbf{C}^T$, $\hat{\boldsymbol{\beta}}_u$ is the unconstrained least squares solution, and

$$\Sigma_{\mathbf{C}\hat{\boldsymbol{\beta}}_u} = \text{var}[\mathbf{C}\hat{\boldsymbol{\beta}}_u].$$

Furthermore, the following identity holds:

$$SSE_r - SSE_f = (\mathbf{C}\hat{\boldsymbol{\beta}}_u - \mathbf{d})^T \left[\mathbf{B}^T\mathbf{B}\right]^{-1} (\mathbf{C}\hat{\boldsymbol{\beta}}_u - \mathbf{d}).$$ ■

Example 6.18 (Regression with a Constrained Coefficient Sum (The Probability Simplex Case)) Suppose we are given a linear regression model

$$\mathbf{Y} = \beta_1\mathbf{x}_1 + \ldots + \beta_q\mathbf{x}_q + \boldsymbol{\epsilon},$$

for which the coefficents $\boldsymbol{\beta}$ represent allocation proportions, so that $\boldsymbol{\beta}$ is a probability distribution and is contained in a probability simplex. This introduces the affine constraint $\mathbf{c}^T\boldsymbol{\beta} = 1$, where $\mathbf{c} = \mathbf{1}_q$, but also the boundary (or inequality) constraint $\beta_j \geq 0$ for all j. The problem of minimizing the sum-of-squares under a boundary constraint requires specialized techniques from numerical analysis, which is beyond the scope of this volume (Stark and Parker, 1995). We can instead, optimistically, enforce the affine constraint alone, hoping that the coefficient estimates are positive, in which case we have our desired estimate.

Two distinct approaches to the problem present themselves, both based on first calculating, then modifying, the unconstrained estimate $\hat{\boldsymbol{\beta}}_u$. First, we may simply normalize $\boldsymbol{\beta}_u$, which gives

$$\hat{\boldsymbol{\beta}}^* = \left(\frac{\hat{\beta}_1}{\sum_{j=1}^q \hat{\beta}_j}, \ldots, \frac{\hat{\beta}_q}{\sum_{j=1}^q \hat{\beta}_j}\right),$$

which clearly satifies the affine constraint.

Interestingly, the constrained least squares method gives a difference solution. The expression for the constrained solution is given in Equation (6.59). The constraint matrix is $\mathbf{C} = \mathbf{1}_{1,q}$, with the complete constraint given as $\mathbf{C}\boldsymbol{\beta} = \mathbf{d} = 1$. From Equation (6.59) we have

$$(\mathbf{X}^T\mathbf{X})^{-1}\mathbf{C}^T = [\alpha_1, \ldots, \alpha_q]^T,$$

where α_j is the sum of the jth row of $(\mathbf{X}^T\mathbf{X})^{-1}$, and

$$\mathbf{C}(\mathbf{X}^T\mathbf{X})^{-1}\mathbf{C}^T = \sum_{j=1}^q \alpha_j.$$

For convenience set $\alpha_j^* = \alpha_j / \sum_{j=1}^q \alpha_j$, so that $\sum_{j=1}^q \alpha_j^* = 1$.

The constrained least squares solution is then

$$\hat{\boldsymbol{\beta}}_c = \hat{\boldsymbol{\beta}}_u + [\alpha_1^*, \ldots, \alpha_q^*]^T(1 - \mathbf{C}\hat{\boldsymbol{\beta}}_u), \tag{6.62}$$

which is a quite different estimator than $\hat{\boldsymbol{\beta}}^*$ (for example, if by chance the elements of $\hat{\boldsymbol{\beta}}_u$ are equal, the same must be true of $\hat{\boldsymbol{\beta}}^*$ but not of $\hat{\boldsymbol{\beta}}_c$).

It is easily verified that $\hat{\boldsymbol{\beta}}_c$ is an unbiased estimator of $\boldsymbol{\beta}$ for which $\mathbf{C}\hat{\boldsymbol{\beta}}_c = 1$ (note that $\hat{\boldsymbol{\beta}}_u$ is unbiased, and we may assume that the true parameter $\boldsymbol{\beta}$ satisfies the affine constraint). ■

6.10 Simultaneous Confidence Intervals

The following situation arises frequently in the inference of linear models, sometimes in very different contexts. We have an estimator $\hat{\boldsymbol{\beta}} = (\hat{\beta}_1, \ldots, \hat{\beta}_q)$ of parameter $\boldsymbol{\beta} = (\beta_1, \ldots, \beta_q)$, with distribution

$$\hat{\boldsymbol{\beta}} \sim N(\boldsymbol{\beta}_0, \Sigma_{\hat{\boldsymbol{\beta}}}). \tag{6.63}$$

Then $[\hat{\boldsymbol{\beta}} - \boldsymbol{\beta}_0]^T \Sigma_{\hat{\boldsymbol{\beta}}}^{-1} [\hat{\boldsymbol{\beta}} - \boldsymbol{\beta}_0] \sim \chi_q^2$ (Theorem 2.2). Suppose $\mathbf{c} \in \mathbb{R}^q$. The obvious estimator for $\eta = \mathbf{c}^T \boldsymbol{\beta}$ is $\hat{\eta} = \mathbf{c}^T \hat{\boldsymbol{\beta}}$. Then $E[\hat{\eta}] = \eta$, $\sigma_{\hat{\eta}}^2 = \text{var}[\hat{\eta}] = \mathbf{c}^T \Sigma_{\hat{\boldsymbol{\beta}}} \mathbf{c}$, and a level $1 - \alpha$ confidence interval is given by $CI = \hat{\eta} \pm z_{\alpha/2} \sigma_{\hat{\eta}}$. We will, however, often encounter applications for which an inference consists of multiple confidence intervals. For example, in the case of simple linear regression, rather than reporting confidence intervals for β_0 and β_1 it may be more useful to report a confidence band

$$CB = \hat{\beta}_0 + \hat{\beta}_1 x \pm z^* \sigma_{\hat{\mu}_x}, \tag{6.64}$$

where the problem of estimating $\mu_x = \beta_0 + \beta_1 x$ was considered in Example 6.7. Clearly, if we wish to report these confidence limits simultaneously at a single confidence level $1-\alpha$ for all x (at least within the range of the observed predictor values) we cannot use $z^* = z_{\alpha/2}$. Fortunately, we may determine a useful and correct value for z^*, and extend the method when MSE replaces σ^2. An example of this type of problem is left to the reader (Problem 6.32).

The basis of the approach is Scheffé's method (Scheffe, 1959). Our approach will be based on the following lemma.

Lemma 6.1 Suppose $\mathbf{x} \in \mathbb{R}^n$, and $\mathbf{A} \in \mathcal{M}_n$ is positive definite. Then

$$\sup_{\mathbf{y} \neq \mathbf{0}} \frac{(\mathbf{x}^T \mathbf{y})^2}{\mathbf{y}^T \mathbf{A} \mathbf{y}} \leq \mathbf{x}^T \mathbf{A}^{-1} \mathbf{x}.$$ ■

Proof. Since $\mathbf{A}$ is positive definite it possesses invertible square root matrix $\mathbf{A}_{1/2}^2 = \mathbf{A}$ (Theorem B.7). Then, applying the Cauchy-Schwarz inequality gives

$$(\mathbf{x}^T \mathbf{y})^2 = (\mathbf{x}^T \mathbf{A}_{1/2}^{-1} \mathbf{A}_{1/2} \mathbf{y})^2 \leq (\mathbf{x}^T \mathbf{A}^{-1} \mathbf{x})(\mathbf{y}^T \mathbf{A} \mathbf{y}),$$

from which the lemma follows. □

Our first result follows.

Theorem 6.12 (Simultaneous Confidence Intervals – χ^2 Distribution) Suppose Equation (6.63) holds. Then all confidence intervals

$$\hat{\eta} \pm (\chi_{q;\alpha}^2)^{1/2} \sigma_{\hat{\eta}}^2$$

hold simultaneously with confidence level $1 - \alpha$, where $\hat{\eta} = \mathbf{c}^T \hat{\boldsymbol{\beta}}$ for any $\mathbf{c} \neq \mathbf{0}$. ■

Proof. We have

$$P\left([\hat{\boldsymbol{\beta}} - \boldsymbol{\beta}_0]^T \Sigma_{\hat{\boldsymbol{\beta}}} [\hat{\boldsymbol{\beta}} - \boldsymbol{\beta}_0] \leq \chi_{q;\alpha}^2 \right) = 1 - \alpha.$$

By Lemma 6.1, noting that $\Sigma_{\hat{\boldsymbol{\beta}}}$ is positive definite, we have

$$[\hat{\boldsymbol{\beta}} - \boldsymbol{\beta}_0]^T \Sigma_{\hat{\boldsymbol{\beta}}}^{-1} [\hat{\boldsymbol{\beta}} - \boldsymbol{\beta}_0] \geq \frac{(\mathbf{c}^T [\hat{\boldsymbol{\beta}} - \boldsymbol{\beta}_0])^2}{\mathbf{c}^T \Sigma_{\hat{\boldsymbol{\beta}}} \mathbf{c}} = \frac{(\hat{\eta} - \eta_0)^2}{\sigma_{\hat{\eta}}^2},$$

hence by assumption

$$P\left(\sup_{\mathbf{c}\neq 0}\frac{(\mathbf{c}^T[\hat{\boldsymbol{\beta}}-\boldsymbol{\beta}_0])^2}{\mathbf{c}^T\Sigma_{\hat{\boldsymbol{\beta}}}\mathbf{c}}\leq\chi^2_{q;\alpha}\right)=1-\alpha. \qquad \square$$

This procedure become very conservative with increasing q. Compare $z_{0.025} = 1.96$ to $(\chi^2_{2;\alpha})^{1/2} = 2.45$, $(\chi^2_{3;\alpha})^{1/2} = 2.80$, $(\chi^2_{4;\alpha})^{1/2} = 3.08$, $(\chi^2_{8;\alpha})^{1/2} = 3.33$. Yet, in some cases this will be the appropriate procedure.

In many applications we will have $\Sigma_{\hat{\boldsymbol{\beta}}} = \sigma^2\mathbf{V}$, where $\mathbf{V}$ is known but σ^2 is unknown. We will often have an unbiased estimate $\hat{\sigma}^2$, for which $\nu\sigma^{-2}\hat{\sigma}^2 \sim \chi^2_\nu$ for some degrees of freedom ν, the exact value of which need not concern us at this point. We also assume $\hat{\boldsymbol{\beta}}$ and $\hat{\sigma}^2$ are independent. If we replace σ^2 with $\hat{\sigma}^2$ we obtain confidence interval $CI = \hat{\eta} \pm t_{\nu;\alpha/2}S_{\hat{\eta}}$, where $S_{\hat{\eta}}$ is the standard error defined by

$$S^2_{\hat{\eta}} = \hat{\sigma}^2\mathbf{c}^T\mathbf{V}\mathbf{c}. \tag{6.65}$$

Then Theorem 6.12 is easily extended to this case, noting that under our stated conditions we must have

$$F = \frac{[\hat{\boldsymbol{\beta}}-\boldsymbol{\beta}_0]^T\mathbf{V}^{-1}[\hat{\boldsymbol{\beta}}-\boldsymbol{\beta}_0]/q}{\hat{\sigma}^2} \sim F_{q,n-q}.$$

Theorem 6.13 (Simultaneous Confidence Intervals – F Distribution) Suppose Equation (6.63) holds, with $\Sigma_{\hat{\boldsymbol{\beta}}} = \sigma^2\mathbf{V}$. Suppose $\hat{\sigma}^2$ is an estimate of σ^2 for which $\nu\sigma^{-2}\hat{\sigma}^2 \sim \chi^2_\nu$ for some degrees of freedom ν, and assume $\hat{\sigma}^2$ and $\hat{\boldsymbol{\beta}}$ are independent. For any estimate $\hat{\eta} = \mathbf{c}^T\hat{\boldsymbol{\beta}}$ of $\eta = \mathbf{c}^T\boldsymbol{\beta}$ let $S_{\hat{\eta}}$ be the standard error defined in Equation (6.65). Then all confidence intervals

$$\hat{\eta} \pm (qF_{q,\nu;\alpha})^{1/2}\, S_{\hat{\eta}}$$

hold with confidence level $1-\alpha$. ■

Proof. The proof follows Theorem 6.12. □

The estimate $\hat{\sigma}^2 = MSE$ (Section 6.5.1) satisfies the conditions of Theorem 6.13. As before, this procedure become very conservative with increasing q. Compare $t_{30;0.025} = 2.04$ to $(2F_{2,30;\alpha})^{1/2} = 2.58$, $(3F_{3,30;\alpha})^{1/2} = 2.96$, $(4F_{4,30;\alpha})^{1/2} = 3.28$, $(5F_{5,30;\alpha})^{1/2} = 3.56$.

We next apply Theorem 6.13 to the problem of constructing confidence bands for simple linear regression introduced earlier in this section.

6.10.1 Confidence Bands for Simple Linear Regression

In this example we will follow through on developing the confidence band defined in Equation (6.64). We are given a simple linear regression model with *iid* normally distributed error terms. From n observations we evaluate $\hat{\beta}_0$, $\hat{\beta}_1$, as well as MSE. The standard error for estimand $\mu_x = \beta_0 + \beta_1 x$ is given in Equation (6.24) as

$$S_{\hat{\mu}_x} = MSE^{1/2}\sqrt{\frac{1}{n}+\frac{(x-\bar{x})^2}{\sum_{i=1}^n(x_i-\bar{x})^2}}$$

By Theorem 6.13 the appropriate critical value in Equation (6.64) will be $z^* = (qF_{q,\nu;\alpha})^{1/2}$, where $q = 2$ and $\nu = n - q = n - 2$, giving confidence band

$$CB = \hat{\beta}_0 + \hat{\beta}_1 x \pm (2F_{2,n-2;\alpha})^{1/2}MSE^{1/2}\sqrt{\frac{1}{n}+\frac{(x-\bar{x})^2}{\sum_{i=1}^n(x_i-\bar{x})^2}}.$$

CB implicitly defines two functions $L(x)$, $U(x)$ which form the envelope $L(x) \leq \hat{\beta}_0 + \hat{\beta}_1 x \leq U(x)$ for all x. The formal inference is then

$$P(L(x) \leq \beta_0 + \beta_1 x \leq U(x) \text{ for all } x) = 1 - \alpha,$$

as stated by Theorem 6.13. See Figure 6.4 for a numerical example of a confidence band. See also Problem 6.32.

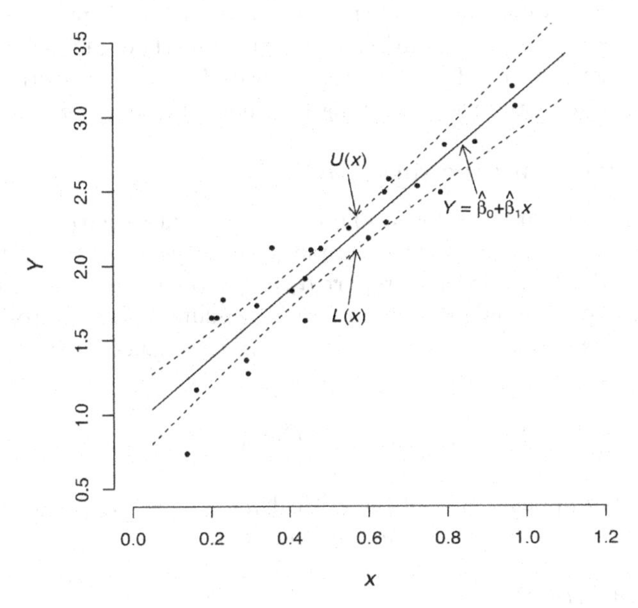

FIGURE 6.4
Confidence bands $(L(x), U(x))$ for Section 6.10.1. In this example $n = 25$, $1 - \alpha = 0.95$, $MSE = 0.046$, $\sum_{i=1}^{n}(x_i - \bar{x})^2 = 1.519$, $\bar{x} = 0.506$, $(\hat{\beta}_0, \hat{\beta}_1) = (0.922, 2.272)$. Observations (Y_i, x_i) are plotted individually.

6.10.2 Multiple Comparsons in ANOVA

We return to the one-way ANOVA model. For q treatments we have $\boldsymbol{\beta} = (\beta_1, \ldots, \beta_k)$, $\hat{\boldsymbol{\beta}} = (\bar{Y}_1, \ldots, \bar{Y}_k)$ and sample sizes $(n_1, \ldots, n_k)$. Then $\hat{\sigma}^2 = MSE$ estimates σ^2. If the null hypothesis of equality of means of Equation (6.6) is rejected, there may be interest in a more detailed summary of the treatments means of $\boldsymbol{\beta}$, often refered to as a *post-hoc* analysis. This is typically centered around comparisons of the form $\Delta_{i,j} = \beta_i - \beta_j$. We may always construct a single level $1 - \alpha$ confidence interval for $\Delta_{i,j}$ of the form

$$\bar{Y}_i - \bar{Y}_j \pm t_{n-k;\alpha/2} \sqrt{MSE \left(\frac{1}{n_i} + \frac{1}{n_j} \right)}. \tag{6.66}$$

However, a *post-hoc* analysis is often more complicated. Suppose $k = 3$, and we are interested in the hypothesis $H : \beta_1 > \beta_2$ and $\beta_1 > \beta_3$. Note that we have not designated H as null or alternative. Suppose we construct confidence intervals (L_1, U_1) for $\beta_1 - \beta_2$ and (L_2, U_2) for $\beta_1 - \beta_3$. If we provisionally accept both confidence intervals as correct we may reach one of three conclusions

(1) If $L_1 > 0$ and $L_2 > 0$ then H is true.

(2) If $U_1 < 0$ or $U_2 < 0$ then H is false.
(3) Otherwise H is unresolved.

Then, as a statistical inference, this conclusion is reported with whatever confidence level applies simulataneously to both confidence intervals.

Thus, we need to define a family-wise error α_{FWE}. Suppose we make m inference statements, to be interpreted as a single inference statement. Then our inference is in error if at least one of the m statements is in error, which occurs with probability no greater than α_{FWE}. We will consider three multiple comparison methods, for which the m inference statements are in the form of confidence intervals for some collection of treatment differences $\Delta_{i,j}$. All are generally used, meaning that none of them is uniformly preferable.

The Bonferroni Correction Procedure (BCP)

First, suppose the error probabilities for m inference statements are $\alpha_1, \ldots, \alpha_m$. By Boole's inequality $\alpha_{FWE} \leq \alpha_1 + \ldots + \alpha_m$. This simple procedure is known as the Bonferroni correction procedure (BCP). In order to report $\alpha_{FWE} = \alpha$ for m confidence intervals, we can construct each to be of confidence level $1 - \alpha/m$ (assuming each is to be of the same confidence level, which is not strictly necessary). The m confidence intervals are then given by

$$\bar{Y}_i - \bar{Y}_j \pm t_{n-k;\alpha/(m2)} \sqrt{MSE\left(\frac{1}{n_i} + \frac{1}{n_j}\right)}$$

Note that the number m appears only in the critical value $t_{n-k;\alpha/(m2)}$, so the effect on the length of the confidence interval can be directly measured.

Tukey's Range Test (HSD)

The second method is known as Tukey's range test or Tukey's honestly significant difference test (HSD test). Suppose we may assume that $n_j = m$, $j = 1, \ldots, k$ for total sample size $n = km$. This method is based on the observation (Tukey, 1949) that the distribution of

$$q = \max_{i,j} \frac{|\bar{Y}_i - \bar{Y}_j|}{\sqrt{MSE/m}} \tag{6.67}$$

depends only on parameters $n - k$ and k, and is refered to as the studentized range distribution. Then if $q_{n-k,k;\alpha}$ is the α critical value for q we can claim that the confidence intervals

$$\bar{Y}_i - \bar{Y}_j \pm \frac{q_{k,n-k;\alpha}}{\sqrt{2}} \sqrt{MSE\left(\frac{1}{n_i} + \frac{1}{n_j}\right)} \tag{6.68}$$

hold for all pairs i, j with $\alpha_{FWE} = \alpha$, where we are assuming $n_i = n_j = m$. If the samples sizes are unequal, the procedure is not exact, but is known to be conservative and is referred to as the Tukey-Kramer pairwise procedure. See, for example, Kutner *et al.* (2004). For more on the studentized range distribution itself see, for example, Copenhaver and Holland (1988).

Scheffé's Method

The third multiple comparison procedure we consider is based on Scheffé's method. To apply Theorem 6.13 note that we have $\boldsymbol{\beta} = (\beta_1, \ldots, \beta_q)$, $\hat{\boldsymbol{\beta}} = (\bar{Y}_1, \ldots, \bar{Y}_q)$, $\hat{\sigma}^2 = MSE$ and $\mathbf{V} = \text{diag}(1/n_1, \ldots, 1/n_q)$. We will consider two versions of the problem. For any linear combination $\hat{\eta} = \mathbf{c}^T\hat{\boldsymbol{\beta}}$ the standard error is given by

$$S_{\hat{\eta}}^2 = \hat{\sigma}^2\mathbf{c}^T\mathbf{V}\mathbf{c} = \hat{\sigma}^2\sum_{j=1}^{q}\frac{c_j^2}{n_j}.$$

So the class of confidence intervals

$$CI = \hat{\eta} \pm \left(qF_{q,\nu;\alpha}\right)^{1/2} S_{\hat{\eta}}, \tag{6.69}$$

are simultaneously correct with probability $1 - \alpha$.

However, Equation (6.69) does not define what is conventionally refered to as Scheffé's method for this application. This relies on the notion of a contrast, which is a linear combination $\mathbf{c}^T\boldsymbol{\beta}$ for which $c_1 + \ldots + c_q = 0$. Because ANOVA models tend to be more concerned with variation of treatments means, analyses are appropriately confined to contrasts. The critcal value universally used in this case is $\left((q-1)F_{q-1,\nu;\alpha}\right)^{1/2}$, which differs from that used in Equation (6.69) in that the numerator degrees of freedom q is replaced by $q - 1$.

The key to understanding this apparent discrepancy is to note that the dimension of the space of all linear combinations is q, while the dimension of the space of all contrasts is $q - 1$. To see this write $\boldsymbol{\beta}^* = (\beta_1 - \beta_q, \beta_2 - \beta_q, \ldots, \beta_{q-1} - \beta_q)$, and $\hat{\boldsymbol{\beta}}^* = (\bar{Y}_1 - \bar{Y}_q, \bar{Y}_2 - \bar{Y}_q, \ldots, \bar{Y}_{q-1} - \bar{Y}_q)$. Then the space of all linear combinations of $\boldsymbol{\beta}^*$ is equivalent to the space of all contrasts of $\boldsymbol{\beta}$. To see this, consider $\eta^* = \mathbf{b}^T\boldsymbol{\beta}^*$ for any nonzero $\mathbf{b} \in \mathbb{R}^{q-1}$. Then

$$\eta^* = \sum_{j=1}^{q-1} b_j(\beta_j - \beta_q) = \sum_{j=1}^{q} c_j\beta_j$$

where $c_j = b_j$, $j = 1, \ldots, q-1$, $c_q = -\sum_{j=1}^{q-1} b_j$. Then η^* is a contrast of $\boldsymbol{\beta}$, since $\sum_{j=1}^{q} c_j = 0$. Conversely, suppose $\eta = \mathbf{c}^T\boldsymbol{\beta}$ is a contrast of $\boldsymbol{\beta}$. Then

$$\eta = \sum_{j=1}^{q} c_j\beta_j = \sum_{j=1}^{q} c_j\beta_j - \beta_q\sum_{j=1}^{q} c_j = \sum_{j=1}^{q-1} c_j(\beta_j - \beta_q)$$

which is a linear combination of $\boldsymbol{\beta}^*$. Thus, we may apply Theorem 6.13 to $\hat{\boldsymbol{\beta}}^*$. Of course, the standard error of a contrast is evaluated the same way as for a linear combination. So, as long as attention is confined to contrasts, we may replace Equation (6.69) with

$$CI = \hat{\eta} \pm \left((q-1)F_{q-1,n-k;\alpha}\right)^{1/2} S_{\hat{\eta}}. \tag{6.70}$$

From the values we have given, the reduction by a single degree of freedom can result in a notable decrease in the critical value used in the confidence intervals.

If we apply the Scheffé's method to the multiple comparison problem, we obtain multiple confidence intervals of the form

$$\bar{Y}_i - \bar{Y}_j \pm \left((q-1)F_{q-1,n-k;\alpha}\right)^{1/2}\sqrt{MSE\left(\frac{1}{n_i} + \frac{1}{n_j}\right)} \tag{6.71}$$

and like Tukey's range test we report all possible comparisons $\beta_i - \beta_j$ with $\alpha_{FWE} = \alpha$.

A Comparison of the Three Multiple Comparison Procedures

It is interesting to note that if we compare the forms of the confidence intervals of the three multiple comparison methods it can be seen that they differ only in the critical values, respectively, $t_{n-k;\alpha/(m2)}$, $q_{k,n-k;\alpha}/\sqrt{2}$ and $((q-1)F_{q-1,\nu;\alpha})^{1/2}$. This allows a direct comparison of the methods. For $k = 2$, we have $q_{k,n-k;\alpha}/\sqrt{2} = ((q-1)F_{q-1,\nu;\alpha})^{1/2}$ (the statistic of Equation (6.67) is equivalent to an F statistic, and for two means β_1, β_2 there is only one (normalized) contrast). Otherwise, we will have in general $q_{k,n-k;\alpha}/\sqrt{2} < ((q-1)F_{q-1,n-k;\alpha})^{1/2}$, so if the goal is to simultaneously estimate all comparisons $\beta_1 - \beta_j$, then Tukey's range test would be preferable to Scheffé's method. On the other hand, if we are interested in resolving contrasts of the form $H : \beta_1 - \beta_2 = \beta_3 - \beta_4$, then only Scheffé's method can be used.

The comparison between the BCP and Tukey's range test will depend on the number of comparisons m for the BCP. For example, if we have $\alpha = 0.05$, $n = 20$ and $k = 4$ treatments, there are up to 6 comparisons which can be made. Tukey's range test reports all comparisons, using critical value $q_{4,20-4;\alpha}/\sqrt{2} = 2.86$. In contrast, the critical values for the BCP will be for $m = 2, 3, 4, 5, 6$, $t_{20-4;\alpha/(m2)} = 2.47, 2.67, 2.81, 2.92, 3.01$. Thus, the BCP will be preferable for $m = 2, 3, 4$, and the Tukey's range test will be preferable for $m = 5, 6$.

6.11 Problems

Problem 6.1 Verify the reciprocal variance weighting rule of Example 6.1.

Problem 6.2 Suppose $\mathbf{X} \in \mathcal{M}_{n,q}$, with $n > q$. Prove that $\mathbf{X}^T\mathbf{X}$ is invertible *iff* $\text{rank}(\mathbf{X}) = q$. **HINT:** Make use of conventional matrix rank identities. Then a square matrix is invertible *iff* it is of full rank.

Problem 6.3 Suppose we are given a sample $\mathbf{X} = (X_1, \ldots, X_n)$ for which $E[X_i] = \theta \in \mathbb{R}$ and for which the covariance matrix of $\mathbf{X}$ is positive definite. Further suppose that the observations X_i can be partitioned into groups $A_1, \ldots, A_m$, and that the groups are mutually independent, while the observations within A_k have any covariance matrix Σ_k. Verify that the BLUE of θ (Definition 6.1) is a weighted linear combination of the BLUEs that would be obtained by considering the groups separately.

Problem 6.4 Suppose we are given a sample $\mathbf{X} = (X_1, \ldots, X_n)$ for which $E[X_i] = \theta \in \mathbb{R}$ and for which the covariance matrix Σ of $\mathbf{X}$ is positive definite. Prove that the variance of the BLUE of θ (Definition 6.1) is equal to $[\mathbf{1}_{n,1}^T \Sigma^{-1} \mathbf{1}_{n,1}]^{-1}$. How can we interpret this quantity?

Problem 6.5 Consider the linear regression model $\mathbf{Y} = \mathbf{X}\boldsymbol{\beta} + \boldsymbol{\epsilon}$, $\boldsymbol{\beta} \in \mathbb{R}^q$, for which $\Sigma_{\mathbf{Y}} = \sigma^2\mathbf{R}$. Let $\hat{\boldsymbol{\beta}}$ be the BLUE of $\boldsymbol{\beta}$, and let $\hat{\boldsymbol{\beta}}^*$ be the BLUE of $\boldsymbol{\beta}$ under the assumption that $\mathbf{R} = \mathbf{I}$ (the ordinary least squares estimate).

(a) Prove that $\hat{\boldsymbol{\beta}}^*$ is unbiased.
(b) Prove that $\text{var}[\hat{\beta}_j] \leq \text{var}[\hat{\beta}_j^*]$ for each $j = 1, \ldots, q$.

Problem 6.6 For a response vector $\mathbf{Y} \in \mathbb{R}^n$ of a linear model, let $\hat{\mathbf{Y}}_f$, $\hat{\mathbf{Y}}_r$ be two vectors of fitted values, with respective residual vectors $\mathbf{e}_f = \mathbf{Y} - \hat{\mathbf{Y}}_f$, $\mathbf{e}_r = \mathbf{Y} - \hat{\mathbf{Y}}_r$. Prove that if $\mathbf{e}_r^T\mathbf{e}_f = \mathbf{e}_f^T\mathbf{e}_f$, then the ANOVA decompositon of Theorem 6.4 holds.

Problem 6.7 Suppose under the conditions of Theorem 6.5 we wanted to test $H_o : \boldsymbol{\mu} \in \mathcal{V}_r$ against $H_a : \boldsymbol{\mu} \in \mathcal{V}_f$. Assume $\sigma^2 = 1$ is known. Both $SSR_{f,r}$ and SSE_r would then have a noncentral χ^2 distribution with the same noncentrality parameter $\Delta = (\boldsymbol{\mu} - \boldsymbol{\mu}_r)^T(\boldsymbol{\mu} - \boldsymbol{\mu}_r)$. Which would be the better choice as a test statistic?

Problem 6.8 Prove that if the constraints of Equation (6.37) hold, equations (6.39) and (6.42) are equivalent.

Problem 6.9 Verify the claim made in Example 6.11. Suppose we accept the constraint $\max_{i,j} |\beta_i - \beta_j| \geq m$, for parameter values $(\beta_1, \ldots, \beta_k)$. Then $\sum_{i=1}^k (\beta_i - \bar{\beta})^2$ is minimized by setting $\beta_1 = \beta_2 + m$, then $\beta_i = \bar{\beta}$, $i = 3, \ldots, k$.

Problem 6.10 Derive the likelihood ratio test (LRT) (Section 5.8) for the ANOVA hypothesis of Section 6.2.2. What is its relationship to the F test given in Theorem 6.7

Problem 6.11 An ANOVA model is analyzed based on data for 3 treatments, of which each have n_j observations. An observation from treatment $j = 1, \ldots, 3$ has distribution $N(\beta_j, \sigma^2)$, the variance being assumed constant. Observations are independent. The treatment means, standard deviations and sample sizes are given in a table below. The sum of squares (SS), mean sum of squares (MSS) and degrees of freedom for treatment and error sources of variation are given in the subsequent table

Treatment j	$\bar{X}_j$	S_j	n_j
1	99.94	12.61	10
2	126.58	13.02	12
3	100.98	17.57	9

Source of Variation	DF	SS	MSS
Treatment	2	5033.37	2516.69
Error	28	5764.66	

Using a Bonferroni multiple comparison procedure, with a family-wise error rate of $\alpha_{FWE} = 0.15$, can we conclude that β_2 is the maximum treatment mean? Repeat using the Tukey range test, and Scheffé's method for contrasts.

Problem 6.12 The monthly power consumption in kwh of samples of $k = 4$ brands of humidifier were monitored, with results given in the table below. Assume the sample for Brand i is from a normally distributed population with mean β_i and fixed variance σ^2.

	1	2	3	4	5	$\bar{X}_i$	S_i	n_i
Brand 1	24.85	22.08	25.91	28.74	21.99	24.71	2.83	5
Brand 2	19.00	14.43	23.73	15.74	23.10	19.20	4.20	5
Brand 3	17.14	13.37	18.64	15.31		16.12	2.28	4
Brand 4	14.38	17.28	14.65	9.13		13.86	3.41	4

(a) Construct an ANOVA table.

(b) Use an F test for null hypothesis $H_o : \beta_i = \beta_j$ for all i,j against $H_a : \beta_i \neq \beta_j$ for some i,j. Use significance level $\alpha = 0.05$. Construct a rejection region, and also give a P-value.

(c) Before the study was carried out, it was conjectured that Brand 1 had the highest mean power consumption. In the study, Brand 1 did have the highest sample mean. However, rejection of the null hypothesis of equal treatment means does not imply by itself that the observed sample mean ordering is the same as the true mean ordering. Construct multiple confidence intervals using an appropriate Bonferroni procedure to determine whether or not Brand 1 has the highest power consumption. Use a familywise error rate of $\alpha_{FWE} = 0.05$.

(d) Use the Tukey range test to construct simultaneous confidence intervals for all paired differences in mean, using $\alpha_{FWE} = 0.05$. Do you reach the same conclusion as in Part (c)?

Problem 6.13 Independent samples for $k = 3$ treatments are summarized in the table below. Assume sample j is from a normally distributed population with mean β_j and fixed variance σ^2.

	1	2	3	4	5	$\bar{X}_i$	S_i	n_i
Treatment 1	13.13	15.16	10.60	16.41	17.99	14.66	2.88	5
Treatment 2	9.04	6.94	9.30	9.00	10.69	8.99	1.34	5
Treatment 3	20.55	16.37	20.20	14.39	21.89	18.68	3.16	5

(a) Construct an ANOVA table.
(b) Use an F test for null hypothesis $H_o : \beta_i = \beta_j$ for all i, j. Use significance level $\alpha = 0.05$.
(c) Is it possible, using any of the three multiple comparison procedures considered in this chapter, to determine the maximum treatment mean with family-wise error rate $\alpha_{FWE} = 0.05$?

Problem 6.14 Prove Theorem 6.6. **HINT:** Make use of the block diagonal structure of $\mathbf{X}^T\mathbf{X}$.

Problem 6.15 Two variables Y and x are believed to have the following relationship: $Y = ax^b$ for two constants a, b. According to a certain conjecture, Y is proportional to the square root of x. In order to resolve this question paired observations $(Y_1, x_1), \ldots, (Y_n, x_n)$ are sampled, where $n = 51$. The simple linear regression model $\log(Y) = \beta_0 + \beta_1 \log(x) + \epsilon$ is fit, with the following output:

```
              Estimate Std. Error   t value     Pr(>|t|)
(Intercept) 2.3137646 0.02512828 92.078108 1.447613e-56
log(x)      0.4985705 0.05774591  8.633867 2.084971e-11
```

Formulate appropriate null and alternative hypotheses for this question in terms of the regression coefficients β_0 and/or β_1. Is there evidence at an $\alpha = 0.05$ significance level with which to reject the conjecture?

Problem 6.16 Consider the matrix representation of the multiple linear regression model $\mathbf{Y} = \mathbf{X}\boldsymbol{\beta} + \boldsymbol{\epsilon}$ where $\mathbf{Y}$ is an $n \times 1$ response vector, $\mathbf{X}$ is a $n \times q$ matrix, $\boldsymbol{\beta}$ is a $q \times 1$ vector of coefficients, and $\boldsymbol{\epsilon}$ is an $n \times 1$ vector of error terms. Let $\hat{\boldsymbol{\beta}}$ be the least squares estimate of $\boldsymbol{\beta}$ (we will assume that $\mathbf{X}^T\mathbf{X}$ is invertible). The vector of residuals is given by $\mathbf{e} = \mathbf{Y} - \hat{\mathbf{Y}}$, where $\hat{\mathbf{Y}} = \mathbf{X}\hat{\boldsymbol{\beta}}$ is the vector of fitted values.

(a) Prove that $\mathbf{e}$ and $\hat{\mathbf{Y}}$ are orthogonal, that is, $\hat{\mathbf{Y}}^T\mathbf{e} = 0$.
(b) Prove that $\mathbf{X}^T\mathbf{e} = \mathbf{0}$ where $\mathbf{0}$ is a $q \times 1$ column vector of zeros.
(c) Prove that if the model contains an intercept (that is, $\mathbf{X}$ contains a column of 1's), then the sum of the residuals is zero.

Problem 6.17 Consider the case of linear regression through the origin, expressed as the model $Y_i = \beta x_i + \epsilon_i$, $i = 1, \ldots, n$ where $\epsilon_i \sim N(0, \sigma^2)$ are *iid* error terms, and $x_1, \ldots, x_n$ are fixed predictor terms. Write explicitly the error sum of squares SSE for this model, where $\hat{\beta}$ is an estimate of β. After verifying that SSE is a second order polynomial in $\hat{\beta}$, determine the least squares estimate of β directly in terms of the observation (Y_i, x_i), $i = 1, \ldots, n$, and derive its variance.

Problem 6.18 Suppose we are given the simple linear regression model $Y = \beta_0 + \beta_1 X + \epsilon$, where $\epsilon \sim N(0, \sigma_\epsilon^2)$. Suppose X is then interpreted itself as a random outcome with distribution $X \sim N(0, 1)$, which is independent of ϵ. What is the relationship between the correlation coefficient ρ of (X, Y) and the parameters of the regression model?

Problem 6.19 For this problem we make use of the `Animals` data set from the `MASS R` package (Venables and Ripley, 2013). Brain size is usually positively associated with body size in animals. Snell's equation of simple allometry is: $E = CS^r$, where E is brain weight, S is body weight, C is the *cephalization factor* and r is the exponential constant. Suppose we are given paired observations of brain and body weight (E_i, S_i), in grams, for a sample of distinct animals indexed by $i = 1, \ldots, n$. The problem is to use this data to estimate r and C.

Consider the transformed data $(Y_i, x_i) = (\log E_i, \log S_i)$, $i = 1, \ldots, n$. Then suppose $n = 28$, and a least squares fit of the model $Y_i = \beta_0 + \beta_1 x_i + \epsilon_i$, where ϵ_i are *iid* normally distributed error terms, yields the following coefficient table:

```
              Estimate Std. Error    t value     Pr(>|t|)
(Intercept) -1.0455742 1.06023599 -0.9861712 3.331381e-01
x            0.5951936 0.09380328  6.3451262 1.016872e-06
```

(a) What is the estimated value of exponential constant r?
(b) Test the null hypothesis $H_o : r = 2/3$ against the two-sided alternative $H_a : r \neq 2/3$ using significance level $\alpha = 0.05$.
(c) What is the estimated value of cephalization factor C?
(d) Construct a 95% confidence interval for the cephalization factor C.

Problem 6.20 Suppose we are given a simple linear regression model with *iid* normal error terms. In Example 6.7 a confidence interval for $\mu_x = \beta_0 + \beta_1 x$ of the form $\hat{\mu}_x \pm t_{n-2;\alpha/2} S_{\hat{\mu}_x}$ was given. Suppose Y_x is a *future* observation from the same model when the independant variable has value x. Construct a prediction interval (L, U) for Y_x with the property that $P(L < Y_x < U) = 1 - \alpha$. This should be based on the same estimates that would be avaliable for the construction of the confidence interval for μ_x.

Problem 6.21 Consider linear regression through the origin (Problem 6.17). Let $\mathbf{Y}^* = (Y_{j_1}, \ldots, Y_{j_m})$ be the subvector of elements Y_i of $\mathbf{Y} = (Y_1, \ldots, Y_n)$ for which $x_i \neq 0$. Show that $\mathbf{Y}^*$ is sufficient for β but not for σ^2.

Problem 6.22 We are given a simple linear regression model $Y = \beta_0 + \beta_1 x$, based on n pairs of independent and dependent variables (Y_i, x_i). Let $\hat{\beta}_i$ be the least squares estimates of β_i, $i = 0, 1$. Suppose we have some choice of the independent variables x_i, subject to the following constraints:

(a) Sample size n is even.
(b) The mean $\bar{x} = n^{-1} \sum_{i=1}^n x_i$ is constrained to equal some fixed number x^*.
(c) We impose the bound $x^* + M \geq x_i \geq x^* - M$, $i = 1, \ldots, n$, for some fixed number $M > 0$.

Show that the variance of $\hat{\beta}_1$ is minimized by setting $x_i = x^* - M$ for half the sample, and $x_i = x^* + M$ for the other half (the selection of independent variables for the purpose of optimizing the efficiency of an inference is refered to as the *design problem*).

Problem 6.23 Show that when the error variables of the linear regression model of Equation (6.44) are *iid* with $\epsilon_1 \sim N(0, \sigma^2)$, then the least squares estimate of $\boldsymbol{\beta}$ is equal to the maximum likelihood estimate.

Problem 6.24 Consider the multiple linear regression model with q predictor variables, $\mathbf{Y} = \mathbf{X}\boldsymbol{\beta} + \boldsymbol{\epsilon}$, where $\boldsymbol{\epsilon}$ is an *iid* sample from $N(0, \sigma^2)$. Assume the columns of $\mathbf{X}$ are linearly independent.

(a) Show that the $q \times 1$ column vector $\mathbf{S} = \mathbf{X}^T \mathbf{Y}$ is a complete sufficient statistic for $\boldsymbol{\beta}$.

(b) Let $\mathbf{A}$ be any $r \times n$ matrix for which $\mathbf{AX} = \mathbf{0}$, where $\mathbf{0}$ is the $r \times p$ matrix of zero elements. Define $\mathbf{V} = \mathbf{AY}$. Show that $\mathbf{V}$ and $\mathbf{S}$ are independent random vectors. **HINT:** Use Basu's theorem (Theorem 3.23).

Problem 6.25 Consider the matrix representation of the multiple linear regression model

$$\mathbf{Y} = \mathbf{X}\boldsymbol{\beta} + \boldsymbol{\epsilon} \tag{6.72}$$

where $\mathbf{Y}$ is an $n \times 1$ response vector, $\mathbf{X}$ is a $n \times q$ matrix, $\boldsymbol{\beta}$ is a $q \times 1$ vector of coefficients, and $\boldsymbol{\epsilon}$ is an $n \times 1$ vector of *iid* error terms with $\epsilon_1 \sim N(0, \sigma^2)$.

(a) Suppose we are given paired observations of the form $(Y_1, u_1), \ldots, (Y_n, u_n)$, where each $u_i \in \{1, 2, 3\}$ is one of three values, and $Y_i \sim N(\mu_k, \sigma^2)$ if $u_i = k$. Assume that the responses Y_i are independent, and that the variance σ^2 is the same for all responses. We decide to express this model as a linear regression model by defining three predictors $\mathbf{x}_1, \mathbf{x}_2, \mathbf{x}_3$, associated with the three outcomes of u_i, using indicator variables, that is,

$$\mathbf{x}_{1i} = I\{u_i = 1\}, \quad \mathbf{x}_{2i} = I\{u_i = 2\}, \quad \mathbf{x}_{3i} = I\{u_i = 3\},$$

for $i = 1, \ldots, n$. Then suppose we attempt to fit the model

$$\mathbf{Y} = \beta_0 + \beta_1 \mathbf{x}_1 + \beta_2 \mathbf{x}_2 + \beta_3 \mathbf{x}_3 + \boldsymbol{\epsilon}, \tag{6.73}$$

We may express this model in the matrix form of Equation (6.72). Derive the matrix $\mathbf{X}^T\mathbf{X}$. Is this matrix invertible? **HINT:** Let n_k be the number of times $x_i = k$, for each $k = 1, 2, 3$.

(b) Show that if any of the four terms associated with coefficients $\beta_0, \ldots, \beta_3$ is deleted from Equation (6.73), then the resulting matrix $\mathbf{X}^T\mathbf{X}$ will be invertible.

(c) In Part (b), four linear regression models are obtained by deleting one of the four terms associated with the coefficients. Show that the least squares fit of each of these will give the same fitted values $\hat{\mathbf{Y}}$, and are therefore equivalent.

Problem 6.26 Consider the matrix representation of the multiple linear regression model $\mathbf{Y} = \mathbf{X}\boldsymbol{\beta} + \boldsymbol{\epsilon}$ where $\mathbf{Y}$ is an $n \times 1$ response vector, $\mathbf{X}$ is a full rank $n \times q$ matrix, $n > q$, $\boldsymbol{\beta}$ is a $q \times 1$ vector of coefficients, and $\boldsymbol{\epsilon}$ is an $n \times 1$ vector of *iid* error terms with $\epsilon_1 \sim N(0, \sigma^2)$. Let $\hat{\beta}_i$ be the least squares estimate of β_i, $i = 1, \ldots, q$. There may be an advantage with respect to the interpretability of the regression coefficients if their estimates are uncorrelated (consider Example 6.16). In this case, the contribution of each predictor to the model can be assessed independently of the other predictors.

(a) Let $\hat{\boldsymbol{\beta}} = [\hat{\beta}_1 \ldots \hat{\beta}_q]^T$ be the vector of least squares coefficient estimates of $\boldsymbol{\beta}$. Show that theses estimates are mutually uncorrelated if and only if $\sum_{i=1}^n x_{ij} x_{ik} = 0$ for each pair $j \neq k$, where $\{\mathbf{X}\}_{i,j} = x_{ij}$.

(b) Consider the simple regression model

$$Y_i = \beta_0 + \beta_1 x_i + \epsilon_i, \ \ i = 1, \ldots, n. \tag{6.74}$$

Suppose the independent variable is transformed to $x_i' = x_i - \bar{x}$, where $\bar{x} = n^{-1} \sum_i x_i$. Then, using the transformed independent variable, consider the alternative model

$$Y_i = \beta_0' + \beta_1' x_i' + \epsilon_i, \ \ i = 1, \ldots, n. \tag{6.75}$$

The values Y_i and ϵ_i are otherwise the same for both models (6.74) and (6.75). Show that the two models are equivalent in the sense that the fitted values $\hat{\mathbf{Y}}$ must be the same. How are the least squares estimates of the coefficients for the respective models related?

(c) Show that for model (6.75) the estimates of the coefficients β_0' and β_1' are uncorrelated.

Problem 6.27 Polynomial regression extends simple linear regression by constructing new predictor variables from powers of a single predictor variable x_i:

$$Y_i = \beta_0 + \beta_1 x_i + \beta_2 x_i^2 + \ldots + \beta_p x_i^p + \epsilon_i. \tag{6.76}$$

Then, we have

$$\mathbf{X} = \begin{bmatrix} 1 & x_1 & x_1^2 & \ldots & x_1^p \\ \vdots & \vdots & \vdots & \vdots & \vdots \\ 1 & x_n & x_n^2 & \ldots & x_n^p \end{bmatrix}$$

The problem with this approach is that $\mathbf{X}$ may possess significant collinearity (Example 6.16). One solution is to use as predictors the linear transformation

$$\mathbf{X}' = \mathbf{XA} = \begin{bmatrix} z_{10} & z_{11} & z_{12} & \ldots & z_{1p} \\ \vdots & \vdots & \vdots & \vdots & \vdots \\ z_{n0} & z_{n1} & z_{n2} & \ldots & z_{np} \end{bmatrix}$$

where $\mathbf{A}$ is a $(p+1) \times (p+1)$ matrix of coefficients. Then $\mathbf{A}$ can be chosen so that $\mathbf{X}'$ is orthogonal. Note that the first column of $\mathbf{X}'$ has been relabelled $j = 0$. The coefficient matrix $\mathbf{A}$ is usually upper triangular, so the transformed predictor variables become

$$\begin{aligned} z_{i0} &= a_{00} \\ z_{i1} &= a_{01} + a_{11} x_i \\ z_{i2} &= a_{02} + a_{12} x_i + a_{22} x_i^2 \\ &\vdots \\ z_{ip} &= a_{0p} + a_{1p} x_i + a_{2p} x_i^2 + \ldots + a_{pp} x_i^p \end{aligned}$$

noting that the first row and column of $\mathbf{A}$ are here labeled $i = j = 0$. The model is now

$$Y_i = \beta_0' z_{i0} + \beta_1' z_{i1} + \beta_2' z_{i2} + \ldots + \beta_p' z_{ip} + \epsilon_i. \tag{6.77}$$

(a) Show that models (6.76) and (6.77) are equivalent in the sense that the fitted values $\hat{\mathbf{Y}}$ must be the same. How are the least squares estimates of the coefficients for the respective models related?

(b) Suppose we wish to choose $\mathbf{A}$ so that the components of $\hat{\boldsymbol{\beta}} = [\hat{\beta}_0 \ldots \hat{\beta}_p]^T$ are uncorrelated. Verify that this is achieved by the choice

$$\mathbf{A} = \begin{bmatrix} 1 & -\bar{x} \\ 0 & 1 \end{bmatrix}$$

for $p = 1$, where $\bar{x} = n^{-1} \sum_i x_i$.

Problem 6.28 This problem refers to the polynomial regression model of Problem 6.27. The `R` function `poly` can be used to construct matrices of the form $\mathbf{X}$ or $\mathbf{X}'$ (note that the first column of one's is not included). Let `x` be a vector of length n representing a single predictor variable. Then `poly(x, 3, raw=TRUE)` produces a matrix of the form $\mathbf{X}$ (that is, the jth column is simply the jth powers of `x`). On the other hand, `poly(x, 3, raw=FALSE)` produces a matrix of the form $\mathbf{X}'$, where the coefficients are chosen so that the columns are orthonormal (i.e. $[\mathbf{X}']^T \mathbf{X}' = \mathbf{I}$, where $\mathbf{I}$ is the $p \times p$ identity matrix).

Generate simulated data for a linear model with the following code. Here `x` is the predictor variable and `y` is the response.

```
> set.seed(12345)
> x = (1:100)/100
> y = rnorm(100,mean=1+5*x^2,sd=1)
> plot(x,y)
```

(a) Write the model explicitly, describing the error term precisely.

(b) Fit the model twice, using the methods given next:

```
> fit1 = lm(y~poly(x,3,raw=T))
> fit2 = lm(y~poly(x,3,raw=F))
```

For each fit report and interpret the `F-statistic` given by the `summary` method for each fit.

(c) Construct a scatter plot of the response variable `y` against predictor variable `x`. Superimpose the fitted values of both fits. Make sure you use distinct plotting characters and colors, so that the two sets of fitted values can be distinguished. For example, you could use the `matplot` function with options `pch=c(2,3)`, `col=c('green','red')` and `add=T`.

(d) Each fit summary reports P-values for rejecting the null hypothesis $H_0 : \beta_j = 0$ for each coefficient. Compare the P-values for the two fits. Do either summaries permit the (correct) conclusion that the mean of the response Y_i is a second order polynomial in the predictor variable x_i? If for either fit the P-values for the coefficients (other than the intercept) are all large (i.e. >0.05) does this contradict the conclusion of Part (b)?

Problem 6.29 Consider the matrix representation of the multiple linear regression model $\mathbf{Y} = \mathbf{X}\boldsymbol{\beta} + \boldsymbol{\epsilon}$ where $\mathbf{Y}$ is an $n \times 1$ response vector, $\mathbf{X}$ is a $n \times q$ matrix, $\boldsymbol{\beta}$ is a $q \times 1$ vector of coefficients, and $\boldsymbol{\epsilon}$ is an $n \times 1$ vector of error terms. The least squares solution is expressed using the coefficient vector $\boldsymbol{\beta}$ which minimizes the error sum of squares $SSE[\boldsymbol{\beta}] = (\mathbf{Y} - \mathbf{X}\boldsymbol{\beta})^T(\mathbf{Y} - \mathbf{X}\boldsymbol{\beta})$.

(a) By setting each partial derivative $\partial SSE[\boldsymbol{\beta}]/\partial \beta_j$ to zero, $j = 1, \ldots, q$, verify that the least squares solution is $\hat{\boldsymbol{\beta}} = (\mathbf{X}^T\mathbf{X})^{-1}\mathbf{X}^T\mathbf{Y}$.

(b) Next, suppose we wish to minimize $SSE[\boldsymbol{\beta}]$ subject to $m < q$ linear constraints on $\boldsymbol{\beta}$, expressed in matrix form as

$$\mathbf{C}\boldsymbol{\beta} - \mathbf{d} = \mathbf{0} \tag{6.78}$$

where $\mathbf{C}$ is an $m \times q$ matrix and $\mathbf{d}$ is an $m \times 1$ column vector and $\mathbf{0}$ is an $m \times 1$ column vector of zeros. This is equivalent to the m linear constraints

$$\begin{aligned} c_{11}\beta_1 + \ldots + c_{1q}\beta_q &= d_1 \\ &\vdots \\ c_{m1}\beta_1 + \ldots + c_{mq}\beta_q &= d_m, \end{aligned}$$

where c_{ij} and d_k are the respective elements of $\mathbf{C}$ and $\mathbf{d}$. Using the Lagrange multiplier method, we can calculate the constrained least squaresconstrained solution $\hat{\boldsymbol{\beta}}_c$ to this problem by minimizing

$$\Lambda = SSE[\boldsymbol{\beta}] + \bar{\lambda}^T(\mathbf{d} - \mathbf{C}\boldsymbol{\beta})$$

with respect to $(\boldsymbol{\beta}, \bar{\lambda})$, where $\bar{\lambda}^T = (\lambda_1, \ldots, \lambda_m)$, while applying constraint (6.78). Verify that

$$\hat{\boldsymbol{\beta}}_c = \hat{\boldsymbol{\beta}}_u + (\mathbf{X}^T\mathbf{X})^{-1}\mathbf{C}^T\left[\mathbf{C}(\mathbf{X}^T\mathbf{X})^{-1}\mathbf{C}^T\right]^{-1}(\mathbf{d} - \mathbf{C}\hat{\boldsymbol{\beta}}_u),$$

where $\hat{\boldsymbol{\beta}}_u = (\mathbf{X}^T\mathbf{X})^{-1}\mathbf{X}^T\mathbf{Y}$ is the unconstrained least squares solution.

Problem 6.30 We are given a simple linear regression model $Y = \beta_0 + \beta_1 x + \epsilon$.

(a) Let $\hat{\beta}_i$ be the least squares estimates of β_i, $i = 0, 1$. Suppose a constant c is added to each response and the model refit. What will be the new least squares estimates of β_i, expressed in terms of the old estimates? Verify your answer analytically.

(b) The coefficient β_0 is referred to as the *intercept term* (where the Y-axis is intercepted by the regression line). It can be interpreted as a summary of the vertical location of the response Y, since the effect of changing the constant c of part (a) is directly observable in β_0. Of course, $\beta_0 = \mu_0$, where $\mu_x = \beta_0 + \beta_1 x$, so we may construct a new intercept μ_x at any x for the same purpose. The least squares estimate will be

$$\hat{\mu}_x = \hat{\beta}_0 + \hat{\beta}_1 x.$$

What analytical criterion can be used to select x, and what would be the resulting optimal choice? **HINT:** Make use of the development of Example 6.7.

Problem 6.31 The least squares solution can be obtained by the transformation $\hat{\mathbf{Y}} = \mathbf{H}_{LS}\mathbf{Y}$ of responses $\mathbf{Y}$, representing a projection onto a vector subspace. Then $\mathbf{H}_{LS}$ is commonly referred to as the "hat matrix". This type of matrix appears in many modeling techniques, both parametric and nonparametric.

(a) The following three modeling techniques yield various forms of fitted vectors $\hat{\mathbf{Y}} = \mathbf{H}'\mathbf{Y}$ as linear transformations of response vector $\mathbf{Y}$. In each case, describe precisely $\mathbf{H}'$, and give its trace.

   (i) We have $\hat{\mathbf{Y}} = (\bar{\mathbf{Y}}, \ldots, \bar{\mathbf{Y}}) \in \mathbb{R}^n$, where $\bar{\mathbf{Y}}$ is the sample mean of the elements of $\mathbf{Y}$.

   (ii) Here, the elements of $\mathbf{Y}$ are assumed sorted in some sequence, either by a time index or some other predictor. We take *moving average*

$$\begin{aligned}
\hat{\mathbf{Y}}_1 &= \frac{\mathbf{Y}_1 + \mathbf{Y}_2}{2} \\
\hat{\mathbf{Y}}_i &= \frac{\mathbf{Y}_{i-1} + \mathbf{Y}_i + \mathbf{Y}_{i+1}}{3}, \quad i = 2, 3, \ldots, n-1, \\
\hat{\mathbf{Y}}_n &= \frac{\mathbf{Y}_{n-1} + \mathbf{Y}_n}{2}
\end{aligned}$$

   (iii) We can define a *saturated model* as one which fits the original reponse vector exactly, that is, $\hat{\mathbf{Y}} = \mathbf{Y}$.

(b) The quantity trace($\mathbf{H}$) is sometimes referred to as the *effective degrees of freedom.* In linear regression it equals the model degrees of freedom (which is equal to the number of regression coefficients). But it has a similar interpretation for other forms of models, and can be taken as an index of model complexity. Order the three "hat matrices" considered in Part (a), along with $\mathbf{H}_{LS}$ (for simple linear regression), by effective degrees of freedom, and describe briefly how this ordering relates to the complexity of the models considered.

Problem 6.32 Suppose we are given a linear regression model

$$Y_i = \beta_0 + \beta_1 x_i + \beta_2 w_i + \beta_3 x_i w_i + \epsilon_i, \quad i = 1, \ldots, n$$

where the error terms are *iid* with $\epsilon_i \sim N(0, \sigma^2)$, σ^2 is unknown. We assume that w_i is an indicator variable, which equals 0 or 1 if the ith observation is from class A or B, respectively.

(a) Write the model as two distinct simple regression models with independent variable x_i for classes A and B

(b) Suppose the model is fit, and we are given values $n = 100$, $MSE = 29.56$, $(\hat{\beta}_0, \hat{\beta}_1, \hat{\beta}_2, \hat{\beta}_3) = (5.833, -3.041, 0.821, 3.101)$. Assume the range of x_1 is $[0, 5]$ and that, we are given the matrix

$$\mathbf{X}^T\mathbf{X} = \begin{bmatrix} 100 & 248.9658 & 153 & 375.9784 \\ 248.9658 & 802.1505 & 375.9784 & 1199.2293 \\ 153 & 375.9784 & 259.0000 & 630.0038 \\ 375.9784 & 1199.2293 & 630.0038 & 1993.3870 \end{bmatrix}$$

Write a computer program which plots simultaneous level $1 - \alpha = 0.95$ confidence bands for the two simple linear regression models of Part (a).

Problem 6.33 Consider the matrix representation of the multiple linear regression model $\mathbf{Y} = \mathbf{X}\boldsymbol{\beta} + \boldsymbol{\epsilon}$, where $\mathbf{Y}$ is an $n \times 1$ response vector, $\mathbf{X}$ is a $n \times q$ matrix, $\boldsymbol{\beta}$ is a $q \times 1$ vector of coefficients, and $\boldsymbol{\epsilon}$ is an $n \times 1$ vector of error terms. Suppose we have matrix $\mathbf{X}^* \in \mathcal{M}_{m,q}$, $m \geq q$, with $\text{rank}(\mathbf{X}^*) = q$, and that the model can be conformally partitioned as

$$\mathbf{Y} = \begin{bmatrix} \mathbf{Y}_1 \\ \mathbf{Y}_2 \\ \vdots \\ \mathbf{Y}_K \end{bmatrix} = \begin{bmatrix} \mathbf{X}^* \\ \mathbf{X}^* \\ \vdots \\ \mathbf{X}^* \end{bmatrix} \boldsymbol{\beta} + \begin{bmatrix} \boldsymbol{\epsilon}_1 \\ \boldsymbol{\epsilon}_2 \\ \vdots \\ \boldsymbol{\epsilon}_K \end{bmatrix} = \mathbf{X}\boldsymbol{\beta} + \boldsymbol{\epsilon},$$

where $K = n/m$ (assumed to be an integer).

(a) Verify that for this model $\text{rank}(\mathbf{X}) = q$.

(b) Let $\hat{\boldsymbol{\beta}}$ be the least squares estimate of the model $\mathbf{Y} = \mathbf{X}\boldsymbol{\beta} + \boldsymbol{\epsilon}$, and let $\hat{\boldsymbol{\beta}}_j$ be the least squares estimate of the model $\mathbf{Y}_j = \mathbf{X}^*\boldsymbol{\beta} + \boldsymbol{\epsilon}_j$, $j = 1, \ldots, K$. Prove that $\hat{\boldsymbol{\beta}}$ is equal to the element-wise sample averages of the estimates $\hat{\boldsymbol{\beta}}_j$.

Problem 6.34 We wish to fit a model of the form $Y_i = g(x_i) + \epsilon_i$, $i = 1, \ldots, n$ where $\epsilon_i \sim N(0, \sigma^2)$ are independent error terms, and $x_i \in [0, 10]$ is a predictor variable. We will assume that $g(x)$ is a step function with two discontinuities at $\xi = 2, 5$. One way to do this is to use the basis functions $b_1(x) = I\{x \leq 2\}$, $b_2(x) = I\{x \in (2, 5]\}$, $b_3(x) = I\{x > 5\}$. Then set $g(x) = \beta_1 b_1(x) + \beta_2 b_2(x) + \beta_3 b_3(x)$ and use least squares regression to estimate β_1, β_2 and β_3 (no intercept term will be added here).

Alternatively, we could use basis functions $h_1(x) = 1$, $h_2(x) = I\{x > 2\}$, $h_3(x) = I\{x > 5\}$, then set $g(x) = \beta_1^* h_1(x) + \beta_2^* h_2(x) + \beta_3^* h_3(x)$ and use least squares regression to estimate β_1^*, β_2^* and β_3^* (again, no intercept term will be added here).

Then suppose, using basis functions $b_1(x), b_2(x), b_3(x)$, least squares estimates $\hat{\beta}_1, \hat{\beta}_2$ and $\hat{\beta}_3$ of coefficients β_1, β_2 and β_3 are obtained. Derive the least squares estimates of β_1^*, β_2^* and β_3^* (using basis functions $h_1(x), h_2(x), h_3(x)$) as a function of $\hat{\beta}_1, \hat{\beta}_2$ and $\hat{\beta}_3$.

Problem 6.35 The standard method for estimating coefficients β_0, β_1 in the simple linear regression model $Y_i = \beta_0 + \beta_1 x_i + \epsilon_i$, $i = 1, \ldots, n$, is to first define the error sum of squares $SSE[\beta_0, \beta_1] = \sum_{i=1}^n (Y_i - \beta_0 - \beta_1 x_i)^2$ as a function of β_0, β_1, then set the least squares estimates $\hat{\beta}_0, \hat{\beta}_1$ to be those values which minimize $SSE[\beta_0, \beta_1]$, that is, $SSE[\hat{\beta}_0, \hat{\beta}_1] = \min_{\beta_0, \beta_1} SSE[\beta_0, \beta_1]$. In ridge regression (as applied to simple linear regression), a penalty term is added to the error sum of squares:

$$\Lambda[\beta_0, \beta_1] = SSE[\beta_0, \beta_1] + \lambda \sum_{i=0,1} \beta_i^2,$$

where $\lambda > 0$ is fixed, and the ridge regression coefficients $\hat{\beta}_0^*, \hat{\beta}_1^*$ are those values which minimize $\Lambda[\beta_0, \beta_1]$, that is, $\Lambda[\hat{\beta}_0^*, \hat{\beta}_1^*] = \min_{\beta_0,\beta_1} \Lambda[\beta_0, \beta_1]$. Assume, for convenience, that $\sum_{i=1}^n x_i = 0$. Derive $\hat{\beta}_0^*$ and $\hat{\beta}_1^*$ as a function of λ, n and summations of the form $Y = \sum_{i=1}^n Y_i$, $XX = \sum_{i=1}^n x_i^2$, $XY = \sum_{i=1}^n x_i Y_i$. See also Problem 6.36.

Problem 6.36 In this problem we consider further the ridge regression model of Problem 6.35. Recall that all eigenvalues of a symmetric matrix $\mathbf{A}$ are real, and $\mathbf{A}$ is invertible if and only if no eigenvalue equals 0. The *condition number* for a symmetric matrix may be taken to be $\kappa(\mathbf{A}) = |\lambda_{max}(\mathbf{A})|/|\lambda_{min}(\mathbf{A})| \geq 1$, where $\lambda_{min}(\mathbf{A}), \lambda_{max}(\mathbf{A})$ are the minimum and maximum eigenvalues in magnitude (alternative definitions exist). If $\kappa(\mathbf{A}) < \infty$, then $\mathbf{A}$ is invertible. However, if there exist eigenvalues of $\mathbf{A}$ which are close to zero (equivalently, $\kappa(\mathbf{A})$ is relatively large), the matrix is referred to as *ill-conditioned*, which may lead to problems for any computation or estimation procedure which relies on the inverse $\mathbf{A}^{-1}$.

In the context of multiple linear regression, multicollinearity occurs when one column of $\mathbf{X}$ is close to a linear combination of other columns In this case $\mathbf{X}^T\mathbf{X}$ will be ill-conditioned, even if it is strictly invertible. This will result in very large variances for at least some components of $\hat{\boldsymbol{\beta}}$ as well as potential inaccuracies in its computation. Such problems can be controlled using regularization, which refers to a process of modifying the objective function in order to stabilize the optimization problem. Ridge regression is one such method, in which the estimates of $\boldsymbol{\beta}$ are those which minimize the objective function

$$\Lambda[\boldsymbol{\beta}] = SSE[\boldsymbol{\beta}] + \lambda \sum_{j=1}^{q} \beta_j^2$$

for a fixed constant $\lambda \geq 0$ (here, $\lambda = 0$ gives the least squares estimate).

(a) By setting each partial derivative $\partial\Lambda[\boldsymbol{\beta}]/\partial\beta_j$ to zero, $j = 1, \ldots, q$, derive a system of q estimating equations

$$\Psi_\lambda(\mathbf{Y}, \boldsymbol{\beta}) = \mathbf{0}$$

which yield the stationary points of $\Lambda[\boldsymbol{\beta}]$ as solutions.

(b) Verify that the Hessian matrix Λ'' of $\Lambda[\boldsymbol{\beta}]$ is positive definite, so that $\Lambda[\boldsymbol{\beta}]$ is strictly convex (Section C.7).

(c) Show that the condition number of Λ'' is given by

$$\kappa(\Lambda'') = \frac{\lambda_{max}(\mathbf{X}^T\mathbf{X}) + \lambda}{\lambda_{min}(\mathbf{X}^T\mathbf{X}) + \lambda},$$

which can be made arbitrarily close to one by making λ large enough.

(d) Derive a close form solution to the ridge regression estimates $\hat{\boldsymbol{\beta}}_\lambda$ of $\boldsymbol{\beta}$. What happens to $\hat{\boldsymbol{\beta}}_\lambda$ as $\lambda \to \infty$?

Problem 6.37 Suppose we have linear models

$$\mathbf{Y} = \mathbf{X}\boldsymbol{\beta} + \boldsymbol{\epsilon}, \quad \mathbf{Y} = \mathbf{X}'\boldsymbol{\beta}' + \boldsymbol{\epsilon},$$

where $\mathbf{y}$ and $\boldsymbol{\epsilon}$ are common to both models, $\mathbf{X}$ is an $n \times q$ matrix of linearly independent columns, and $\mathbf{X}' = \mathbf{X}\mathbf{A}$ for some invertible $q \times q$ matrix $\mathbf{A}$. Then suppose $\hat{\boldsymbol{\beta}}$, $\hat{\boldsymbol{\beta}}'$ are the respective least squares estimates of the coefficients, and denote the fitted values $\hat{\mathbf{Y}} = \mathbf{X}\hat{\boldsymbol{\beta}}$ and $\hat{\mathbf{Y}}' = \mathbf{X}'\hat{\boldsymbol{\beta}}'$.

(a) Prove that the two fitted models are equivalent in the sense that $\hat{\mathbf{Y}} = \hat{\mathbf{Y}}'$.

(b) Prove that the relation $\hat{\boldsymbol{\beta}}' = \mathbf{A}^{-1}\hat{\boldsymbol{\beta}}$ always holds

(c) Suppose we are given the design matrices related by $\mathbf{X}' = \mathbf{X}\mathbf{A}$. Show that we may deduce equation $\mathbf{A} = \left(\mathbf{X}^T\mathbf{X}\right)^{-1}\mathbf{X}^T\mathbf{X}'$ to recover the original transformation.

(d) We wish to fit a polynomial regression model

$$Y_i = \beta_0 + \beta_1 x_i + \beta_2 x_i^2 + \ldots + \beta_q x_i^q + \epsilon_i, \ i = 1, \ldots, n.$$

After some consideration, we use a function within our computing enviroment to linearly transform our original design matrix $\mathbf{X}$ to $\mathbf{X}'$. The method probably uses some system of orthogonal polynomials. We obtain least squares estimates $\hat{\boldsymbol{\beta}}'$ using the matrix $\mathbf{X}'$, but we would like estimates of the original coefficients. Apart from studying the transformation method used to produce $\mathbf{X}'$, how can this be accomplished?

Problem 6.38 Prove Theorem 6.10. **HINT:** Note that $\mathbf{C}_q\mathbf{C}_q^{-1} = \mathbf{I}_q$, so that the inner product of the jth row vector of $\mathbf{C}_q$ and the kth column vector of $\mathbf{C}_q^{-1}$ is zero whenever $j \neq k$.

7

Decision Theory

7.1 Introduction

Much of the theory of statistical inference can be formulated as an optimization problem. If we are to choose the "best estimator" we must define what we mean by "best". Of course, we usually do this by devising some numerical criterion. If the criterion is a single number, the optimization problem is well-defined. However, it will usually be the case for parametric inference that this criterion (which we will eventually call "risk") is a function of the unknown parameter $\theta \in \Theta$. It may then be the case that when we compare two estimators δ_1 and δ_2, we find that δ_1 has smaller risk for $\theta' \in \Theta$, but larger risk for $\theta'' \in \Theta$. Do we then have any basis for choosing between δ_1 and δ_2? Section 7.2 serves as an introduction to the problem, which is the concern of much this chapter. The concepts of admissibility and the "favored hypothesis" are introduced here. It is worth briefly noting at this point that an estimator is "inadmissible" if a better one exists. Surprisingly, a rather large number of estimators that are commonly used are inadmissible, so the topic is not a trivial one (the sample variance S^2 is only one example, Example 7.7).

The problem of prediction is in some ways simpler than the problem of parametric inference. Suppose we are given a random variable $Y \in \mathcal{Y}$, and a more general random observation $\tilde{X} \in \mathcal{X}$. Unlike the parametric inference problem, we know exactly the joint distribution of $\tilde{X}$ and Y. If we observe $\tilde{X}$ but not Y, the prediction problem is to construct a predictor $\delta(\tilde{X})$ which is intended to be close to Y. This problem is introduced in Section 7.3. Since there is no unknown parameter, we are able to present a remarkably general method for determining the optimal predictor $\delta(\tilde{X})$, in the form of the optimal prediction theorem (Theorem 7.1). The method is certainly of interest on its own terms (Example 7.3, Section 7.3.1), but our interest in Theorem 7.1 has more to do with the fact that many decision problems involving unknown parameters can be formulated so as to use essentially the same argument.

The decision problem for parametric inference is introduced in Section 7.4. The elements of a decision problem, such as loss, risk, action space and decision rule are introduced, in the context of the parametric family. The concept of admissibility, introduced by example in Section 7.2, is formally defined.

Probably the most commonly used risk in statistical methdology is the MSE (Section 7.2), which is induced by squared error loss. In much of decision theory the form of the loss or risk is left open, and there are many reasons, both practical and theoretical, to consider risk other than MSE. The various forms of loss and risk commonly used are described in Section 7.5. This section also provides important background for Chapter 9.

In Section 7.6 we consider the problem of determining uniformly minimum risk estimators, which is applied to location and scale parameter models. As will be made clear in Section 7.2 a global uniformly minimum risk estimator will not exist. Therefore, in order to use this approach, the class of estimators must first be restricted. In fact, this approach is quite prominent in the theory of inference. In Chapter 8 it will be shown how to select a minimum risk estimator from the class of unbiased estimators (uniformly minimum

DOI: 10.1201/9781003049340-7

variance unbiased (UMVU) estimators), while in Chapter 9 we introduce the theory of minimum risk equivariant (MRE) estimators. Section 7.6 will serve as something of a preview to Sections 9.2 and 9.3, and it is hoped that the simplicity of the approach will clarify the more mathematically detailed development in Chapter 9.

In Section 7.7 some methods of characterizing admissibility are presented. We present here one of the most important theorems of statistical inference theory, the Rao-Blackwell theorem (Theorem 7.11). This theorem will be stated in an alternative form later as Theorem 8.5. The importance of this theorem to decision theory is as the simple statement that any estimator which is not a mapping of a minimal sufficient statistic is inadmissible with respect to a strictly convex loss function. Theorem 7.15 gives a simple explanation as to why shrinkage estimators, introduced in Section 4.7, play an important role in statistical decision theory.

Section 7.8 introduces Karlin's theorem (Karlin, 1958), which provides sufficient conditions for admissibility for the single parameter exponential family model with respect to MSE risk. This will be applied to a number of important cases. The theorem allows us to confirm that the sample mean from an *iid* normal distribution $N(\mu, \sigma^2)$ is an admissible estimator of μ with respect to MSE risk (Example 7.10). In contrast, the MLE of a scale parameter from a $gamma(\mu, \alpha)$ distribution is inadmissible, while some, but not all, shrinkage estimates of the same estimand are admissible (Example 7.11).

The final two sections of this chapter deal with globally optimal decision rules. While much of statistical theory is concerned with either MLEs or uniformly minimum risk estimators, it is possible to define criteria which permit the definition of procedures which are optimal among *all* alternatives. This involves the construction of a functional which maps a risk function to a single real number, so that all procedures can be ranked without ambiguity.

We consider two such approaches. The Bayesian model was introduced in Sections 3.10, 3.11 and 4.8. Our discussion of Bayesian models in this volume concludes in this chapter, with the introduction of Bayes risk (Sections 7.9-7.10), which is an average risk weighted by a prior distribution. A Bayes procedure is then a decision rule with minimizes Bayes risk. In Section 7.10 we introduce the minimax decision rule, which minimizes the maximum risk over the parameter space. We then show how the Bayes and minimax decision rules yield a connection between admissibility and optimality.

7.2 Ranking Estimators by MSE

If $\delta(\tilde{X})$ is an estimator for $\theta \in \Theta$, the mean squared error is defined as

$$\text{MSE}_\theta[\delta] = E_\theta[(\delta(\tilde{X}) - \theta)^2] = \text{var}_\theta[\delta] + \text{bias}_\theta[\delta]^2,$$

and serves as an error index for δ. The root mean squared error is then $\text{RMSE}_\theta[\delta] = \sqrt{\text{MSE}_\theta[\delta]}$, which has the advantage of being in the same units as θ.

Of course, $\text{MSE}_\theta[\delta]$ is not a single number, but a function over $\theta \in \Theta$, and there will not generally exist a single estimator which uniformly minimizes $\text{MSE}_\theta[\delta]$ over all Θ. Consider the following example.

Example 7.1 (MSE as a Criterion for Comparing Estimators) Suppose $X_1, \ldots, X_n$ is an *iid* sample with $X_1 \sim N(\mu, \sigma^2)$. Consider two estimators for μ:

$$\hat{\mu}_1 = \bar{X}, \quad \hat{\mu}_2 \equiv 0. \tag{7.1}$$

In general, $\hat{\mu}_2$ would be a poor choice of estimator, unless $\mu = 0$, in which case $\hat{\mu}_2$ would be the best possible estimator. And if μ was close enough to zero, $\hat{\mu}_2$ would be preferable to $\hat{\mu}_1$. On the other hand, for all μ sufficiently far from zero $\hat{\mu}_1$ would be preferable to $\hat{\mu}_2$. It is easily verified that the MSE for each estimator is

$$\text{MSE}_\mu[\hat{\mu}_1] = 1/n, \quad \text{MSE}_\mu[\hat{\mu}_2] = |\mu|.$$

Figure 7.1 gives plots of the respective MSEs. Clearly, neither estimator dominates the other. ■

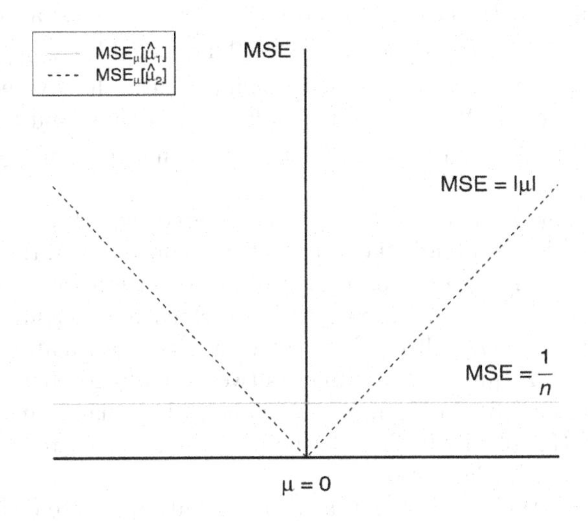

FIGURE 7.1
Plot of $\text{MSE}_\mu[\hat{\mu}_i]$ for estimators proposed in Example 7.1.

Clearly, the estimator $\hat{\mu}_2$ of Example 7.1 anticipates that the true parameter will be at or near $\mu = 0$, and would have very poor properties if this were not the case. Of course, we would not know in advance if this were true. We will refer to this as the "the problem of the favored hypothesis". For any hypothesis $\theta \in \Theta$, we can always find a statistical procedure that is best for that hypothesis. Conversely, no single procedure can be best for all hypotheses $\theta \in \Theta$. In other words, in general there will not exist a decision rule which dominates all others.

Nonetheless, if we are unable to select a single best estimator, we may still be able to eliminate an estimator from consideration. Suppose we add to Example 7.1 the class of estimators $\hat{\mu}_{a,b} = a\bar{X} + b$ (note that $\hat{\mu}_1 = \hat{\mu}_{1,0}$ and $\hat{\mu}_2 = \hat{\mu}_{0,0}$). The MSE as a function of μ is then

$$\text{MSE}_\mu[\hat{\mu}_{a,b}] = a^2/n + ((1-a)\mu - b)^2. \tag{7.2}$$

First, note that if $a > 1$ then

$$\text{MSE}_\mu[\hat{\mu}_{a,b}] \geq a^2/n > \frac{1}{n} = \text{MSE}_\mu[\hat{\mu}_1]. \tag{7.3}$$

In this case $\hat{\mu}_1$ has smaller MSE *for all* μ, so we can eliminate $\hat{\mu}_{a,b}$ from consideration. A similar argument allows the same conclusion when $a < 0$ (Theorem 7.15). Any estimator that can be eliminated from consideration in this manner is refered as inadmissible (Definition 7.1 will give a precise definition of this term).

The situation is somewhat more complicated when $a \in [0,1]$. We can easily find the minimum value of $\text{MSE}_\mu[\hat{\mu}_{a,b}]$ over μ given by Equation (7.2) by setting $\mu = b/(1-a)$, so that

$$\min_\mu \text{MSE}_\mu[\hat{\mu}_{a,b}] = a^2/n. \tag{7.4}$$

If we set $\mu_0 = b/(1-a)$ we may then write

$$\hat{\mu}^*_{a,\mu_0} = \hat{\mu}_{a,b} = a\bar{X} + b = a\bar{X} + (1-a)\mu_0,$$

so that if $a \in [0,1]$, $\hat{\mu}^*_{a,\mu_0}$ is a convex combination of $\bar{X}$ and μ_0 and therefore a shrinkage estimator. Note that μ_0 is the favored hypothesis, and we have just argued that $\text{MSE}_\mu[\hat{\mu}^*_{a,\mu_0}]$ is minimized at $\mu = \mu_0$. Furthermore, if $a \in [0,1)$ then $\text{MSE}_{\mu_0}[\hat{\mu}^*_{a,\mu_0}] < \text{MSE}_{\mu_0}[\bar{X}]$, so that the shrinkage estimator is strictly better than $\bar{X}$ near the favored hypothesis. On the other hand, if $a \neq 1$, then $\text{MSE}_\mu[\hat{\mu}^*_{a,\mu_0}]$ is quadratic in μ so we must have $\text{MSE}_{\mu'}[\hat{\mu}^*_{a,\mu_0}] > \text{MSE}_{\mu'}[\bar{X}]$ for all μ' far enough from μ_0. Thus, the shrinkage estimator does not dominate, nor is dominated by, $\bar{X}$.

One way of dealing with the problem of the favored hypothesis is to restrict attention to procedures which do not permit them. In fact, the conventional theory of inference relies mostly on this approach, with the exception of Bayesian inference. We will see that it is possible to define a class of estimators $\delta \in \mathcal{D}$ in which it is possible to find a uniformly best choice. In Chapter 8 $\mathcal{D}$ will be the class of unbiased estimators, while in Chapter 9 we consider the class of equivariant estimators (we have already defined location and scale equivariance in Section 3.3). In Example 7.1 $\hat{\mu}_1$ is both unbiased and location equivariant while $\hat{\mu}_2$ is neither (Problem 7.9). The best linear unbiased estimator (BLUE) is another example of this approach (Section 6.3).

One problem with this strategy is that a statistical procedure derived in this way may be inadmissible, in the sense that there is procedure outside of $\mathcal{D}$ which is strictly better than any procedure within $\mathcal{D}$. This happens surprisingly often, particularly with unbiased estimators. For example, the sample variance is inadmissible with respect to MSE (Example 7.7), even though it has the uniformly smallest MSE among unbiased estimates of a variance based on a normal *iid* sample. We will also see examples of commonly used MLEs which are inadmissible (Example 7.11). In fact, unbiased scale equivariant estimates of scale parameters are generally inadmissible (Section 7.6).

We conclude this section with another surprising example of inadmissibility.

Example 7.2 (The James-Stein Estimator) The James-Stein estimator provides a well known counterexample to admissibility. Suppose $(X_1,\ldots,X_m) \sim N(\boldsymbol{\mu}, \sigma^2\mathbf{I})$ is a multivariate normal observation, where $\boldsymbol{\mu}^T = (\mu_1,\ldots,\mu_m)$. The maximum likelihood estimate of $\boldsymbol{\mu}$ is $\hat{\boldsymbol{\mu}} = (X_1,\ldots,X_m)$. Before the publication of Stein (1956), it would have been widely assumed that $\hat{\boldsymbol{\mu}}$ should be the obvious choice of estimator for this problem. We can write the mean-squared error of any estimator $\delta(X) = (\delta_1(X),\ldots,\delta_m(X))$ for this problem as

$$\text{MSE}_\theta[\delta] = E_\theta[(\delta_1(X) - \mu_1)^2 + \ldots + (\delta_m(X) - \mu_m)^2],$$

for parameter $\theta = (\mu_1,\ldots,\mu_m,\sigma^2)$. Assuming σ^2 is known, the James-Stein estimator is given by

$$\hat{\boldsymbol{\theta}}_{JS} = \left(1 - \frac{(m-2)\sigma^2}{\hat{\boldsymbol{\mu}}^T\hat{\boldsymbol{\mu}}}\right)\hat{\boldsymbol{\mu}}.$$

This has the form of a shrinkage estimator given by Equation (4.30), with $\hat{\boldsymbol{\theta}} = \hat{\boldsymbol{\mu}}$, $\boldsymbol{\theta}_0 = \mathbf{0} = (0,\ldots,0)$, $\lambda = (m-2)\sigma^2/\hat{\boldsymbol{\mu}}^T\hat{\boldsymbol{\mu}}$. Possibly, $\lambda > 1$. But $E_\theta[\hat{\boldsymbol{\mu}}^T\hat{\boldsymbol{\mu}}] = m\sigma^2 + \boldsymbol{\mu}^T\boldsymbol{\mu}$, so we would expect $\lambda < 1$ with high probability (some modifications of the James-Stein estimator force the bound $\lambda \leq 1$).

It was shown in Stein (1956) that $\text{MSE}_\theta[\hat{\boldsymbol{\theta}}_{JS}] < \text{MSE}_\theta[\hat{\boldsymbol{\mu}}]$ for all $\theta \in \Theta$. Thus, there would be no reason to use the estimator $\hat{\boldsymbol{\mu}}$ in place of $\hat{\boldsymbol{\theta}}_{JS}$, that is, the MLE is inadmissible *wrt* MSE. See also Problem 7.11. ■

7.3 Prediction

This volume is concerned primarily with parametric inference, but it is worth discussing the related but distinct problem of prediction. Suppose we are given a random variable $Y \in \mathcal{Y}$, and a more general random observation $\tilde{X} \in \mathcal{X}$. To fix ideas, we can take $\mathbf{X} = (X_1, \ldots, X_d) \in \mathbb{R}^d$. For the prediction problem there is no unknown parameter θ, so that the joint distribution $(\mathbf{X}, Y)$ is known. The random outcome $\mathbf{X}$, but not Y, is observed. The problem is to construct a predictor $\delta : \mathcal{X} \to \mathcal{Y}$ such that $\delta(\mathbf{X}) \approx Y$.

The prediction problem can be formulated as a decision problem based on the following components:

(i) Two random outcomes $Y \in \mathcal{Y}$, $\mathbf{X} \in \mathcal{X}$. The joint distribution $(\mathbf{X}, Y)$ is known.
(ii) A decision rule $\delta : \mathcal{X} \to \mathcal{Y}$.
(iii) A loss function $L : \mathcal{Y} \times \mathcal{Y} \to \mathbb{R}_{\geq 0}$, with $L(y, y) = 0$ for all $y \in \mathcal{Y}$. Given outcome $y \in \mathcal{Y}$ and decision $d \in \mathcal{Y}$, a loss of $L(y, d)$ is accrued.

If Y is quantitative, then loss will be some distance function on $\mathcal{Y}$. If Y is categorical, then the structure of L is more general, and is typically given in tabular form.

In Section 5.2 we consider the possibility of the randomized test, and we can also consider the randomized decision rule. There are several ways of constructing such a procedure. We may think of it as hierarchical. For example, for a randomized test based on observation $\tilde{X} \in \mathcal{X}$ we would first observe $\delta(\tilde{X})$, then the null hypothesis would be rejected with probability $\delta(\tilde{X})$. We could, equivalently, introduce a randomization quantity U, then interpret a randomized decision rule $\delta(\tilde{X})$, as a *nonrandomized* mapping $\delta^*(\tilde{X}, U)$. In this case, we would have to be careful to recognize that only $\tilde{X}$ had any information relevant to the prediction or inference problem.

The objective is to select $\delta(\mathbf{X})$ which in some sense minimizes loss $L(Y, \delta(\mathbf{X}))$. Of course, this is random, and not observed, so any criteria for the selection of δ must anticipate this form of uncertainty.

If the distribution of $(\mathbf{X}, Y)$ is known, one obvious solution is to measure the predictive ability of δ using the expected value of loss, or risk

$$R(\delta) = E[L(Y, \delta(\mathbf{X}))]. \tag{7.5}$$

In principle, risk can be evaluated for any decision, so the problem of finding a minimum risk decision is well defined. In fact, an optimal decision rule can usually be deduced using a simple device, which we refer to as the optimal predicton theorem. We will later see a number of versions of this method, which follows next.

Theorem 7.1 (Optimal Prediction Theorem) Suppose the pair $(\mathbf{X}, Y)$ has a joint diistribution on space $\mathcal{X} \times \mathcal{Y}$, and we are given loss function $L : \mathcal{Y} \times \mathcal{Y} \to \mathbb{R}_{\geq 0}$. Suppose the following conditions hold:

(i) There exists some $\delta_0 : \mathcal{X} \to \mathcal{Y}$ such that $E[L(Y, \delta_0(\mathbf{X}))] < \infty$;
(ii) For each $\mathbf{x} \in \mathcal{X}$ there exists $d(\mathbf{x})$ which minimizes $E[L(Y, d) \mid \mathbf{X} = \mathbf{x}]$ *wrt* $d \in \mathcal{Y}$.

Then $E[L(Y, \delta(\mathbf{X}))]$ is minimized by $\delta^*(\mathbf{X}) = d(\mathbf{X})$. If the minimizers $d(\mathbf{x})$ are unique for each $\mathbf{x} \in \mathcal{X}$, then $\delta^*(\mathbf{X})$ is also the unique minimizer. ■

Proof. By the law of total probability we may write $E[L(Y, \delta(\mathbf{X}))] = E[\phi_\delta(\mathbf{X})]$ where $\phi_\delta(\mathbf{x}) = E[L(Y, \delta(\mathbf{X})) \mid \mathbf{X} = \mathbf{x}]$. By assumption $\phi_d(\mathbf{x}) \leq \phi_\delta(\mathbf{x})$ for any δ. Furthermore $E[L(Y, d(\mathbf{X}))] = E[\phi_d(\mathbf{X})] \leq E[\phi_{\delta_0}(\mathbf{X})] = E[L(Y, \delta_0(\mathbf{X}))] < \infty$. We conclude $E[L(Y, d(\mathbf{X}))] = E[\phi_d(\mathbf{X})] \leq E[\phi_\delta(\mathbf{X})]$ for any δ. Therefore $\delta^*(\mathbf{X}) = d(\mathbf{X})$ minimizes $E[L(Y, \delta(\mathbf{X}))]$, and must be the unique minimizer if each minimizer $d(\mathbf{x})$ is unique. □

Example 7.3 (The Monty Hall Problem) The Monty Hall Problem is an interesting example of a problem in decision theory, as well as a good example of the often counterintuitive nature of probability theory. The problem is based on the television game show *Let's Make a Deal* (starring Monty Hall). There are three doors. Behind one is a car, and behind the other two are goats. The contestant picks one door. Then, one the other doors is opened, revealing a goat (this can always be done, since there are two goats). The contestant is offered the choice of staying with the original choice, or switching to the one remaining door. The contestant wins whatever is behind the selected door. It has become widely accepted that the best strategy is to switch. This solution can be derived from Theorem 7.1.

We can, without loss of generality, select door 1 first. Let Y be the door with the car, and let X be the door opened by Monty Hall. The decision can be $d \in \{1, 2, 3\}$ the door ultimately chose. If $d = 1$ the contestant stays, otherwise the contestant switches. We admit the possibility that the door opened by Monty Hall is selected by the contestant, assuming that this decision will be rejected as suboptimal.

The loss is $L(Y, d) = I\{Y \neq d\}$, and the joint distribution of (X, Y) is given in the following table:

$P(X = x, Y = y) =$

	$y = 1$	2	3
$x = 1$	0	0	0
2	1/6	0	1/3
3	1/6	1/3	0

To apply Theorem 7.1 for each $x \in \{2, 3\}$ we find $d(x) \in \{1, 2, 3\}$ which minimizes $E[L(Y, d(x)) \mid X = x] = P(Y \neq d(x) \mid X = x)$, then the optimal decision rule is $\delta(X) = d(X)$. For $x = 2$ we have $P(Y \neq 1 \mid X = 2) = 2/3$, $P(Y \neq 2 \mid X = 2) = 1$ and $P(Y \neq 3 \mid X = 2) = 1/3$. The minimum conditional probability is $1/3$, so $d(2) = 3$. By an identical argument $d(3) = 2$. This confirms that the optimal (nonrandomized) strategy is to switch, which attains a risk of $1/3$. That is, the probability of winning the car under this strategy is $2/3$. See also Problems 7.1 and 7.2. ■

7.3.1 Minimum Mean Squared Error (MMSE) Predictors

When the object of a prediction problem Y is quantitative, perhaps the most commonly used loss function is squared error loss $L(y, a) = (y - a)^2$. There are sometimes good reasons to use more robust alternatives. However, the theory that results is both elegant and powerful, and is the obvious starting point for this problem. Thus, our objective is to find predictor $\delta(\mathbf{X})$ which minimizes the risk $\text{MSE}[\delta] = E[(Y - \delta(\mathbf{X}))^2]$.

It is sometimes useful, even necessary, to restrict the class of predictors to some subset of decision rules $\mathcal{D}$, on which the optimization problem is well defined or especially tractable. Any predictor that minimizes $\text{MSE}[\delta]$ over $\delta \in \mathcal{D}$ is the minimum mean squared error predictor (MMSE) for that class. The simplest case would be the class $\delta \equiv c$ for any constant $c \in \mathbb{R}$, which we denote $\mathcal{D}_C$. There will also be interest in linear predictors, which assume the form

$$\delta(\mathbf{X}) = a + \sum_{j=1}^{d} b_j X_j. \tag{7.6}$$

This class will be denoted as $\mathcal{D}_L$. We also define the class of L^2 predictors ($E[\delta(\mathbf{X})^2] < \infty$), which we denote $\mathcal{D}_2$ (here, we would assume $E[Y^2] < \infty$). It may be that $\mathcal{D}$ happens to be a rich enough class to resolve the problem completely, or it may be used to facilitate optimization over a larger class. We will encounter both cases below.

The problem of finding the MMSE predictor on class $\mathcal{D}_C$ is quite straightforward, but has important implications for more general problems. We give the solution in the following theorem.

Theorem 7.2 (Optimal Constant Predictors) Among predictors of the class $\mathcal{D}_C$, either $E[(Y-c)^2] = \infty$, for all $c \in \mathcal{D}_C$, or $E[(Y-c)^2]$ is uniquely minimized by $c = E[Y]$, which is therefore the MMSE predictor within $\mathcal{D}_C$. ■

Proof. The expression $E[(Y-c)^2]$ is a second order polynomial in c, from which the theorem easily follows. □

It turns out that Theorem 7.2, despite its simplicity, leads to a quite definitive resolution of the more general problem. This is achieved by replacing the expectation in predictor $\delta \equiv E[Y]$ with the conditional expectation $\delta(\mathbf{X}) = \mu_Y(\mathbf{X})$ where

$$\mu_Y(\mathbf{x}) = E[Y \mid \mathbf{X} = \mathbf{x}]. \tag{7.7}$$

This result is given in the following theorem, which is a direct application of Theorems 7.1 and 7.2.

Theorem 7.3 (MMSE Predictors in $\mathcal{D}_2$) Suppose $E[Y^2] < \infty$. Then among predictors of the class $\delta \in \mathcal{D}_2$, MSE$[\delta]$ is uniquely minimized by $\delta(\mathbf{X}) = \mu_Y(\mathbf{X})$, as defined in Equation (7.7). ■

Proof. First note that by the total variance identity, if $E[Y^2] < \infty$ then $E[E[Y \mid \mathbf{X}]^2] < \infty$. We may then apply Theorem 7.2 to the problem of minimizing $E[(Y-c)^2 \mid \mathbf{X} = \mathbf{x}]$, which has unique minimizer $\mathbf{c} = E[Y \mid \mathbf{X} = \mathbf{x}] = \mu_Y(\mathbf{x})$. Then take the expectation $\text{MSE}[\mu_Y(\mathbf{X})] = E[E[(Y - \mu_Y(\mathbf{X}))^2 \mid \mathbf{X}]]$. □

It is natural to express the predictor of Theorem 7.3 as

$$Y = \mu_Y(\mathbf{X}) + \epsilon \tag{7.8}$$

where $\epsilon = Y - \mu_Y(\mathbf{X})$ represents prediction error. This allows us to illustrate an important principle of prediction or estimation, in particular, that an optimal procedure will exhaust all relevant information in the data. This principle will be demonstrated in Section 8.2 in the context of parametric estimation.

For the prediction problem, if $\mu_Y(\mathbf{X})$ is the MMSE predictor of Y, then ϵ should represent spurious variation, or noise. Otherwise, we should be able to pool $\mu_Y(\mathbf{X})$ and ϵ to construct a better predictor. Any potential relationship between ϵ and any other predictor may be given in terms of correlation.

Theorem 7.4 (The MMSE Predictor is Uncorrelated with its Error) Suppose $E[Y^2] < \infty$. Let $\mu_Y(\mathbf{X})$ be as defined in (7.7), and ϵ as defined in (7.8). Then $\text{cov}[g(\mathbf{X}), \epsilon] = 0$ for any function $g(\mathbf{X}) \in \mathbb{R}$. ■

Proof. Since $E[\epsilon] = 0$, we may write, using the total expectation identity (1.13), $\text{cov}[g(\mathbf{X}), \epsilon] = E[g(\mathbf{X})\epsilon] = E[E[g(\mathbf{X})\epsilon \mid \mathbf{X}]] = E[g(\mathbf{X})E[\epsilon \mid \mathbf{X}]] = E[g(\mathbf{X})0] = 0$. □

Thus, by Theorem 7.4 ϵ is uncorrelated with any other decision rule. If this were not the case, the MMSE predictor could be improved (see Theorem 8.2).

We next consider an important class of MMSE predictor.

Example 7.4 (Prediction for Multivariate Normal Distributions) The multivariate normal density is described in Section 2.2. Suppose $\mathbf{Z} = (Z_1, \ldots, Z_d) \in \mathbb{R}^d$ is a random vector. Let $\boldsymbol{\mu} \in \mathbb{R}^d$ be the mean vector $E[Z_i] = \mu_i$, and let Σ be a $d \times d$ positive definite matrix, with element (i, j) given by $\sigma_{ij} = \text{cov}[Z_i, Z_j]$. If $\mathbf{Z}$ is multivariate normal we write $\mathbf{Z} \sim N(\boldsymbol{\mu}, \Sigma)$.

The elements of the prediction problem are derived from $\mathbf{Z}$. Consider the partition $\mathbf{Z} = (\mathbf{Y}, \mathbf{X})$, using d_1- and d_2-dimensional subvectors $\mathbf{Y} = (Z_1, \ldots, Z_{d_1})$, $\mathbf{X} = (Z_{d_1+1}, \ldots, Z_{d_1+d_2})$. The mean and covariance of $\mathbf{Z}$ are then partioned by

$$\boldsymbol{\mu} = \begin{bmatrix} \mu_{\mathbf{Y}} \\ \mu_{\mathbf{X}} \end{bmatrix}, \qquad \Sigma = \begin{bmatrix} \Sigma_{\mathbf{YY}} & \Sigma_{\mathbf{YX}} \\ \Sigma_{\mathbf{XY}} & \Sigma_{\mathbf{XX}}, \end{bmatrix} \tag{7.9}$$

where $\mathbf{Y} \sim N(\mu_{\mathbf{Y}}, \Sigma_{\mathbf{YY}})$, $\mathbf{X} \sim N(\mu_{\mathbf{X}}, \Sigma_{\mathbf{XX}})$, and $\Sigma_{\mathbf{YX}} = \Sigma_{\mathbf{XY}}^T$ can be interpreted as the covariance of vectors $\mathbf{Y}$ and $\mathbf{X}$.

Conditional mutivariate normal densities have a convenient form, in particular,

$$\mathbf{Y} \mid \mathbf{X} = \mathbf{x} \sim N(\mu_{\mathbf{Y}|\mathbf{X}}(\mathbf{x}), \Sigma_{\mathbf{Y}|\mathbf{X}}),$$

where

$$\begin{aligned} \mu_{\mathbf{Y}|\mathbf{X}}(\mathbf{x}) &= \mu_{\mathbf{Y}} + \Sigma_{\mathbf{YX}}\Sigma_{\mathbf{XX}}^{-1}(\mathbf{x} - \mu_{\mathbf{X}}), \text{ and} \\ \Sigma_{\mathbf{Y}|\mathbf{X}} &= \Sigma_{\mathbf{YY}} - \Sigma_{\mathbf{YX}}\Sigma_{\mathbf{XX}}^{-1}\Sigma_{\mathbf{XY}}. \end{aligned} \tag{7.10}$$

We then consider the implications of Theorem 7.3 when $(\mathbf{Y}, \mathbf{X})$ jointly possess a multivariate normal density. In this case $d_1 = 1$, and d_2 is the original dimension of $\mathbf{X}$. Simply put, the unique MMSE predictor of $Y \in \mathbb{R}$ based on $\mathbf{X}$ is $\mu_{Y|\mathbf{X}}(\mathbf{X})$ as defined in (7.10), with the actual MSE given by $\Sigma_{Y|\mathbf{X}}$. ■

While prediction involving multivariate normal random vectors has a quite tractable solution, in practice, the calculation of conditional expectations can present considerable technical challenges. For this reason, predictors are sometimes restricted to the linear class $\mathcal{D}_L$. We will express this class in matrix form:

$$\delta(\mathbf{X}) = a + \mathbf{b}^T(\mathbf{X} - \mu_{\mathbf{X}}) \tag{7.11}$$

where $\mathbf{X} \in \mathbb{R}^d$ is a random vector with mean $\mu_{\mathbf{X}}$, and d-dimensional vectors are equivalently $d \times 1$ matrices when used in the context of matrix algebra. Here, a is a real number, while $\mathbf{b} \in \mathbb{R}^d$. This is clearly equivalent to (7.6). The vector $\mu_{\mathbf{X}}$ is explicitly included in (7.11). If it were removed, the constant a could be adjusted to leave the resulting predictor unchanged.

As before, the object is to predict Y. We do not assume that $\mathbf{Z} = (Y, \mathbf{X})$ is a multivariate normal random vectors, but we will assume that the joint mean vector and covariance matrix are partioned as in (7.9).

The predictor MSE is then

$$\text{MSE}[\delta] = E[(Y - a - \mathbf{b}^T(\mathbf{X} - \mu_{\mathbf{X}}))^2].$$

By Theorem 7.2, MSE$[\delta]$ is minimized with respect to a by setting

$$a = E[Y - \mathbf{b}^T(\mathbf{X} - \mu_{\mathbf{X}})] = \mu_Y.$$

Replacing a with this optimal value gives

$$E[(Y - \mu_Y - \mathbf{b}^T(\mathbf{X} - \mu_{\mathbf{X}}))^2] = \Sigma_{YY} - 2\Sigma_{Y\mathbf{X}}\mathbf{b} + \mathbf{b}^T\Sigma_{\mathbf{XX}}\mathbf{b}.$$

It may be verified using matrix calculus that this quantity is minimized by setting $\mathbf{b}^T = \Sigma_{Y\mathbf{X}}\Sigma_{\mathbf{XX}}^{-1}$, giving the best linear MMSE predictor

$$\delta(\mathbf{X}) = \mu_Y + \Sigma_{Y\mathbf{X}}\Sigma_{\mathbf{XX}}^{-1}(\mathbf{X} - \mu_{\mathbf{X}}). \tag{7.12}$$

But this is exactly the same as the minimum MSE predictor when $(Y, \mathbf{X})$ is multivariate normal. Thus, (7.12) will be a good choice in general, and the best choice in the multivariate normal case, using the MSE criterion.

Some examples of MMSE predictors are given in Problems 7.4, 7.5 and 7.6.

7.4 The Structure of Decision Theoretic Inference

We now develop a decision theoretic formulation of parametric inference. We start with an observation $\tilde{X} \in \mathcal{X}$ from p_θ, a member of family $\mathcal{P} = \{p_\theta : \theta \in \Theta\}$, as defined in Section 3.2. We then have a surjective mapping $A : \Theta \to \mathcal{A}$, which defines an action $A(\theta)$, where $\mathcal{A}$ is the action space. The action represents the target of the inference, in response to which a decision $d \in \mathcal{A}$ is made. The objective is to select a decision which in some sense minimizes a loss $L_\theta(d)$. We take $A(\theta)$ itself to be the optimal decision, so that

$$L_\theta(A(\theta)) = 0 \quad \text{and} \quad L_\theta(d) \geq 0, \quad d \in \mathcal{A}. \tag{7.13}$$

As a statistical problem, we observe $\tilde{X} \in \mathcal{X}$ from density p_θ. Since $\tilde{X}$ contains information about θ, we may construct a statistic $\delta(\tilde{X}) \in \mathcal{A}$ with the objective of minimizing expected loss, or risk

$$R(\theta, \delta(\tilde{X})) = E_\theta[L_\theta(\delta(\tilde{X}))]. \tag{7.14}$$

As a practical matter, the assumption that $\delta(\tilde{X}) \in \mathcal{A}$ is sometimes violated. We will, for example, see cases in which $\mathcal{A} = \mathbb{R}_{>0}$, but $\delta(\tilde{X})$ may be negative. Rather than introduce a separate decision space, we will simply deal with such exceptions individually. The cost of doing so seems much less than that of defining a decision space which is distinct from the action space.

Note also that here the loss function is parametrized. In the prediction problem of Section 7.3 loss is defined as a distance function applied to the target and decision rule. In contrast L_θ may be a family of distance functions parametrized by θ. The next example will make this distinction clear.

Example 7.5 (Risk as Mean Squared Error) The mean squared error (MSE) of an estimator $\delta(\tilde{X})$ of estimand $\eta(\theta)$ is the expected value $MSE = E_\theta[(\delta(\tilde{X}) - \eta(\theta))^2]$. This is equivalent to risk under squared error loss $L_\theta(d) = (A(\theta) - d)^2$. Clearly, the loss function satisfies Equation (7.13). First, suppose θ is a single dimensional parameter, for which $\hat{\theta} = \delta(\tilde{X})$ is an estimator. Here, the action is simply $A(\theta) = \theta$, the estimand of $\hat{\theta}$, therefore the action space is $\mathcal{A} = \Theta$. Then $R(\theta, \delta) = E_\theta[(\hat{\theta} - \theta)^2]$.

Then suppose we have a location-scale parameter $\theta = (\mu, \tau) \in \Theta = \mathbb{R} \times \mathbb{R}_{>0}$. Suppose $\hat{\mu} = \delta(\tilde{X})$ is an estimator of μ. Under the decision theoretic formulation, we have action $A(\theta) = \mu \in \mathcal{A} = \mathbb{R}$. We may retain squared error loss $L_\theta(d) = (A(\theta) - d)^2$, but now risk $R(\theta, \delta) = E_\theta[(\hat{\mu} - \mu)^2]$ will depend on τ. Alternatively, we may choose loss $L_\theta(d) = \tau^{-2}(A(\theta) - d)^2$, so that risk will not depend on τ, provided $\hat{\mu}$ is location-scale equivariant in the sense of Definition 3.5. ∎

Following the decision theoretic formulation for the prediction problem of Section 7.3, the parametric decision problem is based on the following components.

(i) A family of models $\mathcal{P} = \{p_\theta : \theta \in \Theta\}$ as defined in Section 3.2.
(ii) An observation $\tilde{X} \in \mathcal{X}$ assumed to have been sampled from some $p_\theta \in \mathcal{P}$.
(iii) An action space $\mathcal{A}$.
(iv) A decision rule $\delta : \mathcal{X} \to \mathcal{A}$.
(v) An optimal action for each $\theta \in \Theta$, in the form $A : \theta \to \mathcal{A}$.
(vi) A parametrized loss function $L_\theta : \mathcal{A} \to \mathbb{R}_{\geq} 0$, $\theta \in \Theta$. We assume $L_\theta(A(\theta)) = 0$.
(vii) For any decision rule δ we have risk function $R(\theta, \delta) = E_\theta[L_\theta(\delta(\tilde{X}))]$, $\theta \in \Theta$.

As for the prediction problem a decision rule may be randomized. So as not to introduce additional notation, a randomized decision rule can be similarly constructed by introducing a randomization quantity U, then expressing a randomized decision rule $\delta(\tilde{X})$, as the non-randomized mapping $\delta^*(\tilde{X}, U)$. In one sense, our observation is now $(\tilde{X}, U)$, but $\tilde{X}$ remains sufficient for θ. Thus, the definition of risk extends naturally to randomized decision rules, and the possibility that an optimal decision rule is randomized must be considered (in fact, this will sometimes be the case).

Decision theory for parametric inference relies on many of the same ideas as for prediction, and the derivation of optimal decision rules will rely on various versions of the optimal prediction theorem (Theorem 7.1). The main difference is that while for the prediction problem risk $R(\delta)$ varies by decision rule δ, for the parametric inference problem risk $R(\theta, \delta)$ varies by both decision rule and parameter θ. Thus, for the prediction problem any two decision rules can always be ordered by preference. In particular, we would prefer δ to δ' if $R(\delta) < R(\delta')$, and we would be indifferent to the choice if $R(\delta) = R(\delta')$.

As demonstrated in Section 7.2, what complicates decision theory for parametric inference is that two decision rules δ, δ' cannot always be ordered unambiguously in the same way, since we may have $R(\theta_1, \delta) < R(\theta_1, \delta')$, but $R(\theta_2, \delta) > R(\theta_2, \delta')$ for distinct parameters θ_1, θ_2. Consider the following example.

Example 7.6 (MSE of Alternative Estimators of a Bernoulli Parameter) Suppose for a certain application, data takes the form (X_1, X_2), where X_i are independent Bernoulli random variables of mean p. The usual unbiased estimator of p is

$$\hat{p}_1 = \frac{X_1 + X_2}{2},$$

which has variance $\sigma_p^2 = p(1-p)/2$. Suppose we also consider as an alternative estimator of p:

$$\hat{p}_2 = \hat{p}_1/2 + 1/4.$$

We can calculate the MSE for each estimator. For $\hat{p}_1$ we have: $\text{var}_p[\hat{p}_1] = p(1-p)/2$; $\text{bias}_p[\hat{p}_1] = E_p[\hat{p}_1] - p = p - p = 0$; $\text{MSE}_p[\hat{p}_1] = p(1-p)/2$. For $\hat{p}_2$ we have: $\text{var}_p[\hat{p}_2] = \text{var}_p[\hat{p}_1/2 + 1/4] = \text{var}_p[\hat{p}_1]/4 = p(1-p)/8$; $\text{bias}_p[\hat{p}_2] = E_p[\hat{p}_2] - p = -p/2 + 1/4$; $\text{MSE}_p[\hat{p}_2] = (p^2 - p + 1/2)/8$.

Figure 7.2 gives plots of the respective MSEs. Clearly, neither estimator improves the other. To be more precise, it is easily verified that the MSE curves intersect at $p = 1/2 \pm \sqrt{3/20} \approx 0.5 \pm 0.387$, and that $\hat{p}_2$ has strictly smaller MSE if $|p - 1/2| < \sqrt{3/20}$, and $\hat{p}_1$ has strictly smaller MSE if $|p - 1/2| > \sqrt{3/20}$. This means that the choice of best estimator depends on the unknown parameter p. ■

In Example 7.6 $\hat{p}_1$ is the conventional sample frequency, and is unbiased. However, $\hat{p}_2$ has smaller MSE over a significant portion of the parameter space. It can also be verified that $\hat{p}_2$ has smaller average risk, and smaller maximum risk (Problem 7.12). From Figure

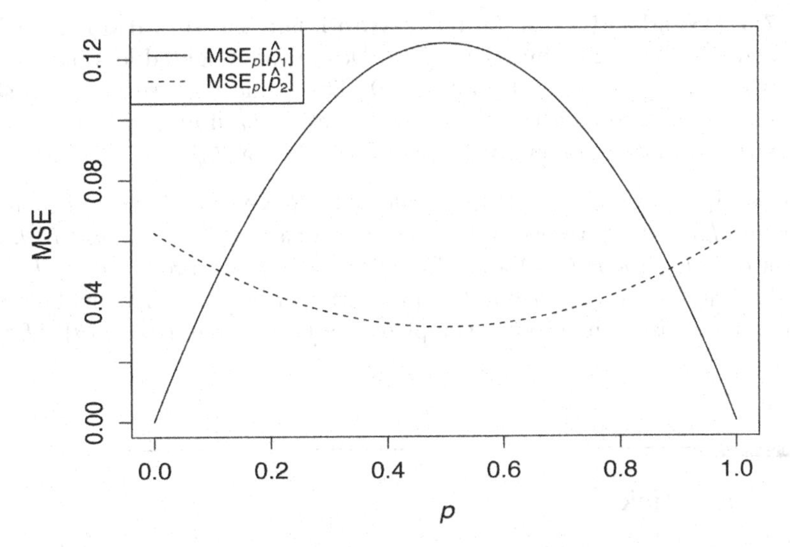

FIGURE 7.2
Plot of MSE for estimators proposed in Example 7.6.

7.2, it is clear that $\hat{p}_2$ reduces the variability of risk in comparison to $\hat{p}_1$, in the sense that it decreases risk near $p = 1/2$ at the cost of increasing risk near $p = 0, 1$. We will see that such considerations play an important role in decision theory.

We may now define precisely our objectives regarding decision theory, the first step of which is to define admissibility.

Definition 7.1 (Admissibility) Suppose for a decision problem we are given two decision rules δ, δ'. We say δ dominates δ' if

$$R(\theta, \delta) \leq R(\theta, \delta'), \ \ \forall \theta \in \Theta,$$

with strict inequality for at least one θ. If strict inequality holds for all θ, then δ uniformly dominates δ'.

A decision rule δ' is inadmissible if it is dominated by at least one other decision rule. A decision rule δ' is admissible if it is not inadmissible. ■

While a universally dominant decision rule will generally not exist (Section 7.2), a great deal of classical inference theory concerns decision rules which are universally dominant *within a restricted class.* This approach is especially prominent in non-Bayesian methods. Accordingly, we offer the following definition.

Definition 7.2 (Uniformly Minimum Risk Decision Rules) Let $\mathcal{D}$ be a class of decision rules with risk $R(\theta, \delta)$ for $\theta \in \Theta$, $\delta \in \mathcal{D}$. Then $\delta^* \in \mathcal{D}$ is a uniformly minimum risk (UMR) decision rule (over $\mathcal{D}$) if for any other $\delta \in \mathcal{D}$, $R(\theta, \delta^*) \leq R(\theta, \delta)$ $\forall \theta \in \Theta$. ■

A class of decision rules $\mathcal{D}$ need not have a UMR decision rule. And as can be seen by Example 7.1 the class $\mathcal{D}$ cannot be too large. However, this approach can yield a powerful theory by defining $\mathcal{D}$ with quite intuitively reasonable constraints.

The admissibility and UMR properties depend on a specific risk/loss function. Thus, δ maybe inadmissible with respect to one type of loss, but not another. We will make use of the following theorem to determine when these properties will be equivalent with respect to distinct loss functions.

Theorem 7.5 (Weighted Loss Functions and the Preservation of Admissibility) Suppose we are given loss function $L_\theta(d)$. Define the weighted loss function $L_\theta^w(d) = w(\theta)L_\theta(d)$, where $0 < w(\theta) < \infty$ for all $\theta \in \Theta$. Then a decision rule is admissible, inadmissible or uniformly minimum risk under loss function L_θ if and only if it is admissible, inadmissible or uniformly minimum risk under loss function L_θ^w. ■

Proof. Suppose δ_1, δ_2 are any two decision rules. Let $R(\theta, \delta)$ and $R^w(\theta, \delta)$ be the risk functions under loss L_θ and L_θ^w respectively. Clearly, we have $R^w(\theta, \delta) = w(\theta)R(\theta, \delta)$, so that all strict risk orderings of $R(\theta, \delta_1)$ and $R(\theta, \delta_2)$ at each θ are preserved by $R^w(\theta, \delta_1)$ and $R^w(\theta, \delta_2)$. But these orderings define the properties of admissibility, inadmissibility and uniformly minimum risk. The proof is completed by noting that $L_\theta = w(\theta)^{-1}L_\theta^w$. □

7.5 Loss and Risk

Some results from decision theory will hold for general loss functions, otherwise we will need to impose certain properties. We will, for example, often refer to convex loss. In other problems we may need to assume invariant properties similar to those discussed in Section 3.3 in the context of location-scale transformations. We consider in this section the various properties of loss and risk which will arise in our subsequent development of the topic.

7.5.1 Optimization of Convex Functions

As is well known, convexity (Definition 1.4) plays an important role in optimization. If a function $f : (a, b) \to \mathbb{R}$ is strictly convex and nonmonotone on (a, b) then it possesses a unique minimum $x^* = \mathrm{argmin}_{x \in (a,b)} f(x)$. It will first be useful to know what transformations of a function preserve convexity.

Theorem 7.6 (Preservation of Convexity and Nonmonotonicity Under Functional Transformations) Suppose $f(x)$ is a convex function on interval $(-\infty, \infty)$. Then the following statements hold.

(i) $f^*(x) = f(ax + b)$ is a convex function of x for any constants $a \neq 0$, b.
(ii) The function $\phi(a) = E[f(X - a)]$ is convex for $a \in (-\infty, \infty)$ for any random variable X, assuming expectations exist.
(iii) If f is strictly convex, then so are f^* and ϕ in parts (i) and (ii).
(iv) If f is nonmonotone, then so are f^* and ϕ in parts (i) and (ii). ■

Proof. (i) By hypothesis the inequalities (1.17) and (1.18) hold after replacing x, x' with $ax+b$ and $ax'+b$. Then $f(p[ax + b] + (1 - p)[ax' + b]) = f(a[px + (1 - p)x'] + b) = f^*(px + (1 - p)x')$. It may then be verified that the defining inequalities hold for f^* using a similar argument for the upper bounds.

(ii) For $a, b \in \mathbb{R}$, $p \in (0, 1)$, we may write

$$\begin{aligned}
\phi(pa + (1 - p)b) &= E[f(X - pa - (1 - p)b)] \\
&\leq E[pf(X - a) + (1 - p)f(X - b)] \\
&= p\phi(a) + (1 - p)\phi(b),
\end{aligned}$$

noting that by part (i) $f(X - a)$ is a convex function of a.

(iii) If f is strictly convex, then inequality can be replaced by strict inequality in the preceding arguments.

(iv) Any nonmonotone convex function on $(-\infty, \infty)$ must have limit ∞ as $x \to \pm\infty$. Nonmonotonicity of f^* and ϕ follows directly. □

We next define properties of a loss function which may lead to a more definitive resolution of a problem.

Definition 7.3 (Classes of Loss Functions) Let $L_\theta(d)$ be a loss function, $d \in \mathcal{A}$, where $\mathcal{A}$ is a convex subset of $\mathbb{R}$.

(i) $L_\theta(d)$ is a (strictly) convex loss function if $L_\theta(d)$ is (strictly) convex in d for all $\theta \in \Theta$.
(ii) $L_\theta(d)$ is a nonmonotone loss function if $L_\theta(d)$ is nonmonotone in d for all $\theta \in \Theta$.
(iii) $L_\theta(d)$ is a location invariant loss function if there exists $\rho(x) \geq 0$ such that $L_\theta(d) = \rho(\mu - d)$, where μ is a location parameter.
(iv) $L_\theta(d)$ is a scale invariant loss function if there exists $\rho(x) \geq 0$ such that $L_\theta(d) = \rho(d/\tau)$, where τ is a scale parameter. ■

We next offer conditions under which risk is convex with respect to location or scale transformations.

Theorem 7.7 (Convex Loss and Risk) Given an observation $\tilde{X}$ from a parametric family Θ, suppose $L_\theta(d)$ is a parametric loss function, $d \in \mathcal{A}$, where $\mathcal{A}$ is a convex subset of $\mathbb{R}$. Then let $R(\theta, \delta) = E_\theta[L_\theta(\delta(\tilde{X})]$ be risk for any decision rule $\delta(\tilde{X})$.

(i) If the function ρ defining either a location or scale invariant loss function $L_\theta(d)$ in Definition 7.3 is (strictly) convex, then $L_\theta(d)$ is a (strictly) convex loss function.
(ii) If $L_\theta(d)$ is a location invariant loss function, and ρ is (strictly) convex and nonmonotone, then $E_\theta[\rho(\delta - a)]$ is (strictly) convex in $a \in \mathcal{A}$ and nonmonotone.
(iii) If $L_\theta(d)$ is a scale invariant loss function, and $\phi(w) = \rho(e^w)$ is (strictly) convex and nonmonotone, then $E_\theta[\rho(\delta/a)]$ is (strictly) convex and nonmonotone in $\log(a)$. ■

Proof. Statements (i)-(ii) follows directly Theorem 7.6. To prove (iii) write $\rho(\delta/a) = \phi(\log(\delta) - \log(a))$. Under the hypothesis, by (ii), $E_\theta[\rho(\delta/a)]$ is a convex mapping of $\log(a)$. □

Remark 7.1 Theorem 7.7 gives conditions under which convexity of the loss function implies convexity of the risk. In the methods introduced below, we will be more interested in optimizing risk, so there is some value in being able to determine when an optimal solution exists, and if it is unique, based on the properties of the loss function.

However, we will encounter important cases in which the assumptions of Theorem 7.7 do not hold. For example, consider scale invariant loss $L(X, d) = \rho(X/d)$, where $\rho(x) = |x-1|^k$, $k \geq 1$. While ρ is convex, $\phi(w) = \rho(e^w)$ is not. However, it satisfies condition (1.9) for loss functions given in Farrell (1964), in particular, that $\phi(0) = 0$, $\phi(w)$ is bounded for $w < 0$, and is convex for $w > 0$. In fact, if $\rho(1) = 0$, is convex, and possesses first and second order derivatives, it may be easily shown that this condition will hold for $\phi(w)$.

It should be pointed out, however, that it in the cases we will consider, it will usually be straightforward to characterize the minimum of a risk function directly, which will suffice for most purposes. ■

7.5.2 Some Important Optimization Problems

The minimization of risk typically reduces to a technical problem of the types considered in the following theorem, which can vary considerably in difficulty.

Theorem 7.8 (Methods for the Minimization of Risk) Suppose X is a random variable and $c \in \mathbb{R}$.

(i) If $E[X^2] < \infty$, then $\phi(c) = E[(X-c)^2]$ is uniquely minimized by $c = E[X]$.
(ii) If $E[|X|] < \infty$, then $\phi(c) = E[|X-c|]$ attains it's minimum if and only if c is any median of X (Definition 1.8).
(iii) If $E[X^2] > 0$, then $\phi(c) = E[(X/c-1)^2]$ is uniquely minimized by $c = E[X^2]/E[X]$.
(iv) If $X > 0$ $wp1$, $E[X] < \infty$ and X possesses a density $f(x) > 0$ on (a,b), then $\phi(c) = E\left[|X/c-1|\right]$ is minimized by any c satisfying

$$\int_0^c xf(x)dx = \int_c^\infty xf(x)dx. \tag{7.15}$$

Any such value is referred to as a scale-median. ∎

Proof. Parts (i) and (iii) are easily proven by writing $\phi(c)$ as a polynomial in c or $1/c$. Note that (iii) holds if $E[X] = 0$ or $E[X^2] = \infty$.

To show (ii), recall from Theorem 1.14 that the medians of X are expressible as a nonempty interval $[m_L, m_U]$. Fix $a < b$, and write

$$\begin{aligned}
|x-b| - |x-a| &= (x-b)I\{x > b\} - (x-b)I\{x \leq b\} \\
&\quad - (x-a)I\{x > a\} + (x-a)I\{x \leq a\} \\
&= (x-b)I\{x > b\} - (x-b)I\{x \leq b\} \\
&\quad - (x-b)I\{x > a\} + (x-b)I\{x \leq a\} \\
&\quad - (b-a)I\{x > a\} + (b-a)I\{x \leq a\} \\
&= -2(x-b)I\{a < x \leq b\} - (b-a)I\{x > a\} \\
&\quad + (b-a)I\{x \leq a\}.
\end{aligned}$$

It follows that

$$\begin{aligned}
E[|X-b|] - E[|X-a|] &= (b-a)\left[P(X \leq a) - P(X > a)\right] \\
&\quad + 2E\left[(b-X)I\{a < X < b\}\right].
\end{aligned} \tag{7.16}$$

Set $a = m_U$ and $b > a$. Then, $P(X \leq a) - P(X > a) \geq 0$ for any median, by Definition 1.8. In addition, $P(X \leq b) > P(X \leq a)$, otherwise b would be a median (see Remark 1.2). This implies the quantity given in (7.16) is strictly positive, so that $E[|X-b|] > E[|X-m_U|]$. A similar argument shows that $E[|X-b|] > E[|X-m_L|]$ when $b < m_L$.

If the median is unique, then the proof is complete. Otherwise, suppose m is in the interior of $[m_L, m_U]$. From Remark 1.2 it follows that $[P(X \leq m) - P(X > m)] = 0$. Furthermore, since $P(m_L < X < m_U) = 0$ we must have $2E\left[(b-X)I\{a < X < b\}\right] = 0$ for any two medians a, b. Then $E[|X-m|]$ is constant for any median m.

To show (iv), first assume $c > 0$. Write

$$\begin{aligned}
E[|X/c-1|] &= c^{-1}E[|X-c|] \\
&= c^{-1}E[(X-c)I\{X > c\} - (X-c)I\{X < c\}] \\
&= P(X < c) - P(X \geq c) + c^{-1}\int_c^\infty xf(x)dx - c^{-1}\int_0^c xf(x)dx.
\end{aligned}$$

The derivative may be calculated by

$$\begin{aligned}
\frac{d}{dc}E[|X/c-1|] &= 2f(c) + \frac{-2c^2f(c) - \left[\int_c^\infty xf(x)dx - \int_0^c xf(x)dx\right]}{c^2} \\
&= \frac{1}{c^2}\left[\int_0^c xf(x)dx - \int_c^\infty xf(x)dx\right].
\end{aligned}$$

Then the solution to (7.15) gives a stationary point, which must be in the interval (a, b). It is easily verified that the second derivative at a stationary point c' is positive provided $f(c') > 0$. We then note that since X is positive, $c < 0$ may be ruled out as a solution to the minimization problem, since we would then have $E[|X/c - 1|] > 1$, which can always be improved following the limit $\lim_{c \to \infty} E[|X/c - 1|] = 1$. □

7.5.3 Classes of Loss and Risk

In Example 7.5 a loss function of the form $L_\theta(d) = L^*(A(\theta), d) = \rho(A(\theta) - d)$ was introduced. Such a loss function is location invariant (Definition 7.3 (iii)), and is invariant to a location transformation on the action space $\mathcal{A} = \mathbb{R}$ in the sense that $L^*(A(\theta) + a, d + a) = \rho(d + a - A(\theta) - a) = L^*(A(\theta), d)$. Location invariant loss functions are commonly induced by $\rho(x) = |x|^k$, $k \geq 1$, giving loss and risk

$$L_\theta(d) = |A(\theta) - d|^k, \quad R(\theta, \delta) = R_k(\theta, \delta) = E_\theta[|\delta(\tilde{X}) - A(\theta)|^k]. \tag{7.17}$$

For $k = 1$, risk is commonly refered to as mean absolute deviation (MAD), and for $k = 2$ mean squared error (MSE).

We may also define a scale invariant loss function $L_\theta(d) = \rho(d/A(\theta))$ (Definition 7.3 (iv)). Here, we assume $\mathcal{A} = \mathbb{R}_{>0}$. Again, we may write $L_\theta(d) = L^*(A(\theta), d)$. Then $L_\theta(d)$ is invariant to a scale transformation on the action space $\mathcal{A}$ in the sense that $L^*(bA(\theta), bd) = \rho((bd)/(bA(\theta))) = L^*(A(\theta), d)$. Scale invariant loss functions are commonly induced by $\rho(x) = |x - 1|^k$, $k \geq 1$, giving loss

$$L_\theta(d) = \left|\frac{d - A(\theta)}{A(\theta)}\right|^k, \quad R(\theta, \delta) = \bar{R}_k(\theta, \delta) = E_\theta\left[\left|\frac{\delta(\tilde{X}) - A(\theta)}{A(\theta)}\right|^k\right]. \tag{7.18}$$

We can characterize the loss defining $R_k(\theta, \delta)$ as absolute error raised to the kth power, and loss defining $\bar{R}_k(\theta, \delta)$ as relative error raised to the kth power. Following Theorem 7.5 the two are clearly realated by the identity

$$\bar{R}_k(\theta, \delta) = R_k(\theta, \delta)/|A(\theta)|^k. \tag{7.19}$$

Since risk is intended to permit a comparison between decision rules, it may seem that, given Equation (7.19), there is no need for separate risks $R_k(\theta, \delta)$ and $\bar{R}_k(\theta, \delta)$, since by Theorem 7.5 a decision rule δ will be admissible or UMR with respect to $\bar{R}_k$ *iff* is it admissible or UMR with respect to R_k. However, the problem of determing UMR decision rules can be considerably simplified if a class of decision rules $\mathcal{D}$ can be defined for which risk is constant. This is the basis for the minimum risk equivariant (MRE) estimators considered in Chapter 9.

7.6 Uniformly Minimum Risk Estimators (The Location-Scale Model)

In this section, we take what might seem to be an improbably simplistic approach to the problem of deriving an optimal estimator for a location or scale parameter. Begin with a single estimator δ_0, then for the location parameter problem we consider all estimators of the form $\delta_0 - b$, where $b \in \mathbb{R}$ is a constant, and δ_0 is assumed to be location equivariant. This defines the class of decision rules $\mathcal{D}$. For a scale parameter τ an equivariant estimator

TABLE 7.1
The following table gives the value c^* which minimizes risk $E[L(X,c)]$ *wrt* c, and the required moment conditions.

$L(X,c) =$	Risk minimizer c^*	Moment conditions
$\lvert X - c \rvert$	c^* is any median of X	$E[\lvert X \rvert] < \infty$
$(X-c)^2$	$c^* = E[X]$	$E[X^2] < \infty$
$\lvert \frac{X}{c} - 1 \rvert$	c^* is any scale median of X	$E[\lvert X \rvert] < \infty$
$\left(\frac{X}{c} - 1\right)^2$	$c^* = E[X^2]/E[X]$	$E[X^2] > 0$
$\frac{X}{c} - 1 - \log\frac{X}{c}$	$c^* = E[X]$	$E[\lvert X \rvert] < \infty$
$\lvert \log\frac{X}{c} \rvert$	c^* is any median of X	$E[\lvert \log(X) \rvert] < \infty$
$\left(\log\frac{X}{c}\right)^2$	$\log c^* = E[\log(X)]$	$E[\log(X)^2] < \infty$

δ_1 of τ^r can be similarly modified to δ_1/a for any constant $a \in \mathbb{R}_{>0}$. The search within $\mathcal{D}$ for a UMR estimator is then reduced to a search over a or b.

Of course, if we accept the sufficiency and equivariance principles of Chapter 3, the approach is at least plausible. We would then select δ_0 or δ_1 to be, in addition to equivariant, a mapping of a minimal sufficient statistic, which may also be a complete statistic. So we will proceed with optimism, assuming such an estimator can be chosen.

The problem of determining the UMR for the location parameter is resolved in the next theorem.

Theorem 7.9 (UMR Estimator for a Location Parameter) Suppose $\mathbf{X} = (X_1, \ldots, X_n) \in \mathcal{X}$ is a sample from a location invariant model $\mathcal{P} = \{p_\theta\}$, $\theta \in \Theta = \mathbb{R}$, $p_\theta(\mathbf{x}) = p_0(\mathbf{x} - \theta)$ for some fixed density p_0. Then let δ_0 be any location equivariant statistic, and define the class of procedures $\mathcal{D}_{\delta_0} = \{\delta_0 - b : b \in \mathbb{R}\}$. Let $L_\theta(d) = \rho(d - \theta)$ be a location invariant loss function, with risk function $R(\theta, \delta)$, and assume $R(\theta', \delta_0) < \infty$ for some fixed θ'. Then the following statements hold:

(i) The risk function $R(\theta, \delta)$ is constant for any $\delta \in \mathcal{D}_{\delta_0}$.
(ii) Suppose there exists b^* which minimizes $E_0[\rho(\delta_0(\mathbf{X}) - b)]$ *wrt* $b \in \mathbb{R}$. Then $\delta_0 - b^*$ is the UMR estimator within $\mathcal{D}_{\delta_0}$. ■

Proof. If $\delta \in \mathcal{D}_{\delta_0}$, then $\delta = \delta_0 - b$ for some $b \in \mathbb{R}$ which *dnd* θ. Then δ is also equivariant. We may therefore write, making use of Theorem 3.2,

$$R(\theta, \delta) = E_\theta[\rho(\delta_0(\mathbf{X}) - b - \theta)] = E_\theta[\rho(\delta_0(\mathbf{X} - \theta) - b)] = E_0[\rho(\delta_0(\mathbf{X}) - b)].$$

The final term of the identity does not depend on θ, which proves statement (i). Statement (ii) follows directly. □

UMR estimators for the scale parameter can be constructed in much the same way.

Theorem 7.10 (UMR Estimator for a Scale Parameter) Suppose $\mathbf{X} = (X_1, \ldots, X_n) \in \mathcal{X}$ is a sample from a scale invariant model $\mathcal{P} = \{p_\theta\}$, $\theta \in \Theta = \mathbb{R}_{>0}$, $p_\theta(\mathbf{x}) = \theta^{-n} p_1(\mathbf{x}/\theta)$ for some fixed density p_1. Then let δ_1 be any scale equivariant estimate of θ^r, and define the class of procedures $\mathcal{D}_{\delta_1} = \{\delta_1/a : a > 0\}$. Let $L_\theta(d) = \rho(d/\theta^r)$ be a scale invariant loss function, with risk function $R(\theta, \delta)$, and assume $R(\theta', \delta_1) < \infty$ for some fixed θ'. Then the following statements hold:

(i) The risk function $R(\theta, \delta)$ is constant for any $\delta \in \mathcal{D}_{\delta_1}$.

(ii) Suppose there exists a^* which minimizes $E_1[\rho(\delta_1(X)/a)]$ *wrt* $a > 0$. Then δ_1/a^* is the UMR estimator within $\mathcal{D}_{\delta_1}$. ■

Proof. The proof is similar to that of Theorem 7.9, and is left to the reader (Problem 7.16). □

Theorems 7.9 and 7.10 form the beginnings of an important principle for the estimation of location and scale parameters. Suppose we have any location equivariant statistic δ and location invariant loss $L_\theta(d) = \rho(d - \theta)$. Then the risk can be improved by subtracting a constant from δ, giving estimator $\delta^* = \delta - b$, where b minimizes $E_0[\rho(\delta - b)]$. Furthermore, such an improvement is easilily implemented. From Theorem 7.8, under MSE risk the improved estimator is $\delta - E_0[\delta]$. and under MAD risk the improved estimator is $\delta - m_0$, where m_0 is any median of δ when the true parameter value is $\theta = 0$.

A similar procedure exists the scale parameter θ. Suppose δ is a scale equivariant estimate of θ^r, and we have scale invariant loss $L_\theta(d) = \rho(d/\theta^r)$. Then the risk can be improved by introducing a multiplicative constant to δ, giving estimator $\delta^* = \delta/a$, where a minimizes $E_1[\rho(\delta/a)]$. Again, such an improvement is easilily implemented. A general form for these constants are given in Table 7.1 for selected loss functions. For relative MAD risk, a is any scale median of δ when the parameter value is $\theta = 1$ (Theorem 7.8). For relative MSE we have improved estimator

$$\delta^* = \frac{E_1[\delta]}{E_1[\delta^2]}\delta. \tag{7.20}$$

Note that by applying Jensen's inequality to Equation (7.20) we can see that any scale equivariant estimator which is admissble must have nonzero bias. On the hand, Stein loss is given by $L_\theta(d) = \rho(d/A)$, $\rho(x) = x - 1 - \log x$, which yields improved estimator

$$\delta^* = \frac{\delta}{E_1[\delta]}, \tag{7.21}$$

which is always unbiased. See Problem 7.19.

This problem will be fully resolved in Sections 9.2 and 9.3. It is worth giving a brief overview at this point. For the location parameter problem it will be shown that given a single equivariant estimator δ_0, *any other* equivariant estimator may be written $\delta = \delta_0 - w(\tilde{Z})$ for some mapping w of a maximal location invariant $\tilde{Z}$ (Definition 3.3). Using much the same argument as Theorem 7.9 it can be shown that, for example, $\delta^* = \delta_0 - E_0[w(\tilde{Z}) \mid \tilde{Z}]$ is UMR for MSE risk among all location equivariant estimators. Similarly,

$$\delta^* = \frac{E_1[\delta_1 \mid \tilde{Z}]}{E_1[\delta_1^2 \mid \tilde{Z}]}\delta_1$$

is UMR for relative MSE risk among all scale equivariant estimators, where δ_1 is any scale equivariant estimator and $\tilde{Z}$ is a maximal scale invariant. Essentialy, for the location or scale parameter problem we may use Table 7.1, conditioning on the maximal invariant.

It should also be noted that this introduces an important interaction between the theories of sufficiency and invariance. By Theorems 3.3 and 3.9 a maximal location or scale invariant $\tilde{Z}$ is also an ancillary statistic with respect to the location or scale parameter. Then if an estimator δ is a mapping of a complete sufficient statistic, by Basu's theorem it is independent of $\tilde{Z}$, and Theorems 7.9 or 7.10 can be used to derive the UMR estimator among the class of equivariant estimators (which will be refered to as the minimum risk equivariant (MRE) estimator in Chapter 9). This topic is investigated in depth in Brown (1968), which can be highly recommended.

We next present an application of Theorem 7.10.

Example 7.7 (Inadmissibility of the Sample Variance) Theorem 7.10 can be used to show that the sample variance is inadmissible. Suppose we are given an *iid* sample of size n from a normal density with unknown mean and variance μ, σ^2. The sample variance is $S^2 = \sum_{i=1}^n (X_i - \bar{X})^2/(n-1)$. Then $(n-1)S^2/\sigma^2 \sim \chi^2_{n-1}$. In addition, $E_\theta[S^2] = \sigma^2$. The variance of any χ^2_ν random variable is $\sigma^2_\nu = 2\nu$, so

$$\text{var}_\theta[S^2] = \frac{\sigma^4}{(n-1)^2}2(n-1) = \frac{2\sigma^4}{n-1}.$$

Consider relative squared error risk. By Theorems 7.8 and 7.10 the UMR estimator of σ^2 among the class $\mathcal{D} = \{aS^2, a > 0\}$ is a^*S^2, where

$$a^* = \frac{E_1[S^2]^2}{E_1[(S^2)^2]} = \frac{\sigma^4}{\sigma^4 + \frac{2\sigma^4}{n-1}} = \frac{1}{1 + \frac{2}{n-1}} = \frac{n-1}{n+1}.$$

Therefore, the estimator

$$\delta^* = \sum_{i=1}^n (X_i - \bar{X})^2/\,(n+1)$$

has strictly smaller relative squared error risk than S^2, and by Theorem 7.5 has strictly smaller squared error risk. In addition, the MLE can be shown to be $\hat{\sigma}^2 = \sum_{i=1}^n (X_i - \bar{X})^2/n$, and is therefore also strictly improved by δ^*. Thus, both S^2 and the MLE are inadmissible with respect to MSE risk. ■

7.7 Some Principles of Admissibility

Definition 7.1 suggests two principles of decision theory. First, if δ dominates δ' we would prefer δ. Second, we would only consider using admissible decision rules. We first consider a few methods by which we may determine whether or not a decision rule is admissible.

We first note that the sufficiency principle can be given some theoretical justification by the Rao-Blackwell theorem (a second version of this theorem will be given as Theorem 8.5). Simply put, under convex loss we may eliminate from consideration decision rules that are not mappings of any minimal sufficient statistic.

Theorem 7.11 (Rao-Blackwell Theorem I) Suppose T is sufficient for Θ. Let δ be an estimator of $\eta(\theta)$. We are given loss function $L_\theta(d)$, assumed to be convex in d for each fixed θ. Suppose δ has finite expectation and risk for all θ. Then $\hat{\eta}(T)$ has no smaller risk than δ, where

$$\hat{\eta}(t) = E_\theta[\delta(\tilde{X}) \mid T = t].$$

Furthermore, if L_θ is strictly convex in d then $\hat{\eta}(T)$ has uniformly smaller risk, unless $\delta(\tilde{X}) \equiv \hat{\eta}(T)$ *wp*1. ■

Proof. By the sufficiency of T, $\hat{\eta}(T)$ *dnd* θ. Applying Jensen's inequality directly, we have, under the assumption of finite risk for δ,

$$E_\theta[L(\eta(\theta), \delta(\tilde{X})) \mid T = t] \geq L(\eta(\theta), E_\theta[\delta(\tilde{X}) \mid T = t]) = L(\eta(\theta), \hat{\eta}(t)). \tag{7.22}$$

Then substitute T for t in the preceding inequality, and take the expected values. Finally, if L is strictly convex in d, the inequality (7.22) is strict, unless $P_\theta(\delta(\tilde{X}) = \hat{\eta}(t) \mid T = t) = 1$. □

We next present a simple application of Theorem 7.11.

Example 7.8 (A Simple Application of the Rao-Blackwell Theorem) Consider the model of Example 7.6. Define a third estimator

$$\hat{p}_3 = \frac{3}{4}X_1 + \frac{1}{4}X_2,$$

which is clearly an unbiased estimate of p. We also know that $X_1+X_2 \sim bin(2,p)$ is minimal sufficient for p. We can verify that, for $i = 1, 2$,

$$\begin{aligned} E_p[X_i \mid X_1 + X_2 = 0] &= 0 \\ E_p[X_i \mid X_1 + X_2 = 1] &= 1/2 \\ E_p[X_i \mid X_1 + X_2 = 2] &= 1, \end{aligned}$$

or, more compactly, $E_p[X_i \mid X_1 + X_2 = x] = x/2$. Applying Theorem 7.11 yields the new predictor

$$\hat{p}_4 = E_p[\hat{p}_3 \mid X_1 + X_2] = \frac{3}{4}(X_1 + X_2)/2 + \frac{1}{4}(X_1 + X_2)/2 = (X_1 + X_2)/2.$$

Then $\text{var}_p[\hat{p}_3] = (5/8)p(1-p)$ and $\text{var}_p[\hat{p}_4] = (1/2)p(1-p)$, verifying the reduction in variance predicted by the Rao-Blackwell theorem. ■

We next consider the randomized decision rule. Since risk is well defined for this class, we must consider the possibility that a nonrandomized decision rule is dominated by a randomized decision rule, and is therefore inadmissible. There are three cases. It may happen that an optimal decision rule must be randomized. This is sometimes the case with minimax decision rules (Section 7.10). We will also see examples for which an optimal nonrandomized decision rule exists, but randomized decision rules also exist which have an identical risk function (see Theorem 9.5 and Remark 9.1). This may occur with, for example, MAD risk and is technically related to the fact that the optimization problems of Theorem 7.8 (ii) and (iv) do not necessarilly have a unique solution. Finally, it may be the case that any randomized decision rule is inadmissible. The next theorem shows that for convex loss, no randomized decision rule can dominate all nonrandomized decision rules, and for strictly convex loss all randomized decision rules are inadmissible.

Theorem 7.12 (Convex Loss and Randomized Decision Rules) Suppose $\delta(\tilde{X}, U)$ is a randomized decision rule, and we are given a convex loss function $L_\theta(d)$. Then there exists a nonrandomized decision rule δ^* with $R(\theta, \delta^*) \leq R(\theta, \delta)$, $\theta \in \Theta$. If $L_\theta(d)$ is strictly convex then $R(\theta, \delta^*) < R(\theta, \delta)$, $\theta \in \Theta$. ■

Proof. For any randomized estimator $\delta(\tilde{X}, U)$ $\tilde{X}$ is sufficient for θ. Then set $\delta^*(\tilde{X}) = E_\theta[\delta(\tilde{X}, U) \mid \tilde{X}]$. Since $\tilde{X}$ is sufficient, $\delta^*(\tilde{X})$ is a valid decision rule. The theorem follows directly from the Rao-Blackwell theorem Theorem 7.11. □

The next theorem follows Theorem 1.7.10 of Lehman and Casella (1998). Essentially, under strictly convex loss, the risk function of any admissible decision rule is unique.

Theorem 7.13 (Uniqueness of Risk for Admissible Decision Rules Under Strictly Convex Loss) Suppose $L_\theta(d)$ is strictly convex, and δ is admissible. If for decision rule δ' $R(\theta, \delta) \equiv R(\theta, \delta')$ then $\delta \equiv \delta'$. ■

Proof. Define new decision rule $\delta^* = (\delta + \delta')/2$. If $\delta \neq \delta'$ then by hypothesis $L_\theta(\delta^*) < (L_\theta(\delta) + L_\theta(\delta^*))/2$, and so $R(\theta, \delta^*) < (R(\theta, \delta) + R(\theta, \delta^*))/2$. But this contradicts the hypothesis that δ is admissible, therefore $\delta \equiv \delta'$. □

It may be useful to known when admissibility is preserved under a transformation of a decision problem. The following theorem considers affine transformations under MSE risk. We will make use of this theorem in Section 7.10.2 (see also Problem 7.25).

Theorem 7.14 (Admissibility is Preserved Under Affine Transformations) If δ is an admissible, inadmissible or UMR estimate of θ under MSE risk, then $\delta^* = a\delta + b$ is an admissible, inadmissible or UMR estimate of $\theta^* = a\theta + b$ for any constants $a \neq 0$, b. ■

Proof. We have $R(\theta^*, \delta^*) = E_{\theta^*}[(\delta^* - \theta^*)^2] = E_\theta[(a\delta + b - a\theta - b)^2] = a^2 R(\theta, \delta)$. The transformation therefore preserves strict ordering of risk functions, which define the admissibility or UMR properties. □

As a practical matter, questions of admissibility often concern estimators of the form $\delta_{a,b}(\tilde{X}) = aT + b$, where $T = T(\tilde{X})$ is an unbiased estimator of $\theta \in \Theta$. The motivation is the construction of Bayesian estimators or shrinkage estimators, in which case we may write

$$\delta_{\lambda,\gamma}(\tilde{X}) = \frac{1}{1+\lambda} T + \frac{\lambda}{1+\lambda} \gamma = aT + b. \tag{7.23}$$

If $\delta_{\lambda,\gamma}$ is a shrinkage estimator we would have $\lambda \geq 0$ and $\gamma \in \Theta$, but it is best to allow at first the constants a, b to assume any value.

If $a = 0$ and $b \in \Theta$ then $\delta_{a,b}$ is admissible, since $R(b, \delta_{0,b}) = 0$. Following Theorem 5.2.6 of Lehmann and Casella (1998) we may rule out admissibility of $\delta_{a,b}$ for various choices of a, b.

Theorem 7.15 (Inadmissibility Cases for $aT + b$) Suppose $\delta_{a,b}(\tilde{X}) = aT(\tilde{X}) + b$ is an estimate of θ, where $E[T(\tilde{X})] = \theta$, and $\text{var}[T] < \infty$. Then $\delta_{a,b}(\tilde{X})$ is inadmissible under MSE risk if (i) $a \notin [0, 1]$; or (ii) $a = 1$, $b \neq 0$. ■

Proof. The MSE risk is given by

$$R(\theta, \delta_{a,b}) = a^2 \text{var}[T] + [(a-1)\theta + b]^2 .$$

(i) If $a > 1$ then

$$R(\theta, \delta_{a,b}) > \text{var}[T] = R(\theta, \delta_{1,0}),$$

so that $\delta_{a,b}$ is inadmissible. If $a < 0$ then

$$R(\theta, \delta_{a,b}) \geq [(a-1)\theta + b]^2 = (a-1)^2[\theta + b/(a-1)]^2 > R(\theta, \delta_{0,-b/(a-1)}).$$

so that $\delta_{a,b}$ is inadmissible.

(ii) If $a = 1$ and $b \neq 0$ then

$$R(\theta, \delta_{a,b}) = \text{var}[T] + b^2 > \text{var}[T] = R(\theta, \delta_{1,0}),$$

so that $\delta_{a,b}$ is inadmissible. □

In effect, Theorem 7.15 states that only unbiased estimators and shrinkage estimators can be admissible under MSE risk.

7.8 Admissibility for Exponential Families (Karlin's Theorem)

The question of admissibility for single parameter exponential families was resolved to a remarkable degree in Karlin (1958). We offer a somewhat more general version of this theorem due to Lehmann and Casella (1998) (Theorem 5.2.14) (proof omitted).

Theorem 7.16 (Karlin's Theorem) Suppose $\tilde{X} \in \mathcal{X}$ has density

$$p_\eta(\tilde{x}) = \beta(\eta) e^{\eta T(\tilde{x})} h(\tilde{x}), \tag{7.24}$$

where $h(\tilde{x}) > 0$ on $\mathcal{X}$. Suppose the natural parameter space $\eta \in \Theta_\eta$ (Equation (3.14)) is an interval with endpoints L, U, which may be finite or infinite. Let $\eta_0 \in \Theta_\eta$ satisfy $L < \eta_0 < U$.

Next consider the estimator $\delta_{\lambda,\gamma}$ of θ given in Equation (7.23), for any $\lambda, \gamma \in \mathbb{R}$. Define the integrals

$$I_U(\eta_U) = \int_{\eta_0}^{\eta_U} \frac{e^{-\gamma\lambda\eta}}{[\beta(\eta)]^\lambda} d\eta \quad \text{and} \quad I_L(\eta_L) = \int_{\eta_L}^{\eta_0} \frac{e^{-\gamma\lambda\eta}}{[\beta(\eta)]^\lambda} d\eta. \tag{7.25}$$

Then a sufficient condition for the admissibility of $\delta_{\lambda,\gamma}$ under MSE risk is that $I_U(\eta_U)$ and $I_L(\eta_L)$ approach ∞ as η_U and η_L approach U and L, respectively. ■

Theorem 7.16 provides a simple, if somewhat unexpected, sufficient condition for the admissibility of $T(\tilde{X})$ itself.

Theorem 7.17 (Corollary to Karlin's Theorem) Under the conditions of Theorem 7.16, if the natural parameter space is $\Theta = (-\infty, \infty)$, then $T(\tilde{X})$ is an admissible estimator for θ under MSE risk. ■

Proof. First note that $T(\tilde{X}) = \delta_{\lambda,\gamma}$ for $\lambda = 0$. In this case, $I_U(\eta_U) = \int_{\eta_0}^{\eta_U} 1 d\eta$, $I_L(\eta_L) = \int_{\eta_L}^{\eta_0} 1 d\eta$. □

In the next three examples, we consider the admissibility of commonly used MLEs. We find that the MLEs for the binomial parameter and the normal mean are admissible. The MLE for the gamma scale parameter is unbiased, and we can conclude from Theorem 7.10 that it is inadmissible under MSE risk (the argument would be essentially the same as that of Example 7.7). However, we will use Karlin's theorem as well as Theorem 7.10 to determine the exact class of admissible shrinkage estimators.

Example 7.9 (Admissibility of the MLE of a Binomial Parameter) The natural parameter for $X \sim bin(n, \theta)$ is $\log(\theta/(1-\theta))$. Expressed in terms of the MLE for θ, $T(X) = X/n$, the parameter becomes $\eta = n\log(\theta/(1-\theta))$, conforming to the notation of Theorem 7.16. The natural parameter space is $\eta \in (-\infty, \infty)$, so by Theorem 7.17 $T(X)$ is admissible. On the other hand, in Brown *et al.* (1992) it is shown that the MLE of the binomial variance is admissible for squared error loss for $n \leq 5$, but inadmissible for $n \geq 6$. ■

Example 7.10 (Admissibility of the MLE of a Normal Mean) Let $\mathbf{X} = (X_1, \ldots, X_n)$ be an *iid* sample with $X_1 \sim N(\mu, \sigma^2)$, and assume σ^2 is known. Following Example 7.9 we can confirm the admissibility of the sample mean (the MLE) $T(\mathbf{X}) = \bar{X}$ as an estimate of μ. The parameter for $\bar{X}$ is $\eta = n\mu/\sigma^2$, the natural parameter space is $\eta \in (-\infty, \infty)$, so by Theorem 7.17 $T(\mathbf{X})$ is admissible. ■

Example 7.11 (For the Gamma Scale Parameter the MLE is Inadmissible, but Shrinkage Estimators are Admissible) Suppose we observe $X \sim gamma(\tau, \alpha)$, and assume α is known. The MLE of τ is $T(X) = X/\alpha$. Then $T(X)$ is unbiased and inadmissible with respect to MSE risk by Theorem 7.10. In fact, any estimator of the form aX is inadmissible if $a \neq 1/(\alpha + 1)$. A similar argument may be used to shown that $aX + b$ is inadmissible if $a > 1/(\alpha + 1)$ and $b > 0$ (Problem 7.24).

Then consider the class of estimators $\delta_{\lambda,\gamma}$ of Equation (7.23). To apply Theorem 7.16, note that the parameter for $T(X)$ is $\eta = -\alpha/\tau$, with natural parameter space $\eta \in (-\infty, 0)$. We then have

$$\beta(\eta) = |\eta|^{\alpha} / (\alpha^{\alpha}\Gamma(\alpha)),$$

conforming to the notation of Equation (7.24). The next step is to evaluate the intergrals of Equation (7.25). Choose any $\eta_0 \in (-\infty, 0)$, then write

$$I_U(\eta) = \int_{\eta_0}^{\eta} \alpha^{\alpha}\Gamma(\alpha) \frac{e^{-\gamma\lambda\eta}}{|\eta|^{\lambda\alpha}} d\eta.$$

As η approaches zero from below, $I_U(\eta)$ diverges if $\lambda\alpha \geq 1$. Then consider the integral

$$I_L(\eta) = \int_{\eta}^{\eta_0} \alpha^{\alpha}\Gamma(\alpha) \frac{e^{-\gamma\lambda\eta}}{|\eta|^{\lambda\alpha}} d\eta.$$

We need to enumerate conditions under which $I_L(\eta)$ diverges as η approaches $-\infty$, noting that after examining $I_U(\eta)$ we may assume that $\lambda\alpha \geq 1$. This means we must have $\gamma \geq 0$. If $\gamma = 0$, we must have $\lambda\alpha \leq 1$. If $\gamma > 0$, $I_L(\eta)$ diverges for any λ.

We finally conclude, therefore, that $aX + b$ is admissible if and only if either $a = 1/(\alpha+1)$ and $b = 0$, or $a \in [0, 1/(\alpha+1)]$ and $b > 0$. ∎

7.9 Bayes Decision Rules

Given a Bayesian model, $f(\tilde{x} \mid \theta)$ with prior π, we informally introduced in Section 4.8 the Bayesian estimator of $\eta(\theta) \in \mathcal{A}$ as $\delta_B(\tilde{X}) = E[\eta(\theta) \mid \tilde{X}]$. More formally, given a proper prior π, a Bayes decision rule $\delta_B \in \mathcal{A}$ is decision rule which minimizes a weighted average of risk, referred to as Bayes risk, defined as

$$R_{\pi}(\delta) = \int_{\Theta} R(\theta, \delta)\pi(\theta) d\nu_{\Theta}(\theta).$$

The assumption that π is a proper prior allows comparison of finite Bayes risk between competing decision rules. However, Bayes risk may also be interpreted as the expected loss under the joint density $f(\tilde{x} \mid \theta)\pi(\theta)$ of $(\theta, \tilde{X})$:

$$\begin{aligned} R_{\pi}(\delta) &= \int_{\Theta} \left[\int_{\mathcal{X}} L_{\theta}(\delta(\tilde{x})) f(\tilde{x} \mid \theta) d\nu_{\mathcal{X}}(\tilde{x}) \right] \pi(\theta) d\nu_{\Theta}(\theta) \\ &= \int_{\Theta} \int_{\mathcal{X}} L_{\theta}(\delta(\tilde{x})) f(\tilde{x} \mid \theta)\pi(\theta) d\nu_{\mathcal{X}}(\tilde{x}) d\nu_{\Theta}(\theta) \\ &= E[L_{\theta}(\delta(\tilde{X}))]. \end{aligned}$$

A Bayes decision rule intended to estimate a parameter will also be refered to as a Bayes estimator of that parameter.

We can reformulate the optimal prediction theorem (Theorem 7.1) to give a general method of deriving such decision rules.

Theorem 7.18 (Evaluation of Bayes Decision Rules) Suppose we are given a Bayesian model $f(\tilde{x} \mid \theta)$, $\pi(\theta)$ and loss function $L_{\theta}(\delta)$. Suppose there exists decision rule $\delta_0 \in \mathcal{A}$ for which $R_{\pi}(\delta_0) < \infty$. Note that we may express Bayes risk as $E[E[L_{\theta}(\delta(\tilde{X})) \mid \tilde{X}]]$. Then suppose for each $\tilde{x} \in \mathcal{X}$ the function

$$h_{\tilde{x}}(d) = E[L_{\theta}(d) \mid \tilde{X} = \tilde{x}], \tag{7.26}$$

is minimized by some $d^*_{\tilde{x}} \in \mathcal{A}$. Then $\delta^*(\tilde{x}) = a^*_{\tilde{x}}$ is the Bayes decision rule *wrt* π. ∎

Proof. The proof is essentially the same as for Theorem 7.1. □

Note that the definition of Bayes risk assumes that π is a proper prior. However, the procedure of Theorem 7.18 is well defined without this restriction, and the minimization of $E[L_\theta(d) \mid \tilde{X} = \tilde{x}]$ with respect to d requires only that the posterior density $\pi(\theta \mid x)$ is well defined. If no assumption on the prior density is made, then we can refer to the decision rule δ^* described in Theorem 7.18 as a generalized Bayes decision rule. However, it must be remembered that a formal Bayes decision rule minimizes Bayes risk, and must therefore itself have finite Bayes risk.

To some degree, Bayes decision rules may be evaluated using the methods introduced in Section 7.6. This is certainly true for location invariant loss functions, as we show in the next example. However, for other forms of loss, the fact that the parameter is random will mean that the two approaches are not entirely comparable.

Example 7.12 (Bayes Decision Rules for Location Invariant Loss) Suppose we wish to derive the generalized Bayes decision rule for an estimand $\eta(\theta)$ under squared error loss. We then minimize $E[(\eta(\theta) - d)^2 \mid \tilde{X} = \tilde{x}]$ *wrt* d. By Theorem 7.8 this is given by $\delta_B(\tilde{X}) = E[\eta(\theta) \mid \tilde{X}]$, the mean of $\eta(\theta)$ under the posterior density $\pi(\theta \mid \tilde{x})$. Thus, the Bayesian estimator introduced in Section 4.8 is in the present context a generalized Bayes decision rule under squared error loss.

Next, we may wish to derive the generalized Bayes decision rule for the same estimand $\eta(\theta)$, but under absolute deviation loss. We then minimize $E[|\eta(\theta) - d| \mid \tilde{X} = \tilde{x}]$ *wrt* d. By Theorem 7.8 this is given by setting $\delta_B(\tilde{X})$ equal to a median of $\eta(\theta)$ under the posterior density $\pi(\theta \mid \tilde{x})$.

In both cases, evaluation of $\delta_B(\tilde{X})$ will depend on the prior π, but it need not be proper. ■

We have seen that the Jeffreys prior (Section 3.11.2) for a location or scale parameter is improper, in which case it may be used to construct generalized Bayes decision rules, but not formal Bayes decision rules.

Example 7.13 (Normal Scale Parameter with Jeffreys Prior) We are given an *iid* sample $\mathbf{X} = (X_1, \ldots, X_n)$, $X_1 \sim N(0, \tau)$. Let $S = \sum_i X_i^2$. Since $\tau^{1/2}$ is a scale parameter, Jeffreys prior for τ is $\pi(\tau) \propto \tau^{-1}$. Under this prior we have posterior density $\pi(\tau \mid \mathbf{x}) \propto \tau^{-\frac{n}{2}-1} e^{-\frac{s}{2\tau}}$, where $s = \sum_i x_i^2$. We may conclude that $\tau \mid \mathbf{X} = \mathbf{x} \sim igamma(2/s, n/2)$. The generalized Bayes estimate under squared error loss $L(\tau, d) = (d - \tau)^2$ is then $\delta_B(\mathbf{X}) = E[\tau \mid \mathbf{X}] = S/(n-2)$, which actually has positive bias. Under scaled squared error loss $L(\tau, d) = (d - \tau)^2/\tau^2$ we have

$$\delta_B^*(\mathbf{X}) = \frac{E[1/\tau \mid \mathbf{X}]}{E[1/\tau^2 \mid \mathbf{X}]} = \frac{S}{n+2},$$

noting that $1/\tau \mid \mathbf{X} = \mathbf{x} \sim gamma(s/2, n/2)$. ■

7.9.1 Bayes Classification

For the classification problem, Θ is a set of categorical values, or labels, which we can give as $\Theta = \{1, \ldots, m\}$. We assume there is a prior distribution $\pi(i) = P(\theta = i)$, and conditional densities $f(\tilde{x} \mid \theta)$ for some observation $\tilde{X} \in \mathcal{X}$. A decision rule takes the form $\delta(\tilde{x}) \in \Theta$, the objective of which is to attain as high a probability as possible for the event $\{\delta(\tilde{X}) = \theta\}$. Accordingly, the appropriate loss function is classification error

$$L_\theta(a) = I\{\theta \neq a\},$$

although in more general decision problems alternative loss functions may be used. To apply Theorem 7.18, set

$$h_{\tilde{x}}(a) = P(\theta \neq a \mid \tilde{X} = \tilde{x}) = 1 - \pi(a \mid \tilde{x}). \tag{7.27}$$

The Bayes decision rule is then

$$\delta^*(\tilde{x}) = \text{argmax}_{a \in \Theta} \pi(a \mid \tilde{x}) = \text{argmax}_{a \in \Theta} f(\tilde{x} \mid a)\pi(a), \tag{7.28}$$

the identity holding since the numerator of the expression $\pi(a \mid x) = f(x \mid a)\pi(a)/f(x)$ is common to all classes. In the context of classification theory, δ^* is conventionally referred to as the Bayes classifier, and is simply the value $a \in \Theta$ which maximizes the posterior density of θ given observation $\tilde{X}$.

This is quite an important result, since it puts a theoretical limit on the accuracy of *any* classifier, assuming common assumptions and criterion. Recently, there has been increased interest in classification algorithms in the context of artificial intelligence, often supported by theories of cognitive science rather than traditional optimization theory. It is sometimes reported that such algorithms outperform humans.

In the view of this author, this should be expected. That Bayes classifiers are optimal is a mathematical certainty, and many widely used classification algorithms (linear/quadratic discriminant analysis, K-nearest neighbor algorithms, for example) are essentially approximate Bayes classifiers, distinguished by the method with which they estimate the conditional densities $f(\tilde{x} \mid \theta)$. It also seems reasonable to suppose that a computer algorithm would be better than a human at both estimating $f(\tilde{x} \mid \theta)$ and consistently implementing a Bayes classifier. Of course, the corollary is that the most intelligent choice of classifier is the Bayes classifier.

The following straightforward example demonstrates the method.

Example 7.14 (Benford's Law) According to Benford's Law (or the *first-digit law*), the frequency distribution of the leading digit in many sets of numerical data is not uniform. Rather, smaller digits tend to occur more frequently, according to the frequencies $p_i = \log_{10}((i+1)/i)$ (the digit '0' is not represented). It may be verified that $p_1 + \ldots + p_9 = 1$. For example, '1' occurs with a frequency of about $p_1 = 0.30$, and '5' or '9' occur with frequencies of about $p_5 = 0.079$ or $p_9 = 0.046$.

This has been observed in accounting documents, and can therefore be used in forensic accounting applications to detect fraud (Bhattarcharya and Kumar, 2008). We may suspect, for example that forged numbers are randomly generated, and therefore each leading digit would appear with equal frequencies $\alpha = 1/9$.

Suppose a classifier is to be developed based on the observed frequencies $\mathbf{X} = (n_1, n_5, n_9)$ of the leading digits that are '1', '5' or '9' in a given document. Let $n = n_1 + n_5 + n_9$. We wish to classify an accounting document as **(A)uthentic** or **(F)orged**. The relative frequencies of these three digits (assuming only these are observed in the data) are

$$\mathbf{P}_A = \left(\frac{0.30}{N}, \frac{0.079}{N}, \frac{0.046}{N}\right), \quad N = 0.30 + 0.079 + 0.046,$$
$$\mathbf{P}_F = (1/3, 1/3, 1/3),$$

for the two classes. For convenience, set $\mathbf{P}_A = (p_1, p_5, p_9)$, $\mathbf{P}_F = (q_1, q_5, q_9)$. Let π_A be the prior probability that a document is authentic. We have

$$P(\mathbf{X} \mid \text{A}) = \binom{n}{n_1, n_5, n_9} p_1^{n_1} p_5^{n_5} p_9^{n_9}$$
$$P(\mathbf{X} \mid \text{F}) = \binom{n}{n_1, n_5, n_9} q_1^{n_1} q_5^{n_5} q_9^{n_9}.$$

Then based on Equation (7.28) we predict that the document is authentic if

$$\frac{p_1^{n_1} p_5^{n_5} p_9^{n_9}}{q_1^{n_1} q_5^{n_5} q_9^{n_9}} \left(\frac{\pi_A}{1 - \pi_A} \right) > 1,$$

and forged if this ratio is less than 1.

A numerical example is instructive. Suppose $\mathbf{X} = (7, 5, 8)$, and take $\pi_A = 1/2$. Then the posterior probability that the document is forged is

$$\begin{aligned} P(\mathrm{F} \mid \mathbf{X}) &= \frac{P(\mathbf{X} \mid \mathrm{F}) \times (1 - \pi_A)}{P(\mathbf{X} \mid \mathrm{A}) \times \pi_A + P(\mathbf{X} \mid \mathrm{F}) \times (1 - \pi_A)} \\ &= \frac{q_1^{n_1} q_5^{n_5} q_9^{n_9} \times (1 - \pi_A)}{p_1^{n_1} p_5^{n_5} p_9^{n_9} \times \pi_A + q_1^{n_1} q_5^{n_5} q_9^{n_9} \times (1 - \pi_A)} \approx 0.999. \end{aligned}$$

■

We next consider how the K-nearest neighbor (KNN) classifier may be interpreted as an approximate Bayes classifier.

Example 7.15 (The K-Nearest Neighbor (KNN) Classifier) Let $(\mathcal{X}, d)$ be a metric space. Suppose have a training data $(Y_i, \tilde{x}_i)$, $i = 1, \ldots, n$, where $Y_i \in \mathcal{A} = \{1, \ldots, m\}$ is an observed class paired with predictor data $\tilde{x}_i \in \mathcal{X}$. Let $n_j = \sum_{i=1}^n I\{Y_i = j\}$. Fix an integer $K \geq 1$, and for any $\tilde{x} \in \mathcal{X}$ define the neighborhood

$$\mathcal{N}_K(\tilde{x}) = \{i : \text{rank of } d(\tilde{x}_i, \tilde{x}) \text{ no greater than } K\},$$

that is, $\mathcal{N}_K(\tilde{x})$ consists of the indices of the K features nearest to $\tilde{x}$ (note that $\tilde{x}$ need not be represented in the data set). Let $n_j(\tilde{x})$ be the number observed classifications $Y_i = j$ in $\mathcal{N}_K(\tilde{x})$. Then a KNN classifier is defined as

$$\begin{aligned} \delta(\tilde{x}) &= j \ \text{ if } j \text{ is the most frequent class in } \mathcal{N}_K(\tilde{x}) \\ &= \mathrm{argmax}_{j \in \mathcal{A}} n_j(\tilde{x}), \end{aligned} \tag{7.29}$$

where ties may be resolved randomly (Fix and Hodges, 1952).

This can be seen as an approximation of a Bayes classifier. Suppose all observations $\tilde{x}_i$, $i \in \mathcal{N}_K(\tilde{x})$ are contained in a neighborhood $E \subset \mathcal{X}$ of $\tilde{x}$. Let ν be some measure on $\mathcal{X}$. Then

$$\frac{K}{n} \approx P(\tilde{x} \in E) = \int_E f(u) d\nu \approx c_E f(\tilde{x}),$$

where $c_E = \int_E d\nu$. We then have approximations

$$\begin{aligned} f(\tilde{x}) &\approx c_E^{-1} \frac{K}{n}, \\ f(\tilde{x}, y) &\approx c_E^{-1} \frac{n_y(\tilde{x})}{n}, \\ f(\tilde{x} \mid y) &\approx c_E^{-1} \frac{n_y(\tilde{x})}{n_y}. \end{aligned} \tag{7.30}$$

This allows us to deduce the "default" prior probabilities $\pi^*(y)$ used by the KNN classifier by the identity $f(\tilde{x}, y) = f(\tilde{x} \mid y) \pi^*(y)$, giving

$$\pi^*(y) \approx n_y / n. \tag{7.31}$$

Then, since

$$f(y \mid \tilde{x}) = f(\tilde{x}, y) / f(\tilde{x}) \approx n_y(\tilde{x}) / K,$$

the KNN classifier $\delta(\tilde{x})$ is an approximate Bayes classifer with prior probabilities (7.31) equal to the prevalence of each class in the sample.

If we wish to base the classifier on *any* prior distribution $\pi(y)$ we have the means to do so, since under the Bayesian model, of the densities of Equation (7.30) exactly one, $f(\tilde{x} \mid y)$, does not depend on $\pi(y)$. Therefore, we can introduce the prior implicitly by setting $f_\pi(\tilde{x}, y) = f(\tilde{x} \mid y)\pi$. The posterior probability is then evaluated as

$$f_\pi(y \mid \tilde{x}) = \frac{f(\tilde{x} \mid y)\pi(y)}{\sum_{j=1}^m f(\tilde{x} \mid j)\pi(j)} \approx \frac{\frac{n_y(\tilde{x})}{n_y}\pi(y)}{\sum_{j=1}^m \frac{n_j(\tilde{x})}{n_j}\pi(j)}$$

and the KNN classifier, under prior distribution $\pi(y)$, becomes

$$\delta_\pi(\tilde{x}) = \text{argmax}_{j \in \mathcal{A}} \frac{n_j(\tilde{x})\pi(j)}{n_j} \approx \text{argmax}_{j \in \mathcal{A}} f_\pi(j \mid \tilde{x}). \tag{7.32}$$

This is equivalent to the conventional KNN classifier (7.29) when $\pi(y)$ is the class prevalence within the data. Note that δ_π of Equation (7.32) is not the same type of adjustment as distance weighting, under which observations in $\mathcal{N}_K(\tilde{x})$ are weighted by their proximity to $\tilde{x}$ (the problem of optimal distance weighting schemes is considered in, for example, Samworth *et al.* (2012)). ∎

7.10 Admissibility and Optimality

A global procedure solves a decision problem over all possible decision rules (we do not consider the possibility of a decision rule which is not Borel measurable, but an optimization theory which anticipates this possibility does exist, for example, in Bertsekas and Shreve (1978)). Such procedures define a functional for a risk function which yields a single value, and on this basis any two decision rules can be compared. A Bayes decision rule minimizes a weighted average of the risk function, and so is in this sense a global procedure.

Another form of this global procedure is to apply the minimax criterion. For each decision rule we determine the maximum risk:

$$R_{max}(\delta) = \sup_{\theta \in \Theta} R(\theta, \delta).$$

We then prefer δ to δ' if $R_{max}(\delta) < R_{max}(\delta')$. Accordingly, we say δ^* is the minimax decision rule if $R_{max}(\delta^*) \leq R_{max}(\delta)$ for any other decision rule δ. We have, for example, noted that in Example 7.6 the decision rule with the smallest average risk also had the smallest maximum risk $R_{max}(\hat{p}_i)$ (Problem 7.12).

7.10.1 Admissibility and Bayes Procedures

We will need to qualify somewhat our definition of admissibility for Bayes decision rules. In particular, δ is inadmissible *a.e.* $[\pi]$ if there exists another decision rule δ' such that $R(\theta, \delta') \leq R(\theta, \delta)$ *a.e.* $[\pi]$, and the set $\{R(\theta, \delta') < R(\theta, \delta)\}$ is of strictly positive measure with respect to π, where π is a density on Θ. Otherwise δ is admissible *a.e.* $[\pi]$.

A fundamental problem of decision theory is to characterize the degree to which the class of admissible procedures overlaps the class of Bayesian procedures. That the overlap is significant seems reasonable, even if the two classes are not identical. It is easy to characterize a large class of Bayes decision rules which are admissible, which we do in the following two theorems.

First, uniqueness implies admissibility, which will hold in general for strictly convex loss.

Theorem 7.19 (Admissibility of Unique Bayes Decision Rules) Suppose δ_B is the unique Bayes decision rule for risk $R(\theta, \delta)$ under prior π. Then δ_B is admissibile *a.e.* $[\pi]$. ■

Proof. Suppose there exists δ' such that $R(\theta, \delta') \leq R(\theta, \delta_B)$ *a.e.* $[\pi]$. Then

$$\int_\Theta R(\theta, \delta')d\pi(\theta) \leq \int_\Theta R(\theta, \delta_B)d\pi(\theta),$$

which means that δ' is also a Bayes decision rule. By hypothesis, δ_B is a unique Bayes decision rule, therefore $\delta' = \delta_B$. □

We will also make used of the following theorem.

Theorem 7.20 (Admissibility of Bayes Decision Rules for Positive Priors) If δ_B is a Bayes estimator *wrt* a proper prior π, and the support of π is Θ, then δ is admissible *a.e.* $[\pi]$. ■

Proof. Suppose for some δ $R(\theta, \delta) \leq R(\theta, \delta_B)$, and $R(\theta, \delta) < R(\theta, \delta_B)$ on a set of positive probability *wrt* π. Then $R_\pi(\delta) < R_\pi(\delta_B)$. □

We may also associate admissible and minimax decision rules.

Theorem 7.21 (Minimax Decision Rules and Admissibility) If an admissible decision rule has constant risk, then it is a minimax decision rule. If the loss is strictly convex, then it is the unique minimax decision rule. ■

Proof. Suppose δ has constant risk $R(\theta, \delta) = c$. Since δ is admissible, we must have $\sup_\theta R(\theta, \delta') \geq c$. Then if δ' is any other minimax decision rule we must have $R(\theta, \delta') = c = R(\theta, \delta)$. If the loss is strictly convex, then by Theorem 7.13 $\delta = \delta'$. □

In Example 7.9 we showed that Karlin's theorem (Theorem 7.16) could be used to show that the MLE of a binomial parameter is admissible. Can we also use Theorem 7.19 or Theorem 7.20 to do this? We offer two approaches to this problem. Only one succeeds, but both are instructive.

Example 7.16 (Admissibility of the MLE of a Binomial Parameter (cont'd)) Recall Example 4.18, in which we developed a Bayes estimator for the binomial parameter based on observation $X \sim bin(n, \theta)$, under a $\pi \sim beta(\alpha, \beta)$ conjugate prior. It was shown that the expected value of the posterior density of θ is $\hat{\theta}_\pi = (X + \alpha)/(n + \alpha + \beta)$. Of course, this is equivalent to a Bayes decision rule under MSE risk.

Then, in Example 4.19 it was noted that the MLE $\hat{\theta} = X/n$ is analytically equal to $\hat{\theta}_\pi$ upon substitution of $\alpha = \beta = 0$. However, the beta density is not defined for these parameters. It seems reasonable to conjecture, nonetheless, that $\hat{\theta}$ must be a Bayes estimate under any prior π for which $P(\theta \in \{0, 1\}) = 1$, this being the limiting case as the hyperparameters α, β approach zero. Certainly, nothing in our definition of a Bayes decision rule precludes such a prior, so we will investigate this conjecture. In this case, under squared error loss, a Bayes decision rule minimizes the Bayes risk

$$R_\pi(\delta) = E[L_\theta(\delta)] = qE[L_0(\delta) \mid \theta = 0] + (1 - q)E[L_1(\delta) \mid \theta = 1] \tag{7.33}$$

where the prior satisfies $\pi(1) = 1 - \pi(0) = q$. If we set $\delta = \hat{\theta}$, then we have

$$P(\delta = 0 \mid \theta = 0) = P(\delta = 1 \mid \theta = 1) = 1. \tag{7.34}$$

Under *any* loss function we have $L_\theta(\theta) = 0$ so we conclude from (7.33) and (7.34) that $R_\pi(\delta) = 0$, so that $\hat\theta$ is a Bayes decision rule.

Can we conclude on this basis that $\hat\theta$ is admissible? Under a prior satisfying $P(\theta \in \{0,1\}) = 1$, any decision rule $\delta(\mathbf{X})$ satisfying $\delta(0) = 0$ and $\delta(1) = 1$ will also have zero Bayes risk. Thus, $\hat\theta$ is not a unique Bayes decision rule and the assumptions of Theorem 7.19 are not satisfied. In addition, the support of the prior for this example is strictly smaller than Θ, so the assumptions of Theorem 7.20 are not satisfied.

Of course, $\hat\theta$ may be a Bayes decision rule under another loss function. If we accept loss function $L_r(\theta,d) = (d-\theta)^2/[\theta(1-\theta)]$, then it may be shown that $\hat\theta$ is the unique Bayes decision rule under prior *beta*$(1,1)$ (i.e. the uniform prior), so $\hat\theta$ is admissible by Theorem 7.19. ∎

7.10.2 Shrinkage Estimates of the Binomial Parameter

We have earlier considered the problem of determining which estimators of the class $\delta_{a,b} = aX + b$ are admissible, and which are not. For example, in Example 7.11 we considered observation $X \sim gamma(\tau,\alpha)$. When α is known, the MLE of τ is $T(X) = X/\alpha$, which is unbiased.

We were able to exhaustively characterize $\delta_{a,b}$ as admissble or inadmissble for all possible choices of $a, b \in \mathbb{R}$, using four separate arguments. By Theorem 7.15 inadmissibilty follows if $a \notin [0,1]$ or $b < 0$. By Theorem 7.10, if $b = 0$, then $\delta_{a,0}$ is inadmissble if $a \neq 1/(\alpha+1)$. A similar argument shows that if $a > 1/(\alpha+1)$ and $b > 0$, then $\delta_{a,b}$. is inadmissible. Finally, from Theorem 7.16 admissibilty follows if $a = 1/(\alpha+1)$ and $b = 0$; or $a \in [0, 1/(\alpha+1)]$ and $b > 0$.

Here, we will undertake a similar analysis for the class $\delta_{a,b} = aT + b$, where $T = X/n$ is an unbiased estimate of the binomial parameter θ based on $X \sim bin(n,\theta)$. Rather than rely on Karlin's theorem, we will make use of the Bayes estimates

$$\hat\theta_B(X) = \frac{X+\alpha}{n+\alpha+\beta} = T\frac{n}{n+\alpha+\beta} + \frac{\alpha}{\alpha+\beta}\frac{\alpha+\beta}{n+\alpha+\beta}$$

obtained using the prior $\theta \sim beta(\alpha,\beta)$. Set $a = n/(n+\alpha+\beta)$, $q = \alpha/(\alpha+\beta)$, $b = (1-a)q$. For any $(a,q) \in (0,1)\times(0,1)$ we may find hyperparameters $\alpha, \beta > 0$ which satisfy these identities. Equivalently, for any (a,b) in the triangular region $E = \{(a,b) : a > 0, b > 0, a+b < 1\}$ $\delta_{a,b}$ is a Bayes decision rule with finite risk, which is also unique under squared error loss. Therefore by Theorem 7.19 all estimators in this class are admissible.

We now need to resolve the problem outside the open region E. We will first consider (a,b) outside of the closure of E, then consider (a,b) on the boundary. By Theorem 7.15 we may conclude that $\delta_{a,b}$ is inadmissible if $a < 0$ or $b < 0$. To resolve the case $a+b > 1$ note that we may write risk

$$R(\theta,\delta_{a,b}) = a^2\theta(1-\theta)/n + (b-(1-a)\theta)^2. \tag{7.35}$$

Then if $a+b > 1$ and $a \in [0,1]$ there exists $c > 0$ such that

$$\begin{aligned} R(\theta,\delta_{a,b}) &= a^2\frac{\theta(1-\theta)}{n} + (c+(1-a)-(1-a)\theta)^2 \\ &= a^2\frac{\theta(1-\theta)}{n} + (c+(1-a)(1-\theta))^2 \\ &> R(\theta,\delta_{a,1-a}), \end{aligned}$$

so that $\delta_{a,b}$ is inadmissible for this case.

At this point we may conclude that $\delta_{a,b}$ is admissible in E, and admissible outside the closure of E. The final step is therefore to resolve the boundary of E. We first use Theorem 7.16 for the case aT, $a \in (0,1]$. Following Example 7.9, for statistic $T(X) = X/n$ the parameter is $\eta = n\log(\theta/(1-\theta))$, for which the natural parameter space is $\eta \in (-\infty, \infty)$. Then, using the notation of Theorem 7.16 we have

$$\beta(\eta) = \left(1 + e^{\eta/n}\right)^{-n}$$

with $\lambda \geq 0$, $\gamma = 0$. To apply Theorem 7.16 we consider the integrals

$$I_U(\eta) = \int_{\eta_0}^{\eta} \left(1 + e^{\eta/n}\right)^{n\lambda} d\eta, \quad I_L(\eta) = \int_{\eta}^{\eta_0} \left(1 + e^{\eta/n}\right)^{n\lambda} d\eta,$$

for any $\eta_0 \in (-\infty, \infty)$. Clearly, for any $\lambda \geq 0$ the integrals $I_U(\eta)$, $I_L(\eta)$ diverge as $\eta \to \infty$, $\eta \to -\infty$, respectively. We therefore conclude that aT is admissible for all $a \in (0,1]$.

Next, consider the case $a = b = 0$. We have $R(0, \delta_{a,b}) = 0$ if and only if $b = 0$ (see Equation (7.35)). Let $R'(\theta, \delta_{a,b}) = 0$ be the derivative of $R(\theta, \delta_{a,b})$ with respect to θ. It is easily seen that $R'(\theta, \delta_{a,0}) = 0$ for $a = 0$ and $R'(\theta, \delta_{a,0}) > 0$ if $a > 0$. It follows that $\delta_{0,0}$ is admissible.

If $a = 0$ and $b \in (0,1)$ then $\delta_{0,b}$ is the only estimator which zero risk at $\theta = b$, and is therefore admissible.

We finally consider the case $a + b = 1$, $a, b \geq 0$. We have already shown that aT is an admissible estimator of θ for any $a \in [0,1]$. By essentially the same argument $\delta^* = a(1-T)$ is an admissible estimator of $1 - \theta$. By Theorem 7.14

$$1 - \delta^* = 1 - a(1-T) = aT + 1 - a$$

is an admissible estimator of θ. This resolves the final case, so we have proven the following theorem.

Theorem 7.22 (Admissible Estimates of the Binomial Parameter) Suppose we are given $X \sim bin(n, \theta)$, and $\delta_{a,b} = aT + b$ is an estimator for θ, where $T = X/n$. Then $\delta_{a,b}$ is admissible under MSE risk if and only if $(a, b) \in \bar{E} = \{(a,b) : a \geq 0, b \geq 0, a + b \leq 1\}$ ∎

The estimators identified by Theorem 7.22 as admissible are precisely the class of shrinkage estimators $\delta = pT + (1-p)\theta_0$, where $p \in [0,1]$ and $\theta_0 \in \Theta$.

There are various approaches to deriving the minimax estimator of θ. By Theorem 7.21 if we can find an admissible estimator with constant risk, then this will be a minimax estimator, which will be unique under MSE risk. Theorem 7.22 identifies all admissible estimators of the form $aT + b$. If one of these has constant risk, we have identified our minimax estimator. The technical problem is to identify coefficients a, b which eliminate dependence on θ of the risk function given in Equation (7.35). We may expand this expression as follows:

$$R(\theta, \delta_{a,b}) = \left[(1-a)^2 - \frac{a^2}{n}\right]\theta^2 + \left[\frac{a^2}{n} - 2(1-a)b\right]\theta + b^2.$$

Of the two solutions for a which force the θ^2 coefficient to zero, one ($a = \sqrt{n}/(1+\sqrt{n})$) does not violate the conditions of Theorem 7.22. It may be verified that $b = (1-a)/2$ forces the θ coefficient to zero. These coefficients satisfy the assumptions of Theorem 7.22, so we have identified the minimax estimator

$$\delta_m = pT + (1-p)(1/2),$$

where $p = \sqrt{n}/(1+\sqrt{n})$, with constant risk

$$R(\theta, \delta_m) = b^2 = 4(1+\sqrt{n})^{-2}.$$

7.11 Problems

Problem 7.1 Verify that the optimal nonrandomized decision rule of Example 7.3 is better that any strictly randomized decision rule.

Problem 7.2 We will alter slightly the rules to the "three doors" game described as the "Monty Hall Problem" (Example 7.3). There are now $m \geq 3$ doors, otherwise the rules are the same. All but one door conceals a goat, the remaining door conceals a car. The contestant picks a door (say, door 1). Monty Hall then opens one of the remaining doors to reveal a goat. The contest can then stick to door 1, or choose another door, winning the car if it is behind the final door chosen. By formulating the game as a decision problem, derive the strategy that maximizes the probability of winning the car.

Problem 7.3 Consider the MMSE predictor problem of Section 7.3.1. Assume $E[Y^2] < \infty$.

(a) Prove that any MMSE predictor $\delta(\mathbf{X})$ must satisfy $E[\delta(\mathbf{X})] = E[Y]$.
(b) Suppose $\delta_1, \delta_2 \in \mathcal{D}_2$ are two predictors of Y. Let $\epsilon = Y - \delta_1(\mathbf{X})$. If $\mathrm{cov}[\epsilon, \delta_2] \neq 0$, show how δ_2 can be used to improve δ_1. You may assume $E[\delta_1(\mathbf{X})] = E[\delta_2(\mathbf{X})] = E[Y]$. **HINT:** See Theorem 8.2.

Problem 7.4 Let $\mathbf{N} = (N_1, \ldots, N_m)$ be a multinomial vector from probabiity distribution $\mathbf{P} = (p_1, \ldots, p_m)$ and sample size n. Assume $m \geq 3$. Derive the MMSE predictor δ of N_1 based only on observation N_2. Is the predictor linear?

Problem 7.5 A random vector $\mathbf{Z} = (Z_1, \ldots, Z_3)$ has mean vector $\boldsymbol{\mu} = (201.3, 250.5, 313.9)$ and covariance matrix

$$\Sigma_{\mathbf{Z}} = \begin{bmatrix} 5.14 & -0.16 & -1.02 \\ -0.16 & 3.10 & 1.36 \\ -1.02 & 1.36 & 21.94 \end{bmatrix}$$

Derive the MMSE linear predictor of Z_1 based on observations (Z_2, Z_3).

Problem 7.6 Let $\mathbf{X} = (X_1, \ldots, X_m)$ be a vector of independent random variables with $X_j \sim pois(\lambda_j)$ $j = 1, \ldots, m$. Assume $m \geq 3$. Suppose we observe $U_1 = X_1 + \ldots + X_m$ and $U_2 = X_1 - X_m$. Find the MMSE linear predictor of X_1 based on observations (U_1, U_2).

Problem 7.7 Suppose we given a joint distribution for $(\mathbf{X}, Y)$, $Y \in \mathbb{R}$. Use Theorem 7.1 and Theorem 7.8 to prove the following.

(a) Suppose $E[|Y|] < \infty$. Then $\delta(\mathbf{X})$ minimizes the mean absolute deviation $MAD = E[|Y - \delta(\mathbf{X})|]$ *iff* $\delta(\mathbf{X})$ is equal to a median of $Y \mid \mathbf{X}$.
(b) Suppose $E[Y^2] < \infty$. Then $\delta(\mathbf{X})$ uniquely minimizes the relative MSE $E[[(Y - \delta(\mathbf{X}))/\delta(\mathbf{X})]^2]$ *iff* $\delta(\mathbf{X}) = E[Y^2 \mid \mathbf{X}]/E[Y \mid \mathbf{X}]$.
(c) Suppose $E[|Y|] < \infty$. Then $\delta(\mathbf{X})$ minimizes the relative mean absolute deviation $MAD = E[|(Y - \delta(\mathbf{X}))/\delta(\mathbf{X})|]$ *iff* $\delta(\mathbf{X})$ is equal to a scale-median of $Y \mid \mathbf{X}$.

Problem 7.8 Suppose we are given simple linear regression model

$$Y = \beta_0 + \beta_1 x + \epsilon, \quad \epsilon \sim N(0, \sigma^2).$$

Suppose, given paired observations (Y_i, x_i), $i = 1, \ldots, n$, we can calculate least squares coefficient estimates $\hat{\beta}_0, \hat{\beta}_1$. We want to predict, for some fixed predictor value x a new observation $Y_x \sim N(\beta_0 + \beta_1 x, \sigma^2)$ from the model which is independent of the observations used in the estimates. We consider two predictors:

$$\bar{Y} = n^{-1} \sum_{i=1}^{n} Y_i \approx Y_x \text{ or } \hat{Y}_x = \hat{\beta}_0 + \hat{\beta}_1 x \approx Y_x.$$

In one sense, $\hat{Y}_x$ is the correct choice, unless $\beta_1 = 0$ (which we don't rule out). One approach is to test against null hypothesis $H_0 : \beta_1 = 0$, then choose $\hat{Y}_x$ if we reject H_0. The other approach is to try to minimize prediction error directly.

Here, the square-error risk is: $MSE(Y') = E[(Y_x - Y')^2]$, where Y' is whatever predictor (that is, $\bar{Y}$ or $\hat{Y}_x$) we are considering.

(a) Express $MSE(Y')$ in terms of σ^2, and the bias and variance of Y'. Assume Y_x and Y' are independent. Note that in this case bias $[Y'] = E[Y'] - E[Y_x]$.
(b) Derive $MSE(Y')$ for $Y' = \bar{Y}$ and $Y' = \hat{Y}_x$.
(c) Give conditions on $\beta_0, \beta_1, \sigma^2, SS_X = \sum_{i=1}^{n}(x_i - \bar{x})^2$ under which $MSE(\bar{Y}) < MSE(\hat{Y}_x)$. Is it possible that $\bar{Y}$ could have smaller prediction error even if $\beta_1 \neq 0$?

Problem 7.9 Verify that the estimator $\hat{\mu}_2$ of Example 7.1 is neither unbiased nor location equivariant.

Problem 7.10 We are given an observation $X \sim bin(n, \theta)$, $\theta \in [0, 1]$. The object is to predict θ using an estimator of the form $\delta_{a,b} = a\hat{\theta} + b$, where $\hat{\theta} = X/n$, and $a, b \in \mathbb{R}$ are two fixed constants which do not depend on θ.

(a) Derive an expression for the MSE of $\delta_{a,b}$ for estimand θ.
(b) Fix some parameter value $\theta_0 \in [0, 1]$. For what values of a, b is the MSE of $\delta_{a,b}$ minimized for $\theta = \theta_0$. **HINT:** You would never seriously consider using this estimator.
(c) Suppose we wish to constrain the parameters a, b defining $\delta_{a,b}$ so that the MSE of $\delta_{a,b}$ is symmetric about $\theta = 1/2$, that is

$$\text{MSE}_\theta[\delta_{a,b}] = \text{MSE}_{1-\theta}[\delta_{a,b}], \;\; \theta \in [0, 1]. \tag{7.36}$$

Derive a linear constraint on a, b which achieves this.
(d) Suppose we consider the problem of Part (b) again. That is, fix some parameter value $\theta_0 \in [0, 1]$, and determine the estimator $\delta_{a,b}$ which minimizes MSE for $\theta = \theta_0$. However, consider only estimators which satisfy the constraint (7.36).

Problem 7.11 Consider the James-Stein estimator for mean vectors $\boldsymbol{\mu}^T = (\mu_1, \ldots, \mu_m) \in \mathbb{R}^m$ (Example 7.2). Suppose we are given $\boldsymbol{\nu} = (\nu_1, \ldots, \nu_m) \in \mathbb{R}^m$. Modify the James-Stein estimator so that it is in the form Equation (4.30), with $\boldsymbol{\nu}$ as favored hypothesis. **HINT:** Subtract $\boldsymbol{\nu}$ from both $\hat{\boldsymbol{\theta}}_{MLE}$ and $\hat{\boldsymbol{\theta}}_{JS}$.

Problem 7.12 In Example 7.6 verify that decision rule $\hat{p}_2$ has both smaller average risk and smaller maximum risk (the average risk can be taken to be the intergral of risk over p).

Problem 7.13 The notion of an unbiased estimator can be generalized by the definition of risk unbiasedness. We say a decision rule δ is risk-unbiased for loss $L_\theta(d)$ if $E_\theta[L_\theta(\delta)] \leq E_\theta[L(\theta', \delta)]$ for all $\theta \neq \theta'$.

(a) Show that for MSE risk, an estimator $\delta(\tilde{X})$ is risk unbiased if it is an estimate of its own mean.
(b) Show that for MAD risk, an estimator $\delta(\tilde{X})$ is risk unbiased if it is an estimate of any median of $\delta(\tilde{X})$.

Problem 7.14 Let $\mathcal{T}$ be a class of estimators that is closed under addition of a constant. Let $L_\theta(d) = \rho(d - \eta(\theta))$ be a location invariant loss function. Prove that a necessary condition for δ to have uniformly minimum risk in $\mathcal{T}$ is that it is risk-unbiased. **HINT:** If δ is not risk unbiased, there must exist θ_1, θ_2 for which $\eta(\theta_1) \neq \eta(\theta_2)$, and $R(\theta_1, \delta) = E_{\theta_1}[\rho(\delta(\tilde{X}) - \eta(\theta_1))] > E_{\theta_1}[\rho(\delta(\tilde{X}) - \eta(\theta_2))]$. Note also that we have not assumed that the estimators in $\mathcal{T}$ are equivariant.

Problem 7.15 Let $\mathcal{T}$ be a class of estimators that is closed under multiplication by a positive constant. Let $L_\theta(d) = \rho(d/\eta(\theta))$ be a scale invariant loss function. Prove that a necessary condition for δ to have uniformly minimum risk in $\mathcal{T}$ is that it is risk-unbiased. **HINT:** If δ is not risk unbiased, there must exist θ_1, θ_2 for which $\eta(\theta_1) \neq \eta(\theta_2)$, and $R(\theta_1, \delta) = E_{\theta_1}[\rho(\delta(\tilde{X})/\eta(\theta_1))] > E_{\theta_1}[\rho(\delta(\tilde{X})/\eta(\theta_2))]$. Note also that we have not assumed that the estimators in $\mathcal{T}$ are scale equivariant.

Problem 7.16 Prove Theorem 7.10. **HINT:** The proof is similar to that of Theorem 7.9.

Problem 7.17 Suppose $X \sim exp(\tau)$. For each of the five scale invariate loss functions of Table 7.1 find the UMR estimator of τ among the class X/a, $a > 0$.

Problem 7.18 Suppose $\mathbf{X} = (X_1, \ldots, X_n)$ is an *iid* sample with $X_1 \sim pareto(1, \alpha)$. Let $S = \sum_{i=1}^n \log(X_i)$.

(a) Find the UMR estimator of α^{-1} *wrt* MSE among the class S/a, $a > 0$.
(b) Find the UMR estimator of α *wrt* MSE among the class S^{-1}/a, $a > 0$.

Problem 7.19 Let δ be a scale equivariant estimator of scale parameter τ. Verify that the estimator of Equation (7.21) is UMR among estimators $a\delta$, $a > 0$, *wrt* Stein loss $L_\theta(d) = \rho(d/\tau)$, $\rho(x) = x - 1 - \log x$.

Problem 7.20 Suppose we may observe a vector of random counts $\mathbf{X} = (X_1, \ldots, X_m)$ which are statistically independent, with $X_i \sim pois(\lambda_i)$. The mean vector is then $\Lambda = (\lambda_1, \ldots, \lambda_m)$. Next, suppose we have a classification problem in which the vector of Poisson counts $\mathbf{X}$ comes from class A or B, defined by respective mean vectors $\Lambda_A = (\lambda_1^A, \ldots, \lambda_m^A)$ or $\Lambda_B = (\lambda_1^B, \ldots, \lambda_m^B)$.

Suppose Λ_A, Λ_B are known. Suppose that the respective classes have prior probabilities π_A, π_B. Show that the Bayes classifier can be constructed, given observation $\mathbf{X}$, from two functions of $\mathbf{X}$ of the form

$$h_A(\mathbf{X}) = a_0 + \sum_{i=1}^m a_i X_i, \quad h_B(\mathbf{X}) = b_0 + \sum_{i=1}^m b_i X_i,$$

with the prediction being A if $h_A(\mathbf{X}) > h_B(\mathbf{X})$ and B if $h_B(\mathbf{X}) > h_A(\mathbf{X})$ (with the prediction made randomly when $h_A(\mathbf{X}) = h_B(\mathbf{X})$).

Problem 7.21 A total of n ancient Roman coins are discovered scattered at an archaeological site. Each coin has a distinctive mark identifying the mint at which the coin was produced. Suppose at the time the coins were produced there existed m mints in operation, and that the exact number m is of interest to historians. We may therefore calculate frequencies N_j, $j = 1, \ldots, m$, equalling the number of coins in the collection from the jth observed mint label, so that $N_1 + \ldots + N_m = n$. If N_+ is the number of distinct mints observed in the sample, we at least know that $m \geq N_+$. Assume, for convenience, that a coin is equally likely to come from any mint, and that the mint assignments are independent. We then interpret $(N_1, \ldots, N_m)$ as a multinomial vector (but for a more subtle interpretation of these frequencies see Nayak (1992)).

(a) Suppose we wish to develop a Bayes classifier to predict m. Assume we are given a prior distribution $\pi_j = P(m = j)$. Write explicitly the posterior distribution, and show how this can be interpreted as a Bayes classifier.

(b) Suppose prior belief favors $m = 15$, and it is known that at least $m = 10$ mints existed. It is then assumed that $m \in [10, 20]$, so $\pi_j = 0$ if $j \notin [10, 20]$, and otherwise

$$(\pi_{10}, \pi_{11}, \ldots, \pi_{19}, \pi_{20}) = K \times (1^2, 2^2, \ldots, 5^2, 6^2, 5^2, \ldots, 2^2, 1^2),$$

for some normalization constant K. Then, suppose there are $n = 8$ coins, with 2 mints represented by 2 coins and 4 mints are represented by 1 coin. Plot the prior and posterior densities for m. What value of m does the Bayes classifier predict?

Problem 7.22 Consider the following classification problem. An object can be classified as type A, B or C. Instead of observing the class, we observe an index $I \in \{1, 2, 3, 4\}$. The conditional probabilities of I given each class are given in the following table:

$I =$	1	2	3	4
A	0.2	0.2	0.3	0.3
B	0.3	0.0	0.0	0.7
C	0.5	0.5	0.0	0.0

Give the Bayes classifier based on observed index I for the following two cases.

(a) The class prior probabilities are $\pi_A = \pi_B = \pi_C = 1/3$.
(b) The class prior probabilities are $\pi_A = 0.25$, $\pi_B = 0.5$, $\pi_C = 0.25$.

Problem 7.23 Suppose we may observe a vector of independent random variables $\mathbf{X} = (X_1, \ldots, X_m)$ with $X_i \sim exp(1/\lambda_i)$. The rate vector is then $\Lambda = (\lambda_1, \ldots, \lambda_m)$. Next, suppose we have a classification problem in which the vector of independent exponentially distributed random variables $\mathbf{X}$ comes from class A or B, defined by respective rate vectors $\Lambda_A = (\lambda_1^A, \ldots, \lambda_m^A)$ or $\Lambda_B = (\lambda_1^B, \ldots, \lambda_m^B)$.

Suppose Λ_A, Λ_B are known. Suppose that the respective classes have prior probabilities π_A, π_B. Show that a Bayes classifier can be constructed, given observation $\mathbf{X} = (X_1, \ldots, X_m)$, from two functions of $\mathbf{X}$ of the form:

$$h_A(X) = a_0 + \sum_{i=1}^{m} a_i X_i \text{ and } h_B(X) = b_0 + \sum_{i=1}^{m} b_i X_i,$$

with the prediction being A if $h_A(\mathbf{X}) > h_B(\mathbf{X})$ and B if $h_B(\mathbf{X}) > h_A(\mathbf{X})$. Express the coefficients a_i, b_i in terms of the parameters Λ_A, Λ_B and the prior probabilities π_A, π_B.

Problem 7.24 Verify the assertions of Example 7.11. Suppose $X \sim gamma(\theta, \alpha)$.

(a) aX is an inadmissible estimate of θ with respect to MSE risk if $a \neq 1/(\alpha + 1)$.
(b) $aX + b$ is an inadmissible estimate of θ with respect to MSE risk if $a > 1/(\alpha + 1)$ and $b > 0$.

Problem 7.25 This problem refers to Theorem 7.22. Suppose we observe $X \sim bin(n, \theta)$, and let $\hat{\theta} = X/n$ be the MLE for θ. Prove that any convex combination $p\hat{\theta} + (1 - p)\theta_1$ is an admissible estimator of any other convex combination $q\theta + (1 - q)\theta_2$, where $p, q \in (0, 1)$, $\theta_1, \theta_2 \in (0, 1)$ are known constants. **HINT:** Make use of Theorem 7.14.

Problem 7.26 Suppose we observe $X \sim bin(n, \theta)$, and let $\hat{\theta} = X/n$ be the MLE for θ. Under MSE risk the risk function for $\hat{\theta} = X/n$ approaches zero as θ approaches 0 or 1, which seems unsatisfactory. One approach would be to consider estimation error relative to standatrd deviation. This can be done using loss function $L_r(\theta, d) = \frac{(d-\theta)^2}{\theta(1-\theta)}$.

(a) Verify that L_r is a strictly convex loss function (Definition 7.3).
(b) Derive the risk function under loss L_r for estimator $\delta_{a,b} = a\hat{\theta} + b$.
(c) Determine conditions on a, b and α, β under which the Bayes risk of $\delta_{a,b}$ under loss L_r and prior $\pi \sim beta(\alpha, \beta)$ is finite.
(d) Suppose we wish to construct a Bayes decision rule under loss L_r and prior $\pi \sim beta(\alpha, \beta)$. Determine conditions on α, β under which a Bayes decision rule exists. Show that when these conditions are met, the Bayes decision rule is given by

$$\delta_\pi(X) = \frac{X + \alpha - 1}{n + \alpha + \beta - 2}.$$

HINT: Use Theorem 7.18. Show that the problem of minimizing risk under L_r loss and prior $beta(\alpha, \beta)$ is equivalent to minimizing risk under squared error loss and prior $beta(\alpha - 1, \beta - 1)$.
(e) Show that $\hat{\theta}$ is the unique minimax estimator of θ under L_r loss. **HINT:** With respect to which prior is $\hat{\theta}$ a Bayes decision rule under L_r loss?

8

Uniformly Minimum Variance Unbiased (UMVU) Estimation

8.1 Introduction

In Chapter 7, we considered a number of general criteria by which to select decision rules for parametric families. In this chapter, we define the class of unbiased estimators δ of an estimand $\eta(\theta)$, that is, estimators satisfying $E_\theta[\delta(X)] = \eta(\theta)$. The goal will be to find the minimum variance estimator in this class. This is an example of a uniformly minimum risk estimator, defined in Section 7.6. In practice, this constraint will often prove to be too restrictive. In addition, many commonly used unbiased estimators are inadmissible (Examples 7.7 and 7.11). However, the theory of unbiased estimators does provide optimal solutions to many fundamental inference problems, as well as considerable insight into problems nominally beyond its scope.

For unbiased estimators, MSE is equal to variance, so a uniformly minimum risk unbiased estimator, where risk is MSE, is commonly referred to as a uniformly minimum variance unbiased (UMVU) estimator (or UMVUE).

The UMVUE is formally defined in Section 8.2, and some of its general properties introduced. The main theoretical results for UMVUEs are presented in Section 8.3, and constitute an important part of the theory of inference. This relies primarily on the idea of sufficiency (Section 3.7) and completeness (Section 3.8). Two main theorems characterize the UMVUE, these are the Rao-Blackwell theorem (Theorem 8.5) and the Lehmann-Scheffé theorem (Theorem 8.6). Another version of the Rao-Blackwell theorem was already presented in Section 7.7, and they can usefully be considered together.

We have opted in this volume not to treat nonparametric methods as a separate topic (by this we mean primarily rank-based methods). This is because it is not really necessary, as these methods can be derived from much the same principles used for parametric inference. We will show that the UMVUE property follows from the existence of a complete sufficient statistic. Recall from Section 3.8 that the completeness property is nested. By this we mean that given a parametric family Θ, and nested subsets $\Theta_0 \subset \Theta_1 \subset \Theta$, if T is a complete statistic on Θ_0, it is also complete on Θ_1 (Theorem 3.18). This suggests the possibility that the order statistics $\mathbf{X}_{ord}$ will be complete and sufficient for a large enough family $\mathcal{P}$. This will be the case if $\mathcal{P}$ consists of all distributions satisfying some general moment or measurability conditions. This is what is refered to as a 'distribution-free' model. Our interest here is in the fact that UMVUEs can be constructed for such models based on the Rao-Blackwell and Lehmann-Scheffé theorems, normally associated with parametric inference. Much of the theory of nonparametric methods can be united under the idea of the U-statistic (Halmos, 1946; Hoeffding, 1948). This topic will be revisited in Sections 9.6 and 12.7. An important example of this method, the rank-based measures of correlation, is considered in Section 8.6.

DOI: 10.1201/9781003049340-8

8.2 Definition of UMVUE's

Unbiasedness can be considered a desirable property for an estimator, all things equal, but we cannot take this as a firm requirement. Sometimes, unbiasedness is difficult to verify, and for a number of problems unbiased estimators do not exist. We may also consider alternative criteria, for example, mean squared error, which is generally given by MSE $= var + bias^2$. In many cases we have a meaningful variance/bias tradeoff, which is usually not resolved by setting $bias = 0$.

Nonetheless, the theory of optimal unbiased estimation is essential to the theory of statistical inference, and is often quite useful even when biased estimators are of interest.

First, it is important to recall that an unbiased estimator need not exist for a given estimand $\eta(\theta)$ (Section 4.2). From Definition 4.2, if one does, that estimand is U-estimable. It will also be important to identify the class of statistics with finite variance over Θ.

Definition 8.1 (Uniformly Finite Variance Estimators) A statistic $\delta \in \mathcal{T}_{stat}$ is of uniformly finite variance if $E_\theta[\delta^2] < \infty$ for all $\theta \in \Theta$. This class will be denoted as $\mathcal{T}_{var}$. ∎

The defninition of the UMVUE follows.

Definition 8.2 (Uniformly Minimum Variance Unbiased (UMVU) Estimators) Suppose $\delta \in \mathcal{T}_{stat} \cap \mathcal{T}_{var}$ and has expectation $\eta(\theta) = E_\theta[\delta(\tilde{X})]$. Then δ is an unbiased estimator of $\eta(\theta)$. If for any other unbiased estimator of $\eta(\theta)$, say δ', we have $\text{var}_\theta[\delta] \leq \text{var}_\theta[\delta']$ for all θ, then δ is a uniformly minimum variance unbiased (UMVU) estimator of $\eta(\theta)$. We use the shorthand UMVUE for uniformly minimum variance unbiased estimator. ∎

For convenience, when there would be no ambiguity, if we say δ is an unbiased estimator, or is a UMVUE, it can be assumed that the estimand is $\eta(\theta) = E_\theta[\delta(\tilde{X})]$.

UMVUEs are Unique

The first thing that can be said about UMVUEs is that they are unique.

Theorem 8.1 (UMVUEs are Unique) If a UMVUE of $\eta(\theta)$ exists, it is the unique UMVUE. ∎

Proof. Suppose $\delta, \delta' \in \mathcal{T}_{var}$ are two UMVUEs of $\eta(\theta)$. Then we must have $\text{var}_\theta[\delta] = \text{var}_\theta[\delta']$. Define a third unbiased estimator $\delta'' = (\delta + \delta')/2$. Then, applying the Cauchy-Schwartz inequality gives

$$\begin{aligned} \text{var}_\theta[\delta''] &= \frac{1}{4}\text{var}_\theta[\delta] + \frac{1}{4}\text{var}_\theta[\delta'] + \frac{1}{2}\text{cov}_\theta[\delta, \delta'] \\ &\leq \frac{1}{4}\text{var}_\theta[\delta] + \frac{1}{4}\text{var}_\theta[\delta'] + \frac{1}{2}\left[\text{var}_\theta[\delta]\text{var}_\theta[\delta']\right]^{1/2} = \text{var}_\theta[\delta]. \end{aligned}$$

However, by hypothesis $\text{var}_\theta[\delta]$ is the best possible variance, in which case we must have equality. Since the mean and variance of δ, δ' are the same, we must have $\delta = \delta'$. □

Recall that in Section 7.3.1 it was argued that an MMSE predictor $\delta(\mathbf{X})$ of Y was optimal in the sense that it exhausted all the information in $\mathbf{X}$ predictive of Y. Mathematically, this idea was expressed as Theorem 7.4, which stated that the error term $\epsilon = Y - \delta(\mathbf{X})$ was uncorrelated with any other predictor of Y. If this were not the case, then ϵ would have additional predictive information which could be used to improve δ. This idea is made clear in the next theorem.

Theorem 8.2 (Correlation is Information) Suppose $X \in \mathcal{T}_{var}$ has variance $\sigma_X^2 > 0$, and $U \in \mathcal{T}_{var}$ has mean 0 and variance $\sigma_U^2 > 0$. Furthermore, suppose the correlation coefficient is ρ. Then there exists constant a such that

$$\mathrm{var}[X + aU] = (1 - \rho^2)\sigma_X^2.$$ ■

Proof. The expression $\mathrm{var}[X + aU]$ is a convex order two polynomial in a, which when minimized completes the proof. □

We give a simple technical application of Theorem 8.2 (compare to the best linear unbiased estimate (BLUE) of Section 6.3).

Example 8.1 (Best Linear Combinations) Suppose $E[X] = E[Y]$, and we want a linear combination $aX + bY$ of the same mean $E[X]$, but of the smallest possible variance. Set $X' = X + a(Y - X)$, and apply Theorem 8.2. See also Problem 8.2. ■

Unbiased Estimates of Zero

We next introduce a new class of statistics.

Definition 8.3 (Unbiased Estimates of Zero) $U \in \mathcal{T}_{stat}$ is an unbiased estimate of zero if $E_\theta[U] = 0$ for all $\theta \in \Theta$. We denote the class of all unbiased estimates of zero as $\mathcal{T}_{zero}$. ■

A class of unbiased statistics is essentially equivalent to the simpler class $\mathcal{T}_{zero}$.

Theorem 8.3 (Characterization of the Class of Unbiased Estimators) The class of all unbiased estimators of $\eta(\theta)$, denoted as $\mathcal{T}_\eta$, may be given by

$$\mathcal{T}_\eta = \{\delta + U : U \in \mathcal{T}_{zero}\}, \tag{8.1}$$

where δ is any single unbiased estimator of $\eta(\theta)$. ■

Proof. Clearly, $\delta + U \in \mathcal{T}_\eta$ for any $U \in \mathcal{T}_{zero}$. Conversely, for any other $\delta' \in \mathcal{T}_\eta$, we may write $\delta' = \delta + U'$ where $U' = \delta' - \delta \in \mathcal{T}_{zero}$. □

Suppose δ is an estimator of $\eta(\theta)$. If it is possible to construct the decomposition $\delta = \delta^* + U$, where U contains no information about θ, then we might expect that δ^* is a better estimator than δ, provided we can remove U. If $U \in \mathcal{T}_{zero}$ then we simply subtract $\delta^* = \delta - U$.

On the other hand, suppose δ and δ' are two distinct unbiased estimates of $\eta(\theta)$. If both contain independent information about θ, then it seems reasonable to suppose that they can be combined to yield a strictly better estimate, in effect combining all information from δ and δ'. In fact, we can see from the proof of Theorem 8.1 that any two distinct unbiased estimators of equal variance can be combined into a strictly better estimator simply by taking the average, and this idea can be extended to more general cases.

This idea is conveniently expressed using the class $\mathcal{T}_{zero}$, since all unbiased estimators can be constructed by selecting a single reference estimator δ, then allowing $\delta + U$ to range over all $U \in \mathcal{T}_{zero}$.

Theorem 8.4 (Correlation of UMVUEs) An estimator $\delta \in \mathcal{T}_\eta$ is UMVU *iff* δ is uncorrelated with all $U \in \mathcal{T}_{zero}$. ■

Proof. Suppose δ is uncorrelated with all $U \in \mathcal{T}_{zero}$. Then $\mathrm{var}_\theta[\delta + U] = \mathrm{var}_\theta[\delta] + \mathrm{var}_\theta[U] \geq \mathrm{var}_\theta[\delta]$. Since any $\delta' \in \mathcal{T}_\eta$ can be written $\delta + U$, δ must be UMVU.

To prove the converse. Suppose δ is UMVU. If for some $\theta_0 \in \Theta$, δ and $U \in \mathcal{T}_{zero}$ possesses nonzero correlation, by Theorem 8.2 there exists a for which $\mathrm{var}_{\theta_0}[\delta + aU] < \mathrm{var}_{\theta_0}[\delta]$, contradicting the hypothesis that δ is UMVU. □

Theorems 8.1, 8.3 and 8.4 suggest that the problem of finding a unique UMVUE is a well defined problem.

8.3 UMVUE's and Sufficiency

The theory of Section 8.2 makes clear the importance of decomposing estimators into informative and uniformative components. However, this approach will not generally lead to a specific construction method for UMVUE's.

In this section we show that a complete sufficient statistic essentially acts as a filter, retaining all the information about $\eta(\theta)$ contained in the data, while removing the uninformative component. Theorem 8.1 states that UMVUE's are unique. Completeness leads to a type of converse statement, in which uniqueness, following from completeness, leads to the UMVU property.

We have already made use of an interesting consequence of the total variance identity, that $E_\theta[Y \mid \tilde{Z}]$ and Y have the same expectation, but $E_\theta[Y \mid \tilde{Z}]$ has no greater variance than Y and has strictly smaller variance unless $\tilde{Z} \mapsto Y$. In the context of estimation, this means that if Y is an unbiased estimator of $\eta(\theta)$ then, provided $E_\theta[Y \mid \tilde{Z}] = h(\tilde{Z})$ is a statistic, $E_\theta[Y \mid \tilde{Z}]$ is also an unbiased estimator of $\eta(\theta)$ with strictly smaller variance, unless $\tilde{Z} \mapsto Y$, in which case the two estimators are identical. Of course, this is precisely what happens when $\tilde{Z}$ is sufficient.

We next present our second version of the Rao-Blackwell theorem. Our first version, Theorem 7.11, could be used here. However, it offers greater clarity to state the theorem in the context of UMVU estimation, while extending the result to biased estimators with the objective of minimizing MSE.

Theorem 8.5 (Rao-Blackwell Theorem II) Suppose $\delta \in \mathcal{T}_{stat}$ estimates $\eta(\theta)$, and $E_\theta[\delta^2] < \infty$ for all θ (δ may be biased). Suppose $S \in \mathcal{T}_{suff}$. Then $\delta_{RB} = E_\theta[\delta \mid S] \in \mathcal{T}_{stat}$, and has strictly smaller MSE than δ for all θ unless $S \mapsto \delta$, in which case the two estimators are identical. ■

Proof. By sufficiency, δ_{RB} is a statistic. If $S \mapsto \delta$ then $\delta = \delta_{RB}$. Otherwise, note that $MSE = bias^2 + var$. The proof is completed by noting that δ_{RB} and δ have equal bias, and δ_{RB} has strictly smaller variance by the total variance identity. □

Note that Theorem 8.5 can be modified by replacing squared error loss $\rho(x) = x^2$ with any other convex loss, as in Theorem 7.11.

The next example introduces the process of "Rao-Blackwellization". Suppose we are given a sample $\mathbf{X} = (X_1, \ldots, X_n)$, and some sufficient statistic $S(\mathbf{X})$. It will sometimes be a good strategy to define a simple statistic based on one observation only, say $\delta(X_1)$, then construct a new statistic by taking the conditional expectation $\delta^*(S(\mathbf{X})) = E[\delta(X_1) \mid S(\mathbf{X})]$.

Example 8.2 (Order Statistics and the Rao-Blackwell Theorem) Suppose $\mathbf{X} = (X_1, \ldots, X_n)$, $n > 1$, is an *iid* sample from a distribution $F \in \mathcal{F}$ on $\mathbb{R}$. Assume F is in L^1. By Example 3.20 the order statistics $\mathbf{X}_{ord} = (X_{(1)}, \ldots, X_{(n)})$ are sufficient for $\mathcal{F}$. Then $\delta_1 = X_1$ is an unbiased estimate of the first moment μ_X. Applying the Rao-Blackwell theorem gives estimator $\delta_{RB} = E_F[X_1 \mid \mathbf{X}_{ord}] = \bar{X}$, the sample mean, since $X_1 \mid \mathbf{X}_{ord}$ is equivalent to a random selection from $X_1, \ldots, X_n$. Clearly, δ_{RB} is also unbiased, and has strictly smaller variance than δ_1. This is a simple example of a U-statistic, which will be introduced in Section 8.5. ■

The Rao-Blackwell theorem states a quite powerful idea. If we are given any estimator $\delta(\tilde{X})$, and we can find some $S \in \mathcal{T}_{suff}$ for which $S \mapsto \delta$ does not hold, then δ can be strictly improved by conditioning on S. However, this possibility by itself does not resolve the problem of deriving a UMVUE.

Suppose we 'Rao-Blackwellize' an estimator δ by conditioning on $S \in \mathcal{T}_{suff}$. This gives an estimator $\delta'(S(\tilde{X}))$. However, if $T \in \mathcal{T}_{min}$ and $T \mapsto S$ does not hold, then $\delta'(S(\tilde{X}))$ can itself be strictly improved by conditioning on T. At this point, no more improvement is possible, at least in this manner, since our current best statistic is a function of T, and $S \mapsto T$ for all $S \in \mathcal{T}_{suff}$ (note that we must always condition on a sufficient statistic, otherwise the resulting mapping will not be a statistic).

If we agree that any UMVU estimator must be a function of a minimal sufficient statistic, we still have not resolved the problem. Suppose we start the 'Rao-Blackwellization' process from distinct initial unbiased estimators δ, δ'. Which estimator $E[\delta \mid T]$, $E[\delta' \mid T]$ is better? If T is complete, the two estimators are the same, which resolves the matter. We have, in effect, proven the following theorem.

Theorem 8.6 (The Lehmann-Scheffé Theorem) Suppose $T \in \mathcal{T}_{comp} \cap \mathcal{T}_{suff}$ and δ is an unbiased estimator of $\eta(\theta)$ for which $T \mapsto \delta$. Assume also that $E_\theta[\delta^2] < \infty$ for all θ. Then δ is the unique UMVUE for $\eta(\theta)$ ■

Proof. Suppose δ' is an unbiased estimator of $\eta(\theta)$ for which $T \mapsto \delta'$. Then $E_\theta[\delta(\tilde{X}) - \delta'(\tilde{X})] = 0$, and since T is complete we must have $\delta = \delta'$. That is, there is only one unbiased estimator of $\eta(\theta)$ that is a function of T. Suppose δ'' is any other unbiased estimator of $\eta(\theta)$ for which $\neg T \mapsto \delta''$. Then by Theorem 8.5, δ has uniformly smaller variance than δ''. □

Note that Theorem 8.6 does not need to state that $T \in \mathcal{T}_{min}$, since by Theorem 3.19 if a minimal sufficient statistic exists, then any complete sufficient statistic is also minimal sufficient.

Theorem 8.6 implies that any function $\delta \in \mathcal{T}_{var}$ of a complete sufficient statistic T is UMVU for estimand $\eta(\theta) = E_\theta[\delta(T)]$, where this is defined. It may also be shown that any statistic T with this property must be complete.

Theorem 8.7 (UMVUEs and Completeness) Suppose for some statistic $T \in \mathcal{T}_{suff}$, $\delta(T)$ is UMVU for any function $\delta \in \mathcal{T}_{var}$. Then T is complete. ■

Proof. Suppose $E_\theta[\delta(T)] = c$, where c *dnd* θ. Then by hypothesis $\delta(T)$ is UMVU for $\eta(\theta) = c$. But $\delta'(T) \equiv c$ is an unbiased estimator of c with uniformly zero variance. Therefore $\delta = \delta'$. □

8.4 Methods of Deriving UMVUEs

Theorems 8.5 and 8.6 yield two methods for determining UMVUEs given a complete sufficient statistic $T(\tilde{X})$. First, if $\delta(\tilde{X}) = \delta(T(\tilde{X}))$ is *any* unbiased estimate of $\eta(\theta)$, then by Theorem 8.6 it is the unique UMVUE, provided it has finite variance (Examples 8.3 and 8.4). If this approach is not practical, we may always 'Rao-Blackwellize' an unbiased estimator δ'. By this, we mean that $\delta(T(\tilde{X})) = E_\theta[\delta'(\tilde{X}) \mid T(\tilde{X})]$ is the UMVUE. This method tends to work well when δ' is selected to be as simple as possible.

We next give three examples of the first approach.

Example 8.3 (UMVUE of the Mean of a Uniform Density on $(0, \theta)$) Suppose $X_1, \ldots, X_n$ is an *iid* sample from $unif(0, \theta)$. From Problem 3.23, $X_{(n)}$ is a complete sufficient statistic for θ. Using the beta function, and noting that θ is a scale parameter, $E_\theta[X_{(n)}] = \theta \int_0^1 nt(1-t)^{n-1}dt = \theta n/(n+1)$. This means $\delta = (1 + 1/n)X_{(n)}/2$ is the UMVUE for $\mu = E_\theta[X_1]$. Here, the sample mean $\bar{X}$ is *not* UMVU. ■

Example 8.4 (UMVUE of the Mean and Variance of a Gamma Density) Suppose $\mathbf{X} = (X_1, \ldots, X_n)$ is an *iid* sample from $gamma(\tau, \alpha)$. Suppose we wish to derive the UMVU estimates of $\mu_X = \alpha\tau$, $\sigma_X^2 = \alpha\tau^2$. The gamma density is an exponential family density, so by Theorem 3.22 $T(\mathbf{X}) = (\sum_i X_i, \sum_i \log(X_i))$ is a complete sufficient statistic for (τ, α). The UMVUEs will therefore be the unique statistics $\delta_{\mu_X}(T(\mathbf{X}))$, $\delta_{\sigma_X^2}(T(\mathbf{X}))$ which are unbiased estimates of their respective estimands.

For μ_X, we clearly have $\delta_{\mu_X} = \bar{X}$. However, while the sample variance S^2 is an unbiased estimate of σ_X^2, it is not a function of the complete statistic. First, suppose α is known. Then

$$E_{(\tau,\alpha)}\left[\left(\sum_{i=1}^n X_i\right)^2\right] = \alpha\tau^2(n + n^2\alpha) = \sigma_X^2(n + n^2\alpha). \tag{8.2}$$

Then the UMVUE is the unique unbiased estimator of σ_X^2 based on the complete sufficient statistic, which must be

$$\delta_{\sigma_X^2} = \frac{\alpha\left(\sum_i X_i\right)^2}{n + n^2\alpha}.$$

The problem is not as straightforward if α is unknown. An expression for the UMVUE can be based on any unbiased estimate of σ_X^2, say $\delta_0(\mathbf{X}) = (X_1 - X_2)^2/2$. Then

$$\delta'_{\sigma_X^2} = E_{(\tau,\alpha)}[\delta_0(\mathbf{X}) \mid T(\mathbf{X})],$$

which of course will not depend on (τ, α). Unfortunately, the author is unable to report a convenient form of this estimator. A more practical approach would be to use the maximum likelihood estimate $\hat{\sigma}_X^2 = \hat{\alpha}\hat{\tau}^2$, where $\hat{\alpha}, \hat{\tau}$ are the MLEs of α, τ. Even here $\hat{\alpha}$ must be calculated numerically (but see Problem 4.17). Then $\hat{\sigma}_X^2$ is a function of the complete sufficient statistic, because the likelihood function itself is. Of course, $\hat{\sigma}_X^2$ will be biased, so the outstanding problem is to ensure that the bias is small, or more precisely, approaches zero quickly enough as $n \to \infty$. This problem will be considered in Chapter 12. ■

Example 8.5 (UMVUEs for the Multinomial Density) A multinomial vector $(n_1, \ldots, n_m)$ of nonnegative counts has density

$$p_\theta(n_1, \ldots, n_m) = \frac{n!}{\prod_j n_j!} \prod_j \theta_j^{n_j},$$

assuming $\sum_j n_j = n$. Suppose $m = 3$, and we impose constraint $(\theta_1, \theta_2, \theta_3) = (\alpha, 2\alpha, 1 - 3\alpha)$, for $\alpha \in (0, 1/3)$. Set

$$h(n_1, n_2, n_3) = \frac{n!}{\prod_j n_j!}$$

We may then write the density in exponential family form:

$$\begin{aligned}
p_\theta(n_1, n_2, n_3) &= \exp\left\{\sum_{j=1}^3 n_j \log(\theta_j)\right\} h(n_1, n_2, n_3) \\
&= \exp\{n_1 \log(\alpha) + n_2 \log(2\alpha) + n_3 \log(1 - 3\alpha)\} h(n_1, n_2, n_3) \\
&= \exp\{n_1 \log(\alpha) + n_2 \log(\alpha) + n_2 \log(2) + \\
&\qquad (n - n_1 - n_2) \log(1 - 3\alpha)\} h(n_1, n_2, n_3) \\
&= \exp\{(n_1 + n_2)[\log(\alpha) - \log(1 - 3\alpha)] + \\
&\qquad n_2 \log(2) + n \log(1 - 3\alpha)\} h(n_1, n_2, n_3).
\end{aligned}$$

The minimal sufficient statistic for α can then be seen to be $T = n_1 + n_2$. Also, T possesses a binomial distribution, and is therefore complete. Since $E_\alpha[T] = 3n\alpha$, $\delta = (n_1 + n_2)/(3n)$ is the UMVUE of α. ∎

The next example demonstrates the 'Rao-Blackwellize' procedure for evaluating a UMVUE (see also Section 12.8.2).

Example 8.6 (UMVUE of the Power of a Binomial) Suppose $X \in bin(n, \theta)$. To find the UMVUE of θ^k, $k = 1, \ldots, n$, write X explicitly as a sum of independent Bernoulli random variables $X = \sum_{i=1}^n U_i$. We make use of the fact that X is sufficient and complete for θ (since it is an exponential family density). Therefore, if for some $\delta = \delta(U_1, \ldots, U_n)$ we have $E_\theta[\delta] = \theta^k$, then $\delta_{RB} = E_\theta[\delta \mid X]$ is the unique UMVUE. We can choose $\delta = \prod_{i=1}^k U_i$ as an unbiased estimator of θ^k. Then set

$$\begin{aligned}
\delta_{RB}(x) &= E_\theta[\delta \mid X = x] \\
&= \frac{P_\theta\left\{U_1 = \ldots U_k = 1,\ \sum_{i=k+1}^n U_i = x - k\right\}}{P_\theta(X = x)} \\
&= \frac{\theta^k \binom{n-k}{x-k} \theta^{x-k}(1-\theta)^{n-x}}{\binom{n}{x}\theta^x(1-\theta)^{n-x}} \\
&= \frac{\binom{n-k}{x-k}}{\binom{n}{x}} \\
&= \frac{x \times (x-1) \times \ldots \times (x-k+1)}{n \times (n-1) \times \ldots \times (n-k+1)} \\
&= \frac{x^{(k)}}{n^{(k)}}.
\end{aligned}$$

Note that this can be interpreted as evaluating to $\delta_{RB}(x) = 0$ if $x < k$. ∎

8.5 Nonparametric Estimation and U-statistics

Most inference problems considered in this volume fall under the heading of "parametric inference". Of course, with very few expections, all inference problems must at some point limit interest to some family of densities $\mathcal{P}$, and in defining a parameteric family all we have required is that $\mathcal{P} = \{p_\theta : \theta \in \Theta\}$ is indexed by parameter θ. For our purposes, we can then take nonparametric inference to mean we are not given a bijection from $\mathcal{P}$ to a finite dimensional subset $\Theta \subset \mathbb{R}^m$.

8.5.1 U-Statistics and the Order Statistics

To fix ideas, we will assume that our observation consists of an *iid* sample $\mathbf{X} = (X_1, \ldots, X_n)$. $X_1 \in \mathbb{R}$, from distribution $F \in \mathcal{F}$. Interest is in some functional $\theta(F)$. This notation emphasizies that we do not have a parametric family in the conventional sense. We may have, for example, $\theta(F) = \mu = E[X_1] = \int x dF$. We then note that the definition of sufficiency, minimal sufficiency and completeness apply without modification to the nonparametric case. The first step to to verify the sufficiency of the order statistics.

Theorem 8.8 (The Order Statistics are Sufficient for *iid* Samples) Suppose $\mathbf{X} = (X_1, \ldots, X_n)$, $X_1 \in \mathbb{R}$ is an *iid* sample from distribution $F \in \mathcal{F}$. Then the order statistic $\mathbf{X}_{ord}$ is sufficient for any family $\mathcal{F}$. ■

Proof. The distribution of $(X_1, \ldots, X_n)$ conditional on $\mathbf{X}_{ord}$ is equivalent to a random permutation of the observed data for any F. This conditional distribution therefore does not depend on $F \in \mathcal{F}$. □

Given Theorem 8.8, the theory of UMVU estimation is therefore available for *iid* samples if $\mathbf{X}_{ord}$ is complete for $\mathcal{F}$. By Theorem 3.18, if there exists a subset $\mathcal{F}_0 \subset \mathcal{F}$ for which $\mathbf{X}_{ord}$ is complete, then it is also complete for $\mathcal{F}$ (the proof of Theorem 3.18 does not require that the parameter spaces are finite dimensional).

One approach is given in Section 4.3 of Lehmann and Romano (2005). It may first be verified that $\mathbf{X}_{ord}$ is equivalent to the sums of powers $T(\mathbf{X}) = \left(\sum_i X_i, \sum_i X_i^2, \ldots, \sum_i X_i^n\right)$. Then suppose $\mathcal{F}$ is the family of all distributions on $\mathbb{R}$ which are absolutely continuous with respect to some common measure ν. It is then possible to define a full rank exponential family with n parameters, absolutely continuous with respect to ν, for which $T(\mathbf{X})$ is minimal sufficient. This in turn implies that $T(\mathbf{X})$, and therefore $\mathbf{X}_{ord}$, is complete for $\mathcal{F}$. See also Halmos (1946) for a direct proof of completeness.

As for parametric families, the Lehmann-Scheffé theorem (Theorem 8.6) may then be applied. There is some flexibility with respect to the choice of $\mathcal{F}$. It need not be restricted to distributions dominated by a common measure ν. As a practical matter, is may be restricted to L^p distributions, in which case the device just used to verify the completeness of $\mathbf{X}_{ord}$ is still applicable.

We now define the U-statistic.

Definition 8.4 (U-statistics) Suppose $\mathbf{X} = (X_1, \ldots, X_n)$, $X_1 \in \mathbb{R}$ is an *iid* sample from distribution $F \in \mathcal{F}$. Then $\theta(F)$ is a homogenous functional over $\mathcal{F}$ of degree k if there exists a real valued function of k variables $\psi(x_1, \ldots, x_k)$ such that $E_F[\psi(X_1, \ldots, X_k)] = \theta(F)$ for all $F \in \mathcal{F}$, and if k is the minimal integer for which this holds. We refer to $\psi(x_1, \ldots, x_k)$ as the kernel of $\theta(F)$. We may assume without loss of generality that a kernel is symmetric (i.e. permutation invariant in its arguments). If not, it may be symmetrized by defining

$$\psi^*(x_1, \ldots, x_k) = \sum_{(\alpha_1, \ldots, \alpha_k) \in P_{k,k}} \frac{\psi(x_{\alpha_1}, \ldots, x_{\alpha_k})}{k!}. \tag{8.3}$$

Then, noting that $(X_1, \ldots, X_k)$ equals in distribution $(X_{\alpha_1}, \ldots, X_{\alpha_k})$ for any permutation $(\alpha_1, \ldots, \alpha_k)$, we must have $E_F[\psi^*(X_1, \ldots, X_k)] = E_F[\psi(X_1, \ldots, X_k)] = \theta(F)$.

Given a kernel $\psi(x_1, \ldots, x_k)$ of $\theta(F)$, the mapping

$$U = \sum_{(\alpha_1, \ldots, \alpha_k) \in P_{n,k}} \frac{\psi(x_{\alpha_1}, \ldots, x_{\alpha_k})}{n^{(k)}} \tag{8.4}$$

is a U-statistic, where $P_{n,k}$ is the set of all ordered selections $(\alpha_1, \ldots, \alpha_k)$ from $\{1, \ldots, n\}$. If $\psi(x_1, \ldots, x_k)$ is permutation invariant in its arguments, then we may write

$$U = \sum_{(\alpha_1, \ldots, \alpha_k) \in C_{n,k}} \frac{\psi(x_{\alpha_1}, \ldots, x_{\alpha_k})}{\binom{n}{k}}. \tag{8.5}$$

where $C_{n,k}$ is the set of all unordered selections $(\alpha_1, \ldots, \alpha_k)$ from $\{1, \ldots, n\}$.

If $\psi(x_1, \ldots, x_k)$ is a kernel for estimand $\theta(F)$, then Equation (8.4) defines an unbiased estimate of $\theta(F)$. ■

A statistic $T(X_1, \ldots, X_n)$ is a mapping of the order statistics *iff* it is symmetric in its arguments. If $\mathbf{X}_{ord}$ is complete and sufficient for $\mathcal{F}$, then since U-statistics are symmetric they are the UMVU estimate of the functional $\theta(F) = E_F[U]$, assuming the variance exists for all $F \in \mathcal{F}$.

This result is extraordinary, despite its apparent simplicity, and the series of papers Halmos (1946) and Hoeffding (1948) represent an important breakthrough. Given the completeness of the order statistics, that there exists a unique symmetric unbiased estimator of a homogenous functional $\theta(F)$, and that this estimator is optimal, might seem counter-intuitive at first. The essential point, however, is the "distribution-free" character of the inference, which in this context simply means that $\mathcal{F}$ is sufficiently large to force $\mathbf{X}_{ord}$ to be complete. Apart from its role in methodology, the theory of U-statistics can be used to resolve more theoretical questions, as we demonstrate in the following example.

Example 8.7 (L-statistics and Unbiasedness) Suppose $\mathbf{X} = (X_1, \ldots, X_n)$, $X_1 \in \mathbb{R}$ is an *iid* sample from distribution $F \in \mathcal{F}$, where $\mathcal{F}$ is the class of all distributions in L^1. Then $\mathbf{X}_{ord}$ is complete and sufficient for $\mathcal{F}$. An L-estimator is a statistic of the form $L = \sum_{i=1}^n a_i X_{(i)}$, that is, a linear combination of the order statistics. These are used in robust statistics, which trade off statistical efficiency for lower sensitivity to extreme observations. A typical example is the $p\%$ trimmed mean, for which the largest and smallest $p\%$ of the observations are removed, then a sample mean calculated for the remainder. It can be written as

$$L = \frac{1}{n-2k} \sum_{i=k+1}^{n-k} X_{(i)}.$$

For estimand $\mu_X = E_F[X_1]$, there cannot be an unbiased L statistic over $\mathcal{F}$ other than the (untrimmed) sample mean. If there was, say L^*, then we would have $E_F[\bar{X} - L^*] = 0$ for all $F \in \mathcal{F}$. But $\bar{X} - L^*$ is a function of $\mathbf{X}_{ord}$, which is complete, so no such L-statistic can exist. ■

Many commonly used statistics are U-statistics, including the sample mean and sample variances, which can therefore be refered to as nonparametric estimates. While the sample mean or variance is not a mapping of a sufficient statistic for many parametric families, they are for a larger nonparametric class $\mathcal{F}$. In Chapter 12 we consider the problem of robust versus efficient information. A statistical procedure δ may be optimal for a parametric family $\mathcal{P}$. but if the model is misspecified, that is, the true density is not in $\mathcal{P}$, then δ may be a poor choice for the problem. Of course, to raise the issue at least implicitly assumes the existence of some functional $\theta(F)$ which is interpretable outside the family $\mathcal{P}$, such as a mean or variance. Therefore, an optimal procedure can always be compared to a nonparametric alternative (see, for example, Section 12.8.2).

Example 8.8 (The Sample Mean and Variance as U-Statistics) Suppose $X_1, \ldots, X_n$, $X_1 \in \mathbb{R}$ is an *iid* sample from distribution $F \in \mathcal{F}$, where $\mathcal{F}$ is the class of all distributions in L^2. Then $\mathbf{X}_{ord}$ is complete and sufficient for $\mathcal{F}$. Next, note that $\mu_X = \int_{-\infty}^{\infty} x dF$ is a homogenous functional of order 1 (Definition 8.4), since $\psi(X_1) = X_1$ is an unbiased estimator of μ_X over all $F \in \mathcal{F}$. The sample mean $\bar{X}$ is an unbiased symmetric estimate of μ_X, and is therefore the UMVUE of μ_X.

In Example 8.3 it was shown that the UMVUE δ^* for μ_X based on an *iid* sample from $unif(0, \theta)$, $\theta \in \Theta = (0, \infty)$ is not the sample mean. However, δ^* is not unbiased over all $F \in \mathcal{F}$, so there is no contradiction. In fact, the theory of U-statistics forces this conclusion, as demonstrated in Example 8.7.

Next, consider the sample variance $S_n^2 = (n-1)^{-1}\sum_{i=1}^n (X_i - \bar{X}_n)^2$, indexed by sample size n. It will be convenient to recognize the linear combination

$$L = X_1 - \bar{X}_n = \frac{n-1}{n}X_1 - \frac{1}{n}X_2 - \ldots - \frac{1}{n}X_n.$$

Since the sample is *iid* we have

$$E_F[L] = \mu_X\left(\frac{n-1}{n} - \frac{1}{n} - \ldots - \frac{1}{n}\right) = 0,$$

$$\text{var}_F[L] = \sigma_X^2\left(\left(\frac{n-1}{n}\right)^2 + \frac{1}{n^2} + \ldots + \frac{1}{n^2}\right) = \sigma_X^2\frac{n-1}{n}.$$

This leads to

$$E_F[S_n^2] = \frac{n}{n-1}\text{var}_F[L] = \sigma_X^2 \tag{8.6}$$

for any $n \geq 2$. Thus, S_n^2 is an unbiased symmetric estimate of σ_X^2, and is therefore UMVU.

However, we may also conclude that σ_X^2 is a homogenous functional of degree 2. By Equation (8.6) we can set kernel $\psi(X_1, X_2) = S_2^2$, so that $E_F[\psi(X_1, X_2)] = \sigma_X^2$ over $F \in \mathcal{F}$. To rule out the possibility that σ_X^2 is a homogenous functional of degree 1, note that X_1^2 is the unique unbiased estimator of σ_X^2 over the family $N(0, \sigma_X^2)$ that is a function of X_1. Therefore, if an order 1 kernel did exist, it would have to be defined by X_1^2. However, X_1^2 is not unbiased over all $F \in \mathcal{F}$, so no kernel of degree 1 can exist for σ_X^2. It is easily verified that $S_2^2 = (X_1 - X_2)^2/2$, so, since S_n^2 is a U-statistic we may immediately conclude

$$S_n^2 = \sum_{i<j}\frac{(X_i - X_j)^2}{2\binom{n}{2}},$$

without the need for an explicit algebraic argument. Again, that S_n^2 is not the UMVUE for σ_X^2 over the family of *gamma*(τ, α) densities (Example 8.4) does not contradict the conclusion that S_n^2 is the UMVUE over $F \in \mathcal{F}$. ■

An additional advantage of the structure of U-statistics is that a general method for evaluating their variance exists, due to Hoeffding (1948). Suppose $\psi(x_1, \ldots, x_m)$ is the kernel of a homogenous functional $\theta(F)$ of degree m. Define the sequence of functions

$$\begin{aligned}
&\psi_c(x_1, \ldots, x_c) \\
&\quad = E_F[\psi(X_1, \ldots, X_c, X_{c+1}, \ldots, X_m) \mid (X_1, \ldots, X_c) = (x_1, \ldots, x_c)] \\
&\quad = E_F[\psi(x_1, \ldots, x_c, X_{c+1}, \ldots, X_m)].
\end{aligned} \tag{8.7}$$

By the law of total expectaton

$$E_F[\psi_c(X_1, \ldots, X_c)] = E_F[\psi(X_1, \ldots, X_m)] = \theta(F). \tag{8.8}$$

Then define

$$\sigma_c^2 = \text{var}_F[\psi_c(X_1, \ldots, X_c)]. \tag{8.9}$$

The following lemma is key to the argument.

Lemma 8.1 Let $X_1, \ldots, X_n$ be an *iid* sample from $F \in \mathcal{F}$. Let $\psi(x_1, \ldots, x_m)$ is the kernel of a homogenous functional $\theta(F)$ of degree m. Suppose $\alpha_1, \ldots, \alpha_m$, $\beta_1, \ldots, \beta_m$, are two selections of m distinct indices from $\{1, \ldots, n\}$, and suppose the two selections possess c indices in common. Then

$$\text{cov}_F[\psi(X_{\alpha_1}, \ldots, X_{\alpha_m}), \psi(X_{\beta_1}, \ldots, X_{\beta_m})] = \sigma_c^2, \tag{8.10}$$

where σ_c^2 is defined by Equation (8.9). ■

Proof. We may assume without loss of generality that the covariance of Equation (8.10) may be written as

$$\begin{aligned}&\text{cov}_F[\psi(X_1,\ldots,X_c,X_{\alpha_{c+1}},\ldots,X_{\alpha_m}),\psi(X_1,\ldots,X_c,X_{\beta_{c+1}},\ldots,X_{\beta_m})]\\&= E_F\left[(\psi(X_1,\ldots,X_c,X_{\alpha_{c+1}},\ldots,X_{\alpha_m})-\theta(F))\right.\\&\qquad\left.\times\,(\psi(X_1,\ldots,X_c,X_{\beta_{c+1}},\ldots,X_{\beta_m})-\theta(F))\right],\end{aligned}$$

where the indices $\alpha_{c+1},\ldots,\alpha_m,\beta_{c+1},\ldots,\beta_m$ are distinct. Thus, conditioning on $(X_1,\ldots,X_c)=(x_1,\ldots,x_c)$ gives

$$\begin{aligned}&E_F\left[(\psi(X_1,\ldots,X_c,X_{\alpha_{c+1}},\ldots,X_{\alpha_m})-\theta(F))\right.\\&\quad\left.\times\,(\psi(X_1,\ldots,X_c,X_{\beta_{c+1}},\ldots,X_{\beta_m})-\theta(F))\mid(X_1,\ldots,X_c)=(x_1,\ldots,x_c)\right]\\&= E_F\left[\psi(x_1,\ldots,x_c,X_{\alpha_{c+1}},\ldots,X_{\alpha_m})-\theta(F)\right]\\&\quad\times E_F\left[\psi(x_1,\ldots,x_c,X_{\beta_{c+1}},\ldots,X_{\beta_m})-\theta(F)\right]\\&= \left(\psi_c(x_1,\ldots,x_c)-\theta(F)\right)^2.\end{aligned}$$

The proof is completed by applying the law of total expectation, noting $E_F[((\psi_c(X_1,\ldots,X_c) - \theta(F))^2F=\sigma_c^2$, following Equations (8.8) and (8.9). □

The main theorem is largely an application of Lemma 8.1

Theorem 8.9 (Variance of a U-statistic) Let $X_1,\ldots,X_n$ be an *iid* sample from $F\in\mathcal{F}$. Let $\psi(x_1,\ldots,x_m)$ a the symmetric kernel of a homogenous functional $\theta(F)$ of degree m. Then the variance of the U-statistic of the form of Equation (8.5) is

$$\begin{aligned}\text{var}_F[U]&=\binom{n}{m}^{-1}\sum_{c=1}^{m}\binom{m}{c}\binom{n-m}{m-c}\sigma_c^2\\&=\sum_{c=1}^{m}c!\binom{m}{c}^2\left[\frac{(n-m)^{(m-c)}}{n^{(m)}}\right]\sigma_c^2,\end{aligned}\tag{8.11}$$

where σ_c^2 is defined by Equation (8.9). It follows that $\text{var}_F[U]=O(1/n)$. ■

Proof. The variance of a U-statistic can be evaluated using the conventional double-summation over all covariances:

$$\text{var}_F[U]=\sum_{(\alpha_1,\ldots,\alpha_m)\in C_{n,m}}\sum_{(\beta_1,\ldots,\beta_m)\in C_{n,m}}\text{cov}_F[\psi(X_1,\ldots,X_{\alpha_m}),\psi(X_1,\ldots,X_{\beta_m})].\tag{8.12}$$

Then Equation (8.11) follows by enumerating the number of terms in Equation (8.12) which equal σ_c^2 according to Lemma 8.1. The details are left to the reader (Problem 8.12).

To complete the proof, note that $\binom{n}{m}^{-1}\sum_{c=1}^{m}\binom{m}{c}\binom{n-m}{m-c}=O(1/n^c)$, and $\sigma_c^2=O(1)$. □

It will be useful to specialize Theorem 8.9 to homogenous functionals of order one and two.

Example 8.9 (Homogenous Functionals of Order $m=1,2$) Assume the conditions of Theorem 8.9 hold. For homogenous functionals of order $m=1$, the kernel may be written $\psi(X_i)$. The U-statistic is therefore a sample mean of *iid* terms,

$$U=\frac{1}{n}\sum_{i=1}^{n}\psi(X_i),$$

so $\text{var}_F[U] = n^{-1}\sigma_1^2$, where $\sigma_1^2 = \text{var}_F[\psi(X_1)]$.

For homogenous functionals of order $m = 2$, Equation (8.11) gives

$$\begin{aligned} \text{var}_F[U] &= 4\left[\frac{(n-2)}{n(n-1)}\right]\sigma_1^2 + 2\left[\frac{1}{n(n-1)}\right]\sigma_2^2 \\ &= \frac{1}{n(n-1)}\left[4(n-2)\sigma_1^2 + 2\sigma_2^2\right]. \end{aligned} \tag{8.13}$$

Then $\psi_1(x_1) = E_F[(x_1 - X_2)^2/2] = (x_1^2 - 2x_1\mu + \mu_2)/2 = (x_1 - \mu)^2/2 + c$, where c is some constant. Then

$$\sigma_1^2 = \text{var}_F[\psi_1(X_1)] = \text{var}_F[(X_1 - \mu)^2/2] = \frac{1}{4}(\bar{\mu}_4 - \sigma_X^4),$$

where $\bar{\mu}_4$ is the fourth central moment of X_1. A similar argument gives $\sigma_2^2 = (\bar{\mu}_4 - \sigma_X^4)/2$. Substituting into Equation (8.13) gives

$$\text{var}_F[U] = \frac{1}{n}(\bar{\mu}_4 - \sigma_X^4).$$ ■

A large sample approximation of the variance of a U-statistic also follows from Theorem 8.9

Example 8.10 (Large Sample Variance of a U-Statistic) Clearly, from Equation (8.11) the contribution of σ_c^2 to $\text{var}_F[U]$ is of order $O(1/n^c)$, so that for large n we have $\text{var}_F[U] \approx m^2\sigma_1^2/n$ (Problem 8.13). ■

8.6 Rank Based Measures of Correlation

Figure 8.1 shows two scatteplots of observed pairs (X_i, Y_i) $i = 1, \ldots, n$. The left plot demonstrates perfect linear correlation, while the right plot exhibits perfect rank correlation. The latter means that the ranks of the observations $X_1, \ldots, X_n$ equal the ranks of the observations $Y_1, \ldots, Y_n$. Clearly, in both cases there is a smooth deterministic relation $Y_i = g(X_i)$ which holds without error. However, while the Pearson moment correlation ρ will equal one for the left plot, it will in general be strictly less than one for increasing but nonlinear deterministic associations such as that shown in the right plot. However, as a matter of interpretation, it might seem reasonable to assign the maximum value of any index of pairwise association to both examples. We next discuss two rank based measures of correlation intended to address this issue.

8.6.1 Kendall's τ

Following Figure 8.1, we first clarify the meaning of perfect correlation (the better term here may be association). Suppose we have paired observations (X_1, Y_1), (X_2, Y_2). Under perfect (positive) association the ranks of the X_i and Y_i observations are equal. Thus, one of the events $A = \{X_1 < X_2\} \cap \{Y_1 < Y_2\}$ or $B = \{X_1 > X_2\} \cap \{Y_1 > Y_2\}$ must occur, noting that $A \cap B = \emptyset$. This means that under perfect positive association $P(A) + P(B) = 1$. On the other hand, under perfect negative association the X_i and Y_i ranks are reversed, so that neither A nor B can occur, that is, $P(A) + P(B) = 0$. Of course, association will usually not be perfect in either direction, so the quantity $P(A) + P(B)$ serves as an index of association,

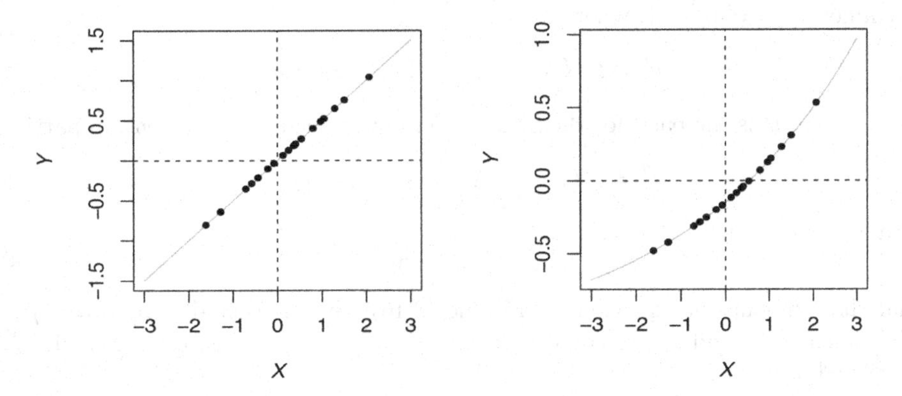

FIGURE 8.1
Scatterplots of observed pairs (X_i, Y_i) $i = 1, \ldots, n$ exhibiting perfect linear correlation (left plot) and perfect rank correlation (right plot).

ranging from $P(A) + P(B) = 0$ (perfect negative association) to $P(A) + P(B) = 1$ (perfect positive association).

Association measures are conventionally standardized to a range of $\rho = [-1, 1]$, with $\rho = 1$ intepretable as perfect positive association, $\rho = -1$ intepretable as perfect negative association, and $\rho = 0$ interpretable as no association. Independence of X_i and Y_i must imply $\rho = 0$, but the converse need not hold.

This convention is easily attained by standardizing the estimand as $\rho = 2(P(A) + P(B)) - 1$. However, the theory of U-statistics now becomes relevant, since ρ can be seen as a homogenous functional of degree $k = 2$, given symmetric kernel

$$\psi((X_1, Y_1), (X_2, Y_2)) = 2[I\{X_1 < X_2, Y_1 < Y_2\} + I\{X_2 < X_1, Y_2 < Y_1\}] - 1,$$

for which $E[\psi((X_1, Y_1), (X_2, Y_2))] = \rho$. Note that a pair of observations defines a single sample, so that the degree of ψ is 2.

The resulting U-statistic is commonly known as Kendall's τ or the Kendall rank correlation coefficient:

$$\begin{aligned} \tau &= \frac{1}{\binom{n}{2}} \sum_{i<j} \psi((X_i, Y_i), (X_j, Y_j)) \\ &= \frac{[\text{Number of concordant pairs}] - [\text{Number of discordant pairs}]}{\binom{n}{2}}. \end{aligned} \tag{8.14}$$

Validating this interpretation is left to the reader (Problem 8.19).

8.6.2 Spearmans Rank Correlation Coefficient

The Spearman rank correlation coefficient (Spearman, 1904) can be simply characterized as a Pearson sample correlation coefficient calculated by replacing the values of (X_i, Y_i) with their ranks (evaluated separately for the X_i and Y_i values). For both scatterplots of Figure 8.1 such a coefficient would equal 1.

Kendall's τ can be understood as a U-statistic for estimand

$$\rho_\tau = 2\eta - 1, \tag{8.15}$$

where

$$\eta = P(\{X_1 < X_2\} \cap \{Y_1 < Y_2\}) + P(\{X_1 > X_2\} \cap \{Y_1 > Y_2\})$$

or, by symmetry, $\rho_\tau = 4\eta' - 1$, where

$$\eta' = P\left(\{X_1 < X_2\} \cap \{Y_1 < Y_2\}\right).$$

A modification of η' is the basis for the Spearman rank correlation coefficient, in particular,

$$\eta'' = P\left(\{X_1 < X_2\} \cap \{Y_1 < Y_3\}\right),$$

which leads to estimand

$$\rho_S = 12\eta'' - 3. \tag{8.16}$$

Note that the estimands of Equations (8.15) and (8.16) are constructed so that $\rho_\tau = \rho_S = 1$ when the ranks of X_i and Y_i are equal *wp*1, and $\rho_\tau = \rho_S = -1$ when the ranks of X_i and Y_i are reversed *wp*1.

8.7 Problems

Problem 8.1 Suppose δ is a UMVU estimator for $g(\theta)$. Let δ_1, δ_2 be any other unbiased estimates of $g(\theta)$ with finite variance. Prove that $\text{cov}[\delta, \delta_1] = \text{cov}[\delta, \delta_2]$. **HINT:** Use Theorem 8.4.

Problem 8.2 Determine the constant a in Example 8.1 which minimizes the variance of X'. Compare to the best linear unbiased estimate (BLUE) of $E[X]$ (Theorem 6.1).

Problem 8.3 Suppose we have *iid* random observations $(X_1, Y_1), \ldots, (X_n, Y_n)$, where (X_1, Y_1) is uniformly distributed on a square defined by sides $(0, a) \times (0, b)$. What is the UMVUE of ab?

Problem 8.4 Let $\mathbf{X} = (X_1, \ldots, X_n)$ be an *iid* sample with $X_1 \sim exp(\tau)$. The object will be to compare estimators for estimand $\eta(\tau) = P_\tau(X_1 > T) = e^{-T/\tau}$, for some fixed $T > 0$.

(a) Determine the UMVUE of $\eta(\tau)$, say $\hat{\eta}_u$.
(b) Show how $P_\tau(\hat{\eta}_u = 0)$ can be evaluated. Will this quantity be nonzero in general?
(c) We can also consider the MLE $\hat{\eta}_{MLE} = \eta(\hat{\tau}) = e^{-T/\hat{\tau}}$, where $\hat{\tau} = n^{-1}\sum_{i=1}^n X_i$, Evaluate numerically the bias, variance and mean squared error (MSE) of $\hat{\eta}_u$ and $\hat{\eta}_{MLE}$ for $\tau = 1$ and all twelve combinations of $T \in (0.5, 1.0, 2.0, 5.0)$ and $n \in (5, 10, 25)$. Possible approaches include simulations; or numerical integration using the R function `integrate()`; or both as a check.
(d) For which pairs (T, n) is the variance smaller for $\hat{\eta}_u$? For which pairs (T, n) is the MSE smaller for $\hat{\eta}_{MLE}$?
(e) Suppose we also include in our comparison the empirical estimator:

$$\hat{\eta}_e = \frac{1}{n}\sum_{i=1}^n I\{X_i > T\}.$$

Without having to do any evaluations or calculations, how will the variance and MSE of $\hat{\eta}_e$ compare with $\hat{\eta}_u$?
(f) Give one reason why $\hat{\eta}_e$ might be used instead of $\hat{\eta}_u$.

Problem 8.5 Suppose we are given an *iid* sample $X_1, \ldots, X_n$, where $X_1 \sim geom(\theta)$.

(a) Verify that the density p_θ for the sample $X_1, \ldots, X_n$ is an exponential family model over $\theta \in (0, 1)$. Give a complete sufficient statistic for θ.

(b) Verify that each of the three estimands θ, $1/\theta$ and $1/\theta^2$ are U-estimable.
(c) Derive the UMVUE of $1/\theta$.
(d) Derive the UMVUE of $1/\theta^2$.
(e) Derive the UMVUE of θ. **HINT:** You can make use of the fact that the sum S of an *iid* sample of n observations from $geom(\theta)$ has a negative binomal distribution, with density

$$P(S = s) = \binom{s-1}{n-1}(1-\theta)^{s-n}\theta^n, \quad s = n, n+1, \ldots$$

Problem 8.6 Let $X_1, \ldots, X_n$ be an *iid* sample with $X_1 \sim pois(\lambda)$.

(a) Derive the distribution of X_1 conditional on $S = \sum_{i=1}^n X_i = s$.
(b) Give the UMVUE for estimand $\eta(\lambda) = P_\lambda(X_1 \geq t)$ for some fixed t.
(c) What is the value of the UMVUE when $S < t$?

Problem 8.7 A multinomial vector $\mathbf{N} = (N_1, \ldots, N_m)$ of nonnegative counts has density

$$p_\theta(n_1, \ldots, n_m) = \frac{n!}{\prod_j n_j!} \prod_j p_j^{n_j},$$

assuming $\sum_j n_j = n$. Suppose $m = 3$, and we impose constraint $(p_1, p_2, p_3) = (1/k_\theta, 2\theta/k_\theta, \theta^2/k_\theta)$, where $\theta > 0$, and k_θ is the appropriate normalization constant.

(a) Express the distribution of $(N_1, \ldots, N_m)$ in exponential family form.
(b) Derive the UMVUE for estimand $g(\theta) = \theta/(1+\theta)$.
(c) By expressing the distribution in terms of its canonical parameter, derive the variance of the UMVUE as a function of θ.

Problem 8.8 Suppose we observe an *iid* sample $X_1, \ldots, X_n$ with $X_1 \sim geom(\theta)$. Derive the UMVU estimator for $\eta(\theta) = P_\theta(X_1 \leq 2) = \theta + \theta(1-\theta)$.

Problem 8.9 Suppose $X_1, \ldots, X_n$, $X_1 \in \mathbb{R}$ is an *iid* sample from distribution $F \in \mathcal{F}$, where $\mathcal{F}$ is the class of all continuous distributions in L^1 which are symmetric about their mean μ. Is $\mathbf{X}_{ord}$ complete for this family? **HINT:** For any $k \neq n/2$ $X_{(n-k)} - \mu$ equals $\mu - X_{(k+1)}$ in distribution.

Problem 8.10 Suppose $\mathbf{X} = (X_1, \ldots, X_n)$, $X_1 \in \mathbb{R}$ is an *iid* sample from distribution $F \in \mathcal{F}$, where $\mathcal{F}$ is the class of all distributions in L^1. Show that for some function $h(n)$, $\delta(X) = h(n)\sum_{i=1}^n (X_i - \bar{X})^3$ is the UMVUE of the third central moment over $F \in \mathcal{F}$, which is a homogenous functional of degree 3.

Problem 8.11 Suppose $X_1, \ldots, X_n, Y_1, \ldots, Y_m$ are independent *iid* samples from respective distributions $F_X, F_Y \in \mathcal{F}$, where $\mathcal{F}$ is a class of all distributions in L^1. Verify that $\bar{X}_n - \bar{Y}_m$ is the UMVU estimator for $\Delta = E_{F_X}[X_1] - E_{F_Y}[Y_1]$.

Problem 8.12 Provide the (largely combinatorical) details for the proof of Theorem 8.9.

Problem 8.13 Show that under the conditions of Theorem 8.9 we have $\mathrm{var}_F[U] \approx m^2\sigma_1^2/n + O(1/n^2)$.

Problem 8.14 Suppose $X_1, \ldots, X_n$, $X_1 \in \mathbb{R}$ is an *iid* sample from distribution $F \in \mathcal{F}$, where $\mathcal{F}$ is the class of all continuous distributions possessing all moments. Let $\mu_X = E_F[X_1]$.

(a) Give the form of the UMVUE of μ_X^k, where k is a positive integer.
(b) Use the result of Problem 8.13 to derive an approximate variance for the estimator of Part (a).
(c) Suppose we allow $\mathcal{F}_q$ to be all distributions F for which $E_F[X_1^q] < \infty$. What does q have to be in order for the approximation of Part (b) to hold?

Problem 8.15 Suppose we are given an *iid* sample $X_1, \ldots, X_n$, and we wish to estimate $P(X_1 > t)$ for some fixed t.

(a) Derive a U-statistic for $P(X_1 > t)$. Of what order is the homogeneous functional for this problem?
(b) Suppose $X_1 \sim unif(\theta)$. Derive the UMVUE for $P(X_1 > 1)$.
(c) Compare the variances of the U-statistic and the UMVUE.

Problem 8.16 Suppose $X \sim bin(n, p)$.

(a) Determine the UMVUE for estimand $p(1-p)$.
(b) Suppose we write $X = \sum_{i=1}^n I_i$, where $\mathbf{U} = (U_1, \ldots, U_n)$ is an *iid* sample with $U_1 \sim bern(p)$. Suppose we consider as an estimator of $\text{var}[U_1] = p(1-p)$ the sample variance of $\mathbf{U}$. How does this estimator compare to the one derived in Part (a)?

Problem 8.17 Suppose we are given an *iid* sample $X_1, \ldots, X_n$ from a parametric family $\mathcal{P}$. For the following families, derive the UMVU estimate of the mean, say, $\hat{\mu}_X$. Compare the variance of $\hat{\mu}_X$ to that of the sample mean $\bar{X}$.

(a) $X_1 \sim pareto(\tau, \alpha)$, α is known.
(b) $X_1 \sim weibull(\tau, \alpha)$, $\alpha > 1$, α is known.
(c) $X_1 \sim unif(0, \theta)$.

Problem 8.18 Suppose we are given an *iid* sample $X_1, \ldots, X_n$ from a parametric family $\mathcal{P}$. For the following families, derive the UMVU estimate of the variance, say $\hat{\sigma}_X^2$. Compare the variance of $\hat{\sigma}_X^2$ to that of the sample variance S^2.

(a) $X_1 \sim gamma(\tau, \alpha)$, α is known.
(b) $X_1 \sim pois(\lambda)$.
(c) $X_1 \sim unif(0, \theta)$.

Problem 8.19 Verify the interpretation of Kendall's τ offered in Equation (8.14).

Problem 8.20 Suppose we are given a sample (X_i, Y_i) $i = 1, \ldots, n$ from a continuous bivariate distribution f_{XY}.

(a) Evaluate the variance of Kendall's τ under the hypothesis that X_i and Y_i are independent. To what degree does this variance depend on f_{XY}?
(b) Give also the large sample approximaton of the variance described in Example 8.10.

Problem 8.21 With respect to Equations (8.15) and (8.16), verify that $\rho_\tau = \rho_S = 1$ when the ranks of X_i and Y_i are equal $wp1$, and $\rho_\tau = \rho_S = -1$ when the ranks of X_i and Y_i are reversed $wp1$.

Problem 8.22 Suppose we are given a sample (X_i, Y_i) $i = 1, \ldots, n$ from a continuous bivariate distribution f_{XY}.

(a) Evaluate the variance of Spearman's rank correlation coefficient under the hypothesis that X_i and Y_i are independent. To what degree does this variance depend on f_{XY}?
(b) Give also the large sample approximation of the variance described in Example 8.10.

9

Group Structure and Invariant Inference

9.1 Introduction

Reading this chapter will require an understanding of elementary group theory, so a brief introduction to the subject is provided in Section D.1. The reader may find this chapter the most mathematically challenging one in this volume, so it should be noted that many of the important ideas of invariance are discussed outside of this chapter. An intuitive introduction to the topic is given in Section 3.3 for location-scale transformations, and some of the implications for inference are discussed in, for example, Sections 7.6 and 10.6. These ideas can be extended to other forms of invariance, such as permutation invariance (Section 10.7). If any chapter in this volume can be described as optional, it is this one, since the reader will still have a good sense of the importance of group invariance to inference. On the other hand, although this may be simply a matter of taste, this chapter may offer a great deal of intellectual satisfaction, and hence stimulate further interest in the theory of inference as a whole.

We have already made the point that invariance plays an important role throughout the theory of inference. In this chapter, and in Chapter 10, we show how optimal statistical procedures can be identified among those constrained to satisfy the invariance (or equivariance) principle of Section 3.13.3. For the problem of estimation, this approach is comparable to the theory of UMVUEs (Chapter 8) in the sense that a uniformly minimum risk estimator (Section 7.6) is identified from among the class of equivariant estimators (see, for example, Definitions 3.3, 3.4 and 3.5). This results in the minimum risk equivariant (MRE) estimator. We will also define invariant hypothesis tests, but a discussion of optimal test procedures will require the Neyman-Pearson lemma (Chapter 10).

In Sections 9.2 and 9.3 we will derive MRE estimators for location and scale parameters (including the Pitman estimators). These sections should be compared to Section 7.6, which in a sense provides a much simpler route to at least some of the solutions offered by the theory of MRE estimation. However, a formal resolution of the MRE problem requires the more advanced mathematical arguments made in this chapter.

The theory of MRE estimation for location or scale parameters serves as a good introduction to the topic of invariance supported by group theory. In Section 9.4 the more general invariant density family is defined. This can be regarded as a generalization of the ideas introduced in Section 3.3, extending the location-scale transformations to any transformation group G acting on a sample space $\mathcal{X}$. See Section D.4 for a formal definition of G.

In Section 9.5 we consider some extended applications of the invariant density family. The invariance structure of the Gauss-Markov theorem (Theorem 6.1) is first described. Section 9.5.2 considers the problem of estimating a scale parameter τ when interest is also in estimating the rate parameter $\lambda = 1/\tau$. We may expect an MRE estimator of λ to be the reciprocal of the MRE estimator of τ, but this requires imposing an additional invariance constraint on the problem, in particular, invariance with respect to the reciprocal transformation. Then the "species sampling" problem of Example 4.4 and Problem 4.8 is

DOI: 10.1201/9781003049340-9

revisited, from the point of view of invariance. This leads to something more than a heuristic solution to the problem.

In Section 9.6 is it shown how the invariance principle of Section 3.13.3 can be applied to hypothesis testing. The main applications considered here are the rank based test procedures induced by the monotone transformation group (the Wilcoxon rank sum and signed rank tests).

9.2 MRE Estimators for Location Parameters

The location invariant model has been introduced in Section 3.3.1 to which readers of this section should refer. In summary, we are given some density $p_0(\mathbf{x})$ on sample space $\mathcal{X} = \mathbb{R}^n$. The support may be strictly smaller than $\mathcal{X}$. Then consider the family of transformations $g_a(x_1, \ldots, x_n) = (x_1 + a, \ldots, x_n + a)$, so that $G_{loc} = \{g_\theta : \theta \in \Theta\}$ is a acting on $\mathcal{X}$, and Θ can be identified as the group $(\mathbb{R}, +)$. If $\mathbf{X}$ has density $p_0(\mathbf{x})$, it is easily verified that the density of $g_\theta \mathbf{X}$ is

$$p_\theta(\mathbf{x}) = p_0(\mathbf{x} - \theta)$$

The parametric family is therefore

$$\mathcal{P} = \{p_0(\mathbf{x} - \theta) : \theta \in \Theta\}.$$

This generates the location parameter family of densities.

Recall the location equivariance property, $S(\mathbf{X} + a) = S(\mathbf{X}) + a$. Similarly, location invariance implies $T(\mathbf{X}+a) = T(\mathbf{X})$, and $M(\mathbf{X})$ is a maximal invariant with respect to G_{loc} *iff* $M(\mathbf{X}) = (X_1 - S(\mathbf{X}), \ldots, X_n - S(\mathbf{X}))$ for some equivariant statistic $S(\mathbf{X})$.

The first step is to construct a compact characterization of the class of location equivariant estimators, which we give in the next theorem.

Theorem 9.1 (The Class of Location Equivariant Estimators) Suppose δ_0 is location equivariant. Then δ is location equivariant *iff* there is location invariant statistic $w(\mathbf{x})$ such that $\delta = \delta_0 - w$. ∎

Proof. First, assume w is invariant. Then $\delta(\mathbf{x}+a) = \delta_0(\mathbf{x}+a) - w(\mathbf{x}+a) = \delta_0(\mathbf{x}) + a - w(\mathbf{x}) = \delta(\mathbf{x}) + a$, so that δ is location equivariant.

Next, assume that δ is location equivariant. Set $w = \delta_0 - \delta$. Then $w(\mathbf{x} + a) = \delta_0(\mathbf{x} + a) - \delta(\mathbf{x} + a) = \delta_0(\mathbf{x}) + a - \delta(\mathbf{x}) - a = \delta_0(\mathbf{x}) - \delta(\mathbf{x}) = w(\mathbf{x})$, so that w is location invariant. The proof follows after noting that $\delta = \delta_0 - w$. □

We first introduce an important consequence of Theorem 9.1. The parametric models considered in this chapter need not have the properties of the exponential family models (the existence of an MGF, for example). So the methods can be quite useful for such cases, and it will be possible to construct estimators which minimize MSE even when the observations themselves do not possess, for example, second order moments. In such cases we may need to make use of the following technical theorem.

Theorem 9.2 (Existence of Moments for Location Equivariant Estimators) Suppose $\mathbf{X}$ is a sample from a location invariant model. Suppose there exists a location equivariant estimator δ_0 for which $E_\theta[|\delta_0|] < \infty$, $\theta \in \Theta$. Let $\tilde{Y} = M(\mathbf{X})$ be a maximal invariant. Then for any other location equivariant estimator δ we have $E_\theta[|\delta| \mid \tilde{Y}] < \infty$, $wp1$. ∎

Proof. By the law of total expectation $E_\theta[|\delta_0| \mid \tilde{Y}] < \infty$, $wp1$. By Theorem 9.1 we may write $\delta(\mathbf{X}) = \delta_0(\mathbf{X}) - w(\tilde{Y})$ for some mapping w, from which the proof follows. □

The following example is of the type anticipated by Theorem 9.2.

Example 9.1 (Location Equivariant Estimators for a Cauchy Location Parameter) Suppose $\mathbf{X} = (X_1, \ldots, X_n)$ is an *iid* sample with $X_1 \sim cauchy(\mu, 1)$. It is easily verified that, for example, $E_\theta[|X_1|] = \infty$. Here, X_1 is a location equivariant estimator. However, the sample median $\hat{m}$ is also a location equivariant estimator (Example 3.7), and we have $E_\theta[|\hat{m}|] < \infty$. Therefore, by Theorem 9.2, $E_\theta[|\delta| \mid \tilde{Y}] < \infty$, $wp1$, for any location equivariant estimator δ. ■

The idea of a location invariant loss function has already been introduced in Definition 7.3 and Theorem 7.7. Here we begin with its intended property. Suppose $\theta \in \mathbb{R}$ is a location parameter. Then a loss function $L_\theta(d)$ is location invariant if $L_\theta(d) = L_{\theta+a}(d+a)$. In the next theorem we recover the form of $L_\theta(d)$ offered as a definition in Definition 7.3.

Theorem 9.3 (Structure of Location Invariant Loss) A loss function $L_\theta(d)$ is location invariant *iff* it assumes the form $L_\theta(d) = \rho(\theta - d)$. ■

Proof. If $L_\theta(d) = \rho(\theta - d)$, invariance follows directly. Conversely, suppose $L_\theta(d) = L_{\theta+a}(d+a)$. Set $a = -d$. Then $L_\theta(d) = L_{\theta-d}(0) = \rho(\theta - d)$. □

We next give a formal statement of the strategy alluded to in Section 7.5.3 which will be used to develop MRE estimators . If we can define a loss function for which risk is constant for any equivariant estimator, then the problem of finding a MRE estimator is well defined under that loss.

Theorem 9.4 (Constant Risk for Location Invariant Loss) Supoose θ is a location parameter, δ is any location equivariant estimator, and $L_\theta(d)$ is location invariant loss function. Then risk $R(\theta, \delta)$, bias $\text{bias}_\theta[\delta]$ and variance $\text{var}_\theta[\delta]$ are constant over $\theta \in \Theta$. ■

Proof. We may write $L_\theta(\delta) = \rho(\delta(\mathbf{X}) - \theta) = \rho(\delta(\mathbf{X} - \theta))$. But by Theorem 3.2 the distribution of $\mathbf{X} - \theta$ *dnd* θ. The proof for bias and variance are left to the reader (Problem 9.2). □

In Section 7.6 we considered the problem of determining the UMR estimator from a class $\{\delta_0 - b, b \in \mathbb{R}\}$ where δ_0 is any location equivariant estimator. However, Theorems 9.1, 9.3 and 9.4 allow us to extend the problem to the entire class of location equivariant estimators by replacing the constant b with a location invariant statistic $w(\mathbf{x})$. Then, using the optimization method introduced in Theorem 7.1, we may derive the minimum risk equivariant (MRE) estimator which is UMR within the class of all location equivariant estimators, with respect to any location invariant loss. The main result of this section follows.

Theorem 9.5 (MRE Estimators for Location Invariant Models) Suppose $\mathbf{X} \in \mathbb{R}^n$ is an observation from a location invariant model, and $L_\theta(d) = \rho(\theta - d)$ is a loss function for location parameter $\theta \in \mathbb{R}$. Let $\tilde{Y} = M(\mathbf{X})$ be any maximal invariant. Suppose at least one equivariant estimator has finite risk. Let δ_0 be any other equivariant estimator.

(i) Assume that for all $\tilde{y} = \tilde{Y}$, there exists a value $w(\tilde{y}) = w^*(\tilde{y}) \in \mathbb{R}$ which minimizes

$$E_0[\rho(\delta_0(\mathbf{X}) - w(\tilde{y})) \mid \tilde{Y} = \tilde{y}]. \tag{9.1}$$

Then $\delta(\mathbf{X}) = \delta_0(\mathbf{X}) - w^*(\tilde{Y})$ is a MRE estimator of θ.

(ii) Suppose δ_0 is a function of a statistic which is complete and sufficient *wrt* θ, and there exists a value $w = w^* \in \mathbb{R}$ which minimizes

$$E_0[\rho(\delta_0(\mathbf{X}) - w)]. \tag{9.2}$$

Then $\delta(\mathbf{X}) = \delta_0(\mathbf{X}) - w^*$ is a MRE estimator of θ.

(iii) If the minimizer of Equation (9.1) or Equation (9.2) exists and is unique, then the MRE estimator of θ is unique.

(iv) If ρ is convex and nonmonotone, then an MRE estimator exists. If ρ is strictly convex and nonmonotone, the MRE estimator is unique. ■

Proof. We prove each statement in turn.

(i) By Theorem 9.3 $L_\theta(d)$ is location invariant. Then suppose we determine an estimator which minimizes risk $R(\theta, \delta)$ for some fixed θ. By Theorem 9.4 risk does not depend on θ, therefore such an estimator will also minimize $R(\theta', \delta)$ for any other $\theta' \in \Theta$. Therefore, we can determine the MRE by miniming $R(0, \delta)$. By Theorem 9.1 any equivariant estimator takes the form $\delta(\mathbf{x}) = \delta_0(\mathbf{x}) - w(\mathbf{x})$, where w is a location invariant statistic. In turn, any invariant statistic may be expressed $w(\mathbf{x}) = w(\tilde{y})$, where $\tilde{y} = M(\mathbf{x})$ is a maximal invariant. Then following the optimal prediction theorem (Theorem 7.1), let $W_{\tilde{y}}$ be the set of values w which minimize $E_0[\rho(\delta_0(\mathbf{X}) - w)) \mid \tilde{Y} = \tilde{y}]$. Select $w^*(\tilde{y}) \in W_{\tilde{y}}$ for each $\tilde{y} \in \mathcal{Y}$. Then $E_0[\rho(\delta_0(\mathbf{X}) - w^*(\tilde{y})) \mid \tilde{Y} = \tilde{y}]$ is minimized for each $\tilde{y}$ so that $\delta(\mathbf{X}) = \delta_0(\mathbf{X}) - w^*(\tilde{Y})$ cannot have larger risk than any other equivariant estimator. Since at least one finite risk equivariant estimator is assumed to exist, $\delta(\mathbf{X}) = \delta_0(\mathbf{X}) - w^*(\tilde{Y})$ is a MRE estimator of θ.

(ii) Any maximal invariant is an ancillary statistic *wrt* θ (Theorem 3.3). If δ_0 is a function of a complete sufficient statistic then by Basu's theorem, $\delta_0(\mathbf{X})$ and $\tilde{Y}$ are independent. Therefore $E_0[\rho(\delta_0(\mathbf{X}) - w(\tilde{y})) \mid \tilde{Y} = \tilde{y}]$ in Equation (9.1) can be replaced by $E_0[\rho(\delta_0(\mathbf{X}) - w)]$. This means the set of minimizers $W_{\tilde{y}}$ of Part (i) is the same for all $\tilde{y}$. We may therefore select a single value $w^* \in W_{\tilde{y}}$, yielding the MRE estimate $\delta_0(\mathbf{x}) - w^*$.

(iii) The uniqueness of the MRE follows directly from the construction $\delta(\mathbf{X}) = \delta_0(\mathbf{X}) - w^*(\tilde{Y})$.

(iv) Part (iv) follows from Theorem 7.6. □

Remark 9.1 (Randomized MRE estimators) Theorem 9.5 (ii) can simplify considerably the problem of determing a MRE estimator. Suppose the conditions of Theorem 9.5 hold. Let $\delta(\mathbf{X})$ be any equivariant estimator of θ that is a function of a complete sufficient statistic. Then there exists an MRE estimate of the form $\delta(\mathbf{X}) - w$ where w is a constant, and the MRE can therefore be derived by the much simpler Theorem 7.9.

If the MRE estimate is unique, then this is clearly also a necessary condition. For example, by Theorem 7.6 if a location invariant loss function is strictly convex and nonmonotonic, then so is the conditional risk used in Theorem 9.5 (i), so the MRE estimate will be unique. However, suppose multiple MRE estimates exist. It will still be the case that $\delta(\mathbf{X}) - w$ is an MRE estimate for any value w in some set W.

Suppose the loss is not strictly convex. Then W may contain multiple minimizers (by hypothesis at least one exists). In the proof of Theorem 9.5 (ii) $w^*(\tilde{y})$ was set to a single value, but this is not strictly necessary, and we could also construct an MRE estimator $\delta(\mathbf{X}) - w^*(\tilde{Y})$, where $w^*(\tilde{Y})$ varies with $\tilde{Y}$. Since here $\tilde{Y}$ and $\mathbf{X}$ are independent, this is essentially a randomized estimator (we could also introduce explicit randomization within the theorem's argument, even without completeness). However, there will always exist an nonrandomized estimator with no worse risk, so the existence of a nonrandomized MRE estimator does not depend on the convexity of the loss function (see also Theorem 7.12). ■

Remark 9.2 (Must δ_0 of Equation (9.1) Have Finite Risk?) It is important to note that while the conditions of Theorem 9.5 state that at least one finite risk equivariant estimator must exist, this need not be the estimator δ_0 which is used in Equation (9.1) or (9.2). This technical detail actually assumes some importance for distributions not possessing a first or second moment. Given that MRE estimation is especially useful for such distributions, it would be limiting to have to assume that δ_0 has finite risk, so it is important to understand this issue. In statement (i) of Theorem 9.5, it is assumed that the optimizing quantity $w^*(\tilde{y})$ is finite. This assumption is not at all trivial, but may be verified using. for example, Theorem 9.2. Above all, it allows us to weaken the assumption that δ_0 is of finite risk. This will be demonstrated in Example 9.4. ∎

The next step is to show how Theorem 9.5 is used to derive specific MRE estimators. Despite the rather indirect definition of the estimators given in Equations (9.1) or (9.2) the final expression will often be quite tractable.

Example 9.2 (Some Specific Location Invariant Loss Functions) Suppose we are given a location invariant model $\mathcal{P}$. Consider loss function $L_\theta(d) = \rho(\theta - d)$ where $\rho(x) = x^2$. Recall from Theorem 7.8 that $E[(X - c)^2]$ is uniquely minimized by $c = E[X]$, assuming $E[X^2] < \infty$. Then suppose at least one equivariant estimator δ_0 of finite risk exists. Since ρ is strictly convex, by Theorem 9.5 a unique MRE estimator of the location parameter exists under mean squared error (MSE) risk, and is given by

$$\delta(\mathbf{X}) = \delta_0(\mathbf{X}) - E_0[\delta_0(\mathbf{X}) \mid \tilde{Y}],$$

where $\tilde{Y} = M(\mathbf{X})$ is any maximal invariant. In addition, if δ_0 is a function of a complete statistic for θ, the MRE estimator becomes

$$\delta(\mathbf{X}) = \delta_0(\mathbf{X}) - E_0[\delta_0(\mathbf{X})].$$

Suppose we next set $\rho(x) = |x|$. By Theorem 7.8 $E[|X - c|]$ is uniquely minimized by any median c of X. Note that ρ is convex, but not strictly convex. Therefore, by Theorem 9.5 a MRE estimator of the location parameter exists under mean absolute deviation (MAD) loss exists and is given by

$$\delta(\mathbf{X}) = \delta_0(\mathbf{X}) - m(\tilde{Y}),$$

where $m(\tilde{Y})$ is any median of $\delta_0(\mathbf{X}) \mid \tilde{Y}$ under density $p_0 \in \mathcal{P}$. Similarly, if δ_0 is a function of a complete statistic for θ, the MRE estimator becomes

$$\delta(\mathbf{X}) = \delta_0(\mathbf{X}) - m,$$

where m is any median of $\delta_0(\mathbf{X})$ under density $p_0 \in \mathcal{P}$. ∎

We next consider some specific examples.

Example 9.3 (MRE Estimators for Exponential Sample with Location Parameter) Let $\mathbf{X} = (X_1, \ldots, X_n)$ be an *iid* sample with $X_1 \sim exp(\theta, \tau)$. We assume τ is known. The statistic $X_{(1)}$ is complete and sufficient for location parameter θ (Section 3.8.3). Also, $\delta_0 = X_{(1)}$ is a location equivariant estimator of θ (any order statistic is location equivariant, Example 3.7). Under density $p_{(0,\tau)}$, $X_{(1)} \sim exp(\tau/n)$. From Theorem 9.5 (ii), following Example 9.2 the MRE estimator for location parameter θ under loss function $L_\theta(d) = (\theta - d)^2$ is

$$\delta^*(\mathbf{X}) = X_{(1)} - E_e[\delta_0(\mathbf{X})] = X_{(1)} - \tau/n.$$

For loss function $L_\theta(d) = |\theta - d|$ we first obtain the median of δ_0 under p_0, which is the solution m to

$$1/2 = P\left\{X_{(1)} > m\right\} = e^{-nm/\tau},$$

or $m = \log(2)/n$. Then the MRE estimator is

$$\delta^*(\mathbf{X}) = X_{(1)} - m = X_{(1)} - \tau \log(2)/n.$$

Clearly, the MRE estimators depend on the loss function. This example is generalized in Problem 9.3. ∎

The next example considers the issue raised in Remark 9.2.

Example 9.4 (MRE Estimators for Cauchy Sample with Location Parameter) We continue with Example 9.1. Let $\mathbf{X} = (X_1, \ldots, X_n)$ be an *iid* sample with $X_1 \sim cauchy(\theta, 1)$. The order statistics are minimal sufficient, so no complete sufficient statistic exists (the difference between any two order statistics is ancillary *wrt* θ). In order to apply Theorem 9.5 with MSE risk we need to verify the existence of at least one finite risk equivariant estimator. The sample median $\hat{m}$ satisfies these conditions. We may then evaluate the MRE estimator as $\delta = \delta_0 - E_0[\delta_0 \mid \tilde{Y}]$ for some maximal invariant $\tilde{Y}$ and any equivariant estimator δ_0. We could use $\delta_0 = \hat{m}$, but we will see that a more convenient alternative will be to set $\delta_0 = X_n$. In fact, this is the approach that will be used below to derive the Pitman estimator, which is analytically tractable, if not trivial.

However, it is worth asking whether or not the fact that $\delta_0 = X_n$ does not have finite risk invalidates this approach (noting that the Cauchy density possesses no finite moments). This is not the case. Theorem 9.5 requires the existence of *any* finite risk equivariant estimator, a condition satisfied by $\hat{m}$. Regarding δ_0, it is assumed in Statement (i) that the optimizing quantity $w^*(\tilde{y})$ is finite. In this example this is equivalent to the assumption that $E_0[X_n \mid \tilde{Y} = \tilde{y}]$ is finite, which holds by Theorem 9.2. ∎

9.2.1 The Pitman Estimator for Location Parameters

While the existence of a complete sufficient simplifies the evaluation of an MRE estimator, it is not needed to verify its existence. Thus, location equivariant estimation is useful for resolving certain estimation problems for which complete sufficient statistics are not avaliable. However, the evaluation of the conditional expectation in Equation (9.1) is often technically challenging (which is not needed when a complete sufficient statistic exists). Fortunately, in the case of squared error loss, an elegant method of deriving the MRE based on Theorem 9.5 exists in the form of the Pitman estimator (Pitman, 1939).

When a complete sufficient statistic is not available we will likely need to evaluate the density of a maximal invariant $M(\mathbf{X}) = (X_1 - S(\mathbf{X}), \ldots, X_n - S(\mathbf{X}))$. Note that this form introduces a constraint. Since $M(\mathbf{X})$ is equivalent to $(X_1 - X_n, \ldots, X_{n-1} - X_n, X_n - X_n) = (X_1 - X_n, \ldots, X_{n-1} - X_n, 0)$, the image of a maximal invariant must be $n-1$ dimensions. We can therefore complete a maximal invariant by setting, for example,

$$M^*(\mathbf{X}) = (Y_1, \ldots, Y_{n-1}, Y_n) = (X_1 - X_n, \ldots, X_{n-1} - X_n, X_n).$$

The density of $M^*(\mathbf{X})$ can be deduced from the density of $\mathbf{X}$ using the transformation method of Section 2.3. If $f_{\mathbf{X}}$ is a density with respect to Lebesgue measure, we note that the determinant of the Jacobian matrix of the transformation is 1. Then the density of $M^*(\mathbf{X})$ may be written

$$f_{M^*}(y_1, \ldots, y_n) = f_{\mathbf{X}}(y_1 + y_n, \ldots, y_{n-1} + y_n, y_n). \tag{9.3}$$

The density of $M(\mathbf{X})$ is then a marginal density of $M^*(\mathbf{X})$ obtainable by integrating through y_n:

$$\begin{aligned} f_M(y_1, \ldots, y_{n-1}) &= \int_u f_{\mathbf{X}}(y_1 + u, \ldots, y_{n-1} + u, u)du \\ &= \int_u f_X(x_1 + u, \ldots, x_{n-1} + u, x_n + u)du, \end{aligned} \tag{9.4}$$

making used of substitution $u \to u + x_n$. See also Problem 9.5.

The Pitman estimator was introduced in Pitman (1939) for both location and scale parameters. We next introduce the location parameter case, then consider the scale parameter in Theorem 9.12.

Theorem 9.6 (Pitman Estimator for Location Parameters) Suppose the conditions of Theorem 9.5 hold, and that $\mathbf{X}$ possesses a density with respect to Lebesgue measure. Then a unique MRE estimator under squared error loss exists and is given by

$$\delta^*(\mathbf{x}) = \frac{\int_{-\infty}^{\infty} u p_0(x_1 - u, \ldots, x_n - u)du}{\int_{-\infty}^{\infty} p_0(x_1 - u, \ldots, x_n - u)du} \tag{9.5}$$

where $p_0 \in \mathcal{P}$. ∎

Proof. In Equations (9.3) and (9.4) set $f_{\mathbf{X}}(\mathbf{x}) = p_0(\mathbf{x})$, then define $\delta_0(\mathbf{X}) = X_n$, which is location equivariant. We make use of maximal invariant

$$\mathbf{Y} = (Y_1, \ldots, Y_{n-1}) = (X_1 - X_n, \ldots, X_{n-1} - X_n),$$

and set $Y_n = X_n$. The conditional density of Y_n given $\mathbf{Y} = \mathbf{y}$ is therefore

$$f_{Y_n|Y=(y_1,\ldots,y_{n-1})}(y_n \mid y_1, \ldots, y_{n-1}) = \frac{p_0(y_1 + y_n, \ldots, y_{n-1} + y_n, y_n)}{\int_{t=-\infty}^{\infty} p_0(y_1 + t, \ldots, y_{n-1} + t, t)dt}.$$

By assumption there exists at least one finite risk equivariant estimator, so Theorem 9.2 applies. We may then evaluate, using Equations (9.3) and (9.4),

$$E_0[X_n \mid \mathbf{Y} = \mathbf{y}] = E_0[Y_n \mid \mathbf{Y} = \mathbf{y}] = \frac{\int_{t=-\infty}^{\infty} t p_0(y_1 + t, \ldots, y_{n-1} + t, t)dt}{\int_{t=-\infty}^{\infty} p_0(y_1 + t, \ldots, y_{n-1} + t, t)dt}.$$

If we express $(y_1, \ldots, y_{n-1})$ in terms of $(x_1, \ldots, x_n)$, then substitute $t = x_n - u$, we may write

$$E_0[X_n \mid \mathbf{Y} = \mathbf{y}] = x_n - \frac{\int_{u=-\infty}^{\infty} u p_0(x_1 - u, \ldots, x_n - u)du}{\int_{u=-\infty}^{\infty} p_0(x_1 - u, \ldots, x_n - u)du}.$$

The Pitman estimator is then $\delta^*(\mathbf{X}) = X_n - E_0[X_n \mid \mathbf{Y}]$, which is equivalent to Equation (9.5). □

We next offer three examples of Pitman estimators for location parameters obtained by applying Theorem 9.6. The first example is for an *iid* sample from an exponential distribution with a location parameter. This problem was already considered in Example 9.3. There, the derivation of the MRE estimator under MSE risk was based on first recognizing the existence of a complete sufficient statistic, then determining its mean. The Pitman estimator will of course solve the same problem. The advantage it has is that the existence of a complete sufficient statistic is not needed. Thus, even when completeness might hold, if resolving the property would be a considerable technical challenge, the Pitman estimator may be used instead. Therefore, it is worth comparing the two approaches.

Example 9.5 (Pitman Estimator for Exponential Sample with Location Parameter) Suppose $\mathbf{X} = (X_1, \ldots, X_n)$ is an *iid* sample with $X_1 \sim exp(\theta, \tau)$. Then $p_{(0,\tau)}$ is the density of $\mathbf{X}$ with $\theta = 0$, where

$$p_{(0,\tau)}(\mathbf{x}) = \tau^{-n} \exp\left\{ -\sum_i x_i/\tau \right\} I\{0 < x_{(1)}\}.$$

The Pitman estimator (Theorem 9.6) is given by

$$\begin{aligned}\delta^*(\mathbf{x}) &= \frac{\tau^{-n}\exp\{-\sum_i x_i/\tau\}\int_{-\infty}^{\infty} u\exp(nu/\tau)I\{u < x_{(1)}\}du}{\tau^{-n}\exp\{-\sum_i x_i/\tau\}\int_{-\infty}^{\infty}\exp(nu/\tau)I\{u < x_{(1)}\}du} \\ &= \frac{\int_{-\infty}^{x_{(1)}} u\exp(nu/\tau)du}{\int_{-\infty}^{x_{(1)}}\exp(nu/\tau)du} \\ &= x_{(1)} - \tau/n.\end{aligned}$$

■

The next example is for an *iid* sample from a uniform distribution with a location parameter. Here, $(X_{(1)}, X_{(n)})$ is minimal sufficient, but not complete, since the distribution of $X_{(n)} - X_{(1)}$ does not depend on the location parameter. Thus, Theorem 9.6 makes more of a contribution than for Example 9.5.

Example 9.6 (Pitman Estimator for Uniform Sample with Location Parameter) Suppose $\mathbf{X} = (X_1, \ldots, X_n)$ is an *iid* sample with $X_1 \sim unif(\theta, \theta+1)$, $\theta \in \mathbb{R}$. Then p_0 is the density of $\mathbf{X}$ with $\theta = 0$, where

$$p_0(\mathbf{x}) = I\{0 < x_{(1)} < x_{(n)} < 1\}.$$

Substitution into Equation (9.5) gives the Pitman estimator

$$\begin{aligned}\delta^*(\mathbf{x}) &= \frac{\int_{-\infty}^{\infty} uI\{u < x_{(1)} < x_{(n)} < 1+u\}du}{\int_{-\infty}^{\infty} I\{u < x_{(1)} < x_{(n)} < 1+u\}du} \\ &= \frac{\int_{x_{(n)}-1}^{x_{(1)}} udu}{\int_{x_{(n)}-1}^{x_{(1)}} du} \\ &= \frac{x_{(1)} + x_{(n)}}{2} - \frac{1}{2}.\end{aligned}$$

■

The Pitman estimator is especially useful in cases where the order statistics are minimal sufficient, as in the next example.

Example 9.7 (Pitman Estimator for Logistic Density with Location Parameter) Suppose $\mathbf{X} = (X_1, \ldots, X_n)$ is an *iid* sample with $X_1 \sim logistic(\theta, 1)$, $\theta \in \mathbb{R}$. Then $p_{(0,1)}$ is the density of $\mathbf{X}$ with $\theta = 0$, where

$$p_{(0,1)} = \prod_{i=1}^{n} \frac{e^{-x_i}}{(1+e^{-x_i})^2}.$$

The order statistics are minimally sufficient (Problem 3.10). The Pitman estimator is then

$$\begin{aligned}\delta^*(\mathbf{x}) &= \frac{\int_{-\infty}^{\infty} u\prod_{i=1}^{n}\frac{e^{-(x_i-u)}}{(1+e^{-(x_i-u)})^2}du}{\int_{-\infty}^{\infty}\prod_{i=1}^{n}\frac{e^{-(x_i-u)}}{(1+e^{-(x_i-u)})^2}du} \\ &= \frac{\int_{-\infty}^{\infty} ue^{nu}\prod_{i=1}^{n}(1+e^{-(x_i-u)})^{-2}du}{\int_{-\infty}^{\infty} e^{nu}\prod_{i=1}^{n}(1+e^{-(x_i-u)})^{-2}du}.\end{aligned}$$

Thus, if no closed form for an MRE can be derived, it can at least be reduced to integrals of a single variable. See also Problem 9.7 for a similar example involving the Cauchy density. ■

9.3 MRE Estimators for Scale Parameters

The scale invariant model has been introduced in Section 3.3.2. A number of problems could be resolved by converting a scale invariant model to a location invariant model using a logarithmic transformation (Farrell, 1964). However, the overall approach for the scale invariant problem so closely resembles that of the location invariant problem, it seems preferable to recognize and rely upon that structure common to both transformations.

As in Section 3.3, we start with some density $p_1(\mathbf{x})$ on sample space $\mathcal{X} = \mathbb{R}^n$. Then consider the family of transformations

$$g_a(x_1, \ldots, x_n) = (ax_1, \ldots, ax_n), \ a > 0.$$

This is clearly a bijective mapping from the scalar multiplication group $(\Theta, \times)$, $\Theta = \mathbb{R}_{>0}$ (Example D.2), so that $G_{sc} = \{g_\theta : \theta \in \Theta\}$ on $\mathcal{X}$ is a transformation group. If $\mathbf{X}$ has density $p_1(\mathbf{x})$, it is easily verified that the density of $g_\theta \mathbf{X}$ is

$$p_\theta(\mathbf{x}) = \theta^{-n} p_1(\mathbf{x}/\theta).$$

The parametric family is therefore

$$\mathcal{P} = \left\{\theta^{-n} p_1(\mathbf{x}/\theta) : \theta \in \Theta\right\}.$$

It is better for scale parameters to permit some flexibility in the definition of the estimand. Here, for some fixed r, we set estimand $\eta(\theta) = \theta^r$. We then have the relation $\eta(a\theta) = a^r \eta(\theta)$. In this case, δ is equivariant if and only if $\delta(a\mathbf{x}) = a^r \delta(\mathbf{x})$, and $L_\theta(d)$ is invariant if and only if $L_\theta(d) = L_{a\theta}(a^r d)$ for all $a \in \Theta$.

Then $M(\mathbf{X}) = (X_1/S(\mathbf{X}), \ldots, X_n/S(\mathbf{X}))$ is a maximal invariant for any scale equivariant estimator $S(\mathbf{X})$ of scale parameter θ. We are, of course, implicitly assuming $S(\mathbf{X}) > 0$. The reader should review the discussion of this issue surrounding Theorem 3.7. Essentially, it is argued that disregarding the case $S(\mathbf{X}) = 0$ is justified. However a more formal approach is possible, in which a maximal invariant is constructed which equals $M(\mathbf{X}) = (X_1/S(\mathbf{X}), \ldots, X_n/S(\mathbf{X}))$ on the subset $\{S(\mathbf{X}) > 0\}$, but which is also well defined on the subset $\{S(\mathbf{X}) = 0\}$. See Problem 9.9.

A compact characterization of scale equivariant estimators closely follows Theorem 9.1 for location equivariant estimators.

Theorem 9.7 (The Class of Scale Equivariant Estimators) Suppose δ_0 is (scale) equivariant. Then δ is equivariant *iff* there is an invariant mapping $w(\mathbf{x})$ such that $\delta = \delta_0/w(\mathbf{x})$. ∎

Proof. First, assume w is invariant. Suppose $\delta(\mathbf{x}) = \delta_0(\mathbf{x})/w(\mathbf{x})$. Then $\delta(ax) = \delta_0(ax)/w(ax) = a^r \delta_0(\mathbf{x})/w(\mathbf{x}) = a^r \delta(\mathbf{x})$, so that δ is equivariant.

Next, assume that δ is equivariant. Set $w = \delta_0/\delta$. Then $w(ax) = \delta_0(ax)/\delta(ax) = [a^r \delta_0(\mathbf{x})]/[a^r \delta(\mathbf{x})] = \delta_0(\mathbf{x})/\delta(\mathbf{x}) = w(\mathbf{x})$. Thus, $\delta = \delta_0/w$, where w is invariant. □

Theorem 9.2 is easily modified for the scale equivariant case:

Theorem 9.8 (Existence of Moments for Scale Equivariant Estimators) Suppose $\mathbf{X}$ is a sample from a scale invariant model, with scale parameter $\theta \in \mathbb{R}_{>0}$. Suppose there exists a scale equivariant estimator δ_0 of θ^r for which $E_\theta[\delta_0^k] < \infty$, $\theta \in \Theta$. Let $\tilde{Y} = M(\mathbf{X})$ be a maximal invariant. Then for any other scale equivariant estimator δ we have $E_\theta[\delta^k \mid \tilde{Y}] < \infty$, *wp*1. ∎

Proof. First note that scale equivariant estimators are nonnegative. By the law of total expectation $E_\theta[\delta_0^k \mid \tilde{Y}] < \infty$, *wp*1. By Theorem 9.7 we may write $\delta(\mathbf{X}) = \delta_0(\mathbf{X})/w(\tilde{Y})$ for some mapping w, from which the proof follows. □

As for the location invariant loss function, scale invariant loss was characterized in Definition 7.3 and Theorem 7.7. We offer this as a separate theorem.

Theorem 9.9 (Structure of Scale Invariant Loss) A loss function $L_\theta(d)$ is scale invariant *iff* it assumes the form $L_\theta(d) = \rho(d/\eta(\theta))$. ■

Proof. If $L_\theta(d) = \rho(d/\eta(\theta))$, then

$$L_{a\theta}(a^r d) = \rho(a^r d/\eta(a\theta)) = \rho(a^r d/[a^r \eta(\theta)]) = \rho(d/\eta(\theta)) = L_\theta(d).$$

Conversely, suppose $L_\theta(d) = L_{a\theta}(a^r d)$ for all $a > 0$. Set $a = 1/\theta$. Then $L_\theta(d) = L_1(d/\eta(\theta))$. □

Finally, the relationship between scale invariant loss, scale equivariant estimation and constant risk follows the location invariant case using essentially the same argument (Theorem 9.4).

Theorem 9.10 (Constant Risk for Scale Invariant Loss) Supoose θ is a scale parameter, δ is any scale equivariant estimator of θ^r, and $L_\theta(d)$ is scale invariant loss function. Then risk $R(\theta, \delta)$ is constant over $\theta \in \Theta$. ■

Proof. The proof is similar to Theorem 9.4. We may write $L_\theta(\delta) = \rho(\delta(\mathbf{X})/\theta^r) = \rho(\delta(\mathbf{X}/\theta))$. But by Theorem 3.8 the distribution of $\mathbf{X}/\theta$ *dnd* θ. □

In Section 7.6 we considered the problem of determining the UMR estimator from a class $\{\delta_0/a, a \in \mathbb{R}_{>0}\}$ where δ_0 is any scale equivariant estimator. Then, similar to Theorem 9.5, Theorems 9.7, 9.9 and 9.10 allow us to extend the problem to the entire class of scale equivariant estimators, and we may derive the minimum risk equivariant (MRE) estimator within this class. The main theorem follows (note that Remarks 9.1 and 9.2 following Theorem 9.5 apply here also).

Theorem 9.11 (MRE Estimators for Scale Invariant Models) Suppose $\mathbf{X} \in \mathbb{R}^n$ is an observation from a scale invariant model, with scale parameter $\theta \in \mathbb{R}_{>0}$, and $L_\theta(d) = \rho(d/\theta^r)$ is a loss function for the problem of estimating θ^r. Let $\tilde{Y}$ be any maximal scale invariant. Suppose at least one equivariant estimator has finite risk. Let δ_0 be any other equivariant estimator.

(i) Assume that for all $\tilde{y} = \tilde{Y}$, there exists a value $w(\tilde{y}) = w^*(\tilde{y})$ which minimizes

$$E_1[\rho(\delta_0(\mathbf{X})/w(\tilde{y})) \mid \tilde{Y} = \tilde{y}]. \tag{9.6}$$

Then $\delta(\mathbf{X}) = \delta_0(\mathbf{X})/w^*(\tilde{Y})$ is a MRE estimator of θ^r.

(ii) Suppose δ_0 is a function of a statistic which is complete *wrt* θ, and there exists a value $w = w^*$ which minimizes

$$E_1[\rho(\delta_0(\mathbf{X})/w)]. \tag{9.7}$$

Then $\delta(\mathbf{X}) = \delta_0(\mathbf{X})/w^*$ is a MRE estimator of θ^r.

(iii) If the minimizer of Equation (9.6) or Equation (9.7) exists and is unique, then the MRE estimator of θ^r is unique.

(iv) If $\phi(w) = \rho(\exp(w))$ is convex and nonmonotone, then an MRE estimator exists. If ϕ is strictly convex and nonmonotone, then the MRE estimator is unique. ∎

Proof. The proof of Theorem 9.11 (i)–(iii) is essentially the same as for Theorem 9.5. Part (iv) is an application of Theorem 7.7 (iii). □

Following Example 9.2, the MRE estimator of Theorem 9.11 assumes a specific form for a given loss function. The cases of relative MAD and MSE risk (Section 7.5.3) are given in the next example.

Example 9.8 (Some Specific Scale Invariant Loss Functions) The reader is encouraged to review Section 7.5.3 for this example. Suppose the conditions of Theorem 9.11 hold. First, we consider relative squared error loss for estimand $\eta(\theta) = \theta^r$, that is,

$$L_\theta(d) = \frac{(\eta(\theta) - d)^2}{\eta(\theta)^2}. \tag{9.8}$$

This loss function is scale invariant, and from Theorem 7.8, the MRE estimator of $\eta(\theta) = \theta^r$ is

$$\delta^*(\mathbf{X}) = \frac{\delta_0(\mathbf{X}) E_1[\delta_0(\mathbf{X}) \mid \tilde{Y}]}{E_1[\delta_0(\mathbf{X})^2 \mid \tilde{Y}]} \tag{9.9}$$

where δ_0 is any other equivariant estimator and $\tilde{Y} = M(\mathbf{X})$ is any maximal scale invariant.

Similarly, relative absolute deviation loss function is given by

$$L_\theta(d) = \frac{|\eta(\theta) - d|}{\eta(\theta)}, \tag{9.10}$$

and is scale invariant. From Theorem 7.8, any MRE estimator of $\eta(\theta) = \theta^r$ takes form

$$\delta^*(\mathbf{X}) = \frac{\delta_0(\mathbf{X})}{w^*(\tilde{Y})} \tag{9.11}$$

where $w^*(\tilde{Y})$ is any scale-median of $\delta_0(\mathbf{X})$ conditional on $\tilde{Y}$ under density $p_1 \in \mathcal{P}$, and δ_0 is any other equivariant estimator.

It is important to note that by Theorem 7.5, the estimator of Equation (9.9) will also be MRE *wrt* squared error loss $L_\theta(d) = (\eta(\theta) - d)^2$, and the estimator of Equation (9.11) will also be MRE *wrt* absolute deviation loss $L_\theta(d) = |\eta(\theta) - d|$. The advantage of the loss functions (9.8) and (9.10) is that they are scale invariant, under which the risk of any scale equivariant estimator is constant. This allows resolution of the optimization problem, but a scale invariant loss need not be the decisive loss of any given decision problem. ∎

We next offer an application of Theorem 9.11.

Example 9.9 (MRE Estimators for Uniform Sample with Scale Parameter) Consider an *iid* sample $\mathbf{X} = (X_1, \ldots, X_n)$ from density $unif(0, \theta)$. The statistic $X_{(n)}$ is complete and sufficient for scale parameter θ (Section 3.8.3). Under p_1 its density is $f(\mathbf{x}) = nx^{n-1}$. Also note that $\delta_0(\mathbf{X}) = X_{(n)}$ is scale equivariant. It is easily verified that

$$E_1[X_{(n)}^r] = \frac{n}{n+r}, \quad r = 1, 2, \ldots.$$

From Theorem 9.11 (ii), following Example 9.8 the MRE estimator for θ under loss function $L_\theta(d) = (\theta - d)^2/\theta^2$ is

$$\delta^*(\mathbf{X}) = \frac{\delta_0(\mathbf{X}) E_1[\delta_0(\mathbf{X})]}{E_1[\delta_0(\mathbf{X})^2]} = \frac{n+2}{n+1} X_{(n)}.$$

Similarly, for loss function $L_\theta(d) = |\theta - d|/|\theta|$ we first obtain the scale median of δ_0 under $p_1 \in \mathcal{P}$, which is the solution c to

$$\int_0^c x f(\mathbf{x}) dx = \int_c^1 x f(\mathbf{x}) dx,$$

equivalently

$$\int_0^c x^n dx = \int_c^1 x^n dx.$$

This has solution $c = (1/2)^{1/(n+1)}$. So the MRE estimator for this particular loss function is

$$\delta^*(\mathbf{X}) = 2^{1/(n+1)} X_{(n)}.$$

As in Example 9.3 the MRE estimators differ by loss function. ■

9.3.1 The Pitman Estimator for Scale Parameters

We next consider the Pitman estimator for a scale parameter, which gives the MRE scale equivariant estimator under relative squared error loss (Example 9.8), and therefore under squared error loss (Theorem 7.5). The approach is similar to that of the location parameter given in Section 9.2.1. The main technical challenge is the evaluation of the density of a maximal scale invariant. This proves to be a somewhat more subtle problem than for the maximal location invariant, particularly when the observations $X_1, \ldots, X_n$ can assume negative values. First note that $|X_n|$ is scale equivariant for scale parameter θ, so

$$M(\mathbf{X}) = (Y_1, \ldots, Y_n) = (X_1/|X_n|, \ldots, X_{n-1}/|X_n|, X_n/|X_n|) \tag{9.12}$$

is a maximal invariant (see also Problem 9.12). We can assume $P(X_n = 0) = 0$, and note that $M(\mathbf{X})$ is a maximal invariant on $\mathcal{X} - \{X_n = 0\}$ (see the discussion of this issue in Section 3.3.2 and Problem 9.9). In this case, note that the last element of $M(\mathbf{X})$, which we denote $W = X_n/|X_n| = \text{sign}(X_n)$, is a discrete random variable assuming values -1 or 1. We may then write

$$M(\mathbf{X}) = (\mathbf{Y}, W)$$

where $\mathbf{Y} = (Y_1, \ldots, Y_{n-1})$. Following Section 9.2.1, we complete $M(\mathbf{X})$ as

$$M^*(\mathbf{X}) = (\mathbf{Y}, X_n, W) = (Y_1, \ldots, Y_{n-1}, X_n, W),$$

the density of which is given by

$$f_{M^*}(y_1, \ldots, x_n, w) = \\ |x_n|^{n-1} f_X(|x_n| y_1, \ldots, |x_n| y_{x-1}, x_n) I\{\text{sign}(x_n) = w\}. \tag{9.13}$$

To derive the density of $M(\mathbf{X}) = (\mathbf{Y}, W)$, we would integrate the density of $M^*(\mathbf{X})$ through X_n, bearing in mind that W is included in the maximal invariant. Interestingly, however, the density of $M(\mathbf{X})$ will not need to be evaluated explicitly in the derivation of the Pitman estimator for a scale parameter, which we give next.

Theorem 9.12 (Pitman Estimator for Scale Parameters) Suppose the conditions of Theorem 9.11 hold. If there exists a finite risk equivariant estimator of scale parameter θ^r, then under loss (9.8) a unique MRE estimator exists and is given by

$$\delta^*(\mathbf{x}) = \frac{\int_0^\infty v^{n+r-1} p_1(vx_1, \ldots, vx_n) dv}{\int_0^\infty v^{n+2r-1} p_1(vx_1, \ldots, vx_n) dv} \tag{9.14}$$

where $p_1 \in \mathcal{P}$. ■

Proof. The proof is similar to Theorem 9.6. First note that $\delta(\mathbf{X}) = |X_n|^r$ is a scale equivariant estimator of θ^r. Then, by Theorem 9.8, for any power $k \leq 2r$,

$$E_1[|X_n|^k|(\mathbf{Y}, W) = (\mathbf{y}, w)] = \frac{\int_{u>0} u^{k+n-1} p_1(uy_1, \ldots, uy_{n-1}, wu) du}{f_M(\mathbf{y}, u)} < \infty.$$

Then take a change in variable $u = v|x_n|$, so that, making use of Equation (9.13),

$$E_1[|X_n|^k|(\mathbf{Y}, W) = (\mathbf{y}, w)] = \frac{\int_{v>0} |x_n|^{k+n} v^{k+n-1} p_1(v|x_n|y_1, \ldots, v|x_n|y_{n-1}, wv|x_n|) dv}{f_M(\mathbf{y}, w)}.$$

Then the MRE estimator is, following Example 9.8,

$$\begin{aligned}\delta(\mathbf{x}) &= |x_n|^r \frac{E_1[|X_n|^r|(Y, W) = (y, w)]}{E_1[|X_n|^{2r}|(Y, W) = (y, w)]} \\ &= |x_n|^r \frac{\int_{v>0} |x_n|^{r+n} v^{r+n-1} p_1(v|x_n|y_1, \ldots, v|x_n|y_{n-1}, wv|x_n|) dv}{\int_{v>0} |x_n|^{2r+n} v^{2r+n-1} p_1(v|x_n|y_1, \ldots, v|x_n|y_{n-1}, wv|x_n| dv}.\end{aligned}$$

Finally, substitute $y_i = x_i/|x_n|$, $i = 1, \ldots, n-1$, noting that $w|x_n| = x_n$, to get

$$\delta(\mathbf{x}) = \frac{\int_{v>0} v^{n+r-1} p_1(vx_1, \ldots, vx_{n-1}, vx_n) dv}{\int_{v>0} v^{n+2r-1} p_1(vx_1, \ldots, vx_{n-1}, vx_n dv}.$$

□

The following two examples demonstrate Theorem 9.12. A version of the first example was already considered in Example 9.9, and we of course obtain the same solution. However, as pointed out in the context of the location parameter problem, evaluation of the Pitman estimator does not rely on any complete sufficient statistic, which can be a practical advantage even when one exists. Thus, both approaches should be kept in mind. See also Problems 9.15 and 9.16.

Example 9.10 (Pitman Estimator for Uniform Sample with Scale Parameter) Suppose $\mathbf{X} = (X_1, \ldots, X_n)$ is an *iid* sample with $X_1 \sim unif(0, \theta)$. Then p_1 is the density of $\mathbf{X}$ with $\theta = 1$, where

$$p_1(\mathbf{x}) = I\{0 < x_{(1)} < x_{(n)} < 1\}.$$

By Theorem 9.12 the Pitman estimator of θ^r is then

$$\begin{aligned}\delta^*(\mathbf{x}) &= \frac{\int_0^\infty v^{n+r-1} I\{0 < vx_{(1)} < vx_{(n)} < 1\} dv}{\int_0^\infty v^{n+2r-1} I\{0 < vx_{(1)} < vx_{(n)} < 1\} dv} \\ &= \frac{\int_0^\infty v^{n+r-1} I\{0 < v < 1/x_{(n)}\} dv}{\int_0^\infty v^{n+2r-1} I\{0 < v < 1/x_{(n)}\} dv} \\ &= \frac{\int_0^{1/x_{(n)}} v^{n+r-1} dv}{\int_0^{1/x_{(n)}} v^{n+2r-1} dv} \\ &= \frac{(n+2r)}{(n+r)} x_{(n)}^r.\end{aligned}$$

■

Example 9.11 (Pitman Estimator for Exponential Density Sample with Scale Parameter) Let $\mathbf{X} = (X_1, \ldots, X_n)$ be an *iid* sample with $X_1 \sim exp(\theta)$. Then p_1 is the density of $\mathbf{X}$ with $\theta = 1$, where $p_1(\mathbf{x}) = \exp(-\sum_i x_i)$. Substitution into Equation (9.14) gives the Pitman estimator

$$\begin{aligned}\delta^*(\mathbf{x}) &= \frac{\int_0^\infty v^{n+r-1} \exp(-v \sum_i x_i)\, dv}{\int_0^\infty v^{n+2r-1} \exp(-v \sum_i x_i)\, dv} \\ &= \frac{\Gamma(n+r)/(\sum_i x_i)^{n+r}}{\Gamma(n+2r)/(\sum_i x_i)^{n+2r}} \\ &= \frac{(\sum_i x_i)^r}{(n+2r-1)^{(r)}},\end{aligned}$$

where $n^{(r)} = n(n-1)\cdots(n-r+1)$ is the partial factorial. ■

9.4 Invariant Density Families

The location invariant, scale invariant and location-scale invariant models of Section 3.3 are examples of invariant density families generated by a transformation group. To a large degree the properties of this class of density families do not depend on the particular transformation group. Essentially, this is the theme we are exploring next.

The invariant density family can be constructed in two complementary ways. Both are based on a transformation group G acting on a sample space $\mathcal{X}$. For many invariant models $\mathcal{X}$ is assumed to be a Lebesgue measure space, but this need not be the case. Recall that any $g \in G$ is a bijection $g : \mathcal{X} \to \mathcal{X}$, and so necessarily possesses an inverse, in particular, the inverse of the bijection (Section D.4). The identity of G is simply the identity transformation. Note that G can be constructed from any collection $\mathcal{C}$ of bijections on $\mathcal{X}$ by taking all finite compositions of elements of $\mathcal{C}$ or their inverses (Theorem D.4). We will always assume that any $g \in G$ is a measurable mapping.

It will often be convenient to identify an index set Γ for G, so that $G = \{g_\gamma : \gamma \in \Gamma\}$. Assuming $g_\gamma \neq g_{\gamma'}$ when $\gamma \neq \gamma'$, there exists a bijection from G to Γ, so by Theorem D.3 there exists a group $(\Gamma, *)$ such that $g_a g_b = g_{a*b}$ for any $a, b \in \Gamma$; g_e is the indentity of G if e is the identity of Γ; and $g_{a^{-1}}$ is the inverse g_a for any $a \in \Gamma$. In practice Γ may be defined first, for example, as an addition group $(\mathbb{R}, +)$ or a scalar multiplication group $(\mathbb{R}_{>0}, \times)$. Although not formally required, the index group Γ will allow greater clarity regarding the invariant family.

An invariant parametric family on $\mathcal{X}$ can then be defined by the pair $(\mathcal{P}, G)$ where $\mathcal{P}$ is a family of densities on $\mathcal{X}$.

Definition 9.1 (Invariant Density Family) Suppose $\mathcal{P} = \{p_\theta : \theta \in \Theta\}$ is a parametric family defined on measure space $\tilde{X} \in \mathcal{X}$. Assume $\theta \in \Theta$ is identifiable (Section 3.2). Let g be a bijection on $\mathcal{X}$, and let $\mathcal{P}'$ be the collection of densities of $g\tilde{X}$ for which $\tilde{X} \sim p \in \mathcal{P}$. If $\mathcal{P} = \mathcal{P}'$ we say that g leaves $\mathcal{P}$ invariant. Then let G be a transformation group on $\mathcal{X}$, and assume that each g leaves $\mathcal{P}$ invariant. Under these conditions $(\mathcal{P}, G)$ is an invariant density family. ■

Any transformation group possesses a maximal invariant $M(\tilde{X})$ (Theorem D.7). Also note that G is required for the definition of an invariant parametric family. It may be that the pairs $(\mathcal{P}, G)$ and $(\mathcal{P}, G')$ both satisfy Definition 9.1 with a common parametric family $\mathcal{P}$ but distinct transformation groups $G \neq G'$ (Problem 9.22).

We offer our first example of an invariant density family outside of the location-scale model.

Example 9.12 (Invariance in the Binomial Model) Suppose $X \sim bin(n, \theta)$, which defines parametric family $\{p_\theta : \theta \in \Theta\}$, where $\Theta = [0, 1]$ and p_θ is the binomial density. Define transformations $g_0 X = n - X$, $g_1 X = X$. Then $G = \{g_0, g_1\}$ is a transformation group acting on $\{0, 1, \ldots, n-1, n\}$. To verify this, first note that G is closed under composition. Then, the identity of G is g_1, and $g_0^{-1} = g_0$, $g_1^{-1} = g_1$. We finally note that if $X \sim bin(n, \theta)$, $g_0 X \sim bin(n, 1-\theta)$ and $g_1 X \sim bin(n, \theta)$, so that each $g \in G$ leaves $\mathcal{P}$ invariant. Thus $(\mathcal{P}, G)$ satisfies Definition 9.1. ∎

If we generate a parametric family $\mathcal{P}$ by applying a transformation group to a single density on $\mathcal{X}$, then we will have an example of Definition 9.1.

Theorem 9.13 (Generating Invariant Density Families) Let p_e be a density on $\mathcal{X}$, and let G be a transformation group acting on $\mathcal{X}$. Suppose $\tilde{X} \sim p_e$, and let $\mathcal{P}$ be the collection of densities p for which $g\tilde{X} \sim p$ for some $g \in G$. Then $(\mathcal{P}, G)$ is an invariant parametric family (Definition 9.1). ∎

Proof. By assumption G is a transformation group on $\mathcal{X}$, so it remains to prove that any $g \in G$ leaves $\mathcal{P}$ invariant. Suppose $\tilde{X}_e \sim p_e$. Then for any $p \in \mathcal{P}$ there exists $g_p \in G$ such that $g_p \tilde{X}_e \sim p$.

Select $g_0 \in G$, and suppose $\tilde{X} = g_p \tilde{X}_e \sim p$. Then $g_0 \tilde{X} = g_0 g_p \tilde{X}_e \in \mathcal{P}$, since $g_0 g_p \in G$. Therefore the transformation g_0 maps $\mathcal{P}$ into $\mathcal{P}$.

We may then write $\tilde{X} = g_0 g_0^{-1} g_p \tilde{X}_e \sim p$. Then $g_0^{-1} g_p \tilde{X}_e \sim p' \in \mathcal{P}$, therefore, for any $p \in \mathcal{P}$ there exists $p' \in \mathcal{P}$ such that g_0 maps p' to p. Therefore, the transformation g_0 maps $\mathcal{P}$ onto $\mathcal{P}$. □

Theorem 9.13 characterizes the approach we have already taken in Sections 3.3, 9.2 and 9.3. However, not all invariant density families can be generated this way (Example 9.12, see also Problem 9.22). In fact, whether or not this is the case will have important implications for invariant inference methods (look ahead, for example, to Theorem 9.16).

9.4.1 Transformation Groups on the Parameter Space

At this point the reader should review Section D.3 on group homomorphisms. Two groups $(G, *)$, $(G', \odot)$ are homomorphic if there exists a mapping $h : G \to G'$ which preserves group operations, in the sense that $h(g_1 * g_2) = h(g_1) \odot h(g_2)$, $g_1, g_2 \in G$ (Definition D.4). If the mapping h is bijective, then $(G, *)$, $(G', \odot)$ are, in addition, isomorphic. For our purposes, two isomorphic groups can be thought of alternative labellings of a single group. Furthermore, if $(G, *)$ is a group, and for some other set G' there exists a bijection $h : G \to G'$, then we may define a binary operation $\odot$ on G' such that $(G, *)$, $(G', \odot)$ are isomorphic (Theorem D.3).

Understanding isomorphisms and homorphisms is crucial to understanding invariant inference. Recall that we define an invariant density family $(\mathcal{P}, G)$ with a transformation group G. If G is indexed by Γ, that is, $G = \{g_\gamma : \gamma \in \Gamma\}$, we assume that G and Γ are bijective, so that we may assume that Γ is also group, and isomorphic to G. Therefore, from the group theoretic point of view, G and Γ are the same group. If we then consider the decision theoretic framework of parametric inference introduced in Section 7.4, we can describe a decision problem as invariant with respect to G if we may induce transformation groups $\bar{G}$ and G^* acting on Θ and action space $\mathcal{A}$, respectively, which are homomorphic to G.

In this section, we consider the problem of constructing the homomorphism $\bar{G}$ on parameter space Θ. We start by verifying that a bijective transformation g on $\mathcal{X}$ induces a bijective transformation $\bar{g}$ on Θ. This is the goal of the next theorem.

Theorem 9.14 (Bijections on the Parameter Space) Let $\mathcal{P} = \{p_\theta : \theta \in \Theta\}$ be a parametric family on $\mathcal{X}$. Assume θ is identifiable. Suppose g is a bijection on $\mathcal{X}$ which leaves $\mathcal{P}$ invariant. Then there exists a unique bijection $\bar{g}$ on Θ such that $g\tilde{X} \sim p_{\bar{g}\theta}$ when $\tilde{X} \sim p_\theta$, equivalently

$$P_\theta(g\tilde{X} \in E) = P_{\bar{g}\theta}(\tilde{X} \in E) \quad \text{and} \quad E_\theta[h(g\tilde{X})] = E_{\bar{g}\theta}[h(\tilde{X})] \tag{9.15}$$

for any $\theta \in \Theta$. ■

Proof. Since g leaves $\mathcal{P}$ invariant, for any $\theta \in \Theta$, we may identify a unique parameter $\bar{g}\theta \in \Theta$ such that $g\tilde{X} \sim p_{\bar{g}\theta}$ when $X \sim p_\theta$. The induced mapping $\bar{g} : \Theta \to \Theta$ is therefore well defined, so it remains to verify that it is bijective. First note that we must have $P_{\bar{g}\theta}(A) = P_\theta(gA)$ for all measurable sets A. Then suppose $\bar{g}\theta = \bar{g}\theta'$. This means $P_\theta(gA) = P_{\theta'}(gA)$. However, since g is bijective, for any measurable set B, there exists A such that $B = gA$. This implies $P_\theta = P_{\theta'}$, equivalently, $\theta = \theta'$. □

Theorem 9.14 establishes that there is a mapping $G \mapsto \bar{G}$, where $\bar{G}$ is a class of bijections on Θ, which satisfies Equation (9.15). The next theorem proves that $(\bar{G}, \circ)$ is a group which is homomorphic to $(G, \circ)$

Theorem 9.15 (Homomorphic Transformation Groups on Θ) Let $(\mathcal{P}, G)$ be an invariant density family, where Γ is the index group for G. For each $\gamma \in \Gamma$, let $\bar{g}_\gamma$ be the bijection on Θ induced by $g_\gamma \in G$, characterized by Theorem 9.14. Define $\bar{G} = \{\bar{g}_\gamma : \gamma \in \Gamma\}$. Then $\bar{G}$ is a transformation group which is homomorphic to G. ■

Proof. By assumption each $\bar{g}_\gamma \in \bar{G}$ is a bijection on Θ, so it remains to verify that $\bar{G}$ is a group, which we do by constructing a homomorphic relationsip to Γ, and hence G.

Consider transformation $\theta' = \bar{g}_a\bar{g}_b\theta$, $a, b \in \Gamma$. If $\tilde{X} \sim p_\theta$ then $g_bX \sim p_{\bar{g}_b\theta}$, and $g_ag_b\tilde{X} \sim p_{\bar{g}_a\bar{g}_b\theta}$. But $g_ag_b = g_{a*b}$, so $\bar{g}_a\bar{g}_b = \bar{g}_{a*b}$, so that $\bar{G}$ is closed under composition. The associativity of $\bar{G}$ follows from the associativity of Γ. Then $\bar{g}_e$ is an identity of $\bar{G}$ where e is the identity of Γ. In addition, any $\bar{g}_a \in \bar{G}$, possesses an inverse $\bar{g}_{a^{-1}} \in \bar{G}$, where a^{-1} is the inverse of $a \in \Gamma$. □

Example 9.13 (Invariance in the Binomial Model (cont'd)) To continue Example 9.12 the induced transformation group acting on $\Theta = (0, 1)$ is given by $\bar{G} = \{\bar{g}_0, \bar{g}_1\}$ where $\bar{g}_0\theta = 1 - \theta$, $\bar{g}_1\theta = \theta$. ■

It is helpful to note in this regard that the axioms of a group (Definition D.1) do not explicitly state that an indentity or an inverse are unique. This is not needed, since uniqueness follows from the axioms. Thus, we need not assume $\bar{g}_a \neq \bar{g}_b$ when $a \neq b$. In other words when $a \neq b$ we may or may not have $p_{\bar{g}_a\theta} \neq p_{\bar{g}_b\theta}$. In terms of group theory, we can say that G and $\bar{G}$ will be homomorphic, but not necessarily isomorphic.

However, a number of important theorems rely on the assumption that $\bar{G}$ is transitive. That will be the case for invariant density families generated when a transformation group acts on a single density (Theorem 9.13). See Definition D.6.

Theorem 9.16 (Transitive Transformation Groups on Θ) Suppose $(\mathcal{P}, G)$ is an invariant parametric family on $\mathcal{X}$. Suppose there exists a density p_e such the collection of densities of $g\tilde{X}$, $g \in G$ is equal to $\mathcal{P}$ when $\tilde{X} \sim p_e$. Then $\bar{G}$, the induced transformation group on Θ, is transitive. ■

Proof. Suppose $\theta, \theta' \in \Theta$. Since G includes the identity transformation, we must have $p_e \in \mathcal{P}$. Denote the corresponding parameter $e \in \Theta$. By hypothesis there exists $g_a, g_b \in G$ such that $\theta = \bar{g}_ae$, $\theta' = \bar{g}_be$. Then $\theta' = \bar{g}_b\bar{g}_a^{-1}\theta$. The proof is completed by noting $\bar{g}_b\bar{g}_a^{-1} \in \bar{G}$. □

We offer a few counterexamples to Theorem 9.16.

Example 9.14 (Invariance in the Binomial Model (cont'd)) The induced transformation group $\bar{G}$ of Example 9.13 is not transitive. ■

Example 9.15 (Transitive Transformation Groups on Θ for the Location-Scale Invariance Model) The induced transformation group $\bar{G}$ for the location, scale and location-scale invariant model are all transitive, since they satisfy the assumptions of Theorem 9.13. However, suppose we are given a parametric family on $\mathcal{X} = \mathbb{R}^n$ which contains all densities of the form $p_{\mu,\sigma}(\mathbf{x}) = \sigma^{-n} p_e((\mathbf{x} - \mu)/\sigma)$, $\mu \in \mathbb{R}$, $\sigma \in \mathbb{R}_{>0}$. Let G_{loc} be the defined by all location transformations on $\mathcal{X}$. Then $(\mathcal{P}, G_{loc})$ satisfies the definition of an invariant density family, but $\bar{G}$ will not be transitive. See Problem 9.22. ■

9.4.2 Invariant Statistics

The invariant statistic was introduced in Section 3.3 in the context of location-scale transformations. The notion extends naturally to any transformation group acting on $\mathcal{X}$, according to the following definition.

Definition 9.2 (Invariant Statistics Under General Transformation Groups) Suppose we are given G on sample space $\mathcal{X}$. Then statistic $U(\mathbf{X})$ on $\mathcal{X}$ is invariant (*wrt* G) if $U(\tilde{X}) = U(g\tilde{X})$ for all $g \in G$. If a statistic $M(\tilde{X})$ is invariant, and $M(\tilde{X}) \mapsto U(\tilde{X})$ for any other invariant statistic $U(\tilde{X})$, then it is a maximal invariant. By Theorem D.7 a maximal invariant always exists. ■

We have seen how the maximal invariant for the location or scale transformation can be used to construct an ancillary statistic (Section 3.3). More generally, suppose $(\mathcal{P}, G)$ is an invariant density family, and let $\bar{G}$ be the transformation group on Θ described in Theorem 9.15. An orbit of $\bar{G}$ is a set of the form $\{\bar{g}\theta : \bar{g} \in \bar{G}\}$, for some fixed $\theta \in \Theta$ (Section D.5). Let $U(\tilde{X})$ be an invariant statistic. From Theorem 9.14 it follows that the distribution of $U(\tilde{X})$ will be the same for all parameters θ within a single orbit of $\bar{G}$. However, if $\bar{G}$ is transitive, then there is only a single orbit Θ, so that the distribution of $U(\tilde{X})$ will not depend on θ. We state this formally in the next theorem.

Theorem 9.17 (Ancillarity of Invariant Statistics Under General Transformation Groups) Suppose the conditions of Theorem 9.15 hold and $U(\tilde{X})$ is invariant *wrt* G. Then the distribution of $U(\tilde{X})$ is the same under densities p_θ and $p_{\bar{g}\theta}$ for any $\theta \in \Theta$, $\bar{g} \in \bar{G}$. In addition, if $\bar{G}$ is transitive, then the distribution of $U(\tilde{X})$ does not depend on θ. ■

Proof. If U is invariant, then by Equation (9.15)

$$P_{\bar{g}\theta}(U(\tilde{X}) \in E) = P_\theta(U(g\tilde{X}) \in E) = P_\theta(U(\tilde{X}) \in E) \tag{9.16}$$

for any measurable set E.

Then suppose $\bar{G}$ is transitive. Fix some parameter $\theta_0 \in \Theta$. Then for any θ there exists $\bar{g} \in \bar{G}$ such $\theta_0 = \bar{g}\theta$. Then by Equation (9.16)

$$P_\theta(U(\tilde{X}) \in E) = P_{\bar{g}\theta}(U(\tilde{X}) \in E) = P_{\theta_0}(U(\tilde{X}) \in E) \quad \theta \in \Theta.$$

Therefore the distribution of $U(\tilde{X})$ does not depend on θ. □

9.4.3 Invariant Pivots

Recall that a pivot is a mapping $\xi(\tilde{X}, \theta)$ the distribution of which does not depend on θ (Section 4.6.1). In Section 4.6.2, we showed that this property was quite natural for location and scale parameters, which is of course related to the group properties of those models. We will next make this explicit.

Definition 9.3 (Invariant Pivot) A pivot is invariant *wrt* transformation group G if $\xi(\tilde{X}, \theta) = \xi(g\tilde{X}, \bar{g}\theta)$ for all $g \in G$. ■

Theorem 9.18 (Properties of Invariant Pivots) Suppose the conditions of Theorem 9.15 hold and the pivot $\xi(\tilde{X}, \theta)$ is invariant *wrt* G.

(i) For any $\theta \in \Theta$, $g \in G$ and measurable set E, $P_\theta\left(\xi(\tilde{X}, \theta') \in E\right) = P_{\bar{g}\theta}\left(\xi(\tilde{X}, \bar{g}\theta') \in E\right)$.
(ii) If $\bar{G}$ is transitive then $P_\theta\left(\xi(\tilde{X}, \theta) \in E\right)$ does not depend on $\theta \in \Theta$.
(iii) If $\bar{G}$ is transitive and commutative then for any $\bar{g} \in \bar{G}$, $P_\theta\left(\xi(\tilde{X}, \bar{g}\theta) \in E\right)$ does not depend on $\theta \in \Theta$. ■

Proof. To prove (i) if $\xi(\tilde{X}, \theta)$ is invariant, then by Equation (9.15)

$$P_\theta(\xi(\tilde{X}, \theta') \in E) = P_\theta(\xi(g\tilde{X}, \bar{g}\theta') \in E) = P_{\bar{g}\theta}(\xi(\tilde{X}, \bar{g}\theta') \in E), \tag{9.17}$$

for any set E. To prove (ii) suppose $\bar{G}$ is transitive. Fix some parameter $\theta_0 \in \Theta$. Then for any θ there exists $\bar{g} \in \bar{G}$ such $\theta = \bar{g}\theta_0$. Then by Equation (9.17)

$$P_{\theta_0}(\xi(\tilde{X}, \theta_0) \in E) = P_{\bar{g}\theta_0}(\xi(\tilde{X}, \bar{g}\theta_0) \in E) = P_\theta(\xi(\tilde{X}, \theta) \in E).$$

To prove (iii) suppose $\bar{G}$ is transitive. As in the previous argument, fix θ_0, then write $\theta = \bar{g}'\theta_0$ for any $\theta \in \Theta$. Then

$$\begin{aligned} P_{\theta_0}(\xi(\tilde{X}, \bar{g}\theta_0) \in E) &= P_{\bar{g}'\theta_0}(\xi(\tilde{X}, \bar{g}'\bar{g}\theta_0) \in E) \\ &= P_{\bar{g}'\theta_0}(\xi(\tilde{X}, \bar{g}\bar{g}'\theta_0) \in E) \\ &= P_\theta(\xi(\tilde{X}, \bar{g}\theta) \in E), \end{aligned}$$

making use of the commutative property. □

9.5 Some Applications of Invariance

Much of the theory of invariant density familes is aimed towards the construction of MRE estimators. However, the idea of invariance on its own (Definition 9.1) can clarify certain inference problems, and even provide solutions to some seemingly intractible or poorly defined problems. In these cases, it is helpful to think of invariance is having a similar objective to sufficiency. In the latter, an observation $\tilde{X} = (S(\tilde{X}), V(\tilde{X}))$ is decomposed into a sufficient statistic $S(\tilde{X})$ and ancillary statistic $V(\tilde{X})$. Then $S(\tilde{X})$ contains all the information in $\tilde{X}$ about the parameter, in the sense that the distribution of $\tilde{X} \mid S(\tilde{X})$ does not depend on parameter θ. The principle of invariance is based on a similar type of decomposition, with the maximal invariant $M(\tilde{X})$ playing the role of ancillary statistic (Theorem 9.17). In this section, and Section 9.6, we consider some extended examples of this point of view.

9.5.1 Invariance and the Gauss-Markov Theorem

Chapter 6 was concerned with linear models and the Gauss-Markov theorem (Theorem 6.1). Recall that we had $\mathbf{Y} = [Y_1, \ldots, Y_n]^T \in \mathcal{Y} = \mathbb{R}^n$, a column vector of random variables, and $\boldsymbol{\beta} \in \mathbb{R}^q$, a q-dimensional parameter that is related to $\mathbf{Y}$ by the equation

$$\mathbf{Y} = \boldsymbol{\mu} + \boldsymbol{\epsilon}, \tag{9.18}$$

$\boldsymbol{\mu} \in \mathcal{M}_{n,1}$, with $E[\boldsymbol{\epsilon}] = 0$, so that $E[\mathbf{Y}] = \mathbf{X}\boldsymbol{\beta}$ and $\text{var}[\mathbf{Y}] = \text{var}[\boldsymbol{\epsilon}] = \Sigma_Y$.

Let Γ be the set of all $n \times n$ nonsingular matrices, which forms a group under matrix multiplication. Then Γ indexes a transformation group G acting on $\mathcal{Y}$ by the evaluation method

$$g_{\mathbf{A}}\mathbf{Y} = \mathbf{A}\mathbf{Y}, \tag{9.19}$$

for any $g_{\mathbf{A}} \in G$, $\mathbf{A} \in \Gamma$, noting that $g_{\mathbf{A}}$ is a bijection on $\mathcal{X}$ (Problem 9.24) Then suppose the density of $\mathbf{Y}$ defined in Equation (9.18) is $p_0(\mathbf{y})$. Then following Theorem 9.13, we can induce an invariant density family $(\mathcal{P}, G)$, where $p \in \mathcal{P}$ *iff* $g_{\mathbf{A}}\mathbf{Y} \sim p$ for some $g_{\mathbf{A}} \in G$. Then $\mathcal{P}$ can be parametrized by the pair $\theta = (\boldsymbol{\mu}, \Sigma)$ of mean vector $\boldsymbol{\mu}$ and covariance matrix Σ of $g_{\mathbf{A}}\mathbf{Y}$, and the parameter space Θ is all such pairs obtainable by any such transformation.

To fix ideas, suppose p_0 possesses mean vector and covariance matrix $\boldsymbol{\mu}_0 = \mathbf{X}\boldsymbol{\beta}_0$ and Σ_0, and set $\theta_0 = (\boldsymbol{\mu}_0, \Sigma_0)$. Then following Theorems 9.14 and 9.15 G induces a transformation group $\bar{G}$ on Θ using the evaluation method $\bar{g}_{\mathbf{A}}\theta_0 = \bar{g}_{\mathbf{A}}(\boldsymbol{\mu}_0, \Sigma_0) = (\mathbf{A}\boldsymbol{\mu}_0, \mathbf{A}\Sigma_0\mathbf{A}^T)$. Since $(\mathcal{P}, G)$ was induced by the method of Theorem 9.13, by Theorem 9.16 $\bar{G}$ is transitive.

We then define a pivot

$$\xi(\mathbf{Y}, \theta) = [\mathbf{Y} - \boldsymbol{\mu}]^T \Sigma^{-1} [\mathbf{Y} - \boldsymbol{\mu}]. \tag{9.20}$$

This pivot is invariant (Definition 9.3), since

$$\xi(g_{\mathbf{A}}\mathbf{Y}, \bar{g}_{\mathbf{A}}\theta) = [\mathbf{A}\mathbf{Y} - \mathbf{A}\boldsymbol{\mu}]^T [\mathbf{A}^T]^{-1}\Sigma^{-1}\mathbf{A}^{-1} [\mathbf{A}\mathbf{Y} - \mathbf{A}\boldsymbol{\mu}] = \xi(\mathbf{Y}, \theta).$$

Set parameter $\theta' = (\mathbf{X}\boldsymbol{\beta}', \Sigma_0)$. By Theorem 9.18 (i), we have

$$P_{\theta_0}(\xi(\mathbf{Y}, \theta') \in E) = P_{g_{\mathbf{A}}\theta_0}(\xi(\mathbf{Y}, \bar{g}_{\mathbf{A}}\theta') \in E). \tag{9.21}$$

We may then find a transformation $g_{\mathbf{A}} \in G$ for which the induced transformation $\bar{g}_{\mathbf{A}} \in \bar{G}$ satisfies $\bar{g}_{\mathbf{A}}(\boldsymbol{\mu}_0, \Sigma_0) = (\mathbf{A}\boldsymbol{\mu}_0, \mathbf{I})$. Then, by Equation (9.21) this means that the problem of minimizing

$$Q(\boldsymbol{\beta}') = [\mathbf{Y} - \mathbf{X}\boldsymbol{\beta}']^T \Sigma_0^{-1} [\mathbf{Y} - \mathbf{X}\boldsymbol{\beta}'] \tag{9.22}$$

with respect to $\boldsymbol{\beta}'$ is statistically identical to the problem of minimizing

$$Q'(\boldsymbol{\beta}') = [\mathbf{Y}' - \mathbf{A}\mathbf{X}\boldsymbol{\beta}']^T [\mathbf{Y}' - \mathbf{A}\mathbf{X}\boldsymbol{\beta}'] \tag{9.23}$$

with respect to $\boldsymbol{\beta}'$, where $\mathbf{Y}' \sim N(\mathbf{A}\boldsymbol{\mu}_0, \mathbf{I})$. Following Section 6.5.2, the minimizer of (9.23) is the ordinary least squares estimate for the transformed linear model $\mathbf{Y}' = \mathbf{A}\mathbf{X}\boldsymbol{\beta} + \mathbf{A}\boldsymbol{\epsilon}$, which is identical to the minimizer of (9.22), which in turn is the generalized least squares estimate for original model of Equation (9.18). Of course, this structure holds whether or not $\mathbf{Y}$ is a multivariate normal random vector.

9.5.2 Invariance *wrt* Reciprocal Transformations

A simple, but perhaps overlooked form of invariance is that with respect to reciprocal transformation. If we use an MRE estimator $\hat{\theta}$ for some scale parameter θ, it would seem natural that $1/\hat{\theta}$ should be the MRE estimator for $1/\theta$, but this requires that a loss function be invariant to reciprocal transformation, in addition to scale transformations.

Suppose we observe $X \sim gamma(\tau, \alpha)$, and wish to estimate τ, while α is known. We will need to assume that $\alpha \geq 3$. Since X is complete for τ as well as scale equivariant, by Theorem 9.11 an MRE estimator will be of the form $\hat{\tau} = X/w$, $w \in \mathbb{R}_{>0}$. The MRE estimator under relative MSE (and therefore under MSE by Equation (7.19) and Theorem 7.5) is then

$$\hat{\tau} = \frac{E_1[X]}{E_1[X^2]} X = \frac{X}{\alpha + 1}.$$

Suppose our intention, however, is to use $\hat{\tau}$ as a point estimate in a future application. If interest is also in the rate parameter $\lambda = 1/\tau$, then the MRE estimator under relative MSE from the class X^{-1}/w would be

$$\hat{\lambda} = \frac{E_1[X^{-1}]}{E_1[X^{-2}]} X^{-1} = \frac{\alpha - 2}{X},$$

where $X^{-1} \sim igamma(\lambda, \alpha)$ (recall that λ is a scale parameter for the inverted gamma distribution).

Note that $\hat{\tau} \neq 1/\hat{\lambda}$, which creates some ambiguity if point estimates of both τ and λ are required. We may therefore opt instead for a loss function which is invariant to reciprocal transformation. In this case, the loss of δ as an estimate of τ would equal the loss of δ^{-1} as an estimate of $\lambda = 1/\tau$. Table 7.1 lists two forms of scale invariant loss function which satisfy this condition, in particular, $L(X, c) = |\log(X/c)|$ and $L(X, c) = \log(X/c)^2$. If we take $L(X, c) = \log(X/c)^2$, then the UMR estimator of τ becomes

$$\hat{\tau}^* = X e^{-E_1[\log(X)]} = X e^{-\psi(\alpha)},$$

where $\psi(\alpha) = \Gamma'(\alpha)/\Gamma(\alpha)$ is the digamma function (see Problem 1.13). It may be shown that $\log(\alpha - 1/2) < \psi(\alpha) < \log(\alpha)$, so that $\hat{\tau} \neq \hat{\tau}^*$, but both estimators are, appropriately, constructed by dividing X by some quantity close to α. Similarly, the MRE estimator of λ becomes

$$\hat{\lambda}^* = X^{-1} e^{-E_1[\log(X^{-1})]} = X^{-1} e^{E_1[\log(X)]} = X^{-1} e^{\psi(\alpha)},$$

so that now $\hat{\tau}^* = 1/\hat{\lambda}^*$, that is, the transformation is reciprocal equivariant. Thus, if the estimation problem is expected to be invariant with respect to the reciprocal transformation, this can be achieved by using a reciprocal invariant loss function.

9.5.3 Invariance and the Species Sampling Problem

Here we revisit the species sampling problem of Example 4.4 (the reader should also review Problem 4.8). This problem involves the estimation of various characteristics of a probability distribution $\mathbf{P} = (p_1, \ldots, p_\nu)$ on the set of labels $\mathcal{S} = \{1, \ldots, \nu\}$ based on an *iid* sample $\mathbf{X} = (X_1, \ldots, X_n)$. We may (tentatively) let $(n_1, \ldots, n_\nu)$ be the corresponding multinomial frequencies. The motivating example, "species sampling", involves, for example, the discovery of new species of beetles within a novel ecosystem. In this sense, the number of distinct species ν clearly becomes an unknown parameter. In fact, it turns out that we should properly regard the entire support $\mathcal{S}$ itself as an unknown parameter, so that the mulitnomial model is not entirely appropriate for this problem. See Bunge and Fitzpatrick (1993) for a review of this problem.

The species sampling problem takes two forms. For the first, the problem is to predict the probability that the next observation X_{n+1} is a new species (Good, 1953; Robbins, 1968). This quantity may be expressed as

$$z_n = \sum_{j=1}^{\nu} p_j I\{n_j = 0\} \tag{9.24}$$

(we use the term "predict" because z is random). Then z_n is interpretable as the unseen support following observation of the nth outcome. Of course, we cannot simply substitute the conventional multinomial estimates $\hat{p}_j = n_j/n$ for p_i in Equation (9.24), since we would always obtain $z \equiv 0$ by its very definition.

In Example 4.4 the problem becomes an estimation problem by constructing a method of moments estimate of $E[z_{n-1}]$, in particular, a_1/n, where a_1 is the number of labels observed exactly once in the sample. Note that z_{n-1} is the unseen support *preceding* the nth outcome, a modifcation of the problem forced by the fact that $E[z_n]$ is not U-estimable. The prediction problem is recovered in Robbins (1968) by the inequality $E\left[(z_{n-1} - a_1/n)^2\right] < 1/n$.

Can a solution to the problem be obtained by attempting to estimate $\mathbf{P}$ directly? The answer given by the multinomial model seems unequivocal, and not helpful, that is, the estimate of the unseen support is always zero. We should note, however that this form of estimate not only predicts z_n (or estimates $E[z_n]$), but also identifies the unseen labels. However obvious this observation seems, it points to the fact that an estimate of $\mathbf{P}$ answers more than is being asked, that is, we seek an inference regarding the unseen support probability, but not the unseen support itself.

This suggests that identifying some form of invariance may allow us to formulate a well defined inference problem. Indeed, in Example 4.4 both the estimate and the estimand are invariant to a permuation of the labels (permutation invariant), yet the multinomial model is still interpretable.

In the second version of the species sampling problem the estimand becomes ν itself. This problem has received considerable attention since Goodman (1949), and Problem 4.8 introduces the well known estimate of ν proposed in Chao (1984). Now, the multinomial model is not relevant, since its definition includes a support set $\{1, \ldots, \nu\}$. While there must be a distribution $\mathbf{P} = (p_1, \ldots, p_\nu)$ for some ν on support $\mathcal{S} = \{1, \ldots, \nu\}$, as well as an *iid* sample $X_1, \ldots, X_n$ from $\mathbf{P}$, we cannot identify any outcome X_i as a particular element of $\mathcal{S}$, since $\mathcal{S}$ is unknown. Of course, for any pair X_i, X_j we can determnine whether or not $X_i = X_j$. The term "species sampling" implies precisely this, that is, that we are discovering $\mathcal{S}$ through sampling, as much as we are estimating frequencies.

We can unify both problems by applying to any estimation the permutation invariance constraint. The multinomial vector $\mathbf{N} = (n_1, \ldots, n_\nu)$ is reduced to a maximal invariant given by the order transformation $\mathbf{N}_{ord} = (n_{(1)}, \ldots, n_{(\nu)})$ of $\mathbf{N}$. However, in the context of this problem it will be natural to introduce the notation $\mathbf{N}_{ord} = \mathbf{S}^\nu = (0, \ldots, 0, s_1, \ldots, s_r)$, where r is the number of distinct labels observed in the sample, and $\mathbf{S} = (s_1, \ldots, s_r)$ is the ordered set of nonnegative frequencies in $\mathbf{N} = (n_1, \ldots, n_\nu)$. Then let $\mathbf{a} = (a_1, \ldots, a_r)$ be the "frequency of frequencies", so that a_j is the number of labels observed exactly j times (this approach is developed in some detail in Nayak (1992)). Then $\mathbf{S}$ and $\mathbf{a}$ are equivalent, and we note that $r = \sum_{j \geq 1} a_j$. A straightforward combinatoric argument gives

$$\begin{aligned} &P\{\mathbf{S}^\nu = (0, \ldots, 0, s_1, \ldots, s_r)\} \\ &\quad = \begin{cases} \binom{r}{a_1, \ldots, a_r} \sum_{k \in A_{\nu,r}} \prod_{i=1}^{r} p_{k_i}^{s_i}; & r \leq \nu \\ 0; & r > \nu \end{cases}, \end{aligned} \tag{9.25}$$

where $A_{\nu,r} = \{\text{all ordered subsets } k = (k_1, \ldots, k_r) \text{ of } \{1, \ldots, \nu\}\}$.

However, it must be pointed out that $\mathbf{S}^\nu$ is not observable if ν is unknown, since we would not observe the number of labels in $\mathcal{S}$ not seen in the sample. We do, of course, observe $\mathbf{S}$, and by applying the factorization theorem to the density of Equation (9.25) we can easily see that $\mathbf{S}$ is sufficient for $\mathbf{P}$. This is true even when ν is unknown, since r is a mapping of $\mathbf{S}$. Therefore the principle of sufficiency and the principle of permutaton invariance both lead to an inference procedure based on $\mathbf{S}$.

The reliance of any solution on the frequency of frequencies $\mathbf{a}$ seems reasonable. If observed labels have been sampled many times, we would expect that most of the support $\mathcal{S}$ has been observed, while the opposite conclusion follows when many species have only been observed once (equivalently, a_1 is large). Accordingly, the estimate of ν proposed in Chao (1984) is

$$\hat{\nu}_{CHAO} = r + \frac{a_1^2}{2a_2}$$

(Problem 4.8, see also Chiu *et al.* (2014)).

To continue with our model, let $\mathcal{P}_\nu$ be the set of distributions on $\{1, \ldots, \nu\}$, and let $\mathcal{P}_\nu^o \subset \mathcal{P}_\nu$ be the subset of distributions for which $p_i > 0$ for all $i \in \{1, \ldots, \nu\}$. Since $\mathbf{S} = (s_1, \ldots, s_r)$ is sufficient, we may evaluate the likelihood for $\mathbf{P} \in \mathcal{P}_\nu$ for any $\nu \in \mathbb{I}_{>0} \cup \{\infty\}$, which is given by

$$H(\mathbf{s}, \mathbf{P}) = \begin{cases} \sum_{k \in A_{\nu,r}} \prod_{i=1}^r p_{k_i}^{s_i}; & \nu \geq r \\ 0; & \nu < r \end{cases}. \tag{9.26}$$

One example of a maximum likelihood solution for Equation (9.26) can be given directly. As pointed out in Nayak (1992), the case $a_1 = r$ (all observed species observed exactly once) yields an MLE under which the next species sampled will be unseen $wp1$ (Problem 9.25).

As another example, consider frequency vector $\mathbf{S} = (1, 1, 2, 5)$. The MLE of $\mathbf{P}$ under constraint $\mathbf{P} \in \mathcal{P}_\nu$, that is, the numerical solutions to the optimization problem

$$\mathbf{P}_\nu^* = \text{argmax}_{\mathbf{P} \in \mathcal{P}_\nu} H(\mathbf{S}, \mathbf{P})$$

are given in Table 9.1 for $\nu = 4, \ldots, 10$. Note that $H(\mathbf{S}, \mathbf{P})$ is a permutation invariant function of $\mathbf{P}$, so an optimal solution $\mathbf{P}_\nu^*$ will not in general be unique, but should rather be interpreted as a class of all permutations of $\mathbf{P}_\nu^*$. Then suppose we have an unconstrained MLE

$$\mathbf{P}^* = \text{argmax}_{\mathbf{P}} H(\mathbf{S}, \mathbf{P}).$$

The MLE of ν associated with $\mathbf{P}^*$ will be ν^*, the number of nonzero probabilities in $\mathbf{P}^*$.

The constrained MLEs $\mathbf{P}_\nu^*$ of Table 9.1 suggest that $\nu^* = 6$. Clearly, $H(\mathbf{S}, \mathbf{P}_6^*) > H(\mathbf{S}, \mathbf{P}_5^*) > H(\mathbf{S}, \mathbf{P}_4^*)$, while the solutions $\mathbf{P}_\nu^*$, $\nu > 6$ evidently contain only 6 nonzero probabilities, and we obtain $\mathbf{P}_\nu^* = \mathbf{P}_6^*$, $\nu > 6$. This conforms to the estimate

$$N_{CHAO} = r + \frac{a_1^2}{2a_2} = 4 + \frac{2^2}{2 \times 1} = 6.$$

We now have a statistical method based on conventional principles of invariance and sufficiency which permits an estimate of ν. What must be emphasized is that Equation (9.26) is permutation invariant in $\mathbf{P}$. Each probability of $\mathbf{P}$ is assigned to each observed species in a symmetric fashion. This means that what is being estimated is not a single multinomial distribution, but merely the ordered probabilities of such a distribution. In other words, no frequency of any MLE $\mathbf{P}_\nu^*$ is assignable to any particular observed species. For the purpose of estimating ν, such an assignment is ancillary information.

TABLE 9.1
Approximate solutions $\mathbf{P}_\nu^*$ for the problem of optimizing $H(\mathbf{S}, \mathbf{P})$ over $\mathcal{P}_\nu$, for $\mathbf{S} = (1, 1, 2, 5)$.

ν	p_1	p_2	p_3	p_4	p_5	p_6	p_7	p_8	p_9	p_{10}
4	0.542	0.153	0.153	0.153						
5	0.550	0.112	0.112	0.112	0.112					
6	0.553	0.089	0.089	0.089	0.089	0.089				
7	0.553	0.089	0.089	0.089	0.089	0.089	0.000			
8	0.553	0.089	0.089	0.089	0.089	0.089	0.000	0.000		
9	0.553	0.089	0.089	0.089	0.089	0.089	0.000	0.000	0.000	
10	0.553	0.089	0.089	0.089	0.089	0.089	0.000	0.000	0.000	0.000

9.6 Invariant Hypothesis Tests

The notion of an invariant test was briefly alluded to in the context of nuisance parameters in Section 3.12. In Section 5.7, location and scale invariance was the basis for the pivot method of defining hypothesis tests. In this section, we introduce the more general invariant hypothesis test. In addition to the definition of an invariant density family (Definition 9.1) we require that the decision problem itself is invariant. This is made clear in the following definition.

Definition 9.4 (Invariant Hypothesis Test) Suppose the conditions of Theorem 9.15 hold for some invariant parametric family $(\mathcal{P}, G)$, given observation $\tilde{X} \in \mathcal{X}$. The problem of testing $H_o : \theta \in \Theta_o$ against alternative $H_a : \theta \in \Theta_a$ is invariant under G if $\bar{g}\Theta_o = \Theta_o$ and $\bar{g}\Theta_a = \Theta_a$ for all $\bar{g} \in \bar{G}$ (Theorem 9.15). If this is the case, then a hypothesis test is invariant if, in addition, the decision rule $\delta(\tilde{X})$ is invariant under G. ∎

Definition 9.4 suffices to guarantee some useful properties for an invariant hypothesis test. If δ is an invariant test then it must be a function of the maximal invariant $M(\tilde{X})$, in which case its distribution is invariant to any transformation $g \in G$. Thus, in contrast with the equivariant estimation problem, in which interest is in θ, for invariant hypothesis testing interest is in a nuisance parameter. Essentially, dependence of the inference on the nuisance parameter is eliminated simply by taking $M(\tilde{X})$ rather than $\tilde{X}$ as the observation.

The problem of determining optimal (uniformly most powerful) hypothesis tests is the subject of Chapter 10, and application specifically to invariant tests is discussed in Section 10.6. However, uniformly most powerful tests need not exist, and the subject of invariant hypothesis testing is usefully considered apart from the question of optimality.

Example 9.16 (Two-sample Test for Location Parameter) Suppose we are given two *iid* samples $X_1, \ldots, X_n$; $Y_1, \ldots, Y_m$, where X_1, Y_1 possess densities $f(x - \theta_0), f(x - \theta_1)$ for some density f. The parameter is $\boldsymbol{\theta} = (\theta_0, \theta_1) \in \Theta = \mathbb{R}^2$. We are interested in the estimand $\Delta(\boldsymbol{\theta}) = \theta_1 - \theta_0$, but not the actual values θ_0, θ_1. The problem should therefore be invariant to the location transformation $g_a(X_1, \ldots, X_n; Y_1, \ldots, Y_m) = (X_1, \ldots, X_n; Y_1, \ldots, Y_m) + a$. Let G be the class of all such transformations. It is easily verified that $(\mathcal{P}, G)$ is an invariant density family, which induces a transformation group $\bar{G}$ of all bijections $\bar{g}_a\boldsymbol{\theta} = (\theta_0 + a, \theta_1 + a)$. The parameter Δ is invariant to the transformation in the sense that

$$\Delta(\bar{g}_a\boldsymbol{\theta}) = \theta_1 + a - \theta_0 - a = \theta_1 - \theta_0 = \Delta(\boldsymbol{\theta}).$$

Therefore, a one-sided test $H_o : \Delta \leq \Delta_0$ against $H_a : \Delta > \Delta_0$; or a two-sided test $H_o : \Delta = \Delta_0$ against $H_a : \Delta \neq \Delta_0$; is an invariant problem. The conventional test statistic for this problem $T = \bar{y} - \bar{x}$, is invariant, since

$$T(g_a(x_1, \ldots, x_n; y_1, \ldots, y_m)) = \bar{y} + a - \bar{x} - a = \bar{y} - \bar{x} = T(x_1, \ldots, x_n; y_1, \ldots, y_m).$$

Thus, any hypothesis test concerning Δ, based on test statistic T is invariant *wrt* location transformations. ■

9.6.1 Rank Statistics and the Monotone Transformation Group

We have considered location transformations $g_a(x_1, \ldots, x_n) = (x_1 + a, \ldots, x_n + a)$ and scale transformations $g_a(x_1, \ldots, x_n) = (ax_1, \ldots, ax_n)$, $a > 0$. In each case, the transformation acts by applying a single monotone mapping $\theta(x)$ to each element of a vector $(x_1, \ldots, x_n) \in \mathbb{R}^n$, which we may write $g_\theta(x_1, \ldots, x_n) = (\theta(x_1), \ldots, \theta(x_n))$. In fact, monotonicity itself defines a transformation group, which we consider in this section. The following theorem is the first step.

Theorem 9.19 (Group Properties of Monotone Transformations) Let $\mathcal{G}$ be the set of all continuous strictly increasing bijective functions $g : \mathcal{I} \to \mathcal{I}$, for some interval $\mathcal{I} \subset \mathbb{R}$. Then $(\mathcal{G}, \circ)$ is a group under composition. ■

Proof. We consider each group axiom (Definition D.1). *Closure.* Suppose $g, h \in \mathcal{G}$. If $x < x'$ then $g(x) < g(x')$, therefore $h(g(x)) < h(g(x'))$. *Associativity.* Function composition is in general associative. *Existence of Identity.* Suppose $g(x) = x$. Then $g \in \mathcal{G}$. If $h \in \mathcal{G}$ then $g \circ h = h \circ g = h$. *Existence of Inverse.* Since $h : \mathcal{I} \to \mathcal{I}$ is assumed to be bijective, it possesses a bijective inverse mapping $h^{-1} : \mathcal{I} \to \mathcal{I}$. If h is strictly increasing and continuous then so is h^{-1}. Then $h \circ h^{-1} = h^{-1} \circ h = x$, which defines the group inverse. □

The monotone function group of Theorem 9.19 allows us to define a transformatiom group similar to the location and scale transformation groups.

Example 9.17 (Monotone Transformation Group) Suppose we have sample space $\mathbf{x} \in \mathcal{X} = \mathcal{I}^n$. Let $(\mathcal{G}, \circ)$ be the monotone function group of Theorem 9.19 on interval $\mathcal{I}$. For any $\theta \in \mathcal{G}$ define transformation $g_\theta : \mathcal{X} \to \mathcal{X}$ by the evaluation method

$$g_\theta(\mathbf{x}) = (\theta(x_1), \ldots, \theta(x_n)).$$

The composition operation $\circ$ is clearly equivalent to $g_{\theta'} \circ g_\theta = g_{\theta' \circ \theta}$. We also have identity $e = g_{\theta_e}$, where $\theta_e(\mathbf{x}) = x$, and inverse $g_\theta^{-1} = g_{\theta^{-1}}$. Then, the class of transformations $G = \{g_\theta : \theta \in \mathcal{G}\}$ is isomorphic to $\mathcal{G}$ (Theorem D.3). ■

We now have another form of invariance, which characterizes inference problems which remain unchanged under monotone transformations. For example, the two scatterplots of Figure 8.1 are equivalent under this form of invariance. The next step is to identify the maximal invariant, which is this case will be the rank statistics.

Theorem 9.20 (Maximal Invariant of the Monotone Transformation Group) Consider the monotone transformation group G of Example 9.17 on the sample space $\mathbf{x} \in \mathcal{X} = \mathcal{I}^n$. Let $\mathcal{N} \subset \mathcal{X}$ be the points $\mathbf{x}$ with at least two identical coordinates. Then the ranks $\mathbf{r} = (r_1, \ldots, r_n)$ of $\mathbf{x} = (x_1, \ldots, x_n)$ is a maximal invariant of G on the sample space $\mathcal{X} - \mathcal{N}$. ■

Proof. The ranks of $\mathbf{x}$ are clearly preserved under any strictly increasing transformation $g\mathbf{x}$ so that $\mathbf{r}$ is invariant *wrt* G.

Next, suppose $\mathbf{x} \in \mathcal{X} - \mathcal{N}$. Let $h(x)$ be a continuous function in $\mathcal{I}$ constructed by setting $h(x_i) = r_i$, $i = 1, \ldots, n$, then applying linear interpolation between these points, and allowing $h(x)$ to have unit slope beyond the range of $x_1, \ldots, x_n$. By construction, $g_h \in G$. Suppose $T(\mathbf{x})$ is invariant under G. Then $T(\mathbf{x}) = T(g_h\mathbf{x}) = T(\mathbf{r})$, that is, $T(\mathbf{x})$ is a function of the ranks. □

Note that the set $\mathcal{N}$ of Theorem 9.20 includes all vectors $\mathbf{x} \in \mathcal{X}$ with ties in rank. For any continuous distribution on $\mathcal{X}$ we would have $P(\mathbf{X} \in \mathcal{N}) = 0$. On the other hand, rank based procedures applied to discrete data are usually modified to accomodate ties.

At this point, we have a transformation group G, and we have identified the rank statistics as the maximal invariant of G. We can then use the method of Theorem 9.13 to construct an invariant density family $(\mathcal{P}, G)$. We will assume that $\mathcal{X}$ is a measurable space paired with Lebesgue measure μ on $\mathbb{R}^n$.

We then identify a single distribution in the form of CDF $F_e(x_1, \ldots, x_n)$, assumed to possess support $\mathcal{X}$. Then, if $\mathbf{X} = (X_1, \ldots, X_n)$ is an observation from F_e, the transformed observation $\mathbf{Y} = g_\theta \mathbf{X} = (\theta(X_1), \ldots, \theta(X_n))$, possesses CDF

$$\begin{aligned} F_\theta(x_1, \ldots, x_n) &= P\left(\theta(X_1) \leq x_1, \ldots, \theta(X_n) \leq x_n\right) \\ &= P\left(X_1 \leq \theta^{-1}(x_1), \ldots, X_n \leq \theta^{-1}(x_n)\right) \\ &= F_e(\theta^{-1}(x_1), \ldots, \theta^{-1}(x_n)). \end{aligned} \tag{9.27}$$

If we let $\mathcal{P} = \{F_\theta : \theta \in \mathcal{G}\}$, then by Theorem 9.13 $(\mathcal{P}, G)$ is an invariant density family. In the next example, we apply Equation (9.27) to the *iid* sample.

Example 9.18 (Monotone Transformations of *iid* Samples) If the observation is an *iid* sample, then the family of densities includes all continuous densities with support on $\mathcal{I}$. Suppose $\mathbf{X} = (X_1, \ldots, X_n)$ is an *iid* sample, $X_1 \sim F_e$, where F_e is a continuous distribution with support on an open inteval $\mathcal{I}$. Suppose $\mathcal{X} = \mathcal{I}^n$ for some interval $\mathcal{I} \subset \mathbb{R}$, and let G be the transformation group of Example 9.17 based on monotone group $\mathcal{G}$ on interval $\mathcal{I}$ (Theorem 9.19). Then, following Equation (9.27), for any $\theta \in \mathcal{G}$ the CDF of $\theta(X_1)$ is $F_\theta(x) = F_e(\theta^{-1}(x))$. Suppose H is any other continuous CDF with support $\mathcal{I}$. Then the inverse mapping restricted to $\mathcal{I}$, $F_e^{-1} : (0,1) \to \mathcal{I}$, is continuous, increasing and bijective. The composition $\theta = F_e^{-1} \circ H$ restricted to $\mathcal{I}$ is well defined, and $\theta \in \mathcal{G}$. Then $F_e(F_e^{-1}(H(x))) = H(x)$. In other words, there exists $\theta \in \mathcal{G}$ such that $H = F_\theta$ for any continuous H with support $\mathcal{I}$. Thus, extending the argument to $\mathcal{I}^n$, $\mathcal{P} = \{F_\theta : \theta \in \mathcal{G}\}$ consists exactly of all continuous CDFs with support $\mathcal{I}$. ∎

9.6.2 Two-Sample Rank Tests

Suppose we are given two *iid* samples $X_1, \ldots, X_n$, $Y_1, \ldots, Y_m$ on support $\mathcal{I}$. We will first assume the samples are mutually independent, with $X_1 \sim F$, $Y_1 \sim H$. We have one-sided test

$$H_o : F = H, \;\; H_a : H(x) \leq F(x), \;\; x \in \mathbb{R}, \;\; F \neq H, \tag{9.28}$$

or two-sided test

$$H_o : F = H, \;\; H_a : H(x) \leq F(x) \text{ or } H(x) \geq F(x), \;\; x \in \mathbb{R}, \;\; F \neq H. \tag{9.29}$$

For the one-sided test, under H_a H is stochastically larger than F (Section 1.5). We can take the parameter space Θ to be all pairs of CDFs (F, H) such that F and H are continuous on support $\mathcal{I}$, and $H \geq_{st} F$ or $F \geq_{st} H$ (which includes the case $F = H$). This defines a parametric family $\mathcal{P}$. We then have the monotone transformation group G of Example 9.17 applied to both samples, $g_\theta(x_1, \ldots, x_n; y_1, \ldots, y_m) = (\theta(x_1), \ldots, \theta(x_n); \theta(y_1), \ldots, \theta(y_m))$. Clearly, g_θ preserves stochastic order, since, for example, $H(x) \leq F(x)$ *iff* $H(\theta(x)) \leq F(\theta(x))$ for any $\theta \in \mathcal{G}$. Then using an argument similar to that of Example 9.18 we can conclude that any g_θ leaves $\mathcal{P}$ invariant, so that $(\mathcal{P}, G)$ is an invariant density family. By essentialy the same argument, each hypothesis in Equation (9.28) or (9.29) is invariant under G.

The maximal invariant under G was identified in Theorem 9.20 which, for the two-sample model, is equivalent to the vector of ranks $(r_1, \ldots, r_n; s_1, \ldots, s_m)$ of the pooled data. From the ranks we may identify a simple sufficient statistic, as shown in the following theorem.

Theorem 9.21 (Sufficient Statistic for Two-Sample Rank Test) The unordered set of ranks $\{r_1, \ldots, r_n\}$ (or the unordered set of ranks $\{s_1, \ldots, s_m\}$) is sufficient for the model defining the hypothesis test of Equation (9.28). ∎

Proof. The distribution of the ranks $(r_1, \ldots, r_n; s_1, \ldots, s_m)$ is symmetric in the first n and last m coordinates. Thus the pair of sets $\{r_1, \ldots, r_n\}$, $\{s_1, \ldots, s_m\}$ is sufficient. However, each of the sets can be determined from the other. □

An immediate consequence of Theorem 9.21 is that any statistic that is a symmetric function of the observed ranks $(R_1, \ldots, R_n)$ (or $(S_1, \ldots, S_m)$) is a function of the sufficient statistic.

Example 9.19 (Sufficient Statistic for Two-Sample Rank Test) To clarify the two-sample rank statistic of Theorem 9.21, suppose $n = 2$, $m = 3$, with $(X_1, X_2) = (9.3, 5.7)$, $(Y_1, Y_2, Y_3) = (10.2, 13.9, 8.4)$. Then $(R_1, R_2; S_1, S_2, S_3) = (3, 1; 4, 5, 2)$. Note that the ranks are calculated without regard to the sample of origin. Then the sufficient statistic identified in Theorem 9.21 is the set $\{1, 3\}$ (or equivalently the set $\{2, 4, 5\}$). ∎

Then let $\bar{G}$ be the transformation group on Θ induced by G (Theorem 9.15). For the invariant hypothesis test we do not expect $\bar{G}$ to be transitive on Θ. Recall that by Definition 9.4 a decision rule $\delta(\tilde{X})$ for an invariant test must be an invariant statistic. If $\bar{G}$ were transitive over the entire parameter space, then by Theorem 9.17 the distribution of $\delta(\tilde{X})$ would be the same in both the null and alternative hypotheses. However, if $\bar{G}$ is transitive over null hypothesis Θ_o, then the invariant test will be similar (Section 5.9), that is, the distribution of $\delta(\tilde{X})$ will be the same for all $\theta \in \Theta_o$. For the two-sample rank test, that this will be the case follows essentially the same argument given in Example 9.18, since $F = H$ under H_o, and therefore $X_1, \ldots, X_n$, $Y_1, \ldots, Y_m$ can be pooled to form a single *iid* sample.

9.6.3 The Wilcoxon Rank-Sum Test

The one- and two-sample tests are ubiquitous in statistical analysis, but are sensitive to model assumptions, and are not robust to outliers. Rank methods are commonly used as alternative procedures which are far less problematic from this point of view. The Wilcoxon rank-sum test (Wilcoxon, 1945) uses the result of Theorem 9.21 to test the hypotheses of Equation (9.28) or (9.29) based on two independent *iid* samples. It is based on the rank-sum statistic $W = \sum_{i=1}^n R_i$, with rejection region $R = \{W > w\}$ for the one-sided test and $R = \{W < w_L\} \cup \{W > w_U\}$ for the two-sided test. Note that W is a mapping of the sufficient statistic identified in Theorem 9.21.

That the rank sum W is equivalent to a multiple sample U-statistic was shown in Mann and Whitney (1947), in which the Mann-Whitney test statistic

$$U = \frac{1}{nm} \sum_{i=1}^{n} \sum_{j=1}^{m} I\{Y_j \leq X_i\}, \tag{9.30}$$

was introduced. We then have estimand $\eta = E[U] = P(Y_1 \leq X_1)$, with $\eta = 1/2$ corresponding to the null hypothesis. Note, however, that the theory for single-sample U-statistics introduced in Section 8.5 would have to be modified for this case (Kowalski and Tu, 2008).

The relationship between W and U emerges as follows. We may express a rank using the identity

$$R_j = \sum_{i=1}^{n} I\{X_i \leq X_j\} + \sum_{i=1}^{m} I\{Y_i \leq X_j\}.$$

Then $W = \sum_{j=1}^{n} R_j = \frac{n(n+1)}{2} + nmU$.

The next example is a typical application of the Wilcoxon rank-sum test.

Example 9.20 (Wilcoxon Rank-Sum Test) Suppose we are given two independent samples of sample sizes $n_1 = 4$, $n_2 = 12$. Let the respective distributions be F_1, F_2. The data is summarized in the following table:

	1	2	3	4	5	6	7	8	9	10	11	12
Sample 1	97.6	99.8	87.4	95.4								
Sample 2	86.5	80.0	70.3	79.5	84.7	84.9	75.7	80.1	69.2	79.3	79.1	70.2
Ranks 1	15	16	13	14								
Ranks 2	12	8	3	7	10	11	4	9	1	6	5	2

We wish to test $H_o : F_1 = F_2$ against $H_a : F_1 \geq_{st} F_2$. By Theorem 9.21 the test may be based on the sum of the ranks of either sample. Since $n_1 < n_2$ it will be more convenient to use the rank sum for the first sample:

$$W = 13 + 14 + 15 + 16 = 58,$$

so that we reject H_o for large values of W. Under the null hypothesis W is equivalent to the sum of four ranks selected randomly from 16 without replacement. In this case, $E_o[W] = n_1(n_1 + n_2 + 1)/2 = 34$, since a randomly chosen rank has mean $(n_1 + n_2 + 1)/2$. It may also be shown that

$$\text{var}_o[W] = \frac{n_1 n_2 (n_1 + n_2 + 1)}{12} = 68. \tag{9.31}$$

See Problem 9.28. If we accept a normal approximation for W, then a level $\alpha = 0.05$ critical value would be

$$w = E_o[W] + z_{0.05}\sqrt{\text{var}_o[W]} = 34 + 1.645\sqrt{68} = 47.57.$$

Since $W = 58 > w$, we reject H_o. It is possible also to determine the exact critical value using a combinatorical argument, which are frequently summarized in tables, or available in a statistical computing platform. It is, of course, not difficult to simulate the rank selection process defining W, and w may be estimated by a simulation program, such as the following `R` code,

```
> x = replicate(10000,sum(sample(16,4)))
> quantile(x,0.95)
95%
 47
```

which quickly offers a reasonable approximation. ■

9.7 Problems

Problem 9.1 Consider a single random variable X with density $f_\theta(x) = I\{x - \theta \in (0, 1/2)\} + I\{x - \theta \in (1, 3/2)\}$ where $\theta \in \mathbb{R}$, $x \in \mathbb{R}$.

(a) Sketch this density for arbitrary θ.
(b) What would be a method of moments estimator of θ?
(c) Derive the equivariant estimator of θ with the uniformly smallest mean squared error.

Problem 9.2 Complete the proof of Theorem 9.4 by proving that if θ is a location parameter; δ is any location equivariant estimator; and $L_\theta(d)$ is location invariant loss function, then $\text{bias}_\theta[\delta]$ and $\text{var}_\theta[\delta]$ are constant over $\theta \in \Theta$.

Problem 9.3 Suppose $X_1, \ldots, X_n$ is an *iid* sample with $X_1 \sim exp(\theta, \tau)$. Assume $\tau > 0$ is known.

(a) Verify that $X_{(1)} = \min_i X_i$ is a complete sufficient statistic for θ.
(b) Suppose we are given a loss function $L_\theta(d) = \rho(\theta - d)$. In addition, assume that for some positive function g, ρ satisfies the identity $\rho(ax) = g(a)\rho(x)$ for any $a > 0$. Prove that an MRE estimator for location parameter θ under this loss function is given by

$$\delta = X_{(1)} - \tau a^*/n,$$

where a^* is any value minimizing $E[\rho(U - a))]$ *wrt* a, where $U \sim exp(1)$.

Problem 9.4 Consider the function

$$h(x; a, b) = \begin{cases} ax; & x \geq 0 \\ \text{-}bx; & x < 0 \end{cases}$$

for any two positive constants $a, b > 0$. Then suppose $\theta \in \Theta = \mathbb{R}$ is a location parameter for model $\mathcal{P}$ with observation $(X_1, \ldots, X_n) \in \mathcal{X}$. Modify Example 9.2 for loss functions $L_\theta(d) = h(d - \theta; a, b)^2$ and $L_\theta(d) = h(d - \theta; a, b)$.

Problem 9.5 Let $\mathbf{X} = (X_1, \ldots, X_n)$ be an *iid* sample from density $f(x)$. Since $X_{(1)}$ is location equivariant we may construct a maximal location invariant by setting $M(\mathbf{X}) = (X_1 - X_{(1)}, \ldots, X_n - X_{(1)})$. The practical problem here is that one of the elements of $M(\mathbf{X})$ will be zero, but the index of that element is random. We can approach the problem by determining the density f_{M^*} of the order statistics $M^*(\mathbf{X}) = (X_{(2)} - X_{(1)}, \ldots, X_{(n)} - X_{(1)})$. The density of $M(\mathbf{X})$ is then equivalent to a symmetric density on $\mathbb{R}^{n-1}$ for which the density of the order statistics is f_{M^*} (Section 2.4). Using this approach show that this density takes the form

$$f_M(w_1, \ldots, w_{n-1}) = n \int_{u \in \mathbb{R}} f(u) \prod_{i=2}^{n} f(w_i + u) d\nu_u, \quad \min w_j \geq 0.$$

Problem 9.6 Suppose we are given the "triangle" density on $\mathbb{R}$, $f(x) = 2xI\{0 < x < 1\}$. We observe an *iid* sample $\tilde{X} = (X_1, \ldots, X_n) \sim p_\theta$ from the location invariant model

$$p_\theta(\tilde{x}) = \prod_{i=1}^{n} f(x_i - \theta), \quad \theta \in \Theta = \mathbb{R}.$$

(a) Show that the order statistics are minimal sufficient for θ.

(b) Derive the MRE estimator for location parameter θ with respect to squared error loss. This may be expressed in terms of integrals of a single variable.
(c) Suppose we are given the density of $X_{(1)}$ conditional on the statistic $Y = (X_1 - X_{(1)}, X_2 - X_{(1)}, \ldots, X_n - X_{(1)})$ when $\theta = 0$, say $f(u \mid y)$. Explain how you would use this density to evaluate the MRE estimator for location parameter θ when the loss function is $L_\theta(d) = |\theta - d|$. Justify your answer. Note that $f(u \mid y)$ itself need not be derived. All that is needed in your answer is an explanation of how it would be used.

Problem 9.7 Suppose $\mathbf{X} = (X_1, \ldots, X_n)$ is an *iid* sample with $X_1 \sim cauchy(\theta, 1)$. Construct a Pitman location estimator as a ratio of two integrals in one variable. Does the fact that the $E_\theta[|X_n|] = \infty$ mean that the argument of Theorem 9.6 cannot be applied to this case? **HINT:** See Example 9.4. For more on estimation problems for Cauchy density samples see Cohen Freue (2007).

Problem 9.8 Suppose $(X_1, \ldots, X_n)$ is an observation from a scale invariant model $p_\theta(\mathbf{x}) = \theta^{-n} p_1(x/\theta)$. Let $(Y_1, \ldots, Y_n)$ be the transformed observation $(\log X_1, \ldots, \log X_n)$. Show that $(Y_1, \ldots, Y_n)$ is an observation from a location invariant model with location parameter $\log \theta$. **HINT:** Use transformation $U_i = \log(X_i/\theta)$, then transformation $Y_i = U_i + \log \theta$.

Problem 9.9 Consider the scale invariant model of Section 9.3. Let $S(\mathbf{X})$ be any scale equivariant estimator of scale parameter θ. Assume $S(\mathbf{X}) \geq 0$. Prove that there exists a maximal scale invariant which equals $M(\mathbf{X}) = (X_1/S(\mathbf{X}), \ldots, X_n/S(\mathbf{X}))$ on the subset $\{S(\mathbf{X}) > 0\}$. **HINT:** Make use of Theorems 3.6, D.7 and D.8.

Problem 9.10 We are given an *iid* sample $\mathbf{X} = (X_1, \ldots, X_n)$ with $X_1 \sim exp(\theta)$. Define statistic $S(\mathbf{X}) = \sum_{i=1}^n X_i / X_{(1)}$, where $X_{(1)} = \min_i X_i$. Prove that $S(\mathbf{X})$ is independent of sample mean $\bar{X} = n^{-1} \sum_1^n X_i$.

Problem 9.11 Suppose $\mathbf{X} = (X_1, \ldots, X_n)$ is a sample from density $f_{\mathbf{X}}(\mathbf{x})$ for which $P(X_i > 0) = 1$ for all i. Define transformation

$$\mathbf{Y} = (Y_1, \ldots, Y_{n-1}) = (X_1/S, \ldots, X_{n-1}/S).$$

where $S = \sum_i X_i$. Show that $\mathbf{Y}$ is a maximal invariant with respect to scale transformations, and that its density has form

$$f_{\mathbf{Y}}(y_1, \ldots, y_{n-1}) = \int_{u>0} u^{n-1} f_{\mathbf{X}}(uy_1, \ldots, uy_{n-1}, u(1 - \sum_{i=1}^{n-1} y_i)) du.$$

Problem 9.12 Suppose $\mathbf{X} = (X_1, \ldots, X_n) \in \mathbb{R}^n$. Consider the mapping $M'(\mathbf{X}) = (Y_1, \ldots, Y_n) = (X_1/X_n, \ldots, X_{n-1}/X_n, X_n/X_n)$.

(a) Is the mapping $M'(\mathbf{X})$ scale invariant on $\{X_n \neq 0\}$?
(b) Does the mapping $M'(\mathbf{X})$ satisfy Definition 3.4 of a maximal scale invariant on $\{X_n \neq 0\}$?
(c) A maximal scale invariant maps to any other scale invariant. Does $M'(\mathbf{X})$ map to the maximal scale invariant $M(\mathbf{X})$ of Equation (9.12)?

Problem 9.13 Define the density function: $f_1(\mathbf{x}) = I\{1 < x < 2\}$, $x \in \mathbb{R}$. Define scale parameter $\theta \in \Theta = \mathbb{R}_{>0}$, and suppose $X_1, \ldots, X_n$ is an *iid* sample from density $f_\theta(x) = \theta^{-1} f_1(x/\theta)$.

(a) Verify that no complete sufficient statistic for θ exists.
(b) Derive the MRE estimator for scale parameter θ with loss function $L_\theta(d) = (\theta - d)^2/\theta^2$.

Problem 9.14 We observe an *iid* sample $\mathbf{X} = (X_1, \ldots, X_n)$ with $X_1 \sim rayleigh(\theta)$. Derive the MRE estimator for scale parameter θ with respect to each of the loss functions given in Equations (9.8) and (9.10).

Problem 9.15 We have an *iid* sample $\mathbf{X} = (X_1, \ldots, X_n)$. Derive the Pitman estimator for estimand θ^r, where θ is a scale parameter, for each of the following models: (a) $X_1 \sim logistic(0, \theta)$; (b) $X_1 \sim cauchy(0, \theta)$. If necessary, the estimator may be given as a ratio of two integrals in one variable. In each case verify that the conditions of Theorem 9.12 hold.

Problem 9.16 Verify that the Pitman estimators of Example 9.11 can also be obtained directly by Theorem 9.11, making use of any available complete sufficient statistic.

Problem 9.17 We are given n *iid* random variables $X_1, \ldots, X_n$. Assume X_1 is distributed as an exponential random variable with mean θ truncated above at $T\theta$. The density of X_1 is then $f(x) = \theta^{-1}e^{-x/\theta}/(1 - e^{-T})$, $x \in (0, T\theta)$ (see also Problem 12.9).

(a) Verify that θ is a scale parameter.
(b) Give the minimal sufficient statistic for θ. Is this statistic also complete?
(c) Derive the Pitman scale estimator of θ. You can express this estimator using F_ν, the CDF of the $gamma(1, \nu)$ distribution.

Problem 9.18 Suppose $X_1, \ldots, X_n$ is an *iid* sample with $X_1 \sim pareto(\theta, \alpha)$. Derive the MRE estimator for scale parameter θ for loss function $L_\theta(\delta) = (\theta - \delta)^2/\theta^2$. Assume α is known.

Problem 9.19 Suppose $X_1, \ldots, X_n$ is an *iid* sample from $exp(\tau)$. Then $S = \sum_i X_i$ is complete sufficient for τ (i.e. due to the exponential family property). The UMVUE for τ is therefore the sample mean $\delta(X) = \bar{X}$. Prove that $\bar{X}$ is an inadmissible estimator of τ with respect to squared error loss.

Problem 9.20 Consider the function $\rho(y) = y - \log y - 1$, $y > 0$.

(a) Verify that $\rho(y)$ is strictly convex on its domain.
(b) Suppose Y is a positive random variable. Determine

$$c^* = \text{argmin}_{c>0} E[\rho(Y/c)]$$

(compare to Theorem 7.8).
(c) Suppose $X_1, \ldots, X_n$ is an *iid* sample from $f_\theta(\mathbf{x})$, where θ is a scale parameter, and $X_1 > 0$, *wp*1. Give the form of the MRE estimator of θ under loss function $L_\theta(d) = \rho(d/\theta)$. Assume δ_0 is any MRE estimator.
(d) Extend Example 9.9 to this loss function, and compare the resulting MRE estimator to those derived using the other two loss functions. See Stein loss (Equation (7.21)).

Problem 9.21 Consider an *iid* sample $(X_1, \ldots, X_n)$ from density $X_1 \sim N(\mu, \sigma^2)$. Assume μ is known. Using the transformed data $(Y_1, \ldots, Y_n) = (X_1 - \mu, \ldots, X_n - \mu)$ derive the MRE estimate for both estimands σ and σ^2 using loss functions (9.8) and (9.10).

Problem 9.22 Let $p_{(0,1)}$ be density on $\mathcal{X} = \mathbb{R}^n$. A parametric family $\mathcal{P}$ contains all densities of the form $p_{(\mu,\tau)}(\mathbf{x}) = \tau^{-n}p_{(0,1)}((\mathbf{x} - \mu)/\tau)$, $\mu \in \mathbb{R}$, $\tau \in \mathbb{R}_{>0}$. Let G_{loc}, G_{sc}, G_{ls} be the transformation groups defined by location, scale, and location-scale transformations on $\mathcal{X}$.

(a) Which of the pairs $(\mathcal{P}, G_{loc})$, $(\mathcal{P}, G_{sc})$, $(\mathcal{P}, G_{ls})$ satisfy the definition of an invariant density family (Definition 9.1)?
(b) Which of the transformation groups G_{loc}, G_{sc}, G_{ls} can be used to generate $\mathcal{P}$ using the method of Theorem 9.13?

Problem 9.23 Suppose $X \in bin(n, \theta)$, for fixed n and unknown θ. The parameter space is restricted to $\theta \in \Theta = \{a, 1-a\}$, where $a \in (0, 1/2)$ is a known constant.

(a) Show that this defines an invariant density family under the transformation class $G = \{X, n-X\}$ acting on the binomial observation X. State clearly the resulting family of densities $\mathcal{P} = \{p_\theta : \theta \in \Theta\}$. Then characterize the equivariant estimator, invariant loss function, and maximal invariant for this family.
(b) Derive the MRE estimator of θ under loss function $L_\theta(d) = (\theta - d)^2$. Verify that the loss function is invariant.
(c) Derive the MRE estimator of θ under loss function $L_\theta(d) = I\{\theta \neq d\}$. Verify that the loss function is invariant.

Problem 9.24 Verify the following claim made in Section 9.5.1. The set Γ of all $n \times n$ non-singular matrices is a group under matrix multiplication. Then Γ indexes a transformation group G acting on $\mathbb{R}^n$ by the evaluation method $g_\mathbf{A}\mathbf{Y} = \mathbf{A}\mathbf{Y}$ for any $g_\mathbf{A} \in G$, $\mathbf{A} \in \Gamma$.

Problem 9.25 Consider the species problem of Section 9.5.3. Justify the claim made in Nayak (1992), that for the case $a_1 = r$ (all observed species observed exactly once), the MLE yields the inference that the next species sampled will be unseen *wp*1.

Problem 9.26 Suppose $\theta \in \Theta = \mathbb{R}$ is a location parameter for model $\mathcal{P}$ on observation $(x_1, \ldots, x_n) \in \mathcal{X}$. The model is then modified so that $\theta \in \Theta_I$ is restricted to be an integer. The model is then invariant to transformations on $\mathcal{X}$ of the form $g_i(x_1, \ldots, x_n) = (x_1 + i, \ldots, x_n + i)$, $i \in \Theta_I$.

(a) Prove that
$$M(\mathbf{x}) = (x_1 - \lfloor x_1 \rfloor, \ldots, x_n - \lfloor x_1 \rfloor)$$
is a maximal invariant.
(b) Prove that $\lfloor \bar{x} \rfloor$ is an equivariant estimator. Then give a general form of the MRE estimator of $\theta \in \Theta_I$ with respect to squared error loss.

Problem 9.27 We are given two independent samples of sample sizes $n_1 = 7$, $n_2 = 10$. The data is summarized in the table below. Suppose $\tilde{\mu}_i$ is the population median of sample i. Perform a two-sided rank sum test for null hypothesis $H_o : F_1 = F_2$. Using both methods of Example 9.20 estimate a P-value.

	1	2	3	4	5	6	7	8	9	10
Sample 1	208.7	211.6	197.6	209.1	198.1	204.1	192.8	-	-	
Sample 2	167.6	162.7	164.4	164.3	172.2	165.9	169.8	157.6	162.9	164.7

Problem 9.28 Prove Equation (9.31). **HINT:** Let R_1, R_2 be two ranks of selected from n without replacement. Evaluate $E[R_1 R_2 \mid R_2 = k]$.

10

The Neyman-Pearson Lemma

10.1 Introduction

The hypothesis test was introduced in Chapter 5, assuming a parametric family $\mathcal{P}$ with parameter space Θ, based on observation $\tilde{X} \in \mathcal{X}$. As in estimation, the principles of sufficiency and of invariance can play an important role (Section 9.6). What is new to this chapter is the central role played by a single theorem, the Neyman-Pearson lemma. We first formulate a hypothesis test as a randomized decision rule $\delta(\tilde{X})$ (Section 10.2). The objective is then to maximize the power among tests of fixed size α. Recall from Chapter 7 that the problem posed by any attempt to rank estimators is that, since risk depends on an unknown parameter, a uniformly best estimator will not exist, unless some additional constraint is imposed. A similar issue arises in testing when the alternative hypothesis Θ_a is composite, in particular, the power of the test will vary over $\theta \in \Theta_a$. However, the resolution of this problem is quite different for hypothesis tests, in comparison to Chapter 7. The Neyman-Pearson lemma (Theorem 10.2) is introduced in Section 10.3, which gives the most powerful test when the null and alternative hypotheses are both simple. On the surface, this result seems to be of limited value, since in practice alternative hypotheses are usually composite. However, it will often be the case that the same test will be most powerful for *all* simple hypotheses *within a composite alternative* Θ_a. In this case we have identified a uniformly most powerful (UMP) test for Θ_a. Section 10.4 introduces the monotone likelihood ratio (MLR) property for parametric families, given which a UMP test will exist (this is derived from likelihood ratio ordering introduced in Section 1.5).

A generalized version of the Neyman-Pearson lemma (Theorem 10.5) is introduced in Section 10.5. Like Theorem 10.2 it yields a most powerful test for simple null and alternative hypothesis, but allows constraints to be introduced. We will show that this expands the utility of the Neyman-Pearson lemma considerably.

The rest of the chapter is devoted to applications of the Neyman-Pearson lemma to the problem of invariant hypothesis testing, which we defined in Section 9.6. In Section 10.6 it is shown how Neyman-Pearson theory can be applied directly to the maximal invariant of a invariant density family. In Section 10.7 we show that the constraints imposed by permutation invariance can be incorporated directly into the generalized version of the Neyman-Pearson lemma, yielding a general method for constructing most powerful permutation invariant tests.

10.2 Hypothesis Tests as Decision Rules

Following Chapter 5, a parameter space Θ is partioned into hypotheses

$$H_o : \theta \in \Theta_o \text{ and } H_a : \theta \in \Theta_a.$$

DOI: 10.1201/9781003049340-10

A rejection region $R \subset \mathcal{X}$ is defined, with the intention that H_o is rejected if $\tilde{X} \in R$. We therefore have acceptance region $A = R^c$. Then the *size* of a rejection region is

$$\sup_{\theta \in \Theta_o} P_\theta(\tilde{X} \in R) = \alpha.$$

We say that a rejection region is of level α if the size does not exceed this number. It will be important to keep this distinction in mind when studying the Neyman-Pearson lemma.

The development of the Neyman-Pearson test will require that a hypothesis test be given as a decision rule. Given rejection region R this will be

$$\delta(\tilde{X}) = I\{\tilde{X} \in R\},$$

so that the action taken if $\delta = 1$ is to reject H_o. However, the development of optimal testing is not really possible unless we allow randomized tests, introduced in Section 5.2. This is expressible as a mapping $\delta : \mathcal{X} \to [0,1]$. Then, given $\tilde{X}$, H_o is rejected with probability $\delta(\tilde{X})$. The rejection probability for a given paremeter θ is then $E_\theta[\delta(\tilde{X})]$. Note that $\delta(\tilde{X}) = I\{\tilde{X} \in R\}$ is simply a special case of a randomized decision rule.

10.3 Neyman-Pearson (NP) Tests

The Neyman-Pearson lemma forms the basis for the theory of optimal statistical hypothesis testing. At first, it might seem restrictive, as both the null and alternative hypotheses are assumed to be simple. However, this offers the advantage that the choice of best hypothesis test can be precisely resolved. If it turns out, as it sometimes will, that the same test is optimal for all hypotheses of a composite alternative interpreted as a simple hypothesis, then this resolution is extended to composite hypotheses.

The optimization problem is to determine the most powerful test of level α. It is based on the elements given in the following definitions.

Definition 10.1 (Most Powerful Test) Suppose we are given a randomized decision rule $\delta : \mathcal{X} \to [0,1]$ for alternatives $H_o : p_0$, $H_a : p_1$, where $p_0, p_1 \in \mathcal{P}$. The size of the test is

$$E_0[\delta(\tilde{X})] = \alpha, \tag{10.1}$$

and the power of the test is

$$E_1[\delta(\tilde{X})] = 1 - \beta. \tag{10.2}$$

Then δ^* is a most powerful (MP) level α test if $E_1[\delta^*(\tilde{X})] \geq E_1[\delta(\tilde{X})]$, for any other level α test δ (of course, δ^* itself is assumed to be level α). ∎

A MP test will assume a specific form, which we refer to as a Neyman-Pearson test.

Definition 10.2 (Neyman-Pearson Test) Suppose we are given two possible densities p_0 and p_1 for an observation $\tilde{X} \in \mathcal{X}$. Assume $\mathcal{X}$ is the union of the support of p_0, p_1. Define the ratio

$$\lambda(\tilde{x}) = \frac{p_1(\tilde{x})}{p_0(\tilde{x})}, \quad \tilde{x} \in \mathcal{X}.$$

A Neyman-Pearson test (NP test) is any test taking the form

$$\delta(\tilde{x}) = \begin{cases} 1 & ; \quad \lambda(\tilde{x}) > k \\ q & ; \quad \lambda(\tilde{x}) = k \\ 0 & ; \quad \lambda(\tilde{x}) < k \end{cases} \tag{10.3}$$

for some $k \in [0, \infty]$ and $q \in [0,1]$. ∎

A number of authors have noted important distinctions between the NP test and the LRT for simple hypotheses, so we will adopt terminology accordingly (Solomon, 1975). This distinction will be seen in Example 10.3 below.

We first need to show that a NP test of any size α may be constructed.

Theorem 10.1 (Existence of NP Test) Suppose we are given simple hypotheses $H_o : p_0$, $H_a : p_1$ for observation $\tilde{X} \in \mathcal{X}$. For any $\alpha \in [0, 1]$ there exists $k \in [0, \infty]$ and $q \in [0, 1]$ such that the NP test defined by (10.3) satisfies (10.1) ■

Proof. If $\alpha = 0, 1$ the theorem is proven by setting $k = \infty, 0$, respectively. Otherwise, assume $\alpha \in (0, 1)$. Define the CDF $F(t) = P_0(\lambda(\tilde{X}) \leq t)$. Under distribution p_0, $\lambda(\tilde{X}) < \infty$ $wp1$, so F is a proper CDF. Denote the left limit of F at t by $F^-(t)$ (so that $P_0(\lambda(\tilde{X}) = t) = F(t) - F^-(t)$). We may then find λ_0 such that $F(\lambda_0) \geq 1 - \alpha \geq F^-(\lambda_0)$. Set $k = \lambda_0$. The size of the test is then $E_0[\delta(\tilde{X})] = P_0(\lambda(\tilde{X}) > \lambda_0) + qP_0(\lambda(\tilde{X}) = \lambda_0) = 1 - F(\lambda_0) + q(F(\lambda_0) - F^-(\lambda_0))$. The proof is completed by noting that for $q = 0$, $E_0[\delta(\tilde{X})] = 1 - F(\lambda_0) \leq \alpha$, and for $q = 1$, $E_0[\delta(\tilde{X})] = 1 - F^-(\lambda_0) \geq \alpha$, therefore there exists q for which (10.1) holds. □

Our first version of the Neyman-Pearson lemma follows.

Theorem 10.2 (Neyman-Pearson Lemma I) Suppose we are given simple hypotheses $H_o : p_0$, $H_a : p_1$ for observation $\tilde{X} \in \mathcal{X}$.

(i) Any test satisfying (10.1) and (10.3) is a most powerful level α test.
(ii) Any most powerful level α test satisfies (10.3) for some $k \in [0, \infty]$. It also satisfies (10.1) unless there exists a test of size $< \alpha$ and power 1. ■

Proof. (i) Suppose δ satisfies (10.1) and (10.3). Let δ' be any level α test. Then for any $\tilde{x} \in \mathcal{X}$, $\delta(\tilde{x}) - \delta'(\tilde{x}) \geq 0$ when $p_1(\tilde{x}) > kp_0(\tilde{x})$, and $\delta(\tilde{x}) - \delta'(\tilde{x}) \leq 0$ when $p_1(\tilde{x}) < kp_0(\tilde{x})$. This means

$$\int_{\mathcal{X}} (\delta(\tilde{x}) - \delta'(\tilde{x}))(p_1(\tilde{x}) - kp_0(\tilde{x}))d\nu_{\mathcal{X}} \geq 0.$$

Decomposing this integral gives

$$E_1[\delta(\tilde{X})] - E_1[\delta'(\tilde{X})] \geq k\left(E_0[\delta(\tilde{X})] - E_0[\delta'(\tilde{X})]\right) \tag{10.4}$$

But $k \geq 0$, δ is a size α test, and δ' is a level α test, therefore, the lower bound of the inequality (10.4) is nonnegative, which completes the proof.

(ii) Next, suppose δ' is a most powerful level α test. By Theorem 10.1 there exists δ satisfying (10.1) and (10.3). Define set

$$S = \{x : p_1(\tilde{x}) \neq kp_0(\tilde{x})\} \cap \{x : \delta'(\tilde{x}) \neq \delta(\tilde{x})\}$$

Suppose $\nu_{\mathcal{X}}(S) > 0$, that is δ' does not satisfy (10.3). Following the argument of part (i), this would imply

$$\int_{x \in S} (\delta(\tilde{x}) - \delta'(\tilde{x}))(p_1(\tilde{x}) - kp_0(\tilde{x}))d\nu_{\mathcal{X}} > 0.$$

However, this contradicts the hypothesis that δ' is most powerful. We finally note that if the size of δ' is strictly less than α, then the power can be increased by increasing the size of the test, unless the power equals 1. □

Although the most powerful test between two simple hypotheses is easily determined, a practical theory of hypothesis testing must eventually be applied to composite hypotheses. Our interest, therefore, is primarily in the uniformly most powerful (UMP) test, which is defined next.

Definition 10.3 (Uniformly Most Powerful (UMP) Test) Suppose we are given a parametric family $\mathcal{P}$ for observation $\tilde{X} \in \mathcal{X}$. Suppose δ is a level α test for the hypotheses: $H_o : \theta \in \Theta_o$ and $H_a : \theta \in \Theta_a$. Then δ is uniformly most powerful (UMP) if for each $\theta \in \Theta_a$ we have $E_\theta[\delta] \geq E_\theta[\delta']$ for any other level α test δ'. ■

The construction of a UMP test can be based on a simple idea. Suppose $\Theta_o = \{\theta_0\}$ is a simple hypothesis, but Θ_a is composite. For each $\theta \in \Theta_a$ we may identify the MP level α NP test δ_θ. Now suppose δ_θ does not depend on θ, that is $\delta_\theta \equiv \delta$. Then δ is UMP. This is summarized by the next theorem.

Theorem 10.3 (Uniformly Most Powerful (UMP) Test) Suppose we are given hypotheses $H_o : \theta = \theta_0$ and $H_a : \theta \in \Theta_a$, where H_o is simple and H_a is composite. If there exists a single test δ which is MP level α for the testing $H_o : \theta = \theta_0$ against $H_a : \theta = \theta_1$ for all $\theta_1 \in \Theta_a$, then δ is UMP level α for testing $H_o : \theta = \theta_0$ and $H_a : \theta \in \Theta_a$. ■

Proof. Because the size of a test is determined only from the null hypothesis, δ is also level α against the composite hypothesis $H_a : \theta \in \Theta_a$. Conversely, any level α test δ' for the composite alternative is level α against a simple alternative in Θ_a. By hypothesis, δ, as a test against the composite alternative, has power no smaller than δ' for any alternative $\theta_1 \in \Theta_a$. □

We demonstrate the idea behind Theorem 10.3 in the next three examples.

Example 10.1 (UMP Test for Normal Means) Let $\mathbf{X} = (X_1, \ldots, X_n)$ be an *iid* sample from $N(\mu, 1)$. Suppose we have simple hypotheses $H_o : \mu = \mu_0$ and $H_a : \mu = \mu_1$. Then to construct the NP test of Definition 10.2 we need the ratio

$$\begin{aligned} \lambda(\mathbf{x}) &= \exp\left\{-\frac{1}{2}\left(\sum_{i=1}^n (x_i - \mu_1)^2 - \sum_{i=1}^n (x_i - \mu_0)^2\right)\right\} \\ &= \exp\left\{(\mu_1 - \mu_0)\sum_{i=1}^n x_i - \frac{n(\mu_1^2 - \mu_0^2)}{2}\right\}. \end{aligned}$$

Note that the values μ_0, μ_1 are considered known. Furthermore, the NP test rejects H_o for large values of $\lambda(\mathbf{X})$. The test therefore can be reduced to

$$\delta(\mathbf{x}) = \begin{cases} 1; & \sum_{i=1}^n x_i > t \\ q; & \sum_{i=1}^n x_i = t \\ 0; & \sum_{i=1}^n x_i < t \end{cases}, \tag{10.5}$$

if $\mu_1 > \mu_0$ and

$$\delta(\mathbf{x}) = \begin{cases} 1; & \sum_{i=1}^n x_i < t \\ q; & \sum_{i=1}^n x_i = t \\ 0; & \sum_{i=1}^n x_i > t \end{cases}, \tag{10.6}$$

if $\mu_1 < \mu_0$.

We first note that because $P_\mu(\sum_{i=1}^n X_i = t) = 0$ for any t, the randomization value q playes no role. Then, the selection of t is constrained by the size α. For the case $\mu_1 > \mu_0$ we set t to be the solution to $P_{\mu_0}(\sum_{i=1}^n X_i > t) = \alpha$, and for $\mu_1 < \mu_0$ we set t to $P_{\mu_0}(\sum_{i=1}^n X_i < t) = \alpha$. Note, then, that for $\mu_1 > \mu_0$ the MP test has rejection region $R = \{\sum_{i=1}^n X_i > t\}$. However, this test does not depend on μ_1, as long as $\mu_1 > \mu_0$. By Theorem 10.3 we have therefore identified the UMP test for hypotheses $H_o : \mu = \mu_0$ and $H_a : \mu > \mu_0$. By essentially the same argument, the UMP test for hypotheses $H_o : \mu = \mu_0$ and $H_a : \mu < \mu_0$ exists, and has rejection region $R = \{\sum_{i=1}^n X_i < t\}$.

We must also note that a UMP test for $H_o : \mu = \mu_0$ and $H_a : \mu \neq \mu_0$ does not exist, since the MP test for simple alternatives is not the same for μ above and below μ_0. ■

Example 10.2 (UMP Test for a Two-Sample Poisson Model with Constraint) Let $X_i \sim pois(\lambda_i)$, $i = 1, 2$ be two independent random variables with a Poisson distribution. Suppose the means λ_1, λ_2 satisfy the constraint

$$\lambda_1 \lambda_2 = \bar{\lambda}. \tag{10.7}$$

Suppose we wish to test $H_o : \lambda_1 = \lambda_2$ against alternative $H_a : \lambda_1 > \lambda_2$, assuming constraint (10.7) holds, and $\bar{\lambda}$ is known.

We can reparametrize by $\lambda_1 = \rho\bar{\lambda}^{1/2}$, $\lambda_2 = \rho^{-1}\bar{\lambda}^{1/2}$. Then the hypotheses are equivalently $H_o : \rho = 1$ against alternative $H_a : \rho > 1$. The density family is

$$\begin{aligned} p_\rho(x_1, x_2) &= \frac{\left[\rho\bar{\lambda}^{1/2}\right]^{x_1} \left[\rho^{-1}\bar{\lambda}^{1/2}\right]^{x_2}}{x_1! x_2!} \exp\left\{\rho\bar{\lambda}^{1/2} + \rho^{-1}\bar{\lambda}^{1/2}\right\} \\ &= \frac{\left[\bar{\lambda}^{1/2}\right]^{x_1+x_2} \left[\rho\right]^{x_1-x_2}}{x_1! x_2!} \exp\left\{\rho\bar{\lambda}^{1/2} + \rho^{-1}\bar{\lambda}^{1/2}\right\} \end{aligned}$$

By the Neyman-Pearson Lemma the MP test for simple hypotheses $H_o : \rho = 1$ against alternative $H_a : \rho = \rho_0$, where $\rho_0 > 1$, rejects for

$$\frac{p_{\rho_0}(X_1, X_2)}{p_1(X_1, X_2)} > k.$$

This is equivalent to

$$X_1 - X_2 > k', \tag{10.8}$$

for some constant k'. But this test does not depend on ρ_0, therefore (10.8) also defines the required UMP test. ■

In the next example, we demonstrate why the randomized test is an important component of the theory of Neyman-Pearson testing.

Example 10.3 (UMP Test for Uniform Samples with Location Parameter) Suppose $\mathbf{X} = (X_1, \ldots, X_n)$ is an *iid* sample from $unif(\theta, \theta + 1)$, $\theta \in \Theta = \mathbb{R}$. Then $p_\theta = I\{\theta < X_{(1)} < X_{(n)} < \theta + 1\}$. Consider hypothetical values $\theta_0 < \theta_1$. For testing against alternative θ_1, the NP test is calculated using set operations on the respective supports of p_{θ_0} and p_{θ_1}, giving

$$\lambda(\mathbf{x}) = \begin{cases} \infty; & \mathbf{x} \in \mathcal{S}_0^c \cap \mathcal{S}_1 \\ 1; & \mathbf{x} \in \mathcal{S}_0 \cap \mathcal{S}_1 \\ 0; & \mathbf{x} \in \mathcal{S}_0 \cap \mathcal{S}_1^c \end{cases}$$

where $\mathcal{S}_0 = \{\theta_0 < x_{(1)} < x_{(n)} < \theta_0 + 1\}$, $\mathcal{S}_1 = \{\theta_1 < x_{(1)} < x_{(n)} < \theta_1 + 1\}$ are the supports of p_{θ_0}, p_{θ_1} respectively. The situation illustrated in Figure 10.1. The respective supports are unit squares with lower left coordinate (θ_i, θ_i), $i = 0, 1$. The test may be represented entirely by the minimal sufficient statistic $(X_{(1)}, X_{(n)})$. Note that since $X_{(1)} < X_{(n)}$, only the region above the diagonal is of interest. The union of the density supports is partitioned into the intersection A_2, and the supports exclusive of the other, A_1 and A_3. This is equivalent to

$$\lambda(\mathbf{x}) = \begin{cases} \infty; & (x_{(1)}, x_{(n)}) \in A_3 \\ 1; & (x_{(1)}, x_{(n)}) \in A_2 \\ 0; & (x_{(1)}, x_{(n)}) \in A_1. \end{cases}$$

Suppose we next proceed a few steps backwards, and specify a rejection region based on intuition. We must reject θ_0 if $X_i > \theta_0 + 1$ for any observation. This occurs *iff* $X_{(n)} > \theta_0 + 1$.

Suppose we also reject θ_0 if $X_{(1)} \geq t_\alpha + \theta_0$, $t_\alpha < 1$, defining rejection region $R_\alpha = \{X_{(1)} \geq \theta_0 + t_\alpha\} \cup \{X_{(n)} \geq \theta_0 + 1\}$. Denote the resulting decision rule $\delta_{R_\alpha} = I\{R_\alpha\}$. We can specify any size $\alpha = P_{\theta_0}(R_\alpha) = (1 - t_\alpha)^n$, or $t_\alpha = 1 - \alpha^{1/n}$, noting that $P_{\theta_0}(X_{(n)} \geq \theta_0 + 1) = 0$. Is δ_{R_α} a NP test? First, note that

$$P_{\theta_0}(A_2) = \begin{cases} (1 + \theta_0 - \theta_1)^n & ; \quad \theta_1 < \theta_0 + 1 \\ 0 & ; \quad \theta_1 \geq \theta_0 + 1 \end{cases}.$$

If randomization is not used, then a NP test can only have sizes 0, $P_{\theta_0}(A_2)$ or 1. Randomization is usually used when discrete distributions do not permit nonrandomized tests to have sizes exactly equal to a specific α. In the current example, although the distribution of $\mathbf{X}$ is not discrete, $\lambda(\mathbf{X})$ itself can assume only values 0, 1 and ∞. Suppose then we fix α and t_α. Consider two cases:

Case 1: $\theta_0 + t_\alpha \leq \theta_1$. In this case $P_{\theta_0}(A_2) \leq \alpha$ Then define the NP test

$$\delta_{NP}(\mathbf{x}) = \begin{cases} 1 & ; \quad \lambda(\mathbf{x}) > 0 \\ \frac{\alpha - P_{\theta_0}(A_2)}{1 - P_{\theta_0}(A_2)} & ; \quad \lambda(\mathbf{x}) = 0 \\ 0 & ; \quad \lambda(\mathbf{x}) < 0 \end{cases}.$$

The size is given by

$$E_{\theta_0}[\delta_{NP}(\mathbf{X})] = P_{\theta_0}(A_2) + \frac{\alpha - P_{\theta_0}(A_2)}{1 - P_{\theta_0}(A_2)}(1 - P_{\theta_0}(A_2)) = \alpha.$$

Interestingly, δ_{R_α} is a NP test. If $\lambda(\mathbf{x}) > 0$ then R_α occurs, and H_o is rejected with probability 1. If $\lambda(\mathbf{x}) = 0$, then $(X_{(1)}, X_{(n)}) \in A_1 \cap A_2^c$. The conditional probability of rejection is

$$P_{\theta_0}(R_\alpha \mid \lambda(\mathbf{X}) = 0) = P_{\theta_0}(R_\alpha \mid (X_{(1)}, X_{(n)}) \in A_1 \cap A_2^c) = \frac{\alpha - P_{\theta_0}(A_2)}{1 - P_{\theta_0}(A_2)}.$$

However, this is exactly the randomization probability used by δ_{NP} when $\lambda(\mathbf{X}) = 0$. Furthermore, by symmetry the distribution of X conditional on A_2 is the same for p_{θ_0} and p_{θ_1}, so that $P_{\theta_0}(R_\alpha \mid \lambda(\mathbf{X}) = 0) = P_{\theta_1}(R_\alpha \mid \lambda(\mathbf{X}) = 0)$. Finally, the case $\lambda(\mathbf{x}) < 0$ plays no role. Thus, δ_{R_α} is equivalent to δ_{NP}. The fact that the randomization procedure is explictly carried out in the testing procedure does not change the fact that with respect to all properties used to characterize an MP test, δ_{R_α} and δ_{NP} are identical.

Case 2: $t_\alpha + \theta_0 > \theta_1$. Here the principle is the same, except that $P_{\theta_0}(A_2) > \alpha$. Then define the NP test

$$\delta_{NP}(\mathbf{x}) = \begin{cases} 1 & ; \quad \lambda(\mathbf{x}) > 1 \\ \frac{\alpha}{P_{\theta_0}(A_2)} & ; \quad \lambda(\mathbf{x}) = 1 \\ 0 & ; \quad \lambda(\mathbf{x}) < 1 \end{cases}.$$

Using a similar argument, it can be shown that δ_{R_α} and δ_{NP} are identical.

Thus, δ_{R_α} is a most powerful level α test for alternatives $H_o : \theta = \theta_0$ and $H_a : \theta = \theta_1$, $\theta_1 > \theta_0$ and therefore a UMP level α test for alternatives $H_o : \theta = \theta_0$ and $H_a : \theta > \theta_0$. ■

10.4 Monotone Likelihood Ratios (MLR)

There is a widely applicable condition under which one-sided UMP tests will exist, which is derived from the likelihood ratio ordering property introduced in Section 1.5.

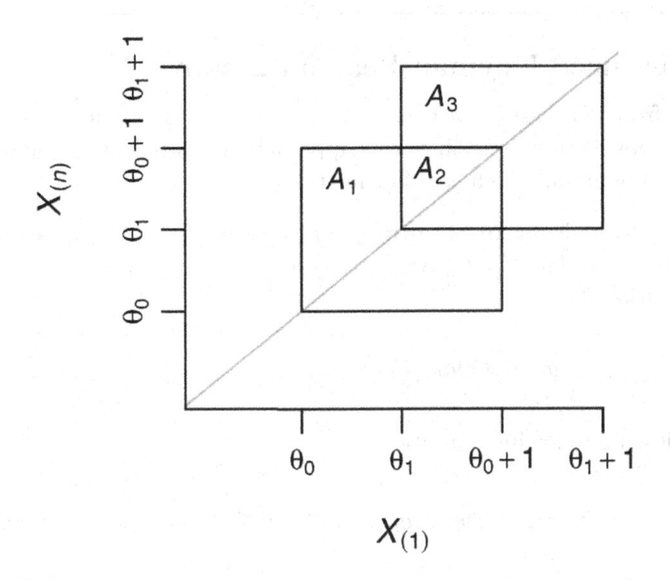

FIGURE 10.1
Geometric representation of NP test for location parameter of a uniform density of Example 10.3.

Definition 10.4 (Monotone Likelihood Ratio (MLR) Property) A parametric family $\mathcal{P}$ for observation $\tilde{X} \in \mathcal{X}$ with $\Theta \subset \mathbb{R}$ is said to have monotone likelihood ratio (MLR) if there exists real valued statistic $T(\tilde{x})$ such that for any $\theta_1 < \theta_2$ the ratio $p_{\theta_2}(\tilde{x})/p_{\theta_1}(\tilde{x})$ is a nondecreasing function of $T(\tilde{x})$. ■

The connection between the MLR property and UMP tests is immediate, and given in the next theorem.

Theorem 10.4 (UMP Tests and the MLR Property) Under the conditions of Definition 10.4, if $\mathcal{P}$ has MLR in $T(\tilde{x})$, then there exists a UMP for testing $H_o : \theta = \theta_0$ against $H_a : \theta > \theta_0$. ■

Proof. The result follows after noting that any most powerful test of $H_o : \theta = \theta_0$ against $H_a : \theta = \theta_1$, $\theta_1 > \theta_0$, assumes the form

$$\delta(\tilde{x}) = \begin{cases} 1; & T(x) > t \\ q; & T(\tilde{x}) = t, \\ 0; & T(\tilde{x}) < t \end{cases} \tag{10.9}$$

where t, q satisfy $P_{\theta_0}(T(\tilde{X}) > t) + qP_{\theta_0}(T(\tilde{X}) = t) = \alpha$. □

The MLR property confers yet one more advantage onto the exponential family model.

Example 10.4 (The MLR property for Exponential Family Models) Suppose we have a one dimensional exponential family model $p_\theta(\tilde{x}) = \exp(\eta(\theta)T(\tilde{x}) - B(\theta))h(\tilde{x})$. If $\eta(\theta)$ is an increasing function of θ, then $p_\theta(\tilde{x})$ has a MLR property for parameter θ and statistic $T(\tilde{x})$. ■

10.5 The Generalized Neyman-Pearson Lemma

Defining UMP tests from NP tests is a powerful method, but is restricted to a relatively small range of applications. An extension of this approach depends on a generalized version of the Neyman-Pearson lemma, which we offer next.

Theorem 10.5 (Neyman-Pearson Lemma II) Suppose $g_1, \ldots, g_{m+1}$ are real valued integrable functions on $\mathcal{X}$, and $b_1, \ldots, b_m$ are finite constants. Let $\mathcal{C}$ be the class of functions $\delta : \mathcal{X} \to [0,1]$ which satisfy

$$\int_{\tilde{x} \in \mathcal{X}} \delta(\tilde{x}) g_i(\tilde{x}) d\nu_{\mathcal{X}}(\tilde{x}) = b_i, \;\; i = 1, \ldots, m. \tag{10.10}$$

Suppose $\delta \in \mathcal{C}$ satisfies the following conditions:

$$\begin{aligned} \delta(\tilde{x}) &= 1 \quad \text{when} \quad g_{m+1}(\tilde{x}) > \sum_{i=1}^{m} k_i g_i(\tilde{x}) \\ \delta(\tilde{x}) &= 0 \quad \text{when} \quad g_{m+1}(\tilde{x}) < \sum_{i=1}^{m} k_i g_i(\tilde{x}), \end{aligned} \tag{10.11}$$

for some constants $k_1, \ldots, k_m$. Then δ maximizes $\int_{\mathcal{X}} \delta(\tilde{x}) g_{m+1}(\tilde{x}) d\nu_{\mathcal{X}}(\tilde{x})$ among decision rules in $\mathcal{C}$. In addition, if the constants $k_1, \ldots, k_m$ are nonnegative, then δ maximizes $\int_{\mathcal{X}} \delta(\tilde{x}) g_{m+1}(\tilde{x}) d\nu_{\mathcal{X}}(\tilde{x})$ among decision rules satisfying constraints

$$\int_{\tilde{x} \in \mathcal{X}} \delta(\tilde{x}) g_i(\tilde{x}) d\nu_{\mathcal{X}}(\tilde{x}) \leq b_i, \;\; i = 1, \ldots, m. \tag{10.12}$$

for constants $b_1, \ldots, b_m$. ■

Proof. A complete treatment of this problem, including existence and sufficient and necessary conditions, can be found in Lehmann and Romano (2005). However, the result is quite similar to Theorem 10.2. To see this, set

$$H(\tilde{x}) = g_{m+1}(\tilde{x}) - \sum_{i=1}^{m} k_i g_i(\tilde{x}).$$

Let δ satisfy (10.11), and let δ' satisfy bounds (10.10). Following Theorem 10.2 we may conclude that $(\delta(\tilde{x}) - \delta'(\tilde{x}))H(\tilde{x}) \geq 0$ for all $\tilde{x} \in \mathcal{X}$. Then

$$0 \leq \int_{\tilde{x} \in \mathcal{X}} (\delta(\tilde{x}) - \delta'(\tilde{x})) H(\tilde{x}) d\nu_{\mathcal{X}}(\tilde{x}) = \int_{\tilde{x} \in \mathcal{X}} \delta(\tilde{x}) g_{m+1}(\tilde{x}) - \int_{\tilde{x} \in \mathcal{X}} \delta'(\tilde{x}) g_{m+1}(\tilde{x}),$$

after applying the constraints of (10.10) or (10.12). □

Of course, Theorem 10.2 is a special case of Theorem 10.5. But while the test size constraint is given as part of the hypothesis in Theorem 10.2, it is nowehere mentioned in the statement of Theorem 10.5 (which, in fact, makes no mention of any probability distribution at all). This means the size constraint must be introduced explicitly. To do this, write

$$E_{\theta_0}[\delta] = \int_{\tilde{x} \in \mathcal{X}} \delta(\tilde{x}) p_{\theta_0}(\tilde{x}) d\nu = \alpha,$$

given simple null hypothesis $H_o : \theta = \theta_o$. Then, in Theorem 10.5 set $g_1(\tilde{x}) = p_{\theta_0}(\tilde{x})$ and $b_1 = \alpha$. If the object is to maximize power at a simple alternative $H_a : \theta = \theta_1$, equal to

$$E_{\theta_1}[\delta] = \int_{\tilde{x}\in\mathcal{X}} \delta(\tilde{x}) p_{\theta_1}(\tilde{x}) d\nu,$$

then we would set $m = 2$ and $g_2(\tilde{x}) = p_{\theta_1}(\tilde{x})$. However, the importance of Theorem 10.5 is that it allows us to introduce additional constraints, and to optimize quantities other than power at a simple alternative (for example, the locally most powerful test of Section 10.5.1). We will see two examples of this approach in the remainder of this section, and a third in Section 10.7.

10.5.1 Locally Most Powerful Tests (LMP)

As a first application of the generalized Neyman-Pearson lemma we introduce the locally most powerful test (LMP). Suppose a UMP one-sided test does not exist. We can reason that any test worth considering will have power close to 1 for alternatives that are far enough away from the null hypothesis. Therefore, we might adopt a strategy of maximizing power near the null hypothesis.

To make this idea more precise, suppose we have hypotheses $H_o : \theta \leq \theta_0$ and $H_a : \theta > \theta_0$. For a test δ, we have rejection probability $\gamma(\theta) = E_\theta[\delta(\tilde{X})]$ for all $\theta \in \Theta$. We then assume that $\gamma(\theta)$ is an increasing function, with $\gamma(\theta_0) = \alpha$ (conditions under which this assumption holds are discussed in Section 5.6). In this case, a LMP test can be defined as one which maximizes the derivative of γ at θ_0 under the constraint $\gamma(\theta_0) = \alpha$. This derivative can be written

$$\left.\frac{d\gamma(\theta)}{d\theta}\right|_{\theta=\theta_0} = \left.\frac{dE_\theta[\delta(\tilde{X})]}{d\theta}\right|_{\theta=\theta_0} = \int_{\tilde{x}\in\mathcal{X}} \delta(\tilde{x}) \left.\frac{\partial p_\theta(\tilde{x})}{\partial\theta}\right|_{\theta=\theta_0} d\nu,$$

assuming we may exchange differentiation and integration. Then, refering to the notation of Theorem 10.5, set

$$g_2(\tilde{x}) = \left.\frac{\partial p_\theta(\tilde{x})}{\partial\theta}\right|_{\theta=\theta_0},$$

noting that we have $m = 1$ constraint, defined by setting $g_1(\tilde{x}) = p_{\theta_0}(\tilde{x})$ and $b_1 = \alpha$. The LMP test is then given by

$$\delta(\tilde{x}) = 1 \text{ when } g_2(\tilde{x}) > kg_1(\tilde{x}), \quad \delta(\tilde{x}) = 0 \text{ when } g_2(\tilde{x}) < kg_1(\tilde{x}),$$

which is equivalent to

$$\delta(\tilde{x}) = 1 \text{ when } \left.\frac{\partial \log p_\theta(\tilde{x})}{\partial\theta}\right|_{\theta=\theta_0} > k, \quad \delta(\tilde{x}) = 0 \text{ when } \left.\frac{\partial \log p_\theta(\tilde{x})}{\partial\theta}\right|_{\theta=\theta_0} < k.$$

Interestingly, this is constructed very much like an unconstrained NP test, except that the alternative density p_{θ_1} is replaced with the derivative of p_θ with respect to θ evaluated at θ_0. This can be justified as follows. The NP test is based on ratio

$$\lambda(\tilde{x}) = \frac{p_{\theta_1}}{p_{\theta_0}} = \frac{p_{\theta_0} + [p_{\theta_1} - p_{\theta_0}]}{p_{\theta_0}} = 1 + \frac{[p_{\theta_1} - p_{\theta_0}]}{p_{\theta_0}}.$$

The NP test can therefore also be based on the ratio $[p_{\theta_1} - p_{\theta_0}]/p_{\theta_0}$. The LMP implements this idea, allowing $\theta_1 \to \theta_0$. For this reason, the LMP test can be interpreted as a large sample method, in the sense that any alternative $\theta \in \Theta_a$ can be assumed to be arbitrarily close to the null θ_0. In fact, the LMP test is equivalent to the score test, a large sample testing method which will be introduced in Section 14.3.1. For the moment, however, we view the LMP test as a possible alternative when a UMP test does not exist. Two examples follow.

Example 10.5 (LMP Test for a Cauchy Sample with Location Parameter) Suppose $\mathbf{X} = (X_1, \ldots, X_n)$ is an *iid* sample from *cauchy*$(\theta, 1)$. The density of $\mathbf{X}$ is then $p_\theta(\mathbf{x}) = \prod_{i=1}^n [\pi(1 + (x_i - \theta)^2)]^{-1}$. It can be shown that a one-sided UMP test does not exist, so we construct a LMP test. The derivative of p_θ can be evaluated as follows:

$$\begin{aligned}
\frac{dp_\theta(\mathbf{x})}{d\theta} &= \sum_{i=1}^n \frac{df_\theta(x_i)}{d\theta} \prod_{j \neq i} f_\theta(x_j) \\
&= \sum_{i=1}^n \frac{2(x_i - \theta)}{\pi(1 + (x_i - \theta)^2)^2} \prod_{j \neq i} f_\theta(x_j) \\
&= \sum_{i=1}^n \frac{2(x_i - \theta)}{(1 + (x_i - \theta)^2)} \prod_{i=1}^n f_\theta(x_j) \\
&= p_\theta(\mathbf{x}) \sum_{i=1}^n \frac{2(x_i - \theta)}{(1 + (x_i - \theta)^2)}.
\end{aligned}$$

Therefore, the LMP test for $H_o : \theta \leq \theta_0$ and $H_a : \theta > \theta_0$ has rejection region

$$R = \left\{ \sum_{i=1}^n \frac{(x_i - \theta_0)}{(1 + (x_i - \theta_0)^2)} > k \right\}. \qquad \blacksquare$$

Example 10.6 (LMP Test for a Weibull Shape Paremeter) Suppose $\mathbf{X} = (X_1, \ldots, X_n)$ is an *iid* sample from *weibull*$(1, \theta)$. The density of the sample is $p_\theta(\mathbf{x}) = \theta^n \left(\prod_i x_i\right)^{\theta - 1} e^{-S_\theta(\mathbf{x})}$ where $S_\theta(\mathbf{x}) = \sum_i x_i^\theta$. The test statistic is based on

$$\begin{aligned}
\frac{1}{p_\theta(\mathbf{x})} \times \frac{dp_\theta(\mathbf{x})}{d\theta} &= \frac{d \log p_\theta(\mathbf{x})}{d\theta} \\
&= \frac{d}{d\theta} \left[n \log(\theta) + (\theta - 1) \left[\sum_i \log(x_i) \right] - S_\theta(\mathbf{x}) \right] \\
&= (n/\theta) + \sum_i \log(x_i)(1 - x_i^\theta).
\end{aligned}$$

So, for a LMP test we reject $H_o : \theta = \theta_0$ in favor of $H_a : \theta > \theta_0$ based on rejection region $\sum_i \log(x_i)(1 - x_i^{\theta_0}) > t$. $\blacksquare$

While the LMP test will be a sound strategy in many applications, it must still be verified that it has sufficient power over the entire range of anticipated alternatives, since nothing in its derivation explicitly guarantees this. We cannot expect, in other words, that optimal power near the null hypothesis will lead to close to optimal power in regions in which we expect power to be close to, say, 90%. This, of course, will usually be the region of interest. Examples of LMP tests which behave poorly in this way are given in Davies (1969). The LMP test is not universally covered in references on the theory of inference, with Cox and Hinkley (1979) being a notable exception. See Example 10.11 below for more on this topic.

10.5.2 Uniformly Most Powerful (UMP) Unbiased Tests

Recall that in Example 10.1 we were able to conclude that a UMP two-sided test for a normal mean cannot exist. Suppose, for the sake of argument, the NP test of Equation (10.5) is used for two-sided alternative $H_a : \mu \neq \mu_0$. The test will be most powerful for alternatives $\mu > \mu_0$. However, for $\mu < \mu_0$, not only will the test not be MP, the power will actually be less than the size α. In other words, this test would not be unbiased, which is a property which can reasonably be expected of any hypothesis test (Section 5.3).

It turns out that the generalized Neyman-Pearson lemma allows us to introduce unbiasedness as a constraint, hence any resulting UMP test can be referred to as the UMP unbiased test. Consider testing $H_o : \theta = \theta_0$ against $H_a : \theta \neq \theta_0$. The size constraint is then

$$\int_{\tilde{x} \in \mathcal{X}} \delta(\tilde{x}) p_{\theta_0}(\tilde{x}) d\nu = \alpha.$$

Using the notation of Theorem 10.5 set $g_1(\tilde{x}) = p_{\theta_0}(\tilde{x})$ and $b_1 = \alpha$.

Next, if the test is unbiased, then the smallest rejection probability is at $\theta = \theta_0$. This forces

$$\left.\frac{d\gamma(\theta)}{d\theta}\right|_{\theta=\theta_0} = \left.\frac{dE_\theta[\delta(\tilde{X})]}{d\theta}\right|_{\theta=\theta_0} = \int_{\mathcal{X}} \delta(\tilde{x}) \left.\frac{\partial p_\theta(\tilde{x})}{\partial \theta}\right|_{\theta=\theta_0} d\nu = 0,$$

assuming we may exchange differentiation and integration (recall that for the LMP test this quantity was maximized, whereas here it is constrained). Then set

$$g_2(\tilde{x}) = \left.\frac{\partial p_\theta(\tilde{x})}{\partial \theta}\right|_{\theta=\theta_0}$$

and $b_2 = 0$. Since the object is to maximize power, we set

$$g_3(\tilde{x}) = p_{\theta_1}(\tilde{x}), \quad \theta_1 \neq \theta_0.$$

Then the problem is to construct a decision rule $\delta \in \mathcal{C}$ of the form

$$\delta(\tilde{x}) = 1 \quad \text{when} \quad g_3(\tilde{x}) > \sum_{i=1}^{2} k_i g_i(\tilde{x}),$$

$$\delta(\tilde{x}) = 0 \quad \text{when} \quad g_3(\tilde{x}) < \sum_{i=1}^{2} k_i g_i(\tilde{x}).$$

To fix ideas, suppose we have a one dimensional exponential family model, with simple hypotheses $H_o : \eta = \eta_0$ against $H_a : \eta = \eta_1$, for $p_\eta(\tilde{x}) = \exp(\eta T(\tilde{x}) - A(\eta))h(\tilde{x})$. This gives

$$\begin{aligned} g_1(\tilde{x}) &= \exp(\eta_0 T(\tilde{x}) - A(\eta_0))h(\tilde{x}) \\ g_2(\tilde{x}) &= (T(\tilde{x}) - A'(\eta_0)) \exp(\eta_0 T(\tilde{x}) - A(\eta_0))h(\tilde{x}) \\ g_3(\tilde{x}) &= \exp(\eta_1 T(\tilde{x}) - A(\eta_1))h(\tilde{x}), \end{aligned}$$

where $A'(\eta)$ is the derivative of $A(\eta)$ with respect to η. We then note that η_0, η_1 are considered known, and may be incorporated into the constants k_1, k_2. In addition, since $h(\tilde{x}) > 0$, the inequalities resolve independently of $h(\tilde{x})$. Thus, if we set

$$H(T(\tilde{x}); k_1, k_2) = e^{(\eta_1 - \eta_0)T(\tilde{x})} - k_1 - k_2 T(\tilde{x})$$

then $H(T(\tilde{x}); k_1, k_2) > 0$ implies $\delta(\tilde{x}) = 1$, and $H(T(\tilde{x}); k_1, k_2) < 0$ implies $\delta(\tilde{x}) = 0$.

Suppose we next set the task of fixing $t_0 < t_1$, then selecting k_1, k_2 so that $H(t_0; k_1, k_2) = H(t_1; k_1, k_2) = 0$. This is equivalent to the linear system of equations

$$\begin{bmatrix} 1 & t_0 \\ 1 & t_1 \end{bmatrix} \begin{bmatrix} k_1 \\ k_2 \end{bmatrix} = \begin{bmatrix} e^{(\eta_1-\eta_0)t_0} \\ e^{(\eta_1-\eta_0)t_1} \end{bmatrix}.$$

As long as $t_0 \neq t_1$, this linear system is nonsingular and we obtain

$$\begin{aligned} k_1 &= \frac{1}{t_1 - t_0}\left[t_1 e^{(\eta_1-\eta_0)t_0} - t_0 e^{(\eta_1-\eta_0)t_1}\right] \\ k_2 &= \frac{1}{t_1 - t_0}\left[-e^{(\eta_1-\eta_0)t_0} + e^{(\eta_1-\eta_0)t_1}\right]. \end{aligned}$$

Finally, it is easily verified that $H(T(\tilde{x}); k_1, k_2)$ is a strictly convex function of $T(\tilde{x})$ (since the second derivative is strictly positive). Therefore,

$$T(\tilde{x}) \notin [t_0, t_1] \implies H(T(\tilde{x}); k_1, k_2) > 0 \implies \delta(\tilde{x}) = 1,$$

and

$$T(\tilde{x}) \in (t_0, t_1) \implies H(T(\tilde{x}); k_1, k_2) < 0 \implies \delta(\tilde{x}) = 0.$$

Then t_0, t_1 are selected to satisfy the constraints (10.10), which do not depend on η. Therefore, a UMP unbiased test exists, provided these constraints can be satisfied. We may then set randomization component $\delta(\tilde{x}) = q_i$ if $T(\tilde{x}) = t_i$, $i = 0, 1$. In particular,

$$\alpha = P_{\eta_0}(T(\tilde{X}) \notin [t_0, t_1]) + q_0 P_{\eta_0}(T(\tilde{X}) = t_0) + q_1 P_{\eta_0}(T(\tilde{X}) = t_1)$$

and

$$\begin{aligned} 0 &= E_{\eta_0}[(T(\tilde{X}) - A'(\eta_0))I\{T(X) \notin [t_0, t_1]\}] \\ &\quad + q_0(t_0 - A'(\eta_0))P_{\eta_0}(T(\tilde{X}) = t_0) \\ &\quad + q_1(t_1 - A'(\eta_0))P_{\eta_0}(T(\tilde{X}) = t_1). \end{aligned}$$

Thus, the two-sided unbiased UMP level α test for null hypothesis $H_o : \eta = \eta_0$ rejects for $T(\tilde{X}) < t_0$ or $T(\tilde{X}) > t_1$ where,

$$\begin{aligned} & P_{\eta_0}(T(\tilde{X}) < t_0) + P_{\eta_0}(T(\tilde{X}) > t_1) = \alpha, \\ & E_{\eta_0}[(T(\tilde{X}) - E_{\eta_0}[T(\tilde{X})])I\{T(\tilde{X}) > t_1\}] \\ & = -E_{\eta_0}[(T(\tilde{X}) - E_{\eta_0}[T(\tilde{X})])I\{T(\tilde{X}) < t_0\}]. \end{aligned} \tag{10.13}$$

In particular, for the two-sided test of a normal mean based on a sample mean, the test $|\bar{X}| > k$ is UMP unbiased. In general, this development is of interest then the distribution of $T(\tilde{X})$ is not symmetric about its mean.

We may verify with the following lemma that a necessary condition for unbiasedness constraint (10.13) is that $t_0 \leq E_{\eta_0}[T(\tilde{X})] \leq t_1$.

Lemma 10.1 Suppose random variable $X \in \mathbb{R}$ has density f *wrt* measure ν and finite mean μ. Suppose the support $\mathcal{S}$ of f is open. If we are given two constants $t_0 < t_1$, $t_0, t_1 \in \mathcal{S}$, for which

$$\int_{x<t_0} (x-\mu) f d\nu + \int_{x>t_1} (x-\mu) f d\nu = 0, \tag{10.14}$$

then we must have $t_0 \leq \mu \leq t_1$. ∎

Proof. We always have $\int_{\mathcal{S}}(x-\mu)f d\nu = 0$, therefore the integral defined in (10.14) is equal to

$$I_{t_0,t_1} = -\int_{x\in[t_0,t_1]}(x-\mu)f d\nu = -\int_{x\in\mathcal{S}}(x-\mu)I\{x\in[t_0,t_1]\}f d\nu.$$

Thus, condition (10.14) is equivalent to $I_{t_0,t_1}=0$. If $t_0>\mu$ or $t_1<\mu$ then the integrand of I_{t_0,t_1} is either nonnegative or nonpositive. In either case, since $t_0,t_1\in\mathcal{S}$, and $\mathcal{S}$ is open, then $I_{t_0,t_1}\neq 0$, which completes the proof. □

An example of a UMP unbiased test follows.

Example 10.7 (UMP Unbiased Test for an Exponential Scale Parameter) Suppose we are given an *iid* sample $\mathbf{X}=(X_1,\ldots,X_n)$, from $X_1\sim exp(\theta)$. We can write the density of $\mathbf{X}$ as $p_\eta(\mathbf{x}) = \exp(\eta T(\mathbf{x}) - A(\eta))h(\mathbf{x})$ where $\eta=-\theta^{-1}$, and $T(\mathbf{x})=\sum_i x_i$. So the null hypothesis $H_o : \eta=\eta_0$ is equivalent to $H_o:\theta=-1/\eta_0$. In addition, $T(\mathbf{X})=T\sim gamma(\theta,n)$. The constraints (10.13) reduce to

$$\begin{aligned} P(T<t_0)+P(T>t_1) &= \alpha \\ -\int_{t=0}^{t_0}(t-n\theta)f_T(t)dt &= \int_{t=t_1}^{\infty}(t-n\theta)f_T(t)dt, \end{aligned}$$

where f_T is the density of T under the H_o. Then consider the right side of the second constraint:

$$\begin{aligned} \int_{t=t_1}^{\infty}(t-n\theta)f_T(t)dt &= \int_{t=t_1}^{\infty}(t-n\theta)\frac{1}{\Gamma(n)\theta}\left(\frac{t}{\theta}\right)^{n-1}\exp(-t/\theta)dt \\ &= n\theta\left(\bar{F}_{n+1}(t_1)-\bar{F}_n(t_1)\right), \end{aligned}$$

where $F_\nu(t)=1-\bar{F}_\nu(t)$ is the CDF of the density $gamma(\theta,\nu)$. We similarly have

$$\begin{aligned} \int_{t=0}^{t_0}(t-n\theta)f_T(t)dt &= \int_{t=0}^{t_0}(t-n\theta)\frac{1}{\Gamma(n)\theta}\left(\frac{t}{\theta}\right)^{n-1}\exp(-t/\theta)dt \\ &= n\theta\left(F_{n+1}(t_0)-F_n(t_0)\right). \end{aligned}$$

The constraint is then reduced to

$$\begin{aligned} F_n(t_0)+\bar{F}_n(t_1) &= \alpha \\ F_{n+1}(t_0)-F_n(t_0) &= F_{n+1}(t_1)-F_n(t_1). \end{aligned}$$

To demonstrate the test, we use parameters $\theta_0=1, n=10$, we allow t_1 to vary above $n\theta_0$. We then solve for fixed t_1 the unbiasedness constraint for t_0, then calculate α. This yields the UMP unbiased tests for the range $\alpha\in[0,1]$. These values are shown in Figure 10.2, scaled for test statistic $\bar{X}$.

The density of $\bar{X}$ is superimposed, rescaled to vertical range $[0,1]$. It might be tempting to argue that if two values of $\bar{X}$ have the same likelihood value for the null hypothesis θ_0, then the inference should be the same, that is, one should be in the rejection region if and only if the other one is. But this is not the case, since the rejection region boundaries do not coincide with the countours of the density. If they did, for large enough α there would be rejection regions which did not include $\bar{X}=\theta_0$. However, by Lemma 10.1 this cannot occur for the UMP unbiased test, since the unbiasedness constraint could not hold.

It must be remembered that the likelihood is a function of a parameter θ and not the data. Furthermore, the likelihood is defined only up to a multiplicative constant. Therefore, the information in the likelihood is discerned by the likelihood ratio of alternative models for fixed data, not for alternative observations for a fixed parameter. ■

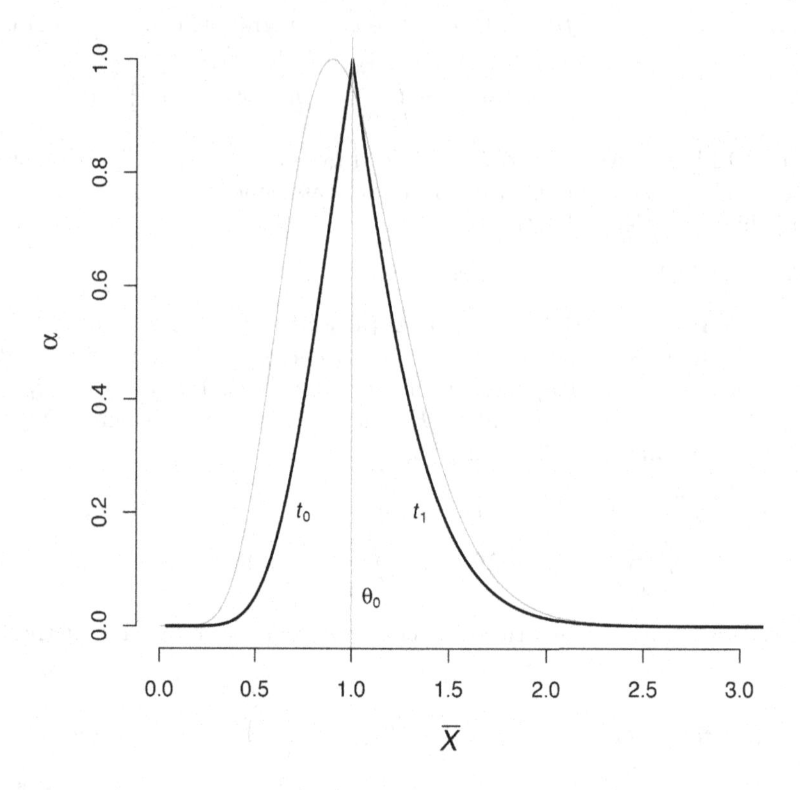

FIGURE 10.2
Boundary points t_0, t_1 defining UMP unbiased test for Example 10.7, standardized for test statistic $\bar{X}$. The vertical axis gives the test size α. The density of $\bar{X}$ under null hypothesis $H_o : \theta_0 = 1$ is superimposed, rescaled to vertical range $[0, 1]$.

10.6 Invariant Hypothesis Tests

Invariant hypothesis tests were defined in Section 9.6. In this section, and Section 10.7 we show how the Neyman-Pearson lemma can be used to derive UMP invariant tests. We will describe two approaches. In this section, we apply Neyman-Pearson theory directly to the maximal invariant. We give a number of examples below. The second is to use the generalized Neyman-Pearson lemma to impose invariance constraints directly. This will be discussed in the subsequent section.

Example 10.8 (Comparison of Two Location Parameter Models) Suppose $\mathbf{X} = (X_1, \ldots, X_n) \in \mathcal{X} = \mathbb{R}^n$ is an *iid* sample from one of two location parameter models $p_j(x_1 - \xi, \ldots, x_n - \xi)$, $j = 0, 1$, where $\xi \in \mathbb{R}$ is an unknown nuisance parameter. We reduce the data to a maximal location invariant on $\mathcal{X}$: $M(\mathbf{X}) = (X_1 - X_n, \ldots, X_{n-1} - X_n) = (Y_1, \ldots, Y_{n-1})$. This density of $M(\mathbf{X})$ may be expressed in terms of the original data as

$$\bar{p}_j = \int_u p_j(x_1 + u, \ldots, x_n + u) d\nu_u, \;\; j = 0, 1,$$

for appropriate measure ν_u. Then, by the Neyman-Pearson lemma (Theorem 10.2), the NP test

$$\delta(\mathbf{x}) = \begin{cases} 1 & ; \quad \frac{\bar{p}_1}{\bar{p}_0} > k \\ q & ; \quad \frac{\bar{p}_1}{\bar{p}_0} = k \\ 0 & ; \quad \frac{\bar{p}_1}{\bar{p}_0} < k \end{cases}$$

will be the MP invariant test for alternatives $H_o : p_0$ versus $H_o : p_1$. See Problem 10.18. ∎

Example 10.9 (Test for Poisson Parameter with Location Nuisance Parameter) Suppose $\mathbf{Y} = (Y_1, \ldots, Y_n)$ is an *iid* sample with $Y_1 \sim pois(\lambda)$. However, the observations take form $X_i = Y_i + \xi$ where ξ is an unknown integer. In this case the density of $\mathbf{X} = (X_1, \ldots, X_n)$ is

$$p_{(\lambda,\xi)}(\mathbf{x}) = \prod_{i=1}^{n} f_\lambda(x_i - \xi).$$

Note that the support of $p_{(\lambda,\xi)}$ is restricted to $\{x_{(1)} \geq \xi\}$.

Suppose interest is in λ, so that ξ is a nuisance parameter. Note that the set of integers is a subgroup of $\mathbb{R}$ under addition (Example D.5), so ξ is interpretable as a location parameter confined to integers. Therefore, as in Example 10.8 we may therefore use maximal invariant $M(\mathbf{X}) = (X_1 - X_n, \ldots, X_{n-1} - X_n) = (Y_1, \ldots, Y_{n-1})$. This density may be expressed in terms of the original data as

$$\begin{aligned} \bar{p}_\lambda(\mathbf{x}) &= \sum_{u \in \mathbb{I}} \prod_{i=1}^{n} f_\lambda(x_i - \xi + u) \\ &= \sum_{u \geq (\xi - x_{(1)})} \prod_{i=1}^{n} f_\lambda(x_i - \xi + u) \\ &= \sum_{u \geq 0} \prod_{i=1}^{n} f_\lambda(x_i - x_{(1)} + u). \end{aligned}$$

As expected, $\bar{p}_\lambda$ does not depend on ξ. It is also naturally expressed in terms of the maximal invariant $M^*(\mathbf{X}) = (X_1 - X_{(1)}, \ldots, X_n - X_{(1)})$, although $M(\mathbf{X})$ and $M^*(\mathbf{X})$ are necessarily equivalent.

At this point, we can apply Neyman-Pearson theory to the density $\bar{p}_\lambda$ to obtain MP invariant tests. Suppose we wish to test $H_o : \lambda = \lambda_0$ versus $H_a : \lambda > \lambda_0$. Then write the density of $M^*(X)$, denoted as $\bar{p}_\lambda$, explicitly as

$$\bar{p}_\lambda(\mathbf{x}) = \sum_{u \geq 0} \prod_{i=1}^{n} \frac{\lambda^{x_i - x_{(1)} + u}}{(x_i - x_{(1)} + u)!} e^{-\lambda}.$$

This density depends on $M^*(\mathbf{X})$ and λ in a rather complex way, so we construct a LMP test. The derivative of $\bar{p}_\lambda(\mathbf{x})$ is given by

$$\begin{aligned} \frac{d\bar{p}_\lambda(\mathbf{x})}{d\lambda} &= \sum_{u \geq 0} \sum_{i=1}^{n} \frac{x_i - x_{(1)} + u - \lambda}{\lambda} \prod_{i=1}^{n} \frac{\lambda^{x_i - x_{(1)} + u}}{(x_i - x_{(1)} + u)!} e^{-\lambda} \\ &= \lambda^{-1}(\bar{S} - n\lambda)\bar{p}_\lambda(\mathbf{x}) + \lambda^{-1} \sum_{u \geq 0} u \prod_{i=1}^{n} f_\lambda(x_i - x_{(1)} + u), \end{aligned}$$

where $\bar{S} = \sum_{i=1}^{n} x_i - x_{(1)}$. Then, following Section 10.5.1, the LMP has rejection region

$$\left(\frac{\bar{S}}{n} - \lambda_0\right) + \frac{\sum_{u \geq 0} u \prod_{i=1}^{n} f_{\lambda_0}(x_i - x_{(1)} + u)}{\bar{p}_{\lambda_0}(\mathbf{x})} > k.$$ ∎

10.7 Permutation Invariant Tests

Location or scale invariance is modeled using homomorphisms from a single group acting on the various elements of an inference problem. Those single groups were, respectively, $(\mathbb{R}, +)$ and $((0, \infty), \times)$. Now consider the permutation group G acting on labels $\{1, \ldots, n\}$ (Example D.4) which acts on a family of densities $\mathcal{P}$ on sample space $\mathcal{X}$ in a way which generates an invariant density family (Section 9.4).

This constraint can arise naturally. In Almudevar (2001), a method for testing relatedness hypotheses based on Mendelian genetic data was developed based on the principle of invariance (Section 3.13.3). Suppose genetic data is collected from a single generation of some species. To fix ideas, we will assume that parental data is not available. Possibly, existence of sibling structure of any kind would be a significant finding. The null hypothesis would be that all sampled individuals are unrelated. Various alternative hypothesis may be considered. The most complex would consist of all partitions into full sibling groups, possibly with additional half-sibling relationship structure imposed on the partition. On the other hand, if sibling structure is expected to be minimal, an alternative hypothesis may be existence of a single sibling pair, all other pairs being unrelated. In either case, it may be appropriate, in the absence of any prior information, for the inference to be invariant with respect to any permutation of the labels.

Suppose we may identify a parameter $\theta_0 \in \Theta$ which is invariant, in the sense that $g\theta_0 = \theta_0$ for all $g \in G$. We take this to be null hypothesis $H_o : \theta = \theta_0$. Then select $\theta' \neq \theta_0$ to define an invariant alternative hypothesis (Definition 9.4)

$$\Theta_a = \{g\theta' : g \in G\}$$

Note that Θ_a need not have the same cardinality $n!$ as G, but if it has cardinality 1, then we have two simple hypothesis, and the matter is easily resolved using a NP test. If this is not the case, then a UMP test will not exist unless the MP tests for H_o against all simple alternatives $\theta \in \Theta_a$ are equivalent. We anticipate that this is not the case, and so we impose invariance by constructing a test subject to the constraint that the power is equal for all alternatives in Θ_a.

This can be done using the generalized Neyman-Pearson lemma. The approach here follows that of Almudevar (2001). We can begin as in Section 10.5.2 with the test size constraint

$$\int_{\tilde{x} \in \mathcal{X}} \delta(\tilde{x}) p_{\theta_0}(\tilde{x}) d\nu = \alpha.$$

This gives $g_1(\tilde{x}) = p_{\theta_0}(\tilde{x})$ and $b_1 = \alpha$. Next, suppose Θ_a contains m simple hypotheses, labeled $\theta_1, \ldots, \theta_m$. To force invariance, we constrain all alternatives to have the same power as θ_1. This introduces $m - 1$ additional constraints:

$$\int_{\tilde{x} \in \mathcal{X}} \delta(\tilde{x}) p_{\theta_1}(\tilde{x}) d\nu - \int_{\tilde{x} \in \mathcal{X}} \delta(\tilde{x}) p_{\theta_j}(\tilde{x}) = 0, \;\; j = 2, \ldots, m,$$

so we set

$$g_j(\tilde{x}) = p_{\theta_1} - p_{\theta_j}$$

and $b_j = 0$ $j = 2, \ldots, m$. Finally, we need only optimize the power for a single alternative, say θ_1, so we set $g_{m+1}(\tilde{x}) = p_{\theta_1}$. The most powerful test then has rejection region

$$p_{\theta_1}(\tilde{x}) > k_1 p_{\theta_0}(\tilde{x}) + \sum_{j=2}^{m} k_j (p_{\theta_1}(\tilde{x}) - p_{\theta_j}(\tilde{x})) \tag{10.15}$$

provided values for $k_1, \ldots, k_m$ can be found which satisfy the m constraints. To do this, set $k_j = -1/m$, $j \geq 2$. The rejection region is now

$$\Lambda(\tilde{x}) = \sum_{j=1}^{m} \frac{p_{\theta_j}(\tilde{x})}{p_{\theta_0}(\tilde{x})} > k. \tag{10.16}$$

The test statistic is therefore the sum of the likelihood ratios used to test H_o against each simple hypothesis in H_a. We only need to note at this point that if group invariance holds, the constraints will also hold, provided the test statistic $\Lambda(\tilde{x})$ is permutation invariant. In this case (10.16) defines the UMP permutation invariant test. The method is demonstrated in the next example.

Example 10.10 (UMP Permutation Invariant Test for Support of a Multinomial Model) A coupon collector expects to see coupons of type $1, \ldots, m$ in frequencies $N_1, \ldots, N_m$ possessing a multinomial density with probabilities $\mathbf{P} = (p_1, \ldots, p_m) = (1/m, \ldots, 1/m)$. This defines null hypothesis H_o. Suppose we consider an alternative hypothesis of the form

$$H_k : k \text{ of the probabilities } (p_1, \ldots, p_m) \text{ are zero, the rest are equal.}$$

Essentially H_k means k of the coupons are no longer being issued.

The problem is clearly permutation invariant. Define statistic

$$T = \sum_{j=1}^{m} I\{N_j = 0\},$$

which is the number of coupons not observed.

We can consider two problems. First, we will derive a MP invariant test for null hypothesis H_o against alternative H_k for some fixed k. We will then consider whether or not this can be used to construct a UMP test for alternative hypothesis

$$H_a : \text{At least one coupon type is not being issued.}$$

Suppose $n = \sum_j N_j$. Consider simple alternative hypothesis

$$\theta_k = (p_1, \ldots, p_{m-k}, p_{m-k+1}, \ldots, p_m) = (1/(m-k), \ldots, 1/(m-k), 0, \ldots, 0),$$

and the null hypothesis θ_0. From the multinomial density we can write

$$\frac{p_{\theta_k}}{p_{\theta_0}} = \frac{(1/(m-k))^{[N_1 + \ldots + N_{m-k}]} \times 0^{[N_{m-k+1} + \ldots + N_m]}}{(1/m)^n}.$$

Interpreting $0^0 = 1$ we have

$$\frac{p_{\theta_k}}{p_{\theta_0}} = \begin{cases} \left[\frac{m}{m-k}\right]^n & ; \quad N_j = 0, \ \ j = m-k+1, \ldots, m \\ 0 & ; \quad \text{otherwise} \end{cases}$$

Let G be the class of all permutations of m labels. Then the alternative hypothesis may be written

$$\Theta_k = \{g\theta_k : g \in G\}.$$

There are $\binom{m}{k}$ simple hypotheses in Θ_k, one for each subset from $\{1, \ldots, m\}$ of size k. Let $\mathcal{S}_{m,k}$ be the collection of all such subsets. Then the MP invariant test statistic is

$$\Lambda = \sum_{\theta' \in \Theta_k} \frac{p_{\theta'}}{p_{\theta_0}} = \left[\frac{m}{m-k}\right]^n \sum_{S \in \mathcal{S}_{m,k}} I\{N_j = 0, \ \ j \in S\}.$$

Then consider again the statistic $T = \sum_{j=1}^{m} I\{N_j = 0\}$. If $T < k$, then all terms of Λ are zero. Otherwise, if $T \geq k$, then there are $\binom{T}{k}$ subsets $S \in \mathcal{S}_{m,k}$ for which $N_j = 0$ for all $j \in S$. Therefore

$$\Lambda = \begin{cases} \left[\frac{m}{m-k}\right]^n \binom{T}{k} & ; \quad T \geq k \\ 0 & ; \quad T < k \end{cases}.$$

Since $\binom{T}{k}$ is increasing in T for fixed k, the MP invariant rejection region is equivalent to

$$T \times I\{T \geq k\} \geq t_\alpha \tag{10.17}$$

for any critical value t_α given by $P_{\theta_0}(TI\{T \geq k\} \geq t_\alpha) = \alpha$.

Next, suppose we wish to test

$$H_o \text{ against } H_a : \text{At least one coupon type is not being issued.} \tag{10.18}$$

Then H_a is defined by alternatives $\cup_{k=1}^{m-1}\Theta_k$ (reasonably, if there is a sample to observe, there must be at least one coupon remaining). Suppose we have rejection region

$$T \geq t_\alpha \tag{10.19}$$

for some critical value t_α given by $P_{\theta_0}(T \geq t_\alpha) = \alpha$. Suppose $k \leq t_\alpha$. Then the tests defined by (10.17) and (10.19) are equivalent. If $k > t_\alpha$ then the test will reject with probability 1 for any alternative in Θ_k. Thus the rejection region (10.19) is a MP permutation invariant test for each alternative hypothesis Θ_k, and is therefore UMP for the alternatives in (10.18). ■

A considerable part of hypothesis testing methodology is concerned with the problem of the homogeneity of several distributions $F_1, \ldots, F_m$. A null hypothesis will typically be $H_o : F_j = F', j = 1, \ldots, m$, where F' is assumed to belong to some class. The alternative hypothesis may then be $H_a : F_i \neq F_j$ for some i, j.

The one-way ANOVA model of Section 6.2.2 is one example. Such tests cannot in general be UMP. To see this, suppose for an ANOVA model we have $k = 3$ treatments, with treatment means (μ_1, μ_2, μ_3). We have hypotheses $H_o : \mu_1 = \mu_2 = \mu_3$ and $H_a : \mu_i \neq \mu_j$ for some i, j. Then any set of treatment means satisfying $\mu_1 = \mu_2 \neq \mu_3$ is in Θ_a. Of course, if we could anticipate this form of alternative we would pool treatments 1 and 2, then do a two-sample test specialized for alternative hypothesis $H_a : \mu_1 = \mu_2 \neq \mu_3$. This test would necessarily be more powerful for this alternative than the conventional F test of Theorem 6.7. We have here another form of the "problem of the favored hypothesis" introduced in Section 7.2. The problem here is that the specialized test is not invariant to permutations of the treatment labels. However, as for the location and scale invariant models, the problem of the favored hypothesis for tests of homogeneity can be voided by imposing this form of invariance (but consider the comments regarding invariance in Section 4.7).

The next example considers a form of test for homogeneity of binomial parameters. It is important to note that permutation invariant tests may be designed to anticipate specific alternative hypotheses.

Example 10.11 (UMP Permutation Invariant Test for Homogeneity of Binomial Parameters) Suppose we are given independent random variables $\mathbf{X} = (X_1, \ldots, X_m)$, with $X_i \sim bin(n_i, p_i)$. We are given a null hypothesis of identical distributions,

$$H_o : (n_i, p_i) = (n, p), \quad i = 1, \ldots, m.$$

The hypothesis test is motivated by the possibility that a subset of the observations differ

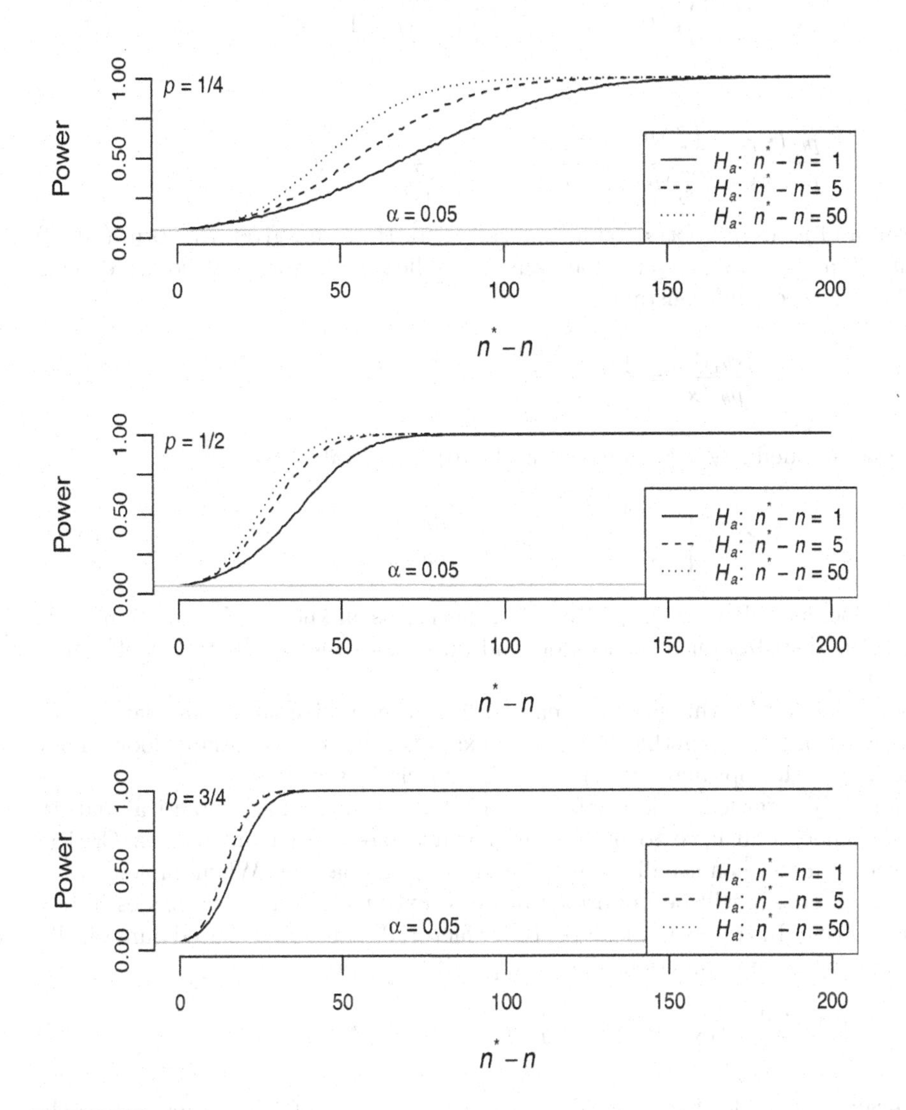

FIGURE 10.3
Power curves for alternatives $n^* - n$ for Example 10.11.

from the distribution $binom(n, p)$ defining the null hypothesis. To fix ideas, suppose for some nonempty subset $I^* \subset \{1, \ldots, m\}$ we have $X_j \sim bin(n^*, p)$ for $j \in I^*$, where $n^* > n$, while the observations remain independent. At this point, an invariant alternative hypothesis is defined by the orbit formed from all permutations of a simple hypothesis θ' (Section D.5). The notion of a locally most powerful test can be applied here, at least by analogy, by selecting θ' not too far from H_o. We can, for example, take $I^* = \{1\}$, then consider the choice of $n^* > n$.

Under the null hypothesis we have density

$$p_{\theta_0}(\mathbf{x}) = \prod_{i=1}^{m} \binom{n}{x_i} p^{x_i}(1-p)^{n-x_i}. \tag{10.20}$$

Suppose alternative θ_j is defined by $I^* = \{j\}$. Then

$$p_{\theta_j}(\mathbf{x}) = \binom{n^*}{x_j} p^{x_j}(1-p)^{n^*-x_j} \prod_{i \neq j} \binom{n}{x_i} p^{x_i}(1-p)^{n-x_i}. \tag{10.21}$$

This gives ratio

$$\frac{p_{\theta_j}(\mathbf{x})}{p_{\theta_0}(\mathbf{x})} = \frac{\binom{n}{x_j} p^{x_j}(1-p)^{n-x_j}}{\binom{n^*}{x_j} p^{x_j}(1-p)^{n^*-x_j}} = \frac{\binom{n}{x_j}}{\binom{n^*}{x_j}}(1-p)^{n^*-n}$$

for $\max_j x_j \leq n$, and $p_{\theta_j}(\mathbf{x})/p_{\theta_0}(\mathbf{x}) = \infty$ for $\max_j x_j > n$. However, given the form (10.16) of the UMP invariant test, we may eliminate any multiplicative factors that do not depend on $\mathbf{x}$, giving, for some positive constant c,

$$\frac{p_{\theta_j}(\mathbf{x})}{p_{\theta_0}(\mathbf{x})} = \begin{cases} c\frac{(n-x_j)!}{(n^*-x_j)!} & ; \quad x_j \leq n \\ \infty & ; \quad x_j > n \end{cases}.$$

Then the test statistic defining rejection region (10.16) is equivalent to

$$\Lambda(\mathbf{x}) = \begin{cases} \sum_{i=1}^m \frac{(n-x_i)!}{(n^*-x_i)!} & ; \quad \max x_i \leq n \\ \infty & ; \quad \max x_i > n \end{cases}. \tag{10.22}$$

Interestingly, Λ does not depend on p. However, since p is unknown H_o is a composite hypothesis, and the distribution of Λ does depend on p. This means the test itself is not independent of p.

Test statistics derived by this method tend to be rather unlike those associated with standard parametric inference problems. In this example, we are essentially looking for outliers. Accordingly, the optimal test statistic is especially sensitive to large values, a property not normally regarded as desirable in inference. This means, as a practical matter, computational methods will have to play an important role in such procedures. Critical values and power estimates can usually be estimated using simulations. When implementing such numerical methods, it will be important to avoid extremely large magnitudes in any numerical computation. In this example, large order factorials can be calculated numerically using a gamma or log gamma function. For example, if the expression

$$\frac{(n-x_j)!}{(n^*-x_j)!} = \exp\left(\log\Gamma(n-x_j+1) - \log\Gamma(n^*-x_j+1)\right)$$

relies on the evaluation of $\log\Gamma(k)$ (avoiding direct calculation of $\Gamma(k)$) then extremely large magnitudes can be avoided. Stirling's approximation can be useful in this type of computation.

To demonstrate the test, we will set $(m, n) = (10, 100)$, and consider a range of p and n^*. Critical values and estimates of power for the test (10.16) can be obtained without too much trouble using simulations. Figure 10.3 shows power curves for values $p = (1/4, 1/2, 3/4)$ and alternatives $n^* - n = 1, 5, 50$.

What we might refer to as the LMP test, based on alternative $H_a : n^* - n = 1$, is clearly a poor choice, exhibiting much lower power over all alternatives above a relatively small value of $n^* - n$. Depending on the parameters, an alternative will not have sufficient power until it exceeds some value. In this example, that threshold varies roughly in the range $n^* - n \in (25, 100)$. Therefore, it would be reasonable to optimize the test for alternatives in that range. Of course, this might be thought of as a refinement, rather than a rejection, of the LMP test method, in that we are maximizing power for the alternative closest to H_o which of interest in any sense.

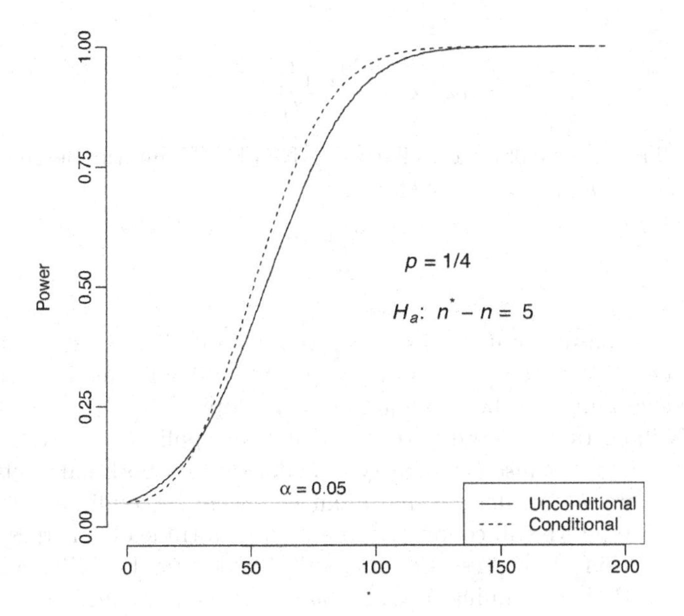

FIGURE 10.4
Power curves for alternatives $n^* - n = 5$, $p = 1/4$, using both conditional and unconditional methods (Example 10.11).

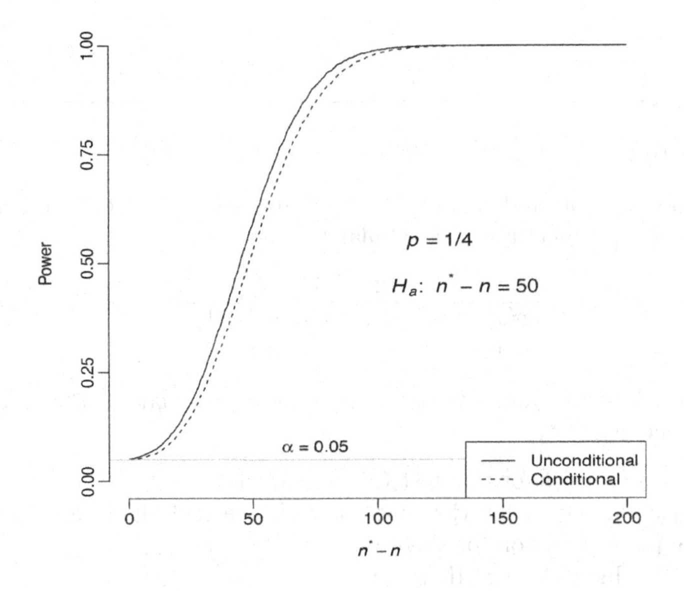

FIGURE 10.5
Power curves for alternatives $n^* - n = 50$, $p = 1/4$, using both conditional and unconditional methods (Example 10.11).

We next modify the test by treating p as a nuisance parameter. Following Section 3.12, $S = \sum_i X_i$ is sufficient for p under all the alternatives considered. The densities (10.20)-(10.21) for the various alternatives can be modified accordingly. Under H_o the density of $\mathbf{X} \mid \{S = s\}$ is given by the multivariate hypergeometric density $multihyper(s, n, \ldots, n)$, given by

$$p_{\theta_0}(\mathbf{x}) = \frac{\prod_{i=1}^{n}\binom{n}{x_i}}{\binom{mn}{s}},$$

where $s = \sum_i x_i$. The density of alternative θ_j is given by the multihypergeometric density *multihyper*$(s, n, \ldots, n, n^*, n, \ldots, n)$, given by

$$p_{\theta_j}(\mathbf{x}) = \frac{\binom{n^*}{x_j}\prod_{i \neq j}\binom{n}{x_i}}{\binom{mn+n^*-n}{s}}.$$

But this yields the same test statistic as that given in Equation (10.22). Therefore, if the test is conditional on $S = s$, the only change is that the critical value is calculated using a multihypergeometric sample, rather than a binomial sample.

However, it is important to note that this is not an application of the conditionality (or ancillarity) principle, because the density of S depends on both parameters of interest (n, n^*), and is therefore not ancillary. If p is unknown, then the conditional test is justified. But, if p is known, should we still condition? In this case, a trade off emerges. Conditioning will reduce the variation of the test statistic $\Lambda(\mathbf{X})$, which tends to increase power. But conditioning on a statistic containing information about the parameter of interest (that is, any statistic with a density dependent on that parameter) will tend to reduce power.

To see this, compare the power curves for the conditional an unconditional tests for $p = 1/4$, $H_a : n^* - n = 5$ (Figure 10.4) and $p = 1/4$, $H_a : n^* - n = 50$ (Figure 10.5). Clearly, neither procedure is uniformly better, although they yield similar power. ■

10.8 Problems

Problem 10.1 Let X be a random variable on sample space $\mathcal{X} = \{1, 2, 3, 4, 5\}$. Assume X either has density $q(x)$ or $p(x)$ given in tabular form:

$x =$	1	2	3	4	5
$p(x)$	0	0.4	0.3	0.2	0.1
$q(x)$	0.4	0.3	0.2	0.1	0

Suppose we wish to test null and alternative hypotheses $H_o : p(x)$ is the true density and $H_a : q(x)$ is the true density .

(a) Give the likelihood ratio for this test (in tabular form).
(b) Give the rejection region for the most powerful test of significance level 0.15, and calculate the power function for this test.
(c) What is the P-value if $X = 2$? If $X = 1$?
(d) Construct a (possibly randomized) test of size $\alpha = 0.15$.

Problem 10.2 Consider two densities on sample space $\{1, 2, 3, 4, 5\}$:

x	1	2	3	4	5
$f_0(x)$	0.2	0.2	0.2	0.2	0.2
$f_1(x)$	0.4	0.3	0.2	0.05	0.05

and suppose X_1, X_2 are independent random variables with density f.

(a) Use the Neyman-Pearson lemma to find the most powerful test of $H_0 : f(x) = f_0(x)$ against $H_1 : f(x) = f_1(x)$ of size $\alpha = 0.12$. Give the rejection region precisely.

(b) What is the power of this test?

Problem 10.3 If a sample of size n is selected randomly *without replacement* from N objects, of which D are defective, then the number of defectives X in the sample has hypergeometric density $P(X = x) = P_D(x) = \binom{N}{n}^{-1}\binom{D}{x}\binom{N-D}{n-x}$, for $\max(0, n + D - N) \leq x \leq \min(n, D)$.

(a) Show that for $d > 0$, $r(x) = P_{D+d}(x)/P_D(x)$ is increasing in x, so that $P_D(x)$ is a monotone likelihood ratio (MLR) model. **HINT**: Show this first for $d = 1$.
(b) Derive a UMP test for the hypotheses $H_o : D \leq D_0$ and $H_a : D > D_0$ for some fixed D_0, based on X.

Problem 10.4 Suppose $X_1, \ldots, X_n$ is an *iid* sample from $N(\theta, \theta)$, $\theta > 0$. Show that a test with rejection region $R = \{\sum_{i=1}^n X_i^2 > K\}$ for some constant K is UMP for testing $H_o : \theta = \theta_0$ against alternative $H_a : \theta > \theta_0$ for any $\theta_0 > 0$. Precisely what distribution would be used to determine critical values for this test?

Problem 10.5 We are given an *iid* sample $X_1, \ldots, X_n$ with $X_1 \sim beta(a, a)$.

(a) Find the form of the most powerful size α test of $H_o : a = 1$ against $H_a : a = a'$, where $a' > 1$. Explain how you would calculate the exact rejection region.
(b) Does a UMP test for $H_o : a = 1$ against $H_a : a > 1$ exist?

Problem 10.6 Suppose we have a single observation $X \sim nb(r, p)$, where r is known.

(a) Find UMP tests for $H_o : p \leq p_0$ against $H_a : p > p_0$, and for $H_o : p \geq p_0$ against $H_a : p < p_0$.
(b) Show how these tests can be inverted to find a level $1 - \alpha$ confidence interval for p.

Problem 10.7 Consider the density function $f_\theta(x) = \theta^{-1}x^{\theta-1}I\{0 < x < 1\}$ for $\theta > 0$. Suppose we wish to test $H_o : \theta \leq 1$ against $H_a : \theta > 1$ based on a single observation X from f_θ.

(a) Find the size α and the power curve (explicitly) for the test which rejects H_0 when $X \geq 1/2$.
(b) Is this a monotone likelihood ratio family?
(c) Does there exist a UMP size α test for $H_o : \theta \leq 1$ against $H_a : \theta > 1$? If so, find it (expressed directly in terms of α). Otherwise, why not?

Problem 10.8 Let $X \sim logistic(\theta, 1)$ be a single observation.

(a) Show that this model possesses the monotone likelihood ratio (MLR) property.
(b) Give the form of a UMP test for hypotheses $H_o : \theta \leq 0$ versus $H_a : \theta > 0$.

Problem 10.9 Suppose $X \sim bin(n, p)$, where p is fixed and n is the unknown parameter. Is this a monotone likelihood ratio family?

Problem 10.10 Suppose $X_1, \ldots, X_n$ is an independent sample with $X_i \sim pois(\lambda\rho^i)$. Assume λ is known and $\rho > 0$.

(a) Derive the MP test for simple hypotheses $H_o : \rho = \rho_0$ and $H_a : \rho = \rho_1$, $\rho_1 > \rho_0$. Show that a UMP test exists for hypotheses $H_o : \rho = \rho_0$ and $H_a : \rho > \rho_0$.
(b) Derive the LMP test for hypotheses $H_o : \rho = \rho_0$ and $H_a : \rho > \rho_0$. Is this the same as the UMP test?

Problem 10.11 Suppose $X_1, \ldots, X_n$ is an *iid* sample with $X_1 \sim N(\theta, \tau(\theta))$, $\theta > 0$, where $\tau(\theta)$ is a nondecreasing function of θ.

(a) Derive the LMP test for hypotheses $H_o : \theta = \theta_0$ and $H_a : \theta > \theta_0$.
(b) Under what conditions on $\tau(\theta)$ will a UMP test exist?

Problem 10.12 Suppose $X_1, \ldots, X_n$ is an *iid* sample with $X_1 \sim logistic(\theta, 1)$. Derive the locally most powerful test for hypotheses $H_o : \theta = \theta_0$ and $H_a : \theta > \theta_0$.

Problem 10.13 The T distribution with $\nu > 0$ degrees of freedom has density

$$f(x) = \frac{\Gamma((\nu+1)/2)}{\sqrt{\nu\pi}\,\Gamma(\nu/2)}\left[1 + \frac{x^2}{\nu}\right]^{-(\nu+1)/2}, \quad x \in \mathbb{R}$$

Let $p_0(\mathbf{x})$ be the density of an *iid* sample of size n from density $f(x)$. Then construct a location parameter model $p_\theta(\mathbf{x}) = p_0(\mathbf{x} - \theta)$, $\theta \in \Theta = \mathbb{R}$. Assume ν is known.

(a) Does a UMP test exist for $H_o : \theta = \theta_0$ against $H_a : \theta > \theta_0$.
(b) Construct a locally most powerful (LMP) size α test for $H_o : \theta = \theta_0$ against $H_a : \theta > \theta_0$.

Problem 10.14 Let $\mathbf{N} = (N_1, N_2, N_3)$ be a multinomial vector for probability distribution $\mathbf{P} = (1/3 + \alpha, 1/3, 1/3 - \alpha)$, $\alpha \in \Theta = (-1/3, 1/3)$.

(a) Give the MP test for hypotheses $H_o : \alpha = 0$ versus $H_a : \alpha = \alpha_0$, $\alpha_0 \neq 0$.
(b) Give the LMP test for hypotheses $H_o : \alpha = 0$ versus $H_a : \alpha > 0$. Does a UMP test exist?
(c) Explain in detail how a critical value would be calculated for the LMP test.
(d) Is N_2 ancillary for α? Explain your answer.
(e) Suppose we condition on N_2. Explain in this case how a critical value would be calculated for the LMP test.

Problem 10.15 Suppose we have an *iid* sample $\mathbf{X} = (X_1, \ldots, X_n)$ from $pareto(\tau, \kappa)$. Set $\theta = (\tau, \kappa)$.

(a) Suppose κ is known. Show that the test $\delta(\mathbf{X}) = I\{X_{(1)} > t_\alpha\}$, is a UMP size α test for hypotheses $H_o : \tau = \tau_0$ and $H_a : \tau > \tau_0$, where t_α is selected to satisfy $E_{\tau_0}[\delta(\mathbf{X})] = \alpha$.
(b) Suppose τ is known. Show that the resulting parametric family has a monotone likelihood ratio with respect to shape parameter κ. Then derive a UMP size α test for hypotheses $H_o : \kappa = \kappa_0$ and $H_a : \kappa > \kappa_0$.
(c) Suppose τ is known. Derive a UMP unbiased size α test for hypotheses $H_o : \kappa = \kappa_0$ and $H_a : \kappa \neq \kappa_0$. **HINT:** Show that $\log(X_1/\tau)$ has an exponential distribution with rate parameter κ.
(d) In general, whenever a two-sided test for $H_o : \theta = \theta_0$ against $H_a : \theta \neq \theta_0$ is based on statistic $T(\mathbf{X})$, which is in some monotone relationship with θ (the MLR property, for example), a common practice is to reject H_o when $T(\mathbf{X}) < q_{\alpha/2}$ or $T(\mathbf{X}) > q_{1-\alpha/2}$, where q_p is the p-quantile of $T(\mathbf{X})$ under H_o. Alternatively, two-sided tests may be symmetric, in the sense that H_o is rejected when $|T(\mathbf{X}) - \tau_T| > K_\alpha$, where τ_T is the mean of $T(\mathbf{X})$ under H_o (which need not be θ_0), and K_α is selected so that the test has size α. Suppose $n = 5$, $\kappa_0 = 10$ and $\alpha = 0.05$. Give a numerical representation of the UMP unbiased test of Part (c). Then derive both the quantile-based and symmetric two-sided tests just described, again using size $\alpha = 0.05$. Is the UMP unbiased test equivalent to either of the alternative tests? Numerical methods may be employed.
(e) Now suppose we wish to test $H_o : \kappa = \kappa_0$ and $H_a : \kappa > \kappa_0$, but scale parameter τ is unknown. How would you construct a UMP scale invariant test? Give a brief outline of your argument.

Problem 10.16 A multinomial sample $\mathbf{N} = (N_1, N_2, N_3)$ of sample size n with $m = 3$ categories is observed. The multinomial probabilities are parametrized as $\mathbf{P} = (p_1, p_2, p_3) = (e^{-2\theta}, 2e^{-\theta}(1 - e^{-\theta}), (1 - e^{-\theta})^2)$, for some $\theta \in \Theta = (0, \infty)$.

(a) Derive a uniformly most powerful (UMP) level α test for null and alternative hypotheses $H_o : \theta = \theta_0$ and $H_a : \theta > \theta_0$, for some fixed $\theta_0 \in (0, 1)$. Explain how this test would be randomized to attain some exact size α.

(b) Describe how a UMP unbiased level α test for null and alternative hypotheses $H_o : \theta = \theta_0$ and $H_a : \theta \neq \theta_0$ would be constructed, for some fixed $\theta_0 \in (0, 1)$. You do not have to derive the test exactly. Rather, give the form of the test as a randomized decision rule $\delta : \mathcal{X} \to \infty$, where $\mathcal{X}$ is the sample space from which the observation is sampled, and $\delta(\mathbf{X})$ is the probability of rejecting H_o given observation $\mathbf{X} \in \mathcal{X}$. State clearly the parameters needed to define δ, and how they would be calculated.

Problem 10.17 Suppose $\mathbf{X} = (X_1, \ldots, X_n)$ is an independent sample from $N(0, \sigma^2)$. Following Example 10.7 show how a UMP unbiased test for hypotheses $H_o : \sigma^2 = \sigma_0^2$ and $H_a : \sigma^2 \neq \sigma_0^2$ can be constructed. Give the rejection region in numerical form for $n = 10$ and $\sigma_0 = 1$.

Problem 10.18 Suppose $\mathbf{X} = (X_1, \ldots, X_n)$ is an *iid* sample with $X_1 \sim f(x - \xi)$, where $\xi \in \mathbb{R}$ is a nuisance parameter. Suppose we wish to test hypothesis

$$H_o : \ f \text{ is the uniform density on } (0, 1),$$

against alternative

$$H_a : \ f \text{ is the symmetric triangular density on } (0, 1).$$

Use the method of Example 10.8 to determine the most powerful location invariant test.

Problem 10.19 Suppose $\mathbf{X} = (X_1, \ldots, X_n)$ is an *iid* sample with $X_1 \sim \tau^{-1} f(x/\tau)$, where $\tau \in \mathbb{R}_{>0}$ is a nuisance parameter. Assume $X_1 > 0$ *wp*1. Suppose we wish to test hypothesis $H_o : f = f_0$ against alternative $H_a : f = f_1$.

(a) Following the method of Example 10.8 give a general form for a MP scale invariant test.

(b) Specialize Part (a) to the case $f_0 \sim exp(\tau)$, $f_1 \sim gamma(\tau, \alpha)$, for some fixed $\alpha \neq 1$.

Problem 10.20 Suppose we are given an *iid* sample $\mathbf{X} = (X_1, \ldots, X_n)$, with $X_1 \sim gamma(\tau, \kappa)$. Suppose we wish to test null and alternative hypotheses $H_o : \kappa = \kappa_0$ and $H_a : \kappa > \kappa_0$ for some fixed $\kappa_0 > 0$. Assume τ is unknown, and is therefore a nuisance parameter. Derive a uniformly most power scale invariant level α test. **HINT:** Suppose we consider transformation

$$\mathbf{Y} = \left(\frac{X_1}{S}, \frac{X_2}{S}, \ldots, \frac{X_n}{S} \right) = (Y_1, \ldots, Y_n).$$

where $S = \sum_{i=1}^n X_i$. Then consider the related one-to-one transformation $\mathbf{Y}^* = (Y_1, \ldots, Y_{n-1}, S)$. Using standard transformation methods determine the density of $\mathbf{Y}^*$, then the marginal density of $(Y_1, \ldots, Y_{n-1})$.

Problem 10.21 Suppose $\mathbf{X} = (X_1, \ldots, X_m)$ are independent normally distributed random variables, with $X_i \sim N(\mu_i, \sigma^2)$. Assume σ^2 is known. Consider null hypothesis $H_o : \mu_i = \mu_o$, $i = 1, \ldots, m$ for some μ_o. Then consider the alternative hypothesis defined by

$$H_a : \mu_j = \mu_a \text{ for some } j, \ \ \mu_i = \mu_o \text{ for all } i \neq j,$$

where $\mu_a \neq \mu_o$. In other words, H_a is the hypothesis that all means μ_i equal μ_o except for one $\mu_j = \mu_a$, where j can be any index $1, \ldots, m$.

(a) Assuming μ_o, μ_a and σ^2 are fixed, derive the MP permutation invariant test for H_o against H_a.
(b) Suppose H_a is modified to allow any $\mu_a > \mu_o$, that is,

$$H_a : \mu_j > \mu_o \text{ for some } j, \quad \mu_i = \mu_o \text{ for all } i \neq j.$$

Assume σ^2 is still fixed. Verify that a UMP invariant test does not exist for H_a. Then verify that for small $\epsilon = \mu_a - \mu_o > 0$ this test is approximately equivalent to $\sum_{j=1}^{m} X_j > k$. The argument need not be rigorous. Just base your argument on a first order Taylor's series approximation of $f(x) = e^x$.

11

Limit Theorems

11.1 Introduction

Large sample, or asymptotic, theory concerns the limiting properties of a statistical methodology as a sample size approaches infinity. This is covered in the final section of this volume, from the current chapter to Chapter 14. In this first chapter on the subject we cover the relevant background in probability theory. More than any other chapter of this volume, this takes the form of a series of technical results, most of which are not directly related to inference. Most of this will be familiar to the reader with a strong background in probability theory.

While there is only one sense in which the sequence of real numbers $1/2, 1/3, 1/4, \ldots$ converges to zero, there are multiple ways of defining the convergence of a sequence of random variables. This is the subject of Section 11.2, while Section 11.3 covers limits of expected values.

Section 11.4 introduces uniform integrability, which allows us to deal with an important technical issue which is sometimes overlooked. Much of asymptotic theory takes the following form. Based on sample $X_1, \ldots, X_n$ we construct estimator $\hat{\theta}_n$ of θ. We imagine the sample size n growing unboundedly, so we have, at least conceptually, a sequence of estimators $\hat{\theta}_n$, $n \geq 1$. Provided certain regularity conditions are met, we may be able to claim that $\sqrt{n}(\hat{\theta}_n - \theta)$ converges in distribution to a normal distribution. Such a statement allows us to conclude that $\sqrt{n}(\hat{\theta}_n - \theta)$ is approximately normally distributed, from which an inference statement can be formed. However, we may also wish to assume that the *moments* of $\sqrt{n}(\hat{\theta}_n - \theta)$ converge to the moments of the limiting distribution, but this need not folllow from convergence in distribution (Example 11.2). The concept of uniform integrability allows a resolution of this problem which serves our purposes.

Suppose we are given a sample $X_1, \ldots, X_n$ from a distribution with mean μ. The law of large numbers (Section 11.5) states that the sample average approaches μ as $n \to \infty$. This holds under a variety of conditions on the sample.

Weak convergence (or convergence in distribution) refers to the convergence not of a sequence of random variables, but of their distributions. The main theorem characterizing this limit process is the Portmanteau theorem, introduced in Section 11.6.

Sections 11.7 and 11.8 extend the modes of convergence of random variables to metric spaces, which allows us to describe convergence of random vectors. Two theorems which are widely used in asymptotic theory are introduced here, the continuous mapping theorem and Slutsky's theorem.

The role played by the moment generating function in the theory of weak convergence is introduced in Section 11.9, while the central limit theory is introduced in Section 11.10.Weak convergence of random vectors is considered in Section 11.11.

DOI: 10.1201/9781003049340-11

11.2 Limits of Sequences of Random Variables

There are multiple ways of defining a limiting process associated with a stochastic model, but these can be regarded as one of two types, those concerning sequences of probability measures, and those concerning sequences of random variables $X_1, X_2, \ldots$ considered jointly. These definitions have certain hierarchical relationships which play an important role in large sample theory.

Notions of the convergence of random variables can be defined in terms of the converegnce of measurable functions. Suppose $X_1, X_2, \ldots$ is a sequence of random variables defined on a probability measure space $(\Omega, \mathcal{F}, P)$. The following definitions of convergence are standard:

Definition 11.1 (Modes of Convergence of Random Variables) Let X and $X_1, X_2, \ldots$ be random variables on a probability measure space $(\Omega, \mathcal{F}, P)$.

(i) If $P(|X_n - X| > \epsilon) \to 0$ for all $\epsilon > 0$, then X_n converges in probability to X. This is denoted as $X_n \overset{i.p.}{\to} X$ (convergence in probability (*i.p.*)).
(ii) If $E[|X_n - X|^p] \to 0$ then X_n converges in L^p to X, denoted as $X_n \overset{L^p}{\to} X$ (convergence in L^p). We assume $p \geq 1$.
(iii) If $P(\lim_{n\to\infty} X_n = X) = 1$ then X_n converges to X with probability one (*wp*1), or almost surely (*a.s.*), denoted as $X_n \overset{wp1}{\to} X$. This is also referred to as strong convergence.
(iv) Let F_X be the CDF of any random variable X. If $\lim_{n\to\infty} F_{X_n}(x) = F_X(x)$ for any point of continuity x of $F_X(x)$, then X_n converges to X in distribution, denoted as $X_n \overset{d}{\to} X$. This is also referred to as weak convergence. ∎

What distinguishes convergence in distribution from the remaining modes is that it is not necessary to assume that the sequence and the limit are defined on a common probability measure space, since the statement concerns the marginal distributions and not the random variables themselves. However, there is no harm in making that assumption, so that it becomes possible to establish precise relationships between the convergence modes. In particular, these modes possess the hierarchical structure summarized in the following theorem.

Theorem 11.1 (Hierarchy of Modes of Convergence) The following implications hold:

(i) $X_n \overset{wp1}{\to} X \Rightarrow X_n \overset{i.p.}{\to} X$,
(ii) $X_n \overset{L^p}{\to} X \Rightarrow X_n \overset{i.p.}{\to} X, \ p \geq 1$,
(iii) $X_n \overset{i.p.}{\to} X \Rightarrow X_n \overset{d}{\to} X$,
(iv) $X_n \overset{L^r}{\to} X \Rightarrow X_n \overset{L^s}{\to} X, \ 1 \leq s < r$. ∎

Proof. (i) Define events $E_n^\epsilon = \{|X_n - X| > \epsilon\}$. If $X_n \overset{wp1}{\to} X$ then for any $\epsilon > 0$, $P(E_n^\epsilon \text{ i.o.}) = 0$. This implies $\lim_{n\to\infty} P(\cup_{m\geq n} E_m^\epsilon) = 0$, from which it follows that $P(E_n^\epsilon) = 0$.

(ii) That $X_n \overset{L^p}{\to} X \Rightarrow X_n \overset{i.p.}{\to} X$ follows a straightforward application of Markov's inequality.

(iii) We may write $P(X_n \leq x) \leq P(X \leq x + \epsilon) + P(X - X_n > \epsilon)$, and since $X_n \overset{i.p.}{\to} X$, taking the limit gives $\limsup_n P(X_n \leq x) \leq P(X \leq x + \epsilon)$. A similar argument leads to the inequality $P(X \leq x - \epsilon) \leq \liminf_n P(X_n \leq x)$. By allowing ϵ to approach zero we may therefore conclude that if x is a point of continuity of $F_X(x)$, then $\lim_n P(X_n \leq x) = P(X \leq x)$.

(iv) That $X_n \xrightarrow{L^r} X \Rightarrow X_n \xrightarrow{L^s} X$ for $1 \leq s < r$ may be verified by an application of Jensen's inequality. □

The hierarchy of Theorem 11.1 is strict. A counterexample to $X_n \xrightarrow{L^r} X \Rightarrow X_n \xrightarrow{L^s} X,\ 1 \leq s < r$ would be any suitable sequence possessing order s but not order r moments.

A counterexample to $X_n \xrightarrow{i.p.} X \Rightarrow X_n \xrightarrow{d} X$ would be any sequence of *iid* random variables with positive variance, taking the limit to be a random variable with the same diistribution as X_1.

Further counterexamples are readily constructed as independent sequences of random variables of the form $P(X_n = x_n) = p_n = 1 - P(X_n = 0)$ with limit $X = 0$.

By Borel-Cantelli lemmas I-II, $X_n \xrightarrow{wp1} X$ *iff* $\sum_n p_n < \infty$ (take $E_n = \{|X_n - X| > \epsilon\}$).

Next, for any $p \geq 1$, $X_n \xrightarrow{L^p} X$ *iff* $x_n p_n \to_n 0$.

Finally $X_n \xrightarrow{i.p.} X$ *iff* $p_n \to_n 0$.

(i) Set $p_n = 1/n$, $x_n = n$. Then X_n converges to X in probabilty but not *wp*1 or in L^p.
(ii) Set $p_n = 1/n^2$, $x_n = n^2$. Then for any $p \geq 1$, X_n converges to X *wp*1 but not in L^p.
(iii) Set $p_n = 1/n$, $x_n = 1$. Then X_n converges to X in L^p but not *wp*1.

The next theorem proves that if a limit $X = c$ is a constant, then weak convergence and convergence *i.p.* are equivalent.

Theorem 11.2 (Equivalence of Weak Convergence and Convergence *i.p.* for Constant Limits) If in Definition 11.1 limit $X = c$ is a constant, then weak convergence and convergence *i.p.* are equivalent. ■

Proof. In general, $X_n \xrightarrow{i.p} X \Rightarrow X_n \xrightarrow{d} X$, so it remains to prove the converse when $X = c$ is a constant. We may write for any $\epsilon > 0$

$$\begin{aligned} P(|X_n - c| > \epsilon) &= P(X_n < c - \epsilon) + P(X_n > c + \epsilon) \\ &\leq F_{X_n}(c - \epsilon) + 1 - F_{X_n}(c + \epsilon). \end{aligned}$$

Suppose $X_n \xrightarrow{d} c$. Every point of $F_X(x)$ is a point of continuity except for $x = c$. Taking the limit then gives

$$\lim_{n\to\infty} P(|X_n - c| > \epsilon) \leq F_X(c - \epsilon) + 1 - F_X(c + \epsilon) = 0,$$

which completes the proof. □

The following stochastic order notation is commonly used.

Definition 11.2 (Stochastic Order) Let X_n, Y_n be sequences of random variables.

(i) If $X_n \xrightarrow{i.p.} 0$ we write $X_n = o_p(1)$.
(ii) We may also say that X_n is *stochastically bounded* if for all $\epsilon >$ there exists $M < \infty$ and $N < \infty$ such that $P(|X_n| \geq M) \leq \epsilon$ for all $n \geq N$ (equivalently, there exists $M < \infty$ such that $\limsup_n P(|X_n| \geq M) \leq \epsilon$). In this case we write $X_n = O_p(1)$.
(iii) We write $X_n = o_p(Y_n)$ if $|X_n|/|Y_n| = o_p(1)$.
(iv) We write $X_n = O_p(Y_n)$ if $|X_n|/|Y_n| = O_p(1)$.
(v) Finally, if $X_n = O_p(Y_n)$ and $Y_n = O_p(X_n)$, we write $X_n \sim_p Y_n$. ■

An informal algerba can then be defined for stochastic order notation.

Lemma 11.1 (Stochastic Algebra) Regarding Definition 11.2, suppose $X_n = O_p(1)$, $Y_n = o_p(1)$, $U_n \overset{d}{\to} U$ and $V_n \overset{i.p.}{\to} V$, where U, V are finite *wp*1. The following statements hold.

(i) $X_n Y_n = o_p(1)$,
(ii) $X_n + Y_n = O_p(1)$,
(iii) $U_n = O_p(1)$,
(iv) $V_n = O_p(1)$. ■

Proof. (i) For any δ we may select $M < \infty$ such that $\limsup_n P(|X_n| > M) \leq \delta$. We may write

$$\begin{aligned} P(|X_n Y_n| > \epsilon) &= P(|X_n Y_n| > \epsilon \wedge |X_n| \leq M) + P(|X_n Y_n| > \epsilon \wedge |X_n| > M) \\ &\leq P(|Y_n| > \epsilon/M) + P(|X_n| > M). \end{aligned}$$

Taking the limit gives $P(|X_n Y_n| > \epsilon) \leq \delta$. The proof is completed by allowing δ to approach zero.

(ii) For any δ we may select $M < \infty$ such that $\limsup_n P(|X_n| > M) \leq \delta$. Fix $\epsilon > 0$. We may write

$$\begin{aligned} P(|X_n + Y_n| > M + \epsilon) &\leq P(|X_n| > M \wedge |Y_n| \leq \epsilon) + P(|Y_n| > \epsilon) \\ &\leq P(|X_n| > M) + P(|Y_n| > \epsilon). \end{aligned}$$

The proof is completed by allowing $n \to \infty$.

(iii) First note that if F_U is the CDF of U, then for $M \in [0, \infty)$, if $\pm M$ are points of continuity we have $P(|U| > M) = 1 - F_U(M) + F_U(\text{-}M)$. For any δ we may find such an M for which $1 - F_U(M) + F_U(\text{-}M) \leq \delta$. Then

$$\lim_n P(|U_n| > M) = 1 - F_U(M) + F_U(\text{-}M) \leq \delta,$$

which completes the proof.

(iv) We need only note that convergence in probability implies convergence in distribution (Theorem 11.1). □

11.3 Limits of Expected Values

The purpose of much of the large sample theory described in this volume is to verify convergence in distribution or convergence in probability of a sequence of statistics to some limit. In turn, many of these results refer explicitly to the marginal moments of some sequence $X_1, X_2, \ldots$, which are assumed to converge to the moments of the limit X. However, we have already seen that convergence *wp*1, in probability or in distribution does not imply convergence of moments, so we need a method of validating this assumption. A counterexample is given in Example 11.2.

There are a number of approaches to this problem, which are to one degree or another dependent on the mode of convergence. The first makes use of standard results from the theory of measure and integration. Suppose we are given a measure space $\mathcal{X}$, and a sequence $f_n : \mathcal{X} \to \mathbb{R}$ which coverges pointwise almost everywhere to a limit f, all measurable functions. Conditions under which $\lim_{n\to\infty} \int f_n d\mu = \int f d\mu$ are given by a number of standard theorems, summarized here in Section C.3. These theorems apply directly to sequences of random variable, which are measurable functions on a probability measure space. It must be noted, however, that convergence of the form $f_n \rightarrow f$ in the context of probability theory is equivalent to $X_n \overset{wp1}{\rightarrow} X$, so these theorems apply only to this mode of convergence. We summarize the main integral limit theorems for random variables in the following theorem. Conditions under which moments converge for sequences of random variables which converge under other modes will be developed in Section 11.4.

Theorem 11.3 (Integral Limit Theorems for Random Variables) Let X_n be a sequence of random variables on a probability measure space $(\Omega, \mathcal{F}, P)$.

(i) **Fatou's Lemma**: If $X_n \geq Y$ *wp*1, where $E[Y]$ is finite, then

$$\liminf_{n\rightarrow\infty} E[X_n] \geq E[\liminf_{n\rightarrow\infty} X_n].$$

(ii) **Monotone Convergence Theorem**: If $X_n \geq Y$ *wp*1, where $E[Y]$ is finite, and $X_n \uparrow X$ *wp*1 then

$$\lim_{n\rightarrow\infty} E[X_n] = E[X].$$

(iii) **Lebesgue Dominated Convergence Theorem** Suppose $X_n \overset{wp1}{\rightarrow} X$. If there exists random variable $Y \geq |X_n|$, $n \geq 1$, *wp*1, for which $E[Y] < \infty$ then

$$\lim_{n\rightarrow\infty} E[X_n] = E[X].$$

(iv) **Bounded Convergence Theorem**: Suppose $X_n \overset{wp1}{\rightarrow} X$. If there exists finite constant $M \geq |X_n|$ *wp*1 then

$$\lim_{n\rightarrow\infty} E[X_n] = E[X].$$

■

Theorem 11.3 can be used to derive conditions under which $E[\sum_{i=1}^{\infty} X_i] = \sum_{i=1}^{\infty} E[X_i]$.

Example 11.1 (Expectations of Random Sums) Suppose a random variable is expressible as a random series $Y = \sum_{i=1}^{\infty} X_i$. If we define the partial sums $Y_n = \sum_{i=1}^{n} X_i$, we take this to mean $Y_n \overset{wp1}{\rightarrow} Y$. It is often important to verify the validity of the series expression for the expectation $E[Y] = \sum_{i=1}^{\infty} E[X_i]$. Clearly, we can write $E[Y_n] = \sum_{i=1}^{n} E[X_i]$, so the problem is resolved by verifying the limit $\lim_{n\rightarrow\infty} E[Y_n] = E[Y]$.

If $X_i \geq 0$ *wp*1, then $Y_n \uparrow Y$, and by the monotone convergence theorem we have $E[Y] = \sum_{i=1}^{\infty} E[X_i]$, whether $E[Y]$ is finite or infinite.

The general case requires an additional condition. Suppose we define $U = \sum_{i=1}^{\infty}|X_i|$ (the terms of the series are nonnegative, so the series is well defined). Using the monotone convergence theorem as for the preceding case, we can conclude that $E[U] = \sum_{i=1}^{\infty} E[|X_i|]$. We then note that $U \geq |Y_n|$, $n \geq 1$. Then, if $E[U] = \sum_{i=1}^{\infty} E[|X_i|] < \infty$, the conditions of the Lebesgue dominated convergence theorem are met, so we may conclude $E[Y] = \sum_{i=1}^{\infty} E[X_i]$. ■

The importance of understanding the assumptions of Theorem 11.3 can be seen in the following example.

Example 11.2 (Counterexample to Theorem 11.3) Suppose $U \sim unif(0, 1)$, and define $X_n = nU^n$, $n \geq 1$. It can be shown that $E[X_n] = n/(n + 1)$, but that $X_n \overset{wp1}{\rightarrow} 0$, since $\lim_{n\rightarrow\infty} nu^n = 0$ for all $u \in (0, 1)$ and $P\{U \in (0, 1)\} = 1$. This also implies $X_n \overset{i.p.}{\rightarrow} 0$ and $X_n \overset{d}{\rightarrow} 0$ (Theorem 11.1). Necessarily, the assumptions of the monotone convergence theorem, the Lebesgue dominated convergence theorem and the bounded convergence theorem do not hold for this example. However, the assumptions of Fatou's lemma do hold. ■

11.4 Uniform Integrability

In this section we continue our discussion of the convergence of moments problem. For any random variable X, if $E[|X|] < \infty$, then $\lim_{K\to\infty} E[|X|I\{|X| > K\}] = 0$ (this follows from the Lebesgue dominated convergence theorem). This condition can be imposed on a collection of random variables, which becomes the uniform integrability property. This is formally defined next.

Definition 11.3 (Uniform Integrability) A collection of random variables $\{X_t\}$, $t \in \mathcal{T}$, is uniformly integrable if $\lim_{K\to\infty} \sup_{t\in\mathcal{T}} E[|X_t|I\{|X_t| > K\}] = 0$. ∎

The uniform integrability property implies $\lim_{n\to\infty} E[X_n] = E[X]$ for sequences satisfying $X_n \overset{wp1}{\to} X$, and can be extended to powers X_n^r. We present without proof the following two theorems (Billingsley (1995) can be recommended as a good reference for this topic).

Theorem 11.4 (Uniform Integrability and Convergence of Moments) Suppose the sequence $X_1, X_2, \ldots$ is uniformly integrable, and possesses limit $X_n \overset{wp1}{\to} X$. Then the limit X is integrable ($E[|X|] < \infty$), with $\lim_{n\to\infty} E[X_n] = E[X]$. ∎

A straightforward condition for uniform integrability is then given in the next theorem.

Theorem 11.5 (Sufficient Conditions for Uniform Integrability) Suppose the sequence $X_1, X_2, \ldots$ possesses limit $X_n \overset{wp1}{\to} X$. Let r be a positive integer. If for some $\epsilon > 0$ we have $\sup_n E[|X_n|^{r+\epsilon}] < \infty$ then $E[|X|^r] < \infty$, the sequence $X_1^r, X_2^r, \ldots$ is uniformly integrable, therefore $\lim_{n\to\infty} E[X_n^r] = E[X^r]$. ∎

See Problem 11.8.

Recall that the technical problem stated in Section 11.3 was to determine conditions under which the limit $\lim_{n\to\infty} E[X_n] = E[X]$ holds for sequences of random variables which converge under the various modes of Definition 11.1. In that section, we offered such conditions for strongly convergent sequences, so we might ask what contribution is made by Theorems 11.4 and 11.5 to the more general problem.

There is a subtle but crucial difference between the assumptions of Theorem 11.4 and Theorem 11.3. Once the existence of a limit $X_n \overset{wp1}{\to} X$ is given, the various statements of Theorem 11.3 place additional conditions on the sequence $X_1, X_2, \ldots$ which must hold $wp1$. For the Lebesgue dominated convergence theorem we must assume that the sequence $|X_n|$ is bounded $wp1$ by another random variable Y which is integrable (the bounded convergence theorem is simply a special case of the Lebesgue dominated convergence). For Fatou's lemma X_n must be bounded below by Y, and for the monotone convergence theorem we must have $X_1 \leq X_2 \leq \ldots wp1$. These conditions are often difficult to claim, and for some proofs their verification is a large part of the argument (for example, Theorems 1.20 and 1.22). It is always possible to set $Y = \sup_n |X_n|$, but the condition $E[Y] < \infty$ would clearly be too restrictive.

In contrast, uniform integrability is defined using the *marginal distributions* of the elements X_n of a sequence. It turns out that this distinction is key to resolving the more general moment convergence problem.

11.4.1 Weak Convergence and Convergence of Moments

Conditions are given in Theorems 11.3 and 11.4 which imply $\lim_{n\to\infty} E[X_n] = E[X]$ for strongly convergent sequences, so the next task is to extend these results to the other modes of convergence. Fortunately, there exists a very elegant (even unexpected) solution to this problem offered by the Skorohod representation theorem, which simply states that for any sequence $X_n \xrightarrow{d} X$, there exists a sequence $X_n \xrightarrow{wp1} X$ with the same marginal distributions. Then, since the definition of uniform integrability places conditions only on the marginal distributions of a sequence, we have our link between convergence *wp*1, in probability, or in distribution, and convergence of moments. It is important to note, however, that this approach cannot guarantee that the conditions of Theorem 11.3 will hold, as discussed above.

Theorem 11.6 (Skorohod's Representation Theorem) Suppose μ_n, $n \geq 1$, μ are probability measures on measurable space $(\mathbb{R}^m, \mathcal{B}^m)$, where $\mathcal{B}^m$ are the Borel sets on $\mathbb{R}^m$. Suppose the sequence μ_n converges weakly to μ. Then there exist random vectors Y_n, Y defined on some common probability measure space $(\Omega, \mathcal{F}, P)$ such that $Y_n \xrightarrow{wp1} Y$, μ_n is the distribution of Y_n, and μ is the distribution of Y. ■

Proof. See Theorem 29.6 of Billingsley (1995). See also Theorem 25.6 of Billingsley (1995) for the (considerably simpler) proof for $\mathbb{R}^1$. □

All that is left is to replace the strongly convergent sequence in Theorem 11.4 with a weakly convergent sequence.

Theorem 11.7 (Convergence of Moments for Weakly Convergent Sequence) Suppose the sequence $X_1, X_2, \ldots$ possesses limit $X_n \xrightarrow{wp1} X$, $X_n \xrightarrow{i.p.} X$ or $X_n \xrightarrow{d} X$. Let r be a positive integer. If for some $\epsilon > 0$ we have $\sup_n E[|X_n|^{r+\epsilon}] < \infty$ then $E[|X|^r] < \infty$, the sequence $X_1^r, X_2^r, \ldots$ is uniformly integrable, therefore $\lim_{n\to\infty} E[X_n^r] = E[X^r]$. ■

Proof. If $X_n \xrightarrow{wp1} X$ the theorem is identical to Theorem 11.5, otherwise, by Theorem 11.1 we may also claim $X_n \xrightarrow{d} X$. In this case, by Theorem 11.6 there exists a sequence and limit $X_n^* \xrightarrow{wp1} X^*$ with the same marginal distrbutions as $X_1, X_2, \ldots$ and X. The proof is completed by noting that the conditions of Theorem 11.5 depend only on the marginal distributions. □

11.5 The Law of Large Numbers

Given partial sums $S_n = \sum_{i=1}^n \epsilon_i$, define the sample means $\bar{S}_n = n^{-1} S_n$. Suppose $E[\epsilon_n] = \mu$ for all n. A strong law of large numbers (SLLN) asserts that $\bar{S}_n \xrightarrow{wp1} \mu$, and may also specify convergence rates. In contrast a weak law of large numbers (WLLN) asserts that $\bar{S}_n \xrightarrow{i.p.} \mu$, and we may also define L^p laws when $\bar{S}_n \xrightarrow{L^p} \mu$. We know that a SLLN or an L^p law implies a WLLN (Theorem 11.1).

Generally, μ can be taken as 0 by replacing ϵ_n with $\epsilon_n - \mu$. Many forms of the SLLN exist. The sequence ϵ_n need not be independent or identically distributed. Generally, some moment condition is required.

Our approach will be based on the following three theorems. See, for example, Durrett (2010) for proofs.

Theorem 11.8 (Kolmogorov's Inequality) Suppose $X_1, \ldots, X_n$ are independent random variables with $E[X_i] = 0$ and $\text{var}[X_i] < \infty$. If $S_k = \sum_{i=1}^k X_i$, $k \leq n$, then

$$P\left(\max_{1\leq k\leq n} |S_k| \geq x\right) \leq x^{-2}\text{var}[S_n].$$

■

The following lemma follows from Kolmogorov's inequality.

Lemma 11.2 Suppose $X_1, X_2, \ldots$ are independent random variables with $E[X_i] = 0$, and let $S_n = \sum_{i=1}^n X_i$. If $\sum_{i=1}^\infty \text{var}[X_i] < \infty$, then the sequence S_n, $n \geq 1$ possesses a limit *wp*1. ■

The next theorem, Kroneker's lemma, does not directly involve a probability measure, but quite relevant to the SLLN.

Theorem 11.9 (Kroneker's Lemma) Suppose we have constants $b_n \uparrow \infty$, and any sequence of real numbers $x_1, x_2, \ldots$. If $\sum_{i=1}^\infty x_i/b_i$ converges, then $\lim_{n\to\infty} b_n^{-1} \sum_{i=1}^n x_i = 0$. ■

We may now derive a SLLN somewhat more general than that stated earlier in this section.

Theorem 11.10 (SLLN for Cumulative Random Sums) Suppose $X_1, X_2, \ldots$ are independent random variables with $E[X_i] = 0$, and let $S_n = \sum_{i=1}^n X_i$. Suppose we have constants $b_n \uparrow \infty$. If $\sum_{i\geq 1} b_i^{-2} E[X_i^2] < \infty$, then $b_n^{-1} S_n \overset{wp1}{\to} 0$. ■

Proof. Under the hypothesis, by Lemma 11.2 $\sum_{i\geq 1} b_i^{-1} X_i$ converges *wp*1. The proof is completed by applying Kroneker's lemma (Theorem 11.9). □

Remark 11.1 A sequence $X_1, X_2, \ldots$ of random variables is a martingale if $E[X_1] = 0$ and $E[X_{n+1} \mid X_1, \ldots, X_n] = 0$, $n \geq 1$. Theorem 11.10 is equivalent to Theorem 2, Section VII.8 of Feller (1971), with the exception that the latter assumes that $X_1, X_2, \ldots$ is a martingale. Since any zero mean independent sequence is also a martingale, the theorem of Feller (1971) is strictly more general. ■

It is quite straightforward to derive a SLLN for a sequence X_n, $n \geq 1$ which is independent but not necessarily *iid*.

Example 11.3 (SLLN for Sample Means) Suppose $X_1, X_2, \ldots$ is an independent sequence with $E[X_i] = \mu$ and $\text{var}[X_i] = \sigma_i^2 < K$, for some finite constant K. Consider the related sequence $Y_i = X_i - \mu$, and set

$$Z_n = n^{-1} \sum_{i=1}^m X_i - \mu = n^{-1} \sum_{i=1}^m Y_i, \quad n \geq 1.$$

To apply Theorem 11.10 note that $E[Y_1] = 0$, $\text{var}[Y_1] = \sigma_i^2 < \infty$, and set set $b_n = n$. Then

$$\sum_{i\geq 1} b_i^{-2} E[Y_i^2] = \sum_{i\geq 1} i^{-2} \sigma^2 \leq K \sum_{i\geq 1} i^{-2} < \infty$$

and therefore $\lim_{n\to\infty} Z_n = 0$ *wp*1, or equivalently, $\lim_{n\to\infty} n^{-1} \sum_{i=1}^n X_i = \mu$ *wp*1. Problem 11.10 shows how the assumptions used for this example can be weakened. ■

In this volume a sequence $\hat{\theta}_n$ of estimators will be taken to be consistent if it converges *i.p.* to the true value θ. Convergence *wp*1 to θ is obviously the stronger statement (Theorem 11.1), but requires more a more technical argument, so the reasonable approach is to first master the weaker form of consistency. The next example is intended to make this distinction clear. However, it also points forward to the fact that weakly consistent estimators are often strongly consistent as well. In addition to the next example, see also Problem 11.14.

Example 11.4 (Strong Convergence of Linear Least Squares Estimates) The following example will help clarify some important issues regarding the convergence properties of sequences of estimators. Let X_n be a sequence of random variables with $E[X_n] = 0$, $\text{var}[X_n] = \sigma_n^2$. Clearly, in order to claim $X_n \to 0$ in any sense the condition $\sigma_n^2 \to 0$ must hold, in which case we may claim $X_n \xrightarrow{L^2} 0$, $X_n \xrightarrow{i.p.} 0$, and $X_n \xrightarrow{d} 0$, by Theorem 11.1.

As for strong convergence, $\sigma_n^2 \to 0$ is clearly a necessary condition, but it is not in general sufficient. To fix ideas, suppose the random variables are independent. Let $F(t) = 1 - \bar{F}(t)$ be a fixed CDF, and suppose σ is a scale parameter, so that $P(|X_n| > t) = F(-t/\sigma_n) + \bar{F}(t/\sigma_n)$. Clearly, $\sigma_n^2 \to 0$ implies $P(|X_n| > t) \to 0$ for any $t > 0$. However, we may always select the sequence σ_n^2 to converge to zero at a slow enough rate to force $\sum_n P(|X_n| > t) = \infty$. Then by Borel-Cantelli lemma II (Theorem 1.5) $P(|X_n| > t \text{ i.o.}) = 1$ and X_n does not converge to 0 *wp*1, even though it converges in the other modes.

To llustrate the subtleties of this issue, consider the linear regression model of Section 6.8. To fix ideas, in model (6.45) set $q = 1$, $\mathbf{x}_1[n] = (x_1, \ldots, x_n)$. The least squares estimate of β_1 is $\hat{\beta}_1[n] = \sum_i x_i y_i / \sum_i x_i^2$ (this is linear regression through the origin, see Problem 6.17). The variance is

$$\text{var}[\hat{\beta}_1[n]] = \sigma^2 \left[\sum_i x_i^2 \right]^{-1}.$$

Clearly, $\sum_{i=1}^n x_i^2 \to_n \infty$ is a necessary and sufficient condition for $\hat{\beta}_1[n] \xrightarrow{i.p.} \beta_1$. What additional conditions are needed to ensure that $\hat{\beta}_1[n] \xrightarrow{wp1} \beta_1$? Given the example above, it might be tempting to assume that strong consistency requires additional conditions on the rate at which $\text{var}[\hat{\beta}_1[n]]$ conveges to zero. This in turn would impose further assumptions on $\mathbf{x}_1[n]$ in addition to $\sum_{i=1}^n x_i^2 \to_n \infty$.

However, this is not the case. One obvious difference between our two examples is that the sequence of estimators $\hat{\beta}_1[n]$ induced by an increasing sample are not independent, but change incrementally with n, and this difference is crucial. We may write

$$\hat{\beta}_1[n] - \beta_1 = \frac{S_n}{\sum_{i=1}^n x_i^2}, \quad S_n = \sum_{i=1}^n x_i \epsilon_i, n \geq 1, \quad S_0 = 0.$$

We may then apply Theorem 11.10 to the sequence $\hat{\beta}_1[n]$. To do so we need to verify the condition

$$\sum_{i=1}^{\infty} \frac{x_i^2}{\left(\sum_{i=1}^n x_i^2\right)^2} < \infty.$$

Here, we may rely on the somewhat counterintuitive Lemma 1 of Taylor (1974), from which it follows that

$$\sum_{j=1}^{n} \frac{x_j^2}{\left(\sum_{i=1}^j x_i^2\right)^2} \leq \frac{2}{x_1^2}, \quad n \geq 1, \tag{11.1}$$

where we assume without loss of generality that $x_1 \neq 0$. The proof is left to the reader (Problem 11.13). Thus, $\sum_{i=1}^n x_i^2 \to_n \infty$ is also a sufficient condition for $\hat{\beta}_1[n] \xrightarrow{wp1} \beta_1$.

Clearly, a sufficient condition for L^2 convergence, and therefore convergence in probability, is

$$(\mathbf{X}^T \mathbf{X})^{-1} \to_n 0, \tag{11.2}$$

which simply states that the covariance matrix of $\hat{\boldsymbol{\beta}}$ approaches zero. That this is also a necessary condition was shown in Drygas (1976), which also gives sufficient conditions for strong convergence. That Equation (11.2) implies strong consistency for normal *iid* error terms was shown in Anderson *et al.* (1976), and in Lai and Robbins (1977) under minimal assumptions on the error terms. ■

Although we usually have $E[|\bar{S}_n|] = O(n^{-1/2})$, it is worth noting that when an SLLN holds, we do not expect $\bar{S}_n = O(n^{-1/2})$ *wp*1. In fact, the rate of convergence of $\bar{S}_n$ for the *iid* case is precisely resolved as the law of the iterated logarithm (LIL).

Theorem 11.11 (Law of the Iterated Logarithm) If $X_1, X_2, \ldots$ is an *iid* sequence with $E[X_i] = 0$, $E[X_i^2] = 1$ then

$$\limsup_{n\to\infty} \frac{\sum_{i=1}^n X_i}{(2n\log\log n)^{1/2}} = 1, \quad wp1.$$ ■

In other words, for L^2 convergence we have rate $E[|\bar{S}_n|] = O(n^{-1/2})$, but for almost sure convergence we have the slightly larger LIL rate $\bar{S}_n = O(2n^{-1/2}(\log\log n)^{1/2})$. There are a variety of conditions beyond those given in Theorem 11.11 under which the LIL holds, and the reader can be referred to a summary of the quite interesting history of the LIL in Durrett (2010).

11.6 Weak Convergence

The definition of convergence in distribution (Definition 11.1) requires careful interpretation, as suggested by the next example.

Example 11.5 (Counterexample of Weak Convergence) Suppose we have a sequence X_n, $n \geq 1$, where $X_n \sim unif(0,n)$. Then for each $x \in \mathbb{R}$ we have CDFs $F_{X_n}(x) \to_n 0$. So a pointwise limit F^* of the CDFs exists, but it is not itself a CDF. ■

The wording of the definition of convergence in distribution assumes that the limit is a proper CDF, so the definition is not formally satisfied by the sequence of Example 11.5. It is therefore worth clarifying this aspect. This can be done by equating convergence in distribution with weak convergence (equivaently, convergence in law or convergence in measure). Any sequence of finite measures μ_n on a metric space converges weakly to μ if any of the equivalent statements of the Portmanteau theorem hold (a good reference for this theorem is Ash and Dolacutuseans-Dade (2000)). We first require the following definition.

Definition 11.4 (Boundary of a Set) Given a metric space Ω, the *boundary* ∂A of $A \subset \Omega$ is the set of all limits of sequences in A which are also limits of sequences in A^c. ■

Example 11.6 (Boundaries of Intervals) If A is any of the intervals (a,b), $[a,b]$, $(a,b]$ or $[a,b)$, then $\partial A = \{a,b\}$. ■

We now state the Portmanteau theorem.

Theorem 11.12 (Portmanteau Theorem) Let Ω be a metric space, and let μ $\mu_1, \mu_2, \ldots$ be finite measures on the Borel sets of Ω. The following conditions are equivalent:

(i) $\lim_{n\to\infty} \int_\Omega f d\mu_n = \int_\Omega f d\mu$ for all bounded continuous $f : \Omega \to \mathbb{R}$.
(ii) $\lim_{n\to\infty} \int_\Omega f d\mu_n = \int_\Omega f d\mu$ for all bounded Lipschitz continuous $f : \Omega \to \mathbb{R}$.
(iii) $\liminf_{n\to\infty} \int_\Omega f d\mu_n \geq \int_\Omega f d\mu$ for all bounded lower semicontinuous $f : \Omega \to \mathbb{R}$.
(iv) $\limsup_{n\to\infty} \int_\Omega f d\mu_n \leq \int_\Omega f d\mu$ for all bounded upper semicontinuous $f : \Omega \to \mathbb{R}$.
(v) $\liminf_{n\to\infty} \mu_n(A) \geq \mu(A)$ for every open set $A \subset \Omega$, and $\lim_{n\to\infty} \mu_n(\Omega) = \mu(\Omega)$.

(vi) $\limsup_{n\to\infty} \mu_n(A) \leq \mu(A)$ for every closed set $A \subset \Omega$, and $\lim_{n\to\infty} \mu_n(\Omega) = \mu(\Omega)$.
(vii) $\lim_{n\to\infty} \mu_n(A) = \mu(A)$ for any Borel set $A \subset \Omega$ for which $\mu(\partial A) = 0$. ■

Under the assumptions of Theorem 11.12, a sequence of measures μ_n *converges weakly* to μ, written $\mu_n \Rightarrow \mu$, if any of the seven statements are satisfied.

Note that statement (*i*) is, nominally, a stronger statement than (*ii*), but they are provably equivalent (a bounded continuous function can be arbitrarily well approximated by a bounded Lipschitz continuouus function). There are, in fact, a variety of statements of this theorem in the literature.

An equivalent definition of convergence in distribution can be given by applying the Portmanteau theorem to sequences of probability measures, that is, a sequence of probability measures converges in distribution to a limiting measure if any of the equivalent statements of Theorem 11.12 are satisfied (a common convention is to use statement (*i*) as the definition). This allows a rigorous definition of convergence of probability measures defined on any metric space.

We'll first ask if the sequence of Example 11.5 satisfies this definition. The Portmanteau theorem requires that if $P_n \overset{d}{\to} P$, and $P_n(\Omega) = 1$, $n \geq 1$ we must have $P(\Omega) = 1$ (this is seen clearly using statement (*i*)). Thus, that sequence is not weakly convergent.

We next need to consider whether or not the definition of convegence in distribution offered in Definition 11.1 (iv) is equivalent to that of weak convergence defined by the Portmanteau theorem.

Theorem 11.13 (Equivalence of Convergence in Distribution and Weak Convergence of Probability Measures) Let X_n be a sequence of random variables measurable *wrt* the Borel sets of $\mathbb{R}$, and let μ_n be the respective probability measures. Then Definition 11.1 (iv) is equivalent to the weak convergence of measures μ_n, as defined by the Portmanteau theorem (Theorem 11.12). ■

Proof. Clearly, statement (vii) of Theorem 11.12 implies Definition 11.1 (iv) (consider Example 11.6). Conversely, let A be an open set in $\mathbb{R}$. Any such open set is a countable union of disjoint open intervals, so we can assume that A itself is an open interval. For any $\epsilon > 0$ we can always find an open interval $B \subset A$ such that $P(B) \geq P(A) - \epsilon$, and such that the endpoints of B are points of continuity of $F_X(x)$. In this case

$$\liminf_{n\to\infty} P_n(A) \geq \liminf_{n\to\infty} P_n(B) = P(B) \geq P(A) - \epsilon,$$

and we have statement (v) of Theorem 11.12 after letting ϵ approach zero. □

A sequence of discrete random variables may converge in distribution to a continuous random variable.

Example 11.7 (Convergence of the Discrete to Continuous Uniform Distribution) Suppose X_n, $n \in \mathbb{I}$, is a discrete random variable with probability mass function $P(X_n = (i-1)/n) = 1/n$, $i = 1, \ldots, n$. The CDF of X_n is then

$$F_{X_n}(x) = \begin{cases} 0 & ; \quad x < 0 \\ i/n & ; \quad x \in [(i-1)/n, i/n),\ i = 1, \ldots, n-1 \\ 1 & ; \quad x \geq (n-1)/n \end{cases}.$$

Suppose $F_U(x)$ is the CDF of $U \sim unif(0,1)$. Then

$$F_U(x) = \begin{cases} 0 & ; \quad x < 0 \\ x & ; \quad x \in [0,1) \\ 1 & ; \quad x \geq 1 \end{cases}.$$

It may be verified that $\sup_{x \in \mathbb{R}} |F_{X_n}(x) - F_U(x)| \leq 1/n$, so that $\lim_{n \to \infty} F_{X_n}(x) = F_U(x)$ for all $x \in \mathbb{R}$. So we can conclude $X_n \xrightarrow{d} U$, according to Definition 11.1 (iv).

We may use Theorem 11.12 to reach the same conclusion. Let $f : [0, 1] \to \mathbb{R}$ be bounded and continuous. Then

$$E[f(U)] = \int_{\mathcal{X}} f(x) dP_U = \int_0^1 f(x) dx,$$

$$E[f(X_n)] = \int_{\mathcal{X}} f(x) dP_{X_n} = n^{-1} \sum_{i=1}^{n} f((i-1)/n).$$

It is easily verified that $\lim_{n \to \infty} E[f(X_n)] = E[f(U)]$, so that statement (i) of Theorem 11.12 holds. See also Problem 11.15. ∎

11.7 Multivariate Extensions of Limit Theorems

In subsequent chapters we will need to extend our definition of the stochastic convergence of random variables to random vectors. It turns out that this is most easily done by interpreting a vector as an element in a metric space. Thus, the topic of this section is in reality stochastic convergence in metric spaces.

A sequence $\tilde{x}_n$ converges to $\tilde{x}$ in a metric space $(\mathcal{X}, d)$ if $\lim_n d(\tilde{x}_n, \tilde{x}) = 0$. We are assuming throughout this volume that a probability measure space is defined on a metric space $(\mathcal{X}, d)$ (usually, d is Euclidean distance in $\mathbb{R}^m$). Thus, a valid definition of convergence *wp*1 or *i.p.* follows the definition of convergence in a metric space. We then have

$$\tilde{X}_n \xrightarrow{i.p} \tilde{X} \text{ iff } \lim_n P\left(d(\tilde{X}_n, \tilde{X}) > \epsilon\right) = 0, \text{ for all } \epsilon > 0, \tag{11.3}$$

and

$$\tilde{X}_n \xrightarrow{wp1} \tilde{X} \text{ iff } P\left(\lim_{n \to \infty} d(\tilde{X}_n, \tilde{X}) = 0\right) = 1. \tag{11.4}$$

That the essential properties of convergence *wp*1 or *i.p.* of Definition 11.1 are preserved by this construction can be seen by first noting that $Y_n = d(\tilde{X}_n, \tilde{X}) \in \mathbb{R}$ is a sequence of random variables. Then Equations (11.3) and (11.4) are equivalent to $Y_n \xrightarrow{i.p.} 0$ and $Y_n \xrightarrow{wp1} 0$, respectively.

As for convergence in distribution, recall that the Portmanteau theorem is defined on a measurable space $(\mathcal{X}, \mathcal{B})$, where $\mathcal{X}$ is a metric space $(\mathcal{X}, d)$, and $\mathcal{B}$ are the Borel sets induced by the metric d. Then given probability measure space $(\Omega, \mathcal{F}, P)$, if $\tilde{X}_n : \Omega \to \mathcal{X}$ is a sequence of measurable mappings, then $\tilde{X}_n \xrightarrow{d} \tilde{X}$ if the induced measures satisfy $P_{\tilde{X}_n} \Rightarrow P_{\tilde{X}}$.

We now have satisfactory definitions of convergence *wp*1, *i.p.* and in distribution for random sequences on metric spaces, although a more convenient characterization of convergence in distribution of random vectors will be offered below in Section 11.11.

Before considering L^p convergence, we will generalize Theorem 11.1 for the three remaining modes. This will require the following lemma:

Lemma 11.3 Suppose $h : \mathcal{X} \to \mathcal{Y}$ is a continuous mapping between metric spaces $(\mathcal{X}, d_x)$, $(\mathcal{Y}, d_y)$. Define the class of subsets

$$H_{\epsilon,\tau} = \{\tilde{x} \in \mathcal{X} : \exists \tilde{y} \in \mathcal{Y} \ni d_x(\tilde{x}, \tilde{y}) < \epsilon, \ d_y(h(\tilde{x}), h(\tilde{y})) > \tau\}$$
$$\epsilon, \tau > 0. \tag{11.5}$$

Then for any fixed τ, $\cap_{\epsilon > 0} H_{\epsilon,\tau} = \emptyset$. ∎

Proof. The lemma follows directly from the definition of continuity for h. □

The main result follows.

Theorem 11.14 (Hierarchy of Modes of Convergence on Metric Spaces) We are given probability measure space $(\Omega, \mathcal{F}, P)$, and metric space $(\mathcal{X}, d)$. Let $\tilde{X}_n : \Omega \to \mathcal{X}$, $n \geq 1$ be a sequence of measurable mappings, and let $\tilde{X}(\omega)$ be another such mapping. Then the following implications hold:

$$\tilde{X}_n \overset{wp1}{\to} \tilde{X} \Rightarrow \tilde{X}_n \overset{i.p.}{\to} \tilde{X}, \quad \tilde{X}_n \overset{i.p.}{\to} \tilde{X} \Rightarrow \tilde{X}_n \overset{d}{\to} \tilde{X}.$$ ■

Proof. $\tilde{X}_n \overset{wp1}{\to} \tilde{X} \Rightarrow \tilde{X}_n \overset{i.p.}{\to} \tilde{X}$ holds by taking $Y_n = d(\tilde{X}_n, \tilde{X})$, limit $Y = 0$, and applying Theorem 11.1.

[$\tilde{X}_n \overset{i.p.}{\to} \tilde{X} \Rightarrow \tilde{X}_n \overset{d}{\to} \tilde{X}$] Let $h : \mathcal{X} \to \mathbb{R}$ be a bounded continuous function. Since h is bounded we must have $\sup_{\tilde{x},\tilde{x}'} |h(\tilde{x}) - h(\tilde{x}')| = M < \infty$. Then we have

$$\begin{aligned} h(\tilde{X}_n) - h(\tilde{X}) &= (h(\tilde{X}_n) - h(\tilde{X}))I\{d(\tilde{X}_n, \tilde{X}) \leq \epsilon\} \\ &\quad + (h(\tilde{X}_n) - h(\tilde{X}))I\{d(\tilde{X}_n, \tilde{X}) > \epsilon\} \\ &= A_1 + A_2. \end{aligned} \tag{11.6}$$

We will consider the terms A_1, A_2 separately. First, A_1 can be further decomposed using the set $H_{\epsilon,\tau}$ defined in Lemma 11.3:

$$\begin{aligned} A_1 &= (h(\tilde{X}_n) - h(\tilde{X}))I\{d(\tilde{X}_n, \tilde{X}) \leq \epsilon\}I\{\tilde{X} \notin H_{\epsilon,\tau}\} \\ &\quad + (h(\tilde{X}_n) - h(\tilde{X}))I\{d(\tilde{X}_n, \tilde{X}) \leq \epsilon\}I\{\tilde{X} \in H_{\epsilon,\tau}\} \\ &= A_{1,1} + A_{1,2}. \end{aligned} \tag{11.7}$$

Using the properties of $H_{\epsilon,\tau}$ we may conclude $A_{1,1} \leq M\tau$, so taking the expectation of (11.7) yields inequality

$$E[A_1] \leq M\tau + MP(\tilde{X} \in H_{\epsilon,\tau}).$$

Similarly, we have

$$E[A_2] \leq MP(d(\tilde{X}_n, \tilde{X}) > \epsilon),$$

so that

$$h(\tilde{X}_n) - h(\tilde{X}) \leq M\tau + MP(\tilde{X} \in H_{\epsilon,\tau}) + MP(d(\tilde{X}_n, \tilde{X}) > \epsilon).$$

Allowing $n \to \infty$ then yields

$$\limsup_n E[h(\tilde{X}_n)] - E[h(\tilde{X})] \leq M\tau + MP(\tilde{X} \in H_{\epsilon,\tau}),$$

since $\tilde{X}_n \overset{i.p.}{\to} \tilde{X}$. By Lemma 11.3, allowing ϵ to approach zero yields

$$\limsup_n E[h(\tilde{X}_n)] - E[h(\tilde{X})] \leq M\tau.$$

But this holds for any $\tau > 0$, so we may conclude

$$\limsup_n E[h(\tilde{X}_n)] - E[h(\tilde{X})] \leq 0.$$

If, in Equation (11.6), $h(\tilde{X}_n) - h(\tilde{X})$ is replaced by $h(\tilde{X}) - h(\tilde{X}_n)$ essentially the same argument leads to the inequality

$$E[h(\tilde{X})] - \liminf_n E[h(\tilde{X}_n)] \leq 0,$$

so that we may conclude that statement (i) of Theorem 11.12 holds. □

Theorem 11.2 establishes for random variables in $\mathbb{R}$ the equivalence of convergence in probability and convergence in distribution when the limit is a constant. We show in the next theorem that this holds also for metric spaces. However, we will use the Portmanteau theorem, and a comparison of the two proof methods is instructive.

Theorem 11.15 (Equivalence of Weak Convergence and Convergence *i.p.* for Constant Limits in Metric Spaces) Under the conditions of Theorem 11.14, if the limit $\tilde{X} = c$ is a constant, then weak convergence and convergence *i.p.* are equivalent. ∎

Proof. In general, $\tilde{X}_n \stackrel{i.p}{\to} \tilde{X} \Rightarrow \tilde{X}_n \stackrel{d}{\to} \tilde{X}$, so it remains to prove the converse when $\tilde{X} = c$ is a constant. Suppose $\tilde{X}_n \stackrel{d}{\to} \tilde{X} = c$. The measure of $\tilde{X}$ is then given by

$$P_{\tilde{X}}(A) = \begin{cases} 1 & ; \quad c \in A \\ 0 & ; \quad c \notin A \end{cases}.$$

Let $A_\epsilon = \{\tilde{x} \in \mathcal{X} : d(\tilde{x}, c) > \epsilon\}$ for some $\epsilon > 0$. The boundary of A_ϵ is $\partial A_\epsilon = \{\tilde{x} \in \mathcal{X} : d(\tilde{x}, c) = \epsilon\}$. Since $c \notin \partial A_\epsilon$ and $c \notin A_\epsilon$ we must have $P_{\tilde{X}}(\partial A_\epsilon) = 0$ and $P_{\tilde{X}}(A_\epsilon) = 0$. Then by statement (vii) of Theorem 11.12 we may conclude that $\lim_n P_{\tilde{X}_n}(A_\epsilon) = 0$, or equivalently, $\tilde{X}_n \stackrel{i.p}{\to} \tilde{X}$. □

L^p Convergence for Random Vectors

The metric space properties of L^p convergence are defined specifically for random vectors. We can extend the definition of L^p norm to random vectors $\mathbf{X} = (X_1, \ldots, X_m) \in \mathbb{R}^m$ by setting

$$\|\mathbf{X}\|_p = \left(\sum_{i=1}^m E[|X_i|^p] \right)^{1/p}$$

for $p \geq 1$. We then say $\mathbf{X}_n \stackrel{L^p}{\to} \mathbf{X}$ if $\|\mathbf{X}_n - \mathbf{X}\|_p \to_n 0$. The remainder of Theorem 11.1 can then be extended to random vectors in $\mathbb{R}^m$.

Theorem 11.16 (Hierarchy of Modes of Convergence Involving L^p Convergence of Random Vectors) We are given probability measure space $(\Omega, \mathcal{F}, P)$. Let $\mathbf{X}_n$, $n \geq 1$ be a sequence of m-dimensional random vectors. Then the following implications hold:

$$\mathbf{X}_n \stackrel{L^p}{\to} \mathbf{X} \Rightarrow \mathbf{X}_n \stackrel{i.p.}{\to} \mathbf{X}, \quad p \geq 1,$$
$$\mathbf{X}_n \stackrel{L^r}{\to} \mathbf{X} \Rightarrow \mathbf{X}_n \stackrel{L^s}{\to} \mathbf{X}, \quad 1 \leq s < r,$$ ∎

Proof. $[\mathbf{X}_n \stackrel{L^p}{\to} \mathbf{X} \Rightarrow \mathbf{X}_n \stackrel{i.p.}{\to} \mathbf{X}, \; p \geq 1]$ Suppose $\mathbf{X}_n \stackrel{L^p}{\to} \mathbf{X}$, where $\mathbf{X} \in \mathbb{R}^m$. Then convergence in probability is equivalent to

$$\lim_{n \to \infty} P(|\mathbf{X}_n - \mathbf{X}| > \epsilon) = 0 \text{ for all } \epsilon > 0.$$

This is implied by $\mathbf{X}_n \stackrel{L^p}{\to} \mathbf{X}$, following a straightforward application of Markov's inequality.

$[\mathbf{X}_n \stackrel{L^r}{\to} \mathbf{X} \Rightarrow \mathbf{X}_n \stackrel{L^s}{\to} \mathbf{X}, \; 0 < s < r]$ It is easily verified that a random vector converges in L^p *iff* all components of the vector converge in L^p. The theorem therefore follows from Theorem 11.1. □

Note that we have omitted the case $p = \infty$. It may be verfied that

$$\lim_{p \to \infty} \left(\sum_{i=1}^m E[|X_i|^p] \right)^{1/p} = \max_i |X_i|,$$

so that the supremum norm is then $\|\mathbf{X}\|_\infty = E[\max_i |X_i|]$. The properties of this norm are essentially the same as any other L^p norm.

11.8 The Continuous Mapping Theorem

It is often necessary to evaluate the asymptotic properties of an expression constructed from multiple limiting processes. The continuous mapping theorem is an important tool in this situation.

Theorem 11.17 (Continuous Mapping Theorem) We are given a probability measure space $(\mathcal{X}, \mathcal{B}, P)$, where $\mathcal{X}$ is a metric space possessing Borel sets $\mathcal{B}$. Let $\mathcal{Y}$ be a second metric space. Suppose we are given a continuous mapping $h : \mathcal{X} \rightarrow \mathcal{Y}$. If $\tilde{X}_n \in \mathcal{X}$, $n \geq 1$ is a random sequence, then

(i) $\tilde{X}_n \overset{d}{\rightarrow} \tilde{X} \Rightarrow h(\tilde{X}_n) \overset{d}{\rightarrow} h(\tilde{X})$,
(ii) $\tilde{X}_n \overset{i.p.}{\rightarrow} \tilde{X} \Rightarrow h(\tilde{X}_n) \overset{i.p.}{\rightarrow} h(\tilde{X})$,
(iii) $\tilde{X}_n \overset{wp1}{\rightarrow} \tilde{X} \Rightarrow h(\tilde{X}_n) \overset{wp1}{\rightarrow} h(\tilde{X})$. ■

Proof. (i) By Theorem 11.12, a sequence of random observations U_n on a metric space $\mathcal{U}$ converges in distribution to U *iff* $\lim_n E[f(U_n)] = E[f(U)]$ for all bounded continuous mappings $f : \mathcal{U} \rightarrow \mathbb{R}$. Then suppose $g : \mathcal{Y} \rightarrow \mathbb{R}$ is bounded and continuous. Then the composite mapping $g \circ h : \mathcal{X} \rightarrow \mathbb{R}$ is also bounded and continuous. Since $\tilde{X}_n \overset{d}{\rightarrow} \tilde{X}$ it follows that $\lim_n E[g(h(\tilde{X}_n))] = E[g(h(\tilde{X}_n))]$, and therefore that $h(\tilde{X}_n) \overset{d}{\rightarrow} h(\tilde{X})$.

(ii) Here we use Lemma 11.3 (see also the proof of Theorem 11.14 for a similar application). Fix $\tau, \epsilon > 0$. Suppose $d_y(h(\tilde{X}_n), h(\tilde{X})) > \tau$. Then either $d_x(\tilde{X}_n, \tilde{X}) > \epsilon$ or $\tilde{X} \in H_{\epsilon,\tau}$ (both may occur). By Boole's inequality this leads to the inequality

$$P(d_y(h(\tilde{X}_n), h(\tilde{X})) > \tau) \leq P(d_x(\tilde{X}_n, \tilde{X}) > \epsilon) + P(\tilde{X} \in H_{\epsilon,\tau}).$$

Since $\tilde{X}_n \overset{i.p.}{\rightarrow} \tilde{X}$, taking the limit as $n \rightarrow \infty$ yields

$$\limsup_n P(d_y(h(\tilde{X}_n), h(\tilde{X})) > \tau) \leq P(\tilde{X} \in H_{\epsilon,\tau}).$$

However, by Lemma 11.3, $P(\tilde{X} \in H_{\epsilon,\tau}) \rightarrow 0$ as $\epsilon \rightarrow 0$, so that $\lim_n P(d_y(h(\tilde{X}_n), h(\tilde{X})) > \tau) = 0$.

(iii) The proof follows by simply noting that by the continuity of h, $\tilde{X}_n \rightarrow \tilde{X} \Rightarrow h(\tilde{X}_n) \rightarrow h(\tilde{X})$. □

The continuous mapping theorem finds a useful expression in Slutsky's theorem.

Theorem 11.18 (Slutsky's Theorem) We are given a probability measure space $(\mathcal{X}, \mathcal{B}, P)$, where $\mathcal{X}$ is a metric space possessing Borel sets $\mathcal{B}$. Suppose $\tilde{X}_n$ is a sequence in $\mathcal{X}$ such that $\tilde{X}_n \overset{d}{\rightarrow} \tilde{X}$. Also, suppose $\tilde{U}_n$ is a sequence in $\mathcal{X}$ such that $\tilde{U}_n \overset{i.p.}{\rightarrow} \tilde{u}$, where $\tilde{u}$ is a constant. Suppose $h : \mathcal{X} \times \mathcal{X} \rightarrow \mathcal{Y}$ is a continuous mapping into metric space $\mathcal{Y}$. Then

$$h(\tilde{X}_n, \tilde{U}_n) \overset{d}{\rightarrow} h(\tilde{X}, \tilde{u}).$$

In particular, if $\mathcal{X} = \mathbb{R}$, then

$$X_n + U_n \xrightarrow{d} X + u,$$
$$U_n X_n \xrightarrow{d} uX,$$
$$X_n / U_n \xrightarrow{d} X/u,$$

where defined. ∎

Proof. Since convergence in probability implies convergence in distribution, the theorem follows a straightforward application of the continuous mapping theorem. □

We next offer a typical application of Slutky's theorem.

Example 11.8 (Asymptotic Normality of the T distribution) In Example 1.31, it was shown that the excess kurtosis, which is the fourth-order cumulant κ_4, of a T_ν distribution approaches zero as $\nu \to \infty$. For the normal distribution, $\kappa_k = 0$ for all $k \geq 3$ (Example 1.30). Of course, this does not prove by itself that the T distribution converges to the normal as $\nu \to \infty$.

Slutsky's theorem offers a simple approach to this problem. Recall that we may construct the T distribution from the ratio $T = Z/\sqrt{W/\nu}$, where $Z \sim N(0,1)$, $W \sim \chi^2_\nu$ and Z and W are indpendent. First note that W is equal in distribution to an *iid* sum of ν random variables of mean one and finite variance. This is all we need to know to conclude that $W/\nu \overset{wp1}{\to} 1$ (Example 11.3). That $T \xrightarrow{d} Z \sim N(0,1)$ follows immediately from Theorem 11.18. ∎

11.9 MGFs, CGFs and Weak Convergence

From Section 1.9.2, we have seen that the MGF is unique, in the sense that two distinct distributions which possess MGFs must possess distinct MGFs.

MGFs often provide an analytically tractible representation of a distribution, so it is worth asking if convergence in distribution is equivalent to convergence of the respective MGFs. The answer is "yes" under general conditions. There are a number of ways to view the problem. Suppose we start with a sequence $F_n(x)$ of CDFs of random variables $X_n \in \mathbb{R}$, $n \geq 1$, and for each an MGF $m_n(t)$ exists on some common interval $t \in (-\alpha, \alpha)$. Furthermore, we may identify a function $m(t)$ such that $\lim_n m_n(t) = m(t) < \infty$ for all $t \in (-\alpha, \alpha)$. It may be proven that this suffices to conclude that there exists a CDF F such that $F_n \Rightarrow F$, and F possesses MGF $m(t)$. See Theorem 2 of Kozakiewicz *et al.* (1947), which otherwise offers a particularly elegant treatment of this topic. Of course, exactly the same argument applies to the CGF.

Example 11.9 (Approximate Normality of the Poisson Distribution) In Example 1.10 it was pointed out that the Poisson arrival process can be modeled as a limiting aggregation of independent arrival processes of general type. Then the number of arrivals X in a Poisson process within a fixed time period has a Poisson distribution. This suggests that $X \sim pois(\lambda)$ can be interpreted as the limit of a random sum, and should therefore be well approximated by a normal distribution for large n. First, the CGF of a Poisson distribution is

$$m_X(t) = \lambda(e^t - 1).$$

To investigate the limiting distribution as $\lambda \to \infty$ it will be necessary to standardize X to have zero mean and unit variance, that is, we define

$$Z_\lambda = \frac{X - \lambda}{\lambda^{1/2}}.$$

If we apply the linear transformation rule for CGFs $c_{a+bX}(t) = at + c_X(bt)$ (Section 1.10.1) we get

$$\begin{aligned}
m_{Z_\lambda}(t) &= -\lambda^{1/2}t + \lambda(e^{t\lambda^{-1/2}} - 1) \\
&= -\lambda^{1/2}t + \lambda\left(\sum_{i=0}^{\infty} \frac{(t\lambda^{-1/2})^i}{i!} - 1\right) \\
&= -\lambda^{1/2}t + t\lambda^{1/2} + \frac{t^2}{2} + \lambda^{-1/2}\frac{t^3}{6} + \ldots \\
&= \frac{t^2}{2} + O(\lambda^{-1/2}).
\end{aligned}$$

This means $\lim_{\lambda\to\infty} m_{Z_\lambda}(t) = t^2/2$ which is the CGF of the $N(0,1)$ distribution. See also Problem 11.18. ■

A central limit theorem (CLT) can be taken to be any theorem giving conditions under which a sequence of random variables or vectors converges weakly to a (possibly multivariate) normal distribution, which we can refer to as asymptotic normality. These are obviously ubiquitous in large sample theory. However, we will find, particularly in Chapters 13 and 14, that the portion of the theory which is concerned with asymptotic normality can be to a large extent separated from the main arguments, and then considered independently.

Example 11.10 (The Central Limit Theorem *via* MGFs) As is well known, the normal distribution is a limiting distribution for a large class of stochastic processes characterized by an additive aggregation of random error. This means under quite general conditions, sequences of estimators $\hat{\theta}_n$ tend to converge weakly to a normal distribution. In addition, since many test statistics are approximately sums of squares of normal random variables, the χ^2_ν distribution frequently appears as the limiting distribution in hypothesis testing.

A CLT is especially easy to construct when the elements of a sequence possess a MGF. Recall that the CGF for the $N(\mu, \sigma^2)$ distribution is

$$c(t) = \mu t + \sigma^2 t^2/2.$$

This yields a remarkable fact. Suppose X_n is a sequence of random variables possessing an MGF. Furthermore assume the following limits hold for finite μ, σ^2:

$$\lim_n E[X_n] = \mu, \quad \lim_n \text{var}[X_n] = \sigma^2, \quad \lim_n \kappa_k[X_n] = 0, \ \ k \geq 3. \tag{11.8}$$

In other words, if the means and variances are convergent, and all cumulants of order $k \geq 3$ approach zero, then the sequence converges to $N(\mu, \sigma^2)$ in distribution.

Suppose $\bar{X}_n = S_n/n$ is the sample mean of an *iid* sample $X_1, X_2, \ldots$ from a distribution of mean zero, which also possesses a MGF. Suppose $\text{var}[X_1] = \sigma^2$. We may normalize $Z_n = n^{1/2}\bar{X}_n = n^{-1/2}S_n$, so that we are trying to determine the limiting distribution of Z_n. We have already seen that the kth order cumulant of an independent sum $S_n = X_1 + \ldots + X_n$ is equal to the sum of the kth order cumulants of the individual terms X_i. We therefore have $\kappa_k[S_n] = n\kappa_k[X_1]$, $k \geq 1$. Then, noting that $E[X_1] = 0$ by assumption, applying the linear transformation rule for CGFs gives

$$\begin{aligned}
\kappa_1[Z_n] &= 0 \\
\kappa_k[Z_n] &= n^{1-k/2}\kappa_k[X_1], \ \ k \geq 2.
\end{aligned}$$

This means that $\kappa_2[Z_n] = \kappa_2[X_1]$ is a constant which does not depend on n. On the other hand, for $k > 2$, $1 - k/2 < 0$, so that $\lim_n \kappa_k[Z_n] = 0$, and we may conclude that

$$\lim_n c_{Z_n}(t) = \sigma^2 t^2/2,$$

equivalently, $Z_n \xrightarrow{d} Z$, where $Z \sim N(0, \sigma^2)$. ∎

11.10 The Central Limit Theorem for Triangular Arrays

We will see in Chapter 12 that it is important to understand the connection between MGFs and the central limit theorem. However, not all distributions of interest possess a MGF. Fortunately, this is far from a necessary condition for a CLT, and there exists a quite rich, and usually quite technical, literature on the subject (Feller (1971); Billingsley (1995); Durrett (2010) can be recommended). Sufficient conditions vary widely, and some CLTs also state rates of convergence or error bounds, for example, the Berry-Esseen theorem (Durrett, 2010).

For our purposes, it should suffice to state the Lyapounov central limit theorem, which offers sufficient conditions similar to those of Theorem 11.5 for uniform integrability (a bound on any absolute moment of order greater than two). We assume we have a triangular array $\{X_{ni}\}$, where for each $n \geq 1$, $X_{n1}, X_{n2}, \ldots, X_{nm_n}$ is a collection of independent random variables for positive integers m_n which approach ∞. Suppose we have

$$E[X_{ni}] = \mu_{ni}, \quad \sigma^2_{ni} = E[X^2_{ni}], \quad s^2_n = \sum_{i=1}^{m_n} \sigma^2_{ni}. \tag{11.9}$$

We give the main theorem without proof (Billingsley, 1995).

Theorem 11.19 (The Lyapounov Central Limit Theorem) Suppose we are given a triangular array of independent random variables for which Equation (11.9) holds. Denote the sums $S_n = \sum_{i=1}^{m_n} X_{ni}$. The Lyapounov condition can be stated

$$\lim_{n\to\infty} s_n^{-(2+\delta)} \sum_{i=1}^{m_n} E[|X_{ni} - \mu_{ni}|^{2+\delta}] = 0, \tag{11.10}$$

for some $\delta > 0$. If (11.10) holds, then $(S_n - E[S_n])/\text{var}[S_n]^{1/2} \xrightarrow{d} N(0, 1)$. ∎

By Theorem 11.19 a CLT for a sample mean of an *iid* sequence requires only the existence of any absolute moment of order greater than 2.

Example 11.11 (CLT for *iid* Random Variables Based on the Lyapounov Condition) Suppose we are given an *iid* independent sequence X_i, $i \geq 1$ for which $E[X_1] = \mu$, $\text{var}[X_i] = \tau$. Assume in addition $E[|X_1 - \mu|^{2+\delta}] = K < \infty$ for some $\delta > 0$. In terms of the triangular array notation of Equation (11.9) we need only write $X_{ni} = X_i$, $i = 1, \ldots, m_n$, where $m_n = n$. This then gives $\sigma^2_{ni} = \tau$ and $s^2_n = n\tau$.

We then need to evaluate the limit (11.10) defining the Lyapounov condition. Given our assumptions we may write

$$s_n^{-(2+\delta)} \sum_{i=1}^{m_n} E[|X_{ni} - \mu|^{2+\delta}] = K\tau^{-(1+\delta/2)} n^{-\delta/2},$$

so that the condition holds. Then by Theorem 11.19 $n^{-1/2}(S_n - \mu)/\sqrt{\tau} \xrightarrow{d} N(0,1)$. See also Problem 11.24. ∎

Of course, the *iid* assumption of Example 11.11 will be too limiting for our purposes. We therefore offer the following corollary to the Lyapounov theorem which places a bound on the standardized absolute moment (Equation (1.46)). The proof makes use of the vector p-norm (Section C.8.1).

Theorem 11.20 (Corollary to the Lyapounov Central Limit Theorem) Suppose we are given a triangular array of independent random variables of Equation (11.9). Let $\dot{\gamma}_k[X] = E[|X - E[X]|^k]/\text{var}[X]^{k/2}$ be the standardized absolute moment of any random variable X. Suppose there exists $\delta > 0$ and finite K such that $\dot{\gamma}_{2+\delta}[X_{ni}] \leq K$ for all elements of $\{X_{ni}\}$. Define the sequence of vectors $\boldsymbol{\tau}_n = (\sigma_{n1}, \ldots, \sigma_{nm_n}) \in \mathbb{R}^{m_n}$, $n \geq 1$. Then Lyapounov's condition of Equation (11.10) holds if $\lim_{n\to\infty} \|\boldsymbol{\tau}_n\|_{2+\delta} / \|\boldsymbol{\tau}_n\|_2 = 0$. ∎

Proof. We may write $s_n^2 = \|\boldsymbol{\tau}_n\|_2^2$. Under the given conditions the limit of Equation (11.10) satisfies the inequality

$$\lim_{n\to\infty} \|\boldsymbol{\tau}_n\|_2^{-(2+\delta)} \sum_{i=1}^{m_n} E[|X_{ni} - \mu_{ni}|^{2+\delta}] \leq \lim_{n\to\infty} \|\boldsymbol{\tau}_n\|_2^{-(2+\delta)} K \sum_{i=1}^{m_n} \sigma_{ni}^{2+\delta}$$

$$= \lim_{n\to\infty} K \left[\|\boldsymbol{\tau}_n\|_{2+\delta} / \|\boldsymbol{\tau}_n\|_2\right]^{2+\delta},$$

from which the theorem follows directly. □

Suppose we are given an independent sequence X_i, $i \geq 1$ for which $E[X_i] = \mu_i$, $\text{var}[X_i] = \sigma_i^2$. Let $\mathbf{a}_n = (a_{n1}, \ldots, a_{nn}) \in \mathbb{R}^n$, $n \geq 1$, be a sequence of vectors of increasing dimension. Many sequences of estimators for an estimand $\theta \in \mathbb{R}$ take the form

$$\hat{\theta}_n = \sum_{i=1}^{n} a_{ni} X_i, \quad n \geq 1. \tag{11.11}$$

This generates a triangular array $\{X_{ni}\}$ by setting $X_{ni} = a_{ni} X_i$. Then suppose the moment condition of Theorem 11.20 holds for the original sequence, that is, for some $\delta > 0$ we have $\dot{\gamma}_{2+\delta}[X_i] \leq K$ for all i. Because the standardized moments are scale invariant, we will also have bound $\dot{\gamma}_{2+\delta}[X_{ni}] \leq K$ for all X_{ni}, so that the assumptions of Theorem 11.20 will hold.

To fix ideas, suppose $\sigma_i^2 \equiv \sigma^2$. Then, using the notation of Theorem 11.20, we have $\boldsymbol{\tau}_n = \sigma \mathbf{a}_n$, so that $(\hat{\theta}_n - E[\hat{\theta}_n])/\text{var}[\hat{\theta}_n]^{1/2} \xrightarrow{N} (0,1)$ if

$$\lim_{n\to\infty} \|\boldsymbol{\tau}_n\|_{2+\delta} / \|\boldsymbol{\tau}_n\|_2 = \lim_{n\to\infty} \|\mathbf{a}_n\|_{2+\delta} / \|\mathbf{a}_n\|_2 = 0. \tag{11.12}$$

Example 11.12 (Asymptotic Normality for Simple Linear Regression) As an example, consider the simple linear regression model $Y_i = \beta_0 + \beta_1 x_i + \epsilon_i$, $i = 1, \ldots, n$, where $E[\epsilon_i] = 0$ (Section 6.2.1). We can consider a limiting case of the model, allowing $\epsilon_1, \epsilon_2, \ldots$ to be an independent sequence of zero-mean random variables with constant variance σ^2 (not assumed to be normally distributed), and imagining some sequence of predictor variables x_i, $i \geq 1$. The BLUE of the coefficient β_1 is

$$\hat{\beta}_1[n] = \frac{\sum_{i=1}^{n}(x_i - \bar{x}_n) Y_i}{\sum_{i=1}^{n}(x_i - \bar{x}_n)^2}, \tag{11.13}$$

noting that the sample mean $\bar{x}_n = n^{-1} \sum_{i=1}^{n} x_i$ depends on n. Comparing Equations (11.11) and (11.12) we can equate the coefficients

$$a_{ni} = \frac{(x_i - \bar{x}_n)}{\sum_{i=1}^{n}(x_i - \bar{x}_n)^2}.$$

Then set $\mathbf{b}_n = ((x_1 - \bar{x}_n), \ldots, (x_n - \bar{x}_n))$. In this case condition (11.12) becomes

$$\lim_{n\to\infty} \|\mathbf{b}_n\|_{2+\delta} \,/\, \|\mathbf{b}_n\|_2 = 0, \tag{11.14}$$

for any $\delta > 0$. In Example 6.7 it was shown that

$$\mathrm{var}[\hat{\beta}_1[n]] = \frac{\sigma^2}{\sum_{i=1}^n (x_i - \bar{x}_n)^2},$$

so we must have, at least, $\sum_{i=1}^n (x_i - \bar{x}_n)^2 \to_n \infty$, or equivalently, $\|\mathbf{b}_n\|_2 \to_n \infty$. From Section 11.4 we may also conclude that this is a necessary and sufficient condition for strong consistency of $\hat{\beta}_1[n]$. However, asymptotic normality need not follow from strong consistency. For example, we may select the predictor variables x_i so that $(x_i - \bar{x}_n) \approx \rho^i$, $\rho > 1$. Clearly, we will have $\|\mathbf{b}_n\|_2 \to_n \infty$, but condition (11.14) will fail, since $\|\mathbf{b}_n\|_k = O(\rho^n)$ for any $k \geq 1$. In addition, suppose each Y_i possesses an MGF, with $\kappa_3[Y_i] \equiv \kappa_3 \neq 0$ then

$$\kappa_3[\hat{\beta}_1[n]] = \kappa_3 \frac{\sum_{i=1}^n (x_i - \bar{x}_n)^3}{\left(\sum_{i=1}^n (x_i - \bar{x}_n)^2\right)^2} = \kappa_3(\|\mathbf{b}_n\|_3 \,/\, \|\mathbf{b}_n\|_2)^3.$$

If Equation (11.14) fails for $\delta = 1$, then asymptotic normality cannot hold. ■

11.11 Weak Convergence of Random Vectors

Because an MGF need not exist, we cannot base a theory of weak convergence on them (although they are worth studying for other reasons). In place of the MGF we can use the characteristic function, which we introduce directly in its multivariate form (compare to the multivariate MGF, Section 2.7). Suppose $\mathbf{X} \in \mathbb{R}^m$ is a random vector. For any $\mathbf{t} \in \mathbb{R}^m$, the characteristic function (CF) is evaluated by

$$\phi_{\mathbf{X}}(\mathbf{t}) = E[e^{i\mathbf{t}^T\mathbf{X}}],$$

where i is the imaginary unit. By Euler's formula we have $e^{ix} = \cos(x) + i\sin(x)$, so that $\phi_{\mathbf{X}}$ maps $\mathbb{R}^m$ to $\mathbb{C}$. The expectation can be evaluated separately for the real and imaginary part of the integrand, so that $\phi_{\mathbf{X}}(\mathbf{t}) = E[\cos(\mathbf{t}^T\mathbf{X})] + iE[\sin(\mathbf{t}^T\mathbf{X})]$. The integrands are bounded, therefore $\phi_{\mathbf{X}}(\mathbf{t})$ always exists. The characteristic function is also invertible. This means it uniquely determines the distribution, so can serve in this sense much the same purpose as the MGF, but with no restriction on $\mathbf{X}$. The role of the characteristic function in the theory of weak convergence is made clear by Lévys continuity theorem (Durrett, 2010).

Theorem 11.21 (Lévy's Continuity Theorem) Let $\mathbf{X}_n \in \mathbb{R}^m$, $n \geq 1$ be a sequence of random vectors with CFs $\phi_{\tilde{X}_n}(\mathbf{t})$.

(i) Suppose $\phi_{\mathbf{X}_n}(\mathbf{t})$ converges pointwise to some function $\phi^*(\mathbf{t})$ which is continuous at $\mathbf{t} = \mathbf{0}$. Then $\phi^*(\mathbf{t})$ is the CF of some random vector $\mathbf{X}^* \in \mathbb{R}^m$ for which $\mathbf{X}_n \xrightarrow{d} \mathbf{X}^*$.

(ii) Suppose $\mathbf{X}^* \in \mathbb{R}^m$ possesses CF ϕ^*. Then $\mathbf{X}_n \xrightarrow{d} \mathbf{X}^*$ *iff* $\phi_{\tilde{X}_n}(\mathbf{t}) \to_n \phi^*(\mathbf{t})$. ■

One important consequence of the uniqueness property of a CF is refered to as the Cramér-Wold device. This states that the distribution of a random vector $\mathbf{X}$ is entirely determined by the distributions of all linear combinations $\mathbf{t}^T\mathbf{X}$. This allows use to easily extend any CLT to the mulivariate case.

Theorem 11.22 (Cramér-Wold Theorem) A sufficient condition for $\mathbf{X}_n \xrightarrow{d} \mathbf{X} \in \mathbb{R}^m$ is that $\mathbf{t}^T\mathbf{X}_n \xrightarrow{d} \mathbf{t}^T\mathbf{X} \in \mathbb{R}$ for any vector $\mathbf{t} \in \mathbb{R}^m$. ∎

Proof. The theorem follows from the continuous mapping theorem (Theorem 11.17) and Lévy's continuity theorem (Theorem 11.21) (Billingsley, 1995). □

The application of Theorem 11.22 is straightforward. Suppose we are given a sequence of random vectors $\mathbf{X}_n = (X_{n1}, \ldots, X_{nm})$, $n \geq 1$. We write the marginal means $\mu_{nj} = E[X_{nj}]$, and normalize by the transformation $\mathbf{Z}_n = \sqrt{n}(X_{n1} - \mu_{n1}, \ldots, X_{nm} - \mu_{nm})$ (this normalization would be appropriate if, for example, each X_{nj} was a sample mean). Next, assume that the sequence of covariance matrices possesses a finite elementwise limit $\Sigma_{\mathbf{Z}_n} \to_n \Sigma$. The main task is to verify that for any nonzero $\mathbf{t} \in \mathbb{R}^m$ we have $\mathbf{t}^T\mathbf{Z}_n \xrightarrow{d} N(0, \sigma_{\mathbf{t}}^2)$ for some variance $\sigma_{\mathbf{t}}^2$. If so, that variance must be $\sigma_{\mathbf{t}}^2 = \mathbf{t}^T\Sigma\mathbf{t}$, so by Theorem 11.22 we must have $\mathbf{Z}_n \xrightarrow{d} N(0, \Sigma)$.

We will make use of the following example in subsequent chapters.

Example 11.13 (Limiting Distribution for the Multinomial Vector) Consider a multinomial random vector $\mathbf{N} = (N_1, \ldots, N_m)$ with frequencies $\mathbf{P} = (\theta_1, \ldots, \theta_m)$, with $\sum_i N_i = n$. A suitable normalizing transformation would be

$$\mathbf{Z}_n = \sqrt{n}(\hat{\theta}_1 - \theta_1, \ldots, \hat{\theta}_{m-1} - \theta_{m-1}),$$

where $\hat{\theta}_j = N_j/n$ (we drop N_m from $\mathbf{N}$ due to the linear constraint). Note that $\sqrt{n}\mathbf{t}^T\mathbf{Z}_n$ can always be expressed as a sum of *iid* zero-mean random variables, so the elements of $\mathbf{Z}_n$ converge weakly to a normal distribution. From Equation (2.4) the covariance matrix of $\mathbf{Z}_n$ does not depend on n, and can be given elementwise by $\{\Sigma\}_{ii} = \theta_i(1 - \theta_i)$ and $\{\Sigma\}_{ij} = -\theta_i\theta_j$, $i \neq j$. By Theorem 11.22 it follows that $\mathbf{Z}_n \xrightarrow{d} N(0, \Sigma)$. ∎

11.12 Problems

Problem 11.1 Suppose $X_1, X_2, \ldots$ is an *iid* sequence of random variables with CDF $F(x) = 1 - \bar{F}(x)$. Let $U_n = \max_{i \leq n} X_i$, and let $F_{U_n}(u) = 1 - \bar{F}_{U_n}(u)$ be its CDF.

(a) Show that $F_{U_n}(u) = F^n(u)$.
(b) Use Boole's inequality to show that $\bar{F}_{U_n}(u) \leq n\bar{F}(u)$.
(c) Evaluate the limit $\lim_{u\to\infty} \bar{F}_{U_n}(u)/(n\bar{F}(u))$.
(d) What does this limit say about the sharpness of Boole's inequality for this application?
(e) What does this limit say about the dependence of the distribution of U_n on the dependence structure of $X_1, \ldots, X_n$? See also Asmussen and Rojas-Nandayapa (2008).

Problem 11.2 Let X be an exponential random variable with mean 1. We will further explore the tail probability bound $P(X \geq t) \leq \inf_{k\geq 1} k!/t^k$, $t > 0$ of Example 1.3, using the following informal approach. One form of Stirling's approximation for the factorial is $\log n! = n \log n - n + O(\log n)$. We may then derive an approximation to the tail probability bound by first applying Stirling's approximation, then minimizing with respect to k as though it were a continuous variable. How does this approximation compare to the exact value of $P(X \geq t)$?

Problem 11.3 Let $X_1, \ldots, X_n$ be an *iid* sample from a uniform density on the unit interval $(0, 1)$. Find a sequence of transformations $Y_n = g_n(X_{(1)})$ for which $Y_n \xrightarrow{d} exp(1)$.

Problem 11.4 Let $X_1, \ldots, X_n$ be an *iid* sample from a uniform density on the unit interval $(0,1)$. Find a sequence of transformations $Y_n = g_n(X_n)$ for which $Y_n \stackrel{d}{\to} exp(1)$.

Problem 11.5 Let $X_n \sim hyper(k, M_n, N_n)$, where $N_n \to_n \infty$ and $M_n/N_n \to_n p \in (0,1)$. Show that $X_n \stackrel{d}{\to} X$, where $X \sim bin(k,p)$. **HINT:** One approach is to verify that $P(X_n = j) \to_n P(X = j)$ for any j.

Problem 11.6 We are given a sequence of independent random variables $X_1, X_2, \ldots$. Suppose X_i has mean 0 and variance σ_i^2, $i = 1, 2, \ldots$, so that the mean, but not the variance, is constant. Find conditions on the sequence σ_i^2 for which X_i converges to zero (a) in probability; (b) in L^2; (c) with probability one.

Problem 11.7 Let $\{X_t\}$, $t \in \mathcal{T}$ be any collection of random variables. Prove the following two statements concerning uniform integrability:

(a) If $\{X_t\}$ is uniformly integrable, then $\sup_{t\in\mathcal{T}} E[|X_t|] < \infty$.
(b) Suppose there exists $Y \geq |X_t|$, $t \in \mathcal{T}$, *wp*1, for which $E[Y] < \infty$. Then $\{X_t\}$ is uniformly integrable.

Problem 11.8 Verify that the sequence of Example 11.2 is not uniformly integrable.

Problem 11.9 Suppose X_n is an orthogonal process, that is, $E[X_n] = 0$, the sequence X_n is uncorrelated, and $\sup_n E[X_n^2] < \infty$. Prove that $\bar{X}_n \stackrel{L^2}{\to} 0$, where $\bar{X}_n = n^{-1}\sum_{i=1}^n X_i$.

Problem 11.10 Suppose Example 11.3 is generalized by allowing $E[X_i] = \mu_i$ to vary with i, and eliminate the assumption that $\text{var}[X_i] = \sigma_i^2$ is bounded by a finite constant. Give conditions on μ_i and σ_i^2 under which $n^{-1}\sum_{i=1}^n X_i \stackrel{wp1}{\to} c$ for some finite constant c.

Problem 11.11 The SLLN of Theorem 11.10 and Example 11.3 requires the existence of a finite second moment. If we are willing to assume that fourth order moments are finite and bounded we may derive a SLLN using a much simpler argument. Suppose $X_1, X_2, \ldots$ is a sequence of independent random variables with zero mean. Show that if $\sup_n E[X_n^4] < \infty$ then $\bar{X}_n \stackrel{wp1}{\to} 0$, where $\bar{X}_n = n^{-1}\sum_{i=1}^n X_i$. **HINT:** Use Markov's inequality to bound $P(|\bar{X}_n| > \epsilon)$ using the fourth order moments, then apply Borel-Cantelli lemma I.

Problem 11.12 Suppose $X_1, X_2, \ldots$ is a sequence of independent random variables with $E[X_n] = 0$ and $\sup_n \text{var}[X_n] < \infty$. We may obtain a weaker version of the law of the iterated logarithm while relaxing somewhat the regularity conditions. Using Theorem 11.10 show that $\bar{X}_n = n^{-1}\sum_{i=1}^n X_i = o(n^{-1/2+\delta})$ *wp*1 for any $\delta > 0$. **HINT**: The proof depends on the careful selection of the constants b_n.

Problem 11.13 In this problem we prove Lemma 1 of Taylor (1974). Let $x_1, x_2, \ldots$ be any sequence of real numbers, and assume $x_1 \neq 0$. Consider the partial sums

$$T_n = \sum_{j=1}^n \frac{x_j^2}{\left(\sum_{i=1}^j x_i^2\right)^2}, \quad n \geq 1.$$

Then $T_n < 2/x_1^2$, $n \geq 1$. **HINT:** Maximize T_n with respect to $x_2, \ldots, x_n$. Note that only the final term of T_n depends on x_n, then use an iterative argument.

Problem 11.14 Let X_n, $n \geq 1$, be an independent sequence of random variables, with $E[X_i] = \mu$, and $\text{var}[X_i] = \sigma_i^2 < \infty$. For each n we may construct a linear unbiased estimated of μ, which may be expressed $\hat{\mu}_n = \sum_{i=1}^n a_{ni}X_i$. Give necessary and sufficient conditions for the existence of such a sequence $\hat{\mu}_n$, $n \geq 1$, which converges (a) weakly to μ; (b) strongly to μ. **HINT:** An obvious candidate for $\hat{\mu}_n$ would be the best linear unbiased estimator (BLUE) of Example 6.1. Then make use of Theorem 11.10 and the argument surrounding Equation (11.1).

Problem 11.15 Verify the result of Example 11.7 using statement (v), then (vi), of Theorem 11.12. **HINT:** It may simplify matters to make use of the characterization of an open set used in the proof of Theorem 11.13.

Problem 11.16 Let (X_n, Y_n), $n \geq 1$ be an *iid* sequence of observations from a bivariate distribution with a finite positive definite covariance matrix.

(a) Decompose the sample variance $S_n^2 = (n-1)\sum_{i=1}^n (X_i - \bar{X}_n)^2$ into a mapping $S_n^2 = H(U_n, W_n, n/(n-1))$, where $U_n = \bar{X}_n = n^{-1}\sum_{i=1}^n X_n$, $W_n = n^{-1}\sum_{i=1}^n X_i^2$. Use the continuous mapping theorem and a SLLN to prove that $S_n^2 \overset{wp1}{\to} \text{var}[X_1]$.
(b) Using an approach similar to Part (a), prove that the sample correlation $r^2 = \sum_{i=1}^n (X_i - \bar{X}_n)(Y_i - \bar{Y}_n) / \left(\sum_{i=1}^n (X_i - \bar{X}_n)^2\right)^{1/2} \left(\sum_{i=1}^n (Y_i - \bar{Y}_n)^2\right)^{1/2}$ converged *wp*1 to the correlation coefficient $\text{cor}[X_1, Y_1]$.

Problem 11.17 Let (X_n, Y_n), $n \geq 1$ be an *iid* sequence of observations from a bivariate distribution with a finite positive definite covariance matrix Σ. Suppose $\mu = E[X_1] = E[Y_1]$, $\sigma^2 = \text{var}[X_1] = \text{var}[Y_1]$, and let ρ be the correlation coefficient.

(a) Suppose Σ is known. Find the best linear unbiased estimator (BLUE) $\hat{\mu}$ of μ (Section 6.3).
(b) Derive the variance $\sigma_{\hat{\mu}}^2$ of $\hat{\mu}$.
(c) Find the limiting distribution of $\sqrt{n}(\hat{\mu} - \mu)/\sigma_{\hat{\mu}}$ as $n \to \infty$. Show how this can be used to derive an approximate level $1 - \alpha$ confidence interval for μ.
(d) Next, suppose Σ is unknown. Let $\hat{\Sigma}$ be an estimate of Σ constructed by using the sample variances and covariance in place of the unknown values. Then let $\hat{\mu}^*$ be an estimate of μ obtained by replacing Σ with $\hat{\Sigma}$ in $\hat{\mu}$. Finally, let $\hat{\sigma}_{\hat{\mu}}^2$ be an estimate of $\sigma_{\hat{\mu}}^2$ obtained by replacing Σ with $\hat{\Sigma}$ in the expression derived in Part (b). Find the limiting distribution of $\sqrt{n}(\hat{\mu}^* - \mu)/\hat{\sigma}_{\hat{\mu}}$ as $n \to \infty$. Show how this can be used to derive an approximate level $1 - \alpha$ confidence interval for μ.

Problem 11.18 Suppose $X \sim gamma(\tau, \alpha)$. Using the method of Example 11.9 show that the distribution of X approaches the normal as $\alpha \to \infty$.

Problem 11.19 Suppose $X \sim bin(n, p)$. Using the method of Example 11.9 show that the distribution of X approaches the normal as $n \to \infty$.

Problem 11.20 Suppose for $n = 1, 2, \ldots$, $X_n \sim bin(n, p_n)$, where $\lambda = np_n$ for a finite constant λ.

(a) By using moment generating functions show that $\{X_n\}$ converges in distribution to a random variable with a Poisson distribution.
(b) Suppose we compute the probability $q_\lambda = P(X \leq 8)$, where X is a Poisson random variable with mean $\lambda = 10$. Then we should have $q_{n,p} \approx q_\lambda$ where $q_{n,p} = P(Y \leq 8)$, for $Y \sim bin(n, p)$ with $p = \lambda/n$, provided n is large enough. To get a sense of how large n should be, write a computer program to construct a plot of $q_{n,p}$ against n, for $n = 10, 11, \ldots, 199, 200$, in each case setting $p = \lambda/n$. Superimpose on the plot a horizontal line at q_λ. Then find the smallest n for which $|q_{n,p} - q_\lambda| \leq 0.01$.

Problem 11.21 Let $\|\cdot\|_p$ be the vector p-norm. Prove that $\|\mathbf{x}\|_r \leq \|\mathbf{x}\|_s$ for any $r \geq s$. **HINT:** Define a new vector $\mathbf{y} \in \mathbb{R}^n$ with elements $y_i = |x_i| / \|\mathbf{x}\|_p$, assuming $\mathbf{x} \neq \mathbf{0}$. Then $\|\mathbf{y}\|_p = 1$.

Problem 11.22 Consider the multiple linear regression model $\mathbf{Y}_n = \mathbf{X}_n\boldsymbol{\beta} + \boldsymbol{\epsilon}_n$ where $\mathbf{Y}_n$ is an $n \times 1$ response vector, $\mathbf{X}_n$ is a $n \times q$ matrix, $\boldsymbol{\beta}$ is a $q \times 1$ vector of coefficients, and $\boldsymbol{\epsilon}_n$ is an $n \times 1$ vector of error terms $\epsilon_1, \ldots, \epsilon_n$. We will assume these error terms are *iid* with mean zero, and ϵ_1 has CGF $c(t)$, but are not necessarily normally distributed.

Then assume for $j = 1, \ldots, q$, we have an indefinite sequence of constants $x_{1j}, x_{2j}, \ldots$, so that we have a sequence of matrices $\mathbf{X}_n$ with element (i,j) equal to x_{ij}. Let $\mathbf{x}_{j,n} = (x_{1j}, \ldots, x_{nj})$. Let $\| \cdot \|_p$ be the vector p-norm. Assume the following conditions hold:

(i) For $j = 1, \ldots, q$,

$$\alpha_j \leq \liminf_{n\to\infty} \frac{\|\mathbf{x}_{j,n}\|_2}{\sqrt{n}} \leq \limsup_{n\to\infty} \frac{\|\mathbf{x}_{j,n}\|_2}{\sqrt{n}} \leq \beta_j$$

for some finite constants $\beta_j \geq \alpha_j > 0$.

(ii) For $j = 1, \ldots, q$, $\lim_{n\to\infty} \frac{\|\mathbf{x}_{j,n}\|_3}{\sqrt{n}} = 0$.

(iii) The minimum eigenvalue of $\mathbf{X}_n^T\mathbf{X}_n$ is bounded below by $\lambda_L n$ for some finite constant $\lambda_L > 0$.

(iv) The maximum eigenvalue of $\mathbf{X}_n^T\mathbf{X}_n$ is bounded above by $\lambda_U n$ for some finite constant $\lambda_U > 0$.

Suppose $\hat{\boldsymbol{\beta}}_n$ is the least squares estimate of $\boldsymbol{\beta}$. Show that $\sqrt{n}(\hat{\boldsymbol{\beta}}_n - \boldsymbol{\beta})$ is approximately normal with covariance matrix Σ_n, and that the elements of Σ_n are of order $O(1)$. **HINT:** For a positive-definite matrix $\mathbf{A}$, we have

$$\lambda_{min}(\mathbf{A}) = \min_{\|x\|=1} \mathbf{x}^T\mathbf{A}\mathbf{x}, \quad \lambda_{max}(\mathbf{A}) = \max_{\|x\|=1} \mathbf{x}^T\mathbf{A}\mathbf{x}.$$

Also, the eigenvalues of $\mathbf{A}^{-1}$ are the reciprocals of the eigenvalues of $\mathbf{A}$. Then show that any diagonal element of $[\mathbf{X}_n^T\mathbf{X}_n]^{-1}$ is greater than $1/\lambda_{max}(\mathbf{X}_n^T\mathbf{X}_n)$, and no element of $[\mathbf{X}_n^T\mathbf{X}_n]^{-1}$ is greater than order $O(1/n)$. Finally, make use of Problem 11.21.

Problem 11.23 Suppose $n_1, n_2, \ldots$ is some known sequence of positive integers. Assume $n_i \geq 3$. Then suppose observations $Y_1, Y_2, \ldots$ come in the form $Y_i = \sum_{j=1}^{n_i} X_{ij}$, where the values X_{ij} are independent and identically distributed random variables with mean $\mu = E[X_{11}]$ and finite central moments $\bar{\mu}_k = E[(X_{11} - \mu)^k]$, $k \geq 1$. Assume μ is known. Note that only the sums Y_i are observed.

(a) For any $n \geq 1$ define the set of partial observations $\mathbf{Y}_n = (Y_1, \ldots, Y_n)$. Show that there exists an unbiased estimator of $\bar{\mu}_3$, say S_n, which is a mapping of $\mathbf{Y}_n$.

(b) Show that if the sequence $n_1, n_2, \ldots$ satisfies the condition $\sum_{i=1}^{\infty} n_i^{-1} = \infty$, the estimators S_n may be constructed so that $\lim_{n\to\infty} \text{var}[S_n] = 0$. Note that if $n_i \approx i^m$, this condition is met if $m \leq 1$, but not if $m > 1$. In general, a consistent estimator of $\bar{\mu}_3$ can exist if n_i approaches ∞, as long as the rate of increase is small enough.

HINT: First show that $\text{var}[Y_i] = Cn_i^m + o(n_i^m)$ for some constants C, m. Then, taking $\text{var}[Y_i] \approx Cn_i^m$, find an approximate best linear unbiased estimator (BLUE) of $\bar{\mu}_3$. Note that the exact value of C doesn't need to be known to apply this method.

Problem 11.24 The CLT conditions of Example 11.11 can be generalized in various ways. Suppose, for example, that we are given an independent sequence X_i, $i \geq 1$ for which $E[X_1] = \mu_i$, $\text{var}[X_i] = \tau_i$ with $\inf_i \tau_i > 0$. Assume in addition that $\sup_i E[|X_i - \mu_i|^{2+\delta}] = K < \infty$ for some $\delta > 0$. Use Theorem 11.19 to prove that $(S_n - E[S_n])/\sqrt{\text{var}[S_n]} \xrightarrow{d} N(0,1)$ where $S_n = \sum_{i=1}^{n} X_i$.

Problem 11.25 Suppose we are given an independent sequence X_i, $i \geq 1$ for which $E[|X_i|^3] < \infty$ for all i. Define sum $S_n = \sum_{i=1}^n X_i$ and let $Z_n = (S_n - E[S_n])/\text{var}[S_n]^{1/2}$. Suppose $\lim_{n\to\infty} \text{var}[S_n] = \infty$.

(a) Suppose there exists finte constant K such that $\text{var}[X_i] \leq KE[|X_i - E[X_i]|^3]$ for all i. Prove that $Z_n \xrightarrow{d} N(0,1)$.
(b) Show that the condition of Part (a) is satisfied when the X_i are bounded random variables.
(c) Suppose $X_i \sim unif(0, \theta_i)$, and $\theta_i \to_i 0$. Give conditions on θ_i under which $Z_n \xrightarrow{d} N(0,1)$.

Problem 11.26 The autoregressive model is frequently used to model correlation in the error terms of a regression model. An autoregressive process of order 1 (usually denoted as $AR(1)$) can be defined by the equation $X_n = \mu + \rho X_{n-1} + \epsilon_n$, $n \geq 1$, where ϵ_n, $n \geq 1$ are uncorrelated, with common mean and variance $E[\epsilon_1] = \mu_\epsilon$, $\text{var}[\epsilon_1] = \sigma_\epsilon^2$. We assume $|\rho| < 1$. For the moment, we only assume that X_0 has finite mean and variance.

(a) Let $\mu_n = E[X_n]$, $\sigma_n^2 = \text{var}[X_n]$, $n \geq 0$. Find iterative relationships between μ_{n+1} and μ_n; and σ_{n+1}^2 and σ_n^2.
(b) A process is stationary if it's joint distribution is invariant *wrt* the index n (i.e. time). In our example, this means $\mu_s \equiv \mu_n$, $\sigma_s^2 \equiv \sigma_n^2$. Thus, to define a stationary process $\mu_{n+1} = \mu_n = \mu_s$ and $\sigma_{n+1}^2 = \sigma_n^2 = \sigma_s^2$ must be solutions to the iterative relationships defined in Part (a). Give all such solutions.
(c) Sometimes a stationary distribution will define a limit towards which a more general class of processes will tend. In our example if $\mu_0 = \mu_s$ and $\sigma_0^2 = \sigma_s^2$ the process will be stationary. Prove that in any other case $\lim_n \mu_n = \mu_s$ and $\lim_n \sigma_n^2 = \sigma_s^2$. **HINT:** Apply the iterative relationships defined in part (b) recursively to express μ_n and σ_n^2 in terms of μ_0 and σ_0^2. This is an example of the Banach fixed point theorem.
(d) Let $\bar{X}_n$ be the sample mean of the sequence $X_1, \ldots, X_n$. Derive a limiting distribution for $\bar{X}_n$. Assume existence of all high enough order moments. **HINT:** First derive the limiting distribution of $\bar{\epsilon}_n$, the sample mean of $\epsilon_1, \ldots, \epsilon_n$.

12

Large Sample Estimation — Basic Principles

12.1 Introduction

In Chapter 11, the mathematical background required for large sample inference was presented. In this chapter, we review some of the basic principles and methods of large sample estimation. Chapter 13 will then describe the role of asymptotic theory in inference problems associated with estimating equations (this includes multivariate maximum likelihood estimation, generalized linear models (GLM), nonlinear regression and generalized estimating equations (GEE)). The volume finishes with the application of asymptotic theory to hypothesis testing (Chapter 14).

Suppose, we have sample $X_1, \ldots, X_n$ from which an estimator $\hat{\theta}_n$ of θ is constructed. As sample size n approaches ∞ we have a sequence of estimators $\hat{\theta}_n$, $n \geq 1$. It may be straightforward to verify that $\sqrt{n}(\hat{\theta}_n - \theta) \xrightarrow{d} Z \sim N(0, \tau)$ for some $\tau > 0$ (Section 11.10). In this case, we refer to τ as the asymptotic variance of $\hat{\theta}_n$ (which means that $\text{var}[\hat{\theta}_n] \approx \tau/n$). There may be, however, reasons to consider basing any inference on a transformation $g(\hat{\theta}_n)$. In this case, we may use the δ-method (Theorem 12.1) to determine the asymptotic distribution of $\sqrt{n}(g(\hat{\theta}_n) - g(\theta))$. This is described in Section 12.2, while the next two sections describe some of the reasons for using the δ-method. Variance stabilizing transformations are described in Section 12.3, and can be used, for example, to eliminate the dependence of an asymptotic distribution on an unknown parameter. It is also possible to use the δ-method to improve a normal approximation by, for example, finding transformations which reduce the skewness of $\hat{\theta}_n$. The reduction of bias is a third application. These methods are covered in Section 12.4.

The Bahadur representation theorem is introduced in Section 12.6, which gives the asymptotic distribution of sample quantiles. Similarly, a central limit theorem for U-statistics, introduced in Section 8.5, is given in Section 12.7.

A large part of Chapters 7, 8 and 9 are devoted to the problem of determining optimal estimators. For large sample estimation this problem assumes a very different character. Recall from Chapter 3 the definition of the regular parametric family (Definition 3.6). This definition enumerates regularity conditions which ensure that Fisher information is well defined (Section 3.5). This leads to one of the fundamental theorems of statistical inference, the information inequality, which is presented in Section 12.8. This gives, for regular families, a lower bound (refered to as the Cramér-Rao lower bound) on the variance of any unbiased estimator. In Chapter 13 we will see that, under certain regularity conditions, this lower bound is attained asymptotically by the MLE.

This theory can be quite useful when considering estimators which may not be optimal, but possess other desirable properties, such as robustnesss (Section 12.8.2). By comparing the variance of an estimator to the Cramér-Rao lower bound we obtain an index, referred to as asymptotic efficiency, which quantifies deviation from optimality. This is the subject of Section 12.9.

DOI: 10.1201/9781003049340-12

12.2 The δ-Method

Suppose we have a sequence of estimators $\hat{\theta}_n$ of $\theta \in \mathbb{R}$, and we may claim $\hat{\theta}_n \xrightarrow{i.p.} \theta$, and that $\sqrt{n}(\hat{\theta}_n - \theta) \xrightarrow{d} Z \sim N(0, \sigma^2)$ for some $\sigma^2 > 0$. By the continuous mapping theorem for any continuous function g we would also have $g(\hat{\theta}_n) \xrightarrow{i.p.} g(\theta)$. This suggests the possibility that there might be an advantage to such a transformation.

First, we would need to find a relationship between the limiting distribution of $\hat{\theta}_n$ and $g(\hat{\theta}_n)$. If we can establish that $\sqrt{n}(g(\hat{\theta}_n) - g(\theta)) \xrightarrow{d} Z \sim N(0, \sigma_g^2)$ for some $\sigma_g^2 > 0$, then we have one basis for comparison. Clearly, there is much more to the theory of normal approximations then establishing convergence in distribution. Asymptotically normal sequences can approach the normal probability distribution at very different rates. In statistics, the normal approximation is used to approximate confidence levels, critical values or tail probabilities. Therefore, an important problem in asymtptotic analysis is to estimate, or improve, the accuracy of an approximation such as

$$P\left(\sqrt{n}(\hat{\theta}_n - \theta)/\sigma \leq z\right) \approx \Phi(z), \tag{12.1}$$

where Φ is the CDF of the standard normal distribution, and n is a fixed value representing a sample size that might be encountered in practice. Thus, it is good practice to consider whether or not there is a transformation g for which

$$P\left(\sqrt{n}(g(\hat{\theta}_n) - g(\theta))/\sigma_g \leq z\right) \approx \Phi(z) \tag{12.2}$$

is a more accurate aproximation then (12.1). For example, the skewness of $\sqrt{n}(\hat{\theta}_n - \theta)$ will approach zero as $n \to \infty$, but for a fixed value of n, it might still be relatively large. In this case, there may be a transformation g for which the skewness of $\sqrt{n}(g(\hat{\theta}_n) - g(\theta))$ is much smaller. In such a case, (12.2) will usually be more accurate than (12.1).

Another reason for considering a transformation arises when the limiting variance depends on the unknown parameter through some functional form $\sigma^2 = \tau(\theta)$ (for example, the binomial or Poisson parametric families). A common approach to this problem is to substitute the estimate $\hat{\theta}$ for θ in the evaluaton of the variance, giving approximation $\hat{\sigma}^2 = \tau(\hat{\theta}_n)$. Of course, this compromises the accuracy of the procedure as a whole. Fortunately, in such cases, we can generally find a variance stabilizing transformation g for which the asymptotic variance σ_g^2 of the sequence $\sqrt{n}(g(\hat{\theta}_n) - g(\theta))$ does not depend on θ. We will see that such a transformation can be derived as the solution to an ordinary differential equation.

We will first verify what seems a reasonable statement, that if $\sqrt{n}(\hat{\theta}_n - \theta) \xrightarrow{d} Z$, we must also have $\hat{\theta}_n \xrightarrow{i.p.} \theta$.

Lemma 12.1 Suppose for a sequence of estimators $\sqrt{n}(\hat{\theta}_n - \theta) \xrightarrow{d} Z$, for any random variable Z for which $|Z| < \infty$ $wp1$. Then $\hat{\theta}_n \xrightarrow{i.p.} \theta$. ∎

Proof. Let $Z_n = \sqrt{n}(\hat{\theta}_n - \theta)$. By Lemma 11.1 we have $Z_n = O_p(1)$. Then $(\hat{\theta}_n - \theta) = Z_n/\sqrt{n}$. Since $1/\sqrt{n} = o_p(1)$, a second application of Lemma 11.1 gives $(\hat{\theta}_n - \theta) = o_p(1)$. □

We will make use of Taylor's remainder theorem (Theorem C.8). Assume $\hat{\theta}$ is an estimate of a parameter θ. We take θ_0 to be the "true" parameter, in the sense that $\hat{\theta}$ is calculated from a sample $\tilde{X}$ from density f_{θ_0}. We wish to consider the transformation $g(\hat{\theta})$, and we will asume g satisfies the regularity conditions required to construct a Taylor's expansion about θ_0:

$$g(\hat{\theta}) = g(\theta_0) + g^{(1)}(\theta_0)(\hat{\theta} - \theta_0) + R(\hat{\theta}; \theta_0), \tag{12.3}$$

where

$$R(\hat{\theta}; \theta_0) = g^{(2)}(\eta(\hat{\theta}))(\hat{\theta} - \theta_0)^2/2, \tag{12.4}$$

and $\eta(x) = px + (1-p)x_0$ for some $p \in [0, 1]$. The main theorem follows.

Theorem 12.1 (The δ-Method) Suppose in Equations (12.1) and (12.2) (i) $\sqrt{n}(\hat{\theta}_n - \theta_0) \xrightarrow{d} Z \sim N(0, \sigma^2)$; and (ii) g satisfies the assumptions of Theorem C.8 in an open neighbourhood of θ_0. Then

$$\sqrt{n}(g(\hat{\theta}_n) - g(\theta_0)) \xrightarrow{d} Z \sim N(0, \sigma_g^2) \quad \text{where } \sigma_g^2 = [g^{(1)}(\theta_0)]^2\sigma^2.$$

■

Proof. We may arrange Equation (12.3) as follows:

$$\sqrt{n}(g(\hat{\theta}) - g(\theta_0)) = \sqrt{n}g^{(1)}(\theta_0)(\hat{\theta} - \theta_0) + \sqrt{n}R(\hat{\theta}; \theta_0). \tag{12.5}$$

The main step is to show that $\sqrt{n}R(\hat{\theta}; \theta_0) = o_p(1)$. First, write

$$\sqrt{n}(\hat{\theta}_n - \theta_0)^2/2 = U_n V_n,$$

where $U_n = \sqrt{n}(\hat{\theta}_n - \theta_0)$ and $V_n = (\hat{\theta}_n - \theta_0)/2$. We will apply Lemma 11.1 several times in the next argument. By hypothesis $U_n \xrightarrow{d} Z$, so $U_n = O_p(1)$. From Lemma 12.1 we may conclude $V_n = o_p(1)$, and therefore $U_n V_n = o_p(1)$. We also have $|\eta(\hat{\theta}_n) - \theta_0| \leq |\hat{\theta}_n - \theta_0|$ from which it follows that $\eta(\hat{\theta}_n) \xrightarrow{i.p.} \eta(\theta_0)$, and by the continuous mapping theorem $g^{(2)}(\eta(\hat{\theta}_n)) \xrightarrow{i.p.} g^{(2)}(\eta(\theta_0))$. We may therefore conclude that $\sqrt{n}R(\hat{\theta}; \theta_0) = o_p(1)$.

By Slutsky's theorem, we may conclude from Equation (12.5) that

$$\sqrt{n}(g(\hat{\theta}) - g(\theta_0)) \xrightarrow{d} g^{(1)}(\theta_0)Z, \;\; Z \sim N(0, \sigma^2),$$

or equivalently $\sqrt{n}(g(\hat{\theta}) - g(\theta_0)) \xrightarrow{d} Z^* \sim N(0, [g^{(1)}(\theta_0)]^2\sigma^2)$. □

We next present a typical application of the δ-method.

Example 12.1 (Asymptotic Distribution of the Estimate of a Binomial Variance) Let $X \sim bin(n, p)$, and let $\hat{p} = X/n$. Then $\sqrt{n}(\hat{p} - p) \xrightarrow{d} N(0, p(1-p))$. We could therefore approximate a level $1 - \alpha$ confidence interval by $\hat{p} \pm z_{1-\alpha/2}S$, where $S^2 = \hat{p}(1-\hat{p})/n$. If the estimand was instead $\theta = p(1-p)$, then following the δ-method we would define estimator $\hat{\theta} = \hat{p}(1-\hat{p})$. A direct application of Theorem 12.1 will give $\sqrt{n}(\hat{\theta} - \theta) \xrightarrow{d} N(0, p(1-p)(1-2p)^2)$.

Notice, however, that the asymptotic variance equals zero for $p = 1/2$ (because the slope of the transformation $g(p)$ at this point is zero). This need not mean that the claimed limit is incorrect, if we interpret the distribution $N(0, 0)$ as a point mass at zero. It simply means that $(\hat{\theta} - \theta)$ converges to 0 at a faster rate with respect to n when $p = 1/2$. To see this, we can increase the expansion of Equation (12.3) by one term to get

$$g(\hat{p}) = g(p_0) + g^{(1)}(p_0)(\hat{p} - p_0) + g^{(1)}(p_0)(\hat{p} - p_0)^2/2 + R(\hat{p}; p_0). \tag{12.6}$$

For the current example this gives at $p_0 = 1/2$

$$g(\hat{p}) = 1/4 - (\hat{p} - p_0)^2/2 + R(\hat{p}; p_0), \tag{12.7}$$

since $g^{(1)}(1/2) = 0$ and $g^{(2)}(1/2) = -2$. Assuming the remainder $R(\hat{p}; p_0)$ becomes negligible compared to the other terms, we have the approximation $n4(1/4 - \hat{p}(1-\hat{p})) \sim \chi_1^2$, based on the approximation $\hat{p} \sim N(1/2, 1/4n)$ (Problem 12.3).

We can say that for $p \neq 1/2$, $[\hat{p}(1-\hat{p}) - p(1-p)] = O_p(1/\sqrt{n})$, but for $p = 1/2$, $[\hat{p}(1-\hat{p}) - p(1-p)] = O_p(1/n)$. ■

It must be remembered that convergence in distribution does not imply convergence of moments, even if the limiting distribution is normal. In particular, the quantities σ^2 or σ_g^2 in Theorem 12.1 need not model any actual variance, either exactly or as an approximation. It is worth noting especially that σ_g^2 depends on the properties of the transformation g only in the neighborhood of $\theta = \theta_0$.

Of course, we will be very much interested in the moments of the sequence and the limit. The analysis presented in this chapter will depend on moments up to the fourth order. Their existence must be verified independently of the δ-method. Fortunately, this may be easily done using the notion of uniform integrability (Section 11.4). In particular, by Theorem 11.7, if our analysis requires moments up to order r, we may impose the condition

$$\sup_n E[|\sqrt{n}(\hat{\theta}_n - \theta_0)|^{r+\epsilon}] < \infty, \quad \epsilon > 0. \tag{12.8}$$

If Equation (12.8) holds for any $\epsilon > 0$, then the kth order moments of $\sqrt{n}(\hat{\theta}_n - \theta_0)$ will converge to the kth order moment of the limit, for any $k \leq r$. In the following discussion, we may assume that this condition is met as needed.

12.3 Variance Stabilizing Transformations

In many statistical models, the variance of an estimator $\hat{\theta}$ of θ depends on the unknown parameter θ. We will write this

$$\text{var}_\theta[\hat{\theta}] = \sigma_{\hat{\theta}}^2 = n^{-1}\tau(\theta).$$

This poses a number of problems. The most obvious of these, that $\sigma_{\hat{\theta}}^2$ is unknown if θ is unknown, is only one. In many statistical models, *homoscedasticity* (constant variance) of multiple estimates or observations is sometimes important, so that statistical estimands can be rendered comparable and additive.

A common heuristic solution is to simply replace the unknown parameter θ with its estimate, that is

$$\sigma_{\hat{\theta}}^2 \approx n^{-1}\tau(\hat{\theta}).$$

However, a general method of deriving approximate variance stabilizing transformations follows from the δ-method. From Theorem 12.1 if, approximately, $\sqrt{n}(\hat{\theta}_n - \theta_0) \sim N(0, \sigma^2)$ then under suitable conditions $\sqrt{n}(g(\hat{\theta}_n) - g(\theta_0)) \sim N(0, \sigma_g^2)$, where $\sigma_g^2 = [g^{(1)}(\theta_0)]^2\sigma^2$. Suppose $\sigma^2 = \tau(\theta)$, and we can find g which satisfies

$$[g^{(1)}(\theta)]^2\tau(\theta) = c, \tag{12.9}$$

for some constant c. We have then, at least as an approximation, eliminated dependence of the variance σ_g^2 of $g(\hat{\theta})$ on θ. Of course, the solution to Equation (12.9) takes the form of the antiderivative

$$g(\theta) = \int_{\theta_0}^{\theta} c^{1/2}/\sqrt{\tau(u)}du, \tag{12.10}$$

for any suitable θ_0. We will examine several parametric models in which this situation arises.

Example 12.2 (Variance Stabilizing Transformations for the Binomial Parameter) If $X \sim bin(n, \theta)$, $\theta \in (0, 1)$, the usual estimator for θ is the sample proportion $\hat{\theta} = X/n$. Then $\sigma_{\hat{\theta}}^2 = n^{-1}\tau(\theta) = n^{-1}\theta(1-\theta)$. Recall the *inverse sine* function arcsin : $[-1,1] \to [-\pi/2, \pi/2]$, where $y = \arcsin(x)$ implies $x = \sin(y)$. Its derivative is

$$\frac{d \arcsin(x)}{dx} = \frac{1}{\sqrt{1-x^2}}, \tag{12.11}$$

so that the transformation

$$g(\hat{\theta}) = \arcsin\left(\sqrt{\hat{\theta}}\right) \tag{12.12}$$

satisfies Equation (12.9).

There has been much earlier investigation into this problem. The use of Equation (12.9) in this application is described in Bartlett (1947), as is the transformation given in Equation (12.12). In fact, many proposed methodologies take the form of refinements to the solution g generated by Equation (12.9), often by introducing a linear transformation of the parameter, yielding something like

$$g^*(\theta) = g(a\theta + b).$$

For example, in Anscombe (1948) a variance stabilizing transformation for the sample proportion $\hat{\theta}$ of the form

$$g_c(\hat{\theta}) = \arcsin\left(\frac{X+c}{n+2c}\right) \tag{12.13}$$

is proposed, and it is argued that $c = 3/8$ is the optimal value when both n and $n - E_\theta[X]$ are large (see Freeman and Tukey (1950); Fisher (1954); Laubscher (1961) for further discussion of this problem). ∎

Example 12.3 (Large Sample Inference for the Odds) Inference for the binomal parameter θ is often formulated in terms of the odds $\eta(\theta) = \theta/(1-\theta)$. If we use estimate $\hat{\eta} = \eta(\hat{\theta})$, we can approximate the large sample variance using the δ-method:

$$\sigma^2_{\hat{\eta}} \approx \left[\frac{d\eta(\theta)}{d\theta}\right]^2 \sigma^2_{\hat{\theta}} = \frac{1}{(1-\theta)^4} \times \frac{\theta(1-\theta)}{n} = \frac{\theta}{n(1-\theta)^3}.$$

The odds itself is often log-transformed. If we apply the δ-method to the transformation $g(\theta) = \log \eta(\theta)$, we obtain approximate variance

$$\sigma^2_{\log \hat{\eta}} \approx \left[\frac{d \log \eta(\theta)}{d\theta}\right]^2 \sigma^2_{\hat{\theta}} = \left[\left(\frac{1}{\theta}\right) + \left(\frac{1}{1-\theta}\right)\right]^2 \times \frac{\theta(1-\theta)}{n} = \frac{1}{n\theta(1-\theta)}.$$

The three forms of inference can be usefully compared using the coefficient of variation. For $\hat{\theta}$, we have

$$\sigma_{\hat{\theta}}/\theta = n^{-1/2}\sqrt{1-\theta}/\sqrt{\theta}.$$

The coefficient of variation diverges to ∞ as $\theta \to 0$, but converges to zero as $\theta \to 1$. Clearly, this form of inference is unreliable near the boundaries of the parameter space.

For the odds estimate $\hat{\eta}$, the coefficient of variation becomes

$$\sigma_{\hat{\eta}}/\eta(\theta) = n^{-1/2}/\sqrt{\theta(1-\theta)}.$$

The coefficient of variation for the odds $\hat{\eta}$, unlike for the sample frequency $\hat{\theta}$, is now symmetric, and diverges to ∞ at θ approaches 0 or 1.

For the log-odds estimate $\log \hat{\eta}$, the coefficient of variation becomes

$$\frac{\sigma_{\log \hat{\eta}}}{\log \eta(\theta)} = \frac{n^{-1/2}}{\sqrt{\theta(1-\theta)} \times \log(\theta/(1-\theta))}.$$

As for the odds, the absolute value of the coefficient of variation is symmetric, and diverges to ∞ at θ approaches 0 or 1.

Figure 12.1 shows plots of the coefficient of variation (absolute value) for the sample frequency $\hat{\theta}$, sample odds $\hat{\eta}$ and sample log-odds $\log \hat{\eta}$. Each plot is given as a function of the binomial parameter θ over the range $[0.001, 0.25]$, and so can be directly compared. For small θ we have $\theta \approx \eta(\theta)$, accordingly, Figure 12.1 shows that the estimation properties of $\hat{\theta}$ and $\hat{\eta}$ are similar. As $\theta \to 0$, the standard deviations of each grow rapidly relative to the estimand.

The situation is similar for the sample log-odds $\log \hat{\eta}$, except that the divergence of the standard deviation does not occur until a much smaller value of θ is reached. Estimation properties of $\log \hat{\eta}$ are therefore better than the other alternatives for small θ (and for θ near 1). ■

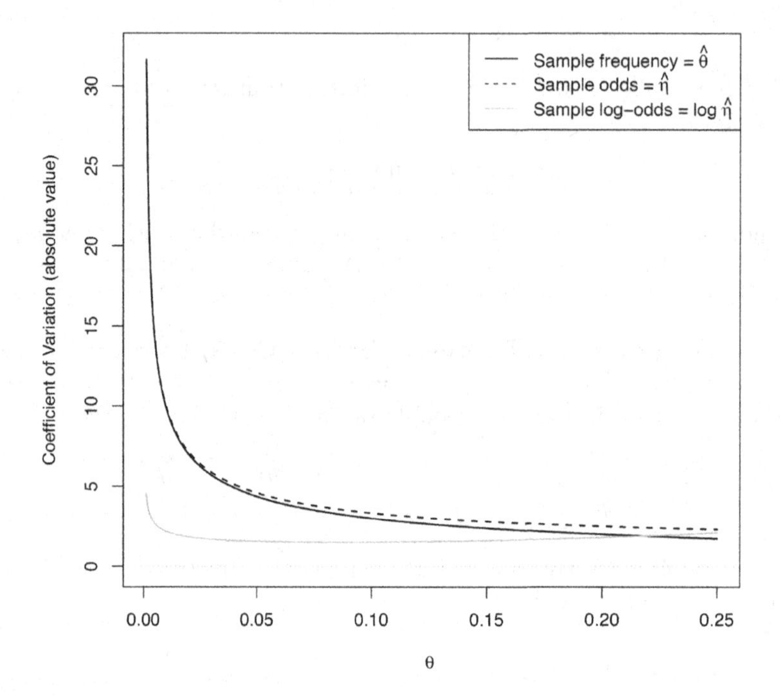

FIGURE 12.1
Coefficient of variation (absolute value) for the sample frequency, sample odds and sample log-odds. Each plot is given as a function of the binomial parameter θ (Example 12.3).

Example 12.4 (Variance Stabilizing Transformations for Scale Parameters) Recall that if a parametric family assumes the form $f_\theta(x) = \theta^{-1} f(x/\theta)$, $\theta \in (0, \infty)$, then θ is a scale parameter. In this case, it may be shown that $E_\theta[X] = \theta\mu_1$, $\text{var}_\theta[X] = \theta^2\sigma_1^2$, where μ_1, σ_1^2 are the mean and variance of density f_1. If we are given an *iid* sample from f_θ, and an estimator $\hat{\theta}$ of θ, we will generally have something like $\text{var}_\theta[\hat{\theta}] = n^{-1}\tau(\theta) = n^{-1}\theta^2\tilde{\sigma}_1^2$. In this case, it is easily verified that the transformation $g(\theta) = \log(\theta)$ satisfies Equation (12.9), and so a log-transformation of an estimate of a scale parameter is commonly employed in practice. Modifications are sometimes introduced, a common example being the introduction of a constant $c > 0$, giving the new transformation $g_c(\theta) = \log(\theta + c)$. This has the effect of bounding the argument to the logarithmic function away from zero. See Bartlett (1947) or Hoyle (1973). ■

Example 12.5 (Log-transformations and Laboratory Assays) The log-transformation is especially important in the analysis of laboratory data. The technology of many assay methods forces an approximate linear relationship between the mean and standard deviation of a measurement, and this is precisely the condition under which the log-transformation eliminates dependence of the variance on the unknown parameter. Of course, the transformation and analysis of laboratory data is itself a significant field of inquiry, and a familiarity with the literature will be important for those working in the field. See, for example, Feng *et al.* (2013) and Bland *et al.* (2013). ■

12.4 The δ-Method and Higher-Order Approximations

The δ-method of Theorem 12.1 is used to derive a limiting distribution. This relies on the first two moments of the estimator $\hat{\theta}_n$. More refined approximations can be obtained by considering higher-order moments. A new technical issue arises, however. This type of analysis requires taking the expectation of the Taylor's series approximation of Theorem 12.1 and requires some care, given the form of the remainder. Conditions under which this approach is valid vary considerably, and are often specialized to a particular application. With this in mind, we offer some general conditions in the next theorem.

Theorem 12.2 (Moment Form of the δ-Method) Let r be any positive integer. Suppose g is a transformation which is $r + 1$ times differentiable. Assume further that $g^{(r+1)}$ is monotone, and that $\sup_n E[[g^{(r+1)}(\hat{\theta}_n)]^2] < \infty$. Then

$$E[g(\hat{\theta}_n)] = g(\theta_0) + \sum_{i=1}^{r} \frac{1}{i!} g^{(i)}(\theta_0) E[(\hat{\theta}_n - \theta_0)^i] + E_n,$$

where $|E_n| \leq ME[(\hat{\theta}_n - \theta_0)^{2(r+1)}]^{1/2}$ for some finite constant M which does not depend on n. ■

Proof. The theorem essentially bounds the remainder term $R_n(\hat{\theta}_n; \theta_0)$ of the expansion

$$g(\hat{\theta}_n) = g(\theta_0) + \sum_{i=1}^{r} g^{(i)}(\theta_0)(\hat{\theta}_n - \theta_0)^i / i! + R_n(\hat{\theta}_n; \theta_0),$$

where, by Theorem C.8,

$$R_n(\hat{\theta}_n; \theta_0) = g^{(r+1)}(\eta(\hat{\theta}_n))(\hat{\theta}_n - \theta_0)^{r+1} / (r+1)!$$

and $\min(\hat{\theta}_n, \theta_0) \leq \eta(\hat{\theta}_n) \leq \max(\hat{\theta}_n, \theta_0)$. Since $g^{(r+1)}$ is monotone we have

$$|R_n(\hat{\theta}_n; \theta_0)| \leq \left(|g^{(r+1)}(\hat{\theta}_n)| + |g^{(r+1)}(\theta_0)| \right) |\hat{\theta}_n - \theta_0|^{r+1} / (r+1)!,$$

and an application of the Hölder inequality gives

$$\begin{aligned} E[|R_n(\hat{\theta}_n; \theta_0)|] &\leq E[g^{(r+1)}(\hat{\theta}_n)^2]^{1/2} E[|\hat{\theta}_n - \theta_0|^{2(r+1)}]^{1/2} / (r+1)! \\ &\quad + |g^{(r+1)}(\theta_0)| E[|\hat{\theta}_n - \theta_0|^{r+1}] / (r+1)! \\ &\leq ME[|\hat{\theta}_n - \theta_0|^{2(r+1)}]^{1/2}, \end{aligned}$$

where M is a finite constant which does not depend on n. □

We next offer four applications of Theorem 12.2.

Example 12.6 (Large Sample Approximation of Variance) Recall that while the δ-method of Theorem 12.1 offers a limiting distribution, it does not formally give an asymptotic approximation of the distribution of $g(\hat{\theta}_n)$. We can, however, use an argument similar to that used in Theorem 12.2 to construct approximations of the form

$$\text{var}[g(\hat{\theta}_n)] = g^{(1)}(\theta_0)^2\text{var}[\hat{\theta}_n] + O(n^{-3/2}), \tag{12.14}$$

for example. Note that a series approximation of the second-order moment of $g(\hat{\theta}_n)$ is required here. See Problem 12.4. ■

Example 12.7 (Large Sample Bias Correction) Suppose $\hat{\theta}_n$ is an unbiased estimate of θ, but we choose to estimate $g(\theta)$ using estimator $g(\hat{\theta}_n)$. Suppose $\sigma_n^2 = \text{var}[\hat{\theta}_n] = O(1/n)$, and that $E[|\theta_n - \theta_0|^6] = o(1/n^2)$. If the assumptions of Theorem 12.2 hold for $r = 2$ we have

$$\begin{aligned} E[g(\hat{\theta}_n)] &= g(\theta_0) + g^{(2)}(\theta_0)\sigma_n^2/2 + o(1/n), \quad \text{or} \\ \text{bias}[g(\hat{\theta}_n)] &= g^{(2)}(\theta_0)\sigma_n^2/2 + o(1/n). \end{aligned} \tag{12.15}$$

Suppose, to take an example, that $\hat{\theta}_n$ is a sequence of sample means for an *iid* sequence with mean θ and variance σ^2. Then $\sigma_n^2 = \sigma^2/n$ and $E[|\theta_n - \theta_0|^6] = O(1/n^3)$, so we have the bias approximation

$$\text{bias}[g(\hat{\theta}_n)] = 2^{-1}g^{(2)}(\theta_0)\sigma^2/n + O(1/n^{3/2}).$$ ■

Equation (12.15) is important for a number of reasons. The accuracy of a large sample normal approximation requires that $|\text{bias}[g(\hat{\theta}_n)]| \ll \text{var}[g(\hat{\theta}_n)]^{1/2}$, and under the conditions of this example we would have (12.15) we have

$$|\text{bias}[g(\hat{\theta}_n)]|/\text{var}[g(\hat{\theta}_n)]^{1/2} = O(n^{-1/2}).$$

It also potentially provides a method of bias correction when using a transformation g. This is demonstrated in the next example.

Example 12.8 (Bias Correction for Transformed Estimates of Scale Parameters) Suppose $\hat{\theta}_n$ is an unbiased estimate of a scale parameter θ, but we choose to estimate $g(\theta) = \log(\theta)$ using $\log(\hat{\theta}_n)$. We have already seen this as a variance stabilizing transformation, in that the approximate variance of $\log\hat{\theta}_n$ does not depend on θ. A similar result holds for the bias.

Suppose, following Example 12.7, $\sigma_n^2 = \text{var}[\hat{\theta}_n] = O(1/n)$, and that $E[|\hat{\theta}_n - \theta_0|^6] = o(1/n^2)$. Then $g^{(2)}(\theta) = -1/\theta^2$ is monotone. To use Theorem 12.2 with $r = 2$, we need to verify that

$$\sup_n E[|\hat{\theta}_n^{-6}|] < \infty. \tag{12.16}$$

Under these conditions the bias approximation of Equation (12.15) holds. Then, if $\hat{\theta}_n$ is a scale equivariant estimator of θ_0 we have $\sigma_n^2 = \theta^2 a_n$, where $a_n = O(1/n)$ is known (that is, it does not depend on the scale parameter). The bias becomes

$$\text{bias}[\log\hat{\theta}_n] = -a_n/2 + O(1/n^{3/2}),$$

the leading term of which does not depend on θ. We therefore have a feasible bias correction method.

As an example, suppose $\hat{\theta}_n$ is the sample mean of an *iid* sequence $X_1, X_2, \ldots$ from $X_1 \sim exp(\theta)$. Here, θ is a scale parameter, and we have $\sigma_n^2 = \theta^2/n$. Then note that $n\hat{\theta}_n \sim gamma(\theta, n)$, so that $1/(n\hat{\theta}_n)$ possesses an inverse gamma distribution, and it may be shown that $E[(n\hat{\theta}_n)^{-6}] = \left[\theta^{-6}(n-1) \times \ldots \times (n-6)\right]^{-1}$ for $n > 6$. It follows that Equation (12.16) holds, as long as we assume $n > 6$. Then the bias of $\log(\hat{\theta}_n)$ as an estimator of $\log(\theta)$ is $E[\log \hat{\theta}_n] - \log \theta_0 = -\left(2n\right)^{-1} + O(1/n^{3/2})$. We may then consider using the bias corrected estimator $\delta = E[\log \hat{\theta}_n] + (2n)^{-1}$. ■

The following example introduces the Anscombe transformation.

Example 12.9 (Variance Stabilizing Transformations for the Poisson Distribution Parameter) A variance stabilizing transformation for the Poisson parameter can be developed in much the same way as for the binomial parameter. In fact papers on this subject tend to consider in one discussion the binomial, Poisson and negative binomial distributions (Bartlett, 1947; Anscombe, 1948; Freeman and Tukey, 1950; Laubscher, 1961; Yu, 2009). See also Hoyle (1973) for a good survey article.

Suppose $X \sim pois(\theta)$, so that $E_\theta[X] = \theta$, and $\hat{\theta} = X$ is an estimator of θ. Then $\text{var}_\theta[\hat{\theta}] = \tau(\theta) = \theta$. It is easily verified that the transformation $g(\theta) = \sqrt{\theta}$ satisfies Equation (12.9). However, as for the binomial parameter, modifications of the transformation generated by Equation (12.9) have been introduced which have better properties. One example is the Anscombe transformation $g^*(\theta) = \sqrt{\theta + 3/8}$ (sometimes $2\sqrt{\theta + 3/8}$ is used, the factor 2 forcing $\text{var}_\theta[g^*(\hat{\theta})] \approx 1$).

We start with an order $r = 4$ expansion of the transformation $g_c(x) = \sqrt{x+c} = \sqrt{(x-\theta) + \theta + c}$ about parameter θ. For convenience, we write $\theta_c = \theta + c$. This gives

$$g_c(x) = \sqrt{\theta_c} + \frac{1}{2}\frac{(x-\theta)}{\theta_c^{1/2}} - \frac{1}{8}\frac{(x-\theta)^2}{\theta_c^{3/2}} + \frac{1}{16}\frac{(x-\theta)^3}{\theta_c^{5/2}} - \frac{5}{128}\frac{(x-\theta)^4}{\theta_c^{7/2}} + R.$$

We can use Theorem 12.2 to bound $E[R]$. For $r = 4$ we need derivative $g_c^{(5)}(x) = (105/32)(x+c)^{-9/2}$. A number of authors (Jones and Zhigljavsky, 2004; Žnidarič, 2009) have investigated the negative moments of the Poisson random variable Y, finding approximations of the form $E[(1/Y^+)^k] = \theta^{-k} + O\left(\theta^{-(k+1)}\right)$, where $Y^+ = Y+1$ or $Y^+ = Y \mid Y \geq 1$. As long as $c > 0$ we may conclude that

$$E[g_c^{(5)}(X)^2] = O(\theta^{-9}) + O(e^{-\theta}) = O(\theta^{-9})$$

since $g_c^{(5)}(x)$ is bounded above and $P(X = 0) = e^{-\theta}$. We conclude directly from Theorem 12.2 that $|E[R]| = O(1/\theta^2)$. Next, we build an approximation of $E[g_c(X)]$ where $X \sim pois(\theta)$, by taking the expectation of each term in the expansion, noting that the first four central moments of X are $\bar{\mu}_1 = 0$, $\bar{\mu}_2 = \theta$, $\bar{\mu}_3 = \theta$, $\bar{\mu}_4 = 3\theta^2 + \theta$. Higher-order moments are of order $\bar{\mu}_k = O(\theta^{k/2})$ for even k, and $\bar{\mu}_k = O(\theta^{(k-1)/2})$ for odd k. This gives

$$\begin{aligned} E_\theta[g_c(X)] &= \sqrt{\theta_c} - \frac{1}{8}\frac{\theta}{\theta_c^{3/2}} + \frac{1}{16}\frac{\theta}{\theta_c^{5/2}} - \frac{5}{128}\frac{3\theta^2+\theta}{\theta_c^{7/2}} + O(1/\theta^2) \\ &= \sqrt{\theta_c} - \frac{1}{8}\frac{(\theta_c - c)}{\theta_c^{3/2}} + \frac{1}{16}\frac{(\theta_c - c)}{\theta_c^{5/2}} - \frac{5}{128}\frac{3(\theta_c-c)^2 + (\theta_c - c)}{\theta_c^{7/2}} \\ &\quad + O(1/\theta_c^2), \end{aligned} \tag{12.17}$$

noting that $O(1/\theta^2) = O(1/\theta_c^2)$. Of course, we may simplify Equation (12.17) by eliminating terms of order $O(\theta_c^{-2})$, which gives

$$E_\theta[g_c(X)] = \sqrt{\theta_c} - \frac{1}{8\theta_c^{1/2}} + \frac{16c - 7}{128\theta_c^{3/2}} + O(\theta_c^{-2}).$$

Taking the square of the preceding approximation, the variance of $g_c(X)$ can then be approximated by

$$\begin{aligned} \text{var}[g_c(X)] &= E[g_c(X)^2] - E[g_c(X)]^2 \\ &= E[X+c] - \theta - c + \frac{1}{4} - \frac{1}{64\theta_c} - \frac{16c-7}{64\theta_c} + O(\theta^{-2}) \\ &= \frac{1}{4} - \frac{8c-3}{32\theta_c} + O(\theta^{-2}). \end{aligned} \tag{12.18}$$

We can conclude that $g_c(x)$ is a variance stabilizing for any $c \geq 0$. However, a higher-order analysis requires $c > 0$, and we may further conclude directly from Equation (12.18) that the approximation error for the variance estimate $\text{var}[g_c(X)] \approx 1/4$ improves from $O(\theta)$ to $O(\theta^{-2})$ upon selection of $c = 3/8$.

A discussion of the Anscombe transformation as a practical inference method requires the resolution of one more issue. Suppose we have an *iid* sample $X_1, \ldots, X_n$, $X_1 \sim pois(\theta)$. We may then apply the transformation to get sample $Y_1, \ldots, Y_n$, where $Y_i = \sqrt{X_i + 3/8}$. Then the sample mean $\bar{Y}$ is an unbiased estimator of $\eta = E[\sqrt{X_1 + 3/8}]$. One advantage of this procedure is that the variance of $\bar{Y}$ does not depend on θ, as an approximation. We may then which to construct an estimate of θ based on $\bar{Y}$ by applying the inverse transform. That is, we may use either $\hat{\theta}_1 = \bar{X}$ or $\hat{\theta}_2 = \bar{Y}^2 - 3/8$ as an estimate of θ. Interestingly, it may be shown that the bias of $\hat{\theta}_2$ does not depend on θ, as an approximation, so we may apply a bias correction. The proof is left to the reader (Problem 12.6). ■

12.4.1 Approximation of Skewness and Skewness Reduction

Suppose we have an *iid* sample $X_1, \ldots, X_n$. By the central limit theorem the sample mean $\bar{X}_n$ will have an approximately normal distribution. In general, however, a normal approximation can be improved by reducing the skewness coefficient SK, since this quantity is zero for the normal density. By the δ-method, a smooth transformation $g(\bar{X}_n)$ will also have an approximately normal distribution. However, similar to the variance stabilizing method of Section 12.3, we may be able to choose a skewness reducing transformation. Our approach will be to construct a higher-order approximation of the third central moment $\bar{\mu}_3[g(\bar{X}_n)]$, and then of the skewness coefficient SK. To do this, we construct an expansion of $\bar{\mu}_3[g(\bar{X}_n)]$. To achieve our purpose, we will need to include all terms of order $O(n^{-2})$ or greater.

To fix ideas, we will assume all order moments of X_1 exist. Then let $\bar{\mu}_k$ be the kth central moment of X_1. We also make use of the notation $\mu = E[X_1]$, $\sigma^2 = \text{var}[X_1]$, $SK = \bar{\mu}_3/\sigma^3$, $KURT = \bar{\mu}_4/\sigma^4$. We may characterize the order of the central moments of $\bar{X}_n$ as

$$|E[(\bar{X}_n - \mu)^k]| = O(n^{-\lceil k/2 \rceil}), \tag{12.19}$$

(see Problem 12.7). We also make use of the standard moment formulas:

$$\begin{aligned} &\bar{\mu}_2[\bar{X}_n] = \bar{\mu}_2/n = O(n^{-1}), \quad \bar{\mu}_3[\bar{X}_n] = \bar{\mu}_3/n^2 = O(n^{-2}), \\ &\bar{\mu}_4[\bar{X}_n] = 3\bar{\mu}_2^2/n^2 + (KURT - 3)\bar{\mu}_2^2/n^3 = O(n^{-2}). \end{aligned} \tag{12.20}$$

We start by constructing an expansion for $g(\bar{X}_n)$ about μ, setting $\bar{U}_n = \bar{X}_n - \mu$ for convenience. This gives

$$\begin{aligned} g(\bar{X}_n) = g(\mu) &+ g^{(1)}(\mu)\bar{U}_n + g^{(2)}(\mu)\bar{U}_n^2/2! \\ &+ g^{(3)}(\mu)\bar{U}_n^3/3! + g^{(4)}(\mu)\bar{U}_n^4/4! + R. \end{aligned} \tag{12.21}$$

The crucial technical problem at this point is to develop a bound for the remainder term

R in (12.21). Suppose the 5th-order derivative $g^{(5)}$ satisfies the bound $|g^{(5)}(x)| \leq M < \infty$, so that $|R| \leq M'|\bar{U}_n^5|$, for some constant $M' < \infty$. Under this assumption we may conclude $|E[R]| = O(n^{-3})$. The simplest way to force this condition is to define the expansion (12.21) on a bounded neighborhood $B_\epsilon(\mu)$ of μ. This is reasonable, since under the SLLN there exists finite N such that $\bar{X}_n \in B_\epsilon(\mu)$ for all $n \geq N$ *wp*1. Alternatively, using the method of Theorem 12.2 we would have the weaker bound $|E[R]| = O(n^{-5/2})$, which is more typical for this type of approximation. Since our intention here is to introduce a general method, it suits our purposes to simply accept the latter bound.

We may write

$$\begin{aligned}\left[g(\bar{X}_n) - E[g(\bar{X}_n)]\right]^3 = \Big\{ & g^{(1)}(\mu)\bar{U}_n + g^{(2)}(\mu)\left(\bar{U}_n^2 - \bar{\mu}_2[\bar{X}_n]\right)/2! \\ & + g^{(3)}(\mu)\left(\bar{U}_n^3 - \bar{\mu}_3[\bar{X}_n]\right)/3! \\ & + g^{(4)}(\mu)\left(\bar{U}_n^4 - \bar{\mu}_4[\bar{X}_n]\right)/4! + (R - E[R])\Big\}^3.\end{aligned}$$

The next step is to expand the right side, taking the expected values term by term, only retaining those of order no smaller than $O(n^{-2})$. While the algebra appears formidable, it is possible to anticipate which terms to include and which to exclude. For example, note that if the term $\left(\bar{U}_n^3 - \bar{\mu}_3[\bar{X}_n]\right)$ is multiplied by any other of the terms, and the expectation then taken, the resulting expression will include only terms of order $o(n^{-2})$. The same must then be true of $\left(\bar{U}_n^4 - \bar{\mu}_4[\bar{X}_n]\right)$ and $R - E[R]$. In fact, we only need to consider the terms

$$\begin{aligned}(g(\bar{X}_n) - E[g(\bar{X}_n)])^3 = & [g^{(1)}(\mu)\bar{U}_n]^3 + \\ & 3[g^{(1)}(\mu)\bar{U}_n]^2 g^{(2)}(\mu)(\bar{U}_n^2 - \bar{\mu}_2[\bar{X}_n])/2! + R'\end{aligned}$$

where all terms in the remainder R' will be of order $o(n^{-2})$ after the expectation is taken. We then have

$$\begin{aligned}E[(g(\bar{X}_n) - E[g(\bar{X}_n)])^3] = & g^{(1)}(\mu)^3\bar{\mu}_3[\bar{X}_n] + \\ & 3g^{(1)}(\mu)^2 g^{(2)}(\mu)\left(\bar{\mu}_4[\bar{X}_n] - \bar{\mu}_2[\bar{X}_n]^2\right)/2! + o(n^{-2}).\end{aligned}$$

Using the expression for $\bar{\mu}_k[\bar{X}_n]$ in Equation (12.20) we then have, after a bit more algebra,

$$E[(g(\bar{X}_n) - E[g(\bar{X}_n)])^3] = \sigma^3 g^{(1)}(\mu)^2[g^{(1)}(\mu)SK + 3g^{(2)}(\mu)\sigma]/n^2 + o(n^{-2}).$$

To approximate the skewness of $g(\bar{X}_n)$ it suffices to divide the approximation of $E[g(\bar{X}_n) - E[g(\bar{X}_n)]]^3$ by a first-order approximation of $\mathrm{var}[g(\bar{X}_n)]$ (Equation (12.14)) raised to the power 3/2. This yields an approximation of the skewness coefficient of $g(\bar{X}_n)$ of the form

$$SK[g(\bar{X}_n)] = \frac{g^{(1)}(\mu)SK + 3g^{(2)}(\mu)\sigma}{|g^{(1)}(\mu)|\sqrt{n}} + O(n^{-3/2}). \tag{12.22}$$

Skewness Reduction

The approximation of Equation (12.22) can be used in a manner similar to the variance stabilizing transformation of Equation (12.9). We can either evaluate the skewness to compare alternative methods, or reduce the skewness directly, if possible, by interpreting the leading term of Equation (12.22) as a differential equation. To do this, set $f(\mu) = g^{(1)}(\mu)$, and solve

$$f(\mu)SK + 3f^{(1)}(\mu)\sigma = 0,$$

which may be alternatively written

$$\frac{d\log f(\mu)}{d\mu} = -\frac{1}{3}\frac{SK}{\sigma}. \tag{12.23}$$

Thus, we can evaluate $\log f(\mu)$ as an antiderivative, then take $g(\mu)$ to be an antiderivative of $f(\mu)$. This approach is demonstrated in the next example.

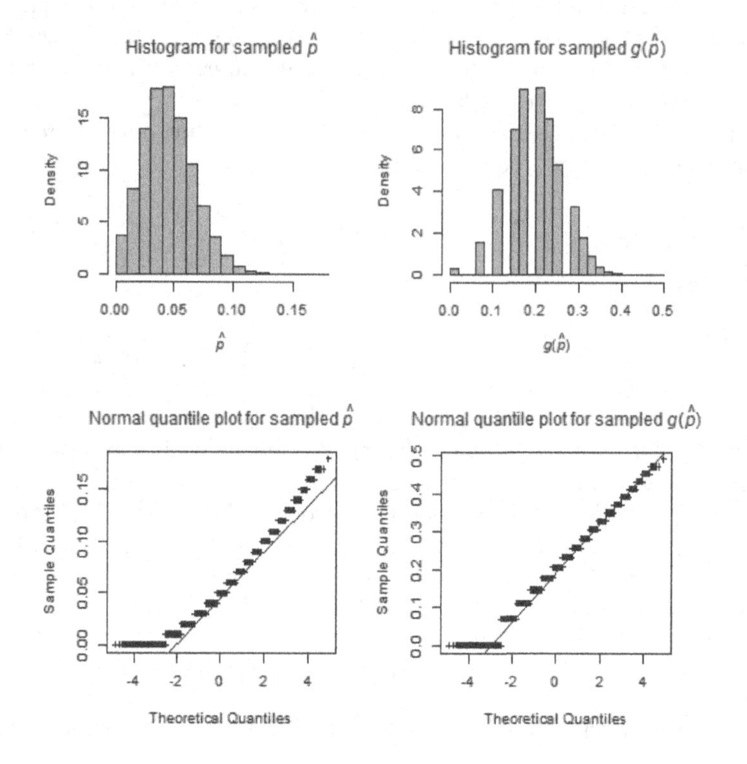

FIGURE 12.2
Histograms and normal quantile plots of $\hat{p}$ and $g(\hat{p})$ (Example 12.10). $N = 10^6$ samples from a $bin(100, 0.1)$ were simulated.

Example 12.10 (Skewness Reduction for the Binomial Distribution) As an example of skewness reduction we will consider the binomial parameter. Let $X \sim bin(n, p)$, and interpret the sample proportion $\hat{p} = X/n$ as a sample mean from an *iid* sample of n Bernoulli random variables $U_i \sim bern(p)$. Then $\mu = p$, $\sigma^2 = \mu(1-\mu)$, $SK = (1-2\mu)/\sqrt{\mu(1-\mu)}$. We can apply Equation (12.23) to solve for $f(\mu) = g^{(1)}(\mu)$:

$$\frac{d \log f(\mu)}{d\mu} = -\frac{1}{3}\frac{SK}{\sigma} = -\frac{1}{3}\frac{(1-2\mu)}{\mu(1-\mu)}.$$

This directly gives a solution $\log f(\mu) = -\frac{1}{3}\log \mu(1-\mu) + c$, where c is any constant, or $f(\mu) = a[\mu(1-\mu)]^{-1/3}$, where $a > 0$ is any positive constant. We can set

$$g(\mu) = a\int_0^{\mu} u^{-1/3}(1-u)^{-1/3}du.$$

Of course, this is just the evaluation method for the CDF $F_{\alpha,\beta}$ of the $beta(\alpha, \beta)$ distribution with $\alpha = \beta = 2/3$. If we set $a^{-1} = B(\alpha, \beta)$, the beta function, as a normalizing constant, we may use transformation $g(\mu) = F_{2/3,2/3}(\mu)$. We next give a numerical demonstration based on $N = 10^6$ simulated replications of $X \sim bin(100, 0.05)$. Histograms and normal quantile plots of $\hat{p}$ and $g(\hat{p})$ are shown in Figure 12.2. The reduction of skewness forced by the transformation is quite apparent in the respective shapes of the histograms, and in the near-linearity of the normal quantile plot for the transformed sample (here, we plot the observed quantiles against the matched quantiles from a $N(0, 1)$ distribution). ■

Example 12.11 (Skewness Reduction for the Location-Scale Model) For a location-scale model SK does not depend on the location-scale parameter. Therefore, the skewness reducing transformation of Equation (12.23) depends on the relationship between the mean μ and variance σ^2, which we may express using the variance function $\sigma^2 = v(\mu)$. The skewness reducing transformation is then the solution to the equation

$$\frac{d \log f(\mu)}{d\mu} = -\frac{1}{3}\frac{SK}{v(\mu)}. \tag{12.24}$$

For a location model (i.e. with no scale parameter) we have $v(\mu) = c$, where c does not depend on μ. In this case, the solution to (12.24) is $f(\mu) = a\exp(-\frac{SK}{3c}\mu)$, where a is any positive constant. As might be expected, the resulting transformation will place less weight on larger observations when skewness is positive ($SK > 0$), and greater weight when skewness is negative. ∎

12.5 The Multivariate δ-Method

The generalization of the δ-method to the multivariate case is straightforward. Suppose a mapping $g : \Theta \to \mathbb{R}$, $\Theta \subset \mathbb{R}^q$ possesses continuous second-order partial derivatives. By the multivariate version of Taylor's remainder theorem (Theorem C.9) the first-order Taylor's approximation in an open neighborhood of $\boldsymbol{\theta}_0 \in \Theta$ takes the form

$$g(\boldsymbol{\theta}) = g(\boldsymbol{\theta}_0) + [g_1^{(1)}(\boldsymbol{\theta}_0), \ldots, g_q^{(1)}(\boldsymbol{\theta}_0)]^T(\boldsymbol{\theta} - \boldsymbol{\theta}_0) + R(\boldsymbol{\theta}; \boldsymbol{\theta}_0),$$

where $g_j^{(1)}(\boldsymbol{\theta}) = \partial g(\boldsymbol{\theta})/\partial\theta_j$, $j = 1, \ldots, q$ are the first-order partial derivatives of g, and the remainder term $R(\boldsymbol{\theta}; \boldsymbol{\theta}_0)$ follows from Equation (C.16). Other than some additional technical detail, the proof of the multivariate case follows Theorem 12.1.

Theorem 12.3 (The Multivariate δ-Method) Suppose for a parameter space $\Theta \subset \mathbb{R}^q$, $\sqrt{n}(\hat{\boldsymbol{\theta}}_n - \boldsymbol{\theta}_0) \xrightarrow{d} \mathbf{Z} \sim N(\mathbf{0}, \Sigma)$, $\boldsymbol{\theta}_0 \in \Theta$, and $g : \Theta \to \mathbb{R}$ satisfies the assumptions of Theorem C.9 in an open neighborhood of $\boldsymbol{\theta}_0$. Then

$$\sqrt{n}(g(\hat{\boldsymbol{\theta}}_n) - g(\boldsymbol{\theta}_0)) \xrightarrow{d} Z \sim N(0, \sigma_g^2),$$

where

$$\sigma_g^2 = \left[g_1^{(1)}(\boldsymbol{\theta}_0), \ldots, g_q^{(1)}(\boldsymbol{\theta}_0)\right]^T \Sigma \left[g_1^{(1)}(\boldsymbol{\theta}_0), \ldots, g_q^{(1)}(\boldsymbol{\theta}_0)\right].$$ ∎

A typical example of the multivariate δ-method follows.

Example 12.12 (Estimate of the Log-Odds Ratio) Consider the 2 × 2 contingency table:

	A	A^c	
B	N_{11}	N_{12}	R_1
B^c	N_{21}	N_{22}	R_2
Total	C_1	C_2	n

The cell counts are assumed to form a multinomial vector $\mathbf{N} = (N_{11}, N_{12}, N_{21}, N_{22})$ from probability distribution $\mathbf{P} = (p_{11}, p_{12}, p_{21}, p_{22})$. The odds ratio is defined as

$$OR = \frac{P(A \mid B)/(1 - P(A \mid B)))}{P(A \mid B^c)/(1 - P(A \mid B^c)))} = \frac{p_{11}p_{22}}{p_{12}p_{21}}.$$

An commonly used approximate level $1-\alpha$ confidence interval for $\log(OR)$ is

$$CI = \log\left(\widehat{OR}\right) \pm z_{\alpha/2}\sqrt{{N_{11}}^{-1} + {N_{12}}^{-1} + {N_{21}}^{-1} + {N_{22}}^{-1}}, \quad (12.25)$$

where $\widehat{OR} = N_{11}N_{22}/N_{12}N_{21}$ (Van Belle *et al.*, 2004). This can be justified using the δ-method.

We take the parameter to be $\boldsymbol{\theta} = (p_{11}, p_{12}, p_{21})$, which is estimated by $\hat{\boldsymbol{\theta}} = (N_{11}/n, N_{12}/n, N_{21}/n)$. Set $p_{22} = p_{22}(\boldsymbol{\theta}) = 1 - p_{11} - p_{12} - p_{21}$. Then $\sqrt{n}(\hat{\boldsymbol{\theta}}_n - \boldsymbol{\theta}) \stackrel{d}{\to} \mathbf{Z} \sim N(0, \Sigma)$, where, following Section 2.1,

$$\Sigma = \begin{bmatrix} p_{11}(1-p_{11}) & -p_{11}p_{12} & -p_{11}p_{21} \\ -p_{11}p_{12} & p_{12}(1-p_{12}) & -p_{12}p_{21} \\ -p_{11}p_{21} & -p_{12}p_{21} & p_{21}(1-p_{21}) \end{bmatrix}.$$

The log-odds ratio is defined by transformation

$$g(\boldsymbol{\theta}) = \log p_{11} - \log p_{12} - \log p_{21} + \log(1 - p_{11} - p_{12} - p_{21}).$$

To apply Theorem 12.3 we need the vector of partial derivatives

$$\mathbf{g}^T = \left[g_{11}^{(1)}(\boldsymbol{\theta}), g_{12}^{(1)}(\boldsymbol{\theta}), g_{21}^{(1)}(\boldsymbol{\theta})\right]^T = \left[\frac{1}{p_{11}} - \frac{1}{p_{22}}, -\frac{1}{p_{12}} - \frac{1}{p_{22}}, -\frac{1}{p_{21}} - \frac{1}{p_{22}}\right].$$

Then $\sqrt{n}(g(\hat{\boldsymbol{\theta}}) - g(\boldsymbol{\theta})) \stackrel{d}{\to} Z \sim N(0, \sigma_g^2)$, where, after some algebra, we have

$$\sigma_g^2 = \mathbf{g}^T\Sigma\mathbf{g} = \frac{1}{p_{11}} + \frac{1}{p_{12}} + \frac{1}{p_{21}} + \frac{1}{p_{22}}.$$

This gives

$$\text{var}[g(\hat{\boldsymbol{\theta}})] \approx n^{-1}\left[\frac{1}{p_{11}} + \frac{1}{p_{12}} + \frac{1}{p_{21}} + \frac{1}{p_{22}}\right] \approx \left[\frac{1}{N_{11}} + \frac{1}{N_{12}} + \frac{1}{N_{21}} + \frac{1}{N_{22}}\right],$$

on which Equation (12.25) is based. See Problem 12.10 for an alternative approach. ∎

12.6 Approximating the Distributions of Sample Quantiles: The Bahadur Representation Theorem

A review of Sections 1.6 and 2.4 is recommended at this point. The definition of quantiles and sample quantiles is subject to varying conventions, but we can accept for the following analysis that based on the quantile function (Definition 1.9)

$$Q(p) = \inf\{x \in \mathbb{R} : F(x) \geq p\}, \;\; p \in (0,1). \quad (12.26)$$

If $F(x)$ possesses an inverse, then $Q(p) = F^{-1}(p)$. Whatever the case, $Q(p)$ is always well defined. Next, suppose $X_1, \ldots, X_n$ is an *iid* sample from a common distribution F. The empirical distribution of the sample is

$$F_n(x) = n^{-1}\sum_{i=1}^{n} I\{X_i \leq x\}. \quad (12.27)$$

Then $F_n(x)$ is the sample proportion of observations $X_i \leq x$. Thus, we have $nF_n(x) \sim bin(x, F(x))$. Furthermore, we can apply the definition of Equation (12.26) to the empirical distribution

$$Q_n(p) = \inf\{x \in \mathbb{R} : F_n(x) \geq p\}, \ \ p \in (0,1). \tag{12.28}$$

We accept $Q_n(p)$ as an estimate of $Q(p)$, although a variety of sample quantile estimates are commonly in use (Hyndman and Fan, 1996).

The problem of approximating the distribution of $Q_n(p)$ has achieved a remarkable unity due to the Bahadur representation theorem, which we state next.

Theorem 12.4 (Bahadur Representation Theorem (Strong Version)) Suppose $X_1, X_2, \ldots$ is an *iid* sequencee from CDF F. Let $\xi = Q(p)$ be a p-quantile, and let $\xi_n = Q_n(p)$ be the sample quantile for $X_1, \ldots, X_n$. Assume (i): $F(x)$ is twice differentiable in a neighborhood of ξ; (ii): the second derivative $F''(x)$ is bounded in a neighborhood of ξ; and (iii): $F'(\xi) = f(\xi) > 0$. Suppose we write

$$\xi_n = \xi - (F_n(\xi) - p)/f(\xi) + R_n. \tag{12.29}$$

We then have $R_n = O\left(n^{-3/4}(\log n)^{1/2}(\log\log n)^{1/4}\right)$ with probability 1. ∎

Proof. See Bahadur (1966). □

The rate $R_n = O\left(n^{-3/4}(\log n)^{1/2}(\log\log n)^{1/4}\right)$ is often simplified, possibly to $R_n = O\left(n^{-3/4}(\log n)^{-3/4}\right)$, which implies the (slightly) faster rate given in Theorem 12.4. In the seminal paper Bahadur (1966) the rate is given as $R_n = O\left(n^{-3/4}\log n\right)$ in the introductory section. To be sure, the Theorem 12.4 rate is identified in that paper, but described as "not a substantial improvement or clarification" of the rate $R_n = O\left(n^{-3/4}\log n\right)$. In Kiefer (1967) the exact convergence rate is obtained, and shown to be $O\left(n^{-3/4}(\log n)^{1/2}(\log\log n)^{1/4}\right)$ with probability 1 (this result is comparable to the law of the iterated logarithm, stated in Theorem 11.11).

The literature commonly refers to *strong* and *weak versions* of the Bahadur representation theorem, Theorem 12.4 being the strong version. The weak version relaxes the assumptions on the CDF $F(x)$, while establishing $\sqrt{n}R_n = o_p(1)$.

Theorem 12.5 (Bahadur Representation Theorem (Weak Version)) Suppose $X_1, X_2, \ldots$ is an *iid* sequencee from CDF F. Let $\xi = Q(p)$ be a p-quantile, and let $\xi_n = Q_n(p)$ be the sample quantile for $X_1, \ldots, X_n$. Assume the first derivative of $F(x)$ exists in a neighborhood of ξ, and satisfies $F'(\xi) = f(\xi) > 0$. Suppose we are given Equation (12.29). We then have $\sqrt{n}R_n = o_p(1)$ ∎

Proof. See Ghosh (1971). □

Remark 12.1 In Ghosh (1971), the main result concerns a more general version of Equation (12.29), in which (among other modifications) ξ_n is allowed to be a p_n sample quantile, where $p_n - p = O(n^{-1/2})$. Here we keep Theorems 12.4 and 12.5 comparable. ∎

An elegant sketch of the proof of Theorems 12.4–12.5 is offered in Bahadur (1966). Define

$$G_n(x) = [F_n(x) - F_n(\xi)] - [F(x) - F(\xi)]. \tag{12.30}$$

Then define a sequence of neighborhoods of ξ, $I_n = (\xi - a_n, \xi + a_n)$, $a_n = O(n^{-1/2}\log n)$, and set $H_n = \sup_{x \in I_n} |G_n(x)|$. If $F'(\xi) = f(\xi) > 0$ then, by definition, $p = F(\xi)$. A similar relationship obviously holds for the sample quantile and sample distribution function,

$$F_n(\xi_n) = p + \epsilon_n. \tag{12.31}$$

Then, using a Taylors series approximation (Section C.9), we have

$$F(x) - F(\xi) = f(\xi)\,(x - \xi) + R(x;\xi). \tag{12.32}$$

Combining Equations (12.30), (12.31) and (12.32) we have, after substituting $x = \xi_n$,

$$\begin{aligned} f(\xi)\,(\xi_n - \xi) &= F(\xi_n) - F(\xi) - R(\xi_n;\xi) \\ &= F_n(\xi_n) - F_n(\xi) - G(\xi_n) - R(\xi_n;\xi) \\ &= p - F_n(\xi) + \epsilon_n - G(\xi_n) - R(\xi_n;\xi), \end{aligned}$$

which can be rewritten as Equation (12.29), the technical details of the proof then verifying that $\sqrt{n}R_n$ approaches zero in some sense.

12.6.1 Normal Approximation for Quantiles

Both the strong and weak version of the Bahadur representation theorem imply that $\sqrt{n}R_n = o_p(1)$ (the reader can verify this for the strong version). Under the conditions of either version we have $nF_n(\xi) \sim bin(n,p)$, so by the central limit theorem, $\sqrt{n}(F_n(\xi) - p) \xrightarrow{d} Z \sim N(0, p(1-p))$, and by Slutsky's theorem

$$\sqrt{n}(\xi_n - \xi) \xrightarrow{d} Z \sim N\left(0, \frac{p(1-p)}{f(\xi)^2}\right), \tag{12.33}$$

provided the assumptions of Theorem 12.5 hold for $F(x)$ in a neighborhood of $x = Q(p)$.

We next offer an application of the Bahadur representation theorem.

Example 12.13 (Normal Approximation for the Interquartile Range (IQR)) The interquartile range (IQR) is a robust measure of the variation of a density, defined as $IQR = \tau_{iqr} = Q(0.75) - Q(0.25)$. The IQR is location invariant. For the normal distribution $N(\mu, \sigma^2)$ we have $\tau_{iqr} = 2\sigma z_{0.25} \approx 1.35\sigma$, where $z_{0.025}$ is the 0.25 critical value from the standard normal distribution $N(0,1)$.

A nonparametric estimator for the IQR is usually used, a natural choice being $\hat{\tau}_{iqr} = Q_n(0.75) - Q_n(0.25)$, The Bahadur representation theorem can used to approximate the distribution of $\hat{\tau}_{iqr}$. However, to do this we need to approximate the joint distribution of $(Q_n(0.25), Q_n(0.75))$. This, in turn, may be expressed in terms of the joint distribution of $(F_n(Q(0.25)), F_n(Q(0.75)))$, following the Bahadur representation theorem. Suppose we have, more generally, $a \leq b$, with $p_a = F(a)$, $p_b = F(b)$. Define random vector $\mathbf{Y} = (F_n(a), F_n(b))$. The mean vector is $E[\mathbf{Y}] = (p_a, p_b)$, so it remains to evaluate the covariance matrix of $\mathbf{Y}$, which will assume the form $\Sigma_{\mathbf{Y}} = n^{-1}\Sigma$. The diagonal elements are given directly by the marginal variances of $(F_n(a), F_n(b))$, in particular, $p_a(1-p_a)/n$, $p_b(1-p_b)/n$. It remains to evaluate the covariance $\text{cov}[F_n(a), F_n(b)]$. To do this, recalling that $a \leq b$, we may write

$$\begin{aligned} & E[F_n(a)F_n(b)] \\ &= \frac{1}{n^2}\sum_{i=1}^{n}\sum_{j=1}^{n} E[I\{X_i \leq a\}I\{X_j \leq b\}] \\ &= \frac{1}{n^2}\left[nE[I\{X_1 \leq a\}I\{X_1 \leq b\}] + n(n-1)E[I\{X_1 \leq a\}I\{X_2 \leq b\}]\right] \\ &= \frac{1}{n^2}\left[nE[I\{X_1 \leq a\}I\{X_1 \leq a\}] + n(n-1)E[I\{X_1 \leq a\}I\{X_2 \leq b\}]\right] \\ &= \frac{1}{n^2}\left[np_a + n(n-1)p_a p_b\right], \end{aligned}$$

which leads to $\text{cov}[F_n(a), F_n(b)] = n^{-1}p_a(1-p_b)$, giving covariance matrix

$$\Sigma_{\mathbf{Y}} = n^{-1}\Sigma, \quad \text{where} \quad \Sigma = \begin{bmatrix} p_a(1-p_a) & p_a(1-p_b) \\ p_a(1-p_b) & p_b(1-p_b) \end{bmatrix}.$$

The Cramér-Wold device allows us to conclude that

$$\sqrt{n}(F_n(a) - p_a, F_n(b) - p_b) \xrightarrow{d} \mathbf{Z} \sim N(0, \Sigma),$$

where $\mathbf{Z}$ is a bivariate normal random vector. We can now construct a normal approximation for $\hat{\tau}_{iqr}$. For convenience set $\xi_p = Q(p)$. We may then write

$$\begin{aligned} \sqrt{n}(\hat{\tau}_{iqr} - \tau_{iqr}) &= \sqrt{n}\left([Q_n(0.75) - Q_n(0.25)] - [Q(0.75) - Q(0.25)]\right) \\ &= \sqrt{n}\left(Q_n(0.75) - Q(0.75)\right) - \sqrt{n}\left(Q_n(0.25) - Q(0.25)\right) \\ &= \sqrt{n}\frac{(F_n(\xi_{0.25}) - 0.25)}{f(\xi_{0.25})} - \sqrt{n}\frac{(F_n(\xi_{0.75}) - 0.75)}{f(\xi_{0.75})} + \epsilon_n, \end{aligned}$$

where $\epsilon_n = o_p(1)$, by Theorem 12.5 (here, ϵ_n is the sum of two remainder terms which are of stochastic order $o_p(1)$). We then need to recognize the linear combination

$$\sqrt{n}(\hat{\tau}_{iqr} - \tau_{iqr}) = \mathbf{a}^T\sqrt{n}(F_n(\xi_{0.25}) - 0.25, F_n(\xi_{0.75}) - 0.75) + \epsilon_n,$$

where $\mathbf{a} = (f(\xi_{0.25})^{-1}, -f(\xi_{0.75})^{-1})$. We therefore conclude that

$$\sqrt{n}(\hat{\tau}_{iqr} - \tau_{iqr}) \xrightarrow{d} \mathbf{Z} \sim N(0, \mathbf{a}^T\Sigma\mathbf{a}),$$

where

$$\begin{aligned} \mathbf{a}^T\Sigma\mathbf{a} &= \frac{0.25(1-0.25)}{f(\xi_{0.25})^2} + \frac{0.75(1-0.75)}{f(\xi_{0.75})^2} - 2 \times \frac{0.25(1-0.75)}{f(\xi_{0.25})f(\xi_{0.75})} \\ &= \frac{1}{16}\left[\frac{3}{f(\xi_{0.25})^2} + \frac{3}{f(\xi_{0.75})^2} - \frac{2}{f(\xi_{0.25})f(\xi_{0.75})}\right]. \end{aligned}$$

■

12.7 A Central Limit Theorem for U-statistics

What makes the theory of U-statistics (Section 8.5) important to the theory of statistical inference is the near universality of their properties, coupled with the fact that a large number of commonly used statistics are U-statistics. One of these properties is the asymptotic normality of any U-statistic with finite variance. Hoeffding (1948) verifies this, using a rather extended argument covering multivariate distributions of multiple U-statistics. However, asymptotic normality follows in a straightforward manner once the resemblance of any U-statistic to an *iid* sum is recognized.

Recall the form of the U-statistic:

$$U = \sum_{(\alpha_1,\ldots,\alpha_k)\in C_{n,k}} \frac{\psi(X_{\alpha_1},\ldots,X_{\alpha_k})}{\binom{n}{k}}, \tag{12.34}$$

where $\psi(x_1,\ldots,x_k)$ is a symmetric kernel of order k. Then Theorem 8.9 presents a general method of evaluating the variance of U, making use of the derived quantities introduced in Equation (8.7):

$$\psi_c(x_1,\ldots,x_c) = E_F[\psi(x_1,\ldots,x_c,X_{c+1},\ldots,X_k)].$$

Suppose we then take the expected value of each term in Equation (12.34) conditional on X_{α_1}. The modified statistic becomes

$$\begin{aligned} U^* &= \sum_{(\alpha_1,\ldots,\alpha_k)\in C_{n,k}} \frac{E_F[\psi(X_{\alpha_1},\ldots,X_{\alpha_k}) \mid X_{\alpha_1}]}{\binom{n}{k}} \\ &= \sum_{(\alpha_1,\ldots,\alpha_k)\in C_{n,k}} \frac{\psi_1(X_{\alpha_1})}{\binom{n}{k}} \\ &= \binom{n-1}{k-1} \sum_{i=1}^{n} \frac{\psi_1(X_i).}{\binom{n}{k}} \\ &= \frac{k}{n} \sum_{i=1}^{n} \psi_1(X_i). \end{aligned} \tag{12.35}$$

All we have done is replace ψ with ψ_1 in Equation (12.34). Asymptotic normality then follows after verifying that U and U^* have the same distribution asymptotically.

Theorem 12.6 (Central Limit Theorem for U-statistics) Suppose $X_1, \ldots, X_n$ is an *iid* sample of random vectors, and $\psi(x_1, \ldots, x_k)$ is a symmetric kernel for a homogenous functional $\theta(F)$ of order k. Define U and U^* by Equations (12.34) and (12.35). Assume $E_F[U^2] < \infty$. Then the following statements hold.

(i) $U - U^* \xrightarrow{L^2} 0$.

(ii) $\sqrt{n}\,(U - \theta(F)) \xrightarrow{d} N(0, k^2\sigma_1^2)$, where $\sigma_1^2 = \text{var}_F[\psi_1(X_1)]$. ■

Proof. The proof is left to the reader (Problem 12.18). □

There are a number of approaches to developing a CLT for the sample variance S^2. In the next example, we first present the problem as an application of Slutsky's theorem (see also Problem 11.16). However, since S^2 is a U-statistic, Theorem 12.6 offers a more direct route.

Example 12.14 (The Asymptotic Normality of the Sample Variance) Let $X_1, \ldots, X_n$ be an *iid* sample with $E[X_1^4] < \infty$, $\mu_X = E[X_1]$ and $\sigma_X^2 = \text{var}[X_1]$. That the sample variance is asymptotically normal can be verified by a decomposition argument. We have

$$S^2 = \sum_{i=1}^{n} (X_i - \bar{X})^2/n - 1 = a_n S_{\mu X}^2 - a_n B_n,$$

and so

$$\sqrt{n}\,(S^2 - \sigma_X^2) = a_n\sqrt{n}\,(S_{\mu X}^2 - \sigma_X^2) + \sqrt{n}a_n B_n + c_n$$

where $S_{\mu X}^2 = n^{-1}\sum_{i=1}^{n}(X_i - \mu_X)^2$, $a_n = n/(n-1)$, $B_n = (\bar{X} - \mu_X)^2$ and $c_n = \sqrt{n}\sigma_X^2/(n-1)$. Then $a_n \to_n 1$, while B_n and c_n can be shown to approach zero in some suitable sense. In addition, $nS_{\mu X}^2$ is an *iid* sum, so the central limit theorem, followed by several applications of Slutsky's theorem, completes the argument.

On the other hand, S^2 is a U-statistic (Example 8.8), so asymptotic normality follows directly from Theorem 12.6, and the variance may be evaluated using the general method for U-statistics (Example 8.9). Using either method we conclude that $\sqrt{n}\,(S^2 - \sigma_X^2) \xrightarrow{d} N(0, \bar{\mu}_4 - \sigma_X^4)$, where $\bar{\mu}_4$ is the fourth-order central moment of X_1. ■

12.8 The Information Inequality

We have defined Fisher information for regular parametric familes in Sections 3.4 and 3.5. Where applicable, it provides an important lower bound on the variance of unbiased estimators. We consider the one-dimensional case $\theta \in \mathbb{R}$.

Theorem 12.7 (The Information Inequality) Suppose we are given a one-dimensional regular parametric family $\Theta \subset \mathbb{R}$ for observation $\tilde{X} \in \mathcal{X}$ (Definition 3.6). If for statistic $T \in \mathcal{T}_{stat}$ we have $E_\theta[T(\tilde{X})] = \eta(\theta)$, with $E_\theta[T^2] < \infty$ for all θ, then $\eta'(\theta) = d\eta(\theta)/d\theta$ exists, and the inequality

$$\text{var}_\theta[T] \geq [\eta'(\theta)]^2 / I(\theta). \tag{12.36}$$

holds for all $\theta \in \Theta$. ∎

Proof. First note that in the following argument exchange of integration and differentiation can be justified following the assumptions of Definition 3.6. We may then write

$$\begin{aligned}
\eta'(\theta) &= \int_{\tilde{x}\in\mathcal{X}} T(\tilde{x}) \frac{\partial p_\theta(\tilde{x})}{\partial \theta} d\nu_{\mathcal{X}} \\
&= \int_{\tilde{x}\in\mathcal{X}} T(\tilde{x}) \frac{\partial \log p_\theta(\tilde{x})}{\partial \theta} p_\theta(\tilde{x}) d\nu_{\mathcal{X}} \\
&= \text{cov}_\theta[T(\tilde{X}), \frac{\partial \log p_\theta(\tilde{X})}{\partial \theta}]
\end{aligned}$$

Applying the Cauchy-Schwartz inequality gives

$$[\eta'(\theta)]^2 \leq \text{var}_\theta[T]\text{var}_\theta[\frac{\partial \log p_\theta(\tilde{X})}{\partial \theta}] = \text{var}_\theta[T] I(\theta),$$

thus verifying (12.36). □

Equation (12.36) is also referred to as the Cramér-Rao lower bound. The following special case follows directly from Theorem 12.7.

Theorem 12.8 (Special Case of the Information Inequality) Under the assumptions of Theorem 12.7, if $E_\theta[T(\tilde{X})] = \theta$ than $\text{var}_\theta[T] \geq I(\theta)^{-1}$. ∎

Proof. Apply Theorem 12.7 with $\eta(\theta) \equiv \theta$. □

One interesting property of the Cramér-Rao lower bound is that the numerator depends only on the estimand of $T(\tilde{X})$, while the denominator $I(\theta)$ depends only on the distribution of $\tilde{X}$. It's importance in the theory of inference lies in the fact that, at least for regular parametric families, it is asymptotically sharp, so that optimality in large sample is precisely defined.

For the remainder of this section, we first specialize the information inequality to exponential family models, which belong to the class of regular families (Section 3.6.3). Section 12.8.2 offers an extended example which explores more fully the role that the Cramér-Rao lower bound might play in large sample inference.

12.8.1 The Information Inequality for Exponential Family Models

Suppose we have an exponential family density of rank 1. From Example 3.16, Fisher's information for the canonical parameter η is $I(\eta) = \text{var}_\eta[T(\tilde{X})]$. Then consider $T(\tilde{X})$ as an estimator of $\theta = h(\eta) = E_\eta[T(\tilde{X})]$. The information for θ, denoted as $I_\theta(\theta)$ is given by the transformation $I(\eta) = I_\theta(h(\eta))\,[h'(\eta)]^2$. However, for the exponential family we always have

$$h(\eta) = E_\eta[T(\tilde{X})] = A'(\eta) \quad \text{and} \quad \text{var}_\eta[T(\tilde{X})] = A''(\eta) = h'(\eta).$$

This gives, after substitution, $I_\theta(\theta) = \text{var}_\theta[T(\tilde{X})]^{-1}$. Thus, the Cramér-Rao lower bound for T is simply $\text{var}_\theta[T(\tilde{X})]$, so that the lower bound of Equation (12.36) is achieved for this case. In fact, it can be shown that under suitable regularity conditions, the Cramér-Rao lower bound is sharp only for the sufficient statistic T from an exponential family density of rank 1 (see Lehmann and Casella (1998)).

12.8.2 Example: An Application of the Information Inequality

If the information inequality is not generally attained (Section 12.8.1) it is worth examining how it might still be applied to a practical inference problem.

Suppose we are given an *iid* sample $\mathbf{X} = (X_1, \ldots, X_n)$, where $X_1 \sim pois(\theta)$. We are interested in finding the UMVUE estimator of

$$P_\theta(X_1 = 0) = \eta(\theta) = e^{-\theta}.$$

Following the "Rao-Blackwellization" approach of Section 8.4, we can use the fact that $S = \sum_{i=1}^n X_i$ is sufficient and complete for θ, since this is an exponential family model (Section 3.8.2). Then $\delta_0 = I\{X_1 = 0\}$ is a unbiased estimator for $\eta(\theta)$. Using the fact that a sum of independent Poisson random variables is also a Poisson random variable, the unique UMVUE may be given by

$$\begin{aligned}
\delta^*(s) &= E_\theta[\delta_0 \mid S = s] \\
&= \frac{P_\theta\{X_1 = 0,\ \sum_{i=2}^n X_i = s\}}{P_\theta\{S = s\}} \\
&= \frac{e^{-\theta}\frac{[(n-1)\theta]^s}{s!}e^{-(n-1)\theta}}{\frac{[n\theta]^s}{s!}e^{-n\theta}} \\
&= ((n-1)/n)^s,
\end{aligned}$$

which gives the unique UMVUE for estimand $\eta(\theta) = P_\theta(X_1 = 0)$. We next derive the Cramér-Rao lower bound. Note that the sample density is

$$p_\theta(\mathbf{x}) = \frac{\theta^s}{\prod_{i=1}^n x_i!}e^{-n\theta}.$$

To calculate the information, set

$$\begin{aligned}
H(\mathbf{x}, \theta) &= \frac{\partial \log p_\theta(\mathbf{x})}{\partial \theta} \\
&= \frac{\partial}{\partial \theta}\left\{s\log(\theta) - n\theta - \log\left(\prod_{i=1}^n x_i!\right)\right\} \\
&= s/\theta - n.
\end{aligned}$$

The information is then $I(\theta) = \text{var}_\theta[H(\mathbf{X}, \theta)] = n/\theta$. The information equality here specializes to

$$\text{var}_\theta[\delta] \geq [\eta'(\theta)]^2 / I(\theta) = \theta e^{-2\theta}/n = \tau_{CR}(\theta).$$

We next evaluate the actual variance of $\delta^*(S)$, in order to compare it to the Cramér-Rao lower bound $\tau_{CR}(\theta)$ just given. First, note that $E_\theta[\delta^*] = e^{-\theta}$. The second moment can be calculated as follows:

$$\begin{aligned}
E_\theta[[\delta^*]^2] &= \sum_{s=0}^{\infty} \left(\frac{n-1}{n}\right)^{2s} \frac{[n\theta]^s}{s!} e^{-n\theta} \\
&= \sum_{s=0}^{\infty} \frac{1}{s!} \left[\left(\frac{n-1}{n}\right)^2 n\theta\right]^s e^{-[(n-1)/n]^2[n\theta]+[(n-1)/n]^2[n\theta]-n\theta} \\
&= e^{[(n-1)/n]^2[n\theta]-n\theta} \\
&= e^{\theta/n-2\theta}.
\end{aligned}$$

This leads to variance

$$\text{var}_\theta[\delta^*] = E_\theta[[\delta^*]^2] - E_\theta[\delta^*]^2 = \left(e^{\theta/n} - 1\right) e^{-2\theta}.$$

To compare $\text{var}_\theta[\delta^*]$ and $\tau_{CR}(\theta)$, note that e^x is a strictly convex function, and $x+1$ is tangent to e^x at $x = 0$, so $e^x \geq x + 1$ for all $x \in \mathbb{R}$, with equality *iff* $x = 0$. Therefore

$$\text{var}_\theta[\delta^*] > \tau_{CR}(\theta), \quad \theta > 0,$$

and the Cramér-Rao lower bound is not attained, as we expect.

If T is any unbiased estimator for which the conditions of Theorem 12.7 hold, we can define an efficiency measure

$$\rho(T) = I(\theta)^{-1}/\text{var}_\theta[T] = \tau_{CR}(\theta)/\text{var}_\theta[T],$$

where $\theta = E_\theta[T]$. By the information inequality $\rho(T) \leq 1$, with $\rho(T) = 1$ indicating maximum efficiency from an information point of view.

To see the implications of this, we'll return our example, for which our efficiency measure will be

$$\rho(\delta^*) = \frac{\tau_{CR}(\theta)}{\text{var}_\theta[\delta^*]} = \frac{\left[\frac{\theta}{n}\right] e^{-2\theta}}{\left(e^{\theta/n} - 1\right) e^{-2\theta}} = \frac{\left[\frac{\theta}{n}\right]}{\left[\frac{\theta}{n} + o\left(\frac{\theta}{n}\right)\right]}$$

where $e^{\theta/n} - 1 = (\theta/n) + o(\theta/n)$ follows from a Taylor's series approximation. This means $\lim_{n\to\infty} \rho(\delta^*) = 1$, that is, as the sample size increases, the variance of δ^* approaches the Cramér-Rao lower bound $\tau_{CR}(\theta)$.

It is worth comparing δ^* to an estimator of $P_\theta(X_1 = 0) = e^{-\theta}$ constructed by direct analogy, in this case, by estimating $P_\theta(X_1 = 0)$ by the proportion of the sample for which $X_i = 0$. In this case we have

$$\delta^{**} = n^{-1} \sum_{i=1}^{n} I\{X_i = 0\}.$$

This is simply an unbiased estimator of a binomial parameter $p = e^{-\theta}$, and so $\text{var}_\theta[\delta^{**}] = n^{-1}e^{-\theta}(1 - e^{-\theta})$. The efficiency is then

$$\rho(\delta^{**}) = \frac{\tau_{CR}(\theta)}{\text{var}_\theta[\delta^{**}]} = \frac{\left[\frac{\theta}{n}\right] e^{-2\theta}}{\frac{1}{n} e^{-\theta}(1 - e^{-\theta})} = \frac{\theta}{e^\theta - 1}.$$

We must have $\rho(\delta^{**}) \leq 1$, and since δ^* is the unique UMVUE, we must also have $\text{var}_\theta[\delta^{**}] < \text{var}_\theta[\delta^*]$, since δ^{**} is clearly not a function of a sufficient statistic. Furthermore, the efficiency $\rho(\delta^*)$ of δ^* depends only weakly on θ (that is, the dependence vanishes as $n \to \infty$), while $\rho(\delta^{**})$ depends quite strongly on θ (but is also independent of n).

However, a practical issue remains. The optimal properties of δ^* depend on a correct specification of the model, that is, that the Poisson model is correct. In fact, the unbiasedness property of δ^* depends on this assumption as well. In contrast δ^{**} is unbiased for any family of distributions. Thus, if some uncertainty regarding the form of the distribution exists, we may prefer to use δ^{**}.

In this case the information inequality is still relevant. We might accept some loss of efficiency in exchange for robustness against model misspecification when using δ^{**}. But we would still wish to ensure that the loss of efficiency is bounded, which could be accomplished using the information inequality. Suppose we anticipate that $p = P_\theta(X_1 = 0)$ will be reasonably large, say $p \geq 0.75$. If the true model was Poisson, then for this bound $\theta = -\log(0.75)$. Noting that $\rho(\delta^{**})$ is a decreasing function of θ, we can conclude that $p \geq 0.75$ implies $\rho(\delta^{**}) \geq -\log(0.75)/(0.25/0.75) \approx 0.863$. This loss of efficiency might be regarded as a reasonable cost to pay for robustness.

12.9 Asymptotic Efficiency

In this section we formalize the notion of efficiency introduced in Section 12.8.2. The asymptotic relative efficiency (ARE) is used to compare estimators from a large sample point of view. Suppose two sequences of estimators $\hat{\theta}_n$ and $\hat{\theta}_n^*$ with a common estimand θ have variances τ_n, τ_n^*. Here, n is the usual sample size parameter. We expect $\tau_n, \tau_n^* \to_n 0$, so a comparison must take the form of a ratio

$$e(\hat{\theta}_n, \hat{\theta}_n^*) = \lim_{n\to\infty} \tau_n^*/\tau_n, \tag{12.37}$$

assuming the limit exists. The limit defined in Equation (12.37) is known as the ARE of $\hat{\theta}_n$ relative to $\hat{\theta}_n^*$. If we write, for convenience, $e(\hat{\theta}, \hat{\theta}^*)$, it is assumed that $\hat{\theta}, \hat{\theta}^*$ represent sequences of estimators that can be indexed by a sample size n. Most asymptotic approximations take the form

$$\sqrt{n}(\hat{\theta}_n - \theta) \xrightarrow{d} N(0, \bar{\tau}), \tag{12.38}$$

so that $\text{var}[\hat{\theta}] \approx \bar{\tau}/n$. Using this convention, Equation (12.37) becomes a ratio of the asymptotic variances

$$e(\hat{\theta}, \hat{\theta}^*) = \bar{\tau}^*/\bar{\tau}. \tag{12.39}$$

It is also possible to define ARE in terms of equivalent sample sizes (Problem 12.25).

The ARE is obviously related to the efficiency measure $\rho(T)$ introduced in Section 12.8.2, except that in the latter case we are not comparing two estimators. Rather, we are comparing one estimator to some theoretically optimal estimator (which may not exist). So it is natural to define the asymptotic efficiency of an estimator $\hat{\theta}$ as

$$\rho(\hat{\theta}) = \lim_n \tau_{CR}/\text{var}[\hat{\theta}].$$

We say that an estimator is asymptotically efficient (or simply efficient) if $\rho(\hat{\theta}) = 1$, in which case $\hat{\theta}$ attains the Cramér-Rao lower bound as $n \to \infty$.

The subject of estimating equations will be discussed in detail in Chapter 13, but it is worth looking briefly ahead, given the insight it offers on the notion of asymptotic efficiency. Suppose an estimator $\hat{\theta}_n$ of $\theta \in \Theta \subset \mathbb{R}$ is the solution to an estimating equation $\Psi_n(\tilde{X}_n, \theta) = 0$, where for any fixed observation $\tilde{X}_n = \tilde{x}_n$, $\Psi_n(\tilde{x}_n, \theta)$ maps Θ to $\mathbb{R}$. We usually assume $E_\theta[\Psi_n(\tilde{x}_n, \theta)] = 0$. Suppose $\Psi_n(\tilde{x}_n, \theta)$ can be normalized so that

$$-\sqrt{n}\Psi_n(\tilde{X}_n, \theta_0) \xrightarrow{d} U_0, \tag{12.40}$$

and $\lim_n E[\Psi_n'(\tilde{X}_n, \theta)] = \tau_0$, where $\Psi_n'(\tilde{x}_n, \theta)$ is the partial derivative of $\Psi_n(\tilde{x}_n, \theta)$ with respect to θ, and θ_0 is the true parameter. It will be shown in Chapter 13 that under certain conditions we will have

$$\sqrt{n}(\hat{\theta}_n - \theta_0) \xrightarrow{d} \tau_0^{-1} U_0. \tag{12.41}$$

We will see that these conditions will be natural to many commonly used statistical methods.

Now suppose $\ell(\theta, \tilde{X}_n)$ is a log-likelihood function, from which we intend to derive the MLE $\hat{\theta}_n$ using a likelihood equation (Section 4.5). In most of our applications the log-likelihood behaves like a cumulative random sum, so we may normalize the estimating equation by taking $\Psi_n(\tilde{X}_n, \theta) = -n^{-1}\partial\ell(\theta, \tilde{X}_n)/\partial\theta$. Under the conditions of Theorem 3.12 we have $E_\theta[\Psi_n(\tilde{X}_n, \theta)] = 0$, $\text{var}_\theta[\Psi_n(\tilde{X}_n, \theta)] = I_n(\theta)/n^2$, and $E_\theta[\Psi_n'(X, \theta)] = I_n(\theta)/n$. We then expect $I_n(\theta) = O(n)$, noting that for the *iid* case we would have $I_n(\theta) = nI_1(\theta)$, where $I_1(\theta)$ is the information for a single observation X_1. Then $\text{var}_\theta[\Psi_n(\tilde{X}_n, \theta)] = O(1/n)$, and Equation (12.40) will hold under general conditions. Next, we may write

$$\tau_0 = \lim_{n\to\infty} E_\theta[\Psi_n'(X, \theta)] = \lim_{n\to\infty} I_n(\theta)/n = \bar{I}(\theta),$$

assuming a finite limit $\bar{I}(\theta)$ exists. Then by Equation (12.41), the limiting variance of $\sqrt{n}(\hat{\theta}_n - \theta_0)$ is $\bar{I}(\theta)^{-1}$, giving asymptotic efficiency

$$\rho(\hat{\theta}_n) = \lim_{n\to\infty} \frac{1/I_n(\theta)}{\bar{I}(\theta)^{-1}/n} = 1.$$

The definition of regularity conditions under which Equation (12.9) holds is a significant technical problem in large sample analysis, which we will consider in Chapters 13 and 14. These are clearly related to the conditions of Definition 3.6 of the regular parametric family, which ensure that the Fisher information is well-defined and interpretable, and which are also stated assumptions for the information inequality (Theorem 12.7).

Suppose we then assume that such regularity conditions hold. It seems that we should be able to conclude at this point that the MLE attains the maximum efficiency asymptotically, and that

$$e(\hat{\theta}, \hat{\theta}_{MLE}) \leq 1 \tag{12.42}$$

where $\hat{\theta}$ and $\hat{\theta}_{MLE}$ both estimate θ. There are, however, a few qualifications. Note that the comparison is meaningful if $\text{var}[\hat{\theta}] = O(1/n)$ and $\text{var}[\hat{\theta}_{MLE}] = O(1/n)$. To fix ideas, suppose we have an *iid* sample $X_1, X_2, \ldots$, and we are estimating parameter θ. The Cramér-Rao lower bound is $\tau_{CR} = 1/(nI(\theta))$ where $I(\theta)$ is the information for a single observation X_1. We may have an estimator $\hat{\theta}$ satisfying $\sqrt{n}(\hat{\theta} - \theta) \xrightarrow{d} N(0, \tau)$, and we already know that $\sqrt{n}(\hat{\theta}_{MLE} - \theta) \xrightarrow{d} N(0, I(\theta)^{-1})$. However, we cannot claim at this point that $\tau \geq I(\theta)^{-1}$, unless $\hat{\theta}$ is an unbiased estimator of θ. The best we can say, by the information inequality, is that

$$\tau \geq \eta'(\theta)^2/I(\theta),$$

where $\eta(\theta) = E_\theta[\hat{\theta}]$. Recall that in Example 12.1 when $\eta'(\theta) = 0$ for a given estimand

$\eta(\theta)$, the variance of the MLE was $O(n^{-2})$ rather than $O(n^{-1})$. In fact, the regularity conditions of Chapters 13 and 14 are concerned as much with precluding this situtation as with validating the information inequality.

Thus, even when the conditions of Theorem 12.7 hold, That $\tau \geq I(\theta)^{-1}$ need not hold is demonstrated in a seminal example known as Hodges' estimator.

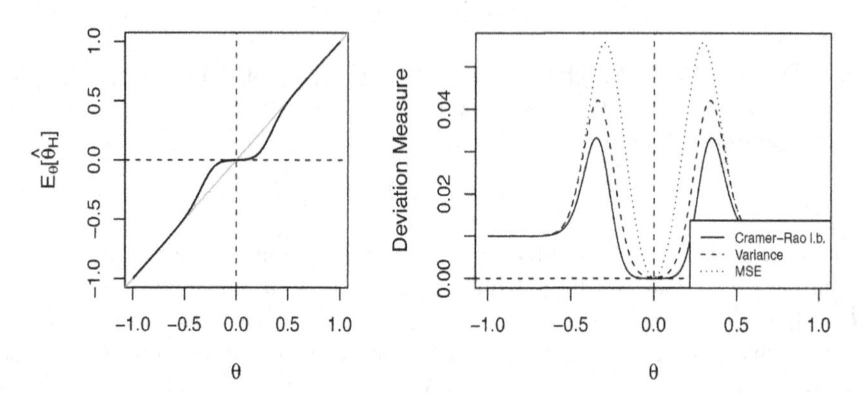

FIGURE 12.3
Expected value $\eta(\theta) = E_\theta[\hat{\theta}_H]$, variance, MSE and Cramér-Rao lower bound τ_{CR} for Hodges' estimator, $n = 100$.

Example 12.15 (Hodges' Estimator) Let $\mathbf{X} = (X_1, \ldots, X_n)$ be an *iid* sample from $N(\theta, 1)$. We have MLE $\hat{\theta}_{MLE} = \bar{X}$, the sample mean. Hodges' estimator is defined as

$$\hat{\theta}_H = \begin{cases} \bar{X} & ; \quad |\bar{X}| \geq n^{-1/4} \\ 0 & ; \quad |\bar{X}| < n^{-1/4} \end{cases},$$

and is a biased but consistent estimator of θ. It is not too difficult to calculate the expected value $\eta(\theta) = E_\theta[\hat{\theta}_H]$, as well as the variance, MSE and Cramér-Rao lower bound τ_{CR}. This is shown in Figure 12.3 for $n = 100$, respectively. The behavior of $\hat{\theta}_H$ near $\theta = 0$ is, predictably, complex. However, the most notable feature is the behavior of $\eta(\theta)$. When bounded away from $\theta = 0$ we have $\eta(\theta) \approx \theta$, however, near $\theta = 0$ the derivative satisfies $\eta'(\theta) < 1$. In addition, it can be verified that $\lim_{n\to\infty} \eta'(0) = 0$. Thus, while we generally expect for the *iid* case $\tau_{CR} = O(1/n)$, for $\hat{\theta}_H$, at $\theta = 0$, we have $\tau_{CR} = o(1/n)$. This means, surprisingly, that the asymptotic efficiency of $\hat{\theta}_H$ relative to $\hat{\theta}_{MLE}$, at $\theta = 0$, is $e(\hat{\theta}_H, \hat{\theta}_{MLE}) = \infty$, which contradicts our expectation expresed in Equation (12.42).

This phenomenon is known as superefficiency, and was not generally expected when introduced in 1953 (note that we are considering only regular families for which Fisher's information is well defined and the information inequality holds). However, it can be shown that under general conditions Equation (12.42) holds, except on a set of Lebesgue measure zero on Θ, so that MLEs can generally be regarded as asymptotically optimal (see Bahadur *et al.* (1964), also Section 6.2 of Lehmann and Casella (1998)). In particular, Hodges' estimator violates Equation (12.42) only at $\theta = 0$. ∎

12.10 Problems

Problem 12.1 Suppose we are given an *iid* sample $\mathbf{X} = (X_1, \ldots, X_n)$, with $\mu_X = E[X_1]$. Let $\bar{X}$ be the sample mean.

(a) Assuming $E[|X_1|^{2k+\epsilon}] < \infty$ for some $\epsilon > 0$, suppose we use $\hat{\theta} = \bar{X}^k$ to estimate $\theta = \mu_X^k$. Give the limiting distribution of $\sqrt{n}(\hat{\theta} - \theta)$.

(b) How can we construct a statistic $h(\mathbf{X})$ for which $\sqrt{n}(\hat{\theta} - \theta)/h(\mathbf{X}) \xrightarrow{d} N(0, 1)$? **HINT:** Use Slutsky's theorem.

Problem 12.2 Consider the sample moment estimator of Problem 12.1. Let S^2 be the sample variance.

(a) For the case $k = 2$, determine an estimator $\hat{\theta}^* = \bar{X}^2 + a_n S^2$ which is unbiased for μ_X^2.

(b) Let $\hat{\theta} = \bar{X}^2$ be the estimator described in Problem 12.1. Derive the bias of $\hat{\theta}$ and the variances of $\hat{\theta}$ and $\hat{\theta}^*$ for finite n. Compare the MSE of the two estimators.

(c) Compare the asymptotic variances of $\hat{\theta}$ and $\hat{\theta}^*$.

Problem 12.3 Using the method of Theorem 12.1, prove the claim of Example 12.1, that for $p = 1/2$, $[n4(1/4 - \hat{p}(1 - \hat{p}))] \xrightarrow{d} \chi_1^2$.

Problem 12.4 Give a heuristic argument to justfy the approximation of Equation (12.14). Using the method of Theorem 12.2 give conditions under which the approximation holds.

Problem 12.5 Suppose $X \sim pois(\theta)$ and $c > 0$. Verify that $E[(X + c)^k] = O(\theta^{-k})$. **HINT:** First, note that $(X + c)^{-1} < X^{-1}$ for any $c > 0$. Then develop an expression for $E[1/X^{(k)} \mid X \geq t]$, for some suitable t, where $X^{(k)}$ is the order k partial factorial of X. Make use of the inequality $1/X^k \leq 1/X^{(k)}$, $X \geq k$, and the fact that $P(X < t) = o(1/\theta^r)$ for any fixed $t, r > 0$.

Problem 12.6 This problem refers to Example 12.9. Suppose we have an *iid* sample $X_1, \ldots, X_n$, $X_1 \sim pois(\theta)$. Apply the Anscombe transformation to get sample $Y_1, \ldots, Y_n$, where $Y_i = \sqrt{X_i + 3/8}$. Then let $\hat{\theta} = \bar{Y}^2 - 3/8$ be an estimate of θ. Show that the bias of $\hat{\theta}$ is $b + O(\theta^{-1})$, where b does not depend on θ.

Problem 12.7 Verify Equation (12.19) for an *iid* sequence $X_1, X_2, \ldots$ with finite kth order moment. **HINT:** What is the largest partition of k labels that contains no singletons?

Problem 12.8 Suppose we are given an *iid* sample of size n from a $rayleigh(\tau)$ distribution, and we wish to estimate the mean μ_X with sample mean $\bar{X}$. Show that there exists a transformation $h(\bar{X}) = \bar{X}^\kappa$ for which the skewness coefficient is of order $O(1/n^{3/2})$. Give κ precisely. **HINT:** Make use of the following identities for the mean, variance and skewness coefficient: $\mu_X = \tau\sqrt{\pi/2}$, $\sigma_X^2 = \tau^2(4 - \pi)/2$, $SK = 2\sqrt{\pi}(\pi - 3)/(4 - \pi)^{3/2}$.

Problem 12.9 We are given n *iid* random variables $X_1, \ldots, X_n$. Assume X_1 is distributed as an exponential random variable with mean θ truncated above at $T\theta$ for some fixed number T. The density of X_1 is then $f_\theta(x) = \theta^{-1}e^{-x/\theta}/(1 - e^{-T})$, $x \in (0, T\theta)$.

(a) Verify the θ is a scale parameter.

(b) Evaluate the skewness coefficient SK of X_1.

(c) Suppose we use sample mean $\bar{X}$ to estimate $\mu = E[X_1]$. Show that there exists a transformation $h(\bar{X})$ for which the skewness coefficient is of order $O(1/n^{3/2})$. Give the transformation exactly.

(d) Find a comparable skewness reducing transformation for the untruncated case ($T = \infty$).

See also Problem 9.17. **HINT:** We may use the indefinte integrals $I_k(x) = \theta^{-1} \int x^k e^{-x/\theta} dx$ for $k = 1, 2, 3$. We can express $I_k(x)$ in terms of $I_{k-1}(x)$ using integration by parts.

Problem 12.10 Repeat Example 12.12 by first expressing the first-order Taylor's series approximation of the log-odds ratio estimate $\log(N_{11}N_{22}/(N_{12}/N_{21}))$ using all four observed frequencies $N_{11}, N_{12}, N_{21}, N_{22}$, then using Theorem 2.1.

Problem 12.11 We are given a multinomial vector $\mathbf{N} = (N_1, N_2, N_3)$, $n = N_1 + N_2 + N_3$, where $E[N_i] = np_i$, $i = 1, 2, 3$. We wish to estimate $\theta = p_1 p_2 - p_3$.

(a) What is the asymptotic distributon of the vector $\mathbf{N} = (N_1, N_2)$? Show how $\mathbf{N}$ would be normalized, and specify the asymptotic distribution as precisely as possible.
(b) Set $\hat{p}_i = N_i/n$, and suppose we use $\hat{\theta} = \hat{p}_1\hat{p}_2 - \hat{p}_3$ to estimate θ. Derive the asymptotic distribution of $\hat{\theta}$.

Problem 12.12 Suppose (X_i, Y_i) $i = 1, \ldots, n$ is an *iid* sample from a bivariate normal distribution, with $E[X_1] = \mu_1$, $E[Y_1] = \mu_2$, $\text{var}[X_1] = \sigma_1^2$, $\text{var}[Y_1] = \sigma_2^2$ and $\text{cov}[X_1, Y_1] = \sigma_{12}$. The correlation coefficient is then $\rho = \sigma_{12}/(\sigma_1^2\sigma_2^2)^{1/2}$. Set $S_X = n^{-1}\sum_{i=1}^n (X_i - \bar{X})^2$, $S_Y = n^{-1}\sum_{i=1}^n (Y_i - \bar{Y})^2$, $S_{XY} = n^{-1}\sum_{i=1}^n (X_i - \bar{X})(Y_i - \bar{Y})$. Then the sample correlation coefficient $r = S_{XY}/\sqrt{S_X S_Y}$ is commonly used as an estimate of ρ. In this problem we will determine an approximate normal distribution for r.

(a) Verify that the distribution of r does not depend on μ_1, μ_2, σ_1^2 or σ_2^2. Therefore we may assume $\mu_1 = \mu_2 = 0$ and $\sigma_1^2 = \sigma_2^2 = 1$.
(b) Under the assumptions of Part (a) determine the limiting distribution of

$$\mathbf{Z}_n = \sqrt{n}\left(n^{-1}\sum_{i=1}^n X_i Y_i - \rho, n^{-1}\sum_{i=1}^n X_i^2 - 1, n^{-1}\sum_{i=1}^n Y_i^2 - 1\right)$$

as $n \to \infty$. Give the covariance matrix of the limiting distribution precisely.
(c) Under the assumptions of Part (a) prove that the limiting distribution of

$$\mathbf{Z}_n^* = \sqrt{n}(S_{XY} - \rho, S_X - 1, S_Y - 1)$$

is the same as that of $\mathbf{Z}_n$ defined in Part (b). **HINT:** Use Slutsky's theorem.
(d) Prove that $\sqrt{n}(r - \rho) \xrightarrow{d} N(0, \sigma_r^2)$. Give σ_r^2 precisely.
(e) Show that $g(\rho) = \text{arctanh}(\rho) = \log((1+\rho)/(1-\rho))/2$ is a variance stabilizing transformation for r. This is known as the Fisher z-transformation, and is described in Fisher (1915) as "not a little attractive", but also not entirely satisfactory in several respects.

Problem 12.13 Inter-rater reliability can be characterized in the following way. A rater's task is to render a binary decision on a sample from a population. For example, the rater may judge whether or not a tissue sample is cancerous. The judgement results in a Yes/No decision. Suppose each of n samples is presented to two raters. There are four possible outcomes (D_1, D_2), where D_i is the decision of rater i. This results in a 2×2 contingency table:

	$D_2 = \text{No}$	$D_2 = \text{Yes}$	
$D_1 = \text{No}$	n_{11}	n_{12}	R_1
$D_1 = \text{Yes}$	n_{21}	n_{22}	R_2
Total	C_1	C_2	n

The cell counts then form a multinomial vector $\mathbf{N} = (N_{11}, N_{12}, N_{21}, N_{22})$ with frequencies $\mathbf{P} = (p_{11}, p_{12}, p_{21}, p_{22})$. The true rate agreement is then $p_A = P(D_1 = D_2) = p_{11} + p_{22}$. The rate of agreement attributable to chance is $P_C = P(D_1 = \text{No}) \times P(D_2 = \text{No}) + P(D_1 = \text{Yes}) \times P(D_2 = \text{Yes}) = (p_{11} + p_{12})(p_{11} + p_{21}) + (p_{21} + p_{22})(p_{12} + p_{22})$. Then Cohen's κ is defined as

$$\kappa = p_A - p_C / (1 - p_C),$$

and is widely used as a measure of inter-rater reliability (Cohen, 1960). Suppose $\hat{\kappa}$ is the MLE of κ. Use the multivariate δ-method to obtain a large sample approximation of the distribution of $\hat{\kappa}$.

Problem 12.14 Recall from Section 4.4 that if F_n is the empirical CDF, the sample quantile function is given by $Q(p; F_n)$. In Gastwirth (1966) the weighted sum of sample quantiles $\hat{\mu} = 0.3Q(1/3; F_n) + 0.4Q(1/2; F_n) + 0.3Q(2/3; F_n)$ was proposed as a robust estimate of a location parameter μ. Compare the asymptotic variance of $\hat{\mu}$ to that of the sample median for an *iid* sample with $X_1 \sim N(\mu, \sigma^2)$.

Problem 12.15 Suppose we are given an *iid* sample with $X_1 \sim N(\mu, \sigma^2)$.

(a) Verify that $Q(1-\alpha) - Q(\alpha) = \sigma(Q_z(1-\alpha) - Q_z(\alpha))$, where $\alpha \in (0, 1/2)$, and Q_z is the quantile function for the $N(0,1)$ distribution.
(b) Suppose we estimate σ using the estimator $\hat{\sigma}_\alpha = (Q(1-\alpha; F_n) - Q(\alpha; F_n))/(Q_z(1-\alpha) - Q_z(\alpha))$ for any $\alpha \in (0, 1/2)$. Find α which minimizes the asymptotic variance of $\hat{\sigma}_\alpha$. This can be done numerically. **HINT:** Make use of Example 12.13.

Problem 12.16 Suppose we are given *iid* sample $\mathbf{X} = (X_1, \ldots, X_n)$ from distribution F. *Gini's mean difference* is defined as $\eta = E[|X_1 - X_2|]$, and was proposed in Gini (1936) as a measure of inequality within F (η may be standardized to be scale invariant, which is then referred to as Gini's coefficient). Suppose we estimate η with the U-statistic $\hat{\eta} = \sum_{i<j} |X_i - X_j| / \binom{n}{2}$. Derive the asymptotic distribution of $\hat{\eta}$.

Problem 12.17 Suppose we are given *iid* sample $\mathbf{X} = (X_1, \ldots, X_n)$ from continuous distribution F. The Wilcoxon one-sample statistic may be expressed as the U-statistic $W = \sum_{i<j} I\{X_i + X_j < 0\} / \binom{n}{2}$. This is used to test against the null hypothesis that the density of F is symmetric about zero (Wilcoxon, 1945). Derive the asymptotic distribution of W under this hypothesis.

Problem 12.18 Prove Theorem 12.6. **HINT:** Use Lemma 8.1 and Theorem 8.9 to show that $E_F[(U - U^*)^2] \rightarrow_n 0$ (see also Problem 8.13). Then apply a suitable central limit theorem to U^*.

Problem 12.19 Let $\mathbf{X} = (X_1, \ldots, X_n)$ be an *iid* sample from *pareto*(τ, α). Assume α is known and that $\alpha > 2$.

(a) Determine the maximum likelihood estimate (MLE) of ν (assuming α is known).
(b) Are the conditions of Theorem 12.7 satisfied for this estimation problem?
(c) Determine the variance of the MLE of τ.
(d) Suppose we estimated τ using a method of moments estimator based on sample mean $\bar{X}$. How would the variance of this estimator compare to that of the MLE as $n \to \infty$?

Problem 12.20 We are given a simple linear regression model $Y_i = \beta_0 + \beta_1 x_i + \epsilon_i$, $i = 1, \ldots, n$, where $\epsilon_i \sim N(0, \sigma_i^2)$ are independent error terms. Assume that the sequences $x_1, x_2, \ldots$ and $\sigma_1^2, \sigma_2^2, \ldots$ consist of repetitions of fixed length sequences $x_1, x_2, \ldots, x_m$ $\sigma_1^2, \sigma_2^2, \ldots, \sigma_m^2$, so that, for example, $x_{m+1} = x_1$, $x_{m+2} = x_2$, and so on. Let $\hat{\beta}_{1:n}$ be the best linear unbiased estimator (BLUE) of β_1, and let $\hat{\beta}^*_{n:1}$ be the ordinary least squares estimate of β_1. What is the asymptotic relative efficiency of $\hat{\beta}^*_{n:1}$ relative to $\hat{\beta}_{n:1}$?

Problem 12.21 Suppose we are given an *iid* sample $\mathbf{X} = (X_1, \ldots, X_n)$, where $X_1 \sim pois(\theta)$.

(a) Show that $\hat{\eta}_S = n^{-1} \sum_{i=1}^n I\{X_i = 1\}$ is an unbiased estimate of $\eta(\theta) = \theta e^{-\theta}$. Derive the variance of $\hat{\eta}_S$.
(b) Derive a UMVU estimator $\hat{\eta}_U$ of estimand $\eta(\theta) = \theta e^{-\theta}$. Derive the variance of $\hat{\eta}_U$.
(c) Derive the Cramér-Rao lower bound τ_{CR} of the variance of unbiased estimates of $\eta(\theta) = \theta e^{-\theta}$.
(d) Evaluate the assymptotic efficiency of $\hat{\eta}_S$ and $\hat{\eta}_U$.

Problem 12.22 Suppose we observe an *iid* random sample $\mathbf{X} = (X_1, \ldots, X_n)$, with $X_1 \sim N(\theta, 1)$, $\theta \in \Theta = \mathbb{R}$.

(a) Find the UMVU estimator of $\eta(\theta) = P_\theta(X_1 \leq 0)$. **HINT:** Make use of the general form for conditional multivariate normal distributions given in Example 7.4. Let $\Phi(x; \mu, \tau)$ be the CDF for the $N(\mu, \tau)$ distribution. Then express your answer in terms of $\Phi(x; \mu, \tau)$, $\mu = \mu(\mathbf{X})$ and $\tau = \tau(\mathbf{X})$.
(b) Derive the Cramér-Rao lower bound for an unbiased estimate of $\eta(\theta)$. Will this be attained by the estimator of Part (a)? Explain briefly your answer.

Problem 12.23 Let $\mathbf{X} = (X_1, \ldots, X_n)$ be an *iid* sample from a gamma distribution $X_1 \sim gamma(\tau, \alpha)$, where the shape parameter α is known and the scale parameter τ is unknown.

(a) Determine the UMVUE, say $\hat{\sigma}_U^2$, of the variance $\sigma_{X_1}^2 = g(\tau) = \alpha\tau^2$. Also derive the variance $\text{var}[\hat{\sigma}_U^2]$ of this estimator.
(b) Derive the Cramér-Rao lower bound τ_{CR} for the variance of an unbiased estimator of $\sigma_{X_1}^2$. Is this attained by the estimator of Part (a)? Either verify your answer directly, or cite a relevant theorem.
(c) Evaluate the asymptotic efficiency of the UMVUE.
(d) Consider as an alternative to the UMVUE the sample variance $S^2 = (n-1)^{-1} \sum_{i=1} (X_i - \bar{X})^2$. Determine $\text{var}[S^2]$ and evaluate the asymptotic efficiency of S^2.

Problem 12.24 Let $\mathbf{X} = (X_1, \ldots, X_n)$ be an *iid* sample from a normal distribution $X_1 \sim N(\mu, \sigma^2)$.

(a) Verify that $T(\mathbf{X}) = \bar{X}^2 - S^2/n$ is an unbiased estimator of μ^2, where $\bar{X} = n^{-1} \sum_i X_i$ and $S^2 = (n-1)^{-1} \sum_i (X_i - \bar{X})^2$ are the sample mean and variance. Determine its variance.
(b) Derive the Cramer-Rao lower bound for the variance of an unbiased estimator of μ^2. Is this attained by the estimator of Part (a)? What is the asymptotic efficiency?

Problem 12.25 The asymptotic relative efficiency (ARE) of Equation (12.37) can be alternatively defined in terms of equivalent sample sizes. Suppose we are given two sequences of estimators $\hat{\theta}_n$ and $\hat{\theta}_n^*$ of estimand θ, which both satisfy limit (12.38). For each n, let $m(n)$ be the sample size for which the variance of $\hat{\theta}_{m(n)}$ is approximately equal to that of $\hat{\theta}_n^*$. Express the ARE $e(\hat{\theta}_n, \hat{\theta}_n^*)$ in terms of the quantities n and $m(n)$.

Problem 12.26 We are given an *iid* sample $\mathbf{X} = (X_1, \ldots, X_n)$ with $X_1 \sim exp(\tau)$. Suppose we wish to estimate τ. We propose two estimators:

(i) $\hat{\tau}_{mean} = \bar{X}$, the sample mean.
(ii) $\hat{\tau}_{med} = g^{-1}(X_q)$, where X_q is the sample q-quantile, and $x_q = g(\tau)$, where x_q is the population q-quantile. The function g is a property of the exponential density.

(a) Derive an expression for the asymptotic relative efficiency (ARE) of $\hat{\tau}_{med}$ relative to $\hat{\tau}_{mean}$ in terms of $q \in (0, 1)$.

(b) What is the largest value of the ARE, and for which value of q is this attained? You may answer Part (b) analytically or numerically.

Problem 12.27 Suppose we have probability distribution $\mathbf{P} = (p_1, \ldots, p_m)$ on support $\mathcal{S} = \{1, \ldots, m\}$, from which we are given *iid* sample $X_1, \ldots, X_n$. The natural choice for estimating $\mathbf{P}$ is the vector of empirical frequences $\hat{\mathbf{P}} = (\hat{p}_1, \ldots, \hat{p}_m)$, where $\hat{p}_j$ is the proportion of observations which equal j. We may not be interested in any particular parameter, but wish to consider estimation of $\mathbf{P}$ as a whole. Suppose we use as a distance measure the L^1 vector norm:

$$\left\|\hat{\mathbf{P}} - \mathbf{P}\right\|_1 = \sum_{j=1}^{m} |p_j - \hat{p}_j|.$$

(a) Prove that $P\left(\left\|\hat{\mathbf{P}} - \mathbf{P}\right\|_1 \leq 2\right) = 1$.

(b) Prove that the bound $E[\left\|\hat{\mathbf{P}} - \mathbf{P}\right\|_1] \leq (m/n)^{1/2}$ holds for all $\mathbf{P}$. **HINT:** First use Jensen's inequality to bound $E[|p_j - \hat{p}_j|]$. Then argue that $E[\left\|\hat{\mathbf{P}} - \mathbf{P}\right\|_1]$ is maximized by the uniform distribution $\mathbf{P}_e = (1/m, \ldots, 1/m)$.

For more on this problem see Section 4.15 of Almudevar (2014). In Blyth (1980) the following bound was reported:

$$E[||\hat{\mathbf{P}} - \mathbf{P}||_1] = n^{-1/2}\sqrt{2/\pi}\sqrt{p(1-p)} + O(n^{-3/2}),$$

which is asymptotically sharp. An exact expression for $\left\|\hat{\mathbf{P}} - \mathbf{P}\right\|_1$ was actually derived by Abraham De Moivre in 1730 (Diaconis and Zabell, 1991).

13

Asymptotic Theory for Estimating Equations

13.1 Introduction

Suppose for parametric family $\mathcal{P}$ with observation $\tilde{X} \in \mathcal{X}$ we have parameter $\boldsymbol{\theta} \in \Theta \subset \mathbb{R}^q$. We have seen that estimators are often derived as solutions to optimization problems. For example, the maximum likelihood estimate is the solution $\hat{\boldsymbol{\theta}}$ to the problem of optimizing some function $\Lambda_{\tilde{X}}(\boldsymbol{\theta}) = \Lambda(\tilde{X}, \boldsymbol{\theta})$ with respect to $\boldsymbol{\theta}$, where the data $\tilde{X}$, once observed, is considered fixed (Section C.7). The least squares estimator can be described in the same way.

Under suitable regularity conditions $\hat{\boldsymbol{\theta}}$ will be a stationary point of $\Lambda_{\tilde{X}}$. Let $\Psi(\tilde{X}, \boldsymbol{\theta})$ be the gradient vector of $\Lambda_{\tilde{X}}$ with respect to $\boldsymbol{\theta}$. Then $\hat{\boldsymbol{\theta}}$ will be a solution with respect to $\boldsymbol{\theta}$ to the estimating equation

$$\Psi(\tilde{X}, \boldsymbol{\theta}) = \mathbf{0} \tag{13.1}$$

where, in this context, $\Psi(\tilde{X}, \boldsymbol{\theta}) = (\Psi_1(\tilde{X}, \boldsymbol{\theta}), \ldots, \Psi_q(\tilde{X}, \boldsymbol{\theta}))$ is referred to as a score function. The term M-estimator is sometimes used to describe any estimator defined this way, although it was originally used in the context of robust estimation (Huber, 2004). We may obtain an estimate $\bar{X}$ of μ by minimizing the sum of squares $\Lambda_{\tilde{X}}(\mu) = \sum_i (X_i - \mu)^2$. While we have seen that this method is optimal with respect to MSE risk it is also susceptible to outliers because of the squared error loss. This type of estimate can be generalized by using objective function $\Lambda_{\tilde{X}}(\mu) = \sum_i \rho(X_i, \mu)$, where ρ is a contrast function, usually a loss function of the form $\rho(x, \mu) = L(x - \mu)$. We could then replace $L(u) = u^2$ with a robust alternative such as $L(u) = |u|$. On the other hand, Huber's Proposal-2 M-estimator is based on a loss function $L(u)$ which is quadratic near $u = 0$ but linear outside a neighborhood of $u = 0$, thus preserving some of the greater efficiency of squared error estimation. The contrast function can be constructed piecewise to be convex and differentiable, so that the M-estimate can be evaluated using estimating equations. Because the asymptotic theory of estimating equations is not related to robustness, we will simply refer to any solution to an estimating equation as an M-estimator.

This process is well known to statisticians, and it is probably the case that most estimation problems carried out in practice can be expressed in this form. It is therefore worth exploring the general structure of this problem, and the extent to which it is shared by various methodologies.

As we will see below, for the most commonly used estimating equations we will have

$$E_{\boldsymbol{\theta}}[\Psi(\tilde{X}, \boldsymbol{\theta})] = \mathbf{0}, \text{ for all } \boldsymbol{\theta} \in \Theta.$$

Under suitable conditions, if $\tilde{X}$ is taken as a sample of unbounded sample size n, we may define a sequence $\hat{\boldsymbol{\theta}}_n$, $n \geq 1$, such that each $\hat{\boldsymbol{\theta}}_n$ is a solution to Equation (13.1), and the estimate is consistent, that is $\hat{\boldsymbol{\theta}}_n \xrightarrow{i.p.} \boldsymbol{\theta}_0$, where $\boldsymbol{\theta}_0$ is the true parameter. Under these (again, ideal) conditions, it can also be claimed that $\sqrt{n}(\hat{\boldsymbol{\theta}}_n - \boldsymbol{\theta}_0) \xrightarrow{d} N(0, \Sigma)$, where Σ is some $q \times q$ covariance matrix.

DOI: 10.1201/9781003049340-13

The study of this inference problem has two quite distinct components. If the ideal conditions to which we just alluded hold, then the limiting distribution $N(0, \Sigma)$ has a remarkably unified form, and in many important applications is directly obtainable from the Fisher information matrix (Sections 3.5 and 12.8). This is why conditions offered in the literature usually include the regular family conditions of Definition 3.6.

The second component concerns those conditions under which the methodology itself is valid. In particular, the existence of a sequence $\hat{\boldsymbol{\theta}}_n$ with the properties just described is by no means guaranteed. These conditions can be quite technical, with little direct relationship to the inference problem itself. While this cannot be avoided in a mathematically rigorous treatment of the subject, they can to a large degree be isolated from a discussion of statistical methodology, which is the approach taken in this volume.

Accordingly, in Section 13.2 we present the main result of the chapter as Theorem 13.1. This gives conditions under which the sequence $\hat{\boldsymbol{\theta}}_n$ is well defined and asymptotically normal. The regularity conditions on which the theorem depends are also presented in Section 13.2, labelled (A1)–(A8), following the usual conventions of large sample theory. These conditions vary considerably in "level". (A1) simply ensures the differentiability of $\Psi(\tilde{X}, \boldsymbol{\theta})$ with respect to $\boldsymbol{\theta}$, without which the required analysis would not be well defined. (A2)–(A4) are applications of either the weak law of large numbers, or the central limit theorem. (A5)–(A6) are somewhat more unintuitive, however, we will show in Section 13.10 how they can be verified for the types of models considered in this chapter. (A7)–(A8) are rather more generous, in that they simply state what a theorem usually sets out to prove. This is why the proof of Theorem 13.1 is so brief. Therefore, a complete treatment of the problem requires proofs that assumptions (A1)–(A6) imply (A7) and (A8), which are presented in Sections 13.8 and 13.9.

Theorem 13.1 is comparable to, for example, Theorem 6.2.1 of Bickel and Doksum (2015) or Theorem 6.5.1 of Lehmann and Casella (1998). The regularity conditions in those references are quite similar to those used in this chapter. However both theorems are confined to the *iid* case. In addition, Theorem 6.5.1 of Lehmann and Casella (1998) is confined to maximum likelihood estimation. Here, we find that by isolating Assumptions (A2)–(A4) (those most model-dependent) this asymptotic approximation method can be easily extended to more general models, using much the same arguments.

Once Theorem 13.1 is given, a large sample theory for quite a wide range of statistical methods can be developed directly. Section 13.3 deals with maximum likelihood estimation. In Section 13.4 we consider a general formulation of the regression model. Suppose we have responses $\mathbf{Y} = (Y_1, \ldots, Y_n)^T$ from a parametric family $p_{\boldsymbol{\theta}} \in \mathcal{P}$, $\Theta \subset \mathbb{R}^q$. Then suppose $\mu_i(\boldsymbol{\theta}) = E_{\boldsymbol{\theta}}[Y_i]$, $i = 1, \ldots, n$ are real valued functions on Θ. Typically, we have an independent variable $\mathbf{x}_i$ such that $\mu_i(\boldsymbol{\theta}) = \mu(\boldsymbol{\theta}, \mathbf{x}_i)$. This model includes as special cases, nonlinear regression, generalized linear models (GLM) and generalized estimating equations (GEE) (J. A. Nelder, 1972; Liang and Zeger, 1986) and quasi-likelihood estimation (Wedderburn, 1974; McCullagh *et al.*, 1983). The conventional estimation methods for these models are based on estimating equations, which, remarkably, all have essentially the same structure, and therefore Theorem 13.1 provides a single asymptotic theory for the entire class. Nonlinear regression, GLMs and GEE models are then considered in Sections 13.5, 13.6 and 13.7, respectively.

The chapter ends with a detailed development of the regularity conditions required for Theorem 13.1 in Sections 13.8, 13.9 and 13.10.

13.2 Consistency and Asymptotic Normality of M-Estimators

The theory of estimating equations tends to depend on large sample approximations, since exact analysis is typically not feasible. We begin with Equation (13.1) defined for a parameter space $\Theta \subset \mathbb{R}^q$. Define $\Psi'(\tilde{X}, \boldsymbol{\theta}) \in \mathcal{M}_q$ elementwise by

$$\left\{\Psi'(\tilde{X}, \boldsymbol{\theta})\right\}_{jk} = \frac{\partial \Psi_j(\tilde{X}, \boldsymbol{\theta})}{\partial \theta_k},$$

assuming the partial derivatives exist. If $\Psi(\tilde{X}, \boldsymbol{\theta})$ is the gradient of an objective function $\Lambda_{\tilde{X}}(\boldsymbol{\theta})$ then $\Psi'(\tilde{X}, \boldsymbol{\theta})$ will be its Hessian matrix (Section C.7). Note, however, that for some methods $\Lambda_{\tilde{X}}(\boldsymbol{\theta})$ is not explicitly given (Example 4.12, Section 13.7), but we will still refer to $\Psi'(\tilde{X}, \boldsymbol{\theta})$ as the Hessian in any case. We will also make use of the three dimensional array of second derivatives $\Psi''(\tilde{X}, \boldsymbol{\theta})$, defined by elements $\{\Psi''(\tilde{X}, \boldsymbol{\theta})\}_{ijk} = \partial^2 \Psi_i(\tilde{X}, \boldsymbol{\theta})/\partial\theta_j \partial\theta_k$. This would also be the array of third order derivatives of any objective function $\Lambda_{\tilde{X}}(\boldsymbol{\theta})$.

We first give an outline of the approach, which will rely on three approximations. The score function $\Psi(\tilde{X}, \boldsymbol{\theta})$, and therefore the Hessian, tend to be constructed from random sums of n observations. We therefore adopt the normalization

$$\bar{\Psi}(\tilde{X}, \boldsymbol{\theta}) = n^{-1}\Psi(\tilde{X}, \boldsymbol{\theta}), \quad \bar{\Psi}'(\tilde{X}, \boldsymbol{\theta}) = n^{-1}\Psi'(\tilde{X}, \boldsymbol{\theta}), \quad \bar{\Psi}''(\tilde{X}, \boldsymbol{\theta}) = n^{-1}\Psi''(\tilde{X}, \boldsymbol{\theta}).$$

Our three approximations are then

$$\begin{aligned} \bar{\Psi}(\tilde{X}, \boldsymbol{\theta}_0) &\approx \mathbf{0}, \\ \bar{\Psi}'(\tilde{X}, \boldsymbol{\theta}_0) &\approx E_{\boldsymbol{\theta}_0}[\bar{\Psi}'(\tilde{X}, \boldsymbol{\theta}_0)] = \Sigma_{\bar{\Psi}'(\boldsymbol{\theta}_0)}, \\ \sqrt{n}\bar{\Psi}(\tilde{X}, \boldsymbol{\theta}_0) &\sim N(\mathbf{0}, \Sigma_{\Psi(\boldsymbol{\theta}_0)}), \end{aligned} \tag{13.2}$$

for some covariance matrix $\Sigma_{\Psi(\boldsymbol{\theta}_0)}$. We will often see that $\Sigma_{\bar{\Psi}'(\boldsymbol{\theta}_0)}$ is positive definite, so $\bar{\Psi}'(\tilde{X}, \boldsymbol{\theta}_0)$ will also be positive definite with probability $1 - \epsilon$. Of course we can replace Equation (13.1) with the equivalent estimating equation $\bar{\Psi}(\tilde{X}, \boldsymbol{\theta}_0) = \mathbf{0}$ by dividing both sides by n.

Suppose we take a Taylor's series approximation

$$\bar{\Psi}(\tilde{X}, \boldsymbol{\theta}) = \bar{\Psi}(\tilde{X}, \boldsymbol{\theta}_0) + \bar{\Psi}'(\tilde{X}, \boldsymbol{\theta}_0)^T[\boldsymbol{\theta} - \boldsymbol{\theta}_0] + o(|\boldsymbol{\theta} - \boldsymbol{\theta}_0|) \tag{13.3}$$

(the details are left for Section 13.9). If a solution $\bar{\Psi}(\tilde{X}, \hat{\boldsymbol{\theta}}) = 0$ exists, upon substitution into Equation (13.3) we have

$$\sqrt{n}(\hat{\boldsymbol{\theta}} - \boldsymbol{\theta}_0) \approx -\left[\bar{\Psi}'(\tilde{X}, \boldsymbol{\theta}_0)\right]^{-1} \sqrt{n}\bar{\Psi}(\tilde{X}, \boldsymbol{\theta}_0). \tag{13.4}$$

Then, if we accept the approximations in Equation (13.2) we have

$$\sqrt{n}(\hat{\boldsymbol{\theta}} - \boldsymbol{\theta}_0) \approx N(0, \Sigma_{\hat{\boldsymbol{\theta}}}), \tag{13.5}$$

where

$$\Sigma_{\hat{\boldsymbol{\theta}}} = \Sigma^{-1}_{\bar{\Psi}'(\boldsymbol{\theta}_0)} \Sigma_{\Psi(\boldsymbol{\theta}_0)} [\Sigma^{-1}_{\bar{\Psi}'(\boldsymbol{\theta}_0)}]^T. \tag{13.6}$$

The main goal of this chapter is to develop a theory to support the approximation given by (13.5)-(13.6). Large sample theory describes what happens as the sample size index n approaches ∞. We therefore have a sequence of observations $\tilde{X}_n$ and score functions $\Psi_n(\tilde{X}_n, \boldsymbol{\theta})$, $n \geq 1$. However, for convenience we may suppress the subscript n, and simply write $\Psi(\tilde{X}, \boldsymbol{\theta})$, while allowing n to play the usual role. The true parameter remains $\boldsymbol{\theta}_0 \in \Theta$.

The first four assumptions follow.

(A1) $\Psi(\boldsymbol{\theta}; X)$ is twice continuously differentiable *wrt* $\boldsymbol{\theta}$, so that the matrix $\Psi'(\tilde{X}, \boldsymbol{\theta})$ and array $\Psi''(\tilde{X}, \boldsymbol{\theta})$ are well defined.

(A2) The limit $\bar{\Psi}(\tilde{X}, \boldsymbol{\theta}_0) \stackrel{i.p.}{\rightarrow} \mathbf{0}$ holds.

(A3) $n^{1/2}\bar{\Psi}(\tilde{X}, \boldsymbol{\theta}_0) \stackrel{d}{\rightarrow} \mathbf{Z} \sim N(\mathbf{0}, \Sigma_{\Psi(\boldsymbol{\theta}_0)})$.

(A4) The limit $\bar{\Psi}'(\tilde{X}, \boldsymbol{\theta}_0) \stackrel{i.p.}{\rightarrow} \Sigma_{\Psi'(\boldsymbol{\theta}_0)}$ exists, with $\det(\Sigma_{\Psi'(\boldsymbol{\theta}_0)}) \neq 0$. $\Sigma_{\Psi'(\boldsymbol{\theta}_0)}$ is either positive definite or negative definite.

(A3) implies (A2) (Lemma 12.1) but it will be useful to refer to these assumptions separately.

Usually, $\Psi(\tilde{X}, \boldsymbol{\theta})$ behaves like a random sum of n terms, so assumptions (A1)–(A4) are direct consequences of the WLLN or the CLT, and so need no further development here. It will usually hold that $E_{\boldsymbol{\theta}}[\bar{\Psi}(\tilde{X}, \boldsymbol{\theta})] = 0$, but we need not assume this if (A2) holds.

Assumption (A5) below is of a similar nature to (A1)–(A4), but will usually requirement more argument to validate. It is essentially a locally uniform stochastic bound on the second derivatives of $\Psi(\tilde{X}, \boldsymbol{\theta})$ near $\boldsymbol{\theta} = \boldsymbol{\theta}_0$. It commonly appears in references on this problem (Theorem 7.8 of Bhattacharya *et al.* (2016) or Theorem 7.3.1 of Lehmann (1999)). (A6) follows from (A5) following an application of the mean value theorem, and essentially captures the motivation for (A5) (a version of this assumption appears in as Assumption (A4) in Section 6.2.1 of Bickel and Doksum (2015) in this context).

(A5) Fix $m_0 > 0$. Then there exists $K_1 < \infty$ such that for any $0 < \tau \leq m_0$

$$P_0\left(\sup_{\boldsymbol{\theta} \in B_\tau(\boldsymbol{\theta}_0)} |\bar{\Psi}''_{jkl}(\tilde{X}, \boldsymbol{\theta})| < K_1\right) \rightarrow_n 1.$$

(A6) Fix $m_0 > 0$. Then there exists $K_0 < \infty$ such that for any $0 < \tau \leq m_0$

$$P_0\left(\sup_{\boldsymbol{\theta} \in B_\tau(\boldsymbol{\theta}_0)} \left\|\bar{\Psi}'(\tilde{X}, \boldsymbol{\theta}) - \bar{\Psi}'(\tilde{X}, \boldsymbol{\theta}_0)\right\|_2 < K_0\tau\right) \rightarrow_n 1.$$

Note that in Assumption (A6) we have made use of the Euclidean matrix norm $\|\mathbf{A}\|_2$ (Section C.8.2).

The remaining assumptions are of a considerably higher level than (A1)–(A6).

(A7) Let $E \subset \Theta$ be any open neighborhood of $\boldsymbol{\theta}_0$. With probability approaching one as $n \rightarrow \infty$ there exists a solution $\hat{\boldsymbol{\theta}}$ of the estimating equation $\Psi(\boldsymbol{\theta}; X) = \mathbf{0}$ in E. As a consequence there is a sequence $\hat{\boldsymbol{\theta}}_n$ with limit $\hat{\boldsymbol{\theta}}_n \stackrel{i.p.}{\rightarrow} \boldsymbol{\theta}_0$.

(A8) Suppose (A7) holds. If $-\sqrt{n}\left[\bar{\Psi}'(\tilde{X}, \boldsymbol{\theta}_0)\right]^{-1}\bar{\Psi}(\tilde{X}, \boldsymbol{\theta}_0) \stackrel{d}{\rightarrow} \mathbf{U}$ then the sequence $\hat{\boldsymbol{\theta}}_n$ described in (A7) satisfies $\sqrt{n}(\hat{\boldsymbol{\theta}}_n - \boldsymbol{\theta}_0) \stackrel{d}{\rightarrow} \mathbf{U}$.

Our assumptions lead naturally to the following theorem.

Theorem 13.1 (Asymptotic Properties of M-estimators) Supposes Assumptions (A3)–(A4) and (A7)–(A8) hold. Then there exists a sequence of estimators $\hat{\boldsymbol{\theta}}_n$ for which the following statements hold:

(i) $P\left(\Psi(\tilde{X}, \hat{\boldsymbol{\theta}}_n) = \mathbf{0}\right) \rightarrow_n 1$,

(ii) $\hat{\boldsymbol{\theta}}_n \stackrel{i.p.}{\rightarrow} \boldsymbol{\theta}_0$

(iii) $\sqrt{n}(\hat{\boldsymbol{\theta}}_n - \boldsymbol{\theta}_0) \stackrel{d}{\rightarrow} \mathbf{Z} \sim N(\mathbf{0}, \Sigma_{\hat{\boldsymbol{\theta}}})$, where $\Sigma_{\hat{\boldsymbol{\theta}}} = \Sigma^{-1}_{\Psi'(\boldsymbol{\theta}_0)}\Sigma_{\Psi(\boldsymbol{\theta}_0)}[\Sigma^{-1}_{\Psi'(\boldsymbol{\theta}_0)}]^T$. ■

Proof. Statements (i)-(ii) follow from (A7). Suppose (A8) holds. By (A3)–(A4) and Slutsky's theorm $\mathbf{U} \sim N(\mathbf{0}, \Sigma_{\hat{\boldsymbol{\theta}}})$. Then statement (iii) follows from the remainder of (A8). □

Theorem 13.1 gives our main result for M-estimators, and, given the generous nature of Assumptions (A7)–(A8) was not too difficult to prove. Of course, the remaining challenge is more formidible. It seems reasonable to accept (A1)–(A5) as practical regularity conditions. and we will eventually show that the regularity conditions of Theorem 13.1 can be replaced by that less generous set of assumptions. We will see, in particular, how (A5) may be verified for the types of models we consider.

For the moment, however, we will present a number of applications of Theorem 13.1, deferring further consideration of regularity conditions to Sections 13.8, 13.9 and 13.10.

13.3 Asymptotic Theory of MLEs

The problem of evaluating MLEs was considered in some detail in Section 4.5, although their distributional properties were not considered. In some cases, the MLE can be determined by setting the gradient of the log-likelihood function to zero, or equivalently, solving the likelihood equations. In this case, the MLE is an M-estimator, with $\Psi(\tilde{X}, \boldsymbol{\theta})$ equal to that gradient. It is this class of MLE we are concerned with in this Chapter.

If the conditions for Theorem 13.1 hold, the asympotic distribution of an MLE is easy to deduce. Following Section 4.5 we are given as objective function the log-likelihood function $\ell(\boldsymbol{\theta}; \tilde{X}) = \log p_{\boldsymbol{\theta}}(\tilde{X})$, with $\boldsymbol{\theta} \in \Theta \subset \mathbb{R}^q$. The score function and Hessian are given elementwise by

$$\Psi_j(\tilde{X}, \boldsymbol{\theta}) = \frac{\partial \ell(\boldsymbol{\theta}; \tilde{X})}{\partial \theta_j}, \quad \Psi'_{jk}(\tilde{X}, \boldsymbol{\theta}) = \frac{\partial^2 \ell(\boldsymbol{\theta}; \tilde{X})}{\partial \theta_j \partial \theta_k}.$$

Assume that $p_{\boldsymbol{\theta}}$ defines a regular family (Definition 3.6). Then by Theorem 3.12

$$E_{\boldsymbol{\theta}}[\Psi(\tilde{X}, \boldsymbol{\theta})] = \mathbf{0}, \quad \mathbf{I}(\boldsymbol{\theta}) = \text{var}_{\boldsymbol{\theta}}[\Psi(\tilde{X}, \boldsymbol{\theta})] = -E_{\boldsymbol{\theta}}[\Psi'(\tilde{X}, \boldsymbol{\theta})],$$

where $\mathbf{I}(\boldsymbol{\theta})$ is the Fisher information matrix. If we let $\bar{\mathbf{I}}(\boldsymbol{\theta}) = n^{-1}\mathbf{I}(\boldsymbol{\theta})$ we have

$$\Sigma_{\Psi(\boldsymbol{\theta})} = \bar{\mathbf{I}}(\boldsymbol{\theta}), \quad \Sigma_{\Psi'(\boldsymbol{\theta})} = -\bar{\mathbf{I}}(\boldsymbol{\theta}), \quad \Sigma_{\hat{\boldsymbol{\theta}}} = \bar{\mathbf{I}}(\boldsymbol{\theta}),$$

so that applying (13.5)-(13.6) gives approximation

$$\sqrt{n}\left[\hat{\boldsymbol{\theta}} - \boldsymbol{\theta}_0\right] \sim N(\mathbf{0}, \bar{\mathbf{I}}(\boldsymbol{\theta}_0)^{-1}). \tag{13.7}$$

It should also be pointed out that a number of options exist for this type of inference. First, note that since $\boldsymbol{\theta}_0$ is unknown, if we are relying on approximation (13.7), we must estimate $\bar{\mathbf{I}}(\boldsymbol{\theta}_0)$, a sensible choice being $\bar{\mathbf{I}}(\boldsymbol{\theta}_0) \approx \bar{\mathbf{I}}(\hat{\boldsymbol{\theta}})$. It is also possible to use the negative Hessian $-\Psi'(\tilde{X}, \hat{\boldsymbol{\theta}})$, again substituting $\hat{\boldsymbol{\theta}}$ for $\boldsymbol{\theta}_0$. This is referred to as the observed Fisher information, in contrast with the expected Fisher information. This is not an attempt to estimate $\bar{\mathbf{I}}(\boldsymbol{\theta}_0)$, but represents a form of conditional inference. The observed Fisher information is interpreted as a ancillary statistic, and this approach represents an application of the conditionality principle (Definition 3.15). Efron and Hinkley (1978) is a seminal paper on this subject.

For a full rank exponential model it is quite straightforward to verify that these conditions hold. Using the natural parametrization we have

$$\Psi(\tilde{X}, \boldsymbol{\eta}) = \mathbf{T} - \nabla A(\boldsymbol{\eta}).$$

Assumption (A1) holds, given the role of $A(\boldsymbol{\eta})$ as a normalization constant. In most applications of this model the elements of $\mathbf{T}$ are random sums, so (A2) and (A4) may follow simply from an application of a WLLN and a CLT. Final, verifying conditions (A3) and (A5) is simplified considerably by the fact that Ψ' and Ψ'' do not depend on $\tilde{X}$, and the matrix of second derivatives of $A(\boldsymbol{\eta})$ define a covariance matrix which will be invertable for a full rank exponential model. If $\boldsymbol{\theta}$ is a smooth 1-1 transformation of the canonical parameter, then asymptotic normality will follow from the multivariate δ-method. This was made precise in Theorems 4.3 and 4.4.

The asymptotic properties of the MLE will be further explored in Section 13.6 on generalized linear models (GLM) and Chapter 14 on large sample hypothesis testing. We finish our discussion on MLEs for the moment by referring to Section 12.9. A sequence of estimators was referred to as asymptotically efficient if the Cramér-Rao lower bound on their variance was attained as $n \to \infty$. But this lower bound is simply the inverse of the Fisher information, so that asymptotic efficiency simply means that the approximation of Equation (13.7) holds in the limit.

Example 13.1 (MLE of the Gamma Scale and Shape Parameters) Suppose $\mathbf{X} = (X_1, \ldots, X_n)$ be an *iid* sample from a $gamma(\tau, \alpha)$ distribution, where τ is a scale parameter and α is a shape parameter. Then X_1 has mean and variance $\mu = \alpha\tau$ and $\sigma^2 = \alpha\tau^2$. Then suppose $\hat{\tau}, \hat{\alpha}$ are the MLEs of (τ, α) Applying Theorem 13.1, the limiting distribution of the MLE vector is then

$$\sqrt{n}[(\hat{\tau}_n, \hat{\alpha}_n) - (\tau, \alpha)] \xrightarrow{d} \mathbf{Z} \sim N(\mathbf{0}, \bar{\mathbf{I}}(\tau, \alpha)^{-1}),$$

where $\mathbf{I}(\tau, \alpha)$ is the Fisher information matrix.

Next, suppose we wish to estimate $\sigma^2 = g(\tau, \alpha) = \alpha\tau^2$ using $\hat{\sigma}^2 = g(\hat{\tau}, \hat{\alpha})$. We can use the multivariate δ-method of Theorem 12.3 to approximate the distribution of $\hat{\sigma}^2$. The gradent matrix of g is the 1×2 matrix $\mathbf{J}_g(a, b) = [2ab \;\; a^2]$. We then have $\sqrt{n}[\hat{\sigma}_n^2 - \sigma^2] \xrightarrow{d} \mathbf{J}_g(\tau, \alpha)\mathbf{Z} = U$ where $U \sim N(0, \sigma_U^2)$, $\sigma_U^2 = [2\tau\alpha \;\; \tau^2]\bar{\mathbf{I}}(\tau, \alpha)^{-1}[2\tau\alpha \;\; \tau^2]^T$. ■

13.4 A General Form for Regression Models

In Chapter 6 we described the linear regression model. More generally, we can take the term regression analysis to refer to the following problem. Consider responses $\mathbf{Y} = (Y_1, \ldots, Y_n)^T$ from a parametric family $p_{\boldsymbol{\theta}}(\mathbf{Y}) \in \mathcal{P}$, $\Theta \subset \mathbb{R}^q$. Then suppose $\mu_i(\boldsymbol{\theta}) = E_{\boldsymbol{\theta}}[Y_i]$, $i = 1, \ldots, n$ are real valued functions on Θ.

The variation in the mean functions can take any form, but the motiviation is often to deduce a functional relationship between a paired vector $\mathbf{x}_i \in \mathbb{R}^q$ and Y_i. In this case we may have $\mu_i(\boldsymbol{\theta}) = f(\boldsymbol{\theta}, \mathbf{x}_i)$ for some function f. The linear regression model is a special case of this model.

Whatever the case, we assume each μ_i possesses all second derivatives. Define the $n \times q$ first derivative array $\dot{\boldsymbol{\mu}}(\boldsymbol{\theta})$ and $n \times q \times q$ second derivative array $\ddot{\boldsymbol{\mu}}(\boldsymbol{\theta})$ by components

$$\{\dot{\boldsymbol{\mu}}(\boldsymbol{\theta})\}_{ij} = \frac{\partial \mu_i(\boldsymbol{\theta})}{\partial \boldsymbol{\theta}_j}, \;\; \{\ddot{\boldsymbol{\mu}}(\boldsymbol{\theta})\}_{ijk} = \frac{\partial^2 \mu_i(\boldsymbol{\theta})}{\partial \boldsymbol{\theta}_j \partial \boldsymbol{\theta}_k}, \;\; i = 1, \ldots, n, \;\; j, k = 1, \ldots, q.$$

In addition, we define the following $n \times 1$ column matrices

$$\begin{aligned}
\boldsymbol{\mu}(\boldsymbol{\theta}) &= [\mu_1(\boldsymbol{\theta}) \cdots \mu_n(\boldsymbol{\theta})]^T, \\
\dot{\boldsymbol{\mu}}_j(\boldsymbol{\theta}) &= [\{\dot{\boldsymbol{\mu}}(\boldsymbol{\theta})\}_{1j} \cdots \{\dot{\boldsymbol{\mu}}(\boldsymbol{\theta})\}_{nj}]^T, \\
\ddot{\boldsymbol{\mu}}_{jk}(\boldsymbol{\theta}) &= [\{\ddot{\boldsymbol{\mu}}(\boldsymbol{\theta})\}_{1jk} \cdots \{\ddot{\boldsymbol{\mu}}(\boldsymbol{\theta})\}_{njk}]^T.
\end{aligned}$$

Then define the $n \times q$ matrix of derivatives $\dot{\boldsymbol{\mu}}(\boldsymbol{\theta})$,

$$\boldsymbol{\Gamma}_{\boldsymbol{\theta}} = \left[\dot{\boldsymbol{\mu}}_1(\boldsymbol{\theta}), \ldots, \dot{\boldsymbol{\mu}}_q(\boldsymbol{\theta})\right]. \tag{13.8}$$

Suppose we now construct an estimating equation by defining the following score function explicitly:

$$\Psi(\mathbf{Y}, \boldsymbol{\theta}) = -{\boldsymbol{\Gamma}_{\boldsymbol{\theta}}}^T \mathbf{W}(\boldsymbol{\theta}) \left[\mathbf{Y} - \boldsymbol{\mu}(\boldsymbol{\theta})\right], \tag{13.9}$$

where $\mathbf{W}(\boldsymbol{\theta})$ is an $n \times n$ matrix, then accept as an estimator for $\boldsymbol{\theta}$ a solution $\hat{\boldsymbol{\theta}}$ to $\Psi(\mathbf{Y}, \boldsymbol{\theta}) = \mathbf{0}$. We have not derived (13.9) according to any optimization principle, but we will see that a large class of statistical optimization problems lead to estimating equations of this form.

The next observation to make is that, following the argument leading to the approximation (13.7), the accuracy of the procedure is determined primarily by the variance of $\Psi(\mathbf{Y}, \boldsymbol{\theta})$ at the true value $\boldsymbol{\theta}_0$. In addition, the discussion of the best linear unbiased estimator (BLUE) in Section 6.3 suggests that the variance of a random sum is minimized by weighting the individual terms by the reciprocal of their variance. Although the model here is more complex, this at least suggests that $\mathbf{W}(\boldsymbol{\theta})$ in (13.9) should be selected to have the same effect (see Example 13.2). The simplest way to do this would be to set $\mathbf{W}(\boldsymbol{\theta})$ equal to the inverse of the covariance matrix of $\mathbf{Y}$, or an estimate if needed. In the examples we consider, this is the solution that optimization principles lead to.

However $\mathbf{W}(\boldsymbol{\theta})$ is derived, we can calculate the Hessian matrix as

$$\begin{aligned} &\left\{\Psi'(\mathbf{Y}, \boldsymbol{\theta})\right\}_{jk} = \\ &\quad \left\{\mathbf{K}(\boldsymbol{\theta})\right\}_{jk} - \left[\ddot{\boldsymbol{\mu}}_{jk}(\boldsymbol{\theta})^T \mathbf{W}(\boldsymbol{\theta}) + \dot{\boldsymbol{\mu}}_j(\boldsymbol{\theta})^T \dot{\mathbf{W}}_k(\boldsymbol{\theta})\right] \left[\mathbf{Y} - \boldsymbol{\mu}(\boldsymbol{\theta})\right], \end{aligned} \tag{13.10}$$

where

$$\mathbf{K}(\boldsymbol{\theta}) = {\boldsymbol{\Gamma}_{\boldsymbol{\theta}}}^T \mathbf{W}(\boldsymbol{\theta}) \boldsymbol{\Gamma}_{\boldsymbol{\theta}}.$$

Suppose $\hat{\boldsymbol{\theta}}$ is the M-estimate solving $\Psi(\mathbf{Y}, \hat{\boldsymbol{\theta}}) = \mathbf{0}$. Following the method of Theorem 13.1, if we match

$$\Sigma_{\Psi(\boldsymbol{\theta})} = n^{-1} {\boldsymbol{\Gamma}_{\boldsymbol{\theta}}}^T \mathbf{W}(\boldsymbol{\theta}) \Sigma_{\mathbf{Y}} \mathbf{W}(\boldsymbol{\theta})^T \boldsymbol{\Gamma}_{\boldsymbol{\theta}}$$

and

$$\Sigma_{\Psi'(\boldsymbol{\theta})} = n^{-1} \mathbf{K}(\boldsymbol{\theta})$$

then

$$\sqrt{n}\left[\hat{\boldsymbol{\theta}} - \boldsymbol{\theta}_0\right] \approx \mathbf{U} \sim N(\mathbf{0}, \Sigma_{\hat{\boldsymbol{\theta}}}) \tag{13.11}$$

where

$$\Sigma_{\hat{\boldsymbol{\theta}}} = n\mathbf{K}(\boldsymbol{\theta})^{-1} {\boldsymbol{\Gamma}_{\boldsymbol{\theta}}}^T \mathbf{W}(\boldsymbol{\theta}) \Sigma_{\mathbf{Y}} \mathbf{W}(\boldsymbol{\theta})^T \boldsymbol{\Gamma}_{\boldsymbol{\theta}} \left[\mathbf{K}(\boldsymbol{\theta})^{-1}\right]^T.$$

When $\mathbf{W}(\boldsymbol{\theta}) = \Sigma_{\mathbf{Y}}^{-1}$, the covariance simplifies to

$$\begin{aligned} \Sigma_{\hat{\boldsymbol{\theta}}} &= n\mathbf{K}(\boldsymbol{\theta})^{-1} {\boldsymbol{\Gamma}_{\boldsymbol{\theta}}}^T \Sigma_{\mathbf{Y}}^{-1} \Sigma_{\mathbf{Y}} \Sigma_{\mathbf{Y}}^{-1} \boldsymbol{\Gamma}_{\boldsymbol{\theta}} \left[\mathbf{K}(\boldsymbol{\theta})^{-1}\right]^T \\ &= n\mathbf{K}(\boldsymbol{\theta})^{-1} {\boldsymbol{\Gamma}_{\boldsymbol{\theta}}}^T \Sigma_{\mathbf{Y}}^{-1} \boldsymbol{\Gamma}_{\boldsymbol{\theta}} \left[\mathbf{K}(\boldsymbol{\theta})^{-1}\right]^T \\ &= n\mathbf{K}(\boldsymbol{\theta})^{-1}. \end{aligned} \tag{13.12}$$

Where needed, we may estimate $\boldsymbol{\theta}$ in Equation (13.12) with $\hat{\boldsymbol{\theta}}$. This form will hold for the methodologies considered in this Chapter.

To a large degree, M-estimators defined by Equation (13.9) share the same structure as the best linear unbiased estimator (BLUE) for linear models described in Chapter 6. This is shown in the next example.

Example 13.2 (On the Relationship between BLUEs and Estimating Equations) The score function of (13.9) can be directly related to the BLUE of $\boldsymbol{\theta} \in \mathbb{R}^q$ for the linear model $\mathbf{Y} = \mathbf{X}\boldsymbol{\theta} + \boldsymbol{\epsilon}$ of Equation (6.2). Since $\boldsymbol{\mu} = \mathbf{X}\boldsymbol{\theta}$ it is easily checked that $\boldsymbol{\Gamma}_{\boldsymbol{\theta}} = \mathbf{X}$. If we set $\mathbf{W}(\boldsymbol{\theta}) = \Sigma_{\mathbf{Y}}^{-1}$, the estimating equation becomes

$$\mathbf{X}^T \Sigma_{\mathbf{Y}}^{-1} [\mathbf{Y} - \mathbf{X}\boldsymbol{\theta}] = \mathbf{0}. \tag{13.13}$$

If $\mathbf{X}$ is of rank q, $\mathbf{X}^T \Sigma_{\mathbf{Y}}^{-1} \mathbf{X}$ will be an invertible $q \times q$ matrix, so Equation (13.13) possesses the unique solution

$$\hat{\boldsymbol{\theta}} = \left[\mathbf{X}^T \Sigma_{\mathbf{Y}}^{-1} \mathbf{X} \right]^{-1} \mathbf{X}^T \Sigma_{\mathbf{Y}}^{-1} \mathbf{Y},$$

which, by the Gauss-Markov theorem (Theorem 6.1) is the BLUE of $\boldsymbol{\theta}$. ■

13.5 Nonlinear Regression

Suppose for the regression model of Section 13.4 the responses are given by

$$Y_i = \mu_i(\boldsymbol{\theta}) + \epsilon_i \tag{13.14}$$

where $\boldsymbol{\epsilon} = (\epsilon_1, \ldots, \epsilon_n)^T$ are zero mean error terms. The linear model is clearly a special case, defined by a linear relationship between $\mu_i(\boldsymbol{\theta})$ and $\boldsymbol{\theta}$ (Example 13.2). When this cannot be assumed, Equation (13.14) is referred to as a nonlinear regression model. As for the linear model, $\mu_i(\boldsymbol{\theta})$ commonly defines a parametrized relationship between the response mean $E[Y_i] = \mu_i(\boldsymbol{\theta})$ and a nonrandom predictor vector $\mathbf{x}_i$ of the form $\mu_i(\boldsymbol{\theta}) = f(\boldsymbol{\theta}, \mathbf{x}_i)$, but this need not be specified at this point.

Recall that for linear models the mean vector $\boldsymbol{\mu}$ is constrained to fall in a vector space $\mathcal{V}$ (Section 6.4), which gives the analysis of this class of models its particular qualities. This structure does not hold exactly for nonlinear models (although $\boldsymbol{\mu}$ is clearly confined to a q-dimensional contour). However, statistical methods for nonlinear regression models do resemble those for linear models, especially for large samples. In fact, a large part of the theory of nonlinear regression models is concerned with the measurement of curvature with respect to the parameter $\boldsymbol{\theta}$ (deviation from linearity), and transformations which approximately linearize models are commonly used (Chapter 4 of Wild and Seber (1989)).

As for linear models, the least squares method is available for nonlinear regression, and is probably the most commonly used procedure. Both the ordinary least squares and generlized least squares criterion can be used (Section 6.5). To define the ordinary least squares criterion, set

$$\Lambda_{\mathbf{Y}}(\boldsymbol{\theta}) = \frac{1}{2} \sum_{i=1}^{n} (Y_i - \mu_i(\boldsymbol{\theta}))^2 = \frac{1}{2} [\mathbf{Y} - \boldsymbol{\mu}(\boldsymbol{\theta})]^T [\mathbf{Y} - \boldsymbol{\mu}(\boldsymbol{\theta})].$$

The gradient and Hessian are is easily calculated to be

$$\Psi(\mathbf{Y}, \boldsymbol{\theta}) = -\boldsymbol{\Gamma}_{\boldsymbol{\theta}}{}^T \left[\mathbf{Y} - \boldsymbol{\mu}(\boldsymbol{\theta}) \right], \ \{\Psi'(y, \boldsymbol{\theta})\}_{jk} = \{\mathbf{K}(\boldsymbol{\theta})\}_{jk} - \ddot{\boldsymbol{\mu}}_{jk}(\boldsymbol{\theta})^T \left[\mathbf{Y} - \boldsymbol{\mu}(\boldsymbol{\theta}) \right].$$

The ordinary least squares estimate of $\boldsymbol{\theta}$ is the solution $\hat{\boldsymbol{\theta}}$ to $\Psi(\mathbf{Y}, \boldsymbol{\theta}) = \mathbf{0}$, which is of the general form given in Equation (13.9). Suppose the model covariance matrix $\Sigma_{\mathbf{Y}}$ is $\sigma^2 \mathbf{I}_n$. If we match $\Sigma_{\Psi(\boldsymbol{\theta})} = n^{-1} \sigma^2 \boldsymbol{\Gamma}_{\boldsymbol{\theta}}{}^T \boldsymbol{\Gamma}_{\boldsymbol{\theta}}$ and $\Sigma_{\Psi'(\boldsymbol{\theta})} = n^{-1} \mathbf{K}(\boldsymbol{\theta})$, then by Equations (13.11)–(13.12) we have the approximation

$$\sqrt{n} \left[\hat{\boldsymbol{\theta}} - \boldsymbol{\theta}_0 \right] \approx \mathbf{U} \sim N(\mathbf{0}, \sigma^2 n \mathbf{K}(\boldsymbol{\theta}_0)^{-1}). \tag{13.15}$$

Then suppose $\Sigma_{\mathbf{Y}} \neq \sigma^2 \mathbf{I}_n$, assuming $\Sigma_{\mathbf{Y}}$ does not depend on $\boldsymbol{\theta}$. In this case the ordinary least squares estimate may still work reasonably well, but it would be more efficient to use the generalized least squares criterion given by

$$\Lambda_{\mathbf{Y}}(\boldsymbol{\theta}) = \frac{1}{2}[\mathbf{Y} - \boldsymbol{\mu}(\boldsymbol{\theta})]^T \Sigma_{\mathbf{Y}}^{-1} [\mathbf{Y} - \boldsymbol{\mu}(\boldsymbol{\theta})].$$

The gradient of $\Lambda_{\mathbf{Y}}(\boldsymbol{\theta})$ then becomes

$$\Psi(\mathbf{Y}, \boldsymbol{\theta}) = -\boldsymbol{\Gamma_\theta}^T \Sigma_{\mathbf{Y}}^{-1} [\mathbf{Y} - \boldsymbol{\mu}(\boldsymbol{\theta})].$$

This assumes the form of the score function of Equation (13.9) with $\mathbf{W}(\boldsymbol{\theta}) = \Sigma_{\mathbf{Y}}^{-1}$, so the approximation of Equation (13.11) holds, that is $\sqrt{n}[\hat{\boldsymbol{\theta}} - \boldsymbol{\theta}_0] \approx \mathbf{U} \sim N(\mathbf{0}, \Sigma_{\hat{\boldsymbol{\theta}}})$, where $\Sigma_{\hat{\boldsymbol{\theta}}} = n\mathbf{K}(\boldsymbol{\theta})^{-1} = n[\boldsymbol{\Gamma_\theta}^T \Sigma_{\mathbf{Y}}^{-1} \boldsymbol{\Gamma_\theta}]^{-1}$.

Wild and Seber (1989) can be especially recommended for this subject. We next offer an example of a seminal nonlinear regression model.

Example 13.3 (The Michaelis-Menten Model) The Michaelis-Menten model with three parameters takes the form

$$Y_i = \mu_i(\boldsymbol{\theta}) + \epsilon_i, \;\; i = 1, \ldots, n, \;\; \text{where } \mu_i(\boldsymbol{\theta}) = \frac{Vx_i}{K + x_i} + C$$

for unknown constants $\boldsymbol{\theta} = (V, K, C)$, known independent variable x_i and unobserved error terms ϵ_i. We assume the error terms are *iid* from distribution $N(0, \sigma^2)$. The ordinary least squares estimates of V, K and C are those values, denoted as $\hat{\boldsymbol{\theta}} = (\hat{V}, \hat{K}, \hat{C})$, which minimize the sum of squares criterion

$$SS(V, K, C) = \sum_{i=1}^{n} \left(Y_i - \frac{Vx_i}{K + x_i} \right)^2.$$

The collection of first order derivatives of $\mu_i(\boldsymbol{\theta})$ can be given as

$$\dot{\boldsymbol{\mu}}_1(\boldsymbol{\theta}) = \begin{bmatrix} \frac{x_1}{K + x_1} & \cdots & \frac{x_n}{K + x_n} \end{bmatrix}^T,$$
$$\dot{\boldsymbol{\mu}}_2(\boldsymbol{\theta}) = \begin{bmatrix} \frac{-Vx_1}{(K + x_1)^2} & \cdots & \frac{-Vx_n}{(K + x_n)^2} \end{bmatrix}^T,$$
$$\dot{\boldsymbol{\mu}}_3(\boldsymbol{\theta}) = [1 \;\; \cdots \;\; 1]^T.$$

Following Equation (13.15) if we set $\boldsymbol{\Gamma_\theta} = [\dot{\boldsymbol{\mu}}_1(\boldsymbol{\theta}), \dot{\boldsymbol{\mu}}_2(\boldsymbol{\theta}), \dot{\boldsymbol{\mu}}_3(\boldsymbol{\theta})]^T$, we then have approximation

$$\sqrt{n}\left[\hat{\boldsymbol{\theta}} - \boldsymbol{\theta}_0\right] \sim N(\mathbf{0}, \Sigma_{\hat{\boldsymbol{\theta}}}),$$

where $\Sigma_{\hat{\boldsymbol{\theta}}} = n\sigma^2 \left[\boldsymbol{\Gamma_\theta}^T \boldsymbol{\Gamma_\theta}\right]^{-1}$. ■

13.6 Generalized Linear Models (GLM)

Recall the exponential family density of one dimension for a random variable $Y \in \mathbb{R}$ (Section 3.6):

$$p_{\boldsymbol{\theta}}(y) = \exp\left[b(\boldsymbol{\theta})y - B(\boldsymbol{\theta})\right] h(y). \tag{13.16}$$

It is sometimes advantageous to introduce a dispersion parameter $\phi > 0$, giving form

$$p_{\boldsymbol{\theta}}(y) = \exp\left[\frac{b(\boldsymbol{\theta})y - B(\boldsymbol{\theta})}{\phi}\right] h(x, \phi). \tag{13.17}$$

This may arise naturally in families for which $\boldsymbol{\theta}$ can be written (μ, ϕ), where $E[Y] = \mu$ and ϕ is a function of a scale parameter. We will see that analysis of the model may be simplified when $b(\boldsymbol{\theta})$ and $B(\boldsymbol{\theta})$ can be expressed as functions of μ alone. We give two examples of this form of parametrization.

Example 13.4 (Parametrization of the Normal Distribution in the Form (μ, ϕ)) Suppose $Y \sim N(\mu, \sigma^2)$. Then the density of Y can assume the form (13.17) with $b(\boldsymbol{\theta}) = \mu$, $B(\boldsymbol{\theta}) = \mu^2/2$ and $\phi = \sigma^2$. ∎

Example 13.5 (Parametrization of the Bernoulli Distribution in the Form (μ, ϕ)) Suppose Y is a Bernoulli random variable of mean μ. Then $b(\boldsymbol{\theta}) = \log(\mu/(1-\mu))$, $B(\boldsymbol{\theta}) = -\log(1-\mu)$ and $\phi = 1$. ∎

When using formulation (13.17), ϕ need not be considered a parameter, so we will now assume that $\boldsymbol{\theta}$ does not include ϕ, and is as a result one-dimensional. At this point, take the canonical representation using parameter $\theta \in \mathbb{R}$ (now written as a real number):

$$p_{\theta}(y) = \exp\left[\frac{\theta y - A(\theta)}{\phi}\right] h(y, \phi). \tag{13.18}$$

The following theorem gives the mean and variance of Y under this parametrization.

Theorem 13.2 (Mean and Variance of Exponential Family Distribution with Dispersion Parameter) Given a density of the form (13.18), we have $E_\theta[Y] = A'(\theta)$ and $\text{var}_\theta[Y] = \phi A''(\theta)$. ∎

Proof. Set $T(Y) = Y/\phi$. By Theorem 3.13 we have $E_\theta[T(Y)] = A'(\theta)/\phi$, equivalently, $E_\theta[Y] = A'(\theta)$. In addition, $\text{var}_\theta[Y] = \phi^2 \text{var}_\theta[T] = \phi^2 A''(\theta)/\phi = \phi A''(\theta)$. □

The essential feature of the general regression model introduced in Section 13.4 is that inference is based directly on estimation of response means, which here take the form

$$\mu(\theta) = E_\theta[Y] = A'(\theta),$$

Assuming $\mu(\theta)$ is invertible, we equivalently write $\theta(\mu') = \mu^{-1}(\mu')$. We also define the variance function $V(\mu) = A''(\theta(\mu))$. This is independent of ϕ, so we must write $\text{var}_\theta[Y] = \phi V(\mu)$, by Theorem 13.2.

Linear Prediction

What is typically of interest in a regression model is the relationship between $\mu_i = E[Y_i]$ and one or several predictor variables. We may be able to associate each Y_i with $\mathbf{x}_i = [x_{i1} \cdots x_{iq}]^T$, a known $q \times 1$ column vector of predictors, collectively given as the $n \times q$ design matrix $\mathbf{X} = [\mathbf{x}_1 \cdots \mathbf{x}_n]^T$. The first column of $\mathbf{X}$ is often identically 1, representing an intercept term, and is added to whatever predictors are available. Then let $\boldsymbol{\beta}$ be an unknown $q \times 1$ column vector. Then we refer to $\eta_i = \boldsymbol{\beta}^T \mathbf{x}_i$ as the linear predictor term. This, of course, is similar to the multiple linear regression model of Section 6.8. However, now we introduce a new idea, which is to associate η_i with μ_i using a link function

$$g(\mu_i) = \eta_i.$$

For example, for the multiple linear regression model of Section 6.8 we wrote

$$\mu_i = \beta_0 + \beta_1 x_{i1} + \ldots + \beta_q x_{iq} = \eta_i. \tag{13.19}$$

Here, the link function is simply the identity $\eta_i = g(\mu_i) = \mu_i$. However, this will not be appropriate for all models. For example, if Y_i is a binary response, then μ_i is confined to the interval $[0, 1]$. If we accept model (13.19), then either we must accept the possibility that estimates of μ_i are outside $[0, 1]$, and therefore not interpretable, or we must somehow introduce constraints to avoid this.

Almost always, the better alternative is to modify the link function $g(\mu)$. To see this, compare Examples 13.4 and 13.5. For the normal distribution, referring to the density form (13.17) we had $b(\theta) = \mu$, while for the binary response we had $b(\theta) = \log(\mu/(1-\mu))$, or the log-odds of μ.

For the normal model (which we associate with Equation (13.19)) μ is both the mean $E[Y]$ and the canonical parameter in (13.18). Suppose we regard the latter property as the more important one. If for the binary response model we proceed by analogy, we would, as for the normal model, set the linear predictor term η equal to the canonical parameter. Then (13.19) would be modified to become

$$\log(\mu_i/(1-\mu_i)) = \beta_0 + \beta_1 x_{i1} + \ldots + \beta_p x_{ip} = \eta_i. \tag{13.20}$$

This defines the logistic regression model.

Does this make sense? First note that the log-odds represents a bijection from the space of probabilities $(0, 1)$ to the set of all real numbers $\mathbb{R}$, eliminating the need for constraints on the set of coefficients $\boldsymbol{\beta}$ (although this eliminates estimates of probabilities equal to 0 or 1, which can arise as a computational issue). However, the advantages go beyond this. The essential feature of model (13.19) is additivity. Suppose means μ and μ' are associated with predictor vectors $\mathbf{x} = [x_1 \cdots x_p]^T$ and $\mathbf{x}' = [x_1' \cdots x_p']^T$. Then the difference in the means is expressible as $\mu' - \mu = \boldsymbol{\beta}^T(\mathbf{x}' - \mathbf{x})$. This is crucial, since the purpose of the model is to infer the relationship between predictors and mean response. Thus, the effect of a perturbation of the predictors on the mean response remains linear with respect to $\boldsymbol{\beta}$, which permits inference methods of considerable analytic precision. It should therefore be remembered that the term "linear" in "linear models" refers to the linearity of the inference problem with respect to $\boldsymbol{\beta}$.

This idea extends quite naturally to model (13.20). Suppose we have the same means and predictors μ, μ' and $\mathbf{x}$, $\mathbf{x}'$ as above. However, instead of the comparison $\mu' - \mu$ we calculate the odds ratio

$$\rho = \frac{\mu'/(1-\mu')}{\mu/(1-\mu)}.$$

Then directly from (13.20) we have $\log(\rho) = \boldsymbol{\beta}^T(\mathbf{x}' - \mathbf{x})$. The inference problem therefore remains linear in $\boldsymbol{\beta}$, and additive models retain the same interpretation as they do in the case of normal linear models.

Accordingly, we have the following definition.

Definition 13.1 (Canonical Link Function) Given model (13.18), $g(\mu)$ is a canonical link function if it is equal to the canonical parameter. Equivalently, the canonical link function may be taken to be $g(\mu) = b(\theta(\mu))$ in Equation (13.17). ■

Example 13.6 (Canonical Link Function for Bernoulli Responses) Continuing Example 13.5, using the canonical parametrization of Equation (13.18), the canonical link function is $\eta = g(\mu) = \log(\mu/(1-\mu))$. ■

Put another way, when an exponential family model is parametrized using a linear prediction term $\eta = \boldsymbol{\beta}^T \mathbf{x}$ introduced to the density using the canonical link function, the response Y has a density of the form

$$p_{\boldsymbol{\beta}}(y) = \exp\left[\frac{\eta y - A(\eta)}{\phi}\right] h(y, \phi). \tag{13.21}$$

Table 13.1 lists the canonical link function and variance function for some commonly used GLMs based on the response distribution.

TABLE 13.1
Canonical link functions and variance functions for some commonly used GLM models. Note that σ^2 is a variance paremeter, while α is a shape parameter.

Response Distribution	Canonical Link Function	Variance Function	Dispersion Parameter
Gaussian	$\eta = \mu$	$V(\mu) = 1$	$\phi = \sigma^2$
Bernoulli	$\eta = \log(\mu/(1-\mu))$	$V(\mu) = \mu(1-\mu)$	$\phi = 1$
Exponential	$\eta = -1/\mu$	$V(\mu) = \mu^2$	$\phi = 1$
Gamma	$\eta = -1/\mu$	$V(\mu) = \mu^2$	$\phi = 1/\alpha$
Poisson	$\eta = \log(\mu)$	$V(\mu) = \mu$	$\phi = 1$

Estimating Equations for Generalized Linear Models

Suppose we are given independent observations $\mathbf{Y}$ paired with design matrix $\mathbf{X}$. Assume that response Y_i has density (13.18) for some parameter θ_i. The dispersion parameter ϕ is constant, and assumed known. Furthermore, assume there is a relation $\theta_i = \theta(\eta_i)$, where $\eta_i = \mathbf{x}_i^T \boldsymbol{\beta}$ are the linear predictor terms, $\boldsymbol{\beta} \in \mathbb{R}^q$. Then the log-likelihood for $\boldsymbol{\beta}$ may be written

$$\ell(\boldsymbol{\beta}, \mathbf{Y}) = \frac{1}{\phi} \sum_i \{Y_i \theta(\eta_i) - A(\theta(\eta_i))\}. \tag{13.22}$$

The partial derivative of $\ell(\boldsymbol{\beta}, \mathbf{Y})$ with respect to β_j is equal to

$$\Psi_j = \frac{1}{\phi} \sum_i \{Y_i - \dot{A}(\theta(\eta_i))\} \dot{\theta}(\eta_i) x_{ij} = \frac{1}{\phi} \sum_i \{Y_i - \mu_i\} \dot{\theta}(\eta_i) x_{ij}, \; j = 1, \ldots, q. \tag{13.23}$$

Equation (13.23) can be written using matrix notation as

$$\Psi(\boldsymbol{\beta}, \mathbf{Y}) = \frac{1}{\phi} \mathbf{X}^T \mathbf{W} [\mathbf{Y} - \boldsymbol{\mu}], \tag{13.24}$$

where $\mathbf{W} = \{w_{ij}\}$ is a diagonal matrix with entries $w_{ii} = \dot{\theta}(\eta_i)$. The solution $\hat{\boldsymbol{\beta}}$ to the estimating equation $\Psi(\mathbf{Y}, \boldsymbol{\beta}) = \mathbf{0}$ is, under general conditions, the MLE based on the log-likelihood function (13.22), so following Equation (13.7) we have approximation

$$\sqrt{n}\left[\hat{\boldsymbol{\beta}} - \boldsymbol{\beta}_0\right] \sim N(\mathbf{0}, \bar{\mathbf{I}}(\boldsymbol{\beta}_0)^{-1}).$$

But $\mathbf{I}(\boldsymbol{\beta}_0)$ is simply the covariance matrix of $\Psi(\mathbf{Y}, \boldsymbol{\beta}_0)$, which is readily evaluated. First note that the covariance matrix $\Sigma_{\mathbf{Y}}$ of $\mathbf{Y}$ is the diagonal matrix with ith diagonal entry $\phi V(\mu_i)$. Then the information matrix is

$$\mathbf{I}(\boldsymbol{\beta}) = \frac{1}{\phi^2} \mathbf{X}^T \mathbf{W} \Sigma_{\mathbf{Y}} \mathbf{W} \mathbf{X} = \frac{1}{\phi} \mathbf{X}^T \mathbf{D} \mathbf{X}$$

where $\mathbf{D} = \{d_{ij}\}$ is a diagonal matrix with diagonal entries $d_{ii} = V(\mu_i)\dot{\theta}(\eta_i)^2$.

It is instructive, at this point, to compare (13.23) to (13.9). Each is a linear transformation of $\mathbf{Y} - \boldsymbol{\mu}$. To continue the comparison, we retrace our steps, writing

$$\Psi_j = \frac{1}{\phi}\sum_i \{Y_i - \dot{A}(\theta(\eta_i))\}\frac{d\theta(\eta_i)}{d\beta_j} = \frac{1}{\phi}\sum_i \{Y_i - \mu_i\}\frac{d\theta(\eta_i)}{d\mu_i}\frac{d\mu_i}{d\beta_j}.$$

This is more compactly expressed in the form (13.9) with $\mathbf{W}(\boldsymbol{\beta}) = -\mathbf{D}(\boldsymbol{\beta})$, where $\mathbf{D}(\boldsymbol{\beta})$ is a diagonal matrix with diagonal entries $\{\mathbf{D}(\boldsymbol{\beta})\}_{ii} = \phi^{-1}d\theta(\eta_i)/d\mu_i$. So, what does this quantity represent? We can write

$$\frac{d\theta(\eta_i)}{d\mu_i} = \left[\frac{d\mu_i}{d\theta(\eta_i)}\right]^{-1} = \left[\frac{d\dot{A}(\theta(\eta_i))}{d\theta(\eta_i)}\right]^{-1} = \left[\ddot{A}(\theta(\eta_i))\right]^{-1} = V(\mu_i)^{-1}.$$

Recall that $\text{var}[Y_i] = \phi V(\mu_i)$, therefore $d_{ii} = 1/\text{var}[Y_i]$. So the components of the score function defining the estimating equation are

$$\Psi_j = \sum_i \frac{(Y_i - \mu_i)}{\phi V(\mu_i)} \times \frac{d\mu_i}{d\beta_j},\ j = 1, \ldots, q, \tag{13.25}$$

which is equivalent to the general form of Equation (13.9), with the weight matrix $W(\boldsymbol{\beta})$ equal to the inverse of the covariance matrix of the responses. Thus, the GLM is simply a special case of the general regression model introduced in Section 13.4.

The Canonical Link Function

Based on canonical density representation (13.18) the form of the canonical link function is given by $\theta = g(\mu)$. It becomes the actual canonical link function when we set $\eta = g(\mu)$, where $\eta = \boldsymbol{\beta}^T\mathbf{x}$ is the linear predictor, which yields density parametrization (13.21). A number of advantages arise when this constraint is used. First, $\dot{\theta}(\eta_i) = 1$, so the score function is simply

$$\Psi(\boldsymbol{\beta}, \mathbf{Y}) = \frac{1}{\phi}\mathbf{X}^T\left[\mathbf{Y} - \boldsymbol{\mu}\right].$$

Next, we can verify that the Hessian Ψ' can be written

$$\Psi'(\boldsymbol{\beta}, \mathbf{Y}) = -\frac{1}{\phi}\mathbf{X}^T\mathbf{D}^*\mathbf{X}, \tag{13.26}$$

where $\mathbf{D}^*$ is a diagonal matrix with diagonal entries

$$d_{ii} = V(\mu_i)\dot{\theta}(\eta_i)^2 - \{Y_i - \mu_i\}\ddot{\theta}(\eta_i). \tag{13.27}$$

But $\dot{\theta}(\eta_i) = 1$, $\ddot{\theta}(\eta_i) = 0$, so that $\mathbf{D}^*$ does not depend on $\mathbf{Y}$, and in fact $\Psi'(\boldsymbol{\beta}, \mathbf{Y}) = -\mathbf{I}(\boldsymbol{\beta})$, so that the log-likelihood is concave *wp*1.

We next offer two examples

Example 13.7 (GLM with Gamma Responses and Canonical Link Function) Suppose $Y \sim gamma(\tau, \alpha)$. This density can be written in exponential family form as

$$f(y) = \exp\{-\tau^{-1}y + \alpha\log(x) - B(\tau, \alpha)\}h(x).$$

We may then reparametrize using the transformation $\mu = \alpha\tau$, giving

$$f(y) = \exp\left\{\frac{-\mu^{-1}y - B^*(\mu, \alpha)}{\phi}\right\}h^*(y),$$

where $\phi = 1/\alpha$ is now the dispersion parameter, and $-1/\mu$ is the canonical parameter. This leads to canonical link function $g(\mu) = -1/\mu$ and variance function $V(\mu) = \mu^2$ (see Table 13.1).

Next, suppose $\mathbf{Y}$ is a vector of n independent responses, with $Y_i \sim gamma(\tau_i, \alpha)$. Given $n \times q$ design matrix $\mathbf{X}$, we construct linear predictor terms $\eta_i = \boldsymbol{\beta}^T \mathbf{x}_i$, $\boldsymbol{\beta} \in \mathbb{R}^q$. The canonical GLM for gamma responses then satisfies

$$E[Y_i] = \mu_i = -1/\boldsymbol{\beta}^T \mathbf{x}_i, \; i = 1, \ldots, n,$$

with estimates $\hat{\boldsymbol{\beta}}$ obtained by solving the estimating equation

$$\Psi(\boldsymbol{\beta}, \mathbf{Y}) = \phi^{-1} \mathbf{X}^T \left[\mathbf{Y} - \boldsymbol{\mu}\right] = \mathbf{0}$$

with respect to $\boldsymbol{\beta}$ (note that only $\boldsymbol{\mu}$ depends on $\boldsymbol{\beta}$). In particular, each element of Ψ is set to zero:

$$\Psi_j = \sum_i x_{ij} \{Y_i + 1/\boldsymbol{\beta}^T \mathbf{x}_i\} = 0, \; j = 1, \ldots, q.$$

The approximate covariance of $\sqrt{n}(\hat{\boldsymbol{\beta}} - \boldsymbol{\beta})$ is then $\bar{\mathbf{I}}(\boldsymbol{\beta})^{-1}$, where

$$\mathbf{I}(\boldsymbol{\beta}) = \phi^{-1} \mathbf{X}^T \mathbf{D} \mathbf{X} = \mathbf{X}^T \mathbf{D}' \mathbf{X}$$

and, for the canonical model, $\mathbf{D}'$ is a diagonal matrix with ith diagonal entry $\alpha V(\mu_i) = \alpha \mu_i^2 = \alpha/\eta_i^2$.

As we did for the logistic regression model, we can ask if the canonical GLM for gamma responses "makes sense". For the former model, our answer depended on whether or not it was reasonable to expect that predictor variables should influence the log-odds of the ressponse means in an additive manner.

A similar analysis can be made for the gamma response. First, the gamma rate parameter may be written $\lambda_i = \alpha/\mu_i$. Next, note that we can, without loss of generality, replace η_i with $-\eta_i$, so that the model becomes

$$\lambda_i = \alpha \eta_i, i = 1, \ldots, n, \tag{13.28}$$

that is, the predictor variables influence the *rate parameter* of the response linearly. Then suppose we are given a Poisson process with rate λ. The time taken to observe α arrivals has distribution $gamma(1/\lambda, \alpha)$. If the response Y_i can be conceived as such a waiting time, then the canonical model assumes a linear relationship between the predictor variables and the arrival rate of the observed Poisson process. Since a Poisson process can itself be thought of as a superposition of separate arrival processes (Example 1.10) its total arrival rate will be the sum of those separate arrival rates. Therefore, λ_i will respond additively to such a superposition process, and will be linearly related to any predictor variable controlling this process in a linearly proportional manner. In this case, the canonical link function would be the suitable choice. ■

It must be stressed that the canonical model need not be the "correct" model. Its suitability is entirely a matter of correct model specification. In Example 13.7 we considered the case of responses Y_i as observations of a Poisson process of rate λ_i. We then have predictor variables $\mathbf{x}_i$ with which to estimate a parametrized functional relationship $\lambda_i = f(\boldsymbol{\beta}, \mathbf{x}_i)$ between λ_i and $\mathbf{x}_i$. However, the exact form of $f(\boldsymbol{\beta}, \mathbf{x}_i)$ is not a statistical question, but scientific one. We may have good reasons to expect the linear relationship $\lambda_i = \boldsymbol{\beta}^T \mathbf{x}_i$ to hold. In this case, if Y_i is the waiting time for α arrivals, then the GLM for gamma responses with the canonical link function is the appropriate choice, given Equation (13.28).

Suppose, on the other hand, that we have the same model, except that Y_i is now the number of arrivals in a fixed time T of the same Poisson process of rate λ_i. We now have $Y_i \sim pois(T\lambda_i)$, so we would use a GLM for Poisson responses. However, the canonical link function for a Poisson response is $g(\mu) = \log \mu$ (Table 13.1), so the relationship between the Poisson process rate λ_i and linear prediction term for the canonical model is

$$\lambda_i = T^{-1} e^{\eta_i}, i = 1, \ldots, n.$$

In this case, the relationship between λ_i and the predictor variables is not linear. Thus, if these two models are intended to model observations of a Poisson process, the only difference being the form of the observation, then the canonical gamma GLM would not be consistent with the canonical Poisson GLM. However, the models could be made consistent by using the identity link function for the Poisson responses. See Problem 13.8.

The next example presents a GLM for exponentially distributed responses in which the rate parameter varies exponentially with predictor variables. This model is noncanonical, but preserves much of the analytical advantages of its canonical counterpart.

Example 13.8 (Negative Exponential Model) The negative exponential model is an example of a noncanonical GLM, described in detail in Knight and Satchell (1996). See also Almudevar (2016). Suppose Y_i are independent exponentially distributed responses with means μ_i given by $\mu_i = e^{\boldsymbol{\beta}^T \mathbf{x}_i}$, $i = 1, \ldots, n$, where $\boldsymbol{\beta} \in \mathbb{R}^q$ and $\mathbf{x}_i = (x_{i1}, \ldots, x_{iq})^T$ are known q-dimensional predictor vectors. Here, the link function is $g(\mu) = \log \mu$, which is not canonical (Table 13.1).

Suppose the true parameter value determining the exact density of Y_i is $\boldsymbol{\beta}_0$. Then the likelihood equation becomes, making use of (13.25)

$$\Psi(\mathbf{Y}, \boldsymbol{\beta}) = \begin{bmatrix} \sum_{i=1}^n \left(\frac{Y_i}{\mu_i} - 1 \right) \times x_{i1} \\ \vdots \\ \sum_{i=1}^n \left(\frac{Y_i}{\mu_i} - 1 \right) \times x_{iq} \end{bmatrix} = 0,$$

with derivative matrix

$$\Psi'(\mathbf{Y}, \boldsymbol{\beta}) = -1 \times \begin{bmatrix} \sum_{i=1}^n x_{i1}^2 \frac{Y_i}{\mu_i} & \cdots & \sum_{i=1}^n x_{i1} x_{iq} \frac{Y_i}{\mu_i} \\ \vdots & \vdots & \vdots \\ \sum_{i=1}^n x_{iq} x_{i1} \frac{Y_i}{\mu_i} & \cdots & \sum_{i=1}^n x_{iq}^2 \frac{Y_i}{\mu_i} \end{bmatrix}. \tag{13.29}$$

Interestingly, Ψ' may still be written in the form

$$\Psi'(\mathbf{Y}, \boldsymbol{\beta}) = -\mathbf{X}^T \mathbf{D} \mathbf{X},$$

where $\mathbf{D}$ is the diagonal matrix with diagonal entry $d_{ii} = Y_i / \mu_i > 0$ $wp1$. Thus, the Hessian matrix of the log-likelihood function is negative definite, despite its dependence on $\mathbf{Y}$. ∎

13.7 Generalized Estimating Equations (GEE)

GLMs are widely used due to their flexibility and statistical efficiency. However, their structure depends very much on the assumption of independence, which makes evaluation of the log-likelihood feasible. Unfortunately, construction of a log-likelihood is challenging when data are not independent, particularly for nonnormal responses. For example, dependence among Bernoulli responses cannot be reduced to a single covariance matrix, and in the worst case the joint distribution of vector of n Bernoullli random variables has a parameter space of $2^n - 1$ dimensions. Model complexity can be reduced by introducing parametric forms of dependence, but there will be multiple ways of doing this (Zeger and Liang, 1986; Cessie and Houwelingen, 1994).

However, Example 13.2 suggests that when constructing a log-likelihood is not feasible, regression model parameters can be estimated using an estimating equation defined by the score function of Equation (13.9). This need not be the gradient of a log-likelihood function, but can be regarded as an extension of the quasi-likelihood method (Example 4.12).

It is possible to simply transfer much of the structure of GLMs to this problem, with the intention of generalizing the model to dependent responses. Accordingly, the response mean $E[Y_i] = \mu_i, i = 1, \ldots, n$ is related to the response variance $\text{var}_{\mu_i}[Y_i] = V(\mu_i)/\phi$, where V and ϕ are the variance function and dispersion parameter introduced in Section 13.6. Similarly, we have linear predictor terms $\eta_i = \boldsymbol{\beta}^T \mathbf{x}_i$, $\boldsymbol{\beta} \in \mathbb{R}^q$, which are related to the response mean by link function $g(\mu_i) = \eta_i$.

In fact, we already showed that the estimating equations which yield the MLEs of a GLM assume the form (13.9) (see Equation (13.25)). If we adopt the parametrization of the GLM the components $\boldsymbol{\Gamma}_{\boldsymbol{\theta}} = \boldsymbol{\Gamma}_{\boldsymbol{\beta}}$ and $\boldsymbol{\mu}(\boldsymbol{\theta}) = \boldsymbol{\mu}(\boldsymbol{\beta})$ of Equation (13.9) would remain the same as for the GLM, depending on $\boldsymbol{\beta}$ but not ϕ. What is new is that $\Sigma_{\mathbf{Y}}$ now models dependence of the responses. Once these components are substituted into Equation (13.9), we have the generalized estimating equation (GEE) (Zeger and Liang, 1986; Liang and Zeger, 1986).

The usual approach in GEE methodology is to write

$$\Sigma_{\mathbf{Y}} = \mathbf{V}(\boldsymbol{\beta}, \phi, \alpha) = \mathbf{D}(\boldsymbol{\beta}, \phi)^{1/2} \mathbf{R}(\alpha) \mathbf{D}(\boldsymbol{\beta}, \phi)^{1/2} \tag{13.30}$$

where $\mathbf{D}(\boldsymbol{\beta}, \phi)$ is a diagonal matrix with diagonal elements $\{\mathbf{D}\}_{ii} = \phi V(\mu_i)$, equal to the marginal response variances, and $\mathbf{R}(\alpha)$ is the correlation matrix of $\mathbf{Y}$. $\mathbf{D}(\boldsymbol{\beta}, \phi)$ must depend on $\boldsymbol{\beta}$, unless $V(\mu)$ does not depend on μ (as may happen). On the other hand, $\mathbf{R}(\alpha)$ is usually assumed to not depend on $\boldsymbol{\beta}$ or ϕ, but may depend on a nuisance parameter α.

This completely defines the GEE

$$\Psi(\mathbf{Y}, \boldsymbol{\beta}) = \boldsymbol{\Gamma}_{\boldsymbol{\beta}}{}^T \left[\mathbf{D}(\boldsymbol{\beta}, \phi)^{1/2} \mathbf{R}(\alpha) \mathbf{D}(\boldsymbol{\beta}, \phi)^{1/2} \right]^{-1} [\mathbf{Y} - \boldsymbol{\mu}(\boldsymbol{\beta})] = \mathbf{0}. \tag{13.31}$$

Fitting the GEE Model

Note that in Equation (13.31) α will probably be, and ϕ may be, unknown. This poses no problem in the case of ϕ, since $\Psi(\mathbf{Y}, \boldsymbol{\beta})$ depends on ϕ only through $\mathbf{D}(\boldsymbol{\beta}, \phi)$, from which it follows that $\Psi(\mathbf{Y}, \boldsymbol{\beta})/\phi$ does not depend on ϕ. This means that the problem of solving the estimating equation does not depend on ϕ, although construction of the standard errors will.

Unfortunately, the same is not true of α. In fact, it expected that the M-estimate should depend on α (after all, evaluation of the best linear unbiased estimator (BLUE) relies on the correct specification of response dependence). Since the M-estimate is not constructed to estimate α, we need an auxilliary estimation procedure.

Suppose we had estimates $\hat{\mu}_i$ of the mean responses, obtained by any means. We could then construct the Pearson residuals

$$e_i = \frac{Y_i - \hat{\mu}_i}{\sqrt{V(\hat{\mu}_i)}} \approx \phi^{1/2} \frac{Y_i - E[Y_i]}{\sqrt{\text{var}[Y_i]}}.$$

Then $E[e_i^2] \approx \phi$. If ϕ is unknown, a rough estimate is given by

$$\hat{\phi} = (n-q)^{-1} \sum_{i=1}^{n} e_i^2,$$

where we adjust for the q degrees of freedom lost to the estimation of $\boldsymbol{\beta}$. Of course, if $\mathbf{R}(\alpha)$ were known, we may use instead $\hat{\phi}' = (n-q)^{-1}\mathbf{e}^T\mathbf{R}(\alpha)^{-1}\mathbf{e}$, where $\mathbf{e}$ is the vector of residuals (McCullagh *et al.*, 1983). We would then expect $(n-q)\hat{\phi}^*/\phi \sim \chi^2_{n-q}$, approximately.

Estimation of α would then depend on the form of $\mathbf{R}(\alpha)$, depending on the fact that $e_i e_j$ is a method of moments estimate of $\mathrm{cor}[Y_i, Y_j]$.

By convention, $\mathbf{R}(\alpha)$ is referred to as the "working correlation matrix". Misspecification of $\mathbf{R}(\alpha)$ results in loss of efficiency but not consistency. As a simple two-step approach we may estimate the coefficients $\boldsymbol{\beta}$ using $\mathbf{R}(\alpha) = \mathbf{I}_n$ (the independence model), then estimate ϕ and α using the Pearson residuals. In the second step $\boldsymbol{\beta}$ is estimated again using working correlation matrix $\mathbf{R}(\hat{\alpha})$, where $\hat{\alpha}$ is the estimate of α obtained from the first step. Alternatively, the two steps can be incorporated into a single iteration of a Gauss-Newton algorithm (this approach is described in detail in Wedderburn (1974)). Following either method, the resulting M-estimate $\hat{\boldsymbol{\beta}}$ will have approximate distribution $\sqrt{n}[\hat{\boldsymbol{\theta}}-\boldsymbol{\theta}] \sim N(\mathbf{0}, n\mathbf{K}(\boldsymbol{\beta}, \phi, \alpha)^{-1})$, where

$$\mathbf{K}(\boldsymbol{\beta}, \phi, \alpha) \approx \boldsymbol{\Gamma}_{\hat{\boldsymbol{\beta}}}^T \mathbf{D}(\hat{\boldsymbol{\beta}}, \hat{\phi})^{-1/2}\mathbf{R}(\hat{\alpha})^{-1}\mathbf{D}(\hat{\boldsymbol{\beta}}, \hat{\phi})^{-1/2}\boldsymbol{\Gamma}_{\hat{\boldsymbol{\beta}}}.$$

Example 13.9 (Exchangeable Correlation for GEEs) Under exchangeable correlation, random variates are partitioned into groups with common pairwise correlation. Between-group correlation is zero. Suppose sample indices $i = 1, \ldots, n$ are partitioned into groups $g = 1, \ldots, n_g$. To fix ideas, assume the dispersion parameter is $\phi = 1$. Let g_i identify the group of sample i. Then for any $i \neq i'$ we have $\mathrm{cor}[y_i, y_{i'}] = \rho \neq 0$ if $g_i = g_{i'}$, and $\mathrm{cor}[Y_i, Y_{i'}] = 0$ otherwise. It is usually expected that $\rho > 0$, but this is not required. Following Liang and Zeger (1986) we have estimate

$$\hat{\rho} = \frac{1}{\phi(N_{pair} - d)} \sum_{g=1}^{n_g} \sum_{\{i<i' : g_i = g_{i'} = g\}} e_i e_{i'}, \tag{13.32}$$

where N_{pair} is the total number of correlated pairs contributing to the sum, and d is the number of estimated parameters. If ϕ is unknown $\hat{\phi}$ can be substituted. ■

13.8 Existence and Consistency of M-Estimators

Clearly, there is no guarantee that a solution $\hat{\boldsymbol{\theta}} \in \mathbb{R}^q$ to Equation (13.1) exists or is unique when it does. Formally, (13.1) defines a point process on $\Theta \subset \mathbb{R}^q$ and not a single random vector (Almudevar *et al.*, 2000; Almudevar, 2016). A resolution of this issue should be regarded as a necessary step in the development of any estimating method of this type.

There are two general approaches to this problem. The first relies on the ideas of Section C.7. Suppose we may interpret $\Psi'(\tilde{X}, \boldsymbol{\theta})$ as a Hessian matrix of an objective function $\Lambda_{\tilde{X}}(\boldsymbol{\theta})$ defined on a convex open set $\boldsymbol{\theta} \in \Theta$. If $\Psi'(\tilde{X}, \boldsymbol{\theta})$ is positive definite then $\Lambda_{\tilde{X}}(\boldsymbol{\theta})$ is a convex mapping, and at most one stationary point exists. If it does, it must be the unique minimizer of $\Lambda_{\tilde{X}}(\boldsymbol{\theta})$. This approach was used in Section 4.5.1 to characterize the MLE of a full-rank exponential family model (here $\Lambda_{\tilde{X}}(\boldsymbol{\theta})$ would be the log-likelihood times -1). If a stationary point does not exist, we may consider minimizing $\Lambda_{\tilde{X}}(\boldsymbol{\theta})$ on the boundary of Θ (see Example 4.9).

However, $\Psi'(\tilde{X},\boldsymbol{\theta})$ will not be positive definite in general (although it's expected value often will be). We therefore rely on the more general approach of verifying that the mapping $\Psi(\tilde{X},\boldsymbol{\theta})$ is invertible near $\boldsymbol{\theta}_0$ with high probability. The first step is to show that with probability $1-\epsilon$ (i) there exists an inverse mapping $\Psi^{-1}(\tilde{X},\boldsymbol{\theta})$ from a neighborhood E of zero to a neighborhood $\Theta_E \subset \Theta$ of $\boldsymbol{\theta}_0$; and that (ii) $\Psi(\tilde{X},\boldsymbol{\theta}) \in E$. It remains to show that ϵ approaches zero as $n \to \infty$. In this section we use this approach to develop conditions under which Assumption (A7) of Section 13.2 holds.

We will make use of Euclidean matrix norm $|||\mathbf{A}|||_2$ for a $q \times q$ matrix $\mathbf{A}$ (Section C.8.2). Denote the open ball $B_\epsilon(\mathbf{x}_0) = \{\mathbf{x} : |\mathbf{x}-\mathbf{x}_0| < \epsilon\}$. We may standardize the score function as

$$\Psi^*(\tilde{X},\boldsymbol{\theta}) = \begin{cases} \Psi'(\tilde{X},\boldsymbol{\theta})^{-1}\Psi(\tilde{X},\boldsymbol{\theta}) & ; \ \det(\Psi'(\tilde{X},\boldsymbol{\theta})) \neq 0 \\ \infty & ; \ \text{otherwise} \end{cases},$$

then define the following quantity:

$$z(\tilde{X},\boldsymbol{\theta},\tau) = \begin{cases} \sup_{\boldsymbol{\theta}' \in B_\tau(\boldsymbol{\theta})} \left|\left|\left| \Psi'(\tilde{X},\boldsymbol{\theta})^{-1}\Psi'(\tilde{X},\boldsymbol{\theta}') - \mathbf{I}_p \right|\right|\right|_2 & ; \ \det(\Psi'(\tilde{X},\boldsymbol{\theta}')) \neq 0 \\ & \quad \forall \boldsymbol{\theta}' \in B_\tau(\boldsymbol{\theta}) \\ \infty & ; \ \text{otherwise} \end{cases}. \tag{13.33}$$

Our main result will be based on Lemma 1 of Almudevar *et al.* (2000), stated here without proof (a useful feature of this lemma is that it makes no reference to any probability measure).

Theorem 13.3 (Almudevar *et al.* (2000)) Suppose for some $\tau > 0$, and $0 < \alpha < 1$ we have $z(\tilde{X},\boldsymbol{\theta}_0,\tau) < \alpha$. Then if $|\Psi^*(\tilde{X},\boldsymbol{\theta}_0)| \leq (1-\alpha)\tau$, there exists a unique solution $\hat{\boldsymbol{\theta}}$ to the equation $\Psi(\tilde{X},\boldsymbol{\theta}) = 0$ which satisfies $|\hat{\boldsymbol{\theta}} - \boldsymbol{\theta}_0| < \tau$. ■

Theorem 13.3 resolves a number of problems inherent in the use of estimating equations. Even in the favorable case of the full rank exponential family model, the MLE cannot be defined without some qualification (the MLE is either the unique solution to the estimating equation, or a solution does not exist). If the conditions of Theorem 13.3 can be met *wp*1, or at least with probability approaching 1, then we may define $\hat{\boldsymbol{\theta}}$ to be the unique solution satisfying $|\hat{\boldsymbol{\theta}} - \boldsymbol{\theta}_0| < \tau$, whether or not other solutions exist.

We will need the following three technical lemmas, which state some consequences of Assumptions (A4) and (A6).

Lemma 13.1 If (A4) and (A6) holds then there exists $K_2 < \infty$ such that the limit

$$P_{\boldsymbol{\theta}_0}\left(\sup_{\boldsymbol{\theta} \in B_\tau(\boldsymbol{\theta}_0)} \left|\left|\left| \bar{\Psi}'(\tilde{X},\boldsymbol{\theta}) - E_{\boldsymbol{\theta}_0}[\bar{\Psi}'(\tilde{X},\boldsymbol{\theta}_0)] \right|\right|\right|_2 < K_2\tau \right) \to_n 1 \tag{13.34}$$

holds for any $0 < \tau \leq m_0$. ■

Proof. By (A4), for any $\tau > 0$,

$$P_{\boldsymbol{\theta}_0}\left(\left|\left|\left| \bar{\Psi}'(\tilde{X},\boldsymbol{\theta}_0) - E_{\boldsymbol{\theta}_0}[\bar{\Psi}'(\tilde{X},\boldsymbol{\theta}_0)] \right|\right|\right|_2 < \tau \right) \to_n 1.$$

By the triangle inequality for norms

$$\begin{aligned} &\sup_{\boldsymbol{\theta} \in B_\tau(\boldsymbol{\theta}_0)} \left|\left|\left| \bar{\Psi}'(\tilde{X},\boldsymbol{\theta}) - E_{\boldsymbol{\theta}_0}[\bar{\Psi}'(\tilde{X},\boldsymbol{\theta}_0)] \right|\right|\right|_2 \\ &\leq \sup_{\boldsymbol{\theta} \in B_\tau(\boldsymbol{\theta}_0)} \left|\left|\left| \bar{\Psi}'(\tilde{X},\boldsymbol{\theta}) - \bar{\Psi}'(\tilde{X},\boldsymbol{\theta}_0) \right|\right|\right|_2 + \left|\left|\left| \bar{\Psi}'(\tilde{X},\boldsymbol{\theta}_0) - E_{\boldsymbol{\theta}_0}[\bar{\Psi}'(\tilde{X},\boldsymbol{\theta}_0)] \right|\right|\right|_2. \end{aligned}$$

Then by (A6), if $\tau \leq m_0$ we may then conclude that Equation (13.34) holds with $K_2 = K_0 + 1$. □

Lemma 13.2 If (A4) and (A6) hold then there exists exists $\tau_3 > 0$ for which

$$P_{\boldsymbol{\theta}_0}\left(\det(\bar{\Psi}'(\tilde{X}, \boldsymbol{\theta})) \neq 0, \forall \boldsymbol{\theta} \in B_{\tau_3}(\boldsymbol{\theta}_0)\right) \rightarrow_n 1. \quad \blacksquare$$

Proof. First note that by (A4) $\det(E_{\boldsymbol{\theta}_0}[\bar{\Psi}'(\tilde{X}, \boldsymbol{\theta}_0)]) \neq 0$. Since a determinant is a polynomial function of the elements of a matrix, if $\det(\mathbf{A}) \neq 0$, then there must be some positive constant ϵ such that $\det(\mathbf{A}') \neq 0$ if $\left|\left|\left|\mathbf{A}' - \mathbf{A}\right|\right|\right|_2 < \epsilon$. Then let $\epsilon > 0$ be such a value for $E_{\boldsymbol{\theta}_0}[\bar{\Psi}'(\tilde{X}, \boldsymbol{\theta}_0)]$. By Lemma 13.1, the lemma holds for any $\tau_3 < \min(\epsilon/K_2, m_0)$. □

Lemma 13.3 If (A4) and (A6) hold, then there exists $K_3 < \infty$ for which $P_{\boldsymbol{\theta}_0}\left(\left|\left|\left|[\bar{\Psi}'(\tilde{X}, \boldsymbol{\theta}_0)]^{-1}\right|\right|\right|_2 < K_3\right) \rightarrow_n 1$. ■

Proof. The argument is similar to that of Lemma 13.2. □

The main theorem follows.

Theorem 13.4 (Existence and Consistency of M-Estimators) If (A2), (A4) and (A6) hold, then there exists an estimator $\hat{\boldsymbol{\theta}}$ such that $P\left(\Psi(\tilde{X}, \hat{\boldsymbol{\theta}}) = 0\right) \rightarrow_n 1$, and for any $\tau > 0$, $P_{\boldsymbol{\theta}_0}\left(|\hat{\boldsymbol{\theta}} - \boldsymbol{\theta}_0| < \tau\right) \rightarrow_n 1$. ■

Proof. Let τ_3, K_3 be the constants of Lemmas 13.2 and 13.3, respectively, and let K_0, m_0 be the constants of (A6). We will define the following events:

$$\begin{aligned}
A &= \left\{\left|\left|\left|[\bar{\Psi}'(\tilde{X}, \boldsymbol{\theta}_0)]^{-1}\right|\right|\right|_2 < K_3\right\}, \\
B_\tau &= \left\{\det(\bar{\Psi}'(\tilde{X}, \boldsymbol{\theta})) \neq 0, \forall \boldsymbol{\theta} \in B_\tau(\boldsymbol{\theta}_0)\right\}, \\
C_\tau &= \left\{\sup_{\boldsymbol{\theta} \in B_\tau(\boldsymbol{\theta}_0)} \left|\left|\left|\bar{\Psi}'(\tilde{X}, \boldsymbol{\theta}) - \bar{\Psi}'(\tilde{X}, \boldsymbol{\theta}_0)\right|\right|\right|_2 < K_0 \tau\right\}, \\
D_\tau &= \left\{|\Psi^*(\tilde{X}, \boldsymbol{\theta}_0)| < \tau\right\}.
\end{aligned}$$

For any $\tilde{X} \in A \cap B_\tau \cap C_\tau$ we have, by submultiplicativity of matrix norms (Equation (C.13)),

$$\begin{aligned}
&\sup_{\boldsymbol{\theta} \in B_\tau(\boldsymbol{\theta}_0)} \left|\left|\left|\bar{\Psi}'(\tilde{X}, \boldsymbol{\theta}_0)^{-1} \bar{\Psi}'(\tilde{X}, \boldsymbol{\theta}) - \mathbf{I}_p\right|\right|\right|_2 \\
&\leq \sup_{\boldsymbol{\theta} \in B_\tau(\boldsymbol{\theta}_0)} \left|\left|\left|\bar{\Psi}'(\tilde{X}, \boldsymbol{\theta}_0)^{-1}\right|\right|\right|_2 \left|\left|\left|\bar{\Psi}'(\tilde{X}, \boldsymbol{\theta}) - \bar{\Psi}'(\tilde{X}, \boldsymbol{\theta}_0)\right|\right|\right|_2 \leq K_3 K_0 \tau
\end{aligned}$$

or equivalently

$$z(\tilde{X}, \boldsymbol{\theta}_0, \tau) < K_3 K_0 \tau.$$

Then select $\tau < \min\{m_0, \tau_3, (2K_3K_0)^{-1}\}$. By Lemma 13.3 $P(A) \rightarrow_n 1$. By Lemma 13.2, $P(B_\tau) \rightarrow_n 1$, by (A6) $P(C_\tau) \rightarrow_n 1$, so we conclude

$$P\left(z(\tilde{X}, \boldsymbol{\theta}_0, \tau) < 1/2\right) \rightarrow_n 1.$$

Finally, by (A2) we have $\bar{\Psi}(\boldsymbol{\theta}_0) \xrightarrow{i.p.} \mathbf{0}$, and by (A4) $[\bar{\Psi}'(\boldsymbol{\theta}_0)]^{-1} \xrightarrow{i.p.} E_{\boldsymbol{\theta}_0}[[\bar{\Psi}'(\boldsymbol{\theta}_0)]^{-1}]$, and therefore $P(D_{\tau/2}) \rightarrow_n 1$, using Slutsky's theorem. The proof is finished by applying Theorem 13.3 with $\alpha = 1/2$. □

We have just shown that Assumptions (A2), (A4) and (A6) together imply Assumption (A7).

13.9 Asymptotic Distribution of $\hat{\theta}_n$

Theorem C.10 gives a first-order Taylor's series expansion for a mapping $f : \mathbb{R}^q \rightarrow \mathbb{R}$. The multidimensional mapping $f : \mathbb{R}^q \rightarrow \mathbb{R}^q$ can be written in vector form as $f(\mathbf{x}) = (f_1(\mathbf{x}), \ldots, f_q(\mathbf{x}))$ using q mappings of the form $f_j : \mathbb{R}^q \rightarrow \mathbb{R}$, $j = 1, \ldots, q$. If we apply Theorem C.10 to each f_j, we get q approximations $f_j(\mathbf{x}) = \hat{f}_j(\mathbf{x}; \mathbf{x}_0) + R_j(\mathbf{x}; \mathbf{x}_0)$, $j = 1, \ldots, q$. This leads to a multidimensional Taylors series approximation about $\mathbf{x}_0$

$$f(\mathbf{x}) = \hat{f}(\mathbf{x}; \mathbf{x}_0) + \mathbf{R}(\mathbf{x}; \mathbf{x}_0)$$

where $\hat{f}(\mathbf{x}; \mathbf{x}_0) = (\hat{f}_1(\mathbf{x}; \mathbf{x}_0), \ldots, \hat{f}_q(\mathbf{x}; \mathbf{x}_0))$ and $\mathbf{R}(\mathbf{x}; \mathbf{x}_0) = (R_1(\mathbf{x}; \mathbf{x}_0), \ldots, R_q(\mathbf{x}; \mathbf{x}_0))^T$. Suppose the approximation is restricted to an open neighborhood E of $\mathbf{x}_0$. Then let

$$R_{max} = \frac{1}{2} \max_{i,j,k} \sup_{\mathbf{x}' \in E} \left| \frac{\partial^2 f_i(\mathbf{x}')}{\partial x_j \partial x_k} \right|.$$

If we apply the bound on the remainder term given by Equation (C.17) we get

$$\max_j |R_j(\mathbf{x}; \mathbf{x}_0)| \leq \frac{1}{2} R_{max} q^2 |\mathbf{x} - \mathbf{x}_0|^2.$$

Using this method to approximate $\bar{\Psi}(\tilde{X}, \boldsymbol{\theta})$ about $\boldsymbol{\theta}_0$ gives

$$\bar{\Psi}(\tilde{X}, \boldsymbol{\theta}) = \bar{\Psi}(\tilde{X}, \boldsymbol{\theta}_0) + \bar{\Psi}'(\tilde{X}, \boldsymbol{\theta}_0)(\boldsymbol{\theta} - \boldsymbol{\theta}_0) + \mathbf{R}(\boldsymbol{\theta}; \boldsymbol{\theta}_0).$$

Confining the approximation to the open ball $\boldsymbol{\theta} \in B_\tau(\boldsymbol{\theta}_0)$, we have bounding constant

$$R_{max} = \frac{1}{2} \max_{i,j,k} \sup_{\boldsymbol{\theta}' \in B_\tau(\boldsymbol{\theta}_0)} \left| \bar{\Psi}''_{ijk}(\tilde{X}, \boldsymbol{\theta}') \right|,$$

so that

$$\max_j |R_j(\boldsymbol{\theta}; \boldsymbol{\theta}_0)| \leq \frac{1}{2} R_{max} q^2 |\boldsymbol{\theta} - \boldsymbol{\theta}_0|^2.$$

If $\bar{\Psi}(\tilde{X}, \hat{\boldsymbol{\theta}}) = 0$ and $\bar{\Psi}'(\tilde{X}, \boldsymbol{\theta}_0)$ is invertible, then

$$\sqrt{n}(\hat{\boldsymbol{\theta}} - \boldsymbol{\theta}_0) = -\sqrt{n}\bar{\Psi}'(\tilde{X}, \boldsymbol{\theta}_0)^{-1}\bar{\Psi}(\tilde{X}, \boldsymbol{\theta}_0) - \sqrt{n}\bar{\Psi}'(\tilde{X}, \boldsymbol{\theta}_0)^{-1}\mathbf{R}(\hat{\boldsymbol{\theta}}; \boldsymbol{\theta}_0).$$

Under Assumption (A5) there exists a finite constant K such that for all small enough $\tau > 0$ the limit $P(R_{max} < K) \rightarrow_n 1$ holds, in which case it follows that

$$|\sqrt{n}\bar{\Psi}'(\tilde{X}, \boldsymbol{\theta}_0)^{-1}\mathbf{R}(\hat{\boldsymbol{\theta}}; \boldsymbol{\theta}_0)| = o_p\left(|\sqrt{n}(\hat{\boldsymbol{\theta}} - \boldsymbol{\theta}_0)|\right).$$

We have proven the following theorem.

Theorem 13.5 (Limiting Distribution of an *M*-estimator) Asssumptions (A4), (A5) and (A7) imply Assumption (A8). ■

By Theorems 13.4 and 13.5 we can now replace the regularity conditions of Theorem 13.1 with (A1)–(A5) (we have already argued that (A5) implies (A6)).

13.10 Regularity Conditions for Estimating Equations

We finally consider conditions under which Assumptions (A1)–(A5) will hold. Suppose we write the score function as

$$\bar{\Psi}(\mathbf{Y}, \boldsymbol{\theta}) = -n^{-1}\Gamma_{\boldsymbol{\theta}}^T \mathbf{W}(\boldsymbol{\theta}) \left[\mathbf{Y} - \boldsymbol{\mu}(\boldsymbol{\theta})\right], \tag{13.35}$$

where $\mathbf{W}(\boldsymbol{\theta})$ is an $n \times n$ matrix, then accept as an estimator for $\boldsymbol{\theta}$ a solution $\hat{\boldsymbol{\theta}}$ to $\bar{\Psi}(\mathbf{Y}, \boldsymbol{\theta}) = \mathbf{0}$. Verifying (A1) is straightforward. Then the elements Ψ_j of $\Psi(\mathbf{Y}, \boldsymbol{\theta})$ are linear combinations of the elements of $[\mathbf{Y} - \boldsymbol{\mu}(\boldsymbol{\theta})]$, and methods such as those introduced in Chapter 11 can be used to define conditions under which (A2) and (A3) hold.

Note that $\Sigma_{\Psi'(\boldsymbol{\theta})} = \bar{\mathbf{K}}(\boldsymbol{\theta}) = n^{-1}{\boldsymbol{\Gamma}_{\boldsymbol{\theta}}}^T \mathbf{W}(\boldsymbol{\theta}) \boldsymbol{\Gamma}_{\boldsymbol{\theta}}$, thus we may verify $\det(\Sigma_{\Psi'(\boldsymbol{\theta}_0)}) \neq 0$ by checking that $\boldsymbol{\Gamma}_{\boldsymbol{\theta}}$ is of rank q and $\mathbf{W}(\boldsymbol{\theta}_0)$ is of rank n. Here, $\boldsymbol{\Gamma}_{\boldsymbol{\theta}}$ plays a role similar to the design matrix $\mathbf{X}$ of multiple linear regression, so a well designed model will ensure that the rows are linearly independent. Otherwise, checking (A4) will be similar to checking (A2).

To consider (A5), we note that the higher-order derivatives of the score function will have the form

$$\bar{\Psi}''_{ijk}(\mathbf{Y}, \boldsymbol{\theta}) = a_{ijk}(\boldsymbol{\theta}) + n^{-1}\mathbf{b}_{ijk}(\boldsymbol{\theta})^T \left[\mathbf{Y} - \boldsymbol{\mu}(\boldsymbol{\theta})\right],$$

where $a_{ijk}(\boldsymbol{\theta}) \in \mathbb{R}$, $\mathbf{b}_{ijk}(\boldsymbol{\theta}) \in \mathcal{M}_{n,1}$ are allowed to depend on $\boldsymbol{\theta}$. Then (A5) will hold if the variances of $\mathbf{Y}$ and the elements of $a_{ijk}(\boldsymbol{\theta}) \in \mathbb{R}$, $\mathbf{b}_{ijk}(\boldsymbol{\theta})$ remain bounded in a neighborhood of $\boldsymbol{\theta}$.

13.11 Problems

Problem 13.1 Suppose we are given an *iid* sample $X_1, \ldots, X_n$ from a parametric family $\mathcal{P}$ with parameter space $\Theta \subset \mathbb{R}^q$. Define $T_j = n^{-1}\sum_i X_i^j$, and suppose $\eta_j(\boldsymbol{\theta}) = E_{\boldsymbol{\theta}}[X_1^j]$, $j = 1, \ldots, q$.

(a) Give a set of estimating equations the solution to which is a method of moments estimator, say $\hat{\boldsymbol{\theta}}_{mom}$ (Section 4.3).
(b) Assuming the conditions of Theorem 13.1 hold, give a general form for the asymptotic distribution of $\hat{\boldsymbol{\theta}}_{mom}$.

Problem 13.2 This example of a nonlinear regression model is discussed in Wild and Seber (1989). We are given parameters $\boldsymbol{\theta} = (\alpha_1, \alpha_2, \gamma) \in \mathbb{R}^3$ and mean function $\mu_i(\boldsymbol{\theta}) = \alpha_1 + \alpha_2 e^{\gamma x_i}$, given predictor variables $x_1, \ldots, x_n$.

(a) Examine the matrix $\mathbf{K}(\boldsymbol{\theta})$ for $\gamma = 0$. Can it be concluded that α_1 and α_2 are unidentifiable under the hypothesis $H_o : \gamma = 0$?
(b) Consider the reparametrization $\boldsymbol{\theta}' = (\beta_1, \beta_2, \gamma)$, giving

$$\mu_i'(\boldsymbol{\theta}') = \beta_1 + \beta_2 \left(\frac{e^{\gamma x_i} - 1}{\gamma}\right), \quad \gamma \neq 0.$$

Note that the second term of $\mu_i'(\boldsymbol{\theta}')$ has a singularity when $\gamma = 0$. Extend $\mu_i'(\boldsymbol{\theta}')$ to $\gamma = 0$ by taking the limit as $\gamma \to 0$. Describe precisely the difference between $\mu_i(\boldsymbol{\theta})$ and $\mu_i'(\boldsymbol{\theta}')$.

(c) Show that the singularity of $\mu_i'(\boldsymbol{\theta}')$ at $\gamma = 0$ can also be removed by replacing $e^{\gamma x_i}$ with a Taylor's series.
(d) Examine the matrix $\mathbf{K}(\boldsymbol{\theta})$ for $\gamma = 0$ for the model $\mu_i'(\boldsymbol{\theta}')$. Can it be concluded that β_1 and β_2 are identifiable under the hypothesis $H_o : \gamma = 0$?

Problem 13.3 Consider the model

$$\mathbf{Y} = \mathbf{X} \begin{bmatrix} \alpha \\ \beta \\ \alpha\beta \end{bmatrix} + \boldsymbol{\epsilon}$$

where $\mathbf{Y}$ is an $n \times 1$ response vector, $\mathbf{X}$ is a $n \times q$ matrix, $\boldsymbol{\epsilon}$ is an $n \times 1$ vector of independent error terms from $N(0, \sigma^2)$. We have parameter $\boldsymbol{\theta} = (\alpha, \beta) \in \Theta = \mathbb{R}^2$.

(a) Can this model be expressed as a linearly constrained multiple linear regression model?
(b) Show that $SSE = (\mathbf{Y} - E_{\boldsymbol{\theta}}[\mathbf{Y}])^T(\mathbf{Y} - E_{\boldsymbol{\theta}}[\mathbf{Y}])$ is a convex function on Θ $wp1$.
(c) Does a least squares estimate of $\boldsymbol{\theta}$ always exist?

Problem 13.4 In a generalized linear model the canonical link function can be deduced by the relation $\eta = g(\mu)$ where η is the canonical parameter of an exponential family density, and μ is the mean of the response. Verify each canonical link function of Table 13.1.

Problem 13.5 Overdispersion occurs when a variance exceeds that predicted by a probabilistic model. In GLMs this can be modeled using dispersion parameters ϕ which exceed the value given in Table 13.1 (the issue is not so relevant to the Gaussian model, however). Suppose a count observation X_i is generated by a Poisson process. We let $T_i \sim pois(\lambda_i)$ be the number of arrivals. However, each arrival contributes m to the total count X_i. What would be the dispersion parameter for this model?

Problem 13.6 For the binomial GLM a response has distribution $Y \sim bin(n,p)$, $\text{var}[Y] = np(1-p)$. For the *beta-binomial model* we assume $\boldsymbol{\theta} \sim beta(\alpha, \beta)$, and $Y^* \mid \boldsymbol{\theta} \sim bin(n, \boldsymbol{\theta})$ (McCulloch *et al.*, 2008). To make the two models comparable assume $p = E[\boldsymbol{\theta}]$ is fixed. We can take the dispersion parameter to be $\phi = \text{var}[Y^*]/\text{var}[Y]$. Evaluate ϕ. Verify that $\phi = 1$ if $n = 1$ and $\phi > 1$ if $n > 1$.

Problem 13.7 Consider a generalized linear model for independent Bernoulli response variables Y_i with mean μ_i, $i = 1, \ldots, n$. Linear predictor terms are given by $\eta_i = \beta_1 x_{i1} + \ldots + \beta_q x_{iq}$, and we have link function $\eta_i = g(\mu_i)$. For the binomial model, the canonical link function is $g_c(\mu) = \log[\mu/(1-\mu)]$. An alternative is the complementary log-log link function $g_{cll}(\mu) = \log(-\log(1-\mu))$. Develop an asymptotic approximation of the distribution of the maximum likelihood estimates of the coefficients $(\beta_1, \ldots, \beta_p)$ for each link function.

Problem 13.8 Consider a generalized linear model for independent Poisson response variables Y_i with mean μ_i, $i = 1, \ldots, n$. Linear predictor terms are given by $\eta_i = \beta_1 x_{i1} + \ldots + \beta_q x_{iq}$, and we have link function $\eta_i = g(\mu_i)$. For the Poisson model, the canonical link function is $g_c(\mu) = \log \mu$. However, the identity link function $g_i(\mu) = \mu$ is also used. Develop an asymptotic approximation of the distribution of the maximum likelihood estimates of the coefficients $(\beta_1, \ldots, \beta_p)$ for each link function.

Problem 13.9 For a certain form of cancer, the probability of recurrence after treatment within two years is to be predicted. The prediction will depend on tumor size x at treatment (in mm). For this purpose, a logistic regression model is fit using observed pairs (Y_i, x_i) obtained from a study, where $y_i = 1$ denotes observed recurrence, and x_i is tumor size. The linear prediction term is $\hat{\eta} = \hat{\beta}_0 + \hat{\beta}_1 x$, where $\hat{\beta}_0 = -4.23$, $\hat{\beta}_1 = 0.23$.

(a) At which tumor size x is the recurrence probability estimated to be 20%?

(b) Suppose $(0.19, 0.27)$ is a 95% confidence interval for $\hat{\beta}_1$. Give a 95% confidence interval for the odds ratio between patients with tumor sizes $x = 10$ and $x = 2$.

Problem 13.10 Suppose a logistic regression model is fit with response $Y =$ `Minor injury during the past year` and one predictor $X =$ `age`. The sample size is $n = 100$ and the range of X in years is $[3.2, 18.1]$. The data was fit using the `R` `glm()` function, and produced the following output:

```
Coefficients:
             Estimate Std. Error z value Pr(>|z|)
(Intercept) -3.89379    1.09305  -3.562 0.000368 ***
x            0.19859    0.09147   2.171 0.029927 *
---
Signif. codes:  0 '***' 0.001 '**' 0.01 '*' 0.05 '.' 0.1 ' ' 1

(Dispersion parameter for binomial family taken to be 1)
```

What is the estimated odds ratio for probability of `Minor injury during the past year` between 15 year old and 5 year old subjects? Give an approximate 95% confidence interval.

Problem 13.11 A logistic regression model is used to predict the probability of a win for team A against team B for a baseball match based on predictor variable $x = w_A - w_B$, where w_A and w_B are the winning percentages for the current season for teams A and B respectively (on a scale of 0 to 100). Thus, we have model

$$P(\text{Team } A \text{ wins}) = \frac{e^{\beta_0 + \beta_1 x}}{1 + e^{\beta_0 + \beta_1 x}}.$$

Suppose the coefficients are estimated to be $\hat{\beta}_0 = 0.00034$, $\hat{\beta}_1 = 0.054$.

(a) What should we expect the value of β_0 to be?

(b) What is the estimated probability that a team with a winning percentage of 60% wins against each of 3 teams with winning percentages 40%, 50% and 55%, respectively?

(c) Suppose games were in realty decided by chance. That is, each team had a 50% chance of winning each game, independent of past winning percentage. What would we expect the coefficients β_0 and β_1 to be? Justify your answer.

Problem 13.12 The autoregressive model is frequently used to model correlation in the error terms of a regression model. A zero mean autoregressive process of order 1 (usually denoted as $AR(1)$) can be defined by the equation

$$\epsilon_n = \rho\epsilon_{n-1} + u_n, n \geq 1,$$

where u_n, $n \geq 1$ are uncorrelated, with zero mean and variance $\text{var}[u_1] = \sigma_u^2$. We assume $|\rho| < 1$. It is recommended that the reader also review Problem 11.26.

(a) Prove that if $E[\epsilon_0] = 0$ and $\text{var}[\epsilon_0] = \sigma_u^2/(1 - \rho^2)$, then $E[\epsilon_n] = 0$ and $\text{var}[\epsilon_n] = \sigma_u^2/(1 - \rho^2)$, $n \geq 1$. This is referred to as a stationary $AR(1)$ process.

(b) Suppose we wish to fit a linear regression model $\mathbf{Y} = \mathbf{X}\boldsymbol{\beta} + \boldsymbol{\epsilon}$, where $\mathbf{Y}$ is an $n \times 1$ response vector, $\mathbf{X}$ is a $n \times q$ matrix, $\boldsymbol{\beta}$ is a $q \times 1$ vector of coefficients, and $\boldsymbol{\epsilon}$ is an $n \times 1$ vector of error terms. Suppose $\boldsymbol{\epsilon}$ is a stationary $AR(1)$ process as described above, and ρ is known. Derive the correlation matrix $\mathbf{R}$ of $\boldsymbol{\epsilon}$ and its inverse. **HINT:** It may be shown that the inverse of an $AR(1)$ correlation matrix is tridiagonal, that is, $\{\mathbf{R}^{-1}\}_{ij} = 0$ if $|i - j| > 1$. Consider the problem of finding constants a, b such that $\rho^k a + \rho^{k+1} b + \rho^{k+2} a = 0$ and $\rho a + b + \rho b = 1$. Then consider $\{\mathbf{r}^{-1}\}_{11}$ and $\{\mathbf{r}^{-1}\}_{nn}$ separately.

(c) Show how $\mathbf{Y}$ and $\mathbf{X}$ can be transformed to $\mathbf{Y}'$ and $\mathbf{X}'$, so that the ordinary least squares estimate of $\boldsymbol{\beta}$ of the model $\mathbf{Y}' = \mathbf{X}'\boldsymbol{\beta} + \boldsymbol{\epsilon}'$ will be the same as the BLUE of $\boldsymbol{\beta}$ for the original model (see Section 9.5.1).

(d) Give the elements of $\mathbf{Y}'$ explicity. **HINT:** A tridiagonal matrix $\mathbf{B}$ can be factored by $\mathbf{B} = \mathbf{LU}$, where $\mathbf{L}$ is lower diagonal, with, $\{\mathbf{L}\}_{ij} = 0$ for $j < i - 1$, and $\mathbf{U}$ is upper diagonal, with, $\{\mathbf{U}\}_{ij} = 0$ for $j < i + 1$.

(e) Suppose ρ is not known. Describe a procedure that might be used to approximate the BLUE of $\boldsymbol{\beta}$.

Problem 13.13 Suppose we wish to fit a model using the GEE method. Suppose the data is assumed to have block exchangeable correlation. This means the n observations are partitioned into K blocks of size m_i, so that $m_1 + \ldots + m_K = n$. Then suppose that any two responses from a common block have pairwise correlation $\rho \neq 0$, while any two responses from different blocks have zero pairwise correlation. Then set $z_i = (y_i - \mu_i)/\tau_i^{1/2}$, and let $\bar{z}_j$ be the sample mean of the z_i values within block j. Let j_i be the block to which the ith response belong.

(a) Show that the estimating equation can be written

$$\Psi(\boldsymbol{\theta}) = \Gamma_{\boldsymbol{\theta}}^T D_{\boldsymbol{\tau}}^{-1/2} \tilde{z}^* \tag{13.36}$$

where the ith element of $\tilde{z}^*$ is $z_i^* = z_i - \alpha_{j_i} \bar{z}_{j_i}$, and α_j is a constant that depends on σ^2, ρ and m_j. **HINT:** See Example 6.8

(b) This type of exchangeable correlation can be modeled by adding a random effect U to a set of independent responses. In this model each block has it's own random effect U_j. How are the random effects predicted using the GEE method.

14

Large Sample Hypothesis Testing

14.1 Introduction

This final chapter presents an application of asymptotic theory to hypothesis testing. For "finite sample inference" the Neyman-Pearson lemma is able to resolve most questions concerning optimality (Chapter 10). However, the approach generally taken for large sample inference is quite different. In this chapter, we will confine attention to the likelihood ratio test (LRT), introduced in Section 5.8, and two related procedures. Under quite general conditions, the limiting distribution of the LRT statistic (expressed as model deviance) is χ^2_q, where q is the difference in degrees of freedom between the full and reduced model. This is the conclusion of Wilks' theorem (Wilks, 1938). Two other tests, the Rao score test (or score test) and the Wald test, can be shown to be asymptotic approximations of the LRT, and have the same limiting distribution. All three are part of conventional statistical methodology, meaning that each has its relative advantages with respect to accuracy and analytical tractability, and should therefore be considered together.

The approach taken here will be based on the estimating equations introduced in Chapter 13, and we similarly begin by enumerating the regularity conditions required by the theory, here labeled (B1)–(B7) (Section 14.2). Unlike Chapter 13, the methods discussed here are based on the likelihood, so regularity conditions are imposed directly on the log-likelihood function and the Fisher information matrix (so we assume that the regular family conditions of Definition 3.6 hold). Nonetheless, a careful study of Assumptions (B1)–(B7) and Assumptions (A1)–(A8) of Chapter 13 will reveal an obvious correspondence.

The tests considered in this chapter all make use of the full and reduced model structure defined for the likelihood ratio test in Section 5.8. The Wald test and the LRT both require evaluation of a MLE, which must therefore be assumed to be consistent. Here, this is Assumption (B6), which can be verified by Theorem 13.4. Similarly, (B7) simply states the asymptotic distribution of the MLE given in Section 13.3, which can be verified by Theorem 13.5.

Section 14.3 introduces the three tests for simple null hypotheses. The extension to composite null hypotheses is given in Section 14.4, and requires some new, and sometimes subtle, arguments. This is the reason for introducing these methods in two stages. Section 14.5 considers the important special case of the test for independence of a contingency table.

Section 14.6 provides a comparison of the three procedures, considering the relative advantages and disadvantages of each. Section 14.7 shows how confidence sets can be constructed from these testing methods. Finally, Section 14.8 discusses the estimation of power. Since each test statistic has an approximate χ^2 distribution, this is largely a question of estimating a noncentrality parameter (Definition 2.2). The fact that each test statistic is an approximation of the other two simplifies the problem, since any approximation of power for one will hold for all three.

DOI: 10.1201/9781003049340-14

14.2 Model Assumptions

Suppose we have parametric family $\mathcal{P} = \{p_{\boldsymbol{\theta}} : \boldsymbol{\theta} \in \Theta\}$ for some open parameter space $\Theta \subset \mathbb{R}^q$. We have log-likelihood function $\ell(\boldsymbol{\theta}; \tilde{x}) = \log p_{\boldsymbol{\theta}}(\tilde{x})$. Here the score function $\Psi(\tilde{X}, \boldsymbol{\theta})$ is the gradient of $\ell(\boldsymbol{\theta}; \tilde{x})$ with respect to $\boldsymbol{\theta}$, and $\Psi'(\tilde{X}, \boldsymbol{\theta})$ is the Hessian. There is a sample size index n, so that $\tilde{X}_n$, $n \geq 1$ defines a hierarchical sequence of data. However, dependence on n will not always be indicated using the conventional subscript notation. It will be used when n plays a particular role. We will make some use of the normalization

$$\bar{\Psi}(\tilde{X}, \boldsymbol{\theta}) = n^{-1}\Psi(\tilde{X}, \boldsymbol{\theta}), \quad \bar{\Psi}'(\tilde{X}, \boldsymbol{\theta}) = n^{-1}\Psi'(\tilde{X}, \boldsymbol{\theta})$$

introduced in Chapter 13, but this will not play as large a role in this chapter.

We can take as a representative model the existence of some indefinite sequence of observations $\tilde{U}_1, \tilde{U}_2, \ldots$, and take $\tilde{X}_n = (\tilde{U}_1, \ldots, \tilde{U}_n)$. Our assumption, informally, is that $\ell(\boldsymbol{\theta}; \tilde{X}_n)$ behaves like a random sum, so that, if we had to, we could write something like $\ell(\boldsymbol{\theta}; \tilde{X}_n) = \sum_{i=1}^n \ell(\boldsymbol{\theta}; \tilde{U}_i)$. However, our development will not rely on this type of explicit form.

We will, at this point, list the assumptions on which our development will depend.

(B1) $\ell(\boldsymbol{\theta}; \tilde{X})$ is twice differentiable *wrt* $\boldsymbol{\theta}$, so that we may define the score vector $\Psi(\tilde{X}, \boldsymbol{\theta})$ of first order partial derivatives and the Hessian matrix $\Psi'(\tilde{X}, \boldsymbol{\theta})$ of second order partial derivatives. We assume $\Psi'(\tilde{X}, \boldsymbol{\theta})$ is continuous in $\boldsymbol{\theta}$.

(B2) $\mathcal{P}$ is a regular parametric family, so that the Fisher information matrix $\mathbf{I}(\boldsymbol{\theta})$ is well defined. We also assume $\mathbf{I}(\boldsymbol{\theta})$ is invertible and therefore positive definite.

(B3) We assume $\mathbf{I}_n(\boldsymbol{\theta}) = O(n)$. We can denote the "average information" by $\bar{\mathbf{I}}(\boldsymbol{\theta}) = \bar{\mathbf{I}}_n(\boldsymbol{\theta}) = n^{-1}\mathbf{I}_n(\boldsymbol{\theta})$. We further assume that the minimum and maximum eigenvalues of $\bar{\mathbf{I}}_n(\boldsymbol{\theta})$, say $\lambda_n^{min}, \lambda_n^{max}$, satisfy

$$0 < \liminf_n \lambda_n^{min} \leq \limsup_n \lambda_n^{max} < \infty.$$

If the average information possesses a limit it will denoted as $\bar{\mathbf{I}}_\infty(\boldsymbol{\theta}) = \lim_n \bar{\mathbf{I}}_n(\boldsymbol{\theta})$. In this case we say $\bar{\mathbf{I}}_\infty(\boldsymbol{\theta})$ exists.

(B4) If $\bar{\mathbf{I}}_\infty(\boldsymbol{\theta})$ exists then $n^{-1}\Psi_n'(\tilde{X}_n, \boldsymbol{\theta}) \xrightarrow{i.p.} \bar{\mathbf{I}}_\infty(\boldsymbol{\theta})$. Otherwise, we assume

$$\bar{\mathbf{I}}_n(\boldsymbol{\theta})^{-1}[n^{-1}\Psi_n'(\tilde{X}_n, \boldsymbol{\theta})] \xrightarrow{i.p.} \mathbf{I}_q.$$

(B5) If $\bar{\mathbf{I}}_\infty(\boldsymbol{\theta})$ exists then the score vector $\Psi(\tilde{X}, \boldsymbol{\theta}_0)$ at any true parameter $\boldsymbol{\theta}_0$ possesses asymptotic distribution $n^{-1/2}\Psi_n(\tilde{X}_n, \boldsymbol{\theta}_0) \xrightarrow{d} N(0, \bar{\mathbf{I}}_\infty(\boldsymbol{\theta}_0))$. Otherwise, we may claim the approximation $n^{-1/2}\Psi_n(\tilde{X}_n, \boldsymbol{\theta}_0) \sim N(0, \bar{\mathbf{I}}_n(\boldsymbol{\theta}_0))$ with error approaching zero as $n \to \infty$.

(B6) Let $B_{\boldsymbol{\theta}_0} \subset \Theta$ be any open subset containing the true parameter $\boldsymbol{\theta}_0$. Then the probability that there exists a solution $\hat{\boldsymbol{\theta}}_n \in B_{\boldsymbol{\theta}_0}$ to the estimating equation $\Psi_n(\tilde{X}_n, \boldsymbol{\theta}) = \mathbf{0}$ approaches 1 as $n \to \infty$. As a consequence, we may define a sequence $\hat{\boldsymbol{\theta}}_n$ that converges in probability to $\boldsymbol{\theta}_0$. We take this sequence to be the MLE of $\boldsymbol{\theta}$.

(B7) If $\bar{\mathbf{I}}_\infty(\boldsymbol{\theta})$ exists then the MLE possesses asymptotic distribution $\sqrt{n}(\hat{\boldsymbol{\theta}}_n - \boldsymbol{\theta}_0) \xrightarrow{d} N(0, \bar{\mathbf{I}}_\infty(\boldsymbol{\theta}_0)^{-1})$ for any true parameter $\boldsymbol{\theta}_0$. Otherwise, we may claim the approximation $\sqrt{n}(\hat{\boldsymbol{\theta}}_n - \boldsymbol{\theta}_0) \sim N(0, \bar{\mathbf{I}}_n(\boldsymbol{\theta})^{-1})$ with error approaching zero as $n \to \infty$.

Some remarks:

(i) Conditons under which these assumptions hold were discussed in Chapter 13 on the asymptotic theory of estimating equations.

(ii) The notion that the log-likelihood behaves like a random sum is largely captured in Assumption (B3). Clearly it covers the *iid* case, but also many interesting models that are neither independent nor identically distributed. In linear models, (B3) will hold when the predictor variables neither converge to a limit nor diverge to ∞. It will also be satisfied under forms of statistical dependence models commonly used, included autoregressive and moving average models and many forms of exchangeable dependence.

(iii) It would be possible to unify the two cases within Assumptions (B5) and (B7) by stating instead that, using (B5) as an example,

$$\frac{1}{\sqrt{n}}\left[\bar{\mathbf{I}}_n^{1/2}(\boldsymbol{\theta}_0)\right]^{-1}\Psi_n(\tilde{X}_n,\boldsymbol{\theta}_0) \xrightarrow{d} N(0,\mathbf{I}_q).$$

where $\bar{\mathbf{I}}_n(\boldsymbol{\theta}_0) = \bar{\mathbf{I}}_n^{1/2}(\boldsymbol{\theta}_0)[\bar{\mathbf{I}}_n^{1/2}(\boldsymbol{\theta}_0)]^T$ (Section B.4.7). Such a formulation does not depend on the existence of $\bar{\mathbf{I}}_\infty(\boldsymbol{\theta}_0)$. However, this would increase the notational burden of what is already a rather technical subject. Furthermore, the existence of $\bar{\mathbf{I}}_\infty(\boldsymbol{\theta}_0)$ is not a natural assumption for, say, linear models, and not really required. On the other hand, as long as (B3) holds, we don't really lose any generality by assuming $\bar{\mathbf{I}}_\infty(\boldsymbol{\theta}_0)$ exists, so we will do so. If for some reason this assumption cannot be made, then the argument can be modifed as suggested above.

14.3 Large Sample Tests for Simple Null Hypotheses

Many large sample hypothesis tests take the form of χ^2 tests, and are often referred to this way. This is a direct consequence of the role played by normal approximations in large sample theory. The basic procedure is quite straightforward. Suppose $\mathbf{Y} = (Y_1,\ldots,Y_q) \sim N(\mathbf{0},\Sigma)$ is a multivariate normal random vector. Assuming Σ^{-1} exists, we can find an invertible matrix $\Sigma_{1/2}$ such that $\Sigma = \Sigma_{1/2}\Sigma_{1/2}^T$ (Section B.4.7). Set $\mathbf{Z} = \Sigma_{1/2}^{-1}\mathbf{Y}$. Then $\mathbf{Z} \sim N(0,\mathbf{I})$. In other words, the elements of $\mathbf{Z}$ are independent, with distribution $N(0,1)$. We will refer to any random vector possessing covariance matrix equal to the identity as an orthogonal random vector. In turn, this allows us to make use of the following quadratic form

$$\mathbf{Y}^T\Sigma^{-1}\mathbf{Y} = \mathbf{Y}^T\left[\Sigma_{1/2}^{-1}\right]^T\Sigma_{1/2}^{-1}\mathbf{Y} = \mathbf{Z}^T\mathbf{Z} \sim \chi^2_q, \tag{14.1}$$

that is, this quadratic form can be expressed as the sum of squares of q independent random variables with distribution $N(0,1)$, which has a χ^2 distribution with q degrees of freedom (see also Section 2.5).

We next introduce the three hypothesis test statistics to be discussed in this chapter.

14.3.1 The Score Test

We can demonstrate how Equation (14.1) can be used to construct a hypothesis test based on a χ^2 approximation. Suppose, following Section 13.10, that the estimating equation $\Psi_n(\tilde{X}_n,\boldsymbol{\theta}) = 0$ has solution $\hat{\boldsymbol{\theta}}_n$ near true parameter value $\boldsymbol{\theta}^* \in \Theta$. Consider a simple null hypothesis $H_o : \boldsymbol{\theta} = \boldsymbol{\theta}_0$, and alternative hypothesis $H_a : \boldsymbol{\theta} \neq \boldsymbol{\theta}_0$. Then $\Psi_n(\tilde{X}_n,\hat{\boldsymbol{\theta}}_n) = 0$, and for large enough n we would expect $\bar{\Psi}_n(\tilde{X}_n,\boldsymbol{\theta}^*) \approx 0$. Put another way, if H_o is true, then we would expect $\bar{\Psi}_n(\tilde{X}_n,\boldsymbol{\theta}_0) \approx 0$.

On the other hand, if H_o was false, we would expect $|\hat{\boldsymbol{\theta}}_n - \boldsymbol{\theta}_0| = d$ to be reasonably large. Assume the normalized Hessian $\bar{\Psi}'_n(\tilde{X}_n, \boldsymbol{\theta})$ is bounded away from zero. In this context, this means that its eigenvalues are bounded away from zero (Section B.4.4). Consider also Assumption (B3). We could then argue that (holding n fixed)

$$|\bar{\Psi}_n(\tilde{X}_n, \hat{\boldsymbol{\theta}}_n) - \bar{\Psi}_n(\tilde{X}_n, \boldsymbol{\theta}_0)| = |\bar{\Psi}_n(\tilde{X}_n, \boldsymbol{\theta}_0)| = O(d).$$

This means observations of the score vector $\Psi_n(\tilde{X}_n, \boldsymbol{\theta}_0)$ with large magnitude can be interpreted as evidence against H_o. We can then use Equation (14.1) to construct a test statistic. If $\Sigma_{\boldsymbol{\theta}}$ is the covariance matrix of $\Psi_n(\tilde{X}_n, \boldsymbol{\theta})$ when $\boldsymbol{\theta}$ is the true parameter, then the statistic

$$X^2_{score} = \Psi_n(\tilde{X}_n, \boldsymbol{\theta}_0)^T \Sigma^{-1}_{\boldsymbol{\theta}_0} \Psi_n(\tilde{X}_n, \boldsymbol{\theta}_0) \tag{14.2}$$

will have an approximate χ^2_q distribution under H_o (Theorem 2.2). Under H_a we will see that, approximately, $X^2_{score} \sim \chi^2_{\Delta;q}$ for noncentrality parameter $\Delta = O(d^2)$ (Section 14.8), and so this is a suitable choice of test statistic. In this case we reject H_o with a significance level of α if $X^2_{score} > \chi^2_{q;\alpha}$, where $\chi^2_{q;\alpha}$ is the α critical value.

When $\Psi_n(\tilde{X}_n, \boldsymbol{\theta})$ is the gradient of a log-likelihood function $\ell(\boldsymbol{\theta}; \tilde{X}) = \log p_{\boldsymbol{\theta}}(\tilde{X})$ this procedure is known as the Rao score test (Rao, 1948) (or simply score test) and it has a more formal justification. Recall that the Neyman-Pearson lemma states that the most powerful test of fixed size α for simple hypotheses $H_o : \boldsymbol{\theta} = \boldsymbol{\theta}_0$ and $H_a : \boldsymbol{\theta} = \boldsymbol{\theta}_0 + \delta$ rejects H_o when $\lambda(\tilde{X}) > k$ and accepts H_o when $\lambda(\tilde{X}) < k$, where k is some constant, and

$$\lambda(x) = p_{\boldsymbol{\theta}_0+\delta}(x)/p_{\boldsymbol{\theta}_0}(x).$$

We can equivalently use a log transformation of $\lambda(x)$,

$$\log \lambda(x) = \ell(\boldsymbol{\theta}_0 + \delta; X) - \ell(\boldsymbol{\theta}_0; X).$$

Assume, for the moment, that $\boldsymbol{\theta} \in \Theta \subset \mathbb{R}$. We then have

$$\log \lambda(x) \approx \delta \Psi_n(\tilde{X}_n, \boldsymbol{\theta}_0).$$

It follows that Equation (14.2) defines approximately a Neyman-Pearson test, or exactly a locally most powerful test (Section 10.5.1).

More generally, for the multivariate case $\boldsymbol{\theta} \in \Theta \subset \mathbb{R}^q$, we have $\Sigma_{\boldsymbol{\theta}} = \mathbf{I}_n(\boldsymbol{\theta})$, so that for the Rao score test Equation (14.2) becomes

$$X^2_{score} = \Psi_n(\tilde{X}_n, \boldsymbol{\theta}_0)^T \mathbf{I}_n(\boldsymbol{\theta}_0)^{-1} \Psi_n(\tilde{X}_n, \boldsymbol{\theta}_0). \tag{14.3}$$

14.3.2 The Likelihood Ratio Test (LRT)

The likelihood ratio test (LRT) was introduced in Section 5.8. Suppose we have parametric family $\{p_{\boldsymbol{\theta}} : \boldsymbol{\theta} \in \Theta\}$ for some parameter space $\Theta \subset \mathbb{R}^q$. We wish to test hypothesis

$$H_o : \boldsymbol{\theta} \in \Theta_0 \text{ against } H_a : \boldsymbol{\theta} \notin \Theta_0,$$

where Θ_0 is a strict subset of Θ. If we denote the likelihood function $\mathcal{L}(\boldsymbol{\theta}; \tilde{X}) = p_{\boldsymbol{\theta}}(\tilde{X})$ the likelihood ratio test (LRT) is based on test statistic

$$\lambda_{lrt}(\tilde{x}) = \frac{\sup_{\boldsymbol{\theta} \in \Theta_0} \mathcal{L}(\boldsymbol{\theta}; \tilde{x})}{\sup_{\boldsymbol{\theta} \in \Theta} \mathcal{L}(\boldsymbol{\theta}; \tilde{x})}.$$

We can transform $\lambda_{lrt}(\tilde{x})$ as

$$X^2_{lrt} = 2\left[\ell(\hat{\boldsymbol{\theta}}; \tilde{X}) - \ell(\hat{\boldsymbol{\theta}}_0; \tilde{X})\right], \tag{14.4}$$

where $\hat{\boldsymbol{\theta}}$ is the MLE of $\boldsymbol{\theta}$, and $\hat{\boldsymbol{\theta}}_0$ is the MLE of $\boldsymbol{\theta}$ restricted to Θ_0.

Assume, for the moment, that $\Theta_0 = \{\boldsymbol{\theta}_0\}$ defines a simple hypothesis. Using the Taylor's series approximation of Theorem C.9, we have the following expansion of $\ell(\boldsymbol{\theta}; \tilde{X})$ about $\hat{\boldsymbol{\theta}}$:

$$\ell(\boldsymbol{\theta}_0; \tilde{X}) = \ell(\hat{\boldsymbol{\theta}}; \tilde{X}) + \Psi(\tilde{X}, \hat{\boldsymbol{\theta}})(\boldsymbol{\theta}_0 - \hat{\boldsymbol{\theta}}) + \frac{1}{2}(\hat{\boldsymbol{\theta}} - \boldsymbol{\theta}_0)^T \Psi'(\tilde{X}, \boldsymbol{\theta}_c)(\hat{\boldsymbol{\theta}} - \boldsymbol{\theta}_0),$$

where $\boldsymbol{\theta}_c = p\boldsymbol{\theta}_0 + (1-p)\hat{\boldsymbol{\theta}}$, for some $p \in [0,1]$. Under general conditions, if $\boldsymbol{\theta}_0$ is the true parameter then as $n \rightarrow \infty$ we will have $n^{-1}\Psi'(\tilde{X}, \boldsymbol{\theta}_c) \overset{i.p.}{\rightarrow} \bar{\mathbf{I}}_\infty(\boldsymbol{\theta}_0)$, assuming $\hat{\boldsymbol{\theta}} \overset{i.p.}{\rightarrow} \boldsymbol{\theta}_0$.

After noting that $\Psi(\tilde{X}, \hat{\boldsymbol{\theta}}) = 0$, since $\hat{\boldsymbol{\theta}}$ is the MLE, we then have approximation

$$X^2_{lrt} = 2\left[\ell(\hat{\boldsymbol{\theta}}; X) - \ell(\boldsymbol{\theta}_0; X)\right] \approx [\sqrt{n}(\hat{\boldsymbol{\theta}}_n - \boldsymbol{\theta}_0)]^T \bar{\mathbf{I}}_\infty(\boldsymbol{\theta}_0)[\sqrt{n}(\hat{\boldsymbol{\theta}}_n - \boldsymbol{\theta}_0)]. \tag{14.5}$$

By Assumption (B7) and Equation (14.1) we may conclude that under H_o we have the limit

$$X^2_{lrt} \overset{d}{\rightarrow} \chi^2_q, \tag{14.6}$$

as predicted by Wilks' theorem (Wilks, 1938).

14.3.3 The Wald Test

A third commonly used large sample test procedure is the Wald test (Wald, 1943). For a simple null hypothesis the test statistic is given by

$$X^2_{wald} = (\hat{\boldsymbol{\theta}} - \boldsymbol{\theta}_0)^T \mathbf{I}(\boldsymbol{\theta}_0)(\hat{\boldsymbol{\theta}} - \boldsymbol{\theta}_0). \tag{14.7}$$

We may write

$$(\hat{\boldsymbol{\theta}}_n - \boldsymbol{\theta}_0)^T \mathbf{I}_n(\boldsymbol{\theta}_0)(\hat{\boldsymbol{\theta}}_n - \boldsymbol{\theta}_0) \approx [\sqrt{n}(\hat{\boldsymbol{\theta}} - \boldsymbol{\theta}_0)]^T \bar{\mathbf{I}}_\infty(\boldsymbol{\theta}_0)[\sqrt{n}(\hat{\boldsymbol{\theta}} - \boldsymbol{\theta}_0)]$$

which is nothing more than the approximation of the LRT statistic given in Equation (14.5), so by the same argument we have under H_o

$$X^2_{wald} \overset{d}{\rightarrow} \chi^2_q. \tag{14.8}$$

In addition, implicit in Assumptions (A4) and (A8) of Section 13.2 is the approximation

$$\sqrt{n}(\hat{\boldsymbol{\theta}} - \boldsymbol{\theta}_0) \approx \sqrt{n}\bar{\mathbf{I}}(\boldsymbol{\theta}_0)^{-1}\bar{\Psi}(\tilde{X}, \boldsymbol{\theta}_0),$$

from which it follows that $X^2_{wald} \approx X^2_{score}$ (see Theorem 13.5 for the formal proof, also the discussion surrounding Equation (13.4)). Thus, X^2_{lrt}, X^2_{score} and X^2_{wald} are mutual approximations of each other.

The next example compares the three tests for an important application.

Example 14.1 (Large Sample Tests for GLMs (Simple Hypotheses)) Recall from Section 13.6 the log-likelihood function for the canonical generalized linear model (GLM):

$$\ell(\boldsymbol{\beta}, \mathbf{Y}) = \frac{1}{\phi}\sum_i \{Y_i\eta_i - A(\eta_i)\}, \;\; i = 1, \ldots, n.$$

The observations Y_i are independent, with mean and variance μ_i, τ_i. We have linear prediction term $\eta_i = \boldsymbol{\beta}^T\mathbf{x}_i$, where $\boldsymbol{\beta} \in \mathbb{R}^q$ is the model parameter, and $\mathbf{x}_i = (x_{i1}, \ldots, x_{iq})^T$ is a fixed vector of predictor variables paired with response Y_i. We also have a positive dispersion parameter ϕ, which is considered fixed. There is a canonical link function $\eta_i = g(\mu_i)$

and variance function $\tau_i = \phi V(\mu_i)$. The components of the gradient and Hessian of $\ell(\boldsymbol{\beta}; \tilde{X})$ are given by

$$\Psi_j = \phi^{-1} \sum_i \{Y_i - \mu_i\} x_{ij}, \quad \Psi'_{jk} = -\frac{1}{\phi} \sum_i V(\mu_i) x_{ij} x_{ik}.$$

We have the more compact representation

$$\Psi(\mathbf{Y}, \boldsymbol{\theta}) = \frac{1}{\phi} \mathbf{X}^T (\mathbf{Y} - \boldsymbol{\mu}), \quad \Psi'(\mathbf{Y}, \boldsymbol{\theta}) = -\frac{1}{\phi^2} \mathbf{X}^T \Sigma_{\mathbf{Y}} \mathbf{X},$$

where $\mathbf{X}$ is the $n \times q$ design matrix with ith row $\mathbf{x}_i^T$, and $\Sigma_{\mathbf{Y}}$ is a diagonal matrix with entries $\{\Sigma_{\mathbf{Y}}\}_{ii} = \phi V(\mu_i)$. We therefore have the information matrix for $\boldsymbol{\beta}$ equal to $\mathbf{I}(\boldsymbol{\beta}) = \phi^{-2} \mathbf{X}^T \Sigma_{\mathbf{Y}} \mathbf{X}$. If $\hat{\boldsymbol{\beta}}$ is the MLE of $\boldsymbol{\beta}$, the LRT, Wald and score test statistics for a simple null hypothesis $H_o : \beta = \beta_0$ are

$$\begin{aligned} X^2_{lrt} &= 2 \left[\ell(\hat{\boldsymbol{\beta}}, \mathbf{Y}) - \ell(\boldsymbol{\beta}_0, \mathbf{Y}) \right], \\ X^2_{score} &= (\mathbf{Y} - \boldsymbol{\mu}_0)^T \mathbf{X} \left[\mathbf{X}^T \Sigma_{\mathbf{Y}} \mathbf{X} \right]^{-1} \mathbf{X}^T (\mathbf{Y} - \boldsymbol{\mu}_0), \text{ and} \\ X^2_{wald} &= \frac{1}{\phi^2} (\hat{\boldsymbol{\beta}} - \boldsymbol{\beta}_0)^T \left[\mathbf{X}^T \Sigma_{\mathbf{Y}} \mathbf{X} \right] (\hat{\boldsymbol{\beta}} - \boldsymbol{\beta}_0). \end{aligned}$$

Simple null hypotheses do not play a big role in the application of linear models, but can be used to construct confidence sets for $\boldsymbol{\beta}$ (Section 14.7) by inverting the hypothesis test. Note that the statistic X^2_{score} does not require the estimate $\hat{\boldsymbol{\beta}}$, so that with a large enough sample size useful inference statements can be made without having to fit the model. The problem will be revisited in Example 14.2. ■

14.3.4 Pearson's χ^2 Test for Goodness-of-Fit

Consider a multinomial random vector $\mathbf{N} = (N_1, \ldots, N_m)$ with frequencies $\mathbf{P} = (\theta_1, \ldots, \theta_m)$, with $\sum_i N_i = n$. We have constraint $\boldsymbol{\theta}_1 + \ldots + \boldsymbol{\theta}_m = 1$, so we reduce the parameter to $\boldsymbol{\theta} = (\theta_1, \ldots, \theta_{m-1})$, and interpret the remaining probability θ_m as a function of $\boldsymbol{\theta}$, that is, $\theta_m = \theta_m(\boldsymbol{\theta}) = 1 - \theta_1 - \ldots - \theta_{m-1}$, and similarly $n_m = n_m(\mathbf{n}) = 1 - n_1 - \ldots - n_{m-1}$, where $\mathbf{n} = (n_1, \ldots, n_{m-1})$. Then exponentiating gives

$$p_{\boldsymbol{\theta}}(n_1, \ldots, n_m) \quad = \quad \exp \left\{ \sum_{j=1}^{m-1} n_j \log(\theta_j) + n_m(\mathbf{n}) \log(\theta_m(\boldsymbol{\theta})) \right\} \frac{n!}{\prod_{i=j}^{m} n_j!}.$$

The log-likelihood function is then

$$\ell(\boldsymbol{\theta}; \mathbf{n}) = \sum_{j=1}^{m-1} n_j \log(\theta_j) + n_m(\mathbf{n}) \log(\theta_m(\boldsymbol{\theta})),$$

from which we get the score vector

$$\Psi_j(\mathbf{n}, \boldsymbol{\theta}) = \frac{n_j}{\theta_j} - \frac{n_m}{\theta_m}, \; j = 1, \ldots, m-1, \tag{14.9}$$

and Hessian

$$\Psi'_{ij}(\mathbf{n}, \boldsymbol{\theta}) = \left\{ \begin{array}{lcl} -\frac{n_j}{\theta_j^2} - \frac{n_m}{\theta_m^2} & ; & i = j \\ -\frac{n_m}{\theta_m^2} & ; & i \neq j \end{array} \right. , \; i, j = 1, \ldots, m-1. \tag{14.10}$$

From the score vector, we obtain directly the MLEs $\hat{\theta}_j = n_j/n$, $j = 1, \ldots, m-1$.

For a multinomial model, the null hypothesis for the conventional Pearson's goodness-of-fit test or (Pearson's χ^2 test) is any simple hypothesis $H_o : \boldsymbol{\theta} = \boldsymbol{\theta}_0$, with rejection region $X^2_{pearson} > \chi^2_{m-1;\alpha}$, where

$$X^2_{pearson} = \sum_{j=1}^{m} \frac{(N_j - n\theta_j)^2}{n\theta_j}.$$

The statistic $X^2_{pearson}$ is commonly presented in the form

$$X^2_{pearson} = \sum_{j=1}^{m} \frac{(O_j - E_j)^2}{E_j},$$

where $O_j = N_j$ is the observed count for outcome j, and $E_j = n\theta_j$ is the expected count under H_o. We will derive the LRT, Wald and Score test statistics, to observe their relation to $X^2_{pearson}$.

Pearson's χ^2 Test and the LRT

For the LRT, we have

$$X^2_{lrt} = 2\left[\ell(\hat{\boldsymbol{\theta}}; \mathbf{N}) - \ell(\boldsymbol{\theta}_0; \mathbf{N})\right] = 2\sum_{j=1}^{m} N_j \log\left(\frac{N_j}{n\theta_j}\right) = 2\sum_{j=1}^{m} O_j \log\left(\frac{O_j}{E_j}\right).$$

If we use a Taylor's series approximation $\log(1+x) = x - x^2/2 + O(x^3)$ we have

$$\log\left(\frac{O_j}{E_j}\right) = \frac{O_j - E_j}{E_j} - \frac{1}{2}\left[\frac{O_j - E_j}{E_j}\right]^2 + O_p(|\hat{\theta}_j - \theta_j|^3),$$

where the order of the error term holds when θ_j is bounded away from zero. Substituting back into the summation, we get the approximation

$$\begin{aligned}
X^2_{lrt} &= 2\sum_{j=1}^{m} O_j \left\{\frac{O_j - E_j}{E_j} - \frac{1}{2}\left[\frac{O_j - E_j}{E_j}\right]^2\right\} + O_p(n \max_j |\hat{\theta}_j - \theta_j|^3) \\
&= 2\sum_{i=1}^{m} \frac{O_j}{E_j}\left\{(O_j - E_j) - \frac{1}{2}\frac{(O_j - E_j)^2}{E_j}\right\} + O_p(n \max_j |\hat{\theta}_j - \theta_j|^3) \\
&= 2\sum_{i=1}^{m} \left[1 + \frac{O_j - E_j}{E_j}\right]\left\{(O_j - E_j) - \frac{1}{2}\frac{(O_j - E_j)^2}{E_j}\right\} + O_p(n \max_j |\hat{\theta}_j - \theta_j|^3) \\
&= 2\sum_{i=1}^{m} (O_j - E_j) + \frac{(O_j - E_j)^2}{E_j} - \frac{1}{2}\frac{(O_j - E_j)^2}{E_j} + O_p(n \max_j |\hat{\theta}_j - \theta_j|^3) \\
&= \sum_{i=1}^{m} \frac{(O_j - E_j)^2}{E_j} + O_p(n \max_j |\hat{\theta}_j - \theta_j|^3),
\end{aligned}$$

after noting that $\sum_j O_j = \sum_j E_j = n$. We may then verify that $O_p(n \max_j |\hat{\theta}_j - \theta_j|^3) = O_p(n^{-1/2})$. Thus, $X^2_{pearson}$ can be interpreted as a quadratic approximation of X^2_{lrt}.

Pearson's χ^2 Test and the Wald test

We next consider the Wald statistic, for which we need the MLEs $\hat{\theta}_j = n_j/n$ and the information matrix $\mathbf{I}(\boldsymbol{\theta})$, the elements of which follow directly from the Hessian matrix given in Equation (14.10), in particular

$$\{\mathbf{I}(\boldsymbol{\theta})\}_{ij} = \begin{cases} \frac{n}{\theta_j} + \frac{n}{\theta_m} & ; \quad i = j \\ \frac{n}{\theta_m} & ; \quad i \neq j \end{cases}. \tag{14.11}$$

This gives

$$\begin{aligned} X^2_{wald} &= (\hat{\boldsymbol{\theta}} - \boldsymbol{\theta}_0)^T \mathbf{I}(\boldsymbol{\theta})(\hat{\boldsymbol{\theta}} - \boldsymbol{\theta}_0) \\ &= n \sum_{j=1}^{m-1} \frac{(\hat{\theta}_j - \theta_j)^2}{\theta_j} + n \sum_{i=1}^{m-1} \sum_{j=1}^{m-1} \frac{(\hat{\theta}_i - \theta_i)(\hat{\theta}_j - \theta_j)}{\theta_k}. \end{aligned}$$

But the double summation may be expressed as a product, giving

$$\begin{aligned} \sum_{i=1}^{m-1} \sum_{j=1}^{m-1} (\hat{\theta}_i - \theta_i)(\hat{\theta}_j - \theta_j) &= \left\{ \sum_{i=1}^{m-1} \hat{\theta}_i - \theta_i \right\}^2 \\ &= \left\{ (1 - \hat{\theta}_m) - (1 - \theta_m) \right\}^2 \\ &= (\hat{\theta}_m - \theta_m)^2. \end{aligned}$$

This means

$$X^2_{wald} = n \sum_{j=1}^{m} \frac{(\hat{\theta}_j - \theta_j)^2}{\theta_j} = \sum_{i=1}^{m} \frac{(O_j - E_j)^2}{E_j} = X^2_{pearson},$$

so that the Pearson χ^2 test is equivalent to the Wald test.

Pearson's χ^2 Test and the Score Test

Finally, we consider the score test. This requires the inverse of the information matrix $\mathbf{I}(\boldsymbol{\theta})^{-1}$. For the multinomial distribution we have $\mathbf{I}(\boldsymbol{\theta})^{-1} = \text{var}[\hat{\boldsymbol{\theta}}] = \text{var}[\mathbf{N}]/n^2$ from which it follows that

$$\{\text{var}[\mathbf{N}]\}_{ij} = \begin{cases} n\theta_j(1 - \theta_j) & ; \quad i = j \\ -n\theta_i\theta_j & ; \quad i \neq j \end{cases}. \tag{14.12}$$

The score vector is given in Equation (14.9), and the terms of $\mathbf{I}(\boldsymbol{\theta})^{-1}$ are given by Equation (14.12) divided by n^2. Evaluating X^2_{score} is then a matter of evaluating the quadratic form of Equation (14.3). This can be simplified by the following observation. Define vectors $\boldsymbol{v} = (\hat{\theta}_1/\theta_1, \ldots, \hat{\theta}_m/\theta_m)$, $\boldsymbol{u} = (\hat{\theta}_m/\theta_m)\mathbf{1} \in \mathcal{M}_{m,1}$. Then

$$X^2_{score} = [\boldsymbol{v} - \boldsymbol{u}]^T \text{var}[\mathbf{N}][\boldsymbol{v} - \boldsymbol{u}] = \boldsymbol{v}^T \text{var}[\mathbf{N}]\boldsymbol{v} = n \left\{ \left[\sum_{i=1}^{m} \frac{\hat{\theta}_i^2}{\theta_i} \right] - 1 \right\},$$

following the method of Theorem 2.1. However, this is equivalent to $X^2_{pearson}$, since

$$X^2_{pearson} = n \sum_{j=1}^{m} \frac{(\hat{\theta}_j - \theta_j)^2}{\theta_j} = n \sum_{j=1}^{m} \left[\frac{\hat{\theta}_j^2}{\theta_j} - 2\hat{\theta}_j + \theta_j \right] = X^2_{score}.$$

Thus, the Pearson χ^2 test is equivalent to both the Wald test and the score test, and is an approximation of the LRT.

14.4 Nuisance Parameters and Composite Null Hypotheses

Suppose a multidimensional parameter can be decomposed into $\boldsymbol{\theta} = (\psi, \xi)$, and we are interested in an inference concerning ψ alone, regarding ξ as a nuisance parameter (Section 3.12). We have already considered this problem in, for example, Sections 5.9 and 10.6. The technical problem, as before, is to design a procedure that does not require knowledge of ξ. For finite sample inference, a number of approaches are available. If ξ is a location or scale parameter, then inference can be based on a maximal invariant, which will have a distribution that does not depend on ξ. If there exists a statistic S which is sufficient for ξ but ancillary to ψ, then the inference can be conditional on S. Large sample theory, on the other hand, is usually based on approximations of the multivariate normal distribution, so the approach usually taken is to use procedures which are marginal with respect to ξ.

So far, we have assumed that $\Theta_0 = \{\boldsymbol{\theta}_0\}$ defines a simple hypothesis. Suppose now we wish to test a hypothesis concerning ψ, say

$$H_o : \psi = \psi_0, \text{ against } H_a : \psi \neq \psi_0.$$

Although the hypothesis concerns a specific value $\psi = \psi_0$, H_o is not a simple hypothesis, since under H_o ξ may take any value compatible with parameter space Θ. So we have $\Theta_0 = \{\boldsymbol{\theta} = (\psi, \xi) \in \Theta : \psi = \psi_0\}$, and we must now consider the problem of extending the LRT, Wald test and score test to composite hypotheses of this form.

14.4.1 The LRT for Composite Hypotheses

Evaluation of the LRT statistic requires the MLE over Θ, and the MLE over Θ_0, which we denote, when well defined,

$$\hat{\boldsymbol{\theta}} = (\hat{\psi}, \hat{\xi}) = \text{argmax}_{\boldsymbol{\theta} \in \Theta} \ell(\boldsymbol{\theta}; \tilde{X}) \text{ and } \hat{\boldsymbol{\theta}}_0 = (\psi_0, \hat{\xi}_0) = \text{argmax}_{\boldsymbol{\theta} \in \Theta_0} \ell(\boldsymbol{\theta}; \tilde{X}).$$

Of course, we generally do not expect $\hat{\xi}$ and $\hat{\xi}_0$ to be equal. The LRT statistic can now be written $X^2_{lrt} = 2\left[\ell(\hat{\boldsymbol{\theta}}; \tilde{X}) - \ell(\hat{\boldsymbol{\theta}}_0, \tilde{X})\right]$. This can be thought of as the difference between two LRTs, where both Θ and Θ_0 define *alternative* hypotheses against a simple null hypothesis $H_o : \boldsymbol{\theta} = \boldsymbol{\theta}_0$, where $\boldsymbol{\theta}_0 \in \Theta_0$. Since we are interested in approximating X^2_{lrt} when H_o is true, we can simply assume $\boldsymbol{\theta}_0 = (\psi_0, \xi_0) \in \Theta_0$ is the true parameter, then verify that any approximate distribution does not depend on ξ_0. This can be written

$$X^2_{lrt} = 2\left[\ell(\hat{\boldsymbol{\theta}}; \tilde{X}) - \ell(\boldsymbol{\theta}_0; \tilde{X})\right] - 2\left[\ell(\hat{\boldsymbol{\theta}}_0; \tilde{X}) - \ell(\boldsymbol{\theta}_0; \tilde{X})\right] = X^2_{full} - X^2_{reduced}.$$

We consider X^2_{full} and $X^2_{reduced}$ in turn.

The first LRT tests $H_o : \boldsymbol{\theta} = \boldsymbol{\theta}_0$ against $H_a : \boldsymbol{\theta} \in \Theta - \{\boldsymbol{\theta}_0\}$, which is simply the LRT described in Section 14.3.2. But it was shown for simple hypotheses that $X^2_{lrt} \approx X^2_{score}$, so we use the approximation

$$X^2_{full} \approx n^{-1}\Psi(\tilde{X}, \boldsymbol{\theta}_0)^T \bar{\mathbf{I}}_\infty(\boldsymbol{\theta}_0)^{-1}\Psi(\tilde{X}, \boldsymbol{\theta}_0)$$

from Equation (14.3).

The test statistic $X^2_{reduced}$ can be approximated in a similar way, once we define precisely the likelihood model. The parameter space Θ indexes a family of densities $p_{\boldsymbol{\theta}} = p_{(\psi,\xi)}$. If $\psi = \psi_0$ is fixed, the parameter is now ξ, and the parameter space equivalent to Θ_0 can be written as

$$\Xi_0 = \{\xi : (\psi_0, \xi) \in \Theta_0\}.$$

If $\Theta \subset \mathbb{R}^m$ and $\psi \in \mathbb{R}^q$, then $\Xi_0 \subset \mathbb{R}^{m-q}$.

Note, then, that the score vector can be partitioned by parameter sets,

$$\Psi(\tilde{X}, \boldsymbol{\theta}) = \left[\Psi_\psi(\tilde{X}, \boldsymbol{\theta}), \Psi_\xi(\tilde{X}, \boldsymbol{\theta})\right],$$

where $\Psi_\psi(\tilde{X}, \boldsymbol{\theta})$ and $\Psi_\xi(\tilde{X}, \boldsymbol{\theta})$ are the partial derivatives of $\ell(\boldsymbol{\theta}; \tilde{X})$ with respect to paramaters in ψ and ξ, respectively . We can similarly define the information matrix $\mathbf{I}^\xi(\boldsymbol{\theta}) = \text{var}_{\boldsymbol{\theta}}[\Psi_\xi(\tilde{X}, \boldsymbol{\theta})]$ for $\boldsymbol{\theta} \in \Theta_0$. Thus, for the problem of testing $H_o : \boldsymbol{\theta} = \boldsymbol{\theta}_0$ against $H_a : \boldsymbol{\theta} \in \Theta_0 - \{\boldsymbol{\theta}_0\}$, we have score vector $\Psi_\xi(\tilde{X}, \psi_0, \xi)$ and information $\mathbf{I}^\xi(\psi_0, \xi) \approx n\bar{\mathbf{I}}^\xi_\infty(\psi_0, \xi)$, which can be regarded as functions of ξ rather than $\boldsymbol{\theta}$. We have now defined a LRT with a simple null hypothesis, so we have the approximation

$$X^2_{reduced} \approx \frac{1}{n}\Psi_\xi(\tilde{X}, \psi_0, \xi_0)^T \bar{\mathbf{I}}^\xi_\infty(\boldsymbol{\theta}_0)^{-1} \Psi_\xi(\tilde{X}, \psi_0, \xi_0).$$

Combining X^2_{full} and $X^2_{reduced}$ gives approximation

$$\begin{aligned} X^2_{lrt} \approx & \frac{1}{n}\Psi(\tilde{X}, \boldsymbol{\theta}_0)^T \bar{\mathbf{I}}_\infty(\boldsymbol{\theta}_0)^{-1} \Psi(\tilde{X}, \boldsymbol{\theta}_0) \\ & - \frac{1}{n}\Psi_\xi(\tilde{X}, \psi_0, \xi_0)^T \bar{\mathbf{I}}^\xi_\infty(\boldsymbol{\theta}_0)^{-1} \Psi_\xi(\tilde{X}, \psi_0, \xi_0). \end{aligned} \tag{14.13}$$

To be consistent with our earlier approximation for the simple null hypothesis case, we would like to show that $X^2_{lrt} \sim \chi^2_q$, approximately. The normalized quadratic model of Equation (14.1) applies here, so we may write

$$X^2_{lrt} \approx \sum_{i=1}^m u_i^2 - \sum_{i=q+1}^m v_i^2,$$

where $(u_1, \ldots, u_m)$ and $(v_{q+1}, \ldots, v_m)$ are each orthogonal normal random vectors. If it were possible to show that $\sum_{i=q+1}^m u_i^2 = \sum_{i=q+1}^m v_i^2$ we could immediately conclude that $X^2_{lrt} \sim \chi^2_q$. We cannot make this argument, unfortunately, but we can notice that if $\Psi(\tilde{X}, \boldsymbol{\theta}_0)$ itself was an orthogonal normal random vector then our anticipated result would hold. This is because $\mathbf{I}(\boldsymbol{\theta}) = \text{var}[\Psi(\tilde{X}, \boldsymbol{\theta}_0)]$ and $\Psi_\xi(\tilde{X}, \psi_0, \xi_0)$ is a subvector of $\Psi(\tilde{X}, \boldsymbol{\theta}_0) = [\Psi_\psi(\tilde{X}, \psi_0, \xi_0), \Psi_\xi(\tilde{X}, \psi_0, \xi_0)]$. As a consequence

$$\begin{aligned} X^2_{lrt} &\approx \Psi(\tilde{X}, \boldsymbol{\theta}_0)^T \mathbf{I}(\boldsymbol{\theta}_0)^{-1} \Psi(\tilde{X}, \boldsymbol{\theta}_0) - \Psi_\xi(\tilde{X}, \psi_0, \xi_0)^T \mathbf{I}_\xi(\boldsymbol{\theta}_0)^{-1} \Psi_\xi(\tilde{X}, \psi_0, \xi_0) \\ &= \Psi(\tilde{X}, \boldsymbol{\theta}_0)^T \Psi(\tilde{X}, \boldsymbol{\theta}_0) - \Psi_\xi(\tilde{X}, \psi_0, \xi_0)^T \Psi_\xi(\tilde{X}, \psi_0, \xi_0) \\ &= \Psi_\psi(\tilde{X}, \boldsymbol{\theta}_0)^T \Psi_\psi(\tilde{X}, \boldsymbol{\theta}_0) + \Psi_\xi(\tilde{X}, \boldsymbol{\theta}_0)^T \Psi_\xi(\tilde{X}, \boldsymbol{\theta}_0) \\ &\quad - \Psi_\xi(\tilde{X}, \psi_0, \xi_0)^T \Psi_\xi(\tilde{X}, \psi_0, \xi_0) \\ &= \Psi_\psi(\tilde{X}, \boldsymbol{\theta}_0)^T \Psi_\psi(\tilde{X}, \boldsymbol{\theta}_0) \sim \chi^2_q. \end{aligned}$$

Although this might seem like too restrictive an assumption, it becomes relevant after noting that the LRT is invariant to reparametrization. However, the score and the information matrix do depend on the parametrizaton. Thus, the next task is to define a transformation $\boldsymbol{\eta} = \boldsymbol{\eta}(\boldsymbol{\theta})$ under which the resulting score vector $\Psi^{\boldsymbol{\eta}}(\tilde{X}, \boldsymbol{\eta})$ is orthogonal. To do this, set

$$\boldsymbol{\eta}(\boldsymbol{\theta}) = \mathbf{M}(\boldsymbol{\theta} - \boldsymbol{\theta}_0),$$

where $\mathbf{M} = \mathbf{I}(\boldsymbol{\theta})^{1/2}$, $\mathbf{I}(\boldsymbol{\theta})^{1/2}[\mathbf{I}(\boldsymbol{\theta})^{1/2}]^T = \mathbf{I}(\boldsymbol{\theta})$. Note that $\mathbf{I}(\boldsymbol{\theta})^{1/2}$ can be chosen to be lower triangular (Section B.4.7). Denote the transformed parameter space $\Theta^{\boldsymbol{\eta}} = \boldsymbol{\eta}(\Theta)$, and the equivalent null hypothesis $\Theta^{\boldsymbol{\eta}_0} = \boldsymbol{\eta}(\Theta_0)$. We can always assume that ψ represents the first q coordinates of $\boldsymbol{\theta}$. Then, since $\mathbf{M}$ is lower triangular, we can claim that $\eta_1 = \ldots = \eta_q = 0$ *iff* $\psi = \psi_0$ for any $\boldsymbol{\theta} = (\psi, \xi)$. This means that

$$\Theta_0^{\boldsymbol{\eta}} = \{\boldsymbol{\eta} \in \Theta^{\boldsymbol{\eta}} : \eta_1 = \ldots = \eta_q = 0\}.$$

At this point we may approximate the LRT statistic using the new parametrization. Noting that the inverse transformation is $\boldsymbol{\theta}(\boldsymbol{\eta}) = \boldsymbol{\theta}_0 + \mathbf{M}^{-1}\boldsymbol{\eta}$, the score vector is now

$$\Psi^{\boldsymbol{\eta}}(\tilde{X}, \boldsymbol{\eta}) = \nabla_{\boldsymbol{\eta}} \ell(\boldsymbol{\eta}; \tilde{X}) = \nabla_{\boldsymbol{\eta}} \ell(\boldsymbol{\theta}(\boldsymbol{\eta}); \tilde{X}) = \mathbf{J} \nabla_{\boldsymbol{\theta}} \ell(\boldsymbol{\theta}(\boldsymbol{\eta}); \tilde{X}) = [\mathbf{M}^{-1}]^T \Psi(\tilde{X}, \boldsymbol{\theta}).$$

where $\mathbf{J} = \frac{\partial \boldsymbol{\theta}}{\partial \boldsymbol{\eta}}$ is the Jacobian matrix of the transformation $\boldsymbol{\theta}(\boldsymbol{\eta})$. Then $\Psi^{\boldsymbol{\eta}}(\tilde{X}, \boldsymbol{\eta})$ has covariance matrix

$$\begin{aligned}
\text{var}[\Psi^{\boldsymbol{\eta}}(\tilde{X}, \boldsymbol{\eta})] &= \mathbf{I}(\boldsymbol{\theta})^{1/2} \text{var}_{\boldsymbol{\theta}}[\Psi(\tilde{X}, \boldsymbol{\eta})] \left[\mathbf{I}(\boldsymbol{\theta})^{1/2}\right]^T \\
&= \mathbf{I}(\boldsymbol{\theta})^{1/2} \mathbf{I}(\boldsymbol{\theta})^{-1} \left[\mathbf{I}(\boldsymbol{\theta})^{1/2}\right]^T \\
&= \mathbf{I}_m,
\end{aligned}$$

where $\mathbf{I}_m$ is the $m \times m$ identity matrix. Given that the null hypothesis is defined entirely by setting constant the first q coordinates of $\boldsymbol{\eta}$, we need only identify the subvector $\Psi_2^{\boldsymbol{\eta}}(\tilde{X}, \boldsymbol{\eta})$, defined as the final $m - q$ coordinates of $\Psi^{\boldsymbol{\eta}}(\tilde{X}, \boldsymbol{\eta}) = [\Psi_1^{\boldsymbol{\eta}}(\tilde{X}, \boldsymbol{\eta}), \Psi_2^{\boldsymbol{\eta}}(\tilde{X}, \boldsymbol{\eta})]$. Then $\Psi_2^{\boldsymbol{\eta}}(\tilde{X}, \boldsymbol{\eta})$, is also an orthogonal random vector, and therefore has covariance $\mathbf{I}_2^2(\boldsymbol{\eta}) = \mathbf{I}_{m-q}$. We then reevaluate Equation (14.13) with the new parametrization, which gives the approximation

$$\begin{aligned}
X_{lrt}^2 &\approx \Psi^{\boldsymbol{\eta}}(\tilde{X}, \boldsymbol{\eta}_0)^T \mathbf{I}^{\boldsymbol{\eta}}(\boldsymbol{\eta}_0)^{-1} \Psi^{\boldsymbol{\eta}}(\tilde{X}, \boldsymbol{\eta}_0) - \Psi_2^{\boldsymbol{\eta}}(\tilde{X}, \boldsymbol{\eta}_0)^T \mathbf{I}_2^{\boldsymbol{\eta}}(\boldsymbol{\eta}_0)^{-1} \Psi_2^{\boldsymbol{\eta}}(\tilde{X}, \boldsymbol{\eta}_0) \\
&= \Psi^{\boldsymbol{\eta}}(\tilde{X}, \boldsymbol{\eta}_0)^T \Psi^{\boldsymbol{\eta}}(\tilde{X}, \boldsymbol{\eta}_0) - \Psi_2^{\boldsymbol{\eta}}(\tilde{X}, \boldsymbol{\eta}_0)^T \Psi_2^{\boldsymbol{\eta}}(\tilde{X}, \boldsymbol{\eta}_0) \\
&= \Psi_1^{\boldsymbol{\eta}}(\tilde{X}, \boldsymbol{\eta}_0)^T \Psi_1^{\boldsymbol{\eta}}(\tilde{X}, \boldsymbol{\eta}_0) + \Psi_2^{\boldsymbol{\eta}}(\tilde{X}, \boldsymbol{\eta}_0)^T \Psi_2^{\boldsymbol{\eta}}(\tilde{X}, \boldsymbol{\eta}_0)] \\
&\quad - \Psi_2^{\boldsymbol{\eta}}(\tilde{X}, \boldsymbol{\eta}_0)^T \Psi_2^{\boldsymbol{\eta}}(\tilde{X}, \boldsymbol{\eta}_0) \\
&= \Psi_1^{\boldsymbol{\eta}}(\tilde{X}, \boldsymbol{\eta}_0)^T \Psi_1^{\boldsymbol{\eta}}(\tilde{X}, \boldsymbol{\eta}_0) \sim \chi_q^2,
\end{aligned}$$

thus confirming Wilks' theorem for composite null hypotheses.

14.4.2 The Score Test for Composite Hypotheses

Again, we have $\boldsymbol{\theta} = (\psi, \xi)$, and we wish to test null hypothesis $H_o : \psi = \psi_0$. The score test can be generalized to composite null hypotheses using a simple principle. We can retain the original form of the statistic, except that $\boldsymbol{\theta}_0$ is replaced by $\hat{\boldsymbol{\theta}}_0 = (\psi_0, \hat{\xi}_0)$, the MLE of $\boldsymbol{\theta}$ under the null hypothesis H_o. This gives

$$X_{score}^2 = \Psi_n(\tilde{X}_n, \hat{\boldsymbol{\theta}}_0)^T \mathbf{I}(\hat{\boldsymbol{\theta}}_0)^{-1} \Psi_n(\tilde{X}_n, \hat{\boldsymbol{\theta}}_0). \tag{14.14}$$

Recall that the score vector is partitioned as $\Psi(\tilde{X}, \boldsymbol{\theta}) = [\Psi_\psi(\tilde{X}, \boldsymbol{\theta}), \Psi_\xi(\tilde{X}, \boldsymbol{\theta})]$. Note, then, that $\hat{\boldsymbol{\theta}}_0$ would be evaluated by solving equation

$$\Psi_\xi(\tilde{X}, \psi_0, \xi) = 0$$

with respect to ξ, holding $\psi = \psi_0$ fixed. This means $\Psi(\tilde{X}, \hat{\boldsymbol{\theta}}_0) = [\Psi_\psi(\tilde{X}, \hat{\boldsymbol{\theta}}_0), 0]$, so X_{score}^2 becomes

$$\begin{aligned}
X_{score}^2 &= [\Psi_\psi(\tilde{X}, \hat{\boldsymbol{\theta}}_0), 0]^T \mathbf{I}(\hat{\boldsymbol{\theta}}_0)^{-1} [\Psi_\psi(\tilde{X}, \hat{\boldsymbol{\theta}}_0), 0] \\
&= \Psi_\psi(\tilde{X}, \hat{\boldsymbol{\theta}}_0)^T \left[\mathbf{I}(\hat{\boldsymbol{\theta}}_0)^{-1}\right]_{\psi\psi} \Psi_\psi(\tilde{X}, \hat{\boldsymbol{\theta}}_0) \qquad (14.15)
\end{aligned}$$

where $\left[\mathbf{I}(\hat{\boldsymbol{\theta}}_0)^{-1}\right]_{\psi\psi}$ is the block submatrix of $\mathbf{I}(\hat{\boldsymbol{\theta}}_0)^{-1}$ corresponding to the dimensions of ψ.

Suppose we are given the block form of a matrix,

$$\mathbf{M} = \begin{bmatrix} \mathbf{A} & \mathbf{B} \\ \mathbf{C} & \mathbf{D} \end{bmatrix},$$

where $\mathbf{A}$ and $\mathbf{D}$ are square matrices. The Schur complement of block $\mathbf{D}$ is $\mathbf{A} - \mathbf{B}\mathbf{D}^{-1}\mathbf{C}$, assuming $\mathbf{D}^{-1}$ exists. Then $(\mathbf{A} - \mathbf{B}\mathbf{D}^{-1}\mathbf{C})^{-1}$ is the upper left block submatrix of $\mathbf{M}^{-1}$. This means if we let

$$\mathbf{I}_{\psi|\xi}(\boldsymbol{\theta}) = \mathbf{I}_{\psi\psi}(\boldsymbol{\theta}) - \mathbf{I}_{\psi\xi}(\boldsymbol{\theta})\mathbf{I}_{\xi\xi}(\boldsymbol{\theta})^{-1}\mathbf{I}_{\xi\psi}(\boldsymbol{\theta}) \tag{14.16}$$

then

$$\left[\mathbf{I}(\boldsymbol{\theta})^{-1}\right]_{\psi\psi} = \mathbf{I}_{\psi|\xi}(\boldsymbol{\theta})^{-1}$$

and the score test statistic can be written

$$X^2_{score} = \Psi_\psi(\tilde{X}, \hat{\boldsymbol{\theta}}_0)^T \mathbf{I}_{\psi|\xi}(\hat{\boldsymbol{\theta}}_0)^{-1} \Psi_\psi(\tilde{X}, \hat{\boldsymbol{\theta}}_0).$$

However, if $\Psi(\tilde{X}, \boldsymbol{\theta})$ was Gaussian, we can recognize $\mathbf{I}_{\psi|\xi}(\boldsymbol{\theta})$ as the conditional covariance of $\Psi_\psi(\tilde{X}, \boldsymbol{\theta}) \mid \Psi_\xi(\tilde{X}, \boldsymbol{\theta}) = 0$, which is how we must interpret $\Psi_\psi(\tilde{X}, \hat{\boldsymbol{\theta}}_0)$ (Example 7.4). Thus $\bar{\mathbf{I}}_{\psi|\xi}(\hat{\boldsymbol{\theta}}_0) = n^{-1}\mathbf{I}_{\psi|\xi}(\hat{\boldsymbol{\theta}}_0)$ is the asymptotic variance of $\Psi_\psi(\tilde{X}, \hat{\boldsymbol{\theta}}_0)$, so we have the quadratic form (14.1) as an asymptotic approximation, which leads to

$$X^2_{score} \xrightarrow{d} \chi^2_q.$$

Example 14.2 (Score Test for Logistic Regression Gradient) Large sample tests for GLMs were constructed in Example 14.1 for simple null hypotheses. Here we demonstrate how the score test for that case can be modified for a composite hypothesis.

Suppose we are given a random vector $\mathbf{Y} = (Y_1, \ldots, Y_n)^T$ of independent observations $Y_i \sim bern(\mu_i)$. Let $\mathbf{x} = (x_1, \ldots, x_n)^T$ be a vector of independent variables. For a simple logistic regression model (Section 13.6) we have

$$\mu_i = \frac{e^{\beta_0 + \beta_1 x_i}}{1 + e^{\beta_0 + \beta_1 x_i}}, \quad i = 1, \ldots, n,$$

where $\boldsymbol{\beta} = (\beta_0, \beta_1)^T$ is the parameter to be estimated (set dispersion parameter $\phi = 1$). Then let $\mathbf{X} = [\mathbf{1}\ \mathbf{x}] \in \mathcal{M}_{n,2}$ be the design matrix. Logistic regression is a GLM with a canonical link function, so the score function is

$$\Psi(\mathbf{Y}, \boldsymbol{\beta}) = \mathbf{X}^T[\mathbf{Y} - \boldsymbol{\mu}], \tag{14.17}$$

$\boldsymbol{\mu} = \mathbf{X}\boldsymbol{\beta}$, with information matrix $\mathbf{I}(\boldsymbol{\beta}) = \mathbf{X}^T\Sigma_{\mathbf{Y}}(\boldsymbol{\beta})\mathbf{X}$, where $\Sigma_{\mathbf{Y}}(\boldsymbol{\beta})$ is the $n \times n$ diagonal matrix with entries $\{\Sigma_{\mathbf{Y}}\}_{ii} = \mu_i(1 - \mu_i)$.

Next, suppose we wish to test null hypothesis $H_o : \beta_1 = 0$. Comparing the notation of the logistic regression model to that used in this section, we set $\boldsymbol{\theta} = (\beta_0, \beta_1)$. The nuisance parameter is $\xi = \beta_0$, and the parameter of interest is $\psi = \beta_1$, with null hypothesis $H_o : \psi = 0$. Under H_o, the model is easily fit, the constrained MLE derived from $\hat{\mu}_i \equiv \hat{p}_0 = \bar{Y}$ (under H_o, $Y_i \sim bern(p_0)$ for some p_0). This leads to $\hat{\boldsymbol{\theta}}_0 = (\log(\hat{p}_0/(1 - \hat{p}_0)), 0)$. Then

$$\mathbf{I}(\hat{\boldsymbol{\theta}}_0) = \hat{p}_0(1 - \hat{p}_0)\begin{bmatrix} n & \sum_i x_i \\ \sum_i x_i & \sum_i x_i^2 \end{bmatrix}. \tag{14.18}$$

We next identify the elements of Equation (14.15). From Equation (14.17) we have

$$\Psi_{\beta_1}(\mathbf{Y}, \hat{\boldsymbol{\theta}}_0) = \sum_{i=1} x_i(Y_i - \hat{p}_0). \tag{14.19}$$

Applying, for example, the Schur complement method to Equation (14.18) we can write the submatrix of Equation (14.16) as

$$\mathbf{I}_{\beta_1|\beta_0}(\hat{\boldsymbol{\theta}}_0) = \hat{p}_0(1-\hat{p}_0)SS_{\mathbf{x}},$$

where $SS_{\mathbf{x}} = \sum_{i=1}(x_i - \bar{x})^2$. This gives the score statistic

$$X^2_{score} = \frac{\left[\sum_{i=1} x_i(Y_i - \hat{p}_0)\right]^2}{\hat{p}_0(1-\hat{p}_0)SS_{\mathbf{x}}}, \tag{14.20}$$

which under H_o will have an approximate χ^2_1 distribution.

Of course, the form of X^2_{score} given in Equation (14.15) was obtained simply by replacing $\boldsymbol{\theta}_0$ with $\hat{\boldsymbol{\theta}}_0$ in Equation (14.3), which is the form of X^2_{score} for the simple null hypothesis $H_o : \boldsymbol{\theta} = \boldsymbol{\theta}_0$. To do the same thing for this example, we first note that

$$\Psi(\mathbf{Y}, \hat{\boldsymbol{\theta}}_0) = \begin{bmatrix} \Psi_{\beta_0}(\mathbf{Y}, \hat{\boldsymbol{\theta}}_0) \\ \Psi_{\beta_1}(\mathbf{Y}, \hat{\boldsymbol{\theta}}_0) \end{bmatrix} = \begin{bmatrix} 0 \\ \sum_{i=1} x_i(Y_i - \hat{p}_0) \end{bmatrix}. \tag{14.21}$$

Then applying Equations (14.18) and (14.21) directly to Equation (14.14) yields Equation (14.20), using essentially the same calculation needed to apply the form of Equation (14.15) to this problem. See also Problem 14.5. ■

14.4.3 The Wald Test for Composite Hypotheses

Generalization of the Wald test to composite hypotheses $H_o : \psi = \psi_0$ is largely a matter of evaluating the MLE $\hat{\boldsymbol{\theta}} = (\hat{\psi}, \hat{\xi})$ over the full model Θ, then marginalizing to the dimensions of ψ. This means the test is based on $\hat{\psi}$ rather than $\hat{\psi}_0$. Then, since $\mathbf{I}(\boldsymbol{\theta})^{-1}$ is the approximate covariance of $\boldsymbol{\theta}$, we have, assuming $\boldsymbol{\theta}_0$ is the true parameter,

$$\text{var}_{\boldsymbol{\theta}_0}[\hat{\psi}] \approx \left[\mathbf{I}(\boldsymbol{\theta}_0)^{-1}\right]_{\psi\psi} = \mathbf{I}_{\psi|\xi}(\boldsymbol{\theta}_0)^{-1}$$

following the discussion in Section 14.4.2. We then have approximation $\boldsymbol{\theta} \approx \hat{\boldsymbol{\theta}}$, then accepting the quadratic form (14.1) as an asymptotic approximation, the Wald statistic becomes

$$X^2_{wald} = [\sqrt{n}(\hat{\psi} - \psi_0)]^T \bar{\mathbf{I}}_{\psi|\xi}(\hat{\boldsymbol{\theta}})[\sqrt{n}(\hat{\psi} - \psi_0)] \sim \chi^2_d. \tag{14.22}$$

A typical application follows.

Example 14.3 (Wald Test for Logistic Regression Gradient) Here we derive the Wald test for the problem considered in Example 14.2. In contrast to the score test, the Wald test is constructed from the unconstrained MLE $\hat{\boldsymbol{\theta}} = (\hat{\beta}_0, \hat{\beta}_1)$. Like the score test, we require the matrix block $\mathbf{I}_{\psi|\xi}(\hat{\boldsymbol{\theta}})$, except that it is evaluated using the unconstrained MLE, which will differ from $\mathbf{I}_{\psi|\xi}(\hat{\boldsymbol{\theta}}_0)$ evaluated at the constrained MLE $\hat{\boldsymbol{\theta}}_0$ as required by the score test. Then the Wald test statistic becomes

$$X^2_{wald} = \hat{\beta}_1^2 \mathbf{I}_{\beta_1|\beta_0}(\hat{\boldsymbol{\theta}}),$$

which under H_o will have an approximate χ^2_1 distribution. Usually, $\bar{\mathbf{I}}_{\psi|\xi}(\hat{\boldsymbol{\theta}})$ will not have a convenient closed form representation as does $\bar{\mathbf{I}}_{\psi|\xi}(\hat{\boldsymbol{\theta}}_0)$. Whether or not this is a disadvantage depends on the application (Section 14.6). ■

14.5 Pearson's χ^2 Test for Independence in Contingency Tables

Consider an $r \times c$ contingency table (Section 3.9). Each cell i, j of the table (i is the row, j is the column) labels an event C_{ij}. A single trial results in the occurrence of exactly one event, each with probability $p_{ij} = P(C_{ij})$. Then n independent trials result in the counts N_{ij} of the number of occurrences of C_{ij}. The array $\mathbf{N} = \{N_{ij}\}$ can be interpreted as a multnomial random vector which happens to employ double subscripting.

There is often interest in determining whether or not the row and column events are independent. In particular, the row and column events are

$$A_i = \cup_{j=1}^{c} C_{ij}, \ i = 1, \dots, r$$
$$B_j = \cup_{i=1}^{r} C_{ij}, \ j = 1, \dots, c.$$

The row and column probabilities are given by

$$\alpha_i = P(A_i) = \sum_{j=1}^{c} p_{ij}, \ i = 1, \dots, r,$$
$$\beta_j = P(B_j) = \sum_{i=1}^{r} p_{ij}, \ i = 1, \dots, c.$$

Then $C_{ij} = A_i \cap B_j$, and independence of the rows and columns means that $P(C_{ij}) = P(A_i)P(B_j)$. This can be expressed as the null hypothesis

$$H_o : p_{ij} = \alpha_i \beta_j, \ i = 1, \dots, r, \ j = 1, \dots, c. \tag{14.23}$$

Since $\mathbf{N}$ is a multinomial random vector, the dimension of the unconstrained parameter space Θ is the number of outcome minus one, equal to $rc - 1$. The parameter itself is $\boldsymbol{\theta} = \{p_{ij} : i < r \text{ or } j < c\}$.

The hypothesis H_o cannot be written in the form $H_o : \psi = \psi_0$ where ψ is a subset of parameters. However, if we use the score test, we do not need to create a new parametrization which allows this, we need only determine the MLE of $\boldsymbol{\theta}$ under H_o (this will be shown in Section 14.6). We first note that under H_o, all parameters p_{ij} are functions of the row and column probabilities α_i, β_j. It is easily verified that the MLEs of α_i, β_j under H_o are $\hat{\alpha}_i = N_{i \cdot}/n$, $\hat{\beta}_j = N_{\cdot j}/n$, where $N_{i \cdot}, N_{\cdot j}$ denote the row and column totals. Therefore, the MLE of $\boldsymbol{\theta}$ under H_o is $\hat{\boldsymbol{\theta}}_0 = \{\hat{\alpha}_i \hat{\beta}_j : i < r \text{ or } j < c\}$.

We have shown earlier that for a simple null hypothesis $H_o : \boldsymbol{\theta} = \boldsymbol{\theta}_0$ based on a multinomial random vector $\mathbf{N}$ with probabilites $\boldsymbol{\theta}$, the score statistic is equivalent to the Pearson χ^2 statistic

$$X^2_{pearson} = \sum_{j=1}^{m} \frac{(O_j - E_j)^2}{E_j}, \tag{14.24}$$

where $O_j = N_j$ is the observed count for outcome j, and $E_j = n\theta_j$ is the expected count under H_o. Since the score test is transformation invariant (Section 14.6), all we need to do to test a composite null hypothesis such as that of Equation (14.23) is to replace the parameters in Equation (14.24) (which define a simple hypothesis) with $\hat{\boldsymbol{\theta}}_0$, the MLE of $\boldsymbol{\theta}$ under the composite null hypothesis of interest. In our example this becomes

$$X^2_{pearson} = \sum_{i=1}^{r} \sum_{j=1}^{c} \frac{(O_{ij} - E_{ij})^2}{E_{ij}}, \tag{14.25}$$

where $O_{ij} = n_{ij}$ is the observed count for cell i, j, and $E_{ij} = n\hat{\alpha}_i\hat{\beta}_j$ is the expected count under H_o. The final step is to determine the number of constraints defining H_o. One way to do this is to interpret the row and column probabilities α_i, β_j as nuisance parameters. Since we have constraints $\sum_i \alpha_i = \sum_j \beta_j = 1$, the total number of nuisance parameters is $r+c-2$. There are $rc-1$ parameters in total. The number of remaining parameters (which need not be explicitly identified) is therefore $rc-1-r-c+2 = (r-1)(c-1)$. Fixing these parameters defines H_o, so we have

$$X^2_{pearson} \xrightarrow{d} \chi^2_{(r-1)(c-1)},$$

which is, of course, the well known Pearson's χ^2 test for independence in contingency tables.

14.6 A Comparison of the LR, Wald and Score Tests

The LRT, Wald test and score test are all commonly used in practice, and have competing advantages and disadvantages that typically depend on a particular inference problem. All three have a common asymptotic model based on the normalized quadratic form of Equation (14.1), and as a consequence all have a χ^2_q asymptotic distribution with q degrees of freedom, where q is the number of constraints in $\boldsymbol{\theta}$ needed to define the null hypothesis H_o relative to H_a. However, in practice, for fixed sample sizes the power and accuracy can vary considerably, and a choice for a specific application should be based on reports specific to that application. For example, Wild and Seber (1989) compares the LR, Wald and score tests for nonlinear regression models, while McCulloch *et al.* (2008) contains a comparison of the Wald test and LRT for GLMs.

The computational burdens can also be compared. Both computing an MLE and matrix inversion can be considered. The LRT requires evaluation of both the full and reduced MLE, the Wald test requires the full MLE, while the score test requires only the reduced MLE (thus, for a simple null hypothesis, the score test does not need to evaluate an MLE).

On the other hand, the Wald test may be more convenient in a multivariate application when multiple hypothesis tests are anticipated. Most statistical software packages will output the MLE vector of all parameters of a model, as well as an estimated covariance matrix. Then, using the Wald statistic X^2_{wald}, any hypothesis test for a subset of parameters can be implemented simply by marginalizing the MLE vector and covariance matrix.

The LRT requires no matrix inversion. Both the Wald and score tests make use of the Schur complement

$$\mathbf{I}_{\psi|\xi}(\boldsymbol{\theta}) = \mathbf{I}_{\psi\psi}(\boldsymbol{\theta}) - \mathbf{I}_{\psi\xi}(\boldsymbol{\theta})\mathbf{I}_{\xi\xi}(\boldsymbol{\theta})^{-1}\mathbf{I}_{\xi\psi}(\boldsymbol{\theta}).$$

Evaluation of the information matrix $\mathbf{I}(\boldsymbol{\theta})$ is usually straightforward, so the challenge, both computational and analytical, is in the computation of inverses. The Schur complement requires the inversion of the submatrix $\mathbf{I}_{\xi\xi}(\boldsymbol{\theta})$ of $\mathbf{I}(\boldsymbol{\theta})$, unless the null hypothesis is simple. Then the score test, but not the Wald test, requires inversion of $\mathbf{I}_{\psi|\xi}(\boldsymbol{\theta})$. However, it should be noted that $\mathbf{I}^{-1}(\boldsymbol{\theta})$ is, approximately or exactly, the covariance of $\hat{\boldsymbol{\theta}}$. Thus, if $\hat{\boldsymbol{\theta}}$ has a tractable analytic form it may be possible to avoid matrix inversion by setting $\mathbf{I}^{-1}(\boldsymbol{\theta}) = \text{var}_{\boldsymbol{\theta}}[\hat{\boldsymbol{\theta}}]$. In general, it may simplify matters to remember that when the test is given in quadratic form $X^2 = \mathbf{Z}^T\Sigma^{-1}\mathbf{Z}$, then Σ will be the covariance matrix of $\mathbf{Z}$.

One further important point of comparison concerns invariance to parametric transformation. Suppose we have a one-to-one parametric transformation $\boldsymbol{\eta} : \Theta \to \Theta_{\boldsymbol{\eta}}$. We can, in principle, use either parametrization to develop a test statistic, resulting in either $X^2_{\boldsymbol{\theta}}$ or $X^2_{\boldsymbol{\eta}}$. If $X^2_{\boldsymbol{\theta}} = X^2_{\boldsymbol{\eta}}$ then the test is transformation invariant. As we discussed in Section 14.4.1

maximum likelihood estimation, and therefore the LRT is transformation invariant. That the score test is also transformation invariant can be easily shown. Let $\mathbf{J}$ be the Jacobian matrix for the transformation $\boldsymbol{\eta}(\boldsymbol{\theta})$. Then the score vectors under the two parametrizations are related by

$$\Psi(\tilde{X}, \boldsymbol{\eta}) = \mathbf{J}^T \Psi(\tilde{X}, \boldsymbol{\theta})$$

and the information matrices are relate by

$$\mathbf{I}(\boldsymbol{\eta}) = \mathbf{J}^T \mathbf{I}(\boldsymbol{\theta}) \mathbf{J}.$$

This means, for the score statistic we have

$$\begin{aligned} X^2_{\boldsymbol{\eta}} &= \Psi(\tilde{X}, \boldsymbol{\eta})^T \mathbf{I}(\boldsymbol{\eta})^{-1} \Psi(\tilde{X}, \boldsymbol{\eta}) \\ &= \Psi(\tilde{X}, \boldsymbol{\theta})^T \mathbf{J} \left[\mathbf{J}^T \mathbf{I}(\boldsymbol{\theta}) \mathbf{J} \right]^{-1} \mathbf{J}^T \Psi(\tilde{X}, \boldsymbol{\theta}) \\ &= \Psi(\tilde{X}, \boldsymbol{\theta})^T \mathbf{J} \mathbf{J}^{-1} \mathbf{I}(\boldsymbol{\theta})^{-1} [\mathbf{J}^T]^{-1} \mathbf{J}^T \Psi(\tilde{X}, \boldsymbol{\theta}) \\ &= \Psi(\tilde{X}, \boldsymbol{\theta})^T \mathbf{I}(\boldsymbol{\theta})^{-1} \Psi(\tilde{X}, \boldsymbol{\theta}) \\ &= X^2_{\boldsymbol{\theta}}. \end{aligned}$$

However, the Wald test is not transformation invariant (except asymptotically), and this can be an important consideration. A hypothesis may be naturally expressed in a certain parametrization, but the test statistic may be more easily evaluated in another. In this case, the LRT or score test statistic can be evaluated using the more tractable parametrization without altering its intepretation. This approach was used in Section 14.4.1.

In addition, in our development we have assumed that the null hypothesis can be defined simply by constraining a subvector of $\boldsymbol{\theta}$ to be constant. In our development, this idea was expressed by partitioning the parameter $\boldsymbol{\theta} = (\psi, \xi)$, and defining null hypotheses of the form $H_o : \psi = \psi_0$. Of course, this will not always be the case. For example, we may have parameter $\boldsymbol{\theta} = (\boldsymbol{\theta}_1, \boldsymbol{\theta}_2)$ and null hypothesis $H_o : \boldsymbol{\theta}_1 = \boldsymbol{\theta}_2$. However, if we use a transformation invariant test, the asymptotic approximations developed earlier will hold as long as *there exists* a parametric transformation for which the null hypothesis can be written in the form $H_o : \psi = \psi_0$ under the new parametrization. For example, for parameter $\boldsymbol{\theta} = (\boldsymbol{\theta}_1, \boldsymbol{\theta}_2)$ and null hypothesis $H_o : \boldsymbol{\theta}_1 = \boldsymbol{\theta}_2$ we can define the new parameters $\boldsymbol{\eta}_1 = \boldsymbol{\theta}_1$ and $\boldsymbol{\eta}_2 = \boldsymbol{\theta}_2 - \boldsymbol{\theta}_1$, so that H_o is now equivalent to $H_o : \boldsymbol{\eta}_2 = 0$. Of course, the reparametrization need not be carried out, since evaluation of X^2_{lrt}, or X^2_{score} given in the form of Equation (14.14) does not depend on any particular structure of Θ_0. Under any one-to-one transformation, the dimension of the null hypothesis will be the same, say, q, so we will retain the approximation χ^2_q distribution.

14.7 Confidence Sets

The LR, Wald and score tests may be used in the same way to construct confidence sets for a subvector ψ of $\boldsymbol{\theta} = (\psi, \xi)$. In general, confidence sets can be constructed by "inverting a hypothesis test". Suppose we include ψ_0 in a confidence set $C_{1-\alpha}$ *iff* the null hypothesis $H_o : \psi = \psi_0$ is not rejected at a significance level α. If the true parameter is $\boldsymbol{\theta}^* = (\psi^*, \xi^*)$, then $H_o : \psi = \psi^*$ will be rejected with probability no greater than α. Then

$$P_{(\psi,\xi)}(\psi \in C_{1-\alpha}) \geq 1 - \alpha$$

so that $C_{1-\alpha}$ is a true level $1-\alpha$ confidence set. For the LR, Wald and score tests, this type of confidence set is easily constructed by setting

$$C_{1-\alpha} = \left\{ \psi_0 : X^2 \leq \chi^2_{q};\alpha \right\}, \tag{14.26}$$

where $\chi^2_{q;\alpha}$ is the α critical value of the χ^2_q distribution, and q is the number of constraints on $\boldsymbol{\theta}$ defining the null hypothesis H_o.

14.8 Estimating Power for Approximate χ^2 Tests

Power analyses for tests based on the T, F or χ^2 distributions usually depend on the noncentrality parameter (Definition 2.2). We then note that the χ^2 statistics derived from the LRT, score test and Wald test are mutual approximations, so a noncentrality parameter derived from one test will hold for the others. Here, we will use the Wald test.

When assessing the asymptotic power of a test, the alternative hypothesis is usually scaled by n, that is, for a simple hypothesis we consider testing $H_o : \boldsymbol{\theta} = \boldsymbol{\theta}_0$, against $H_a : \boldsymbol{\theta} = \boldsymbol{\theta}_0 + \boldsymbol{\delta}/\sqrt{n}$, for $\boldsymbol{\theta}_0 \in \mathbb{R}^q$, $\boldsymbol{\delta} \in \mathbb{R}^q$. For convenience, set $\boldsymbol{\theta}_n = \boldsymbol{\theta}_0 + \boldsymbol{\delta}/\sqrt{n}$. Suppose we create a new statistic of the form

$$X^2_a = n(\hat{\boldsymbol{\theta}} - \boldsymbol{\theta}_0)^T \bar{\mathbf{I}}(\boldsymbol{\theta}_n)(\hat{\boldsymbol{\theta}} - \boldsymbol{\theta}_0) = [\sqrt{n}(\hat{\boldsymbol{\theta}} - \boldsymbol{\theta}_n) - \boldsymbol{\delta}]^T \bar{\mathbf{I}}(\boldsymbol{\theta}_n)[\sqrt{n}(\hat{\boldsymbol{\theta}} - \boldsymbol{\theta}_n) - \boldsymbol{\delta}].$$

It follows that under H_a we have approximation

$$X^2_a \sim \chi^2_{\Delta;q}, \quad \text{where} \quad \Delta = \boldsymbol{\delta}^T \bar{\mathbf{I}}(\boldsymbol{\theta}_n)\boldsymbol{\delta}. \tag{14.27}$$

But we usually have $\boldsymbol{\theta}_n \overset{i.p.}{\to} \boldsymbol{\theta}_0$, so that under suitable regularity conditions $\bar{\mathbf{I}}(\boldsymbol{\theta}_n) \to \bar{\mathbf{I}}(\boldsymbol{\theta}_0)$, yielding approximation $X^2_{wald} \approx X^2_a$. Thus, Equation (14.27) gives an approximate centrality parameter for the three tests. See problems 14.10 and 14.11.

14.9 Problems

Problem 14.1 Let $\mathbf{X} = (X_1, \ldots, X_n)$ be an *iid* sample with $X_1 \sim pois(\lambda)$.

(a) What is the LRT statistic for hypothesis $H_o : \lambda = \lambda_0$ against $H_a : \lambda \neq \lambda_0$? Express the statistic as an approximate χ^2 statistic.

(b) What is the score test for the same hypotheses? Show that this is equivalent to the Wald test.

(c) Express the LRT and score tests as functions of the MLE $\hat{\lambda} = \bar{X}$. Setting $\lambda_0 = 20$, and $n = 10$, create a plot of both statistics with horizontal axis $\hat{\lambda}$. Given that the Poissson distribution has positive skewness, what advantage might the LRT have over the score test?

(d) When an approximate χ^2 test has rejection region $X^2 > \chi^2_\alpha$, then α is the nominal significance level. Bearing in mind that X^2 need not have an exact χ^2 distribution, the actual significance level is $\alpha_a = P_{\boldsymbol{\theta}_0}(X^2 > \chi^2_\alpha)$. Create a program that simulates the random vector $\mathbf{X}$ for the values given in Part (c), then calculates the LRT and score test statistics. Replicating the simulation $N = 100,000$ times, estimate α_a for each test the value of α_a when $\alpha = 0.05$. What will the margin of error of the estimates be, approximately?

(e) How can we charactize a test for which $\alpha_a < \alpha$, or $\alpha_a > \alpha$.

(f) By modifying your program, estimate the power of each test for the alternative $\lambda = 23$. Is it possible to conclude on the basis of this simulation study that one test is more powerful than the other for this particular alternative? Repeat the question for alternative $\lambda = 17$. **HINT:** We mean, in particular, a more powerful level $\alpha = 0.05$ test.

(g) When the null hypothesis is simple, it is always possible to estimate the critical value need to ensure that $\alpha_a = \alpha$ by simulating the test statistic under the null hypothesis. Using this method, for each test estimate X^2_α for which $P_{\boldsymbol{\theta}_0}(X^2 > X^2_\alpha) = \alpha$. Repeat Part (f) using this critical value in place of χ^2_α. Is it possible to say at this point which test is more powerful?

Problem 14.2 Let $\mathbf{X} = (X_1, \ldots, X_n)$ be an *iid* sample with $X_1 \sim pois(\lambda)$. Recall the Anscombe transformation $g(x) = \sqrt{x + 3/8}$ of Example 12.9. Suppose we wish to test hypothesis $H_o : \lambda = \lambda_0$ against $H_a : \lambda \neq \lambda_0$. If we set statistic $T(\mathbf{X}) = n^{-1}\sum_{i=1} g(X_i)$ we can construct an approximate χ^2 statistic

$$X^2 = n(T(\mathbf{X}) - g(\lambda_0))^2/\sigma_g^2,$$

where σ_g^2 is the approximate variance of $g(X)$ (recall that this does not depend on λ).

(a) Using the methodology of Problem 14.1, estimate the actual significance level α_a of the test which rejects for $X^2 > \chi^2_\alpha$, for $\alpha = 0.05$, assuming $\lambda_0 = 20$, and $n = 10$. Then estimate the power for alternatives $\lambda = 17, 23$.

(b) From Example 12.9 we saw that the bias of $g(X)$ as an estimate of $g(\lambda)$ was approximately $-1/(8g(\lambda))$. Because critical values are determined from the null distribution, we should be able to improve the accuracy of the test statistic X^2 by correcting for bias at λ_0. Create a new statistic X^2_a which does this, and repeat Part (a). What do you find?

(c) Using the methodology of Problem 14.1, Part (g), estimate for each test statistic the critical value X^2_α for which $\alpha_a = \alpha$. Then estimate again the power for alternatives $\lambda = 17, 23$ using these critical values. What do you find?

(d) Does bias correction improve the accuracy of the reported signficance level when critical values from the χ^2 distribution are used? Does bias correction improve the power?

Problem 14.3 Suppose $\mathbf{Y}_1, \ldots, \mathbf{Y}_m$ are m independent samples of size n_j from distribution $pois(\lambda_j)$, for $j = 1, \ldots, m$. Suppose we wish to test hypothesis $H_o : \lambda_1 = \ldots = \lambda_m$ against alternative $H_a : \lambda_i \neq \lambda_j$ for some pair i, j.

(a) Derive the approximate X^2 statistic for the LRT and score test for this hypothesis, and give its approximate distribution (see also Problem 5.17).

(b) Suppose for $m = 4$, $n_1 = \ldots = n_4 = 1$, the single observations from the four samples are observed to be $10, 21, 7, 16$. Estimate the P-values for rejecting H_o for each test.

(c) Suppose we accept the constraint $n_1 = \ldots = n_m = n$ on the sample sizes. Show precisely how we could estimate the sample size parameter n needed to reject an alternative $(\lambda_1, \ldots, \lambda_m)$ with power $1 - \beta$ for a size α test.

Problem 14.4 A probability distribution on the positive integers $1, 2, 3, \ldots, N$ conforms to Zipf's law if the PMF satisfies $p_i \propto 1/i$. It is observed in many types of data. For example, if the frequencies of word occurrences are given in decreasing order, then Zipf's Law is often observed, where p_i is the frequency of the ith most common word. Suppose to test this idea, the frequencies of the four most commonly used words in a certain text are observed to be $162, 77, 63, 47$.

(a) Formulate null and alternative hypotheses to test whether or not thc counts conform to Zipf's Law.
(b) Use an approximate Pearson's χ^2 statistic to carry out the test. What is the P-value?
(c) Yate's correction (Yates, 1934) is often applied to the Pearson's χ^2 statistic, and takes the form

$$X^2 = \sum_i (|O_i - E_i| - 0.5)^2/E_i.$$

Suppose $X \sim bin(n,p)$, and $X^* \sim N(\mu, \sigma^2)$, where $E[X^*] = E[X]$, $\text{var}[X^*] = \text{var}[X]$. Write a simple computer program that accepts numbers (n, p, x) and returns $P(X \leq x)$, $P(X^* \leq x)$ and $P(X^* + 1/2 \leq x)$. Try your program on, for example, $x = 0, 1, \ldots, 10$ for $n = 10$, $p = 1/2$. How does the output motivate Yate's correction?
(d) Repeat Part (b) using Yate's correction. What happens to the P-value?
(e) Does Yate's correction always decrease the P-value?

Problem 14.5 Suppose in Example 14.2 we are able to deduce that $X^2_{score} = K\Psi_{\beta_1}(\mathbf{Y}, \hat{\boldsymbol{\theta}}_0)^2$, where $\Psi_{\beta_1}(\mathbf{Y}, \hat{\boldsymbol{\theta}}_0) = \sum_{i=1} x_i(Y_i - \hat{p}_0)$, the remaining problem being to determine K. Show how K can be deduced by evaluating $\text{var}[\Psi_{\beta_1}(\mathbf{Y}, \hat{\boldsymbol{\theta}}_0)]$.

Problem 14.6 We are given a random vector $\mathbf{Y} = (Y_1, \ldots, Y_n)$ of independent observations $Y_i \sim pois(\mu_i)$. Let $\mathbf{x} = (x_1, \ldots, x_n)$ be a vector of independent variables. For a generalized linear model with Poisson responses and canonical link function, we have $\mu_i = e^{\beta_0 + \beta_1 x_i}$, $i = 1, \ldots, n$. Following Example 14.2 derive the score test statistic for testing $H_o : \beta_1 = 0$ against $H_a : \beta_1 \neq 0$.

Problem 14.7 Consider the negative exponential model of Example 13.8 with predictor term $\eta_i = \beta_0 + \beta_1 x_i$, $i = 1, \ldots, n$. Derive the score test statistic for testing $H_o : \beta_1 = 0$ against $H_a : \beta_1 \neq 0$.

Problem 14.8 Suppose a certain genetic locus possesses two alleles a, A with population frequencies $p_a = 1 - p_A$. Under Hardy-Weinberg equilibrium a genotype Y is a pair of alleles selected independently, so that $P(Y = (aa)) = p_a^2$, $P(Y = (AA)) = p_A^2$, $P(Y = (aA)) = 2p_a p_A$, noting that the pair is unordered. Departure from Hardy-Weinberg equilibrium can be modeled in various ways. A general method is to use the disequilibrium coefficient D_{uv}, which is assigned to each genotype in the following way:

$$P(Y = (uu)) = p_u^2 + D_{uu}, \quad P(Y = (uv)) = 2p_u p_v - 2D_{uv}, \quad u \neq v.$$

(a) Suppose we assume that the marginal allele probabilities cannot depend on any disequilibrium coefficient. Show that for the two allele case this implies $D_{AA} = D_{aa} = D_{aA}$.
(b) Suppose we observe a multinomial random vector $\mathbf{N} = (N_{aa}, N_{AA}, N_{aA})$ of genotype frequencies. Derive a Pearson's χ^2 test statistic $X^2_{pearson}$ for null and alternative hypotheses $H_o : D_{AA} = 0$ and $H_a : D_{AA} \geq 0$. What is its approximate distribution?
(c) Show that $X^2_{pearson}$ is equivalent to $X^2 = n\hat{D}^2_{AA}/(\hat{p}^2_A(1 - \hat{p}_A)^2)$ where $\hat{D}_{AA}, \hat{p}_A$ are the MLEs of D_{AA}, p_A, respectively (Weir, 1996).

Problem 14.9 Suppose $(X_1, \ldots, X_n)$ is an *iid* sample with $X_1 \sim N(\tau, \tau)$, $\tau > 0$, and we wish to test hypotheses $H_o : \tau = \tau_0$ and $H_a : \tau \neq \tau_0$.

(a) Derive the approximate χ^2 statistic for the LRT, score test and Wald test. Verify that each statistic is a function of $S = \sum_i X_i^2$.
(b) Construct a superimposed plot of each statistic as a function of S. Use $\tau_0 = 2$, $n = 10$, and allow $S \in (0.1, 15)$.
(c) Suppose each test was to be expressed as a rejection region of the form $R = \{S \notin (t_1, t_2)\}$. How would the tests compare?

Problem 14.10 Suppose $\boldsymbol{\theta} \in \mathbb{R}^q$ are the frequencies of a multinomial random vector. Suppose we use Pearson's χ^2 statistic $X^2_{pearson}$ to test simple hypothesis $H_o : \boldsymbol{\theta} = \boldsymbol{\theta}_0$ against $H_a : \boldsymbol{\theta} \neq \boldsymbol{\theta}_0$. Show that under the alternative hypothesis $H_a : \boldsymbol{\theta} = \boldsymbol{\theta}_0 + \boldsymbol{\delta}/\sqrt{n}$, $\boldsymbol{\delta} = (\delta_1, \ldots, \delta_q)$ we have the approximate distribution $X^2_{pearson} \sim \chi^2_{\Delta;q-1}$, where

$$\Delta = \sum_{j=1}^{q} \delta_j^2/\theta_j.$$

HINT: First, argue that we must have $\delta_1 + \ldots + \delta_q = 0$. Then the Pearson's χ^2 and Wald test statistics are equivalent.

Problem 14.11 Consider the nonlinear regression model $Y_i = \mu_i(\boldsymbol{\theta}) + \epsilon_i$, $i = 1, \ldots, n$, where $\boldsymbol{\theta} \in \mathbb{R}^q$, and the error terms ϵ_i are *iid* with $\epsilon_i \sim N(0, \sigma^2)$. Assume σ^2 is unknown. Let $\boldsymbol{\Gamma}_{\boldsymbol{\theta}}$ be the matrix of partial derivatives of $\mu_i(\boldsymbol{\theta})$ defined in Section 13.5. Assume rank$(\boldsymbol{\Gamma}_{\boldsymbol{\theta}}) = q$ for all $\boldsymbol{\theta}$ near $\boldsymbol{\theta}_0$.

(a) Show that the score test for hypothesis $H_o : \boldsymbol{\theta} = \boldsymbol{\theta}_0$ is given by

$$X^2 = \frac{[\mathbf{Y} - \boldsymbol{\mu}(\boldsymbol{\theta}_0)]^T \boldsymbol{\Gamma}_{\boldsymbol{\theta}_0} (\boldsymbol{\Gamma}_{\boldsymbol{\theta}_0}^T \boldsymbol{\Gamma}_{\boldsymbol{\theta}_0})^{-1} \boldsymbol{\Gamma}_{\boldsymbol{\theta}_0}^T [\mathbf{Y} - \boldsymbol{\mu}(\boldsymbol{\theta}_0)]}{[\mathbf{Y} - \boldsymbol{\mu}(\boldsymbol{\theta}_0)]^T [\mathbf{Y} - \boldsymbol{\mu}(\boldsymbol{\theta}_0)]/n}.$$

(b) Prove that under H_o the distribution of X^2/q is exactly $F_{q,n}$.
(c) Show that under alternative hypothesis $H_a : \boldsymbol{\theta}_n = \boldsymbol{\theta}_0 + \boldsymbol{\delta}/\sqrt{n}$ we have approximate distribution $X^2 \sim \chi^2_{\Delta;q}$ where $\Delta = \sigma^{-2} \boldsymbol{\delta}^T \boldsymbol{\Gamma}_{\boldsymbol{\theta}_0}^T \boldsymbol{\Gamma}_{\boldsymbol{\theta}_0} \boldsymbol{\delta}/n$ is a noncentrality parameter.

See Wild and Seber (1989) for a discussion of refinements of the score test for this problem.

Problem 14.12 Suppose we are given paired random vectors $\mathbf{X} = (X_1, \ldots, X_n)$, $\mathbf{Y} = (Y_1, \ldots, Y_n)$, where (X_i, Y_i) has a bivariate normal density with means μ_x, μ_y, variances σ_x^2, σ_y^2 and correlation coefficient ρ. Assume all parameters are unknown. Derive the LRT statistic for testing $H_o : \rho = 0$.

Problem 14.13 Suppose $\mathbf{Y}_1, \ldots, \mathbf{Y}_m$ are m independent samples of size n_j from distribution $N(\mu_j, \sigma_j^2)$, for $j = 1, \ldots, m$. Suppose we wish to test hypothesis $H_o : \sigma_1^2 = \ldots = \sigma_m^2$ against alternative $H_a : \sigma_i^2 \neq \sigma_j^2$ for some pair i, j.

(a) Derive the LRT statistic for this hypothesis, and give its approximate distribution.
(b) Let S_j^2 be the sample variance of the sample $\mathbf{Y}_j$, $j = 1, \ldots, m$, and let $S_p^2 = (N - k)^{-1} \sum_{j=1}^{m} (n_i - 1) S_j^2$ be the pooled variance, where $N = n_1 + \ldots + n_m$. Bartlett's test is commonly used for this problem, which is based on statistic

$$X^2_{bartlett} = \kappa^{-1}(N - m)\log(S_p^2) - \sum_{j=1}^{m} (n_j - 1)\log(S_j^2)$$

where $\kappa = 1 + [3(m-1)]^{-1} \sum_{j=1}^{m} (n_j - 1)^{-1} - (N - m)^{-1}$ is a correction factor (Bartlett, 1937). How does the LRT statistic differ from $X^2_{bartlett}$?

A

Parametric Classes of Densities

The following table gives notational conventions for the distributions introduced in Section 1.4ons for commonly used continuous distributions. The density f_X, mean and variance μ_X, σ_X^2 and the moment generating function $m(t)$ are given, as well as formula for higher-order moments. The symbol $\mathcal{S}_X$ refers to the support of the density, which by convention is taken to be an open set. In most cases, a density $f_X(x)$ is nonnegative on the support $x \in \mathcal{S}_X$. The exception is for the binomial (or Bernoulli) density when $p = 0$ or $p = 1$; the Poisson density when $\lambda = 0$, and the negative binomial (or geometric) when $p = 1$. In this cases the support consists of a single outcome. Some densities make use of the gamma function $\Gamma(\alpha)$ and the beta function $B(\alpha, \beta)$ (Section C.5). For integers $1 \leq r \leq n$ the partial factorial is defined as $n^{(r)} = n(n-1)\cdots(n-r+1)$.

Note that the Bernoulli, χ^2, exponential and geometric densities are special cases of other parametric classes listed in this table. This is indicated in their entries.

Bernoulli: $X \sim bern(p)$ equivalent to $X \sim bin(1,p)$ (see **Binomial**)

Beta: $X \sim beta(\alpha,\beta),\ \ \alpha,\beta > 0,\ \ \mu_X = \frac{\alpha}{\alpha+\beta}, \sigma_X^2 = \frac{\alpha\beta}{(\alpha+\beta)^2(\alpha+\beta+1)}$

$f_X(x) = x^{\alpha-1}(1-x)^{\beta-1}/B(\alpha,\beta),\ \ \mathcal{S}_X = (0,1)$

$m(t) = 1 + \sum_{k=1}^{\infty}\left(\prod_{j=0}^{k-1}\frac{\alpha+r}{\alpha+\beta+r}\right)(t^k/k!),\ \ t \in \mathbb{R}$

$E[X^k] = \Gamma(\alpha+\beta)\Gamma(\alpha+k)/(\Gamma(\alpha+\beta+k)\Gamma(\alpha)),\ \ k > 0$

Binomial: $X \sim bin(n,p),\ \ p \in [0,1], n \in \mathbb{I}_{>0},\ \ \mu_X = np, \sigma_X^2 = np(1-p)$

$f_X(x) = \binom{n}{x}p^x(1-p)^{n-x},\ \ \mathcal{S}_X = \{0,1,\ldots,n\}$

$m(t) = (p\exp(t) + (1-p))^n,\ \ t \in \mathbb{R}$

$E[X^{(k)}] = p^k\frac{n!}{(n-k)!},\ \ k \in \{1,2,\ldots,n\}$

Cauchy: $X \sim cauchy(\mu,\tau),\ \ \mu \in \mathbb{R}, \tau \in \mathbb{R}_{>0}$, Moments are undefined

$f_X(x) = \left[\pi(1 + [(x-\mu)/\tau]^2)\right]^{-1},\ \ \mathcal{S}_X = \mathbb{R},\ \ m(t)$ undefined

Moments are undefined

$\boldsymbol{\chi^2}$**:** $X \sim \chi_\nu^2$ equivalent to $X \sim gamma(\nu/2, 1/2)$ (see **Gamma**)

Double Exponential: $X \sim DE(\mu,\tau),\ \ \mu \in \mathbb{R}, \tau \in \mathbb{R}_{>0},\ \ \mu_X = \mu, \sigma_X^2 = 2\tau^2$

$f_X(x) = \frac{\exp(-|x-\mu|/\tau)}{2\tau},\ \ \mathcal{S}_X = \mathbb{R},\ \ m(t) = \frac{\exp(\mu t)}{(1-(t\tau)^2)},\ \ |t| < 1/\tau$

$E[(X-\mu)^k] = \tau^k k!, k$ is even; $E[(X-\mu)^k] = 0, k$ is odd

---

Exponential: $X \sim exp(\tau)$ equivalent to $X \sim gamma(\tau, 1)$ (see **Gamma**)

DOI: 10.1201/9781003049340-A

Gamma: $X \sim gamma(\tau, \alpha),\ \ \tau, \alpha > 0,\ \ \mu_X = \alpha\tau, \sigma_X^2 = \alpha\tau^2$

$f_X(x) = x^{\alpha-1}\frac{\exp(-x/\tau)}{\tau^\alpha\Gamma(\alpha)},\ \ \mathcal{S}_X = \mathbb{R}_{>0},\ \ m(t) = (1-t\tau)^{-\alpha},\ \ t < 1/\tau$

$E[X^k] = \tau^k\Gamma(\alpha+k)/\Gamma(\alpha),\ \ k > 0$

Geometric: $X \sim geom(p)$ equivalent to $X \sim nb(1,p)$ (see **Negative Bin.**)

Inverse Gamma: $X \sim igamma(\xi, \alpha),\ \ \xi, \alpha > 0$

$\mu_X = \xi/(\alpha-1), \alpha > 1,\ \ \sigma_X^2 = \xi^2/[(\alpha-1)^2(\alpha-2)]$

$f_X(x) = \frac{\xi^\alpha\exp(-\xi/x)}{\mu x^\alpha\Gamma(\alpha)},\ \ \mathcal{S}_X = \mathbb{R}_{>0},\ \ m(t)$ undefined

$E[X^k] = \xi^k/[(\alpha-1)\cdots(\alpha-k)],\ \ \alpha > k;\ \ = \infty,\ \ \alpha \le k$

Logistic: $X \sim logistic(\mu, \tau),\ \ \mu \in \mathbb{R}, \tau \in \mathbb{R}_{>0},\ \ \mu_X = \mu, \sigma_X^2 = \tau^2\pi^2/3$

$f_X(x) = \frac{\exp(-(x-\mu)/\tau)}{\tau(1+\exp(-(x-\mu)/\tau))^2},\ \ \mathcal{S}_X = \mathbb{R}$

$m(t) = e^{\mu t}B(1-\tau t, 1+\tau t),\ \ |t| < 1/\tau$

$E[(X-\mu)^k] = \tau^k\int_0^1 \log(u/(1\text{-}u))^k du, k$ is even; $E[(X-\mu)^k] = 0, k$ is odd

---

Negative Binomial: $X \sim nb(r,p),\ \ p \in (0,1],\ \ r \in \mathbb{I}_{>0}$

$\mu_X = r/p,\ \ \sigma_X^2 = rp/(1-p)^2$

$f_X(x) = \binom{x-1}{r-1}p^r(1-p)^{x-r},\ \ \mathcal{S}_X = \{r, r+1, \ldots\}$

$m(t) = \left(\frac{p\exp(t)}{1-(1-p)\exp(t)}\right)^r,\ \ t < -\log(1-p)$

$E[(X-r)^{(k)}] = (r+k)^{(k)}[(1-p)/p]^k, k \ge 1$

Normal: $X \sim N(\mu, \sigma^2),\ \ \mu \in \mathbb{R}, \sigma \in \mathbb{R}_{>0},\ \ \mu \in \mathbb{R},\ \ \mu_X = \mu, \sigma_X^2 = \sigma^2$

$f_X(x) = \frac{\exp\left(-(x-\mu)^2/(2\sigma^2)\right)}{(2\pi\sigma^2)^{1/2}},\ \ \mathcal{S}_X = \mathbb{R},\ \ m(t) = \exp(\mu t + \sigma^2t^2/2),\ \ t \in \mathbb{R}$

$E[(X-\mu)^k] = \sigma^k\frac{k!}{2^{k/2}(k/2)!}, k$ is even; $E[(X-\mu)^k] = 0, k$ is odd

---

Pareto: $X \sim pareto(\tau, \alpha),\ \ \tau, \alpha > 0$

$\mu_X = \frac{\alpha\tau}{(\alpha-1)}, \alpha > 1,\ \ \sigma_X^2 = \frac{\alpha\tau^2}{(\alpha-1)^2(\alpha-2)}, \alpha > 2$

$f_X(x) = \tau^{-1}\alpha/(x/\tau)^{\alpha+1},\ \ \mathcal{S}_X = (\tau, \infty),\ \ m(t)$ undefined

$E[X^k] = \alpha\tau^k/(\alpha-k),\ \ \alpha > k;\ \ = \infty,\ \ \alpha \le k$

Poisson: $X \sim pois(\lambda),\ \ \lambda > 0,\ \ \mu_X = \lambda, \sigma_X^2 = \lambda$

$f_X(x) = \frac{\lambda^x}{x!}\exp(-\lambda),\ \ \mathcal{S}_X = \mathbb{I}_{\ge 0},\ \ m(t) = \exp(\lambda(\exp(t)-1)),\ \ t \in \mathbb{R}$

$E[X^{(k)}] = \lambda^k,\ \ k \ge 1$

Rayleigh: $X \sim rayleigh(\tau)$ equivalent to $X \sim weibull(\tau, 2)$ (see **Weibull**)

Weibull: $X \sim weibull(\tau, \alpha),\ \ \tau, \alpha > 0$

$\mu_X = \tau\Gamma(1+1/\alpha),\ \ \sigma_X^2 = \tau^2\left[\Gamma(1+2/\alpha) - \Gamma(1+1/\alpha)^2\right]$

$f_X(x) = \tau^{-1}(x/\tau)^{\alpha-1}\exp\left(-(x/\tau)^\alpha\right),\ \ \mathcal{S}_X = \mathbb{R}_{>0}$

$m(t) = \sum_{n=0}^\infty \frac{(t\tau)^n}{n!}\Gamma\left(1+\frac{n}{\alpha}\right),\ \ \alpha \ge 1$

$E[X^k] = \tau^k\Gamma(1+k/\alpha),\ \ k > 0$

B

Topics in Linear Algebra

B.1 Numbers

The set of (finite) real numbers is denoted as $\mathbb{R}$, and the set of extended real numbers is denoted as $\bar{\mathbb{R}} = \mathbb{R} \cup \{-\infty, \infty\}$. The corresponding restrictions to nonnegative real numbers are written $\mathbb{R}_{\geq 0} = [0, \infty)$ and $\bar{\mathbb{R}}_{\geq 0} = \mathbb{R}_{\geq 0} \cup \{\infty\}$. The strictly positive real numbers will be denoted as $\mathbb{R}_{>0} = (0, \infty)$, with extension $\bar{\mathbb{R}}_{>0} = \mathbb{R}_{>0} \cup \{\infty\}$. We use standard notation for open, closed, left closed and right closed intervals (a, b), $[a, b]$, $[a, b)$, $(a, b]$.

The set of (finite) integers will be denoted as $\mathbb{I} = \{\ldots, -2, -1, 0, 1, 2, \ldots\}$. The terms "natural" or "whole" number are not consistently defined, so we apply the same notational conventions used above to $\mathbb{I}$, so that $\bar{\mathbb{I}} = \mathbb{I} \cup \{-\infty, \infty\}$, $\mathbb{I}_{\geq 0} = \{0, 1, 2, \ldots\}$, $\bar{\mathbb{I}}_{\geq 0} = \bar{\mathbb{I}}_{\geq 0} \cup \{\infty\}$, $\mathbb{I}_{>0} = \{1, 2, 3, \ldots\}$, $\bar{\mathbb{I}}_{>0} = \mathbb{I}_{>0} \cup \{\infty\}$.

A rational number is any real number expressible as a well defined ratio of integers. The set of all finite rational numbers is denoted as $\mathbb{Q}$. Let $\mathbb{C}$ denote the set of complex numbers $z = a + bi$, $a, b \in \mathbb{R}$, where the imaginary unit i is the solution to $i^2 = -1$. Addition and multiplication in $\mathbb{C}$ is intuitive, so that if $z = a + bi$, $z' = c + di$ we have $z + z' = (a + c) + (b + d)i$ and $zz' = (ac - bd) + (ad + bc)i$. The conjugate of $z = a + bi$ is defined as $\bar{z} = a - bi$.

A field is a set $\mathbb{K}$ of any mathematical objects which can be added, subtracted, multiplied and divided in essentially the same manner as real numbers. In particular, $\mathbb{K}$ is closed under these operations, and possesses additive and multiplicative identities corresponding to 1 and 0 respectively. $\mathbb{R}$, $\mathbb{C}$ and $\mathbb{Q}$ are fields, but $\mathbb{I}$ is not.

For integers $1 \leq r \leq n$ the partial factorial is given by $n^{(r)} = n(n-1)\cdots(n-r+1)$. Then $n^{(n)} = n!$ is the full factorial. For $0 \leq m \leq n$ the binomial coefficient is $\binom{n}{m} = n!/(m!(n-m)!)$.

B.1.1 Polynomials

We can take the class of polynomials in x with coefficients in $\mathbb{K}$ to be denoted $\mathbb{K}[x]$. An order n polynomial in $\mathbb{K}[x]$ is written

$$p_n(x) = \sum_{i=0}^{n} a_i x^i$$

where $a_i \in \mathbb{K}$. The properties of x need not be specified, as long as the powers $1, x, \ldots, x^n$ are distinct. The set of roots of p_n is given by $\{x : p_n(x) = 0\}$.

A field is algebraically closed if every nonconstant polynomial in $\mathbb{K}[x]$ contains at least one root in $\mathbb{K}$. The fundamental theorem of algebra states that the field of complex numbers is algebraically closed, that is, any complex polynomial has at least one complex root. Suppose z_1 is a complex root of $p_n(z) \in \mathbb{C}[x]$. It can be verfied that the factorization

$$p_n(z) = p_{n-1}(z)(z - z_1)$$

DOI: 10.1201/9781003049340-B

holds. Then $p_{n-1}(z)$ also has at least one complex root, say z_2. By induction we always have the factorization

$$p_n(z) = c\prod_{i=1}(z - z_i).$$

where $c \in \mathbb{C}$. This factorization is unique. Thus, an order n complex polynomial has n roots, although the roots need not be unique. Associated with each root z' is its multiplicity m, the largest m admitting the factorization $p_n(z) = p_{n-m}(z)(z - z')^m$.

The field of real numbers is not algebraically closed since, for example, $p_2(x) = x^2+1$ does not have a real root. However, since $\mathbb{R} \subset \mathbb{C}$, an order n polynomial with real coefficients is an element of $\mathbb{C}[x]$, and therefore has n (possibly complex) roots with multiplicities totaling n.

B.2 Equivalence Relations

Suppose we are given a set $\mathcal{X}$. A relation $\mathcal{R}$ is a collection of ordered pairs of elements from $\mathcal{X}$. If $(\tilde{x}, \tilde{y}) \in \mathcal{R}$ we write $\tilde{x} \sim \tilde{y}$, conversely, the statement $\tilde{x} \sim \tilde{y}$ implies $(\tilde{x}, \tilde{y}) \in \mathcal{R}$. A certain type of relation, the equivalence relation, will play a recurring role throughout this volume.

Definition B.1 (Equivalence Relation) A relation $\sim$ on a set $\mathcal{X}$ is an equivalence relation if it satisfies the following three axioms for any $\tilde{x}, \tilde{y}, \tilde{z} \in \mathcal{X}$:

Reflexivity: $\tilde{x} \sim \tilde{x}$.
Symmetry: If $\tilde{x} \sim \tilde{y}$ then $\tilde{y} \sim \tilde{x}$.
Transitivity: If $\tilde{x} \sim \tilde{y}$ and $\tilde{y} \sim \tilde{z}$ then $\tilde{x} \sim \tilde{z}$. ■

Given an equivalence relation, an equivalence class is any set of the form $E_{\tilde{x}} = \{\tilde{y} \in \mathcal{X} \mid \tilde{y} \sim \tilde{x}\}$. If $\tilde{y} \in E_{\tilde{x}}$ then $E_{\tilde{y}} = E_{\tilde{x}}$. Each element $\tilde{x} \in \mathcal{X}$ is in exactly one equivalence class, so $\sim$ induces a partition of $\mathcal{X}$ into equivalence classes. In Euclidean space, 'is parallel to' is an equivalence relation, while 'is perpendicular to' is not.

Example B.1 (Equivalence Relation on $\mathbb{I}$) Let n be any positive integer. Define a relation for $i, j \in \mathbb{I}$ as $i \sim j$ *iff* $i - j$ is a multiple of n. Reflexivity is satisfied by taking 0 to be an integer multiple of n. Symmetry holds, since if k is a multiple of n so is $-k$. To test transitivity, suppose $i, j, k \in \mathbb{I}$, with $i - j = a$, $k - j = b$. If a and b are multiples of n then so is $j - i = (k - j) + (j - i) = a + b$. Hence, transitivy holds. For example, if $n = 3$, there are 3 equivalence classes, $\{\ldots, -6, -3, 0, 3, 6, \ldots\}$, $\{\ldots, -5, -2, 1, 4, 7, \ldots\}$ and $\{\ldots, -4, -1, 2, 5, 8, \ldots\}$. These form a partition of $\mathbb{I}$. ■

Mathematically, an equivalence relation is equivalent to a partition. See the following example.

Example B.2 (Equivalence Classes and Partitions) An equivalence relation can be defined by a partition. Suppose we have partition $\mathcal{X} = \{1, 2, 3, 4, 5, 6\} = \{1, 2, 3\} \cup \{4, 5\} \cup \{6\}$. Include $i \sim j$ in a relation $\mathcal{R}$ *iff* i, j are in the same partition subset. Then

$$\begin{aligned}\mathcal{R} =& \{1 \sim 1, 2 \sim 2, 3 \sim 3, 1 \sim 2, 2 \sim 1, 1 \sim 3, 3 \sim 1, 2 \sim 3, 3 \sim 2\} \\ &\cup \{4 \sim 4, 5 \sim 5, 4 \sim 5, 5 \sim 4\} \cup \{6 \sim 6\}\end{aligned}$$

is an equivalence relation. ■

B.3 Vector Spaces

We introduce here the vector space. The formal definiton is fairly technical (Fraleigh, 1999) but the following informal definition will suite our purposes.

Definition B.2 (Vector Spaces) Let $\mathcal{V}$ be a set of some objects (vectors) for which addition is well defined. Let $\mathcal{F}$ be a field, and assume scalar multiplication of a vector is well defined, that is $a\boldsymbol{v} \in \mathcal{V}$ for any $a \in \mathbb{K}$, $\boldsymbol{v} \in \mathcal{V}$. In particular, $a\boldsymbol{v} = \boldsymbol{v}$ for $a = 1 \in \mathbb{K}$. Then $\mathcal{V}$ is a vector space if for any $\boldsymbol{u}, \boldsymbol{v} \in \mathcal{V}$ and $a, b \in \mathbb{K}$ $a\boldsymbol{u} + b\boldsymbol{v} \in \mathcal{V}$. The elements of $\mathcal{V}$ are referred to as vectors. ■

We are relying on the conventional notion of 'linear combination', under which distributive and associative properties hold, for example, $a\boldsymbol{u} + b\boldsymbol{u} = (a + b)\boldsymbol{u}$, $a\boldsymbol{u} + a\boldsymbol{v} = a(\boldsymbol{u} + \boldsymbol{v})$, $a(b\boldsymbol{u}) = (ab)\boldsymbol{u}$. Addition of vectors is also assumed to be commutative. Note that $\mathcal{V}$ must possess an additive identity $\boldsymbol{0} = 0\boldsymbol{u}$ for any $\boldsymbol{u} \in \mathcal{V}$ (a field must contain a zero), for which $\boldsymbol{u} = \boldsymbol{u} + \boldsymbol{0} = \boldsymbol{0} + \boldsymbol{u}$. Any vector $\boldsymbol{u}$ possesses an inverse $-\boldsymbol{u}$, for which $\boldsymbol{u} + (-\boldsymbol{u}) = (-\boldsymbol{u}) + \boldsymbol{u} = \boldsymbol{0}$. In particular, $-\boldsymbol{u} = a\boldsymbol{u}$ where $a = -1 \in \mathbb{K}$.

B.3.1 Finite Dimensional Vector Spaces

Elements $\mathbf{x}_1, \ldots, \mathbf{x}_m$ of $\mathbb{R}^n$ are linearly independent if $\sum_{i=1}^m a_i \mathbf{x}_i = 0$ implies $a_i = 0$ for all i. Equivalently, no $\mathbf{x}_i$ is a linear combination of the remaining vectors. The span of a set of vectors $\tilde{x} = (\mathbf{x}_1, \ldots, \mathbf{x}_m)$, denoted as $\text{span}(\tilde{x})$, is the set of all linear combinations of vectors in $\tilde{x}$, which must be a vector space. Suppose the vectors in $\tilde{x}$ are not linearly independent. This means that, say, $\mathbf{x}_m$ is a linear combination of the remaining vectors, and so any linear combination in $\text{span}(\tilde{x})$ including $\mathbf{x}_m$ may be replaced with one including only the remaining vectors, so that $\text{span}(\tilde{x}) = \text{span}(\mathbf{x}_1, \ldots, \mathbf{x}_{m-1})$. The dimension of a vector space $\mathcal{V}$ is the minimum number of vectors whose span equals $\mathcal{V}$. Clearly, this equals the size of any set of linearly independent vectors which span $\mathcal{V}$. Any such set of vectors forms a basis for $\mathcal{V}$. Any vector space has a basis.

Suppose $\mathcal{V}$ is the vector space spanned by a basis $\tilde{x} = (\mathbf{x}_1, \ldots, \mathbf{x}_m)$, and that $\mathcal{V}' \subset \mathcal{V}$ is also a vector space. Then any vector $\mathbf{x} \in \mathcal{V}'$ is a linear combination of vectors from $\tilde{x}$, which must therefore be true of any set of vectors forming a basis for $\mathcal{V}'$. This means that the dimension of $\mathcal{V}'$ cannot be greater than that of $\mathcal{V}$, and that $\mathcal{V}' = \mathcal{V}$ if the dimensions are equal. We say $\mathcal{V}'$ is a vector subspace of $\mathcal{V}$. Trivially, $\mathcal{V}$ is a vector subspace of itself, but the term is usually applied to subspaces of strictly smaller dimension.

The inner product (scalar or dot product) is a binary operation defined for vectors $\mathbf{x}, \mathbf{y} \in \mathbb{R}^m$ evaluated by $\mathbf{x} \circ \mathbf{y} = \sum_{i=1}^m x_i y_i$. If $\mathbf{x} \circ \mathbf{y} = 0$ then $\mathbf{x}$ and $\mathbf{y}$ are orthogonal. Then $\mathbf{x}$ is orthogonal to vector space $\mathcal{V}$ if it is orthogonal to every vector in $\mathcal{V}$. This holds *iff* $\mathbf{x}$ is orthogonal to each vector of a basis defining $\mathcal{V}$. Two vector spaces $\mathcal{V}$, $\mathcal{V}'$ are orthogonal if $\mathbf{x} \circ \mathbf{y} = 0$ for all $\mathbf{x} \in \mathcal{V}$, $\mathbf{y} \in \mathcal{V}'$. A set of vectors $(\mathbf{x}_1, \ldots, \mathbf{x}_n)$ is orthogonal if $\mathbf{x}_i \circ \mathbf{x}_j = 0$, $i \neq j$. They are orthonormal if, in addition, $\mathbf{x}_i \circ \mathbf{x}_i = 1$, $i = 1, \ldots, n$.

B.4 Matrices

Let $\mathcal{M}_{n,m}(\mathbb{K})$ denote the set of $n \times m$ matrices $\mathbf{A}$ with elements in $\mathbb{K}$. The square matrices are denoted as $\mathcal{M}_n(\mathbb{K}) = \mathcal{M}_{n,n}(\mathbb{K})$. As it happens, most matrices used in this volume will be real-valued, so the shorthand $\mathcal{M}_{n,m} = \mathcal{M}_{n,m}(\mathbb{R})$ will usually be used.

We will use various notational forms for matrices and their elements, as necessitated by the complexity of their representation. The i,jth element of $\mathbf{A}$ may be written $\mathbf{A}_{i,j} \in \mathbb{K}$, or simply $\mathbf{A}_{ij}$. We also may use the notation $\mathbf{A}_{i,j} = \{\mathbf{A}\}_{i,j} = \{\mathbf{A}\}_{ij}$ when $\mathbf{A}$ is replaced by a more complex expression. It will sometimes be more convenient to reserve a separate symbol for a matrix element, in which case we write $\mathbf{A} = \{a_{i,j}\} = \{a_{ij}\}$, so that $\mathbf{A}$ is a collection of double indexed elements a_{ij}, $i = 1,\ldots,m$, $j = 1,\ldots,n$. In this volume, matrices will be usually be written using bold font. This notation will also be used to represent multidimensional arrays. For example $\{\mathbf{A}\}_{ijk}$ represents an object of elements $\{a_{ijk}\}$ referenced using three independent indices.

We will often find it convenient to regard a vector $\mathbf{x} \in \mathbb{R}^n$ as a special case of a matrix. In this case $\mathbf{x}$ will be a column vector $\mathbf{x} \in \mathcal{M}_{n,1}$. As a matrix, a vector will be written in bold font, with lower case characters reserved for vectors. We may then interpret a matrix in $\mathcal{M}_{m,n}$ as an ordered set of m row vectors or n column vectors.

Linear combinations of matrices of common dimension are well defined. If $\mathbf{A}_j \in \mathcal{M}_{n,m}(\mathbb{K})$, $c_j \in \mathbb{K}$, $j = 1,\ldots,k$, then we may write $\mathbf{B} = c_1\mathbf{A}_1 + \ldots + c_k\mathbf{A}_k$, where $\{\mathbf{B}\}_{i,j} = c_1 \{\mathbf{A}_1\}_{i,j} + \ldots + c_k \{\mathbf{A}_k\}_{i,j}$. Thus, any $\mathcal{M}_{n,m}(\mathbb{K})$ is a vector space.

For $\mathbf{A} \in \mathcal{M}_{n,k}(\mathbb{K})$, $\mathbf{B} \in \mathcal{M}_{k,m}(\mathbb{K})$ we always understand matrix multiplication to mean that $\mathbf{C} = \mathbf{AB}$ possesses elements $\mathbf{C}_{ij} = \sum_{k'=1}^{k} \mathbf{A}_{ik'}\mathbf{B}_{k'j}$. Matrix multiplication is generally not commutative. Repeated multiplication of square matrices can be represented using power notation, that is, $\mathbf{A}^2 = \mathbf{AA}$, which may be extended iteratively to $\mathbf{A}^k = \mathbf{AA}^{k-1}$.

The identity matrix will be written $\mathbf{I}_n \in \mathcal{M}_n$. All elements of the matrices $\mathbf{0}_{n,m} \in \mathcal{M}_{n,m}$ and $\mathbf{1}_{n,m} \in \mathcal{M}_{n,m}$ equal 0 and 1, respectively. When the dimension of a matrix is easily deduced, we may write instead $\mathbf{I}, \mathbf{0}$ or $\mathbf{1}$.

We say that an ordered set of matrices are conformal when they possess the dimensions required for a given matrix expression involving addition or multiplication to be well defined (the matrices then conform).

The transpose $\mathbf{A}^T \in \mathcal{M}_{m,n}(\mathbb{K})$ of a matrix $\mathbf{A} \in \mathcal{M}_{n,m}(\mathbb{K})$ has elements $\mathbf{A}^T_{ji} = \mathbf{A}_{ij}$. We say $\mathbf{A} \in \mathcal{M}_n$ as a symmetric matrix if $\mathbf{A} = \mathbf{A}^T$. Then $(\mathbf{A}_1\mathbf{A}_2\cdots\mathbf{A}_{k-1}\mathbf{A}_k)^T = \mathbf{A}_k^T\mathbf{A}_{k-1}^T\cdots\mathbf{A}_2^T\mathbf{A}_1^T$ for conformal matrices.

It is important to note that the transpose operation is usually reserved for real matrices. When $\mathbf{A} \in \mathcal{M}_{n,m}(\mathbb{C})$ the conjugate matrix is written $\bar{\mathbf{A}}$, and is the element-wise complex conjugate of $\mathbf{A}$. The conjugate transpose (or Hermitian adjoint) of $\mathbf{A}$ is $\mathbf{A}^* = \bar{\mathbf{A}}^T$.

The diagonal of a matrix $\mathbf{A} \in \mathcal{M}_n(\mathbb{K})$ is the vector of diagonal elements $\text{diag}(\mathbf{A}) = (\mathbf{A}_{11},\ldots,\mathbf{A}_{nn})$. Then $\mathbf{A}$ is a diagonal matrix if the only nonzero elements are on the diagonal.

In this volume, vector spaces are usually composed of column vectors, and analyzed using matrix algebra. In general, for any matrix $\mathbf{X}$, we will take $\mathcal{V}_{\mathbf{X}}$ as the vector space spanned by the column vectors of $\mathbf{X}$. Note that any vector space $\mathcal{V} \in \mathbb{R}^n$ can be constructed this way by setting the column vectors of $\mathbf{X}$ equal to any set of basis vectors. If $\mathbf{a}, \mathbf{b} \in \mathcal{M}_{n,1}$ are two column vectors, their inner product may be written $\mathbf{a} \circ \mathbf{b} = \mathbf{a}^T\mathbf{b}$.

A matrix $\mathbf{Q} \in \mathcal{M}_n(\mathbb{R})$ is orthogonal if $\mathbf{Q}^T\mathbf{Q} = \mathbf{QQ}^T = \mathbf{I}$. Equivalently, $\mathbf{Q}$ is orthogonal if and only: (i) it's column vectors are orthonormal; (ii) it's row vectors are orthonormal; (iii) it possesses inverse $\mathbf{Q}^{-1} = \mathbf{Q}^T$. Similary, a matrix $\mathbf{Q} \in \mathcal{M}_n(\mathbb{C})$ is unitary if $\mathbf{Q}^*\mathbf{Q} = \mathbf{QQ}^* = \mathbf{I}$.

B.4.1 Determinant, Trace and Rank of a Matrix

Suppose $\mathbf{A} \in \mathcal{M}_{n,m}$ and let $\alpha \subset \{1,\ldots,n\}$, $\beta \subset \{1,\ldots,m\}$ be any two nonempty subsets of indices. Then $\mathbf{A}[\alpha,\beta] \in \mathcal{M}_{|\alpha|,|\beta|}$ is the submatrix of $\mathbf{A}$ obtained by deleting all elements except for $\mathbf{A}_{i,j}$, $i \in \alpha$, $j \in \beta$. If $\mathbf{A} \in \mathcal{M}_n$ then $\mathbf{A}[\alpha,\alpha]$ is a principal submatrix.

The determinant associates a scalar with $\mathbf{A} \in \mathcal{M}_m$ through the recursive formula

$$\det(\mathbf{A}) = \sum_{i=1}(-1)^{i+j}\mathbf{A}_{i,j}\det(\mathbf{A}^{i,j}) = \sum_{j=1}(-1)^{i+j}\mathbf{A}_{i,j}\det(\mathbf{A}^{i,j})$$

where $\mathbf{A}^{i,j} \in \mathcal{M}_{m-1}$ is the matrix obtained by deleting the ith row and jth column of $\mathbf{A}$. Note that in the respective expressions any j or i may be chosen, yielding the same number, although the choice may have implications for computational efficiency. As is well known, for $\mathbf{A} \in \mathcal{M}_1$ we have $\det(\mathbf{A}) = \mathbf{A}_{1,1}$ and for $\mathbf{A} \in \mathcal{M}_2$ we have $\det(\mathbf{A}) = \mathbf{A}_{1,1}\mathbf{A}_{2,2} - \mathbf{A}_{1,2}\mathbf{A}_{2,1}$. In general, $\det(\mathbf{A}^T) = \det(\mathbf{A})$, $\det(\mathbf{AB}) = \det(\mathbf{A})\det(\mathbf{B})$, $\det(\mathbf{I}) = 1$ which implies $\det(\mathbf{A}^{-1}) = 1/\det(\mathbf{A})$ when the inverse exists.

The trace of a matrix $\mathbf{A} \in \mathcal{M}_n$, denoted as trace($\mathbf{A}$), is defined as the sum of the diagonal elements. We will make use of the following trace identities.

Theorem B.1 (Trace Identities) Suppose $\mathbf{A} \in \mathcal{M}_n$, $\mathbf{A}' \in \mathcal{M}_n$, $\mathbf{B} \in \mathcal{M}_{m,n}$, $\mathbf{C} \in \mathcal{M}_{n,m}$. The following statements hold.

(i) $\mathrm{trace}(\mathbf{A}) = \mathrm{trace}(\mathbf{A}^T)$.
(ii) $\mathrm{trace}(\mathbf{A} + \mathbf{A}') = \mathrm{trace}(\mathbf{A}) + \mathrm{trace}(\mathbf{A}')$.
(iii) $\mathrm{trace}(\mathbf{BC}) = \mathrm{trace}(\mathbf{CB})$. ∎

The rank of a matrix $\mathbf{A} \in \mathcal{M}_{m,n}$, denoted as rank($\mathbf{A}$), is the largest number of column vectors which form a linearly independent set. It may be shown that $\mathrm{rank}(\mathbf{A}) = \mathrm{rank}(\mathbf{A}^T)$, so we may replace column vectors with row vectors in the definition. The following theorem lists some important rank identities and inequalities (Horn and Johnson, 1985).

Theorem B.2 (Rank Identities and Inequalities) Let $\mathbf{A} \in \mathcal{M}_{m,n}$. The following statements hold.

(i) $\mathrm{rank}(\mathbf{A}) \leq \min(m, n)$.
(ii) For any $\mathbf{A} \in \mathcal{M}_{m,n}$, $\mathrm{rank}(\mathbf{A}) = \mathrm{rank}(\mathbf{A}^T)$.
(iii) For any $\mathbf{A} \in \mathcal{M}_{m,n}$, $\mathrm{rank}(\mathbf{A}^T\mathbf{A}) = \mathrm{rank}(\mathbf{A})$.
(iv) If $\mathbf{B} \in \mathcal{M}_{k,m}$ is of rank m, then $\mathrm{rank}(\mathbf{BA}) = \mathrm{rank}(\mathbf{A})$.
(v) If $\mathbf{C} \in \mathcal{M}_{n,k}$ is of rank n, then $\mathrm{rank}(\mathbf{AC}) = \mathrm{rank}(\mathbf{A})$. ∎

If $\mathrm{rank}(\mathbf{A}) = \min(m, n)$, then $\mathbf{A}$ is a full rank matrix. The rank of a diagonal matrix is equal to the number of nonzero diagonal entries.

B.4.2 Singular and Nonsingular Matrices

A large number of algorithms are associated with the problem of determining a solution $\mathbf{x} \in \mathcal{M}_{n,1}(\mathbb{K})$ to the linear systems of equations $\mathbf{Ax} = \mathbf{b}$ for some fixed $\mathbf{A} \in \mathcal{M}_n(\mathbb{K})$, $b \in \mathcal{M}_{n,1}(\mathbb{K})$. Any matrix $\mathbf{A}^{-1}$ for which $\mathbf{AA}^{-1} = \mathbf{A}^{-1}\mathbf{A} = \mathbf{I}$ is the inverse matrix of $\mathbf{A}$. It need not exist, but if it does it is unique. Then $\mathbf{x} = \mathbf{A}^{-1}\mathbf{b}$ is the unique solution to $\mathbf{Ax} = \mathbf{b}$.

Theorem B.3 (Characterization of Nonsingular Matrices) The following statements are equivalent for $\mathbf{A} \in \mathcal{M}_n(\mathbb{K})$, and a matrix satisfying any one is referred to as nonsingular or invertible, any other matrix then being singular.

(i) $\mathbf{A}^{-1}$ exists.
(ii) The columns vectors of $\mathbf{A}$ are linearly independent.
(iii) The row vectors of $\mathbf{A}$ are linearly independent.
(iv) $\det(\mathbf{A}) \neq 0$.
(v) $\mathrm{rank}(\mathbf{A}) = n$

(vi) $\mathbf{Ax} = \mathbf{b}$ possesses a unique solution for any fixed $\mathbf{b}$.
(vii) $\mathbf{x} = \mathbf{0}$ is the only solution of $\mathbf{Ax} = \mathbf{0}$. ■

Clearly, if $\mathbf{A}$ is nonsingular so is $\mathbf{A}^{-1}$ (since $\mathbf{A}$ is the inverse of $\mathbf{A}^{-1}$). In addition, an inverse of a real matrix must be real. To see this suppose, by contradiction, that $\mathbf{A}(\mathbf{B} + i\mathbf{C}) = \mathbf{I}$, where $\mathbf{A}, \mathbf{B}, \mathbf{C}$ are square real-valued matrices, and $\mathbf{A}$ is nonsingular. Then if $\mathbf{A}^{-1} = (\mathbf{B} + i\mathbf{C})$ we must have $\mathbf{A}^{-1}\mathbf{C} = \mathbf{0}$, which by Theorem B.3 implies $\mathbf{C} = \mathbf{0}$.

B.4.3 Eigenvalues and Spectral Decomposition

For $\mathbf{A} \in \mathcal{M}_n(\mathbb{R})$, $\mathbf{x} \in \mathcal{M}_{n,1}(\mathbb{C})$, and $\lambda \in \mathbb{C}$ we may define the eigenvalue equation

$$\mathbf{Ax} = \lambda\mathbf{x}, \tag{B.1}$$

and if the pair $(\lambda, \mathbf{x})$ is a solution for which $\mathbf{x} \neq \mathbf{0}$, then λ is an eigenvalue of $\mathbf{A}$ and $\mathbf{x}$ is an associated eigenvector of λ. Any such solution $(\lambda, \mathbf{x})$ may be called an eigenpair. Clearly, if $\mathbf{x}$ is an eigenvector, so is any nonzero scalar multiple. Let R_λ be the set of all eigenvectors $\mathbf{x}$ associated with λ, If $\mathbf{x}, \mathbf{y} \in R_\lambda$ then $a\mathbf{x} + b\mathbf{y} \in R_\lambda$, so that R_λ is a vector space. The dimension of R_λ is known as the geometric multiplicity of λ. We may refer to R_λ as an eigenspace (or eigenmanifold). In general, the spectral properties of a matrix are those pertaining to the set of eigenvalues and eigenvectors.

The eigenvalue equation (B.1) may be written $(\mathbf{A} - \lambda\mathbf{I})\mathbf{x} = 0$. However, by Theorem B.3 this has a nonzero solution if and only if $\mathbf{A} - \lambda\mathbf{I}$ is singular, which occurs if and only if $p_{\mathbf{A}}(\lambda) = \det(\mathbf{A} - \lambda\mathbf{I}) = 0$. By the construction of a determinant, $p_{\mathbf{A}}(\lambda)$ is an order n polynomial in λ, known as the characteristic polynomial of $\mathbf{A}$. The set of all eigenvalues of $\mathbf{A}$ is equivalent to the set of solutions to the characteristic equation $p_{\mathbf{A}}(\lambda) = 0$. The multiplicity of an eigenvalue λ as a root of $p_{\mathbf{A}}(\lambda)$ is referred to as its algebraic multiplicity. A simple eigenvalue has algebraic multiplicity 1. The geometric multiplicity of an eigenvalue can be less, but never more, than the algebraic multiplicity. A matrix with equal algebraic and geometric multiplicities for each eigenvalue is a nondefective matrix, and is otherwise a defective matrix.

We denote the set of all eigenvalues as $\sigma(\mathbf{A})$ (this includes multiplicities). An important fact is that $\sigma(\mathbf{A}^k)$ consists exactly of the eigenvalues $\sigma(\mathbf{A})$ raised to the kth power, since if $(\lambda, \mathbf{x})$ solves $\mathbf{Ax} = \lambda\mathbf{x}$, then $\mathbf{A}^2\mathbf{x} = \mathbf{A}\lambda\mathbf{x} = \lambda\mathbf{Ax} = \lambda^2\mathbf{x}$, and so on. A quantity of particular importance is the *spectral radius* $\rho(\mathbf{A}) = \max\{|\lambda| \mid \lambda \in \sigma(\mathbf{A})\}$. There is sometimes interest in ordering the eigenvalues by magnitude. If there exists an eigenvalue $\lambda_1 = \rho(\mathbf{A})$, this is referred to as the principal eigenvalue, and any associated eigenvector is a principal eigenvector.

The following theorem gives a number of important identities involving eigenvalues.

Theorem B.4 (Eigenvalue Identities) Suppose for square matrix $\mathbf{A} \in \mathcal{M}_n$ we have eigenvalues $\lambda_1, \ldots, \lambda_n$. The following identities hold:

(i) $\det(\mathbf{A}) = \prod_{i=1}^n \lambda_i$,
(ii) $\text{trace}(\mathbf{A}) = \sum_{i=1}^n \lambda_i$. ■

Note that in Theorem B.4 the eigenvalues are represented in the same frequency as their algebraic multiplicities. Although eigenvalues of a real matrix can be complex, its trace and determinant are always real.

B.4.4 Symmetric and Positive Semidefinite Matrices

Matrices $\mathbf{A}, \mathbf{B} \in \mathcal{M}_n$ are similar if there exists a nonsingular matrix $\mathbf{S}$ for which $\mathbf{B} = \mathbf{S}^{-1}\mathbf{AS}$. Similarity is an equivalence relation (Section B.2). A matrix is diagonalizable if it is similar to some diagonal matrix $\mathbf{D}$. Diagonalization offers a number of advantages. We always have diagonal factorization $\mathbf{B} = \mathbf{S}^{-1}\mathbf{D}\mathbf{S}$, from which it follows that $\mathbf{B}^k = \mathbf{S}^{-1}\mathbf{D}^k\mathbf{S}$, which is easily evaluated and studied. More generally, diagonalization can make apparent the behavior of a matrix interpreted as a transformation. Suppose in the factorization $\mathbf{B} = \mathbf{S}^{-1}\mathbf{D}\mathbf{S}$ $\mathbf{D}$ is diagonal. Then the action of $\mathbf{B}$ on a vector is decomposed into $\mathbf{S}$ (a change in coordinates), $\mathbf{D}$ (elementwise scalar multiplication) and $\mathbf{S}^{-1}$ (the inverse change in coordinates). It can be shown that a matrix is diagonalizable if and only if it is nondefective.

Real symmetric matrices are encountered frequently in the theory of inference, especially as covariance matrices (Equation (2.1)) and projection matrices (Section B.4.5). They are always diagonalizable and have the important property that all eigenvalues are real-valued (this is not true of all real-valued matrices). Furthermore, a diagonal factorization can be constructed using orthogonal matrices.

Theorem B.5 (Spectral Decomposition of Symmetric Matrices) A matrix $\mathbf{A} \in \mathcal{M}_n$ is symmetric *iff* there exists an orthogonal matrix $\mathbf{Q} \in \mathcal{M}_n$ and diagonal matrix $\mathbf{\Lambda} \in \mathcal{M}_n$ for which $\mathbf{A} = \mathbf{Q}\mathbf{\Lambda}\mathbf{Q}^T$. ∎

Clearly, the matrices $\mathbf{\Lambda}$ and $\mathbf{Q}$ of Theorem B.5 may be identified with the eigenvalues and eigenvectors of $\mathbf{A}$, with n eigenvalue equations given by the respective columns of $\mathbf{A}\mathbf{Q} = \mathbf{Q}\mathbf{\Lambda}$. An important implication of this is that all eigenvalues of a symmetric matrix are real, and eigenvectors may be selected to be orthonormal.

For column vector $\mathbf{x} \in \mathcal{M}_{n,1}$ and symmetric matrix $\mathbf{A} \in \mathcal{M}_n$ we have quadratic form $\mathbf{x}^T\mathbf{A}\mathbf{x}$, which is interpretable either as a 1×1 matrix, or as a scalar, as convenient. If $\mathbf{A}$ is not symmetric, it can be replaced by $\bar{\mathbf{A}} = (\mathbf{A} + \mathbf{A}^T)/2$, since $\mathbf{x}^T\mathbf{A}\mathbf{x} = (\mathbf{x}^T\mathbf{A}\mathbf{x})^T = \mathbf{x}^T\mathbf{A}^T\mathbf{x}$. But $\bar{\mathbf{A}}$ is symmetric, so without loss of generality we may assume that the matrix defining a quadratic form is symmetric.

The utility of matrix algebra to the theory of inference depends a great deal on the following definition.

Definition B.3 (Positive Semidefinite Matrices) Let $\mathbf{A} \in \mathcal{M}_n$ be symmetric. Then $\mathbf{A}$ is a positive semidefinite matrix if $\mathbf{x}^T\mathbf{A}\mathbf{x} \geq 0$ for all nonzero column vectors $\mathbf{x}$. If $\mathbf{x}^T\mathbf{A}\mathbf{x} > 0$, then $\mathbf{A}$ is a positive definite matrix. If the inequality is replaced by $\mathbf{x}^T\mathbf{A}\mathbf{x} \leq 0$ then $\mathbf{A}$ is a negative semidefinite matrix, and if $\mathbf{x}^T\mathbf{A}\mathbf{x} < 0$ then $\mathbf{A}$ is a negative definite matrix. Clearly, if $\mathbf{A}$ is positive (semi)definite, $-\mathbf{A}$ is negative (semi)definite. ∎

It is not hard to verify that $\mathbf{B}^T\mathbf{B}$ is positive semidefinite for any $\mathbf{B} \in \mathcal{M}_{n,m}$. We will make use of the following properties of positive semidefinite matrices.

Theorem B.6 (Eigenvalues of Positive Semidefinite Matrices) A symmetric matrix $\mathbf{A} \in \mathcal{M}_n$ is positive semidefinite *iff* all of its eigenvalues are nonnegative. In addition, $\mathbf{A}$ is positive definite *iff* all of its eigenvalues are positive. The rank of $\mathbf{A}$ equals the number of nonzero eigenvalues. As a consequence, a positive definite matrix is nonsingular. ∎

Theorem B.7 (Roots of Positive Semidefinite Matrices) If $\mathbf{A} \in \mathcal{M}_n$ is positive semidefinite then for any positive integer k there exists a unique positive semidefinite matrix $\mathbf{B} \in \mathcal{M}_n$ of the same rank such that $\mathbf{A} = \mathbf{B}^k$.

Conversely, if $\mathbf{B} \in \mathcal{M}_n$ is positive semidefinite, then for any positive integer k, $\mathbf{A} = \mathbf{B}^k$ is a positive semidefinite matrix of the same rank.

As a consequence, $\mathbf{A}$ is positive semidefinite *iff* there exists a positive semidefinite matrix $\mathbf{B}$ of the same rank such that $\mathbf{A} = \mathbf{B}^2$. In addition, $\mathbf{B}$ is the unique matrix with those properties. ∎

B.4.5 Idempotent Matrices and Projection Matrices

Idempotent matrices are introduced in Section 2.5, and play an important role in a certain class of inference problems (Chapter 6).

Definition B.4 (Idempotent Matrices) A square matrix $\mathbf{H} \in \mathcal{M}_n$ idempotent if $\mathbf{H} = \mathbf{HH}$. ■

We will make use of the following properties of an idemopotent matrix.

Theorem B.8 (Properties of Idempotent Matrices) Suppose $\mathbf{H} \in \mathcal{M}_n$ is an idempotent matrix. Then the following statements hold:

(i) $\mathbf{H}$ is diagonalizable.
(ii) The eigenvalues of $\mathbf{H}$ are zero or one. The number of eigenvalues equal to one is equal to rank($\mathbf{H}$). Therefore, if $\mathbf{H}$ is symmetric it is positive semidefinite.
(iii) trace($\mathbf{H}$) = rank($\mathbf{H}$). ■

Idempotent matrices arise naturally from the following construction.

Theorem B.9 (Construction of Idempotent Matrices) Suppose $\mathbf{B} \in \mathcal{M}_{n,q}$ has rank q. Then $\mathbf{B}^T\mathbf{B}$ is nonsingular, and $\mathbf{H} = \mathbf{B}(\mathbf{B}^T\mathbf{B})^{-1}\mathbf{B}^T$ is a symmetric idempotent matrix of rank q. ■

Under the conditions of Theorem B.9, $\mathbf{H} = \mathbf{B}(\mathbf{B}^T\mathbf{B})^{-1}\mathbf{B}^T$, $\mathbf{B} \in \mathcal{M}_{n,q}$, is a projection matrix. Suppose $\hat{\mathbf{x}} = \mathbf{Hx}$. Then $\hat{\mathbf{x}} = \mathbf{B}(\mathbf{B}^T\mathbf{B})^{-1}\mathbf{B}^T\mathbf{x} = \mathbf{Bb}$ for some $\mathbf{b} \in \mathcal{M}_{q,1}$, so that $\hat{\mathbf{x}} \in \mathcal{V}_\mathbf{B} \subset \mathbb{R}^n$ for any $\mathbf{x}$. If we write the residual

$$\mathbf{e} = \mathbf{x} - \hat{\mathbf{x}} = (\mathbf{I} - \mathbf{H})\mathbf{x}$$

then, since $\mathbf{H}$ is idempotent, $\mathbf{e}^T\hat{\mathbf{x}} = \mathbf{x}^T(\mathbf{I}-\mathbf{H})\mathbf{Hx} = \mathbf{x}^T(\mathbf{H}-\mathbf{H})\mathbf{x} = \mathbf{0}$, that is, $\mathbf{e}$ is orthogonal to $\hat{\mathbf{x}}$. We say $\hat{\mathbf{x}}$ is the orthogonal projection of $\mathbf{x}$ onto $\mathcal{V}_\mathbf{B}$, which is a vector subspace of dimension q. Similarly, $(\mathbf{I} - \mathbf{H})$ is a projection matrix onto $\mathcal{V}_\mathbf{B}^\perp \subset \mathbb{R}^n$, the vector space orthogonal to $\mathcal{V}_\mathbf{B}$ of dimension $n - q$. We say that $\mathbf{x} = \hat{\mathbf{x}} + \mathbf{e}$ is an orthogonal decomposition of $\mathbf{x}$.

Suppose $\mathbf{x}' \in \mathcal{V}_\mathbf{B}$. Then $\mathbf{x}' = \mathbf{Hx}'$, and the mapping $\mathbf{H} : \mathbb{R}^n \to \mathcal{V}_\mathbf{B}$ is surjective. A projection matrix must be unique. To see this suppose $\mathbf{H}_1, \mathbf{H}_2$ are two projection matrices onto $\mathcal{V}_B$. Then $\mathbf{H}_1\mathbf{y} = \mathbf{H}_2\mathbf{y}$ for all $\mathbf{y} \in \mathbb{R}^n$.

B.4.6 Loewner Ordering

Loewner ordering of symmetric matrices is worth noting, since it plays an important role in the Gauss-Markov theorem (Theorem 6.1).

Definition B.5 (Loewner Ordering) Suppose $\mathbf{A}, \mathbf{B} \in \mathcal{M}_n$ are symmetric matrices. Then $\mathbf{A} \geq_L \mathbf{B}$ if $\mathbf{A} - \mathbf{B}$ is positive semidefinite. For strict ordering, we write $\mathbf{A} >_L \mathbf{B}$ if $\mathbf{A} - \mathbf{B}$ is positive definite. This ordering is refered to as Loewner ordering. ■

The importance of Loewner ordering can be seen when comparing quadratic forms. Suppose $\mathbf{A} \geq_L \mathbf{B}$. Then $\mathbf{x}^T\mathbf{Ax} - \mathbf{x}^T\mathbf{Bx} = \mathbf{x}^T(\mathbf{A} - \mathbf{B})\mathbf{x} \geq 0$, with $\mathbf{x}^T\mathbf{Ax} > \mathbf{x}^T\mathbf{Bx}$ for strict ordering $\mathbf{A} >_L \mathbf{B}$.

B.4.7 Matrix Factorization

We will often need to find a factorization $\mathbf{A} = \mathbf{A}_{1/2}\mathbf{A}_{1/2}^T$ of a positive definite matrix $\mathbf{A}$, where $\mathbf{A}_{1/2}$ is refered to as a square root matrix of $\mathbf{A}$. Such a square root matrix is not unique, but by Theorem B.7 a unique *symmetric* square root matrix exists which is also positive definite. Thus, we could always take the symmetric square root matrix as a canonical solution. However, there may be reasons to consider alteratives, typically for computational reasons.

For example, for any nonsingular square matrix $\mathbf{A}$ a LU-decomposition takes the form $\mathbf{A} = \mathbf{LU}$, where $\mathbf{L}$ is lower triangular and $\mathbf{U}$ is upper triangular, and both are nonsingular. With one slight caveat, we can say that an LU-decomposition always exists.

Theorem B.10 (*LU*-Decomposition) Suppose $\mathbf{A}$ is a square nonsingular matrix. Then there exists a factorization

$$\mathbf{A} = \mathbf{PLU}, \tag{B.2}$$

where $\mathbf{P}$ is a permutation matrix, $\mathbf{L}$ is a nonsingular lower triangular and $\mathbf{U}$ is a nonsingular upper triangular matrix. ■

The existence of a permutation matrix in the factorization means that the rows of $\mathbf{A}$ may have to be reordered in order for an LU-decomposition to exist. If $\mathbf{A}$ is positive definite, then something more can be said, in particular, that there exists a unique lower triangular matrix $\mathbf{L}$ such that $\mathbf{A} = \mathbf{LL}^T$ (the Choleski decomposition method is an efficient algorithm for computing this factorization (Chambers, 1977)).

B.5 Dimension of a Subset of $\mathbb{R}^d$

The dimension of a finite dimensional vector space equals the size of any basis. We also wish to define the dimension of subsets other than vector spaces. We first give conditions under which the dimension of $E \subset \mathbb{R}^d$ is d.

Definition B.6 (Dimension of a Subset of $\mathbb{R}^d$) A subset $E \subset \mathbb{R}^d$ has dimension d if every point $\mathbf{x}$ in the interior O_E is contained in a neighborhood $B_\epsilon(\mathbf{x})$ also contained in O_E (Definition C.2). ■

Many subsets of interest can be generated by affine transformations of the form

$$L(\mathbf{x}) = \mathbf{Ax} + \mathbf{b}, \;\; \mathbf{x} \in \mathbb{R}^d \tag{B.3}$$

where $\mathbf{A} \in \mathcal{M}_d$ and $\mathbf{b} \in \mathbb{R}^d$. Suppose E can be expressed

$$E = \{L(\mathbf{x}) : \mathbf{x} \in E^*\} \tag{B.4}$$

where $E^* \subset \mathbb{R}^d$ is a set of dimension d. Then the dimension of E is equal to the rank of $\mathbf{A}$. Furthermore, a countable union $E = \cup_{i \geq 1} E_i$ of sets of dimension $d' \leq d$ is also of dimension d'.

Example B.3 (Dimension of a Probability Simplex) Suppose we are given a subset $E \subset \mathbb{R}^3$ defined by

$$E = \{(p_1, p_2, p_3) \in \mathbb{R}^3 : p_1 + p_2 + p_3 = 1, \;\; \min_i p_i > 0\}.$$

This can be recognized as a probability simplex in $\mathbb{R}^3$. It can be visualized as a triangle with vertices $(0,0,1),(0,1,0),(0,0,1)$. Although it is a subset of $\mathbb{R}^3$, it has zero volume, since it is essentially a two dimensional object. But, E is neither a vector space, nor a subset of any two dimensional vector space. We can, however, represent E using (B.3) and (B.4), setting

$$\mathbf{A} = \begin{bmatrix} 1 & 0 & 0 \\ 0 & 1 & 0 \\ -1 & -1 & 0 \end{bmatrix}, \quad \mathbf{b} = (0,0,1),$$

$$E^* = \{(p_1,p_2,p_3) \in \mathbb{R}^3 : p_1 \geq 0, \ p_2 \geq 0, \ p_1 + p_2 \leq 1\}.$$

Clearly, E^* is of dimension 3, while $\mathbf{A}$ is of rank 2, since the first two columns are linearly independent, but all three columns are not. This verifies that E is of dimension 2. ∎

The dimension of a Euclidean subset is a topological property. Suppose E_1, E_2 are two Euclidean subsets. Suppose there exists a bijective mapping $g : E_1 \rightarrow E_2$ such that g and its inverse g^{-1} are continuous. Then E_1 and E_2 are homeomorphic, or topologically equivalent. In particular, they have the same dimension.

C

Topics in Real Analysis and Measure Theory

C.1 Metric Spaces

A metric can be thought of as a generalization of the notion of Euclidean distance, allowing more flexible notions of distance, while retaining the most important properties.

Definition C.1 (Metric Space) Suppose we have a set of objects $\mathcal{X}$ and a real-valued mapping $d : \mathcal{X} \times \mathcal{X} \to \mathbb{R}$ operating on $\tilde{x}, \tilde{y} \in \mathcal{X}$. Then d is a metric if

(i) $d(\tilde{x}, \tilde{y}) \geq 0$ (nonnegativity);
(ii) $d(\tilde{x}, \tilde{y}) = 0$ if and only if $\tilde{x} = \tilde{y}$ (identifiabilty);
(iii) $d(\tilde{x}, \tilde{y}) = d(\tilde{y}, \tilde{x})$ (symmetry);
(iv) $d(\tilde{x}, \tilde{y}) \leq d(\tilde{x}, \tilde{z}) + d(\tilde{z}, \tilde{y})$ for any $\tilde{z} \in \mathcal{X}$ (triangle inequality).

The pair $(\mathcal{X}, d)$ is refered to as a metric space. ■

Two example of metric spaces follow.

Example C.1 (Euclidean Metric) For $\mathbf{x}, \mathbf{y} \in \mathbb{R}^m$ the Euclidean metric is given by $d(\mathbf{x}, \mathbf{y}) = \sqrt{(x_1 - y_1)^2 + \ldots (x_m - y_m)^2}$. This is interpretable as the distance between two points in $\mathbb{R}^m$. That d satisfies the nonnegativity, identifiability and symmetry axioms of Definition C.1 is clear. A proof that d satisfies the triangle inequality in $\mathbb{R}^2$ is given in Euclid's Elements. It states the intuitive notion that the total distance of travel on a straight line between $\mathbf{x}$ and $\mathbf{y}$ can never be decreased by travelling from first to $\mathbf{z}$ then to $\mathbf{y}$. ■

Example C.2 (Discrete Metric) For $\tilde{x}, \tilde{y} \in \mathcal{X}$ the discrete metric is given by $d(\tilde{x}, \tilde{y}) = I\{\tilde{x} \neq \tilde{y}\}$, which satisfies all the axioms of Definition C.1. Under the discrete metric all subsets of $\mathcal{X}$ are both open and closed (Definition C.2). A sequences $\tilde{x}_1, \tilde{x}_2, \ldots$ converges to $\tilde{x}_0$ in the discrete metric *iff* for some N $\tilde{x}_i = \tilde{x}_0$ for all $i > N$. ■

A metric induces the notion of an open set.

Definition C.2 (Open and Closed Sets) We are given a metric space $(\mathcal{X}, d)$. An (open) ball is any subset of the form $B_\epsilon(\tilde{x}) = \{\tilde{y} \in \mathcal{X} : d(\tilde{x}, \tilde{y}) < \epsilon\}$. A subset $E \subset \mathcal{X}$ is open if for each $\tilde{x} \in E \ \exists \ \epsilon > 0 \ni B_\epsilon(\tilde{x}) \subset E$. Equivalently, E is open *iff* it is union of open balls. The interior of a set B is the largest open set contained in B. The closure of B is the smallest closed set containing B. ■

Any class of sets satisfying the axioms of a topology define a system of open sets.

Definition C.3 (Topological Space) Let $\mathcal{O}$ be a collection of subsets of a set Ω. Then $(\Omega, \mathcal{O})$ is a topological space (or $\mathcal{O}$ is a topology on Ω) if the following axioms hold:

(i) $\Omega \in \mathcal{O}$ and $\emptyset \in \mathcal{O}$,
(ii) if $A, B \in \mathcal{O}$ then $A \cap B \in \mathcal{O}$,
(iii) for any collection of sets $\{A_t\}$ in $\mathcal{O}$ (countable or uncountable) we have $\cup_t A_t \in \mathcal{O}$.

DOI: 10.1201/9781003049340-C

Any element of $\mathcal{O}$ is referred to as open set, and the complement of any set in $\mathcal{O}$ is a closed set. Note also that a set can be both open and closed, in particular $\emptyset$ and Ω. ■

The class of open sets $\mathcal{O}$ of a metric space satisfies the axioms of a toplogy.

C.2 Measure Theory

The following definitions support the development of a measure offered in Section 1.2. The definition of a measure on a set Ω requires a system of subsets known as a σ-field.

Definition C.4 (σ-field) Let $\mathcal{F}$ be a collection of subsets of Ω. Then $\mathcal{F}$ is a σ-field (or σ-algebra) if the following axioms hold:

(i) $\Omega \in \mathcal{F}$,
(ii) if $E \in \Omega$ then $E^c \in \Omega$,
(iii) if $E_1, E_2, \ldots \in \Omega$ then $\cup_i E_i \in \Omega$.

If $\mathcal{F}$ is a σ-field on Ω, then $(\Omega, \mathcal{F})$ is known as a measurable space. ■

We offer a general definition of the countably additive measure. All measures used in this volume will be of this type.

Definition C.5 (Countably Additive Measure) A set function $\mu : \mathcal{F} \to \bar{\mathbb{R}}_{\geq 0}$, where $\mathcal{F}$ is a σ-field on Ω, is a measure if $\mu(\emptyset) = 0$, and it is countably additive, that is for any countable collection of disjoint sets $E_1, E_2, \ldots$ we have $\sum_i \mu(E_i) = \mu\left(\cup_j E_j\right)$. Then $(\Omega, \mathcal{F}, \mu)$ is a measure space. A measure μ on a σ-field is a finite measure if $\mu(\Omega) < \infty$. We then say μ is a σ-finite measure if there exists a countable collection of subsets $E_i \in \mathcal{F}$ such that $\cup_i E_i = \Omega$ with $\mu(E_i) < \infty$. ■

We will require the following two definitions associated with a measure space.

Definition C.6 (Almost Everywhere) Suppose we are given a measure μ on Ω. Any property which refers to Ω is said to hold almost everywhere *wrt* μ if it holds except on a set of measure zero. This is abbreviated *a.e.* $[\mu]$, or simply *a.e.* when the context is clear. ■

Definition C.7 (Absolute Continuity) Let ν and μ be two measures on a measurable space $(\Omega, \mathcal{F})$. If $\mu(E) = 0 \Rightarrow \nu(E) = 0$ for all $E \in \mathcal{F}$, then ν is absolutely continuous with respect to μ. This is written $\nu \ll \mu$, and we also say ν is dominated by μ. If $\nu \ll \mu$ and $\mu \ll \nu$ then ν and μ are equivalent. Conversely ν and μ are singular if there exists $E \in \mathcal{F}$ for which $\nu(E) = \mu(E^c) = 0$, also written $\nu \perp \mu$. ■

Any measure used in this volume will be one of the following four types.

Example C.3 (Probability Measure) A probability measure is any measure P (Definition C.5) on a measurable space $(\Omega, \mathcal{F})$ for which $P(\Omega) = 1$. Then $(\Omega, \mathcal{F}, P)$ is a probability measure space. ■

Example C.4 (Borel Sets and Borel Measures) Given a metric space $(\mathcal{X}, d)$, let τ be the class of all open subsets of $\mathcal{X}$ (recall that the definition of an open set depends on d). Then τ is a topology (Definition C.3). The Borel sets $\mathcal{B}_{\mathcal{X}}$ is the class of all subsets obtainable from countably many set operations on elements of τ, including union, intersection and complementation. It may be show that $\mathcal{B}_{\mathcal{X}}$ is the smallest σ-field containing τ. The measurable space $(\mathcal{X}, \mathcal{B}_{\mathcal{X}})$ is refered to as a Borel space. Then a Borel measure is a measure defined on a Borel space. ■

Example C.5 (Lebesgue Measure) Lebesgue measure μ can be defined from a Borel measure defined on the Euclidean metric space $\mathcal{X} = \mathbb{R}^m$. To define μ, let $R \in \mathbb{R}^m$ be a rectangle with sides defined by intervals I_j of length a_j, $j = 1, \ldots, m$, then set $\mu(R) = \prod_{j=1}^m a_j$. Note that it doesn't matter whether or not the intervals are open or closed. It may be proven that there is only one measure μ defined on $\mathcal{B}$ for which $\mu(R)$ is equal to the volume of R for all rectangles.

It is then possible to extend μ further in the following way. Suppose $\mathcal{N}$ is the class of all null sets, that is $A \in \mathcal{N}$ *iff* $A \subset E$ for some $E \in \mathcal{B}$ for which $\mu(E) = 0$. Then $\mathcal{B} \cup \mathcal{N}$ is a σ-field, and μ can be extended to $\mathcal{B} \cup \mathcal{N}$ by setting $\mu(A) = 0$ for any $A \in \mathcal{N}$. A measure which is defined on its null sets is a complete measure. Then $\mathcal{M} = \mathcal{B} \cup \mathcal{N}$ defines the class of Lebesgue measurable sets, and is strictly larger than $\mathcal{B}$. Note that if $E \subset \mathbb{R}^m$ is of zero volume in $\mathbb{R}^m$ then $\mu(E) = 0$ (for example, a two-dimensional plane in $\mathbb{R}^3$). ∎

Example C.6 (Counting Measure) Suppose S is a countable set. The process of counting is additive in the same sense as volume, and so can be used to define a measure. A single element of S is assigned a volume of 1. Now, suppose $S \subset \Omega$, and let $\mu(E)$ equals the number of elements of S in E. Note that Ω, or E itself, need not be discrete. We can also say that $S \subset \Omega$ are the elements $\omega \in \Omega$ for which $\mu(\{\omega\}) > 0$. The remaining elements of Ω have measure 0. ∎

Given two measurable spaces $(\mathcal{X}, \mathcal{F}_{\mathcal{X}})$, $(\mathcal{Y}, \mathcal{F}_{\mathcal{Y}})$, a mapping $f : \mathcal{X} \to \mathcal{Y}$ is measurable if $f^{-1}(E) \in \mathcal{F}_{\mathcal{X}}$ for all $E \in \mathcal{F}_{\mathcal{Y}}$. Note that the measurablility property refers to both $\mathcal{X}$ and $\mathcal{Y}$.

Throughout most of this volume a function f will be some mapping from a measurable space Ω to a metric space $\mathcal{X}$. In addition, we will assume that f is a measurable mapping from Ω to the Borel sets $\mathcal{B}_{\mathcal{X}}$. If Ω is a Borel space then f is Borel measurable. If Ω is the class of Lebesgue measurable sets on $\mathbb{R}^m$ then f is Lebesgue measurable. Since the Borel sets on $\mathbb{R}^m$ are also Lebesgue measurable, a Borel measurable function on $\mathbb{R}^m$ is also Lebesgue measurable, but the converse does not hold.

It is important to be aware of the notion of measurability, although it will not play a large role in our theory. This is because almost all functions we would consider using are measurable. First note that any continuous function $f : \mathbb{R} \to \mathbb{R}$ is Borel measurable. It suffices to note that any preimage $f^{-1}(E)$ of an open set $E \subset \mathbb{R}$ is open. Then if f is continuous everywhere except on a finite set (i.e. piecewise continuous) it is not much harder to argue that $f^{-1}(E)$ is a Borel set when E is open. This covers most functions encountered in statistical theory.

In the context of a measure space, the notion of the equality of two functions is often usefully reduced to equivalence classes of functions. In particular, given a measure space $(\Omega, \mathcal{F}, \mu)$, following Definition C.6 we might find it useful to consider f and g equal if $f(\tilde{x}) = g(\tilde{x})$ except for $\tilde{x} \in E$ for which $\mu(E) = 0$, that is $f = g$ *a.e.* $[\mu]$.

C.3 Integration

Suppose we have measure space $(\mathcal{X}, \mathcal{F}, \mu)$ and a measurable function $f : \mathcal{X} \to \mathbb{R}$. We say f is a simple function if there is a finite measurable partition $A_1, \ldots, A_m$ of $\mathcal{X}$ and finite constants $a_1, \ldots a_m$ such that $f(\tilde{x}) = \sum_{i=1}^m a_i I\{\tilde{x} \in A_i\}$. We should have little disagreement in defining the integral of a simple function as

$$\int_{\mathcal{X}} f d\mu = \sum_{i=1}^m a_i \mu(A_i).$$

Then, if f is the limit of a sequence of simple functions, the integral of f should be the limit of the integrals of the sequence. Accordingly, for nonnegative f we may define

$$\int_{\mathcal{X}} f d\mu = \sup\left\{ \int_{\mathcal{X}} s d\mu \mid \text{all simple functions } s\text{: } 0 \leq s \leq f \right\}. \tag{C.1}$$

If $E \in \mathcal{F}$ then we let $\int_E f d\mu = \int_{\mathcal{X}} f I_E d\mu$. This defines the Lebesgue integral, which is the standard method of constructing integrals on measure spaces. This contrasts with the Riemann integral, which uses a similar method, except that step functions replace simple functions as the approximators. A function is Riemann integrable with respect to Lebesgue measure on a bounded interval if and only if it is continuous $a.e.[\mu]$, and equals the Lebesgue integral when this is the case (for example, Ash (1972) Theorem 1.7.1).

The definition is extended to general functions by the decomposition $f = f^+ - f^-$, where $f^+ = fI\{f > 0\}$ and $f^- = -fI\{f < 0\}$ are nonnegative, by evaluating

$$\int_{\mathcal{X}} f d\mu = \int_{\mathcal{X}} f^+ d\mu - \int_{\mathcal{X}} f^- d\mu.$$

In this case, we say $\int_{\mathcal{X}} f d\mu$ exists, or is well-defined, if $\int_{\mathcal{X}} |f| d\mu < \infty$.

C.3.1 Convergence of Integrals

Suppose we are given measure space $(\mathcal{X}, \mathcal{F}, \mu)$. If f_n is a sequence of measurable real-valued functions there will often be a need to relate the convergence properties of f_n to those of $\int_{\mathcal{X}} f_n d\mu$. The following theorems are therefore of considerable importance.

Theorem C.1 (Fatou's Lemma) For any sequence $f_n \geq 0$ $a.e.[\mu]$ $\liminf_{n\to\infty} \int_{\mathcal{X}} f_n d\mu \geq \int_{\mathcal{X}} \liminf_{n\to\infty} f_n d\mu$. ■

In many statements of Fatou's lemma the sequence f_n is assumed to possess a limit, but this does not strengthen the conclusion.

Theorem C.2 (Monotone Convergence Theorem) For any sequence $f_n \geq 0$ for which $f_n \uparrow f$ $a.e.[\mu]$ $\lim_{n\to\infty} \int_{\mathcal{X}} f_n d\mu = \int_{\mathcal{X}} f d\mu$. ■

Theorem C.3 (Dominated Convergence Theorem) Let f_n be a sequence with limit f $a.e.[\mu]$. If there exists function $h \geq |f_n|$ $a.e.[\mu]$ for which $\int_{\mathcal{X}} h d\mu < \infty$ then $\lim_{n\to\infty} \int_{\mathcal{X}} f_n d\mu = \int_{\mathcal{X}} f d\mu$. ■

An important special case of the Dominated Convergence Theorem is often stated independently:

Theorem C.4 (Bounded Convergence Theorem) Let f_n be a sequence with limit f $a.e.[\mu]$. Suppose μ is a finite measure, and there exists constant $M \geq |f_n|$ $a.e.[\mu]$. Then $\lim_{n\to\infty} \int_{\mathcal{X}} f_n d\mu = \int_{\mathcal{X}} f d\mu$. ■

C.4 Exchange of Integration and Differentiation

The exchange of integration and differentiation is an operation which arises in several problems in statistical inference. We therefore need to know when this is justified. One version of this problem leads to the well known Leibniz's rule of calculus:

Theorem C.5 (Leibniz's Rule) We are given function $f(\theta, x)$ of $x \in \mathbb{R}$, $\theta \in \Theta \subset \mathbb{R}$. Assume $f(\theta, x)$ is continuous *wrt* in (θ, x), and $\partial f(\theta, x)/\partial\theta$ exists and is continuous *wrt* (θ, x). Then the following equality holds

$$\frac{d}{d\theta}\left[\int_{a(\theta)}^{b(\theta)} f(\theta, x)dx\right] =$$
$$f(\theta, b(\theta))\frac{db(\theta)}{d\theta} - f(\theta, a(\theta))\frac{da(\theta)}{d\theta} + \int_{a(\theta)}^{b(\theta)} \frac{\partial f(\theta, x)}{\partial\theta} dx,$$

for differentiable $a(\theta), b(\theta)$, where the integrals exist. ■

The limits in the equality of Theorem C.5 are assumed finite. This is too restrictive for the applications we will encounter in this volume. However, extension to unbounded integration limits requires additional regularity conditions. We will rely on a version of Leibniz's rule extended to integration over general measures. This is an application of the dominated convergence theorem (Theorem C.3).

Theorem C.6 (Exchange of Integration and Differentiation I) Suppose $\Theta \subset \mathbb{R}$ is an open set, and $(\mathcal{X}, \mathcal{F}, \mu)$ is a measure space. Suppose we are given function $f : \Theta \times \mathcal{X} \to \mathbb{R}$. Suppose we can define the function $F : \Theta \to \mathbb{R}$ as $F(\theta) = \int_\Omega f(\theta, x)d\mu$. Furthermore, suppose the following conditions hold:

(i) The mapping $x \mapsto f(\theta, x)$ is measurable and integrable for all $\theta \in \Theta$.
(ii) The mapping $\theta \mapsto f(\theta, x)$ is differentiable for all $x \in \mathcal{X}$.
(iii) $\left|\frac{\partial f(\theta,x)}{\partial\theta}\right| \leq g(x)$ for all θ, x, for some integrable $g : \mathcal{X} \to \mathbb{R}$.

Then $F(\theta)$ is differentiable, and

$$\frac{dF(\theta)}{d\theta} = \int_\Omega \frac{\partial f(\theta, x)}{\partial\theta} d\mu. \tag{C.2}$$

for any $\theta \in \Theta$. ■

Condition (iii) of Theorem C.6 might seem restrictive. However, we may apply the theorem to each θ separately, setting Θ to be any neighborhood of θ. In this case, an obvious choice for the bounding function would be

$$g(x) = \sup_{\theta' \in (\theta-\epsilon, \theta+\epsilon)} \left|\left\{ \left.\frac{\partial f(\theta, x)}{\partial\theta}\right|_{\theta'} \right\}\right|$$

for any $\epsilon > 0$. This leads to a straightforward condition under which Theorem C.6 holds.

Theorem C.7 (Exchange of Integration and Differentiation II) Suppose $\Theta \subset \mathbb{R}$ is an open set, and $(\mathcal{X}, \mathcal{F}, \mu)$ is a measure space. Suppose we are given function $f : \Theta \times \mathcal{X} \to \mathbb{R}$. Suppose we can define the function $F : \Theta \to \mathbb{R}$ as $F(\theta) = \int_\Omega f(\theta, x)d\mu$. Furthermore, suppose the following conditions hold:

(i) The mapping $x \mapsto f(\theta, x)$ is measurable and integrable for all $\theta \in \Theta$.
(ii) The mapping $\theta \mapsto f(\theta, x)$ is differentiable for all $x \in \mathcal{X}$.
(iii) $\int_\Omega \left|\left\{ \left.\frac{\partial f(\theta,x)}{\partial\theta}\right|_{\theta'} \right\}\right| d\mu < \infty$ for all $\theta' \in \Theta$.
(iv) For all $x \in \mathcal{X}$, $\frac{\partial f(\theta,x)}{\partial\theta}$ is monotone in θ.

Then $F(\theta)$ is differentiable, and

$$\frac{dF(\theta)}{d\theta} = \int_\Omega \frac{\partial f(\theta, x)}{\partial\theta} d\mu. \tag{C.3}$$

for any $\theta \in \Theta$. ■

C.5 The Gamma and Beta Functions

The gamma and beta functions play an important role in a number of inference problems, so it is worth reviewing their properties. First, the gamma function is defined by the definite integral

$$\Gamma(t) = \int_{x=0}^{\infty} x^{t-1}e^{-x}dx, \quad t > 0. \tag{C.4}$$

It can be shown that $\Gamma(t+1) = t\Gamma(t)$, and since $\Gamma(1) = 1$ it follows by a simple induction argument that

$$\Gamma(n) = (n-1)! \tag{C.5}$$

for integers $n = 1, 2, 3, \ldots$. The gamma function can therefore be thought of as a generalization of the factorial. In addition, we have $\Gamma(1/2) = \sqrt{\pi}$, which leads to the recursive relation $\Gamma(n+1/2) = (n-1+1/2)\Gamma(n-1+1/2)$, so that

$$\begin{aligned}\Gamma(n+1/2) &= (n-1+1/2)(n-2+1/2) \times ... \times (1+1/2)(1/2)\sqrt{\pi} \\ &= \frac{(2n-1)!!}{2^n}\sqrt{\pi}\end{aligned} \tag{C.6}$$

for $n = 1, 2, 3, \ldots$, where $n!!$ is the double factorial

$$n!! = \begin{cases} n(n-2) \times \ldots \times 4 \times 2 & ; \quad n \text{ is even} \\ n(n-2) \times \ldots \times 3 \times 1 & ; \quad n \text{ is odd} \end{cases}.$$

Similarly, the beta function is defined by the definite integral

$$B(\alpha, \beta) = \int_{u=0}^{1} u^{\alpha-1}(1-u)^{\beta-1}du, \quad \alpha, \beta > 0.$$

It can be shown that we always have $B(\alpha, \beta) = \Gamma(\alpha)\Gamma(\beta)/\Gamma(\alpha+\beta)$.

C.6 Stirling's Approximation of the Factorial

The factorial $n!$ can be accurately approximated using series expansions. See, for example, Feller (1968) (Chapter 2). Stirling's approximation for the factorial is given by $s_n = (2\pi)^{1/2}n^{n+1/2}e^{-n}$, $n \geq 1$, and if we set $n! = s_n\rho_n$, we have

$$e^{1/(12n+1)} < \rho_n < e^{1/(12n)}. \tag{C.7}$$

The approximation is quite sharp, guaranteeing that (a) $\lim_{n\to\infty} n!/s_n = 1$; (b) $1 < n!/s_n < e^{1/12} < 1.087$ for all $n \geq 1$; (c) $(12n+1)^{-1} < \log(n!) - \log(s_n) < (12n)^{-1}$ for all $n \geq 1$.

C.7 The Gradient Vector and the Hessian Matrix

Positive definite matrices (Section B.4.4) play an important role in the theory of inference, so it is worth singling out some specific properties:

(i) If $\mathbf{A}$ is positive definite (positive semidefinite) then -$\mathbf{A}$ is negative definite (negative semidefinite).
(ii) If $\mathbf{A}$ is positive definite then $\mathbf{x}^T\mathbf{A}\mathbf{x} > 0$ for any $p \times 1$ column vector $\mathbf{x}$.
(iii) The kth leading principal minor of an $p \times p$ matrix $\mathbf{A}$ is the determinant of the submatrix defined by the first k rows and columns. Then $\mathbf{A}$ is positive definite if and only if all leading principle minors are positive.
(iv) A matrix $\mathbf{A}$ is positive definite if and only if it can be written as $\mathbf{A} = \mathbf{B}^T\mathbf{B}$ for some $n \times p$ matrix $\mathbf{B}$ with linearly independent columns.

A positive definite matrix must have strictly positive diagonal entries, but this is not sufficient. By Property (ii) a 2×2 real symmetric matrix is positive definite if and only if both diagonal entries are positive *and* the determinant is positive.

Let $F(\mathbf{x})$ be a twice-differentiable real-valued function of q variables, $\mathbf{x} \in \mathbb{R}^q$. The gradient of F is the vector of first-order partial derivatives

$$\mathbf{F}'(\mathbf{x}) = \nabla F(\mathbf{x}) = \left(\frac{\partial F(\mathbf{x})}{\partial x_1}, \ldots, \frac{\partial F(\mathbf{x})}{\partial x_q}\right) \tag{C.8}$$

The Hessian matrix is defined as a matrix of partial second order derivatives $\mathbf{F}''(\mathbf{x}) \in \mathcal{M}_q$, defined elementwise as

$$\{\mathbf{F}''(\mathbf{x})\}_{jk} = \frac{\partial^2 F(\mathbf{x})}{\partial x_j \partial x_k}, \tag{C.9}$$

which is clearly a real symmetric matrix.

Let $\mathbf{x}_0$ be a stationary (or critical) point of F (that is, $\nabla F(\mathbf{x}_0) = \mathbf{0}$). If $\mathbf{F}''(\mathbf{x}_0)$ is positive definite, then $\mathbf{x}_0$ is a local minimum, and uniquely so in a small enough neighborhood. If the Hessian matrix is negative definite, then $\mathbf{x}_0$ is a locally unique maximum. This must be the case, since a local minimum of F would be a local maximum of $-F$ and a positive definite matrix multiplied by -1 is negative definite. If the Hessian matrix possesses both positive and negative eigenvalues but no nonzero eigenvalues (is of full rank) then $\mathbf{x}_0$ is a saddlepoint, and is not a local extrema. When zero-valued eigenvalues exist, multiple cases are possible. Furthermore, if the domain E of F is convex and if the Hessian matrix is positive semidefinite on the interior of E then F is a convex function. It is strictly convex if the Hessian matrix is in addition positive definite. A strictly convex function possesses at most one local minimum, which must also be a global minimum.

C.8 Normed Vector Spaces

A metric introduces a notion of distance to any set $\mathcal{X}$, while a norm adds a notion of magnitude to a vector space. For a vector $\boldsymbol{u} = (u_1, \ldots, u_m)$ in Euclidean space, this magnitude is taken to be $|\boldsymbol{u}| = \sqrt{u_1^2 + \ldots u_m^2}$. Since subtraction of vectors is defined on a vector space we can define Euclidean distance between vectors $\boldsymbol{u}$ and $\boldsymbol{v}$ as $|\boldsymbol{u} - \boldsymbol{v}|$, which satisfies the axioms of a metric. In much the same way that the metric generalizes Euclidean distance, the norm generalizes Euclidean magnitude.

Definition C.8 (Normed Vector Space) Let $\mathcal{V}$ be a vector space. A real-valued mapping $\|\cdot\| : \mathcal{V} \to \mathbb{R}$ is a norm if for any vectors $\boldsymbol{u}, \boldsymbol{v} \in \mathcal{V}$ and scalar $a \in \mathbb{R}$ the following axioms hold:

(i) $\|a\boldsymbol{u}\| = |a| \, \|\boldsymbol{u}\| \geq 0$ (absolute scalability);

(ii) $\|\boldsymbol{u}\| = 0$ implies $\boldsymbol{u} = \mathbf{0}$ (identifiabilty);
(iii) $\|\boldsymbol{u} + \boldsymbol{v}\| \leq \|\boldsymbol{u}\| + \|\boldsymbol{v}\|$ (triangle inequality).

Note that $\mathbf{0}$ is the zero vector, or additive identity of a vector space. It is not necessary to state that $\|\mathbf{0}\| = 0$ since this is implied by Axiom (i). In addition, that $\|\boldsymbol{u}\| \geq 0$ follows from Axiom (i) ($\|\boldsymbol{u}\| = \|-\boldsymbol{u}\|$) and Axiom (iii) (set $\boldsymbol{v} = -\boldsymbol{u}$). Then $(\mathcal{V}, \|\cdot\|)$ is a normed vector space. ■

If $(\mathcal{X}, \|\cdot\|)$ is a normed vector space, then $(\mathcal{X}, d)$ is a metric space, where $d(\boldsymbol{u}, \boldsymbol{v}) = \|\boldsymbol{u} - \boldsymbol{v}\|$, which in turn induces a topology on $\mathcal{X}$.

C.8.1 Vector p-Norms

The p-norm for a vector $\mathbf{x} \in \mathbb{R}^n$ is defined as $\|\mathbf{x}\|_p = \left(\sum_{i=1}^n |x_i|^p\right)^{1/p}$, $p \geq 1$. In particular $|\mathbf{x}| = \|\mathbf{x}\|_2$ is the Euclidean norm. It may be shown that $\|\cdot\|_p$ satisfies the axioms of Definition C.8.

C.8.2 Matrix Norms

The space of all matrices $\mathcal{M}_n$ is a vector space under the usual linear combination $c\mathbf{A} + d\mathbf{B}$, $\mathbf{A}, \mathbf{B} \in \mathcal{M}_n$, $c, d \in \mathbb{R}$. Then if $\|\cdot\|_\alpha$ is some vector norm on $\mathbb{R}^n$, the quantity

$$\|\mathbf{A}\|_\alpha = \max_{\|\mathbf{x}\|_\alpha = 1} \|\mathbf{A}\mathbf{x}\|_\alpha \tag{C.10}$$

defines a matrix norm for any $\mathbf{A} = \{a_{ij}\} \in \mathcal{M}_n$. It may be verified that $\|\cdot\|_\alpha$ satisfies the axioms of Definition C.8. However, a matrix norm is also submultiplicative. Directly from Equation (C.10) we have the inequality

$$\|\mathbf{A}\mathbf{x}\|_\alpha \leq \|\mathbf{A}\|_\alpha \|\mathbf{x}\|_\alpha . \tag{C.11}$$

Then if $\mathbf{A}, \mathbf{B} \in \mathcal{M}_n$ we may write, applying inequality (C.11) twice,

$$\|\mathbf{A}\mathbf{B}\mathbf{x}\|_\alpha \leq \|\mathbf{A}\|_\alpha \|\mathbf{B}\mathbf{x}\|_\alpha \leq \|\mathbf{A}\|_\alpha \|\mathbf{B}\|_\alpha \|\mathbf{x}\|_\alpha . \tag{C.12}$$

It follows by constraining $\|\mathbf{x}\|_\alpha = 1$ in (C.12) that

$$\|\mathbf{A}\mathbf{B}\|_\alpha \leq \|\mathbf{A}\|_\alpha \|\mathbf{B}\|_\alpha , \tag{C.13}$$

which defines the submultiplicativity property.

As an example, if $\|\cdot\|_\alpha = \|\cdot\|_p$ (any vector p-norm of Section C.8.1) then $\|\mathbf{A}\|_p$ is the p-matrix norm, which evaluates to

$$\|\mathbf{A}\|_p = \max_{\|\mathbf{x}\|_p = 1} \|\mathbf{A}\mathbf{x}\|_p. \tag{C.14}$$

Next, suppose $\mathbf{A}$ is positive semidefinite. Then we may write $\mathbf{A} = \mathbf{Q}^T \mathbf{\Lambda} \mathbf{Q}$, where $\mathbf{Q}$ is an orthogonal matrix, and $\mathbf{\Lambda}$ is a diagonal matrix of the eigenvalues of $\mathbf{A}$. Let $\lambda_{min}(\mathbf{A})$ and $\lambda_{max}(\mathbf{A})$ be the smallest and largest eigenvalues of $\mathbf{A}$ (all of which are nonnegative real numbers). Then

$$\min_{\|\mathbf{x}\|_2 = 1} \mathbf{x}^T \mathbf{A} \mathbf{x} = \min_{\|\mathbf{x}\|_2 = 1} \mathbf{x}^T \mathbf{Q}^T \mathbf{\Lambda} \mathbf{Q} \mathbf{x} = \min_{\|\mathbf{x}\|_2 = 1} \mathbf{x}^T \mathbf{\Lambda} \mathbf{x} = \lambda_{min}(\mathbf{A})$$

by the orthogonality of $\mathbf{Q}$, and by essentially the same argument we have

$$\max_{\|\mathbf{x}\|_2 = 1} \mathbf{x}^T \mathbf{A} \mathbf{x} = \lambda_{max}(\mathbf{A}).$$

We refer to $\|\mathbf{A}\|_2$ as the Euclidean matrix norm (or spectral norm), which by Equation (C.14) can be evaluated as

$$\|\mathbf{A}\|_2 = \max_{\|\mathbf{x}\|_2=1} |\mathbf{x}^T\mathbf{A}^T\mathbf{A}\mathbf{x}|^{1/2} = \lambda_{max}^{1/2}(\mathbf{A}^T\mathbf{A}),$$

where $\lambda_{max}(\mathbf{A}^T\mathbf{A})$ is the maximum eigenvalue of $\mathbf{A}^T\mathbf{A}$ (since $\mathbf{A}^T\mathbf{A}$ is positive semidefinite, $\lambda_{max}(\mathbf{A}^T\mathbf{A})$ is real and nonnegative).

Now suppose $\mathbf{A}$ is positive definite. Since the eigenvalues of $\mathbf{A}^2$ are the squares of the eigenvalues of $\mathbf{A}$, and since for a symmetric matrix $\mathbf{A}^T = \mathbf{A}$, we may also conclude

$$\lambda_{min}(\mathbf{A}) = \min_{\|\mathbf{x}\|_2=1} \|\mathbf{A}\mathbf{x}\|_2 \text{ and } \lambda_{max}(\mathbf{A}) = \max_{\|\mathbf{x}\|_2=1} \|\mathbf{A}\mathbf{x}\|_2,$$

for any positive semidefinite matrix. This then implies $\|\mathbf{A}\|_2 = \lambda_{max}(\mathbf{A})$. Chapter 5 of Horn and Johnson (1985) can be highly recommended as a reference for the subject of matrix norms.

C.9 Taylor's Remainder Theorem

Suppose $f : \mathbb{R} \to \mathbb{R}$ is n times differentiable. Let $f^{(j)}(x)$ be the jth order derivative of $f(x)$. The nth order Taylor's polynomial about x_0 is defined as

$$\hat{f}_n(x; x_0) = \sum_{i=1}^{n} \frac{f^{(i)}(x_0)}{i!}(x - x_0)^i, \tag{C.15}$$

and the remainder term is given by $R_n(x; x_0) = f(x) - \hat{f}_n(x; x_0)$. The use of $\hat{f}_n(x; x_0)$ to approximate $f(x)$ is made precise by Taylor's remainder theorem:

Theorem C.8 (Taylor's Remainder Theorem for Functions of One Variable) Suppose $f(x)$ is $n + 1$ times continuously differentiable on (a, b). Given the approximation of Equation (C.15) let $R_n(x; x_0) = f(x) - \hat{f}_n(x; x_0)$ be the remainder term. Then for each $x, x_0 \in (a, b)$ there exists $\eta(x) \in (a, b)$ such that

$$R_n(x; x_0) = \frac{f^{(n+1)}(\eta(x))}{(n + 1)!}(x - x_0)^{n+1}$$

and $\eta(x) = px + (1 - p)x_0$ for some $p \in [0, 1]$. ■

The Taylor's series approximation of multivariate functions takes much the same form as Equation (C.15) and Theorem C.8, but requires the use of tensors or multidimensional arrays for terms above the order of two. The following two special cases will serve our purposes.

Theorem C.9 (Taylor's Remainder Theorem for Multivariate Functions I) Suppose $F(\mathbf{x})$ is a twice continuously differentiable real-valued function of q variables on an open set $E \subset \mathbb{R}^q$, with gradient and Hessian $\mathbf{F}'(\mathbf{x}), \mathbf{F}''(\mathbf{x})$. Then for any $\mathbf{x}, \mathbf{x}_0 \in E$ there exists $\boldsymbol{\eta}(\mathbf{x}) \in E$ such that

$$F(\mathbf{x}) = F(\mathbf{x}_0) + \mathbf{F}'(\mathbf{x}_0)^T(\mathbf{x} - \mathbf{x}_0) + \frac{1}{2}(\mathbf{x} - \mathbf{x}_0)^T\mathbf{F}''(\boldsymbol{\eta}(\mathbf{x}))(\mathbf{x} - \mathbf{x}_0) \tag{C.16}$$

and $\boldsymbol{\eta}(\mathbf{x}) = p\mathbf{x} + (1 - p)\mathbf{x}_0$ for some $p \in [0, 1]$. ■

Theorem C.10 (Taylor's Remainder Theorem for Multivariate Functions II) Suppose $F(\mathbf{x})$ is a twice continuously differentiable real-valued function of q variables on an open set $E \subset \mathbb{R}^q$, with gradient and Hessian $\mathbf{F}'(\mathbf{x}), \mathbf{F}''(\mathbf{x})$. Suppose for any $\mathbf{x}, \mathbf{x}_0 \in E$ we write

$$F(\mathbf{x}) = F(\mathbf{x}_0) + \mathbf{F}'(\mathbf{x}_0)^T(\mathbf{x} - \mathbf{x}_0) + R(\mathbf{x}; \mathbf{x}_0).$$

Then

$$|R(\mathbf{x}; \mathbf{x}_0)| \leq \sup_{\{\mathbf{x}' \in E : |\mathbf{x}' - \mathbf{x}_0| \leq |\mathbf{x} - \mathbf{x}_0|\}} \frac{1}{2} \left\| \mathbf{F}''(\mathbf{x}') \right\|_2 |\mathbf{x} - \mathbf{x}_0|^2, \tag{C.17}$$

where $\|\cdot\|_2$ is the Euclidean matrix norm (Section C.8.2). ■

Formal proofs of Theorems C.8-C.10 can be found in, for example, Rudin (1991).

D

Group Theory

D.1 Definition of a Group

Consider the pair $(G, *)$, where G is a set on which a binary operation $*$ is defined. If $a, b \in G$ then $c = a * b$ is refered to as a product. For example $(\mathbb{R}, +)$ represents addition of real numbers. A group is such a pair which satisfies the following axioms:

Definition D.1 (Axioms of a Group) A group is a pair $(G, *)$ where G is a set and $*$ is a binary operation on G satisfying the axioms:
Closure: If $a, b \in G$ then $a * b \in G$,
Associativity: For all $a, b, c \in G$ we have $(a * b) * c = a * (b * c)$,
Existence of identity: There exists $e \in G$ such that for all $a \in G$ we have $a * e = e * a = a$,
Existence of inverse: For each $a \in G$ there exists $b \in G$ such that $a * b = b * a = e$. ■

Definition D.2 (Commutative Group) A commutative (Abelian) group is a group $(G, *)$ which satisfies the additional axiom:
Commutativity: For any $a, b \in G$ we have $a * b = b * a$. ■

Note that the axioms of Definition D.1 do not explicitly state that an identity or inverse must be unique. However, uniqueness follows from the axioms. First, any inverse in a group $(G, *)$ must be unique. To see this, suppose b, c are both inverses of $a \in G$. Then $b = b * e = b * (a * c) = (b * a) * c = e * c = c$. The identity must also be unique, since it is its own inverse. As a consequence, when verifying that $(G, *)$ is a group, it is not necessary to explicitly verify uniqueness of the identity or inverse.

The degree to which group theory plays a role in the theory of inference can vary considerably by volume. Lehmann and Casella (1998) and Lehmann and Romano (2005) are especially notable in this respect, and contain some background material on group theory which emphasizes statistical applications. More general introductions to group theory can be found in, for example, Rose (1978); Rotman (1995); Fraleigh (1999).

Example D.1 (Addition) Set $G = \mathbb{R}$ with $*$ being the addition operation. Clearly, closure and associativity are satisfied. The identity is $e = 0$ and the inverse of a is $a^{-1} = -a$. Thus, $(\mathbb{R}, +)$ is a group. Addition is also commutative, but groups need not possess this property. We don't need to formally define subtraction as a separate operation. What we think of as $a - b$ would be written $a + b^{-1}$. ■

Example D.2 (Scalar Multiplication) Set $G = \mathbb{R}_{>0}$ with $*$ being the scalar multiplication operation. Clearly, closure and associativity are satisfied. The identity is $e = 1$ and the inverse of a is $a^{-1} = 1/a$. Thus, $(\mathbb{R}_{>0}, \times)$ is a group. It is also commutative. Similar to subtraction within the addition group, we do not formally define division. Instead, we set $a/b = a * b^{-1}$. ■

DOI: 10.1201/9781003049340-D

Example D.3 (Vector Spaces) That a vector space $\mathcal{V} \subset \mathcal{R}^n$ is also a group under vector addition $(x_1, \ldots, x_n) + (y_1, \ldots, y_n) = (x_1 + y_1, \ldots, x_n + y_n)$ follows from the definition given in Definition B.2. A vector space is also a commutative group. ■

Example D.4 (Permutations) A permutation of n distinct objects, which can be labeled $\mathcal{X} = \{1, \ldots, n\}$, is any ordered selection of all labels. For example, $(3, 2, 4, 1)$ and $(2, 1, 3, 4)$ are distinct permutations of 4 labels. There are $n!$ distinct permutations. The set of all permutations possesses the type of symmetry characteristic of a group. However, in order to define this set as a group a binary operation is needed. Let σ be a bijective mapping of the labels (of which there are $n!$). Then define transformation: $g_\sigma(i_1, i_2, i_3, i_4) = (\sigma(i_1), \sigma(i_2), \sigma(i_3), \sigma(i_4))$. Thus the permutation $(3, 2, 4, 1)$ is constructed by applying the bijection $\sigma(1) = 3$, $\sigma(2) = 2$, $\sigma(3) = 4$, $\sigma(4) = 1$ to the ordering $(1, 2, 3, 4)$, giving $g_\sigma(1, 2, 3, 4) = (3, 2, 4, 1)$. Since there is a one-to-one mapping between the set of permutations and the set of bijections σ we can take the permutation group to be the set of all σ paired with the binary operation of composition.

An alternative way of representing σ is given by Cauchy's two-line notation. Two permutations are aligned vertically, where σ maps an element of the top permutation to the one below it:

$$\sigma = \begin{pmatrix} 1 & 2 & 3 & 4 \\ \sigma(1) & \sigma(2) & \sigma(3) & \sigma(4) \end{pmatrix}, \tag{D.1}$$

or in our example,

$$\sigma = \begin{pmatrix} 1 & 2 & 3 & 4 \\ 3 & 2 & 4 & 1 \end{pmatrix},$$

(note that in this expression the symbol σ is used in two different forms, although denoting essentially the same object). A product is then the composition of two bijections, for example

$$\begin{pmatrix} 1 & 2 & 3 & 4 \\ 3 & 2 & 4 & 1 \end{pmatrix} * \begin{pmatrix} 1 & 2 & 3 & 4 \\ 3 & 1 & 2 & 4 \end{pmatrix} = \begin{pmatrix} 1 & 2 & 3 & 4 \\ 2 & 1 & 4 & 3 \end{pmatrix}.$$

Similarly, the inverse is obtained by exchanging the rows, for example

$$\begin{pmatrix} 1 & 2 & 3 & 4 \\ 3 & 2 & 4 & 1 \end{pmatrix}^{-1} = \begin{pmatrix} 3 & 2 & 4 & 1 \\ 1 & 2 & 3 & 4 \end{pmatrix} = \begin{pmatrix} 1 & 2 & 3 & 4 \\ 4 & 2 & 1 & 3 \end{pmatrix}.$$ ■

D.2 Subgroups

A subgroup is a subset of a group that is itself a group. A formal definition follows.

Definition D.3 (Subgroup) Let $(G, *)$ be a group. Then $H \subset G$ is a subgroup of G if it forms a group under the same operation $*$. If $H \neq G$ then H is a proper subgroup. Note that $H = \{e\}$ satisfies the group axioms, and so is the trivial subgroup. ■

A subgroup can be generated by any subset $E \subset G$. This is shown in the next theorem.

Theorem D.1 (Generated Subgroups) Let $(G, *)$ be a group and let E be any subset of G. Let H_E be the set of all finite products of elements of E and their inverses. Then H_E is a subgroup. ■

A subgroup generated by a single element of G is known as a cyclic group. These are always commutative.

Example D.5 (Addition Subgroup) Following Example D.1, we will apply Theorem D.1 to $E = \{1\} \subset G$. The subgroup H_E is generated from all finite additions and subtractions of 1, that is $H_E = \mathbb{I}$ is a subgroup. This is an example of a cyclic group. ■

Example D.6 (Scalar Multiplication Subgroup) Following Example D.2, let H be the set of all positive rational numbers. Then H is closed under multiplication, contains the identity $e = 1$, and the inverse of a positive rational number is also a positive rational number. Therefore, H is a subgroup of $(\mathbb{R}_{>0}, \times)$. ■

Example D.7 (Vector Space Subgroup) Following Example D.3, suppose a vector space $\mathcal{V} \subset \mathbb{R}^n$ is spanned by basis vectors $(x_1, \ldots, x_m)$. Then a vector space spanned by any subset of those basis vectors is a subgroup of $\mathcal{V}$. ■

Example D.8 (Permutation Subgroup) Following Example D.4, given labels $\{1, 2, 3, 4\}$ consider the single permutation

$$\sigma = \begin{pmatrix} 1 & 2 & 3 & 4 \\ 4 & 1 & 2 & 3 \end{pmatrix}.$$

Following Theorem D.1, repeated application of σ to the identity $(1, 2, 3, 4)$ generates a subgroup $\{(1, 2, 3, 4), (4, 1, 2, 3), (3, 4, 1, 2), (2, 3, 4, 1)\}$. This is also a cyclic group. ■

D.3 Group Homomorphisms

One notable feature of Example D.4 is that two distinct groups were defined; first, the set of all orderings of n labels, and second, the set of all bijections on n labels. There was a bijection between these groups defined by Equation (D.1), so that an element of one group could always be unambiguously associated with exactly one element of the other. Thus, although each group contains elements which are distinct types of mathematical objects, their group properties are identical, that is, they are essentially the same group.

This idea proves to be important in many applications, and can be expressed in terms of homomorphisms, which are mappings between groups which preserve group structure.

Definition D.4 (Homomorphism) We are given two groups $(G, *)$, $(G', \odot)$. A group homomorphism is a mapping $h : G \to G'$ which satifies

$$h(g_1) \odot h(g_2) = h(g_1 * g_2) \tag{D.2}$$

for any $g_1, g_2 \in G$. If h is in addition bijective, then it is a group isomorphism. If an homomorphism (isomorphism) exists, the groups are called homomorphic (isomorphic). ■

A homomorphism is a mapping which is compatible with group structure, while isomorphic groups are essentially equivalent.

Theorem D.2 (Identity and Inverse of a Homomorphism) Suppose we are given the homomorphism of Definition D.4.

(i) If e is the identity element of G, then $h(e)$ is the identity element of G'.
(ii) For any $g \in G$, $h(g^{-1}) = h(g)^{-1}$. ■

Suppose we are given group $(G, *)$ and a set G' for which there exists a bijective mapping $h : G \to G'$. From the point of view of group theory, G' is simply a relabeling of G, and so should inherit the group properties of G. The key to this idea is that h is bijective, so with each $g' \in G'$ we may associate a unique $g = h^{-1}(g') \in G$. The binary operation $g_1' \odot g_2' = h(g_1 * g_2)$ is therefore well defined, and satisfies Equation (D.2). We can summarize this idea in the following theorem.

Theorem D.3 (Induced Isomorphisms) Suppose we are given group $(G, *)$, any set G' and a bijective mapping $h : G \to G'$. Define the binary operation $\odot$ on G' by:

$$g_1' \odot g_2' = h(h^{-1}(g_1') * h^{-1}(g_2')), \ \ g_1', g_2' \in G'. \tag{D.3}$$

Then $(G', \odot)$ is a group which is isomorphic to $(G, *)$. If $(G, *)$ is commutative then so is $(G', \odot)$. ■

If we are given group $(G, *)$, a set G' and a surjective but not injective mapping $h : G \to G'$, then the binary operation of Equation (D.3) need not be well defined. In this case, we would need to know that for any $g_1', g_2' \in G'$, $h(g)$ is constant for all $g = g_1 * g_2$ for which $g_1 \in h^{-1}(\{g_1'\})$, $g_2 \in h^{-1}(\{g_2'\})$.

Example D.9 (Isomorphism Between Addition and Multiplication) Consider the groups $(G, *) = (\mathbb{R}, +)$ and $(G', *) = (\mathbb{R}_{>0}, \times)$ of Examples D.1 and D.2. These groups are isomorphic. Define the mapping $h : G \to G'$ by the function $h(x) = e^x$, $h^{-1}(a) = \log(a)$. This is a bijection, so the assumptions of Theorem D.3 hold. Condition (D.2) of Definition D.2 follows. For any $x, y \in G$ we have $h(x) \times h(y) = e^x \times e^y = e^{x+y} = h(x + y)$, and for $a, b \in G'$, $h^{-1}(a) + h^{-1}(b) = \log(a) + \log(b) = \log(a + b) = h^{-1}(a + b)$. As predicted by Theorem D.2: (*i*) the identity of G maps to the identity of G', since $e^0 = 1$; (*ii*) the inverse of $x \in G$, denoted as x^{-1}, can be evaluated by $e^{x^{-1}} = h(x^{-1}) = 1/e^x = e^{-x}$, equivalently, $x^{-1} = -x$. ■

A homomorphism that is not an isomorphism often represents a reduction of the dimension of a group, typically a partial representation of an element of the group, or a projection onto a lower dimension.

Example D.10 (Vector Space Homomorphism) Consider the vector space $\mathcal{V} = \mathbb{R}^2$. Define the mapping $h : \mathcal{V} \to \mathbb{R}$ by the evaluation method $h(\mathbf{x}) = x_2 - x_1$, for any $\mathbf{x} = (x_1, x_2) \in \mathcal{V}$. We wish to show that h is a homorphism from $\mathcal{V}$ to $(\mathbb{R}, +)$. This may be done by directly verifying condition (D.2). Suppose $\mathbf{x}, \mathbf{y} \in \mathcal{V}$. Then $h(\mathbf{x}) + h(\mathbf{y}) = (x_2 - x_1) + (y_2 - y_1) = (x_2 + y_2) - (x_1 + y_1) = h(\mathbf{x} + \mathbf{y})$. Therefore, h defines a homomorphism between groups $\mathcal{V}$ and $(\mathbb{R}, +)$. ■

Example D.11 (Permutation Homomorphism) Consider the permutation bijection of Example D.8

$$\sigma_1 = \begin{pmatrix} 1 & 2 & 3 & 4 \\ 4 & 1 & 2 & 3 \end{pmatrix}.$$

This generated a cyclic subgroup H. Define a mapping h of a permutation p by $h(p) = \{\sigma p : \sigma \in H\}$. For example, $h(3, 2, 4, 1) = \{(3, 2, 4, 1), (1, 3, 2, 4), (4, 1, 3, 2), (2, 4, 1, 3)\}$. It is important to note that the definition of h is consistent in the sense that $p \in h(p)$, since H contains the identity. Furthermore, $h(p)$ can be represented by any element of $h(p)$, that is, $h(p) = h(p')$ *iff* $p' \in h(p)$. We then note that for any two permutations p_1, p_2, we have the identity $h(p_1 * p_2) = h(p_1' * p_2')$ for any pair $p_1' \in h(p_1)$, $p_2' \in h(p_2)$. Thus condition (D.2) can be used to define a product $h(p_1) \odot h(p_2)$. That $G' = \{h(p) : p \in G\}$ is a group under operation $\odot$ can then be verified from this identity. ■

D.4 Transformation Groups

The type of group of particular interest in statistical inference is the transformation group. Suppose we are given a class G of bijective transformations on a domain $\mathcal{X}$, that is, mappings $g \in G$ of the form $g : \mathcal{X} \to \mathcal{X}$. We may then interpret composition as a binary operation $\circ$ on G. Thus, if for $g_1, g_2 \in G$ we set $g_3 = g_2 \circ g_1$, this is understood to mean $g_3(x) = g_2(g_1(x))$ for all $x \in \mathcal{X}$.

Definition D.5 (Transformation Group) Let G be a class of bijective transformations on a domain $\mathcal{X}$. Suppose $e \in G$, where $e(x) \equiv x$ is the identity transformation. Then G is a transformation group if $(G, \circ)$ is a group, where $\circ$ denotes composition. Transformations, and hence transformation groups, may or may not be commutative. ∎

A few facts are important to remember when determing whether or not G is a transformation group. Compositions of bijections are themselves bijections, and for the composition operation, associativity holds in general. The identity transformation will be the identity of the group. Since g is bijective it possesses an inverse g^{-1}. This makes the construction of a transformation group straightforward. The following theorem is similar to Theorem D.1.

Theorem D.4 (Generated Transformation Groups) Given set $\mathcal{X}$, suppose $\mathcal{C}$ is a class of bijections $g : \mathcal{X} \to \mathcal{X}$. Let G be the class of compositions of a finite number of transformations from $\mathcal{C}$ and their inverses. Then G is a transformation group. ∎

Note that a surjection $g : \mathcal{X} \to \mathcal{X}$ is bijective if $\mathcal{X}$ is finite, but not in general. For example, $f(x) = \log(x)I\{x > 0\}$ defines a surjection on $\mathbb{R}$, but not a bijection.

A transformation group can be associated with the idea of invariance. Suppose we have transformation $g : \mathcal{X} \to \mathcal{X}$. The subset $\mathcal{X}' \subset \mathcal{X}$ is invariant under g is the image satisfies the identity $g\mathcal{X}' = \mathcal{X}'$. For example, let $\mathcal{X}$ be the set of rotationally oriented squares, and let G be the set of rotation transformations g_θ of θ radians about the center. Any subset of $\mathcal{X}$ is invariant under $g_{\pi/2}$. For any element $x \in \mathcal{X}$, the set $\mathcal{X}' = \{x, g_{\pi/4}x\}$ is invariant under $g_{\pi/4}$. Finally, $\mathcal{X}$ itself is invariant under any $g_\theta \in G$.

In the next two examples we consider a class of transformation groups that play an important role in the theory of inference. They are related in that they are all subgroups of the affine transformation group acting on $\mathbb{R}^n$.

Example D.12 (Affine Transformations on $\mathbb{R}^n$) Suppose $\mathcal{X} = \mathbb{R}^n$. Define the pair $\boldsymbol{\theta} = (\mathbf{A}, \mathbf{b})$, where $\mathbf{A} \in \mathcal{M}_n$, $\mathbf{b} \in \mathcal{M}_{n,1}$ and $\mathbf{A}$ is invertible, and let Θ_{aff} be the set of all such pairs. Then define the affine transformation $g_{\boldsymbol{\theta}} : \mathcal{X} \to \mathcal{X}$ by the evaluation method $g_{\boldsymbol{\theta}}\mathbf{x} = \mathbf{A}\mathbf{x} + \mathbf{b}$. Since $\mathbf{A}$ is invertible $g_{\boldsymbol{\theta}}$ is bijective. We wish to verify that $G_{aff} = \{g_{\boldsymbol{\theta}} : \boldsymbol{\theta} \in \Theta_{aff}\}$ is a transformation group. Select $\boldsymbol{\theta}_1, \boldsymbol{\theta}_2 \in \Theta_{aff}$, denoted as $\boldsymbol{\theta}_1 = (\mathbf{A}, \mathbf{b})$, $\boldsymbol{\theta}_2 = (\mathbf{C}, \mathbf{d})$. The composition operation $\circ$ is evaluated by

$$g_{\boldsymbol{\theta}_1} \circ g_{\boldsymbol{\theta}_2}\mathbf{x} = \mathbf{A}\left[\mathbf{C}\mathbf{x} + \mathbf{d}\right] + \mathbf{b} = g_{\boldsymbol{\theta}_3}\mathbf{x}, \tag{D.4}$$

where $\boldsymbol{\theta}_3 = (\mathbf{AC}, \mathbf{Ad} + \mathbf{b}) \in \Theta_{aff}$ (the product of invertible matrices is invertible). Thus, G_{aff} is closed under composition.

We can then define the equivalent binary operation $\odot$ on Θ_{aff}

$$(\mathbf{A}, \mathbf{b}) \odot (\mathbf{C}, \mathbf{d}) = (\mathbf{AC}, \mathbf{Ad} + \mathbf{b}).$$

The task is now to verify that $(\Theta_{aff}, \odot)$ is a group. First note that $\odot$ is not commutative. Let $e = (\mathbf{I}, \mathbf{0}) \in \Theta_{aff}$. Then $(\mathbf{A}, \mathbf{b}) \odot e = (\mathbf{AI}, \mathbf{A0} + \mathbf{b}) = (\mathbf{A}, \mathbf{b})$ and $e \odot (\mathbf{A}, \mathbf{b}) = (\mathbf{IA}, \mathbf{Ib} + \mathbf{0}) = (\mathbf{A}, \mathbf{b})$, so that e is an identity.

To verify the existence of an inverse, set $\boldsymbol{\theta}_1 = (\mathbf{A}, \mathbf{b}) \in \Theta_{aff}$ and $\boldsymbol{\theta}_2 = (\mathbf{A}^{-1}, -\mathbf{A}^{-1}\mathbf{b})$. Note that $\boldsymbol{\theta}_2 \in \Theta_{aff}$. Then $\boldsymbol{\theta}_1 \odot \boldsymbol{\theta}_2 = (\mathbf{A}\mathbf{A}^{-1}, -\mathbf{A}\mathbf{A}^{-1}\mathbf{b}\mathbf{d} + \mathbf{b}) = e$, and $\boldsymbol{\theta}_2 \odot \boldsymbol{\theta}_1 = (\mathbf{A}^{-1}\mathbf{A}, \mathbf{A}^{-1}\mathbf{b} - \mathbf{A}^{-1}\mathbf{b}) = e$, so that $\boldsymbol{\theta}_2$ is the inverse of $\boldsymbol{\theta}_1$.

To verify associativity, set $\boldsymbol{\theta}_i = \Theta_{aff}$, $i = 1, 2, 3$, denoted as $\boldsymbol{\theta}_i = (\mathbf{A}_i, \mathbf{b}_i)$. Then

$$\begin{aligned}\boldsymbol{\theta}_1 \odot [\boldsymbol{\theta}_2 \odot \boldsymbol{\theta}_3] &= (\mathbf{A}_1, \mathbf{b}_1) \odot (\mathbf{A}_2\mathbf{A}_3, \mathbf{A}_2\mathbf{b}_3 + \mathbf{b}_2)\\ &= (\mathbf{A}_1\mathbf{A}_2\mathbf{A}_3, \mathbf{A}_1\mathbf{A}_2\mathbf{b}_3 + \mathbf{A}_1\mathbf{b}_2 + \mathbf{b}_1)\end{aligned}$$

and

$$\begin{aligned}[\boldsymbol{\theta}_1 \odot \boldsymbol{\theta}_2] \odot \boldsymbol{\theta}_3 &= (\mathbf{A}_1\mathbf{A}_2, \mathbf{A}_1\mathbf{b}_2 + \mathbf{b}_1) \odot (\mathbf{A}_3, \mathbf{b}_3)\\ &= (\mathbf{A}_1\mathbf{A}_2\mathbf{A}_3, \mathbf{A}_1\mathbf{A}_2\mathbf{b}_3 + \mathbf{A}_1\mathbf{b}_2 + \mathbf{b}_1).\end{aligned}$$

We have just shown directly that associativity holds, so that $(\Theta_{aff}, \odot)$ is a group. Clearly, G_{aff} and Θ_{aff} are bijective. Furthermore, g_e is the identity transformation on $\mathcal{X}$, and $g_{\boldsymbol{\theta}_1} \circ g_{\boldsymbol{\theta}_2} = g_{\boldsymbol{\theta}_1 \odot \boldsymbol{\theta}_2}$. Therefore G_{aff} is a transformation group. ■

Example D.13 (Location-Scale Transformation on $\mathbb{R}^n$) Suppose $\mathcal{X} = \mathbb{R}^n$. For any pair of real numbers $\theta = (\mu, \sigma) \in \Theta_{ls} = \mathbb{R} \times \mathbb{R}_{>0}$ define the location-scale transformation $g_\theta : \mathcal{X} \to \mathcal{X}$ by the evaluation method $g_\theta \mathbf{x} = \mu + \sigma\mathbf{x} = (\mu + \sigma x_1, \ldots, \mu + \sigma x_n)$. Let $G_{ls} = \{g_\theta : \theta \in \Theta_{ls}\}$ be the class of location-scale transformations. We can verify that G_{ls} is a transformation group by showing that it is a subgroup of the affine transformation group G_{aff} (Example D.12). We can recognize g_θ as a special case of the affine transformation $g_{\boldsymbol{\theta}}$ of where $\boldsymbol{\theta} = (\sigma\mathbf{I}, \mu\mathbf{1})$. Define the subset

$$E = \{g_{\boldsymbol{\theta}} : \boldsymbol{\theta} = (\sigma\mathbf{I}, \mu\mathbf{1}),\ (\mu, \sigma) \in \Theta_{ls}\} \subset G_{aff}.$$

It is easily verified that G_{ls} is the set of all finite compositions of elements of E and their inverses, so that G_{ls} is a subgroup of G_{aff} by Theorem D.1 (in fact, $E = G_{ls}$).

We can similarly define for any $\mu \in \Theta_{loc} = \mathbb{R}$ the location transformation $g_\mu : \mathcal{X} \to \mathcal{X}$ by the evaluation method $g_\mu \mathbf{x} = \mu + \mathbf{x} = (\mu + x_1, \ldots, \mu + x_n)$, and also the scale transformation $g_\sigma \mathbf{x} = \sigma\mathbf{x} = (\sigma x_1, \ldots, \sigma x_n)$, for any $\sigma \in \Theta_{sc} = \mathbb{R}_{>0}$. Using an argument similar to the preceding one, we may verify that $G_{loc} = \{g_\mu : \mu \in \Theta_{loc}\}$ and $G_{sc} = \{g_\sigma : \sigma \in \Theta_{sc}\}$ are subgroups of G_{aff}. ■

Any group $(\Theta, *)$ has an equivalent representation as a transformation group on Θ.

Theorem D.5 (Transformation Group Isomporphisms) Any group $(\Theta, *)$ is isomorphic to a transformation group on Θ. ■

Theorem D.5 is proven by the following argument. For any $\theta \in \Theta$ define transformation g_θ by the evaluation method $g_\theta \theta' = \theta * \theta'$. Set $G = \{g_\theta : \theta \in \Theta\}$. Clearly, $g_{\theta_1} \neq g_{\theta_2}$ unless $\theta_1 = \theta_2$. Therefore $g_\theta : \Theta \to G$ is a bijection, and by Theorem D.3 $(G, \odot)$ is a group under binary operation $g_{\theta_1} \odot g_{\theta_2} = g_{\theta_1 * \theta_2}\theta = (\theta_1 * \theta_2) * \theta = \theta_1 * (\theta_2 * \theta) = g_{\theta_1}(g_{\theta_2} * \theta) = (g_{\theta_1} \circ g_{\theta_2})\theta$, so that $\odot$ is equivalent to composition.

D.5 Orbits and Maximal Invariants

A transformation group may or may not be transitive, according to the following definition.

Definition D.6 (Transitive Transformation Groups) A transformation group $(G, \circ)$ acting on $\mathcal{X}$ is transitive if for any $x, y \in \mathcal{X}$ there exists $g \in G$ such that $y = gx$. ■

Suppose we are given a transformation group $(G, \circ)$ on $\mathcal{X}$. Define the relation

$$x \sim y \text{ } \textit{iff} \text{ there exists } g \in G \text{ such that } x = gy. \tag{D.5}$$

The group properties of G can be used to show that Equation (D.5) defines an equivalence relation (Section B.2).

Theorem D.6 (Equivalence Classes of Transformation Groups) For any transformation group $(G, \circ)$ on $\mathcal{X}$ Equation (D.5) defines an equivalence relation. ■

By Theorem D.6 Equation (D.5) defines an equivalence relation. An induced equivalence class is referred to as an orbit. The orbit of $x \in \mathcal{X}$ may be thought of as the set of all points obtained by applying all transformations in G to x. If G is transitive, then $\mathcal{X}$ itself is the only orbit. Otherwise, the set of equivalence classes can be thought of as a reduction of $\mathcal{X}$. This idea is naturally expressed as a mapping on $\mathcal{X}$ that is equivalent to the set of all orbits.

Definition D.7 (Invariants and Maximal Invariants) A mapping $M : \mathcal{X} \to \mathcal{H}$ is group invariant (or invariant) if $M(x) = M(gx)$ for all $g \in G$ and $x \in \mathcal{X}$. An invariant mapping is a maximal invariant if $M(x) = M(y)$ implies $x \sim y$. Equivalently, $M(x) = M(y)$ if and only if x and y belong to the same orbit. ■

A maximal invariant of a transformation group always exists.

Theorem D.7 (Existence of Maximal Invariants) Any transformation group G on $\mathcal{X}$ possesses a maximal invariant. ■

We will sometimes need to identify transformation groups on subsets $\mathcal{X}' \subset \mathcal{X}$ of a domain.

Theorem D.8 (Transformation Groups on Subsets of $\mathcal{X}$) Suppose we have transformation group G on $\mathcal{X}$. Suppose there exists a subset $\mathcal{X}' \subset \mathcal{X}$ on which each transformation $g \in G$ is a bijection. Then G is also a transformation group on $\mathcal{X}'$. ■

The class of all invariant mappings can be characterized as all mappings of any single maximal invariant.

Theorem D.9 (Equivalence of Maximal Invariants) Given a transformation group G on $\mathcal{X}$ with maximal invariant M, a mapping H is invariant if and only if there exists a mapping $M \mapsto H$. It follows that all maximal invariants are equivalent. ■

Example D.14 (Order Permutation on $\mathbb{R}^n$) Suppose $\mathcal{X} = \mathbb{R}^n$. Let Θ be the permutation group acting on n distinct labels. For any $\sigma \in \Theta$ define the transformation on $\mathcal{X}$, $g_\sigma(x_1, \ldots, x_n) = (x_{\sigma(1)}, \ldots, x_{\sigma(n)})$. Then $G = \{g_\sigma : \sigma \in \Theta\}$ is a transformation group on $\mathcal{X}$. This transformation group is not transient. A maximal invariant is given by the order transformation $(x_{(1)}, \ldots, x_{(n)})$, where $x_{(k)}$ is the kth ranked element of x.

Given any mapping $h(x)$ on $\mathcal{X}$, an invariant mapping can be created by $H(x) = \sum_{\sigma \in \Theta} h(g_\sigma x)$ or $H(x) = \prod_{\sigma \in \Theta} h(g_\sigma x)$, as appropriate. Sums or products of the form $H(x) = \sum_{i=1}^n h(x_i)$ or $H(x) = \prod_{i=1} h(x_i)$ are invariant mappings. ■

Bibliography

Agresti, A. (2018). *An Introduction to Categorical Data Analysis*. John Wiley & Sons.

Almudevar, A. (2001). Most powerful permutation invariant tests for relatedness hypotheses using genotypic data. *Biometrics*, **57**(4), 1080–1088.

Almudevar, A. (2014). *Approximate Iterative Algorithms*. CRC Press.

Almudevar, A. (2016). Higher order density approximations for solutions to estimating equations. *Journal of Multivariate Analysis*, **143**, 424–439.

Almudevar, A., Field, C., and Robinson, J. (2000). The density of multivariate M-estimates. *Annals of Statistics*, **28**(1), 275–297.

Anderson, T., Taylor, J. B., *et al.* (1976). Strong consistency of least squares estimates in normal linear regression. *The Annals of Statistics*, **4**(4), 788–790.

Anscombe, F. J. (1948). The transformation of Poisson, binomial and negative-binomial data. *Biometrika*, **35**(3/4), 246–254.

Ash, R. B. (1972). *Real Analysis and Probability*. Academic Press, Orlando, Florida, 1st edition.

Ash, R. B. and Dolacutuseans-Dade, C. A. (2000). *Real Analysis and Probability*. Academic Press, San Diego, 2nd edition.

Asmussen, S. and Rojas-Nandayapa, L. (2008). Asymptotics of sums of lognormal random variables with Gaussian copula. *Statistics and Probability Letters*, **78**(16), 2709–2714.

Bahadur, R. R. (1966). A note on quantiles in large samples. *The Annals of Mathematical Statistics*, **37**(3), 577–580.

Bahadur, R. R. *et al.* (1964). On Fisher's bound for asymptotic variances. *The Annals of Mathematical Statistics*, **35**(4), 1545–1552.

Bartlett, M. S. (1937). Properties of sufficiency and statistical tests. *Proceedings of the Royal Society of London. Series A, Mathematical and Physical Sciences*, **160**(901), 268–282.

Bartlett, M. S. (1947). The use of transformations. *Biometrics*, **3**(1), 39–52.

Bertsekas, D. P. and Shreve, S. E. (1978). *Stochastic Optimal Control: The Discrete-Time Case*. Academic Press, NY.

Bhattacharya, R., Lin, L., and Patrangenaru, V. (2016). *A Course in Mathematical Statistics and Large Sample Theory*. Springer, NY.

Bhattarcharya, S. and Kumar, K. (2008). Forensic accounting and Benford's law. *IEEE Signal Processing Magazine*, **25**(2), 152.

Bickel, P. J. and Doksum, K. A. (2015). *Mathematical Statistics: Basic Ideas and Selected Topics, Volume I*. CRC Press, Boca Raton.

Billingsley, P. (1995). *Probability and Measure*. John Wiley and Sons, NY, 3rd edition.

Birnbaum, A. (1962). On the foundations of statistical inference. *Journal of the American Statistical Association*, **57**(298), 269–306.

Bland, J., Altman, D., and Rohlf, F. (2013). In defence of logarithmic transformations. *Statistics in Medicine*, **32**(21), 3766–3768.

Blyth, C. R. (1980). Expected absolute error of the usual estimator of the binomial parameter. *The American Statistician*, **34**(3), pp. 155–157.

Box, G. E., Hunter, W. H., Hunter, S., *et al.* (1978). *Statistics for experimenters*. John Wiley and Sons, NY.

Box, J. F. (1980). R. A. Fisher and the design of experiments, 1922-1926. *The American Statistician*, **34**(1), 1–7.

Brazzale, A. R., Davison, A. C., and Reid, N. (2006). *Applied asymptotics*. Cambridge University Press, Cambridge, to appear.

Brown, L. (1968). Inadmissibility of the usual estimators of scale parameters in problems with unknown location and scale parameters. *The Annals of Mathematical Statistics*, **39**(1), 29–48.

Brown, L. D., Chow, M., and Fong, D. K. (1992). On the admissibility of the maximum-likelihood estimator of the binomial variance. *Canadian Journal of Statistics*, **20**(4), 353–358.

Bunge, J. and Fitzpatrick, M. (1993). Estimating the number of species: A review. *Journal of the American Statistical Association*, **88**(421), 364–373.

Casella, G. and Berger, R. L. (2002). *Statistical Inference*. Duxbury, Pacific Grove, 2nd edition.

Cessie, S. L. and Houwelingen, J. C. V. (1994). Logistic regression for correlated binary data. *Journal of the Royal Statistical Society. Series C (Applied Statistics)*, **43**(1), 95–108.

Chambers, J. (1977). *Computational Methods for Data Analysis*. John Wiley & Sons.

Chao, A. (1984). Nonparametric estimation of the number of classes in a population. *Scandinavian Journal of Statistics*, **11**(4), 265–270.

Chipman, J. S. and Rao, M. (1964). Projections, generalized inverses, and quadratic forms. *Journal of Mathematical Analysis and Applications*, **9**(1), 1–11.

Chiu, C.-H., Wang, Y.-T., Walther, B. A., and Chao, A. (2014). An improved nonparametric lower bound of species richness via a modified Good–Turing frequency formula. *Biometrics*, **70**(3), 671–682.

Clopper, C. J. and Pearson, E. S. (1934). The use of confidence or fiducial limits illustrated in the case of the binomial. *Biometrika*, **26**(4), 404–413.

Cochran, W. G. (1934). The distribution of quadratic forms in a normal system, with applications to the analysis of covariance. In *Mathematical Proceedings of the Cambridge Philosophical Society*, volume 30, pages 178–191. Cambridge University Press.

Cochran, W. G. (1972). Some effects of errors of measurement on linear regression. *Proceedings of the Berkeley Symposium on Mathematics Statistics and Probabability*, **6**(1), 527–540.

Cohen, J. (1960). A coefficient of agreement for nominal scales. *Educational and Psychological Measurement*, **20**(1), 37–46.

Cohen Freue, G. V. (2007). The Pitman estimator of the Cauchy location parameter. *Journal of Statistical Planning and Inference*, **137**(6), 1900–1913.

Copenhaver, M. D. and Holland, B. (1988). Computation of the distribution of the maximum studentized range statistic with application to multiple significance testing of simple effects. *Journal of Statistical Computation and Simulation*, **30**(1), 1–15.

Cox, D. R. and Hinkley, D. V. (1979). *Theoretical Statistics.* CRC Press, London.

Davies, R. B. (1969). Beta-optimal tests and an application to the summary evaluation of experiments. *Journal of the Royal Statistical Society, Series B*, **31**(3), 524–538.

Deming, W. E. (1943). *Statistical adjustment of data.* Wiley, NY.

Diaconis, P. and Zabell, S. (1991). Closed form summation for classical distributions: variations on a theme of De Moivre. *Statistical Science*, **6**(3), 284–302.

Diaconis, P., Holmes, S., and Montgomery, R. (2007). Dynamical bias in the coin toss. *SIAM Review*, **49**(2), 211–235.

Drygas, H. (1976). Weak and strong consistency of the least squares estimators in regression models. *Zeitschrift für Wahrscheinlichkeitstheorie und Verwandte Gebiete*, **34**(2), 119–127.

Dubins, L. and Savage, L. (1976). *Inequalities for Stochastic Processes (How to Gamble if You Must).* Dover Publications, NY, 2nd edition.

Durrett, R. (2010). *Probability: Theory and Examples.* Cambridge University Press, NY, 4th edition.

Efron, B. and Hinkley, D. V. (1978). Assessing the accuracy of the maximum likelihood estimator: Observed versus expected Fisher information. *Biometrika*, **65**(3), 457–483.

Evans, M. *et al.* (2013). What does the proof of Birnbaum's theorem prove? *Electronic Journal of Statistics*, **7**, 2645–2655.

Farrell, R. H. (1964). Estimators of a location parameter in the absolutely continuous case. *The Annals of Mathematical Statistics*, **35**(3), 949–998.

Feller, W. (1968). *Probability Theory and Its Applications, Volume 1.* John Wiley and Sons, NY, 3rd edition.

Feller, W. (1971). *Probability Theory and Its Applications, Volume 2.* John Wiley and Sons, NY, 2nd edition.

Feng, C., Wang, H., Lu, N., and Tu, X. M. (2013). Log transformation: Application and interpretation in biomedical research. *Statistics in Medicine*, **32**(2), 230–239.

Field, C. A. and Ronchetti, E. (1990). *Small Sample Asymptotics.* IMS Lecture Notes-Monograph Series.

Fisher, R. A. (1915). Frequency distribution of the values of the correlation coefficient in samples from an indefinitely large population. *Biometrika*, **10**(4), 507–521.

Fisher, R. A. (1925). *Statistical Methods for Research Workers.* Oliver & Boyd, Edinburgh & London.

Fisher, R. A. (1935). *The Design of Experiments.* Oliver & Boyd, Edinburgh & London.

Fisher, R. A. (1954). The analysis of variance with various binomial transformations. *Biometrics*, **10**(1), 130–139.

Fix, E. and Hodges, J. (1952). Discriminatory analysis: Nonparametric discrimination: Consistency properties. *USAF School of Aviation Medicine, Project no. 21-29-004.*

Fraleigh, J. B. (1999). *A First Course in Abstract Algebra.* Addison-Wesley, Reading, 6th edition.

Freeman, M. F. and Tukey, J. W. (1950). Transformations related to the angular and the square root. *The Annals of Mathematical Statistics*, **21**(4), 607–611.

Fuller, W. A. (2009). *Measurement Error Models*, volume 305. John Wiley & Sons, NY.

Gałecki, A. and Burzykowski, T. (2013). *Linear Mixed-Effects Models Using R.* Springer, NY.

Gastwirth, J. L. (1966). On robust procedures. *Journal of the American Statistical Association*, **61**(316), 929–948.

Gentleman, R., Carey, V., Huber, W., Irizarry, R., and Dudoit, S. (2006). *Bioinformatics and Computational Biology Solutions using R and Bioconductor*. Springer, NY.

Ghosh, B. K. (1970). *Sequential Tests of Statistical Hypotheses.* Addison-Wesley, Reading.

Ghosh, B. K. (1973). Some monotonicity theorems for χ^2, F and T distributions with applications. *Journal of the Royal Statistical Society, Series B*, **35**(3), 480–492.

Ghosh, J. K. (1971). A new proof of the Bahadur representation of quantiles and an application. *The Annals of Mathematical Statistics*, **42**(6), 1957–1961.

Gilmour, S. G. (1996). The interpretation of Mallows's c_p-statistic. *Journal of the Royal Statistical Society: Series D (The Statistician)*, **45**(1), 49–56.

Gini, C. (1936). On the measure of concentration with special reference to income and statistics. *Colorado College Publication, General Series*, **208**(1), 73–79.

Good, I. J. (1953). The population frequencies of species and the estimation of population parameters. *Biometrika*, **40**(3-4), 237–264.

Goodman, L. A. (1949). On the estimation of the number of classes in a population. *The Annals of Mathematical Statistics*, **20**(4), 572–579.

Graybill, F. A. and Marsaglia, G. (1957). Idempotent matrices and quadratic forms in the general linear hypothesis. *The Annals of Mathematical Statistics*, **28**(3), 678–686.

Halmos, P. R. (1946). The theory of unbiased estimation. *The Annals of Mathematical Statistics*, **17**(1), 34–43.

Halsey, L., Curran-Everett, D., Vowler, S., and Drummond, G. (2015). The fickle P value generates irreproducible results. *Nature Methods*, **12**, 179–85.

Hoeffding, W. (1948). A class of statistics with asymptotically normal distribution. *The Annals of Mathematical Statistics*, **19**(3), 293–325.

Horn, R. A. and Johnson, C. R. (1985). *Matrix Analysis*. Cambridge University Press, Cambridge.

Hoyle, M. H. (1973). Transformations: An introduction and a bibliography. *International Statistical Review / Revue Internationale de Statistique*, **41**(2), 203–223.

Huber, P. J. (2004). *Robust Statistics*. John Wiley & Sons, NY.

Hyndman, R. J. and Fan, Y. (1996). Sample quantiles in statistical packages. *The American Statistician*, **50**(4), 361–365.

J. A. Nelder, R. W. M. W. (1972). Generalized linear models. *Journal of the Royal Statistical Society, Series A*, **135**(3), 370–384.

James, G. S. (1952). Notes on a theorem of Cochran. *Mathematical Proceedings of the Cambridge Philosophical Society*, **48**(3), 443–446.

Jeffreys, H. (1946). An invariant form for the prior probability in estimation problems. *Proceedings of the Royal Society, Series A*, **186**(1007), 453–461.

Jones, C. and Zhigljavsky, A. A. (2004). Approximating the negative moments of the Poisson distribution. *Statistics & Probability Letters*, **66**(2), 171–181.

Karlin, S. (1958). Admissibility for estimation with quadratic loss. *The Annals of Mathematical Statistics*, **29**(2), 406–436.

Kiefer, J. (1967). On Bahadur's representation of sample quantiles. *The Annals of Mathematical Statistics*, **38**(5), 1323–1342.

Kleiber, C. and Kotz, S. (2003). *Statistical Size Distributions in Economics and Actuarial Sciences*. John Wiley & Sons, NY.

Knight, J. L. and Satchell, S. E. (1996). The exact distribution of the maximum likelihood estimators for the linear regression negative exponential model. *Journal of Statistical Planning and Inference*, **50**(1), 91–102.

Kolmogorov, A. N. (1933). *Grundbegriffe der Wahrscheinlichkeitrechnung, Ergebnisse Der Mathematik; translated as Foundations of Probability (1950)*. Chelsea Publishing Company, NY.

Kowalski, J. and Tu, X. M. (2008). *Modern Applied U-statistics*. John Wiley and Sons, NY.

Kozakiewicz, W. *et al.* (1947). On the convergence of sequences of moment generating functions. *The Annals of Mathematical Statistics*, **18**(1), 61–69.

Kutner, M. H., Nachtsheim, C. J., and Neter, J. (2004). *Applied Linear Statistical Models*. McGraw-Hill Irwin, Boston, 5th edition.

Lai, T. L. and Robbins, H. (1977). Strong consistency of least squares estimates in regression models. *Proceedings of the National Academy of Sciences*, **74**(7), 2667–2669.

Laubscher, N. F. (1961). On stabilizing the binomial and negative binomial variances. *Journal of the American Statistical Association*, **56**(293), 143–150.

Lehmann, E. and Casella, G. (1998). *Theory of Point Estimation.* Springer, NY, 2nd edition.

Lehmann, E. L. (1999). *Elements of Large-Sample Theory.* Springer, NY.

Lehmann, E. L. and Romano, J. P. (2005). *Testing Statistical Hypotheses.* Springer, NY, 3rd edition.

Lehmann, E. L. and Scheffé, H. (1950). Completeness, similar regions, and unbiased estimation: Part I. *Sankhyā: The Indian Journal of Statistics (1933–1960)*, **10**(4), 305–340.

Lehmann, E. L. and Scheffé, H. (1955). Completeness, similar regions, and unbiased estimation: Part II. *Sankhyā: The Indian Journal of Statistics (1933–1960)*, **15**(3), 219–236.

Liang, K.-Y. and Zeger, S. L. (1986). Longitudinal data analysis using generalized linear models. *Biometrika*, **73**(1), 13–22.

Mallows, C. L. (2000). Some comments on C_p. *Technometrics*, **42**(1), 87–94.

Mann, H. B. and Whitney, D. R. (1947). On a test of whether one of two random variables is stochastically larger than the other. *The Annals of Mathematical Statistics*, **18**(1), 50–60.

Mayo, D. G. *et al.* (2014). On the Birnbaum argument for the strong likelihood principle. *Statistical Science*, **29**(2), 227–239.

McCullagh, P. (1984). Tensor notation and cumulants of polynomials. *Biometrika*, **71**(3), 461–476.

McCullagh, P. *et al.* (1983). Quasi-likelihood functions. *The Annals of Statistics*, **11**(1), 59–67.

McCulloch, C. E., Searle, S. R., and Neuhaus, J. M. (2008). *Generalized, Linear, and Mixed Models.* Wiley, NY, 2nd edition.

Nayak, T. K. (1992). On statistical analysis of a sample from a population of unknown species. *Journal of Statistical Planning and Inference*, **31**(2), 187–198.

Newman, M. (2005). Power laws, Pareto distributions and Zipf's law. *Contemporary Physics*, **46**(5), 323–351.

Pearce, S. (1992). Introduction to Fisher (1925) Statistical Methods for Research Workers. In *Breakthroughs in Statistics, Volume II*, pages 59–65. Springer, NY.

Pinheiro, J. and Bates, D. (2009). *Mixed-Effects Models in S and S-PLUS.* Springer, NY.

Pitman, E. J. G. (1939). The estimation of the location and scale parameters of a continuous population of any given form. *Biometrika*, **30**(3/4), 391–421.

Rao, C. R. (1948). Large sample tests of statistical hypotheses concerning several parameters with applications to problems of estimation. In *Mathematical Proceedings of the Cambridge Philosophical Society*, volume 44, pages 50–57. Cambridge University Press.

Richards, S. (2012). A handbook of parametric survival models for actuarial use. *Scandinavian Actuarial Journal*, **2012**(4), 233–257.

Robbins, H. E. (1968). Estimating the total probability of the unobserved outcomes of an experiment. *The Annals of Mathematical Statistics*, **39**(1), 256–257.

Robins, J. M., Rotnitzky, A., and Zhao, L. P. (1994). Estimation of regression coefficients when some regressors are not always observed. *Journal of the American statistical Association*, **89**(427), 846–866.

Rohatgi, V. K. and Székely, G. J. (1989). Sharp inequalities between skewness and kurtosis. *Statistics & Probability Letters*, **8**(4), 297–299.

Rose, J. S. (1978). *A Course on Group Theory*. Cambridge University Press, Cambridge.

Ross, S. M. (1996). *Stochastic Processes*. John Wiley and Sons, NY, 2nd edition.

Rotman, J. J. (1995). *An Introduction to the Theory of Groups*. Springer, NY.

Rudin, W. (1991). *Functional analysis*. McGraw-Hill Science, NY.

Samworth, R. J. *et al.* (2012). Optimal weighted nearest neighbour classifiers. *The Annals of Statistics*, **40**(5), 2733–2763.

Satterthwaite, F. E. (1946). An approximate distribution of estimates of variance components. *Biometrics Bulletin*, **2**(6), 110–114.

Scheffe, H. (1959). *The Analysis of Variance*. John Wiley & Sons, NY.

Smirnov, N. (1948). Table for estimating the goodness of fit of empirical distributions. *The Annals of Mathematical Statistics*, **19**(2), 279–281.

Solomon, D. L. (1975). A note on the non-equivalence of the Neyman-Pearson and generalized likelihood ratio tests for testing a simple null versus a simple alternative hypothesis. *The American Statistician*, **29**(2), 101–102.

Spearman, C. (1904). The proof and measurement of association between two things. *The American Journal of Psychology*, **15**(1), 72–101.

Stark, P. B. and Parker, R. L. (1995). Bounded-variable least squares: an algorithm and applications. *Computational Statistics*, **10**, 129–129.

Stein, C. (1956). Inadmissibility of the usual estimator for the mean of a multivariate normal distribution. Proc. 3rd Berkeley Sympos. Math. Statist. Probability 1, 197–206 (1956).

Sun, Y., Baricz, Á., and Zhou, S. (2010). On the monotonicity, log-concavity, and tight bounds of the generalized Marcum and Nuttall Q-functions. *IEEE Transactions on Information Theory*, **56**(3), 1166–1186.

Tate, R. F. and Klett, G. W. (1959). Optimal confidence intervals for the variance of a normal distribution. *Journal of the American Statistical Association*, **54**(287), 674–682.

Taylor, J. B. (1974). Asymptotic properties of multiperiod control rules in the linear regression model. *International Economic Review*, **15**(2), 472–484.

Tian, Y. and Styan, G. P. (2006). Cochran's statistical theorem revisited. *Journal of Statistical Planning and Inference*, **136**(8), 2659–2667.

Tukey, J. W. (1949). Comparing individual means in the analysis of variance. *Biometrics*, **5**(2), 99–114.

Žnidarič, M. (2009). Asymptotic expansion for inverse moments of binomial and Poisson distributions. *Open Statistics and Probability Journal*, **1**, 7–10.

Van Belle, G., Fisher, L. D., Heagerty, P. J., and Lumley, T. (2004). *Biostatistics: A Methodology for the Health Sciences.* John Wiley & Sons, NY, 2nd edition.

Venables, W. N. and Ripley, B. D. (2013). *Modern Applied Statistics with S.* Springer, NY, 4th edition.

Verbeke, G. (1997). *Linear Mixed Models for Longitudinal Data.* Springer, NY.

Wald, A. (1943). Tests of statistical hypotheses concerning several parameters when the number of observations is large. *Transactions of the American Mathematical Society*, **54**(3), 426–482.

Wasserstein, R. L. and Lazar, N. A. (2016). The ASA statement on P-values: Context, process, and purpose. *The American Statistician*, **70**(2), 129–133.

Wedderburn, R. W. M. (1974). Quasi-likelihood functions, generalized linear models, and the Gauss—Newton method. *Biometrika*, **61**(3), 439–447.

Weir, B. S. (1996). *Genetic Data Analysis II.* Sinauer Associates, Sunderland.

Welch, B. L. (1947). The generalisation of student's problems when several different population variances are involved. *Biometrika*, **34**(1-2), 28–35.

Whittle, P. (2000). *Probability via Expectation.* Springer, NY, 4th edition.

Wilcoxon, F. (1945). Individual comparisons by ranking methods. *Biometrics Bulletin*, **1**(6), 80–83.

Wild, C. and Seber, G. (1989). *Nonlinear Regression.* Wiley, NY.

Wilks, S. S. (1938). The large-sample distribution of the likelihood ratio for testing composite hypotheses. *The Annals of Mathematical Statistics*, **9**(1), 60–62.

Wold, H. O. A. and Whittle, P. (1957). A model explaining the Pareto distribution of wealth. *Econometrica*, **25**(4), 591–595.

Yates, F. (1934). Contingency tables involving small numbers and the χ^2 test. *Supplement to the Journal of the Royal Statistical Society*, **1**(2), 217–235.

Yates, F. (1951). The influence of statistical methods for research workers on the development of the science of statistics. *Journal of the American Statistical Association*, **46**(253), 19–34.

Ye, Z.-S. and Chen, N. (2017). Closed-form estimators for the gamma distribution derived from likelihood equations. *The American Statistician*, **71**(2), 177–181.

Yu, G. (2009). Variance stabilizing transformations of Poisson, binomial and negative binomial distributions. *Statistics & Probability Letters*, **79**(14), 1621–1629.

Zeger, S. L. and Liang, K.-Y. (1986). Longitudinal data analysis for discrete and continuous outcomes. *Biometrics*, **42**(1), 121–130.

Index

Printed in the United States
by Baker & Taylor Publisher Services